Media resources included with student textbooks will enhance their understanding of biology!

Thank you for considering *Essential Biology* for your non-majors biology course. At Benjamin Cummings our goal is to help students succeed by supporting a variety of learning styles. Media provides a powerful tool for visualizing concepts and an opportunity for active learning. In fact, the integration of text and media is one of the important features of *Essential Biology.*

Therefore, every new student copy of *Essential Biology* includes:

- A 12-month, pre-paid subscription to the **Essential Biology Place Website**
- The **Essential Biology Place CD-ROM**

The **Essential Biology Place** contains many opportunities to help students master material and prepare for exams, including:

- Over 150 interactive exercises (at least one for every main concept in the book)
- 400 chapter quiz questions
- practice test questions
- annotated web links for every chapter
- and much more! (See the back of this page for a complete list of features.)

You can also use these resources in teaching your course.

> **To help students get the most out of these media resources, the text includes Media Flags that describe what they will find on the website and CD-ROM.**

s another p
PCR is ofte
noresis. Wit
ed from a t

Just how reliable is DNA fingerprinting? In bility of two people having identical DNA finge chance in 100,000 and one in a billion. The exac

Use DNA fingerprinting to analyze evidence from a crime scene in Web/ CD Activity 11F.

many markers are compared
markers are in the population
data on the frequencies of var
population as a whole and in
are enabling forensic scientist
rate statistical calculations. Tl
can arise from insufficient statistical data, huma
DNA fingerprints are now accepted as compelli
and scientists alike. In fact, many argue that DN

As you consider *Essential Biology* for adoption, Benjamin Cummings is pleased to offer you six months of unlimited access to the **Essential Biology Place Website.** Here's how to activate your complimentary subscription:

1. Go to www.essentialbiology.com

2. Click "Launch."

3. Click "Register Here."

4. Enter your pre-assigned access code exactly as it appears below.

WSEB-STOWS-EYING-COWED-TRASH-MOLES

5. Click "Submit."

6. Complete the on-line registration form to create your own personal user ID and password.

7. Once your personal user ID and password are confirmed, go to www.essentialbiology.com, type in your new user ID and password, and click "Enter."

Your access code can only be used once to establish your subscription, which is not transferable. *Write down your new ID and password. You will need them to access the site.*

If you choose to adopt *Essential Biology*, you will receive a complimentary subscription to the **Essential Biology Place** for the duration of the adoption. Additionally, all new student copies of the book will include a CD-ROM and a 12-month subscription to the **Essential Biology Place.** For students who do not buy a new copy of the textbook, please ask your local bookstore to order the **Essential Biology Place Website Subscription** (ISBN 0-8053-6631-8) and the **Essential Biology Place CD-ROM and Website Subscription** (ISBN 0-8053-6630-X). Students can also purchase subscriptions to the **Essential Biology Place Website** on-line at www.essentialbiology.com.

Turn the page to learn more…

0-8053-7422-1

Features of the Essential Biology Place

Feature	Description	Website	CD-ROM*
Objectives	Lists the key concepts students should understand after studying each chapter	✔	✔
Activities	Over 130 interactive activities, including animations, virtual labs, review exercises, and videos; at least one activity for every major topic in a chapter	✔	✔
The Process of Science	Over 20 extensive exercises that encourage experimentation, exploration, and analysis	✔	✔
Quiz	20 multiple-choice questions per chapter; a total of 400 questions	✔	✔
Evolution Link	Links to websites plus questions that reinforce the connection between evolution and each chapter	✔	Link to website
Web Links	Links to relevant websites with descriptions of the sites, organized by chapter	✔	Link to website
News	Recent developments in biology, organized by chapter	✔	Link to website
Test Flight	A database of multiple-choice questions that can be used to construct practice tests from more than one chapter	✔	Link to website
Glossary	Definitions of boldface terms plus selected audio pronunciations	✔	✔
Networking	A chat room and discussion message board exclusively for users of *Essential Biology*	✔	Link to website
Syllabus Manager	A tool for instructors to create a complete on-line syllabus that students can access from the **Essential Biology Place Website**	✔	Link to website
Instructor Resources	Art, animations, and tables for instructors to use in lectures plus an electronic version of the *Instructor's Guide to Text and Media* and links to photo resources. Requires an instructor's access code.	✔	Link to website

*The **Essential Biology Place Website** and **CD-ROM** contain the same core content. For web-based features, the CD-ROM provides a link to the Essential Biology Place Website, which requires a live internet connection and an access code.

System Requirements

WINDOWS
166 MHz Intel Pentium processor or greater
Windows 95, 98, NT4, 2000
32 MB RAM installed
800 × 600 screen resolution
4× CD-ROM drive
Browser: Internet Explorer 5.0 or Netscape Communicator 4.7
Plug-ins: Shockwave Player 8, Flash Player 4, QuickTime 4

MACINTOSH
120 MHz PowerPC
OS 8.1 or higher
24 MB RAM available
800 × 600 screen resolution, thousands of colors
4× CD-ROM drive
Browser: Internet Explorer 5.0 or Netscape Communicator 4.7
Plug-ins: Shockwave Player 8, Flash Player 4, QuickTime 4

How to Start the Essential Biology Place CD-ROM

Windows: Insert the CD in the CD-ROM drive. The first time that you run the program, the installer will search for, and if necessary, install plug-ins and a web browser. From then on, the program will launch automatically after you load the CD into the drive. Click "Launch" on the splash screen to begin. To open the Readme file, which has more information, double-click on My Computer on the desktop, select (but do not double-click) the caterpillar icon, choose Explore in the File menu, and double-click on Readme.txt.

Macintosh: After inserting the CD in the CD-ROM drive, double-click on "EB Place Installer." The installer will search for, and if necessary, install plug-ins and Netscape Communicator 4.7. Double-click on the caterpillar icon to start the program. Click "Launch" on the splash screen to begin. The Readme file, with further information, is located on the CD-ROM. Double-click on the Readme file to open it.

How to Contact Technical Support

For technical support, please visit www.awl.com/techsupport, send an email to online.support@pearsoned.com (for website questions) or send an email to media.support@pearsoned.com (for CD-ROM questions) with a detailed description of your computer system and the technical problem. You can also call our tech support hotline at 800-677-6337, Monday - Friday, 8:00 A.M. - 5:00 P.M. CST.

How to Contact Us

We invite your comments and suggestions via email at question@awl.com, or contact your Benjamin Cummings sales representative.

Important: Please read the License Agreement, located on the launch screen, before using the Essential Biology Place Website or CD-ROM. By using the Essential Biology Place Website or CD-ROM you indicate that you have read, understood, and accepted the terms of this agreement.

ESSENTIAL

Biology

Neil A. Campbell
University of California, Riverside

Jane B. Reece
Palo Alto, California

An imprint of Addison Wesley Longman, Inc.
San Francisco   Boston   New York
Capetown   Hong Kong   London   Madrid   Mexico City
Montreal   Munich   Paris   Singapore   Sydney   Tokyo   Toronto

Executive Editor: Erin Mulligan

Senior Editor: Beth Wilbur

Senior Text and Media Developmental Editor: Pat Burner

Senior Developmental Editor: Shelley Parlante

Project Editor: Evelyn Dahlgren

Publishing Assistant: Aaron Gass

Associate Multimedia Producer: Maureen Kennedy

Media Developmental Editor and Lead Artist: Russell Chun

Managing Editor, Production: Wendy Earl

Production Editor: Leslie Austin

Art and Design Manager: Bradley Burch

Senior Art Supervisor: Donna Kalal

Illustrations: Precision Graphics

Text and Cover Designer: Carolyn Deacy

Photo Researcher: Stephen Forsling

Copy Editor: Janet Greenblatt

Proofreader: Martha Ghent

Indexer: Katherine Pitcoff

Marketing Manager: Josh Frost

Senior Project Editor, Marketing: Kirsten Watrud

Prepress/Manufacturing Supervisor: Vivian McDougal

Compositor and Prepress: GTS Graphics

Cover Printer: The Lehigh Press, Inc.

Printer: Von Hoffmann Press, Inc.

On the cover: Spicebush swallowtail caterpillar. © Jeff Lepore/Photo Researchers, Inc.

Library of Congress Cataloging-in-Publication Data

Campbell, Neil A., 1946–
    Essential biology / Neil A. Campbell, Jane B. Reece.
      p. cm.
    Includes bibliographical references (p.).
    ISBN 0-8053-7393-4 (pbk. : alk. Paper)
    1.   Biology. I. Reece, Jane B. II. Title.

QH308.2.C343 2001
570—dc21
                                           00-065587

3 4 5 6 7 8 9 10—VHP—04 03 02

Benjamin Cummings
1301 Sansome Street
San Francisco, CA 94111

To Rochelle and Allison, with love
N. A. C.

To Paul, with love
J. B. R.

About the Authors

Neil A. Campbell has taught general biology for 30 years and is a coauthor, with Jane Reece and Larry Mitchell, of *Biology*, Fifth Edition, and *Biology: Concepts and Connections*, Third Edition. His enthusiasm for sharing the fun of science with students stems from his own undergraduate experience. He began at Long Beach State College as a history major, but switched to zoology after general education requirements "forced" him to take a science course. Following a B.S. from Long Beach, he earned an M.A. in Zoology from UCLA and a Ph.D. in Plant Biology from the University of California, Riverside. He has published numerous articles on how certain desert plants survive in salty soil and how the sensitive plant (Mimosa) and other legumes move their leaves. His diverse teaching experiences include courses for non-biology majors at Cornell University, Pomona College, and San Bernardino Valley College, where he received the college's first Outstanding Professor Award in 1986. Dr. Campbell is currently a visiting scholar in the Department of Botany and Plant Sciences at UC Riverside.

Jane B. Reece has worked in biology publishing since 1978, when she joined the editorial staff of Benjamin Cummings. She is a coauthor, with Neil Campbell and Larry Mitchell, of *Biology*, Fifth Edition, and *Biology: Concepts and Connections*, Third Edition. Her education includes an A.B. in Biology from Harvard University, an M.S. in Microbiology from Rutgers University, and a Ph.D. in Bacteriology from the University of California at Berkeley. At UC Berkeley, and later as a postdoctoral fellow in genetics at Stanford University, her research focused on genetic recombination in bacteria. She taught biology at Middlesex County College (New Jersey) and Queensborough Community College (New York). During her 12 years as an editor at Benjamin Cummings, Dr. Reece played major roles in a number of successful textbooks including *Microbiology: An Introduction*, by G. J. Tortora, B. R. Funke, and C. L. Case, and *Molecular Biology of the Gene*, Fourth Edition, by J. D. Watson et al. She was also a coauthor of *The World of the Cell*, Third Edition, with W. M. Becker and M. F. Poenie.

Of Related Interest from the Benjamin Cummings Series in the Life Sciences

General Biology

N. A. Campbell, L. G. Mitchell, and J. B. Reece
Biology: Concepts & Connections, Third Edition (2000)

N. A. Campbell, J. B. Reece, and L. G. Mitchell
Biology, Fifth Edition (1999)

R. A. Desharnais and J. R. Bell
BiologyLabs On-Line (2000)

J. Dickey
Laboratory Investigations for Biology (1995)

J. B. Hagen, D. Allchin, and F. Singer
Doing Biology (1996)

R. Keck and R. Patterson
Biomath: Problem Solving for Biology Students (2000)

A. E. Lawson and B. D. Smith
Studying for Biology (1995)

J. Lee
The Scientific Endeavor (2000)

J. G. Morgan and M. E. B. Carter
Investigating Biology, Third Edition (1999)

J. A. Pechenik
A Short Guide to Writing about Biology, Fourth Edition (2001)

G. I. Sackheim
An Introduction to Chemistry for Biology Students, Sixth Edition (1999)

R. M. Thornton
The Chemistry of Life CD-ROM (1998)

Biotechnology

D. Bourgaize, T. Jewell, and R. Buiser
Biotechnology: Demystifying the Concepts (2000)

Cell Biology

W. M. Becker, L. J. Kleinsmith, and J. Hardin
The World of the Cell, Fourth Edition (2000)

Biochemistry

R. F. Boyer
Modern Experimental Biochemistry, Third Edition (2000)

C. K. Mathews, K. E. van Holde, and K. G. Ahern
Biochemistry, Third Edition (2000)

Genetics

R. J. Brooker
Genetics: Analysis and Principles (1999)

J. P. Chinnici and D. J. Matthes
Genetics: Practice Problems and Solutions (1999)

R. P. Nickerson
Genetics: A Guide to Basic Concepts and Problem Solving (1990)

P. J. Russell
Fundamentals of Genetics, Second Edition (2000)

P. J. Russell
Genetics, Fifth Edition (1998)

Human Biology

M. D. Johnson
Human Biology: Concepts and Current Issues (2001)

Molecular Biology

M. V. Bloom, G. A. Freyer, and D. A. Micklos
Laboratory DNA Science (1996)

J. D. Watson, N. H. Hopkins, J. W. Roberts, J. A. Steitz, and A. M. Weiner
Molecular Biology of the Gene, Fourth Edition (1987)

Microbiology

G. J. Tortora, B. R. Funke, and C. L. Case
Microbiology: An Introduction, Seventh Edition (2001)

Anatomy and Physiology

E. N. Marieb
Essentials of Human Anatomy and Physiology, Sixth Edition (2000)

E. N. Marieb
Human Anatomy and Physiology, Fifth Edition (2001)

E. N. Marieb and J. Mallatt
Human Anatomy, Third Edition (2001)

Ecology and Evolution

C. J. Krebs
Ecological Methodology, Second Edition (1999)

C. J. Krebs
Ecology: The Experimental Analysis of Distribution and Abundance, Fifth Edition (2001)

J. W. Nybakken
Marine Biology: An Ecological Approach, Fifth Edition (2001)

E. R. Pianka
Evolutionary Ecology, Sixth Edition (2000)

R. L. Smith
Ecology and Field Biology, Sixth Edition (2001)

R. L. Smith and T. M. Smith
Elements of Ecology, Fourth Edition Update (2000)

Plant Ecology

M. G. Barbour, J. H. Burk, W. D. Pitts, F. S. Gilliam, and M. W. Schwartz
Terrestrial Plant Ecology, Third Edition (1999)

Preface

A decade ago in his popular book, *Megatrends 2000*, futurist John Naisbitt predicted the coming of the Age of Biology. It was a safe forecast. Biology was already emerging as the new millennium's central science—the concourse where all the natural sciences meet and intersect with the humanities and social sciences. But the *scale* of biology's impact is probably more surprising to most scientists. In addition to illuminating life more brightly than ever, modern biology is remodeling medicine, agriculture, forensics, conservation science, anthropology, psychology, sociology, and almost every other "-ology" that's paying attention. This is the best time ever to take a biology course!

It is a privilege for us to be able to help instructors share the story of life with students during this golden age of biology. We see our responsibility as science educators to be especially important in communicating with students who are *not* biology majors, because their attitudes about science and scientists are likely to be shaped by a single, required science course—*this* course. And because biology and society are so interwoven today, we believe that non-majors courses are the most important classes biology departments teach. Many of our colleagues around the country apparently agree, for the most enthusiastic and innovative among them are drawn to the adventure of teaching non-majors. It's among the liveliest and most creative communities in all of science education.

As important and inspirational as modern biology is, these are also the most challenging times to teach and learn the subject. The same discovery explosion that makes biology exhilarating today also threatens to suffocate our students under an avalanche of information. With each of its many subfields bustling in research activity, biology grows larger every year, while the academic semester stays the same size. Something has to give.

In this era of ever-expanding biology, we have no choice but to make our courses less encyclopedic. Leading the way in this movement are the many thoughtful instructors who have opted to cover fewer main biology topics rather than compromise depth in the most important areas. To support this trend, we have created this new biology textbook, which we call *Essential Biology*. Yes, it is a shorter biology text than most, but we did not achieve this brevity by trying to fit all of biology into less space. Instead, we focused on just four core topics: cells, genes, evolution, and ecology. By exploring these four areas and fitting them together, students can synthesize a coherent view of life. In the context of these four main topics, students will encounter diverse organisms and their evolutionary adaptations. However, we have not included separate units on the anatomy and physiology of plants, animals, and other organisms. This enabled us to keep *Essential Biology* manageable in size—and, well, *essential*—without being superficial in the areas we chose to cover. We take the "less is more" mantra in education today to mean fewer topics, not more dilute explanations.

Even with a decision to cover fewer major topics, there is still the potential for biology to collapse into a formless pile of terms and factoids. An integrated view of life depends on a theme that cuts across all topics, and that theme is evolution. Understanding how evolution accounts for both the unity and diversity of life makes biology whole. Evolution will continue to provide this form no matter how big and complex biology becomes. Every chapter of *Essential Biology* connects to this evolutionary theme of life.

We named our new book *Essential Biology* partly because this look at life is relatively brief, selective, and integrated. But the title has a second meaning. It announces an emphasis on concepts and applications that are *essential* for students to make biologically informed decisions throughout their lives—to evaluate certain health and environmental issues, for example. From ethical and safety concerns surrounding genomics to debates about global warming, students will find biology in the news every day. Biology *is* more essential than ever in a general education, and we've tried to spotlight this central place of biology in modern culture.

Long after students have forgotten most of the specific content of their college courses, they will be left with general impressions that will influence their interests, opinions, values, and actions. We hope this textbook, and its supporting media, will help students fold biological perspectives into their personal worldviews. Please let us know how we are doing and how we can improve the next edition of *Essential Biology*.

Neil Campbell
Department of Botany
    and Plant Sciences
University of California
Riverside, CA 92521

Jane Reece
Benjamin Cummings Publishing
1301 Sansome St.
San Francisco, CA 94111

Acknowledgments

As the authors of *Essential Biology*, we're spoiled! We've managed to surround ourselves with the very best publishing professionals and biology colleagues. Though we must take sole responsibility for any of the textbook's shortcomings, its merits reflect the contributions of many associates who helped us build a new kind of textbook and who created integral media activities that complement the lessons.

First, we thank senior editor Beth Wilbur. When the going got tough—and it always does at some point during the development of a textbook—Beth joined the team. She brought a fresh and objective perspective, a collaborative nature, adaptability and humor, and a rational sense of priorities. Her follow-through and confidence-building leadership were reassuring. And speaking of leadership, it was executive editor Erin Mulligan who enlisted us to write *Essential Biology* and who recruited Beth Wilbur as our editor. We appreciate Erin's confidence in us and value her brand of publishing partnership with authors. Of course, publishing excellence flows from the top, and we are grateful to Linda Davis, president of Benjamin Cummings, for the example she provides to her entire company. Benjamin Cummings authors are very fortunate.

Especially important in the creation of *Essential Biology* were three team members: Pat Burner, Shelley Parlante, and Russell Chun. Senior developmental editors Pat and Shelley worked with us and the production team to make *every* paragraph and *every* illustration work better for students. Their commitment to quality and clarity shows on every page. Shelley was responsible for the excellent developmental work on the genetics chapters and helped us refine our vision for the whole book. Russell, as multimedia developmental editor, is a gifted artist and animator. He shaped the book's art program and brought to life the concepts on the *Essential Biology Place* Website and CD-ROM. Pat, in addition to the extensive and topnotch development she did on three-fourths of the book, was also Russell's partner in developing the media. The commitment that Pat, Shelley, and Russell made to *Essential Biology* inspired the whole team.

Several others played key roles in developing and refining the text, graphics, and media of this project. Associate multimedia producer Maureen Kennedy had a big impact on the planning of the CD-ROM and website. Senior art supervisor Donna Kalal not only kept the book's illustrations and photos moving through the pipeline, but her keen eye also helped assure the quality of the art program. Photo researcher Stephen Forsling worked very hard to find just the right pictures to fit each topic and make the book look great. Copy editor Janet Greenblatt provided excellent editorial counsel and proofreader Martha Ghent was vigilant, assuring quality on every page. Project editor Evelyn Dahlgren managed the editorial schedules, coordinated development of the supplements package, monitored the entire project's budget, and helped out in many ways with the media. Katherine Pitcoff created the book's very functional index. And publishing assistant Aaron Gass tirelessly supported the whole team and played an especially important role in coordinating the media initiatives.

The production team transformed our manuscripts and drawings into a real book. Wendy Earl, the managing editor for production, was heroic in mobilizing the entire production effort throughout our very tight schedule. Production editor Leslie Austin managed the book schedules and the production team with expertise and good humor. Art and design manager Bradley Burch and text and cover designer Carolyn Deacy made *Essential Biology* beautiful as well as functional. Finally, manufacturing supervisor Vivian McDougal worked miracles to manage the printing of the bound book you now hold.

"Marketing" is an alien concept to most of us in the biology community. But for what we try to do as authors, "market" translates as "the students and instructors we are trying to serve." Marketing manager Josh Frost and senior project editor for market development Kirsten Watrud kept us focused on the needs of students and instructors. Kirsten initiated valuable contacts with biology instructors and students, and coordinated class testing of our material. We also thank the Addison Wesley field staff for representing *Essential Biology* on campuses. These representatives will also tell us what you like (and don't like) about this book and media, which will help us improve the next edition.

Numerous colleagues in the biology community also contributed to *Essential Biology*. For Units One, Three, and Four, Ed Zalisko of Blackburn College synthesized the lists of intriguing facts that introduce the chapters and wrote many of the summaries. Arlene Larson of the University of Colorado, Denver, read the entire book to help assure biological accuracy. Although Larry Mitchell did not work directly on this textbook, his past collaboration with us on our other books is certainly reflected in *Essential Biology*. At the end of these acknowledgments you'll find a list of the many instructors who provided valuable information about their courses, reviewed chapters, and/or conducted class tests of *Essential Biology* with their students. We thank them for their efforts and support.

Most of all, we thank our families and friends who continue to tolerate our obsession with doing our best for science education. We two have worked together on various projects for more than 20 years. It's still as much fun as ever!

Neil Campbell
Jane Reece

Reviewers and Advisors

Tammy Adair, *Baylor University*, Estrella Ang, *University of Pittsburgh at Greensburg*, David Arieti, *Oakton Community College*, Barbara J. Backley, *Elgin Community College*, Linda Barham, *Meridian Community College*, Rudi Berkelhamer, *University of California, Irvine*, Karyn Bledsoe, *Western Oregon University*, Lisa Boggs, *Southwestern Oklahoma State University*, James Botsford, *New Mexico State University*, Cynthia Bottrell, *Scott Community College*, Robert Boyd, *Auburn University*, B. J. Boyer, *Suffolk County Community College*, Patricia Brewer, *University of Texas, San Antonio*, Evert Brown, *Casper College*, Mary H. Brown, *Lansing Community College*, Richard D. Brown, *Brunswick Community College*, Carol T. Burton, *Bellevue Community College*, Rebecca Burton, *Alverno College*, Warren R. Buss, *University of Northern Colorado*, Miguel Cervantes-Cervantes, *Lehman College, City University of New York*, William H. Coleman, *University of Hartford*, James Conkey, *Truckee Meadows Community College*, Karen A. Conzelman, *Glendale Community College*, James T. Costa, *Western Carolina University*, Paul Decelles, *Johnson County Community College*, Galen DeHay, *Tri County Technical College*, Cynthia L. Delaney, *University of South Alabama*, Jean DeSaix, *University of North Carolina, Chapel Hill*, Edward Devine, *Moraine Valley Community College*, Dwight Dimaculangan, *Winthrop University*, Deborah Dodson, *Vincennes Community College*, Don Dorfman, *Monmouth University*, Terese Dudek, *Kishawaukee College*, Virginia Erickson, *Highline Community College*, Marirose T. Ethington, *Genesee Community College*, Jean Everett, *College of Charleston*, Phillip Fawley, *Westminster College*, Joseph Faryniarz, *Naugatuck Valley Community College*, Dennis M. Forsythe, *The Citadel*, Suzanne S. Frucht, *Northwest Missouri State University*, Edward G. Gabriel, *Lycoming College*, Anne M. Galbraith, *University of Wisconsin, La Crosse*, Gail Gasparich, *Towson University*, Sharon L. Gilman, *Coastal Carolina University*, Mac Given, *Neumann College*, Patricia Glas, *The Citadel*, Ralph C. Goff, *Mansfield University*, Marian R. Goldsmith, *University of Rhode Island*, Curt Gravis, *Western State College of Colorado*, Larry Gray, *Utah Valley State College*, Robert S. Greene, *Niagara University*, Ken Griffin, *Tarrant County Junior College*, Paul Gurn, *Naugatuck Valley Community College*, Peggy J. Guthrie, *University of Central Oklahoma*, Blanche C. Haning, *North Carolina State University*, Lazlo Hanzely, *Northern Illinois University*, Linda Hensel, *Mercer University*, Phyllis C. Hirsch, *East Los Angeles College*, Carl Huether, *University of Cincinnati*, Celene Jackson, *Western Michigan University*, Richard J. Jensen, *Saint Mary's College*, Tari Johnson, *Normandale Community College*, Greg Jones, *Santa Fe Community College*, Tracy L. Kahn, *University of California, Riverside*, Arnold J. Karpoff, *University of Louisville*, Valentine Kefeli, *Slippery Rock University*, Cheryl Kerfeld, *University of California, Los Angeles*, Henrik Kibak, *California State University, Monterey Bay*, Kerry Kilburn, *Old Dominion University*, Peter Kish, *Oklahoma School of Science and Mathematics*, Robert Kitchin, *University of Wyoming*, Michael E. Kovach, *Baldwin-Wallace College*, Gary Kwiecinski, *The University of Scranton*, Lynn Larsen, *Portland Community College*, Barbara Liedl, *Central College*, Harvey Liftin, *Broward Community College*, Maria P. MacWilliams, *Seton Hall University*, Michael Howard Marcovitz, *Midland Lutheran College*, Angela M. Mason, *Beaufort County Community College*, Roy B. Mason, *Mt. San Jacinto College*, John Mathwig, *College of Lake County*, Mary Anne McMurray, *Henderson Community College*, Patricia S. Muir, *Oregon State University*, Lois H. Peck, *University of the Sciences, Philadelphia*, John S. Peters, *College of Charleston*, Bill Pietraface, *State University of New York, Oneonta*, Philip Ricker, *South Plains College*, Laurel Roberts, *University of Pittsburgh*, Maxine Losoff Rusche, *Northern Arizona University*, Leba Sarkis, *Aims Community College*, Neil Schanker, *College of the Siskiyous*, Robert Schoch, *Boston University*, John Richard Schrock, *Emporia State University*, Brian W. Schwartz, *Columbus State University*, Wayne Seifert, *Brookhaven College*, Patty Shields, *George Mason University*, Cahleen Shrier, *Azusa Pacific University*, Jed Shumsky, *Drexel University*, Jeffrey Simmons, *West Virginia Wesleyan College*, Anu Singh-Cundy, *Western Washington University*, Thomas Smith, *Armstrong Atlantic State University*, Marshall D. Sundberg, *Emporia State University*, Sharon Thoma, *Edgewood College*, Sumesh Thomas, *Baltimore City Community College*, Betty Thompson, *Baptist University*, Bruce L. Tomlinson, *State University of New York, Fredonia*, Bert Tribbey, *California State University, Fresno*, Stephen M. Wagener, *Western Connecticut State University*, Harold Webster, *Pennsylvania State University, DuBois Campus*, Ted Weinheimer, *California State University, Bakersfield*, Joanne Westin, *Case Western Reserve University*, Quinton White, *Jacksonville University*, Leslie Y. Whiteman, *Virginia Union University*, Dwina Willis, *Freed Hardeman University*, David Wilson, *University of Miami*

Class-Test Instructors

Donna Bivans, *East Carolina University*, Reggie Cobb, *Nash Community College*, Pat Cox, *University of Tennessee, Knoxville*, Ade Ejire, *Johnston Community College*, Carl Estrella, *Merced College*, Dianne Fair, *Florida Community College*, Lynn Fireston, *Ricks College*, Patricia Glas, *The Citadel*, Consetta Helmick, *University of Idaho*, Richard Hilton, *Towson University*, Howard Hosick, *Washington State*, Greg Jones, *Santa Fe Community College*, Joyce Kille-Marino, *College of Charleston*, Laurel Roberts, *University of Pittsburgh*, Mike Runyan, *Lander University*, Travis Ryan, *Furman University*, Walter Saviuk, *Daytona Beach Community College*, William Smith, *Brevard Community College, Melbourne, Autralia*, Virginia Vandergon, *California State University, Northridge*, Dave Webb, *St. Clair County Community College*, Wayne Whalley, *Utah Valley State College*, Michael Womack, *Macon State College*

How to Use *Essential Biology*

CHAPTER 3

The Molecules of Life

 Americans consume an average of 140 pounds of

sugar per person per year. Most of the sugar in

soft drinks comes from corn. Cellulose, found

in plant cell walls, is the most abundant organic

 compound on Earth. A typical cell in your body has

about 3 meters of DNA. Cells are

composed of 70–95% water.

Know where you're headed . . .

Chapter outline. Use the chapter outline as your road map.

. . . and why.

Opening "snapshots." Get a feel for what you'll learn from the intriguing facts at the start of each chapter.

39

Overview: Our Dependence on Plants

Plants and other photosynthetic organisms convert the energy of sunlight to the chemical energy of sugar and other organic compounds. It is the source of all our food. And we depend on plants for more than our food. You are probably wearing underwear, jeans, or other clothing made of another product of photosynthesis, cotton. Most of our homes are framed with lumber, which is wood produced by photosynthetic trees. Even the text you are now reading is printed on paper, still another material that can be traced to photosynthesis in plants. But mostly, photosynthesis is all about feeding the biosphere.

Discover more ways we are linked to plants in Web/CD Activity 6A.

Plants are **autotrophs,** which means "self-feeders" in Greek. The term is a bit misleading in its implication that plants do not require nutrients—they do. But those nutrients are entirely inorganic: carbon dioxide from the air, and water and minerals from the soil. From that inorganic diet, plants can make all their own organic molecules, including carbohydrates, lipids, proteins, and nucleic acids. That is the definition of an autotroph: an organism that makes all its own organic matter from inorganic nutrients. In contrast, humans and other animals are **heterotrophs,** meaning "other-feeders." We heterotrophs cannot make organic molecules from inorganic ones. That is why we must eat. Heterotrophs depend on autotrophs for their organic fuel and material for growth and repair.

The most common form of autotrophic nutrition is photosynthesis, which uses light energy to drive the synthesis of organic molecules. Almost all plants are photosynthetic, and so are certain groups of protists and bacteria (Figure 6.2).

Start with the big picture . . .

Overview. Each chapter starts with an overview that gives you a context for the details to come.

take it one step at a time,

Numbered steps. Follow biological processes easily using numbered steps in art, text, and legends.

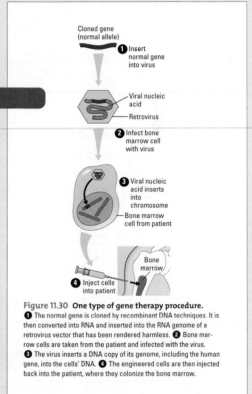

Figure 11.30 One type of gene therapy procedure.
❶ The normal gene is cloned by recombinant DNA techniques. It is then converted into RNA and inserted into the RNA genome of a retrovirus vector that has been rendered harmless. ❷ Bone marrow cells are taken from the patient and infected with the virus. ❸ The virus inserts a DNA copy of its genome, including the human gene, into the cells' DNA. ❹ The engineered cells are then injected back into the patient, where they colonize the bone marrow.

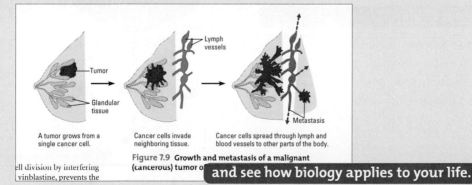

A tumor grows from a single cancer cell.

Cancer cells invade neighboring tissue.

Cancer cells spread through lymph and blood vessels to other parts of the body.

Figure 7.9 Growth and metastasis of a malignant (cancerous) tumor o

ell division by interfering
vinblastine, prevents the

and see how biology applies to your life.

Health and environmental applications. In every chapter, *Essential Biology* makes connections to topics that will be meaningful to you and your family long after this course is done.

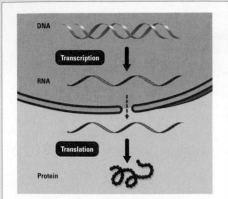

Figure 9.14 The flow of genetic information in a eukaryotic cell: a review. A sequence of nucleotides in the DNA is transcribed into a molecule of RNA in the cell's nucleus (purple area). The RNA travels to the cytoplasm (blue-green area), where it is translated into the specific amino acid sequence of a protein.

Consistent colors and symbols. Consistent colors and symbols— the aqua cell interior with purple nucleus, the blue DNA helix, the red RNA ribbon, the purple protein, and many others—help you recognize biological players that appear throughout *Essential Biology.*

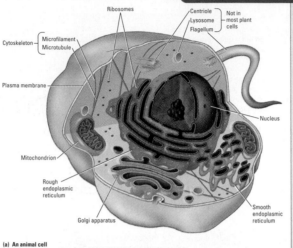

Figure 4.6 Panoramic view of idealized animal cell and plant cell. For now, the labels are just words, but these organelles will come to life as we take a closer look at how each of these parts works. To keep from getting lost on our tour of cells, we'll carry miniature versions of these overview diagrams as our road maps, with "you are here" highlighted.

(a) An animal cell

visualize the processes of life,

Art that explains main ideas. Use the illustrations to reinforce what you read in the text. Photos, illustrations, and captions combine to explain key ideas.

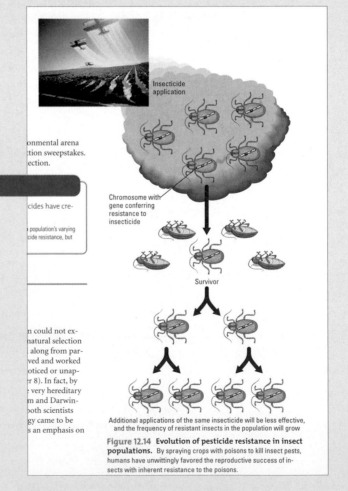

Figure 12.14 Evolution of pesticide resistance in insect populations. By spraying crops with poisons to kill insect pests, humans have unwittingly favored the reproductive success of insects with inherent resistance to the poisons.

and know where you are.

Orientation diagrams. Orientation diagrams show you where you are in a complex process, and where this topic fits into the bigger picture.

Stage 1: Glycolysis The word *glycolysis* means "splitting of sugar." That is exactly what happens (Figure 5.14). Glycolysis breaks the six-carbon glucose in half, forming two three-carbon molecules. These molecules then donate high-energy electrons to NAD^+, the electron carrier. Glycolysis also makes some ATP directly when enzymes transfer phosphate groups from fuel molecules to ADP (Figure 5.15). What remains of the fractured glucose at the end of glycolysis are two molecules of pyruvic acid. The pyruvic acid still holds most of the energy of glucose, and that energy is harvested in the Krebs cycle.

Glycolysis | Krebs Cycle | Electron Transport

ATP   ATP   ATP

Watch the process of glycolysis in Web/CD Activity 5E.

Natural Selection in Action: The Evolution of Pesticide-Resistant Insects

Natural selection and the adaptive evolution it causes are observable phenomena. A classic and unsettling example is the evolution of pesticide resistance in hundreds of insect species.

Pesticides are poisons used to kill insects that are pests in crops, swamps, backyards, and homes. Examples are DDT, now banned in many countries, and malathion. These chemical weapons against insects have proved to be double-edged swords. We have used pesticides to control insects that eat our crops, transmit diseases such as malaria, or just annoy us around the house or campground. But widespread use of these poisons, which are not specific for the intended targets, has also produced some colossal environmental problems, which we'll examine in Chapter 17. Our focus here is the evolutionary outcome of introducing these chemicals into the environments of insects.

Whenever a new type of pesticide is used to control agricultural pests, the story is usually the same[. ...] small amount of the poison[...] But subsequent sprayings a[...] increasing the amount of th[...] mention environmental) co[...] to a different pesticide until[...]

Just what is happen[...] relatively f[ew ...]

Make changes in a virtual environment and observe the effects on a population of leafhoppers in Web/CD Activity 12C/ The Process of Science.

[...]now[...] [t]he case[...] that destro[...] of the inse[...] viduals to[...] for pestici[...] tion of pe[...] lation incr[...] in its environment.

This example of insect a[...] points about natural selecti[...] process of editing than it is a[...]

Watch, practice, and learn with multimedia,

Multimedia for every topic. Media flags in the text guide you to media activities—at least one for every main topic—that let you see animations, perform virtual experiments, and more.

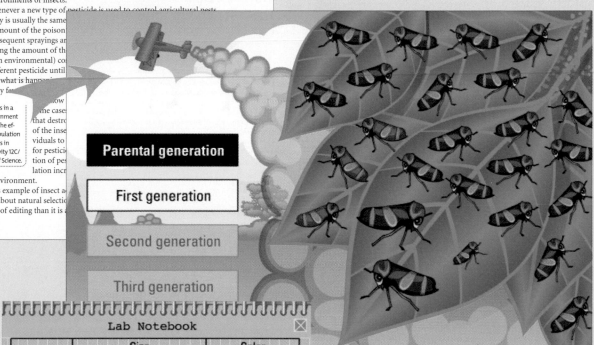

Parental generation

First generation

Second generation

Third generation

Lab Notebook

	Size			Color	
	Large	Medium	Small	Black	Orange
Parental generation					
First generation					
Second generation					
Third generation					
Conclusions:					Check answer

using the CD-ROM or website—your choice.

Core content in both formats. You can access more than 150 media activities, 400 chapter quiz questions, and a glossary with audio pronunciations using either the Essential Biology Place Website or the CD-ROM included in this book. In addition, web links, news, practice test questions, and more are available on the website. For more information, see the insert at the beginning of the book.

CheckPoint with every topic. A CheckPoint at the end of each main section of the chapter lets you assess your understanding.

R allele. In contrast to the Hardy-Weinberg equilibrium of a nonevolving population, we now have the changing gene pool of an evolving population.

One of the products of the modern synthesis was a definition of evolution that is based on population genetics: *Evolution is a generation-to-generation change in a population's frequencies of alleles.* Because this describes evolution on the smallest scale, it is sometimes referred to more specifically as **microevolution.**

CheckPoint

1. What is the smallest biological unit that can evolve?
2. Define microevolution.
3. Which term in the Hardy-Weinberg formula ($p^2 + 2pq + q^2 = 1$) corresponds to the frequency of individuals who have no alleles for the disease PKU?
4. Which of the following variations in a human population is the best example of polymorphism: height, ABO blood group, number of fingers, or math proficiency?
5. Which process, mutation or sexual recombination, results in most of the generation-to-generation variability in human populations?

Answers: 1. A population **2.** Microevolution is a change in a population's frequencies of alleles. **3.** p^2 **4.** Blood group **5.** Sexual recombination

Evolution Link. As the capstone of each chapter, an Evolution Link helps you connect the chapter's subject to the overarching theme of evolution.

chromosome, usually on the X chromosome **3.** All female offspring will be heterozygous ($X^+ X$), with red eyes; all male offspring will be white-eyed ($X^c Y$). **4.** $\frac{1}{4}$ ($\frac{1}{2}$ chance the child will be male $\times \frac{1}{2}$ chance that he will inherit the X carrying the disease allele)

Evolution Link: The Telltale Y Chromosome

The Y chromosome of human males is only about one-third the size of the X chromosome and carries only $\frac{1}{100}$ as many genes. As mentioned earlier, most of the Y genes seem to function in maleness and male fertility and are not present on the X. In prophase I of meiosis, only two tiny regions of the X and Y chromosomes can cross over (recombine). Crossing over requires that the DNA in the recombining regions line up and match very closely, and for the human X and Y chromosomes, this can only happen at their tips.

Nevertheless, biologists believe that X and Y were once a fully homologous pair, having evolved from a pair of autosomes about 300 million years ago. Since that time, four major episodes of change, the most recent about 40 million years ago, have rearranged pieces of the Y chromosome in a way that prevents the matching required for recombination with the X. Much of the reshuffling seems to have resulted from inversions (see Figure 7.24c).

Figure 8.35 A Lemba tribesman. DNA sequences from the Y chromosomes of Lemba men suggest that the Lemba are descended from ancient Jews.

Chapter Review. The summary gives you a quick review of the chapter along with a guide to the resources on the website and CD-ROM.

Chapter Review

Summary of Key Concepts

Overview: Why and How Genes Are Regulated

- Gene expression—the flow of information from genes to proteins—is subject to control, mainly by the turning on and off of genes. Cells become specialized in structure and function because only certain genes of the genome are expressed. In eukaryotic cells, there are multiple possible control points in the pathway of gene expression. However, the most important control point in both eukaryotes and prokaryotes is at gene transcription.
- **Web/CD Activity 10A** *Control of Gene Expression Overview*

Gene Regulation in Bacteria

- **The *lac* Operon** In prokaryotes, genes for enzymes with related functions are often controlled together by being grouped into regulatory units called operons. Regulatory proteins bind to control sequences in the DNA and turn operons on or off in response to environmental changes. The *lac* operon, for example, produces enzymes that break down lactose only when lactose is present.
- **Web/CD Activity 10B** *The lac Operon in E. Coli*

- **Other Kinds of Operons** While the repressor for the *lac* operon is innately active, some operons, such as those that affect the synthesis of certain amino acids, have repressors that are innately *inactive*. The activation of still other operons depends on activator proteins.

Self-Quiz

1. Which of the following shows the effects of a density-dependent limiting factor?
   a. A forest fire kills all the pine trees in a patch of forest.
   b. Early rainfall triggers the explosion of a locust population.
   c. Drought decimates a wheat crop.
   d. Silt from logging kills half the young salmon in a stream.
   e. Rabbits multiply, and their food supply begins to dwindle.

2. With regard to its percent increase, a population that is growing logistically
   a. grows fastest when density is lowest.
   b. has a high intrinsic rate of increase.
   c. grows fastest at an intermediate population density.
   d. grows fastest as it approaches carrying capacity.
   e. is always slowed by density-independent factors.

3. Pine trees in a forest tend to shade and kill pine seedlings that sprout
ne

Self-Quiz in book and media. Try the self-quiz questions at the end of each chapter, and go to the website or CD-ROM for twenty additional questions on each chapter. Test Flight on the website lets you prepare for a test on multiple chapters by building a practice test from a database of hundreds of questions.

QUIZ Chapter 7

1. Which statement regarding meiosis is correct?

○ Crossing over occurs during prophase II.
○ DNA is duplicated once and centromeres are split once
○ Haploid cells become diploid during this process.
○ Centromeres are split and sister chromatids separate during anaphase I.
○ DNA is duplicated twice and cytokinesis occurs once.

2. Which of the following is a function of meiosis?

○ asexual reproduction
○ reduction of the chromosome number of daughter cells to half that of the parent cell
○ Production of a mature organism from a zygote
○ replacement of dead or dying cells
○ growth and development of an individual organism

The Process of Science

1. A population of snails has recently become established in a new region. The snails are preyed on by birds that break the snails open on rocks, eat the soft bodies, and leave the shells. The snails occur in both striped and unstriped forms. In one area, researchers counted both live snails and broken shells. Their data are summarized here:

	Striped	Unstriped
Living	264	296
Broken	486	377

Based on these data, which snail form is more subject to predation by birds? Predict how the frequencies of striped and unstriped individuals might change over the generations.

2. Explore the Galápagos Islands and other places Darwin visited in *The Voyage of the Beagle.* Conduct virtual experiments on an evolving population of leafhoppers in *Effects of Environmental Changes on a Population.* Both activities are available in The Process of Science section of Chapter 1 on the website and CD-ROM.

Biology and Society

To what extent are humans in a technological society exempt from natural selection? Explain your answer.

Process of Science questions. The Process of Science questions challenge you to think scientifically about the questions of biology.

Biology and Society questions. Biology and Society questions invite you to apply the biology you've learned to evaluate ethical or policy-related issues you're likely to read about or vote on at the polls.

Supplements for the Instructor

Benjamin Cummings Digital Library for *Essential Biology*
(0-8053-7395-0)
Features virtually all the illustrations from the book (approximately 400 images), tables, and selected animations for lecture presentation. Edit labels, import illustrations and photos from other sources, and export figures into other programs, including PowerPoint.

Instructor's Guide to Text and Media (0-8053-7402-7)
Edward J. Zalisko, Blackburn College
Provides objectives, lecture outlines, and references to relevant supplements and media. Includes tips for using the extensive *Essential Biology* media package, with many suggestions for first-time instructors.

Transparency Acetates (0-8053-7404-3)
Includes virtually all of the art and tables from the text.

Test Bank (0-8053-7396-9)
Eugene J. Fenster, Longview Community College
Features factual, conceptual, and application-oriented multiple-choice questions.

TestGen EQ (0-8053-7412-4)
Allows the instructor to view and edit electronic questions from the Test Bank, export the questions to tests, and print them in a variety of formats.

Course Management Systems
Includes WebCT and Blackboard for *Essential Biology*. For more information, please visit http://cms.awlonline.com/.

The Essential Biology Place Website
www.essentialbiology.com
In addition to the items for students listed below, the website includes these materials for instructors:

- Syllabus Manager—Instructors can create a complete on-line syllabus, including assignments, projects, and due dates.
- Instructor Resources—Art, animations, and tables for instructors to use in lecture, plus an electronic version of the *Instructor's Guide to Text and Media* and links to photo resources.

BiologyLabs On-Line Instructor's Resource Site
http://biologylab.awlonline.com
Features answers for each lab in **BiologyLabs On-Line** (see description below). The instructor's site also has additional background material, including suggestions for tailoring the labs and assignments to different levels of difficulty.

Supplements for the Student

The Essential Biology Place Website and CD-ROM
(Website 0-8053-6631-8; CD-ROM 0-8053-6630-X)
Includes more than 150 media activities, 400 chapter quiz questions, hundreds of practice test questions, and annotated web links.

- Activities—More than 130 interactive exercises and animations help students understand complex biological processes.
- The Process of Science—Activities that engage students in scientific inquiry.
- Quizzes—Twenty multiple-choice questions per chapter encourage students to test their understanding of each chapter.
- Evolution Link—Links to websites, plus questions that reinforce the connection between evolution and each chapter.
- Test Flight—Students build tests from a database of multiple-choice questions, allowing them to test themselves on more than one chapter at a time.
- Glossary—Definitions of boldface terms with selected audio pronunciations.

Go to www.essentialbiology.com to explore the media for *Essential Biology*. See the insert before Chapter 1 for more information.

BiologyLabs On-Line (0-8053-7443-4)
http://biologylab.awlonline.com
Robert A. Desharnais, California State University, Los Angeles, and Jeffrey R. Bell, California State University, Chico
BiologyLabs On-Line allows students to learn biological principles by designing and conducting simulated experiments on-line. The labs are available for sale separately or packaged together in a combined 10-pack.

Student Study Guide (0-8053-7403-5)
Edward J. Zalisko, Blackburn College
Includes study tips, content-organizing tables, multiple-choice questions, matching questions, true/false questions, questions about text figures, analogy questions, key terms, word roots, and crossword puzzles.

Laboratory Investigations for Biology (0-8053-0922-5)
Jean Dickey, Clemson University
Includes 20 carefully designed labs in investigative, traditional, and observational formats. This lab manual can be customized—ask your sales representative for details.

Brief Contents

Contents

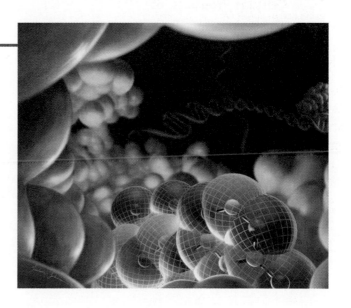

CHAPTER 4

A Tour of the Cell 58

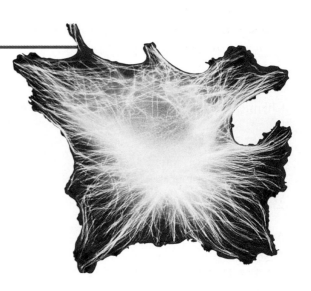

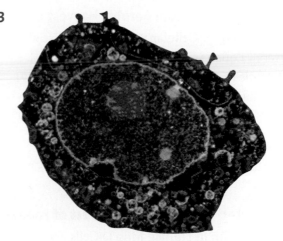

CHAPTER 5

Cellular Respiration: Harvesting Chemical Energy 83

CHAPTER 6

Photosynthesis: Converting Light Energy to Chemical Energy 103

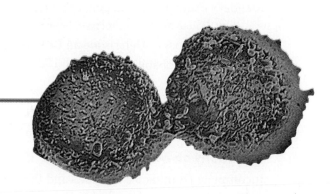

UNIT TWO Genetics

CHAPTER 7

The Cellular Basis of Reproduction and Inheritance 120

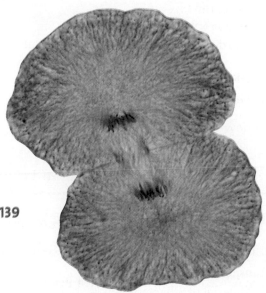

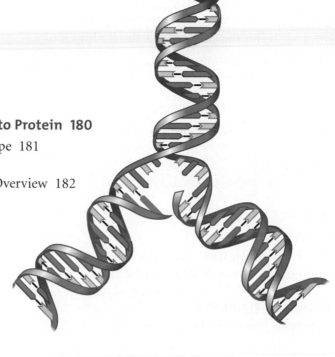

CHAPTER 10

The Control of Gene Expression 198

CHAPTER 11

DNA Technology and the Human Genome 219

UNIT THREE Evolution

CHAPTER 12

How Populations Evolve 250

CHAPTER 13

How Biological Diversity Evolves 276

CHAPTER 14

The Evolution of Microbial Life 303

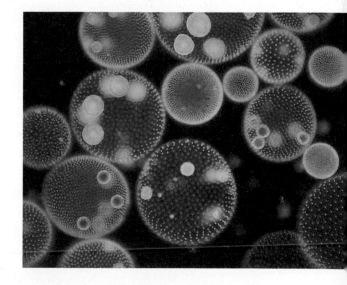

UNIT FOUR   Ecology

CHAPTER 17

The Biosphere: Earth's Diverse Environments 392

CHAPTER 18

Population Ecology 418

Essential Biology

Introduction: Biology Today

Biologists have identified about 1.5 million

species of living organisms.  All organ-

isms share a common chemical language for their genetic

material, DNA.  In 2000, scientists announced a

rough draft for the sequence of all 3 billion letters in the

 DNA of a human. The human population of

Earth doubled in the past 40 years, passing the 6 billion

mark in 1999.

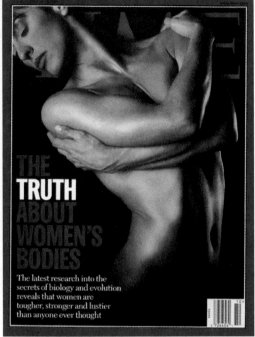

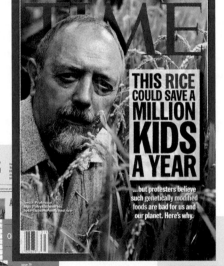

We are living in the golden age of biology. The largest and best-equipped community of scientists in history is beginning to solve biological puzzles that once seemed unsolvable. We are moving ever closer to understanding how a single cell becomes a plant or animal; how plants trap solar energy and store that energy in food; how organisms network in biological communities such as forests and coral reefs; and how the great diversity of life on Earth evolved from the first microbes. Exploring life has never been more exhilarating. Welcome to the big adventure of the twenty-first century!

Modern biology is as important as it is inspiring, with exciting break-throughs changing our very culture. Genetics and cell biology are revolutionizing medicine and agriculture. Molecular biology is providing new tools for anthropology, helping us trace the origin and dispersal of early humans. Ecology is helping us evaluate environmental issues, such as the causes and consequences of global warming. Neuroscience and evolutionary biology are reshaping psychology and sociology. Biology is even entering the legal system, as terms such as *DNA fingerprinting* work their way into our vocabulary. These are just a few examples of how biology is weaving into the fabric of society as never before. It is no wonder that biology is daily news, as illustrated in Figure 1.1.

So, whatever your reasons for taking this course, even if only to meet your college's science requirement, you'll soon discover that this is the best time ever to study biology. To help you get started, this first chapter of *Essential Biology* surveys the scope of biology, introduces evolution as the theme that unifies all of biology, and sets the study of life in the broader context of science as a process of inquiry.

Figure 1.1 A small sample of biology in the news.

The Scope of Biology

Biology is the scientific study of life. It's a huge subject that gets bigger every year because of the great discovery explosion. We can think of biology's enormous scope as having two major dimensions. First, life is structured on a size scale ranging from the molecular to the global. The second dimension of biology's scope stretches across the enormous diversity of life on Earth, now and throughout life's history.

Life at Its Many Levels

Take a tour that goes from outer space to inside an atom in The Process of Science activity *Hyperspace Tour,* available on the *Essential Biology Place* website and CD-ROM.

In *Essential Biology,* we will probe life all the way down to the submicroscopic scale of molecules such as DNA, the chemical responsible for inheritance. At the other extreme of biological size and complexity, our exploration will take us all the way to the global scale of the entire biosphere, which consists of all the environments on Earth that support life—soil, oceans, lakes and other bodies of water, and the inner atmosphere. Figure 1.2 takes us into this world of life through a series of views that zooms in a thousand times with each step. We start by approaching the biosphere from space. A thousand times closer, we can recognize Manhattan, with its Central Park. Another thousand-power jump puts us in a wooded area of the park. The next scale change takes us up close to the beautiful creature that graces the cover of this textbook. It's the caterpillar (larval stage) of an insect called the spicebush swallow-tail butterfly. The "eyespot" markings may trick some predators such as birds into perceiving the caterpillar as a snake instead of something to eat. Our next size change vaults us into the caterpillar to see that its body consists of microscopic units called cells. In each caterpillar cell, you can see a nucleus, the part of the cell that contains all the genes that this insect inherited from its parents. And our last thousand-power change zooms us into the nucleus, where we can behold the molecular architecture of the DNA that makes up the genes. From the interactions within the biosphere to the molecular machinery within cells, biologists are investigating life at its many levels. Let's take a closer look here at just two biological levels near opposite ends of the size scale: ecosystems and cells.

A satellite view of Earth

Approaching Central Park (the red rectangle)

A Central Park woodland

Spicebush swallowtail caterpillar

Cells Nucleus within cell

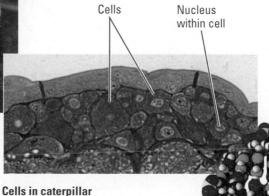

Cells in caterpillar

DNA

Figure 1.2 Zooming in on life. Each scale change in this series takes us a thousand times closer. Biologists explore life at levels ranging from the biosphere to the molecules that make up cells.

Ecosystems Life does not exist in a vacuum. Each organism (living thing) interacts continuously with its environment, which includes other organisms as well as nonliving components of the environment. The roots of a tree, for example, absorb water and minerals from the soil. Leaves take in a gas called carbon dioxide from the air. Chlorophyll, the green pigment of the leaves, absorbs sunlight, which drives the plant's production of sugar from carbon dioxide and water. This food production is called photosynthesis. The tree also releases oxygen to the air, and its roots help form soil by breaking up rocks. Both organism and environment are affected by the interactions between them. The tree also interacts with other living things, including microorganisms (microscopic organisms) in the soil that are associated with the plant's roots and animals that eat its leaves and fruit. We are, of course, among those animals, as is our bookcover caterpillar.

Ecology is the branch of biology that investigates these relationships between organisms and their environments. Ecologists use the term *ecosystem* for the biological level consisting of all the organisms and nonliving factors affecting life in a particular area, such as a wooded glen or grassland.

The dynamics of any ecosystem depend on two main processes. One is the cycling of nutrients. For example, minerals that plants take up from the soil will eventually be recycled to the soil by microorganisms that decompose leaf litter and other organic refuse. The second major process in an ecosystem is the flow of energy from sunlight to producers and then on to consumers. Producers are photosynthetic organisms, such as plants; consumers are the organisms, such as animals, that feed on plants, either directly (by eating plants) or indirectly (by eating animals that ate plants). Figure 1.3 depicts this solar powering of ecosystems.

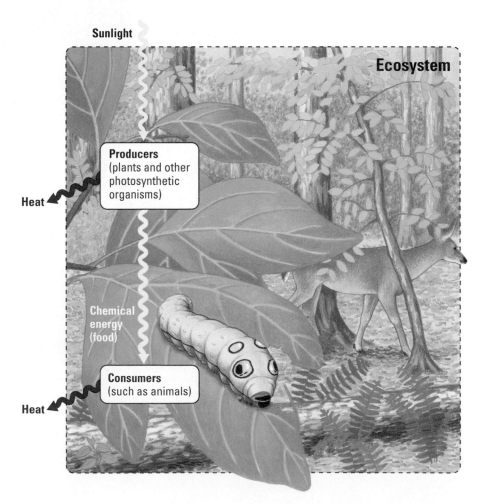

Figure 1.3 Energy flow in an ecosystem. Living is work, and work requires that organisms obtain and use energy. Most ecosystems are solar powered. The energy that enters an ecosystem as sunlight exits as heat, which all organisms dissipate to their surroundings whenever they perform work.

The biosphere is enriched by a great variety of ecosystems. A tropical rain forest in South America is an ecosystem. Very different from tropical forests are the ecosystems we call deserts, such as those of the southwestern United States. A coral reef, such as the Great Barrier Reef off the eastern coast of Australia, is an ecosystem, and so is any small pond that may exist on your campus or in your city. Even a woodland patch in New York's Central Park qualifies as an ecosystem, small and artificial as it is by forest standards and disrupted as it is by human visitors. The fact is, humans are organisms that now have some presence, often disruptive, in all ecosystems. And the collective clout of 6 billion humans and their machines impacts the entire biosphere, which is the global ecosystem, the sum of dynamic processes in all ecosystems. For example, our fuel-burning, forest-chopping actions are changing the atmosphere and the planet's climate in ways that we do not yet fully understand, though we already know that this global vandalism jeopardizes the diversity of life on Earth.

Cells and Their DNA   Let's downsize now from ecosystems to cells. The cell has a special place in the hierarchy of biological organization. It is the lowest level of structure that can perform *all* activities required for life, including the capacity to reproduce.

All organisms are composed of cells. They occur singly as a great variety of unicellular organisms, mostly microscopic. And cells are also the subunits that make up the tissues and organs of plants, animals, and other multicellular organisms. In either case, the cell is the organism's basic unit of structure and function. The ability of cells to divide to form new cells is the basis for all reproduction and for the growth and repair of multicellular organisms, including humans.

We can distinguish two major kinds of cells: prokaryotic and eukaryotic. The prokaryotic cell is much simpler and usually much smaller than the eukaryotic cell. The cells of the microorganisms we commonly call bacteria are prokaryotic. All other forms of life, including plants and animals, are composed of eukaryotic cells. In contrast to the prokaryotic cell, the eukaryotic cell is subdivided by internal membranes into many different functional compartments, or organelles. For example, the nucleus, the largest organelle in most eukaryotic cells, houses the DNA, the heritable material that directs the cell's many activities. Prokaryotic cells also have DNA, but it is not packaged within a nucleus (Figure 1.4).

Though very different in structural complexity, prokaryotic and eukaryotic cells have much in common at the molecular level. Most importantly, all cells use DNA as the chemical material of genes, the units of inheritance that transmit information from parents to offspring. Of course, bacteria and humans inherit different genes, but that information is encoded in a chemical language common to all organisms. This language of life has an alphabet of just four letters, DNA's four molecular building blocks, their chemical names abbreviated A, G, C, and T (Figure 1.5).

An average-sized gene may be hundreds or thousands of chemical letters long. That gene's meaning to the cell is written in its specific sequence of these letters, just as the message of a sentence is encoded in its arrangement of letters selected from the 26 letters of the English alphabet. One gene may be translated as "Build a blue pigment in a colored bacterial cell." Another gene may mean "Make human insulin in this cell." Insulin, a hormone secreted into the bloodstream by cells of an organ called the pancreas, helps regulate your body's use of sugar as a fuel. In people who have certain forms of the disease diabetes, insulin is in short supply. The loss of

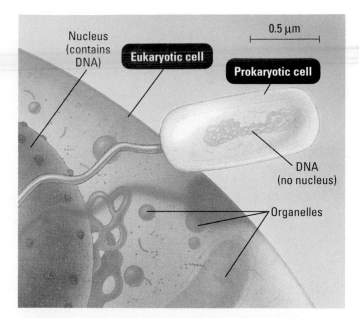

Figure 1.4 Two main kinds of cells: prokaryotic and eukaryotic. The scale bar represents 0.5 micrometers (μm). There are 1000 μm in a millimeter (mm). (You can review metric measurements in Appendix A.)

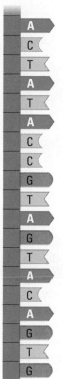

Figure 1.5 The language of DNA. This diagram uses simple shapes and letters to symbolize the four kinds of chemical building blocks that are chained together in DNA. A gene is a segment of DNA composed of hundreds or thousands of these building blocks, of which we see only a short stretch here. Each gene encodes information in its specific sequence of the four chemical letters that are universal in the language of life.

Figure 1.6 DNA technology in the drug industry. This employee of a biotechnology company is monitoring a tank of microorganisms that are genetically engineered to produce an ingredient of hepatitis B vaccine.

normal regulation of blood sugar can be debilitating or even deadly. Many people with diabetes now bring their sugar levels back into balance by injecting themselves with insulin produced by genetically engineered bacteria. The engineering involves transplanting a gene for insulin production from a human cell into a bacterial cell. Each time the bacterial cell reproduces, it copies its new insulin gene, which is inherited by offspring along with the bacterial genes. Soon there is a massive colony of insulin-producing bacterial cells. This was one of the earliest products of a new biotechnology that has transformed the pharmaceutical industry and extended millions of lives (Figure 1.6). And it is only possible because biological information is written in the universal chemical language of DNA.

The entire "book" of genetic instructions that an organism inherits is called its genome. The nucleus of each human cell packs a genome that is 3 billion chemical letters long. In the summer of 2000, scientists tabulating the sequence of these letters announced a "rough draft" of the human genome (see Figure 1.1). The press and world leaders acclaimed this international achievement as the greatest scientific triumph ever. But unlike past cultural zeniths, such as the landing of Apollo astronauts on the moon, the sequencing of the human genome is more a commencement than a climax. As the quest continues, biologists will learn the functions of thousands of genes and how their activities are coordinated in the development of an organism. It is a striking example of human curiosity about life at its many levels.

Life in Its Diverse Forms

Figure 1.7 A small sample of biological diversity.
Shown here are just some of the tens of thousands of species in the butterfly and moth collection at the National Museum of Natural History in Washington, D.C. As diverse as the species are, they are all variations on a common anatomical theme. One of biology's major goals is to explain how such diversity arises while also accounting for characteristics common to different species.

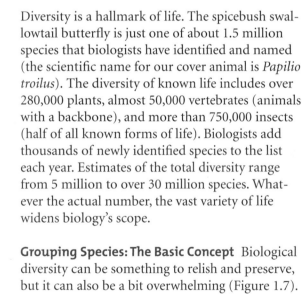

Diversity is a hallmark of life. The spicebush swallowtail butterfly is just one of about 1.5 million species that biologists have identified and named (the scientific name for our cover animal is *Papilio troilus*). The diversity of known life includes over 280,000 plants, almost 50,000 vertebrates (animals with a backbone), and more than 750,000 insects (half of all known forms of life). Biologists add thousands of newly identified species to the list each year. Estimates of the total diversity range from 5 million to over 30 million species. Whatever the actual number, the vast variety of life widens biology's scope.

Grouping Species: The Basic Concept Biological diversity can be something to relish and preserve, but it can also be a bit overwhelming (Figure 1.7). Confronted with complexity, humans are inclined to categorize diverse items into a smaller number of groups. Grouping species that are similar is natural for us. We may speak of squirrels and butterflies, though we recognize that many different species belong to each group. We may even sort groups into broader categories, such as rodents (which include squirrels) and insects (which include butterflies). Taxonomy, the branch of biology that names and classifies species, formalizes this hierarchical ordering according to a scheme you will learn about in Chapter 13. Here we consider only kingdoms and domains, the broadest units of classification.

The Three Domains of Life Until the past decade, most biologists divided the diversity of life into five main groups, or kingdoms. (The most familiar two are the plant and animal kingdoms.) But new methods, such as comparisons of DNA among organisms, has led to an ongoing reassessment of the number and boundaries of kingdoms. Various classification schemes are now based on six, eight, or more kingdoms. But as the debate continues on the kingdom level, there is broader consensus that the kingdoms of life can now be assigned to three even higher levels of classification called domains.

To help you understand kingdoms and domains, go to Web/CD Activity 1A, *Classification Schemes.*

The three domains are named Bacteria, Archaea, and Eukarya (Figure 1.8). The first two domains, Bacteria and Archaea, recognize two very different groups of organisms that have prokaryotic cells. In the five-kingdom system, these prokaryotes were combined in a single kingdom. But newer evidence suggests that the organisms known as archaea are actually more closely related to eukaryotes than they are to bacteria.

All the eukaryotes (organisms with eukaryotic cells) are grouped into at least four kingdoms in the domain Eukarya. Kingdom Protista consists of eukaryotic organisms that are generally single-celled—for example, the microscopic protozoans, such as the amoebas. Many biologists extend the boundaries of the kingdom Protista to include certain multicellular forms, such as seaweeds, that seem to be closely related to the unicellular protists. (Other biologists split the protists into multiple kingdoms.) The remaining three kingdoms—Plantae, Fungi, and Animalia—consist of multicellular eukaryotes. These three kingdoms are distinguished partly by their modes of nutrition. Plants produce their own sugars and other foods by photosynthesis. Fungi are mostly decomposers that absorb nutrients by breaking down dead organisms and organic wastes, such as leaf litter and animal feces. Animals obtain food by ingestion, which is the eating and digesting of other organisms. It is, of course, the kingdom to which we belong.

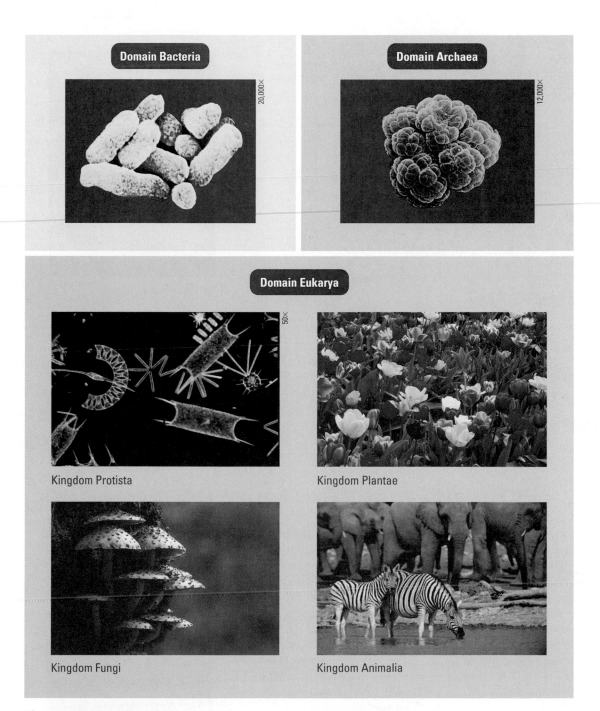

Figure 1.8 **The three domains of life.**

Unity in the Diversity of Life If life is so diverse, how can biology have any unifying themes? What, for instance, can a tree, a mushroom, and a human possibly have in common? As it turns out, a great deal! Underlying the diversity of life is a striking unity, especially at the lower levels of biological organization. We have already seen one example: the universal genetic language of DNA. That fundamental fact connects all kingdoms of life, even uniting prokaryotes such as bacteria with eukaryotes such as humans. And among eukaryotes, unity is evident in many of the details of cell structure (Figure 1.9). Above the cellular level, however, organisms are so variously adapted to their ways of life that describing and classifying biological diversity remains an important goal of biology (see Figure 1.7). What can account for this combination of unity and diversity in life? The scientific explanation is the biological process called evolution.

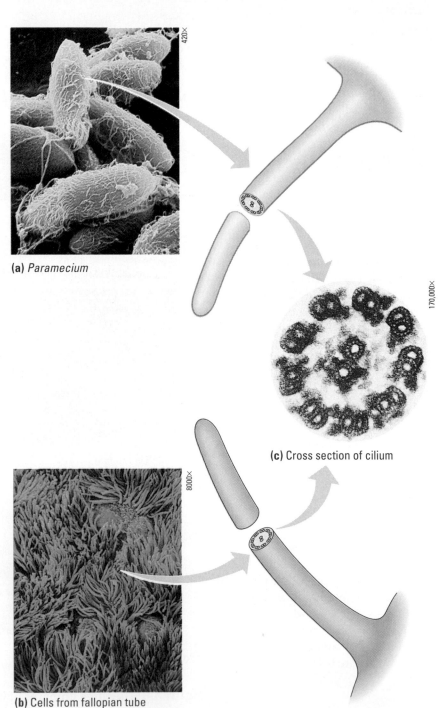

(a) *Paramecium*

(c) Cross section of cilium

(b) Cells from fallopian tube

Figure 1.9 An example of unity underlying the diversity of life: the universal architecture of eukaryotic cilia. Cilia, extensions of cells that function in locomotion, occur in eukaryotic organisms as diverse as the protozoan *Paramecium* and the human. (a) The cilia of *Paramecium* propel the cell through water. (b) The cells that line the fallopian tubes (oviducts) of human females are also equipped with cilia. These cilia sweep egg cells from a woman's ovary to her uterus. (c) A powerful instrument called an electron microscope reveals a common pattern of tubules in the cilia of *Paramecium,* humans, and a diversity of other eukaryotes. This micrograph (photo taken with a microscope) represents this universal architecture in a cross-sectional view of a cilium (singular).

Evolution: Biology's Unifying Theme

The history of life, as documented by fossils and other evidence, is a saga of a restless Earth billions of years old, inhabited by a changing cast of living forms (Figure 1.10). Life evolves. Just as an individual has a family history, each species is one twig of a branching tree of life extending back in time through ancestral species more and more remote. Species that are very similar, such as the brown bear and the polar bear, share a common ancestor that represents a relatively recent branch point on the tree of life (Figure 1.11). But through an ancestor that lived much farther back in time, all bears are also related to squirrels, humans, and all other mammals. Hair and milk-producing mammary glands are just two of a long list of uniquely mammalian traits. It is what we would expect if all mammals descended from a common ancestor, a proto-typical mammal. And mammals, birds, reptiles, and all other vertebrates share a common ancestor even more ancient. Evidence of a still broader relationship can be found in such similarities as the matching machinery of all eukaryotic cilia (see Figure 1.9). Trace life back far enough, and there are only fossils of the primeval prokaryotes that inhabited Earth over 3 billion years ago. We can recognize some of their vestiges in our own cells. All of life is connected. And the basis for this kinship is evolution, the process that has transformed life on Earth from its earliest beginnings to the extensive diversity we see today. Evolution is the theme that unifies all of biology.

Figure 1.10 Digging into the past. Paleontologist (fossil specialist) Robert Bakker displays the leg bone of a dinosaur unearthed in Wyoming. The fossil record supports other evidence that life has changed dramatically over Earth's long history.

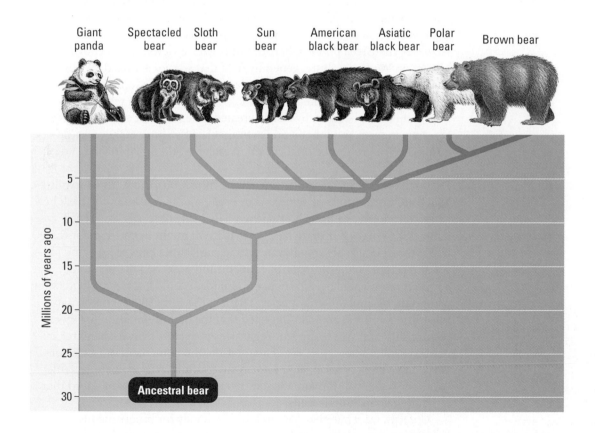

Figure 1.11 An evolutionary tree of bears. The tree is based on both the fossil record and evaluation of relationships among living bears by comparing genetic information.

The Darwinian View of Life

The evolutionary view of life came into focus in 1859 when British biologist Charles Darwin published *The Origin of Species* (Figure 1.12). His book developed two main points. First, Darwin marshaled the available evidence in support of the evolutionary view that species living today descended from ancestral species (we'll examine some of the evidence for evolution in Chapter 12). Darwin called this process "descent with modification." It is an insightful euphemism for "evolution," as it captures the duality of life's unity (descent) and diversity (modification). In the Darwinian view, for example, the diversity of bears is based on different modifications of a common ancestor from which all bears descended. As the second main point in *The Origin of Species,* Darwin proposed a mechanism for descent with modification. He called this process natural selection. Before we examine how natural selection works and how Darwin derived the idea, let's place the Darwinian revolution in its historical setting.

Darwin's Cultural and Scientific Context

The Darwinian view of life contrasts sharply with one that sees a relatively young Earth populated by millions of unrelated species. *The Origin of Species* was truly radical for its time. Not only did it challenge prevailing scientific views; it also shook the deepest roots of Western culture.

The Idea of Fixed Species Like many concepts in science, the basic idea of biological evolution can be traced back to the ancient Greeks. About 2500 years ago, the Greek philosopher Anaximander promoted the idea that life arose in water and that simpler forms of life preceded more complex ones. However, the Greek philosopher Aristotle, whose views had an enormous impact on Western culture, generally held that species are fixed, or permanent, and do not evolve. Judeo-Christian culture fortified this idea with a literal interpretation of the biblical book of Genesis, which tells the story of each form of life being individually created. The idea that all living species are static in form and inhabit an Earth that is only about 6000 years old dominated the cultural climate of the Western world for centuries.

Lamarck and Adaptive Evolution In the mid-1700s, the study of fossils, which are the imprints or remnants of organisms that lived in the past, began to take form as a branch of science. The study of fossils led French naturalist Georges Buffon to suggest that Earth might be much older than 6000 years. He also observed some telling similarities between specific fossils and certain living animals. In 1766, Buffon proposed the possibility that a species represented by a particular fossil form could be an ancient version of a group of similar living species. Then, in the early 1800s, French naturalist Jean Baptiste Lamarck suggested that the best explanation for this relationship of fossils to current organisms is that life evolves. Lamarck explained evolution as a process of adaptation, the refinement of characteristics that equip organisms to perform successfully in their environments. An example of evolutionary adaptation is the powerful beak of a bird that feeds by cracking tough seeds.

Unfairly, we remember Lamarck today mainly for his erroneous view of how adaptations evolve. He proposed that by using or not using its body parts, an individual develops certain characteristics, which it passes on to its offspring. Lamarck's proposal is known as the *inheritance of acquired characteristics.* For example, this idea would view the strong beaks of seed-cracking birds as the cumulative result of ancestors exercising their beaks

Figure 1.12 Charles Darwin (1809–1882). Darwin and his son William posed for this photograph in 1842. Darwin was already one of Britain's most renowned naturalists, but it was his publication of *The Origin of Species* 17 years later that guaranteed his immortality as the most influential scientist in the development of modern biology. He is buried next to Isaac Newton in London's Westminster Abbey.

during feeding and passing that acquired beak power on to offspring. However, there is no evidence for inheritance of acquired characteristics. A carpenter who builds up strength and stamina through a lifetime of pounding nails with a heavy hammer will not pass enhanced biceps on to children. The incompatibility of modern genetics with Larmarck's idea that acquired characteristics can be inherited obscures the important fact that Lamarck helped set the stage for Darwin by proposing that adaptations evolve as a result of interactions between organisms and their environments.

The Voyage of the *Beagle* Charles Darwin was born in 1809 on the same day as Abraham Lincoln. (In that same year, Larmarck published some of his ideas on evolution.) Even as a boy, Darwin's consuming interest in nature was evident. When he was not reading nature books, he was in the fields and forests fishing, hunting, and collecting insects. His father, an eminent physician, could see no future for a naturalist and sent Charles to the University of Edinburgh to study medicine. But Charles, only 16 years old at the time, found medical school boring and distasteful. He left Edinburgh without a degree and then enrolled at Christ College at Cambridge University, intending to become a minister. At Cambridge, Darwin became the protégé of the Reverend John Henslow, a professor of botany. Soon after Darwin received his B.A. degree in 1831, Professor Henslow recommended the young graduate to Captain Robert FitzRoy, who was preparing the survey ship *Beagle* for a voyage around the world. It was a tour that would have a profound effect on Darwin's thinking and eventually on the thinking of the entire world.

Darwin was 22 years old when he sailed from Great Britain with HMS *Beagle* in December 1831. The main mission of the voyage was to chart poorly known stretches of the South American coastline (Figure 1.13).

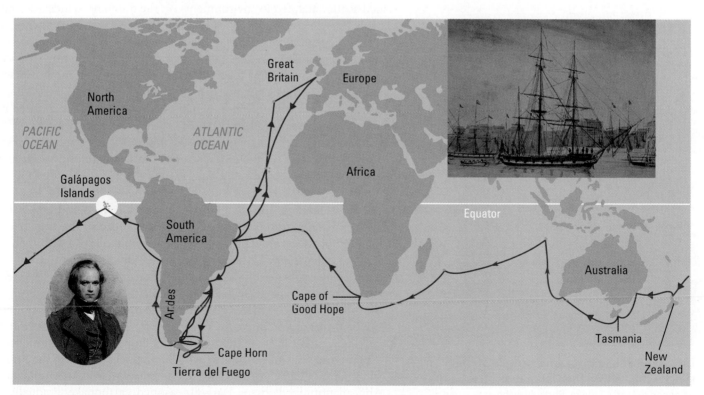

Figure 1.13 The voyage of the *Beagle*. The two insets show the ship and a young Darwin.

While the crew of the ship surveyed the coast, Darwin spent most of his time on shore, observing and collecting thousands of specimens of the native plants and animals of South America. As the ship worked its way around the continent, Darwin observed the various adaptations of organisms that inhabited such diverse environments as the Brazilian jungles, the grasslands of the Argentine pampas, the desolate and frigid lands of Tierra del Fuego near Antarctica, and the towering heights of the Andes.

In spite of their unique adaptations, the plants and animals throughout the continent all had a definite South American stamp, very distinct from the life-forms of Europe. That in itself may not have surprised Darwin. But the plants and animals living in temperate regions of South America seemed more closely related to species living in tropical regions of that continent than to species living in temperate regions of Europe. And the South American fossils Darwin found, though clearly different species from modern ones, were distinctly South American in their resemblance to the living plants and animals of that continent. Despite growing up in the generally antievolutionist climate of the Victorian era, Darwin had a questioning mind. Could his observations mean that the contemporary species owed their South American features to descent from ancestral species on that continent?

Figure 1.14 A marine iguana, an example of the unique species inhabiting the Galápagos. These reptiles dive into the ocean to feed on algae. Partially webbed feet and a flattened tail are two of the adaptations that make marine iguanas such good swimmers.

For a quick tour of the Galápagos Islands, go to Web/CD Activity 1B. To join Darwin at more stops on his voyage, visit The Process of Science section of the website or CD-ROM.

Darwin was particularly intrigued by the geographic distribution of organisms on the Galápagos Islands. These are relatively young volcanic islands about 900 kilometers (540 miles) off the Pacific coast of South America (see Figure 1.13). Most of the animals of the Galápagos live nowhere else in the world, but they resemble species living on the South American mainland (Figure 1.14). It is as though the islands had been colonized by plants and animals that strayed from the mainland and then diversified as they adapted to environments on the different islands. Among the birds Darwin collected on the Galápagos were 13 species of finches. Some were unique to individual islands, while others were distributed on two or more islands that were close together. The unique adaptations of these birds included beaks modified for feeding on certain kinds of foods. Darwin did not appreciate the full significance of the finches he collected until years after returning to Britain. Since then, biologists have applied modern methods of comparing species to reconstruct the evolutionary history of Darwin's finches. The branching of this evolutionary tree traces the "descent with modification" of the 13 finch species from a common South American ancestor (Figure 1.15).

The New Geology During the *Beagle*'s long sails between ports, Darwin, in spite of his seasickness, managed to do a lot of reading. He was strongly influenced by the recently published *Principles of Geology*, by Scottish geologist Charles Lyell. The book presented the case for an ancient Earth sculpted by gradual geological processes that continue today. In essence, the key to the past is the present. The gradual erosion of a river bed can add up over the millennia to a deep river-carved canyon. A mighty mountain range can be thrust up millimeter by millimeter by earth-moving earthquakes occurring sporadically over thousands or millions of years. Darwin actually experienced such an earthquake while doing field studies in the Andes Mountains of Chile. He also collected fossils of marine (sea) snails at high Andean altitudes. Perhaps, Darwin reasoned, earthquakes gradually lifted the rock bearing those marine fossils from the seafloor. In the context of such experiences, Lyell began to make sense to Darwin.

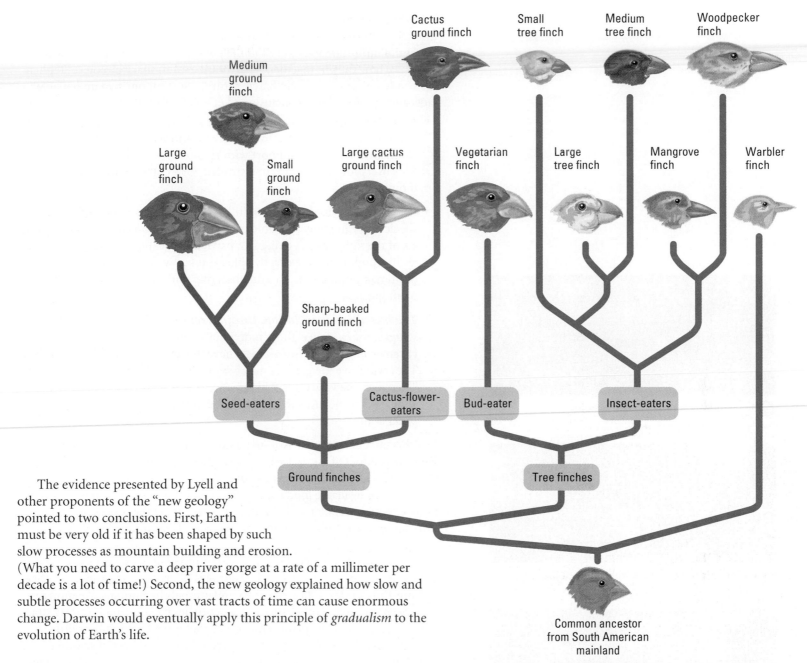

Figure 1.15 labels (clockwise):

Cactus ground finch
Small tree finch
Medium tree finch
Woodpecker finch
Medium ground finch
Large ground finch
Small ground finch
Large cactus ground finch
Vegetarian finch
Large tree finch
Mangrove finch
Warbler finch
Sharp-beaked ground finch

Seed-eaters
Cactus-flower-eaters
Bud-eater
Insect-eaters

Ground finches
Tree finches

Common ancestor from South American mainland

The evidence presented by Lyell and other proponents of the "new geology" pointed to two conclusions. First, Earth must be very old if it has been shaped by such slow processes as mountain building and erosion. (What you need to carve a deep river gorge at a rate of a millimeter per decade is a lot of time!) Second, the new geology explained how slow and subtle processes occurring over vast tracts of time can cause enormous change. Darwin would eventually apply this principle of *gradualism* to the evolution of Earth's life.

Figure 1.15 An evolutionary tree for the 13 species of Galápagos finches. These finch species are all very closely related as descendants of a common ancestor that managed to reach the islands from the South American mainland. Note the diversity of beaks, which are adapted to certain food sources on the different islands.

Natural Selection

Soon after returning to Great Britain in 1836, Darwin started reassessing all that he had observed during the five-year voyage of the *Beagle*. He began to think of adaptation to the environment and the origin of new species as closely related processes. If some geographic barrier—an ocean separating islands, for instance—isolates two populations of a single species, the populations could diverge more and more in appearance as each adapted to local environmental conditions. Over many generations, the two populations could become dissimilar enough to be designated separate species—13 species in the case of the Galápagos finches, with their beak adaptations and other refinements suited to their environments (see Figure 1.15). Darwin realized that explaining such adaptation was the key to understanding descent with modification, or evolution. This focus on adaptation helped Darwin envision his concept of natural selection as the mechanism of evolution.

Darwin's Inescapable Conclusion Darwin synthesized the concept of natural selection from two observations that by themselves were neither profound nor original. Others had the pieces of the puzzle, but Darwin could see how they fit together. As Harvard evolutionary biologist Stephen Jay Gould has put it, Darwin based natural selection on "two undeniable facts and an inescapable conclusion." Let's look at his logic:

- *Fact #1: Overproduction and struggle for existence.* Any population of a species has the potential to produce far more offspring than the environment can possibly support with resources such as food and shelter. This overproduction leads to a struggle for existence among the varying individuals of a population.

- *Fact #2: Individual variation.* Individuals in a population of any species vary in many heritable traits. (A population consists of individuals of the same species living in a particular location.) No two individuals in a population are exactly alike. You know this variation to be true of human populations; careful observers find variation in populations of all species.

→ *The inescapable conclusion: Unequal reproductive success.* In the struggle for existence, those individuals with traits best suited to the local environment will, on average, have the greatest reproductive success: They will leave the greatest number of surviving, fertile offspring. Therefore, the very traits that enhance survival and reproductive success will be disproportionately represented in succeeding generations of a population.

It is this unequal reproductive success that Darwin called natural selection. And the product of natural selection is adaptation, the accumulation of favorable variations in a population over time. Examples are beaks well equipped for available food sources and markings—such as those of our book cover's caterpillar—that reduce predation. As the process of adaptation, natural selection is also the mechanism of evolution (Figure 1.16).

Observing Artificial Selection Darwin found convincing evidence for the power of unequal reproduction in examples of artificial selection, the selective breeding of domesticated plants and animals by humans. We humans have been modifying other species for centuries by selecting breeding stock with certain traits. The plants and animals we grow for food bear little resemblance to their wild ancestors (Figure 1.17). The power of selective breeding is especially apparent in our pets, which have been bred more for fancy than for utility. For example, people in different cultures have customized hundreds of dog breeds as different as German shepherds and chihuahuas, all descendent from one ancestral species of wild dog. Darwin could see that in artificial selection, humans were substituting for the environment in screening the heritable traits of populations.

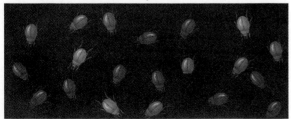

❶ Population with varied inherited traits

❷ Elimination of individuals with certain traits

❸ Reproduction of survivors

❹ Increasing frequency of traits that enhance survival and reproductive success

Figure 1.16 Natural selection. ❶ This imaginary beetle population has colonized a locale where the soil has been blackened by a recent brush fire. Initially, the population varies extensively in the coloration of individuals, from very light gray to charcoal. ❷ For hungry birds that prey on the beetles, it is easiest to spot the beetles that are lightest in color. ❸ The selective predation favors the survival and reproductive success of the darker beetles. Thus, genes for dark color are passed along to the next generations in greater frequency than genes for light color. ❹ Generation after generation, the beetle population adapts to its environment through natural selection.

Cabbage

Brussels sprouts

Cauliflower

Kale

Broccoli

Wild mustard

Kohlrabi

Figure 1.17 An example of artificial selection.
These vegetables (and more) all have a common ancestor in one species of wild mustard. Humans have customized the plant by selecting different parts of the plant to accentuate as food.

Observing Natural Selection If artificial selection could achieve so much change so rapidly, Darwin reasoned, then natural selection should be capable of considerable adaptation of species over hundreds or thousands of generations. There are many examples of natural selection in action. For instance, in Chapter 12, we'll see how resistance to antibiotics by disease-causing bacteria evolves by natural selection. Here, let's take a look at what is perhaps the best-known example of natural selection in action: the evolution of camouflaging coloration in a British moth.

The English peppered moth lives throughout the midland region of England, which is heavily industrialized. There are two varieties of the moth, differing in coloration. The form for which the peppered moth is named is light with splotches of pigment. The other variety is uniformly dark. The moths rest on trees and rocks that are often encrusted by light-colored lichens, which are combinations of fungi and algae. Against this background, light moths are camouflaged, but the dark moths are very conspicuous to birds and other predators (Figure 1.18). Before the Industrial Revolution, dark peppered moths were rare, presumably because most were eaten before they could pass their genes for dark color on to offspring. But industrial pollution darkened the landscape in the late 1800s, mainly by killing the lichens that had covered the rocks and the dark bark of trees. Against this blackened background, it was the dark moths, not the light ones, that were now camouflaged. The relative numbers of dark moths compared to light ones began to increase. By the turn of the century, the moth population in some regions of industrial England consisted almost entirely of dark moths. The same evolutionary phenomenon occurred in hundreds of other species of moths in polluted regions of Europe and North America. (Recent evidence suggests that there are factors in addition to predation by birds that may be involved in the relative success of the light versus dark moths.)

Learn more about evolution, biology's unifying theme, in the Evolution Link on the *Essential Biology Place* website.

Note that adaptation of these moths to their changing environment does not support Lamarck's notion of the inheritance of acquired characteristics.

Figure 1.18 Two varieties of the English peppered moth against light and dark backgrounds.

The environment did not *create* favorable characteristics, but only screened the heritable variations among individuals of a population. Note also from the case of the English peppered moth that natural selection picks traits that work best in the *present* local environment. In the past few decades, light moths have made a strong comeback as pollution controls enabled the countryside to return to natural hues. In Chapter 12, you will learn more about how natural selection works.

Darwin's Long Delay By the early 1840s, Darwin had composed a long essay describing the major features of natural selection. He realized that his evolutionary ideas and evidence would cause a social furor, however, and he delayed publication of his essay. Then, in the mid-1850s, Alfred Wallace, a British naturalist doing fieldwork in Indonesia, developed a concept of natural selection identical to Darwin's. When Wallace sent Darwin a manuscript describing his own ideas on natural selection, Darwin thought, "All my originality will be smashed." However, in 1858, two of Darwin's colleagues presented Wallace's paper and excerpts from Darwin's earlier essay together to the scientific community. With the publication in 1859 of *The Origin of Species,* the short version of his essay, Darwin presented the world with an avalanche of evidence and a strong, logical argument for evolution. He also explained natural selection as the mechanism of descent with modification.

Darwin's book fueled an explosion in biological research and knowledge that continues today. In every chapter of *Essential Biology,* we will link to evolution—the unifying theme of biology.

The Process of Science

Darwin helped make biology a science by seeking natural rather than supernatural causes for the unity and diversity of life. But what *is* science? And how do we tell the difference between science and other ways we try to make sense of nature?

The word *science* is derived from a Latin verb meaning "to know." Science is a way of knowing. It developed from our curiosity about ourselves and the world around us. This basic human drive to understand is manifest in two main scientific approaches: discovery science and hypothesis-driven science. Most scientists practice a combination of these two forms of inquiry.

Discovery Science

Science seeks natural causes for natural phenomena. This limits the scope of science to the study of structures and processes that we can observe and measure, either directly or indirectly with the help of tools such as microscopes that extend our senses. This dependence on observations that other people can confirm demystifies nature and distinguishes science from belief in the supernatural. Science can neither prove nor disprove that angels, ghosts, deities, or spirits, both benevolent and evil, cause storms, rainbows, illnesses, and cures, for such explanations are outside the bounds of science.

Verifiable observations and measurements are the data of discovery science (Figure 1.19). In our quest to describe nature accurately, we discover its structure. In biology, discovery science enables us to describe life at its many levels, from ecosystems down to cells and molecules. Darwin's careful

Figure 1.19 Careful observation and measurement provide the raw data for science. Cornell University's Eloy Rodriguez collects a sample of chemicals extracted from a plant. Analysis of the sample will enable Dr. Rodriguez to determine the types and amounts of different chemical substances in the plant extract.

description of the diverse plants and animals he collected in South America is an example of discovery science, sometimes called descriptive science. A more recent example is the sequencing of the human genome; it's not really a set of experiments, but a detailed dissection and description of the genetic material.

Discovery science can lead to important conclusions based on a type of logic called inductive reasoning. An inductive conclusion is a generalization that summarizes many concurrent observations. "All organisms are made of cells" is an example. That induction was based on two centuries of biologists discovering cells in every biological specimen they observed with microscopes. The careful observations of discovery science and the inductive conclusions they sometimes produce are fundamental to our understanding of nature.

Hypothesis-Driven Science

The observations of discovery science engage inquiring minds to ask questions and seek explanations. Ideally, such investigation consists of what is called the scientific method. As a formal process of inquiry, the scientific method consists of a series of steps, but few scientists adhere rigidly to this prescription. While it would be misleading to reduce science to a stereotyped method, we *can* identify the key element of the method that drives most modern science. It is called hypothetico-deductive reasoning, or more simply hypothesis-driven science (Figure 1.20).

A hypothesis is a tentative answer to some question—an explanation on trial. It is usually an educated guess. We all use hypotheses in solving everyday problems. Let's say, for example, that your flashlight fails during a camp-out. That's an observation. The question is obvious: Why doesn't the flashlight work? A reasonable hypothesis based on past experience is that the batteries in the flashlight are dead.

The *deductive* in hypothetico-deductive reasoning refers to the use of deductive logic to test hypotheses. Deduction contrasts with induction, which, remember, is reasoning from a set of specific observations to reach a general conclusion. In deduction, the reasoning flows in the opposite direction, from the general to the specific. From general premises, we extrapolate to the specific results we should expect if the premises are true. If all organisms are made of cells (premise 1), and humans are organisms (premise 2), then humans are composed of cells (deductive prediction about a specific case).

In the process of science, the deduction usually takes the form of predictions about what outcomes of experiments or observations we should expect *if* a particular hypothesis (premise) is correct. We then test the hypothesis by performing the experiment to see whether or not the results are as predicted. This deductive testing takes the form of "*If . . . then*" logic:

- *Observation:* My flashlight doesn't work.
- *Question:* What's wrong with my flashlight?
- *Hypothesis:* The flashlight's batteries are dead.
- *Prediction: If* this hypothesis is correct,
- *Experiment:* and I replace the batteries with new ones,
- *Predicted result: then* the flashlight should work.

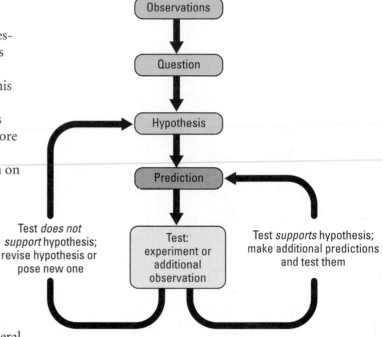

Figure 1.20 The scientific method. Although science rarely conforms exactly to this protocol of steps, inquiry usually involves posing and testing hypotheses.

Let's say the flashlight *still* doesn't work. We can test an alternative hypothesis if new flashlight bulbs are available (Figure 1.21). We could also blame the dead flashlight on campground ghosts playing tricks, but that hypothesis is untestable and therefore outside the realm of science.

A Case Study in the Process of Science

The snakelike appearance of the caterpillar on the front cover of this book is an example of an evolutionary adaptation called mimicry. It is a reasonable hypothesis that such deception reduces the caterpillar's risk of being

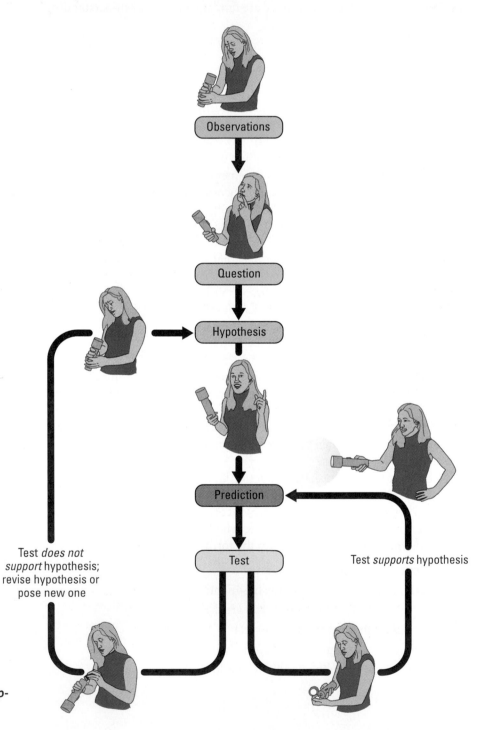

Figure 1.21 Applying the scientific method to a campground problem.

Observations

Question

Hypothesis

Prediction

Test *does not support* hypothesis; revise hypothesis or pose new one

Test

Test *supports* hypothesis

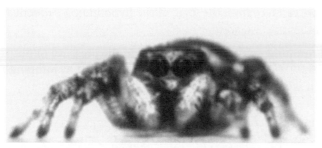
Front view of zebra spider

Figure 1.22 The art of deception: flies that mimic their predators, jumping spiders. Wing markings that resemble spider legs coupled with wing waving that imitates the territorial displays of a zebra spider may reduce the odds of a snowberry fly becoming spider food. Scientists have tested this hypothesis with simple but elegant experiments.

Back view of snowberry fly

eaten, but no one has published a test of this hypothesis. However, two research teams have independently reported clever experiments testing the protective advantage of mimicry in certain fly species that seem to imitate jumping spiders.

This spider mimicry is an unusual case because the flies are apparently imitating one of their predators, the jumping spider. Instead of using webs to net food, jumping spiders capture their prey by stalking them catlike and then pouncing. Flies of various species are favorite foods. Jumping spiders defend their territories against other members of their population by waving their legs, a display that usually causes the trespasser to flee (something like fist waving by territorial humans). Spider-mimicking flies have markings on their wings that look like spider legs, and when approached by jumping spiders, the flies wave their wings (Figure 1.22). The markings and behavior are suggestive of jumping spiders' territorial posturing. But does the mimicry actually turn jumping spiders away?

Biologists tested the hypothesis by measuring the behavior of spiders placed in clear containers with flies. A research team at one university used a black dye to mask the wing markings on some of the putative spider mimics. The scientists reported that jumping spiders pounced on these altered flies more frequently than they did on untreated flies with normal wing markings (Figure 1.23). This is an example of a controlled experiment. Such an experiment is designed to compare an experimental group (flies with their wing markings masked, in this case) with a control group (untreated flies with normal wing markings). Ideally, a control group and an experimental group differ only in the one variable an experiment is designed to test—in our example, the effect of the flies' leglike wing markings on the behavior of jumping spiders. This provides a basis for comparison, enabling researchers to draw conclusions from the effects of their experimental manipulation. Without the normal flies as a control group, there would be no way to tell whether it was absence of wing markings or some other factor that was increasing the pouncing response of the spiders. Perhaps, for example, jumping spiders are just famished at the time of day the experiments were performed.

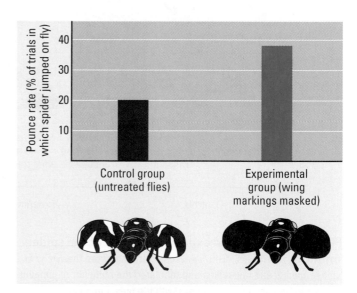

Figure 1.23 One experimental test of the spider mimic hypothesis. The histogram compares how frequently zebra spiders jumped on normal snowberry flies (control group) versus flies with their wing markings masked by black dye (experimental group). Each group consisted of trials with over 30 fly-spider pairs in clear containers. (Data from Monica H. Mather and Bernard D. Roitberg, "A Sheep in Wolf's Clothing: Tephritid Flies Mimic Spider Predators," *Science*, vol. 236, page 309 (April 17, 1987). Copyright © 1987 American Association for the Advancement of Science.)

In this research, we can recognize the logic of the hypothetico-deductive process:

- **Observations:** 1. Jumping spiders wave their legs in the presence of potential competitors. 2. Certain fly species, when approached by jumping spiders, wave their wings, which have markings that resemble spider legs.
- **Question:** What is the function of the flies' wing markings and waving behavior?
- **Hypothesis:** The markings and wing waving increase survival of the flies by causing jumping spiders to flee.
- **Prediction:** *If* this hypothesis is correct,
- **Experiment:** and the flies' wing markings are masked with a dye,
- **Predicted result:** *then* jumping spiders should pounce on the experimental flies more often than they do on control flies with normal wing markings.

The experimental tests supported the hypothesis in this case.

A second research team at another university also found support for the hypothesis in experiments of a different design. These scientists actually performed wing transplants on the flies, using scissors and glue. The flies can wave their wings and even fly normally after such surgery. The fly surgeons transplanted wings between the spider mimics and houseflies, which are about the same size but lack the wing markings and waving behavior. Spider mimics with housefly wings could wave, but the absence of markings resulted in their being eaten as frequently as houseflies. Houseflies with the wings of the spider mimics had the markings, but not the waving behavior, and the spiders nailed them about as often as they did normal houseflies. The researchers also transplanted wings between spider mimics to be sure the surgery didn't affect the results; the mimicry worked almost as well for those with wing transplants as it did for normal mimics with their own wings. The results support the hypothesis that *both* the wing markings *and* the wing-waving behavior contribute to the survival of the mimics in the presence of jumping spiders (Figure 1.24).

Our case study reinforces the important point that scientists must test their hypotheses. Without such testing, ideas about nature, such as speculations on the function of mimicry, are "just so" stories. And explaining that something is true just because "it's so" is not very convincing.

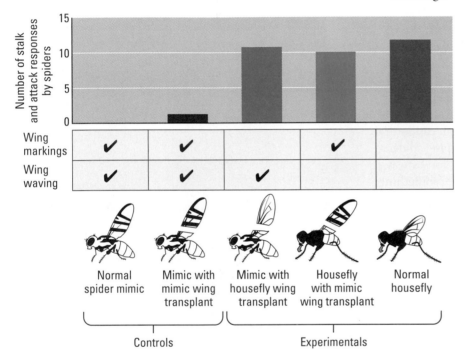

Figure 1.24 A second experimental test of the spider mimic hypothesis. The researchers transplanted the wings of spider mimics and houseflies and tabulated the behavior of jumping spiders placed in clear containers with the flies. Each bar in the histogram represents the number of "stalk and attack" responses of spiders in 20 separate trials, each with a different spider and fly. (Adapted with permission from Erick Greene, Larry J. Orsak, and Douglas W. Whitman, "A Tephritid Fly Mimics the Territorial Displays of Its Jumping Spider Predators," *Science*, vol. 236, page 310 (April 17, 1987). Copyright © 1987 American Association for the Advancement of Science.)

Practice developing hypotheses and designing experiments in Web/CD Activity 1C, *Investigating Acid Rain.*

The Culture of Science

As in our case study of research on spider mimics, it is not unusual that several scientists are asking the same questions. Such convergence contributes to the progressive and self-correcting qualities of science. Scientists build on what has been learned from earlier research, and they pay close attention to contemporary scientists working on the same problem. Scientists share information through publications, seminars, meetings, and personal communication. The internet has added a new medium for this exchange of ideas and data.

Both cooperation and competition characterize the scientific culture (Figure 1.25). Scientists working in the same research field subject one another's work to careful scrutiny. It is common for scientists to check on the conclusions of others by attempting to repeat experiments. This obsession with evidence and confirmation helps characterize the scientific style of inquiry. Scientists are generally skeptics.

We have seen that science has two key features that distinguish it from other styles of inquiry: (1) a dependence on observations and measurements that others can verify and (2) the requirement that ideas (hypotheses) are testable by experiments that others can repeat.

Theories in Science

Many people associate facts with science, but accumulating facts is not what science is primarily about. A telephone book is an impressive catalog of factual information, but it has little to do with science. It is true that facts, in the form of verifiable observations and repeatable experimental results, are the prerequisites of science. What really advances science, however, is some new theory that ties together a number of observations that previously seemed unrelated. The cornerstones of science are the explanations that apply to the greatest variety of phenomena. People like Newton, Darwin, and Einstein stand out in the history of science not because they discovered a great many facts, but because their theories had such broad explanatory power.

What is a scientific theory, and how is it different from a hypothesis? Compared to a hypothesis, a theory is much broader in scope. This is a hypothesis: "Mimicking jumping spiders is an adaptation that helps protect some flies from predation." But *this* is a theory: "Adaptations such as mimicry evolve by natural selection."

Because theories are so comprehensive, they only become widely accepted in science if they are supported by an accumulation of extensive and varied evidence. This use of the term *theory* in science for a comprehensive explanation supported by abundant evidence contrasts with our everyday usage, which equates theories more with speculations or hypotheses. Natural selection qualifies as a scientific theory because of its broad application and because it has been validated by a continuum of observations and experiments. (For one example, you can jump ahead to the Evolution Link at the end of Chapter 18.)

Scientific theories are not the only way of "knowing nature," of course. A comparative religion course would be a good place to learn about the diverse legends that tell of a supernatural creation of Earth and its life. Science and religion are two very different ways of trying to make sense of nature. Art is still another way. A broad education should include exposure to these different ways of viewing nature. Each of us synthesizes our worldview by integrating our life experiences and multidisciplinary education. As a science textbook and part of that broad education, *Essential Biology* showcases life in the scientific context of evolution, the one theme that continues to hold all of biology together no matter how big and complex the subject becomes.

Figure 1.25 Science is a social process. Here, New York University plant biologist Gloria Coruzzi (left) mentors one of her students in the methods of molecular biology.

Figure 1.26 DNA technology and the law. Forensic technicians can use traces of DNA extracted from a blood sample or other body fluid collected at a crime scene to produce a molecular "fingerprint." The stained bands visible in this photograph represent fragments of DNA, and the pattern of bands varies from person to person. The legal applications of DNA technology have become very public in the past decade. It is just one example of biology's prominent role in society today.

Science, Technology, and Society

Science and technology are interdependent. New technologies, such as more powerful microscopes and computers, advance science. And scientific discoveries can lead to new technologies. In most cases, technology applies scientific discoveries to the development of new goods and services. For example, it was 50 years ago that two scientists, James Watson and Francis Crick, discovered the structure of DNA through the process of science. Their discovery eventually led to a variety of DNA technologies, including the genetic engineering of microorganisms to mass-produce human insulin and the use of DNA fingerprinting for investigating crimes (Figure 1.26). Perhaps Watson and Crick envisioned that their discovery would someday inform new technologies, but that probably did not motivate their research, nor could they have predicted exactly what the applications would be. The direction technology takes depends less on the curiosity that drives basic science than it does on the current needs of humans and the changing climate of culture.

Technology has improved our standard of living in many ways, but it is a double-edged sword. Technology that keeps people healthier has enabled the population to grow more than tenfold in the past three centuries, to double to 6 billion in just the past 40 years. The environmental consequences are sometimes devastating. Acid rain, deforestation, global warming, nuclear accidents, toxic wastes, and extinction of species are just a few of the repercussions of more and more people wielding more and more technology. Science can help us identify such problems and provide insight about what course of action may prevent further damage. But solutions to these problems have as much to do with politics, economics, culture, and the values of societies as with science and technology. Now that science and technology have become such powerful functions of society, every thoughtful citizen has a responsibility to develop a reasonable amount of scientific and technological literacy.

We wrote this book to help students who are not biology majors develop an appreciation for the science of life and apply that understanding to evaluate such social issues as health and environmental quality. We believe that such a biological perspective is essential for any educated person, which is why we named our book *Essential Biology*. We hope we serve you well.

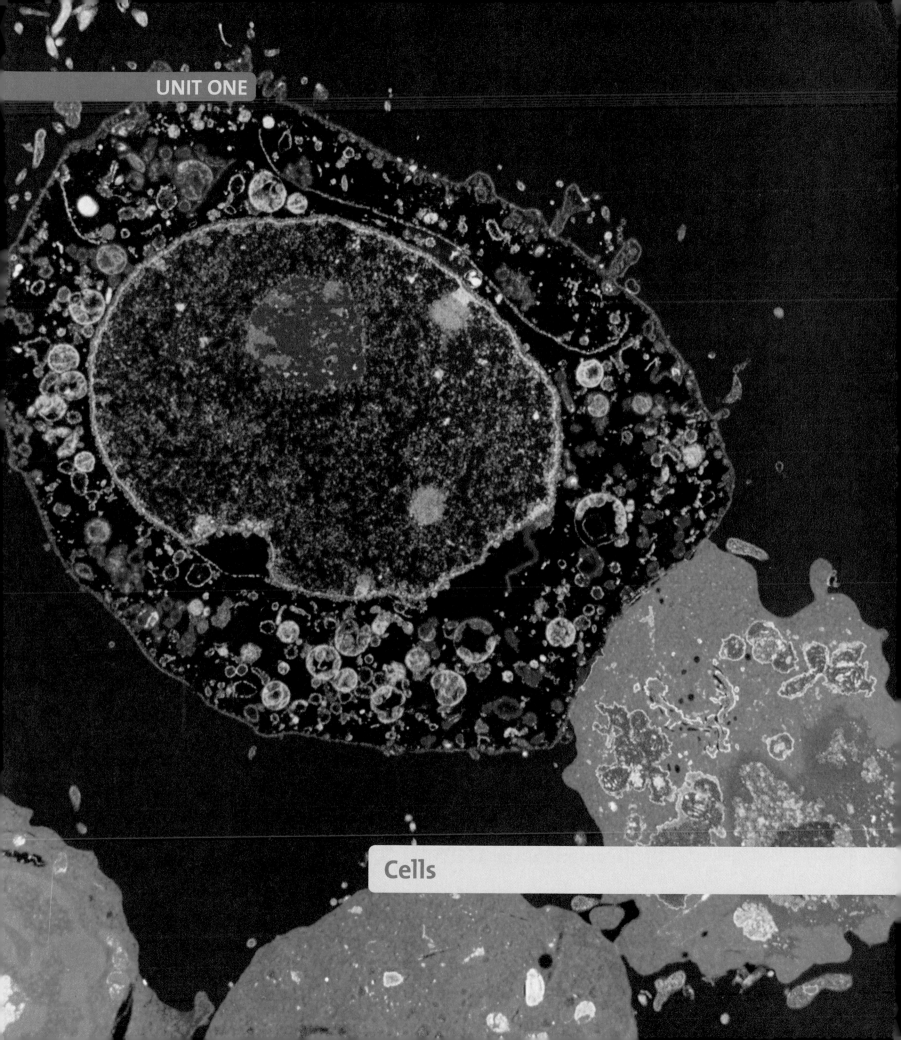

Cells

Essential Chemistry for Biology

Rain in the eastern United States can be more acidic

than vinegar.  About 65% of your weight is

oxygen atoms.  Because of surface tension,

some insects can walk on water.  A person

can produce up to 12 liters (over 3 gallons) of sweat per

day.  The iron in a multivitamin pill is the same

element as the iron in a train or a ship.

Your body is a chemical system. True, it's one of the most elegant and complex of all chemical collections, but it is chemical nevertheless. The genes you inherited from your parents are chemicals, and so are the muscle proteins that move your eyes as you read this sentence. Your ability to see these words results from light interacting with chemicals located in the inner lining of your eyeballs. And your mind's processing of this information depends on chemical signals between cells in your brain. You nourish those brain cells and all other cells of your body with chemicals derived from food. In fact, at its most basic level, life is all about chemicals and how they interact. In this chapter, you will learn some essential chemistry that you will be able to apply throughout your study of life.

Overview: Tracing Life Down to the Chemical Level

Life at the chemical level is mostly invisible to us because it occurs at the microscopic level of cells. So let's start with a macroscopic scene, such as the African savanna in Figure 2.1, and work our way down to smaller and smaller levels of biological organization. You learned about some of these layers of life in Chapter 1, but a review will place the chemical level in context.

The savanna is an example of what biologists call an ecosystem. An **ecosystem** consists of all the organisms living in a particular area, as well as all the nonliving, physical components of the environment that affect the organisms, such as water, air, soil, and sunlight. The ecosystem and all the organizational levels below it form a hierarchy, with each level building on the ones below it. Below the ecosystem level, all the organisms in a savanna are collectively called a **community.** Below the community, an interacting group of individuals of one species—zebras in our example—is called a **population.** Below population in the hierarchy is the **organism,** an individual living thing.

Life's hierarchy continues to unfold within the individual organism. The zebra's body consists of several **organ systems,** such as a nervous system, a digestive system, and a circulatory system, the example illustrated in Figure 2.1. Each organ system consists of **organs.** For instance, the main organs of the circulatory system are the heart and the blood vessels. As we continue downward through the hierarchy, each organ is made up of several different **tissues,** each of which consists of a group of similar cells. A **cell** is life's basic unit of structure and function. Each tissue has specific functions, performed by the cells that compose the tissue.

Finally, we reach the chemical level in the hierarchy. Each cell consists of an enormous number of chemicals that cooperate to give the cell the properties we recognize as life. The chemical illustrated in Figure 2.1 is DNA (see Chapter 1). It is the chemical of inheritance, the substance of genes. The DNA you received from your parents programmed your development from a single cell and continues to direct the chemical activities of all your cells. DNA is an example of a **molecule,** a cluster of even smaller chemical units called **atoms.** In the computer graphic in Figure 2.1, which illustrates only a tiny segment of one DNA molecule, each of the spheres represents an atom.

Play *The Levels of Life Card Game* in Activity 2A, available on the website and CD-ROM.

Ecosystem
African savanna

Community
All organisms in savanna

Population
Herd of zebras

Organism
Zebra

Organ system
Circulatory system

Organ
Heart

Tissue
Heart muscle tissue

Cell
Heart muscle cell

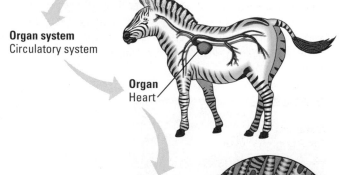

Molecule
DNA

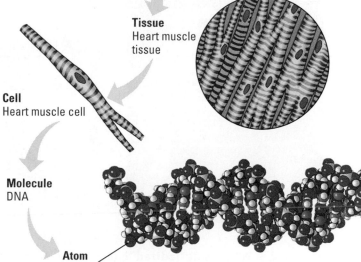

Atom
Oxygen atom

Figure 2.1 Tracing life down to the chemical level: The hierarchy of biological organization.

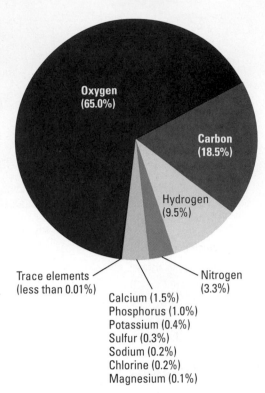

Figure 2.2 **Chemical composition of the human body by weight.** Note that oxygen, carbon, hydrogen, and nitrogen make up 96.3% of the weight of the human body. (These percentages include water, H_2O.) The trace elements, which make up less than 0.01%, include boron, chromium, cobalt, copper, fluorine, iodine, iron, manganese, molybdenum, selenium, silicon, tin, vanadium, and zinc.

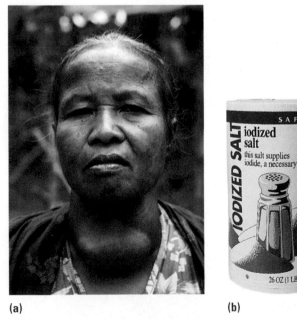

(a)   (b)

Figure 2.3 **Why is salt "iodized"?** (a) Goiter, an enlargement of the thyroid gland, shown here in a Malaysian woman, can occur when a person's diet does not include enough iodine, a trace element. (b) Goiter has been reduced in many nations by adding iodine to table salt (iodized salt). Unfortunately, goiter still affects many thousands of people in developing nations.

Some Basic Chemistry

Taking any biological system apart, we eventually end up at the chemical level. Let's now *begin* at this basic biological level by exploring the chemistry of life. Imagination will help. Try picturing yourself small enough to actually crawl into molecules and climb around on their atoms. The illustrations in this chapter and the activities on the *Essential Biology Place* CD-ROM and website will help you in this exploration at the molecular level.

Matter: Elements and Compounds

Humans and other organisms and everything around them are all made of matter, the physical "stuff" of the universe. Defined more formally, **matter** is anything that occupies space and has mass.

Matter is composed of chemical elements. A chemical **element** is a substance that cannot be broken down to other substances. There are 92 naturally occurring elements. Examples of elements are gold, copper, carbon, and oxygen. Each element has a symbol, the first letter or two of its English, Latin, or German name. For instance, the symbol for gold, Au, is from the Latin word *aurum;* the symbol O stands for the English word *oxygen.* Note that neither a rock nor a human body qualifies as an element, because both humans and rocks can be decomposed (not a good experiment to try) to the many elements that are combined in "human matter" or "rock matter."

About 25 of the 92 elements are essential to life. Four of these elements—oxygen (O), carbon (C), hydrogen (H), and nitrogen (N)—make up about 96% of the human body, which is typical of living matter in chemical composition (Figure 2.2). Calcium (Ca), phosphorus (P), potassium (K), sulfur (S), and a few other elements account for most of the remaining 4%. The **trace elements,** listed below Figure 2.2, are also essential. They are called *trace* elements not because they are of marginal importance (you can't live without them), but because your body requires them in only trace (very small) amounts. Some trace elements, such as iron, are needed by all forms of life. Others are required only by certain species. The average human, for example, needs about 0.15 milligram (mg) of the trace element iodine each day. A deficiency of iodine prevents normal functioning of the thyroid gland and results in goiter, an abnormal enlargement of the thyroid gland (Figure 2.3). Too much iodine can also cause goiter.

Elements can combine to form compounds. The DNA in Figure 2.1 is an example of a **compound,** a substance containing two or more elements in a fixed ratio. Compounds are much more common than pure elements. In fact, few elements exist in a pure state in nature. Many compounds consist of only two elements. For instance, table salt (sodium chloride, NaCl) has equal parts of the elements sodium (Na) and chlorine (Cl). In contrast, most of the compounds in living organisms contain at least three or four different elements, mainly carbon, hydrogen, oxygen, and nitrogen.

Atoms

Each element consists of one kind of atom, which is different from the atoms of other elements. An **atom,** named from a Greek word meaning "indivisible," is the smallest unit of matter that still retains the properties of an element. In other words, the smallest amount of the element carbon is one carbon atom. And that's a very small "piece" of carbon. It would take about a million of those carbon atoms to stretch across the period printed at the end of this sentence.

The Structure of Atoms Physicists have split the atom into more than a hundred types of subatomic particles. However, to understand basic atomic structure, we only have to consider three: protons, electrons, and neutrons. A **proton** is a subatomic particle with a single unit of positive electrical charge (+). An **electron** is a subatomic particle with a single unit of negative electrical charge (−). A third type of subatomic particle, the **neutron,** is electrically neutral (has no electrical charge).

Let's look at the structure of an atom of the element helium (He). You're already familiar with helium—it's the "lighter-than-air" gas that can be used to make balloons rise. Each atom of helium has 2 neutrons and 2 protons tightly packed as the atom's central core, or **nucleus.** Two electrons, much smaller than the neutrons and protons, orbit the nucleus at nearly the speed of light (Figure 2.4). The attraction between the negatively charged electrons and the positively charged protons keeps the electrons near the nucleus.

Elements differ in the number of subatomic particles in their atoms. All atoms of a particular element have the same unique number of protons. This is the element's **atomic number.** Thus, an atom of helium, with 2 protons, has an atomic number of 2. Note that in this atom, the atomic number is also the number of electrons. When an atom has an equal number of protons and electrons, its net electrical charge is zero and so the atom is neutral.

An atom's **mass number** is the sum of the numbers of protons and neutrons in its nucleus. For helium, the mass number is 4. Mass is a measure of the amount of matter in an object. A proton and a neutron are almost identical in mass, and for convenience we use the mass of a single proton or neutron as one unit of mass. An electron has very little mass—only about $\frac{1}{2000}$ the mass of a proton. So an atom's mass is essentially equal to its total number of protons + neutrons—the atom's mass number.

See the structure of more atoms in Web/CD Activity 2B.

Isotopes Some elements have variant forms called isotopes. The different **isotopes** of an element have the same numbers of protons and electrons but different numbers of neutrons. Table 2.1 shows the numbers of subatomic particles in the three isotopes of carbon. Carbon-12 (abbreviated ^{12}C), with 6 neutrons and 6 protons (mass number 12), makes up about 99% of all naturally occurring carbon. Most of the other 1% consists of carbon-13, with 7 neutrons and 6 protons (mass number 13). A third isotope, carbon-14, with 8 neutrons and 6 protons (mass number 14), occurs in minute quantities. Notice that all three isotopes have 6 protons—otherwise, they would not be carbon. Both carbon-12 and carbon-13 are stable isotopes, meaning their nuclei remain permanently intact. The isotope carbon-14, on the other hand, is unstable, or radioactive. A **radioactive isotope** is one in which the nucleus decays, giving off particles and energy.

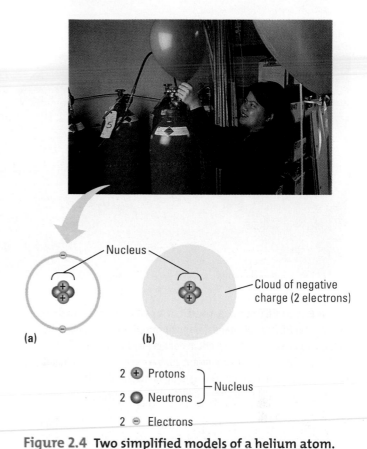

2 ⊕ Protons ⎫
　　　　　　　⎬ Nucleus
2 ● Neutrons ⎭

2 ⊖ Electrons

Figure 2.4 Two simplified models of a helium atom.
Both models are highly diagrammatic, and we use them only to show the basic components of atoms. The atomic nucleus consists of tightly packed neutrons and protons, 2 each in the case of helium. The electrons orbit the nucleus, with electrical attraction between the negatively charged electrons and the positively charged protons in the nucleus holding the atom together. **(a)** This model shows the number of electrons in the atom. **(b)** This model, slightly more realistic, shows the electrons as a spherical cloud of negative charge surrounding the nucleus. Neither model is drawn to scale. In real atoms, the electron cloud is much bigger compared to the nucleus. If the electron cloud was the size of a football stadium, the nucleus would only be the size of a fly on the field.

Table 2.1	Isotopes of Carbon		
	Carbon-12	**Carbon-13**	**Carbon-14**
Protons	6 ⎫ Mass	6 ⎫ Mass	6 ⎫ Mass
Neutrons	6 ⎬ number 12	7 ⎬ number 13	8 ⎬ number 14
Electrons	6	6	6

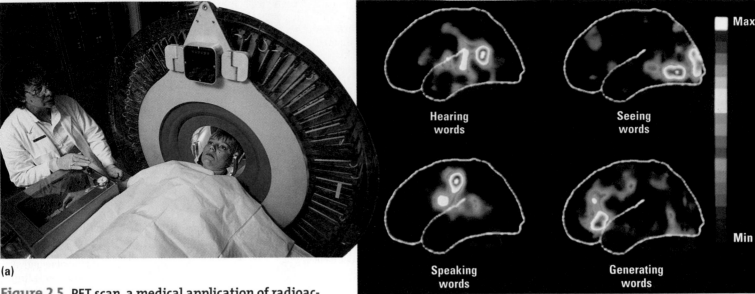

(a)

(b)

Figure 2.5 PET scan, a medical application of radioactivity. (a) PET, which stands for positron-emission tomography, detects locations of intense chemical activity in the body. The patient is first injected with a nutrient such as sugar that is labeled with a radioactive isotope. The isotope emits subatomic particles called positrons. The positrons collide with electrons made available by chemical reactions in the body. An instrument called the PET scanner detects the energy released by these collisions and maps the chemical "hot spots," regions of an organ that are most chemically active at the time. (b) A computer that is linked to the scanner translates the energy data into an anatomical image. The color of the image varies with the amount of the isotope present in an area. For example, these four PET scans of one patient's brain show chemical activity under four different conditions, all related to language. (These are side views, with the front of the brain at the left.) In addition to enabling physicians to diagnose brain disorders, PET can also help detect certain heart problems and cancers.

Figure 2.6 The Tokaimura nuclear accident. In 1999, workers at the Tokaimura nuclear power plant accidentally triggered a chain reaction that leaked radiation into the surroundings. A mile from the plant, radiation levels peaked at 15,000 times higher than normal environmental levels. The accident exposed 46 workers in the power plant to very dangerous radiation doses. About 300,000 people in nearby towns were monitored for radioactive contamination, as shown here, and will continue to be monitored for possible long-term damage.

Radioactive isotopes have many uses in biological research and medicine. Living cells cannot distinguish radioactive isotopes from nonradioactive isotopes of the same element. Consequently, organisms take up and use compounds containing radioactive isotopes in the usual way. This makes radioactive isotopes useful as tracers—biological spies, in effect—for monitoring the fate of atoms in living organisms.

Medical diagnosis is an important application of radioactive tracers. Certain kidney disorders, for example, can be diagnosed by injecting tiny amounts of radioactive isotopes into a patient's blood and then measuring the amount of radioactive tracer passed in the urine. Health professionals also use radioactive tracers in combination with sophisticated imaging instruments such as PET scanners, which can monitor chemical processes as they actually occur in the body (Figure 2.5). In most diagnostic uses of radioactive tracers, the patient receives only a tiny amount of an isotope that decays completely in minutes or hours.

Though radioactive isotopes have many beneficial uses, uncontrolled exposure to them can harm living organisms by damaging cellular molecules, especially DNA. The explosion of a nuclear reactor at Chernobyl, Ukraine, in 1986 released large amounts of radioactive isotopes into the environment, killing 30 people within a few weeks. The survivors have suffered increased rates of thyroid cancer and of birth defects in their children, and thousands may be at an increased risk of cancers in the future. A 1999 nuclear accident at the Tokaimura power plant, about 150 kilometers (km) upwind from Tokyo, Japan, renewed concerns about the biological risks of radioactivity (Figure 2.6).

Natural sources of radiation can also pose a threat. Radon, a radioactive gas, may be a cause of lung cancer. Radon can contaminate buildings in regions where underlying rocks naturally contain uranium, a radioactive substance. Homeowners can buy a radon detector or hire a company to test their homes to ensure that radon levels are safe.

Electron Arrangement and the Chemical Properties of Atoms Of the subatomic particles we have discussed, it is mainly electrons that determine how an atom behaves when it encounters other atoms. Electrons vary in the amount of energy they possess. The farther an electron is from the nucleus, the greater its energy. Electrons in an atom occur only at certain energy levels, called electron shells. Depending on their atomic number, atoms may have one, two, or more electron shells, with electrons in the outermost shell having the highest energy. Each shell can accommodate up to a specific number of electrons. The innermost shell is full with only 2 electrons. So in atoms with more than 2 electrons, the remainder are found in shells farther from the nucleus. For example, the second and third shells can each hold up to 8 electrons.

The number of electrons in the outermost shell determines the chemical properties of an atom. Atoms whose outer shells are not full tend to interact with other atoms—that is, to participate in chemical reactions. Figure 2.7 shows the electron shells of four biologically important elements. Because the outer shells of all four atoms are incomplete, these atoms react readily with other atoms. The hydrogen atom is highly reactive because it has only 1 electron in its single electron shell, which can accommodate 2 electrons. Atoms of carbon, nitrogen, and oxygen are also highly reactive because their outer shells, which can hold 8 electrons, are incomplete. In contrast, the helium atom in Figure 2.4 has a single, first-level shell that is full with 2 electrons. As a result, helium is chemically inert (unreactive).

Review electron arrangement in Web/CD Activity 2C; build atoms in Activity 2D.

Chemical Bonding and Molecules

How does a chemical reaction enable an atom to fill its outer electron shell? When two atoms with incomplete outer shells react, each atom gives up or acquires electrons so that both partners end up with completed outer shells. Atoms do this either by transferring or sharing outer electrons. These interactions usually result in atoms staying close together, held by attractions called **chemical bonds.**

Ionic Bonds Table salt is an example of how electron transfer can bond atoms together. The two ingredients of table salt are the elements sodium (Na) and chlorine (Cl). When a sodium atom donates 1 electron to a chlorine atom, the electron transfer results in both atoms having full outer shells of electrons (Figure 2.8). Before the electron transfer, each of these atoms was electrically neutral because its proton and electron numbers balanced. The electron transfer moves one unit of negative charge from sodium to chlorine. The atoms are now **ions,** the term for atoms that are electrically charged as a result of gaining or losing electrons. Loss of an electron gives the sodium ion a charge of $+1$, while chlorine's gain of an electron gives it a charge of -1. The sodium ion (Na^+) and chloride ion (Cl^-) are held together by an **ionic bond,** the attraction between oppositely charged ions. Table salt, or sodium chloride (NaCl), fits our definition of a compound, a substance consisting of two or more elements.

See how ions are made in Web/CD Activity 2E.

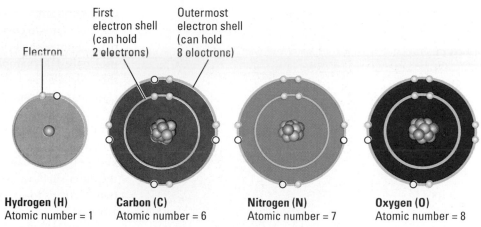

First electron shell (can hold 2 electrons)

Outermost electron shell (can hold 8 electrons)

Electron

Hydrogen (H) Atomic number = 1

Carbon (C) Atomic number = 6

Nitrogen (N) Atomic number = 7

Oxygen (O) Atomic number = 8

Figure 2.7 Atoms of the four elements most abundant in life. All four atoms are chemically reactive because their outermost electron shells are incomplete. The small empty circles in these diagrams represent unfilled "spaces" in the outer electron shells.

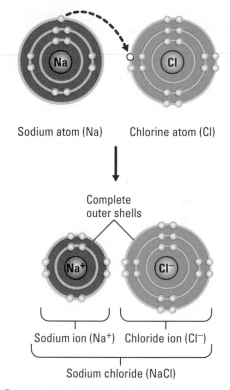

Sodium atom (Na)   Chlorine atom (Cl)

Complete outer shells

Sodium ion (Na^+)   Chloride ion (Cl^-)

Sodium chloride (NaCl)

Figure 2.8 Electron transfer and ionic bonding. Notice that sodium has only 1 electron in its outer shell, whereas chlorine has 7. When these atoms collide, the chlorine atom strips sodium's outer electron away. In doing so, chlorine fills its outer shell with 8 electrons. Sodium, in losing 1 electron, ends up with only two shells, the outer shell now having a full set of 8 electrons. Because electrons are negatively charged particles, the electron transfer between the two atoms moves one unit of negative charge from sodium to chlorine. The atoms are now ions, which are electrically charged atoms. Two ions with opposite charges attract each other; when the attraction holds them together, it is called an ionic bond.

Covalent Bonds In contrast to the complete transfer of electrons that leads to ionic bonds, a **covalent bond** forms when two atoms *share* one or more pairs of outer-shell electrons. Atoms held together by covalent bonds form a **molecule.** For example, a covalent bond connects the two hydrogen atoms in the molecule H_2, a gas present in the atmosphere. The subscript 2 in H_2 indicates that a hydrogen molecule consists of two hydrogen atoms (Figure 2.9).

The number of covalent bonds an atom can form is equal to the number of additional electrons needed to fill its outer shell. Note in Figure 2.9 that hydrogen (H) can form one covalent bond, oxygen (O) can form two, and carbon (C) can form four. The single covalent bond in H_2 completes the outer shells of both hydrogen atoms. In contrast, an oxygen atom needs two electrons to complete its outer shell. In an O_2 molecule, the two oxygen atoms share two pairs of electrons, forming a double covalent bond.

Watch covalent bonds being made in Web/CD Activity 2F.

Though H_2 and O_2 are molecules, neither qualifies as a compound because these molecules are each composed of only one element. An example of a molecule that is also a compound is methane (CH_4), a common gas produced by certain bacteria. You can see in Figure 2.9 that each of the four hydrogen atoms in this molecule shares one pair of electrons with the single carbon atom. Note that water, H_2O, is also a compound.

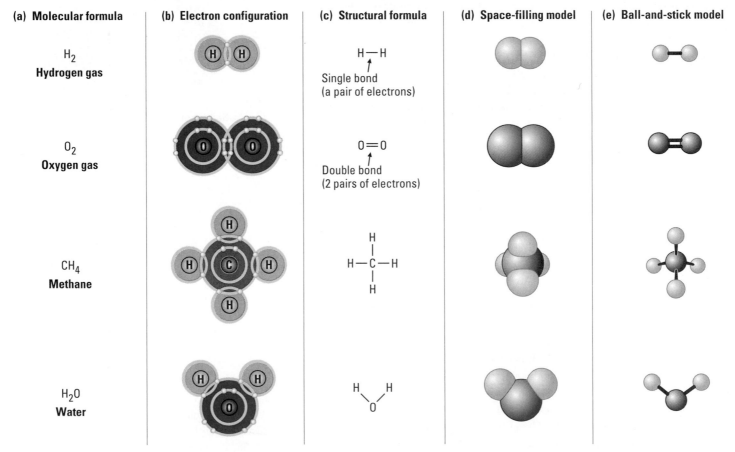

(a) Molecular formula	(b) Electron configuration	(c) Structural formula	(d) Space-filling model	(e) Ball-and-stick model
H_2 **Hydrogen gas**		H—H Single bond (a pair of electrons)		
O_2 **Oxygen gas**		O=O Double bond (2 pairs of electrons)		
CH_4 **Methane**		H—C—H (with H above and below)		
H_2O **Water**		H H \ O /		

Figure 2.9 Alternative ways to represent molecules. **(a)** A molecular formula, such as H_2 or CH_4, only tells you the number of each kind of atom in a molecule. **(b)** You can see from the electron configuration that each atom completes its outer shell by sharing electrons with its covalent partner. **(c)** In a structural formula, each line represents a covalent bond, a pair of shared electrons. Hydrogen always forms one covalent bond: oxygen, two; and carbon, four. The double covalent bond (double lines) in O_2 consists of two pairs of shared electrons. **(d)** A space-filling model, in which the color-coded balls symbolize atoms, shows the shape of a molecule. **(e)** In a ball-and-stick model, the "balls" represent atoms and the "sticks" represent the bonds between the atoms.

Chemical Reactions

The chemistry of life is dynamic. Your cells are constantly rearranging molecules by breaking existing chemical bonds and forming new ones. Such changes in the chemical composition of matter are called **chemical reactions.** A simple example is the reaction between oxygen gas and hydrogen gas that forms water (this is an explosive reaction, which, fortunately, does not occur in your cells):

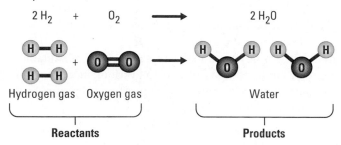

Let's translate the chemical shorthand: Two molecules of hydrogen gas (2 H_2) react with one molecule of oxygen gas (O_2) to form two molecules of water (2 H_2O). The arrow indicates the conversion of the starting materials, the **reactants** (2 H_2 and O_2), to the **products** (2 H_2O).

Notice that the same *numbers* of hydrogen and oxygen atoms are present in reactants and products, although they are grouped differently. Chemical reactions cannot create or destroy matter, but only rearrange it. These rearrangements usually involve the breaking of chemical bonds in reactants and the forming of new bonds in products. You can see those rearrangements in our ball-and-stick models for water formation; the "sticks" represent bonds between the atoms, which are shown as "balls."

The water molecule we have built here is a good conclusion to this section on basic chemistry. Water is a substance so important in biology that we take a closer look at its life-supporting properties in the next section.

1. What four elements are most abundant in living matter?

2. Why is water classified as a compound, but O_2 is not?

3. A nitrogen atom has 7 protons, and the most common isotope of nitrogen has 7 neutrons. A radioactive isotope of nitrogen has 8 neutrons. What is the atomic number and mass number of this radioactive nitrogen?

4. Why are radioactive isotopes useful as tracers in research on the chemistry of life?

5. Explain what holds together the ions in a crystal of table salt (NaCl).

6. What is chemically nonsensical about this structure?

$$H—C{=}C—H$$

7. Fill in the blanks with the correct numbers in the following chemical process:

$$C_6H_{12}O_6 + \_\_ O_2 \rightarrow \_\_ CO_2 + \_\_ H_2O$$

Answers: 1. Carbon, oxygen, hydrogen, and nitrogen **2.** Because water consists of two different elements **3.** Atomic number = 7; mass number = 15 **4.** Organisms incorporate radioactive isotopes of an element into their molecules just as they do the nonradioactive isotopes, and researchers can detect the presence of the radioactive isotopes. **5.** Ionic bonds, the electrical attraction between the Na^+ ions and Cl^- ions, holds the ions together. **6.** Each carbon atom has only three covalent bonds instead of the required four. **7.** $C_6H_{12}O_6 + \underline{\textbf{6}} O_2 \rightarrow \underline{\textbf{6}} CO_2 + \underline{\textbf{6}} H_2O$

Figure 2.10 **A watery world.** Three-quarters of Earth's surface is submerged in water. Although most of this water is in liquid form, water is also present on Earth as ice and vapor (including clouds). Water is the only common substance to exist in the natural environment in all three physical states of matter: solid, liquid, and gas. These three states of water are visible in this view of Earth from space.

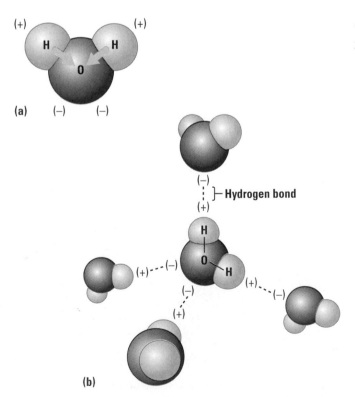

(a)

(–) (–)

(–) —} Hydrogen bond
(+)

(+)---(–)   (–)

(+)---(–)

(+)

(b)

Water and Life

Life on Earth began in water and evolved there for 3 billion years before spreading onto land. Modern life, even land-dwelling life, remains tied to water. You have personal experience with this dependence each time you seek watery fluids to quench your thirst and replenish your body's water content. Inside your body, your cells are surrounded by a fluid that is mostly water, and your cells themselves range from 70% to 95% in water content. Water is the biological medium here on Earth and possibly on other planets throughout the universe.

The abundance of water is a major reason Earth is inhabitable (Figure 2.10). Water is so common that it is easy to overlook its extraordinary behavior. We can trace water's unique life-supporting properties to the structure and interactions of its molecules.

The Structure of Water

Studied in isolation, the water molecule is deceptively simple. Its two hydrogen atoms are joined to the oxygen atom by single covalent bonds:

However, the electrons of the covalent bonds are not shared equally between the oxygen and hydrogen. Oxygen pulls electrons much more strongly than does hydrogen. This results in the electrons of the covalent bonds spending a disproportionate amount of their time in the neighborhood of a water molecule's oxygen atom. The unequal distribution of negatively charged electrons, combined with the V shape of the molecule, makes a water molecule polar. A **polar molecule** has opposite charges on opposite ends. In the case of water, the oxygen end of the molecule has a slight negative charge, while the two hydrogen atoms are slightly positive (Figure 2.11a).

Review the polarity of water in Web/CD Activity 2G.

The polarity of water results in weak electrical attractions between neighboring water molecules. The molecules tend to orient such that the hydrogen atom of one molecule is near the oxygen atom of an adjacent water molecule. These weak attractions are called **hydrogen bonds** because of the role the slightly positive hydrogen atoms play in the cohesion between molecules (Figure 2.11b).

Figure 2.11 **Water, a polar molecule.** (a) The arrows indicate the stronger pull that oxygen has, compared to its hydrogen partners, on the shared electrons of covalent bonds. The roughly V-shaped water molecule is polar, its oxygen end slightly negative and its hydrogen ends slightly positive in charge. (b) The charged regions of the polar water molecules are attracted to oppositely charged areas of neighboring molecules. Each molecule can hydrogen-bond to a maximum of four partners. (The dashed lines symbolize hydrogen bonds.) At any instant in liquid water at 37°C (human body temperature), only about 15% of the molecules are bonded to four partners. But so many water molecules are involved in some degree of hydrogen bonding at any moment that liquid water is much more cohesive than almost any other liquid.

Water's Life-Supporting Properties

The polarity of water molecules and the hydrogen bonding that results explains most of water's life-supporting properties. We explore four of those properties here: the cohesive nature of water; the ability of water to moderate temperature; the biological significance of ice floating; and the versatility of water as a solvent.

Cohesion of Water Water molecules stick together as a result of hydrogen bonding. Hydrogen bonds between molecules of liquid water last for only a few trillionths of a second, yet at any instant, many of the molecules are hydrogen-bonded to others. This tendency of molecules to stick together, called **cohesion,** is much stronger for water than for most other liquids. The cohesion of water is important in the living world. Trees, for example, depend on cohesion to help transport water from their roots to their leaves (Figure 2.12).

To see an animation of water moving up a tree, check out Web/CD Activity 2H.

Related to cohesion is *surface tension,* a measure of how difficult it is to stretch or break the surface of a liquid. Hydrogen bonds give water unusually high surface tension, making it behave as though it were coated with an invisible film (Figure 2.13).

How Water Moderates Temperature If you have ever burned your finger on a metal pot while waiting for the water in it to boil, you know that water heats up much more slowly than metal. In fact, because of hydrogen bonding, water has a better ability to resist temperature change than most other substances. Earth's giant water supply keeps temperatures within limits that permit life.

Temperature and heat are related, but different. A swimmer crossing San Francisco Bay has a higher temperature than the water, but the bay contains far more heat because of its immense volume. **Heat** is the amount of energy associated with the movement of the atoms and molecules in a body of matter. **Temperature** measures the intensity of heat—that is, the average speed of molecules rather than the total amount of heat energy in a body of matter.

When water is heated, the heat energy first disrupts hydrogen bonds and then makes water molecules move faster. The temperature of the water doesn't go up until the water molecules start moving faster. Because heat is first used to break hydrogen bonds rather than raise the temperature, water absorbs and stores a large amount of heat while warming up only a few degrees. Conversely, when water cools, hydrogen bonds form, a process that releases heat. Thus, water can release a relatively large amount of heat to the surroundings while the water temperature drops only slightly.

A large body of water can store a huge amount of heat from the sun during warm periods. And in cold conditions, heat given off from the gradually cooling water can warm the air. That's why coastal areas generally have milder climates than inland regions. Water's resistance to temperature change also stabilizes ocean temperatures, creating a favorable environment for marine life.

Another way that water moderates temperature is by **evaporative cooling.** When a substance evaporates, the surface of the liquid remaining behind cools down. This occurs because the molecules with the greatest energy (the "hottest" ones) tend to vaporize first. It's as if the five fastest runners on your college soccer team left school, lowering the average speed of the remaining team. On a global scale, surface evaporation cools tropical

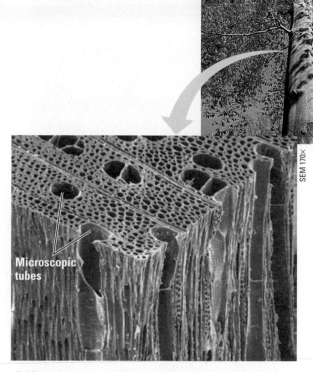

Figure 2.12 Cohesion and water transport in plants. The evaporation of water from leaves pulls water upward from the roots through microscopic tubes in the trunk of the tree. Because of cohesion, the pulling force is relayed through the tubes all the way down to the roots. As a result, water rises against the force of gravity.

Figure 2.13 A water strider walking on water. The cumulative strength of hydrogen bonds between water molecules allows this insect to walk on ponds without breaking the surface.

Figure 2.14 Sweating as a mechanism of evaporative cooling.

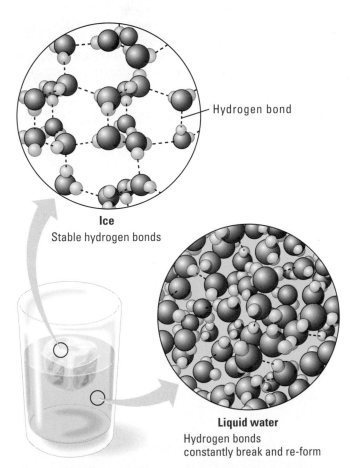

Ice
Stable hydrogen bonds

Hydrogen bond

Liquid water
Hydrogen bonds
constantly break and re-form

Figure 2.15 Why ice floats. Compare the spaciously arranged molecules in the ice crystal with the tightly packed molecules in liquid water. The more stable hydrogen bonding in ice holds the molecules at "arm's length," resulting in ice being less dense than liquid water. The expansion of water as it freezes can crack boulders when the water is in crevices, and it can also break the water pipes of an unheated house in winter.

oceans. On the scale of individual organisms, evaporative cooling prevents some land-dwelling creatures from overheating. It's why sweating helps you maintain a constant body temperature, even when exercising on a hot day (Figure 2.14). And the old expression "It's not the heat that's so bad, it's the humidity" has its basis in the difficulty of evaporating water into air that is already saturated with water vapor.

The Biological Importance of Ice Floating When most liquids get cold, their molecules move closer and closer together. If it's cold enough, the liquid becomes a solid. Water, however, behaves differently. When water molecules get cold enough, they move apart, forming the solid state we call ice. A chunk of ice has fewer molecules than an equal volume of liquid water, so ice floats because it is less dense than liquid water. Like water's other life-supporting properties, floating ice is a consequence of hydrogen bonding. In contrast to the short-lived hydrogen bonds in liquid water, those in ice are longer lasting, with each molecule bonded to four neighbors. As a result, ice is a spacious crystal (Figure 2.15).

How does the fact that ice floats help support life on Earth? Imagine what would happen if ice sank. All ponds, lakes, and even the oceans would eventually freeze solid. During summer, only the upper few inches of the oceans would thaw. Instead, when a deep body of water cools, the floating ice insulates the liquid water below, allowing life to persist under the frozen surface.

Water as the Solvent of Life You know from experience that you can dissolve sugar or salt in water. The resulting concoction is a **solution,** a liquid consisting of a homogeneous mixture of two or more substances. The dissolving agent is called the **solvent,** and a substance that is dissolved is a **solute.** When water is the solvent, the result is called an **aqueous solution** (from the Latin *aqua,* water).

The fluids of organisms are aqueous solutions. Water can dissolve an enormous variety of solutes necessary for life. It is the solvent inside all cells, in blood, and in plant sap. Figure 2.16 illustrates how water can dissolve ionic compounds such as table salt. In addition to dissolving salts, water can also dissolve many polar molecules, such as sugars. Recall that water and other polar molecules have localized regions of positive and negative charge due to unequal electron sharing between atoms. Water molecules can cling to the charged regions of polar molecules, dissolving them.

Acids, Bases, and pH

In the aqueous solutions within organisms, most of the water molecules are intact. However, some of the water molecules actually break apart (dissociate) into ions. The ions formed are called hydrogen ions (H^+) and hydroxide ions (OH^-). For the proper functioning of chemical processes within organisms, the right balance of H^+ ions and OH^- ions is critical.

A chemical compound that donates H^+ ions to solutions is called an **acid.** One example of a strong acid is hydrochloric acid (HCl), the acid in your stomach. In solution, HCl dissociates completely into H^+ and Cl^- ions. A **base** (or alkali) is a compound that accepts H^+ ions and removes them from solution. Some bases, such as sodium hydroxide (NaOH), do this by adding OH^- ions, which combine with H^+ to form H_2O.

To describe the acidity of a solution, we use the **pH scale** (pH stands for potential hydrogen). The scale ranges from 0 (most acidic) to 14 (most basic). Each pH unit represents a tenfold change in the concentration of H^+ (Figure 2.17). For example, lemon juice at pH 2 has 100 times more H^+ than an equal amount of tomato juice at pH 4. Pure water and aqueous solutions that are neither acidic nor basic are said to be neutral; they have a pH of 7. They do contain some H^+ and OH^- ions, but the concentrations of the two kinds of ions are equal. The pH of the solution inside most living cells is close to 7.

Experiment with the pH scale in Web/CD Activity 2I.

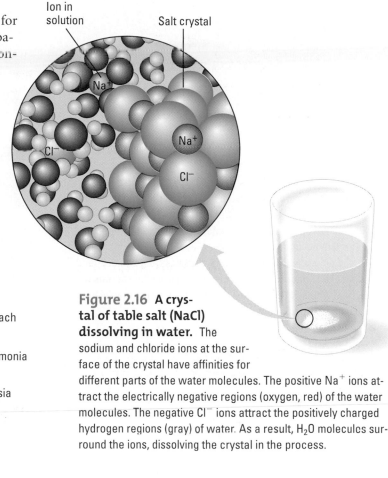

Figure 2.16 A crystal of table salt (NaCl) dissolving in water. The sodium and chloride ions at the surface of the crystal have affinities for different parts of the water molecules. The positive Na^+ ions attract the electrically negative regions (oxygen, red) of the water molecules. The negative Cl^- ions attract the positively charged hydrogen regions (gray) of water. As a result, H_2O molecules surround the ions, dissolving the crystal in the process.

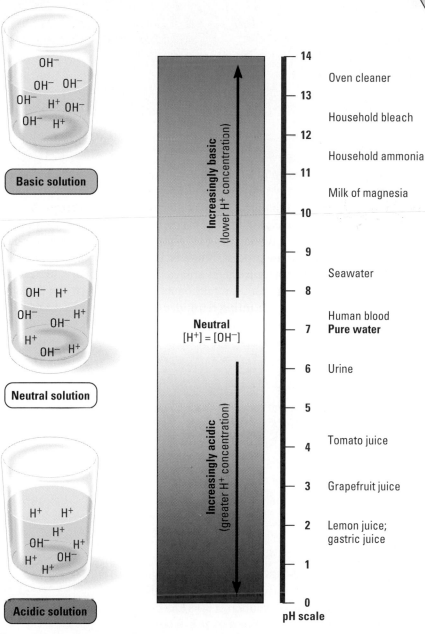

Basic solution

Neutral solution

Acidic solution

Increasingly basic (lower H^+ concentration)

Neutral $[H^+] = [OH^-]$

Increasingly acidic (greater H^+ concentration)

14
13 — Oven cleaner
— Household bleach
12
— Household ammonia
11
— Milk of magnesia
10
9
— Seawater
8
— Human blood
7 **Pure water**
6 — Urine
5
4 — Tomato juice
3 — Grapefruit juice
2 — Lemon juice; gastric juice
1
0

pH scale

Figure 2.17 The pH scale. A solution having a pH of 7 is neutral, meaning that its H^+ and OH^- concentrations are equal. The lower the pH below 7, the more acidic the solution, or the greater its excess of H^+ ions compared to OH^- ions. The higher the pH above 7, the more basic the solution, or the greater the deficiency of H^+ ions relative to OH^- ions.

Figure 2.18 The effects of acid precipitation on a forest. Acid fog and acid rain may have contributed to the death of many of the spruce and fir trees on Mt. Mitchell in North Carolina. Acid precipitation results from water in the atmosphere reacting with certain pollutants spewed as exhaust from automobiles, factories, and power plants. The acid rain, snow, or fog can descend on land or lakes hundreds of miles downwind from the sources of pollution. Rain with a pH between 2 and 3—more acidic than vinegar—has been recorded in the eastern United States.

Even a slight change in pH can be harmful to organisms because the molecules in cells are very sensitive to H^+ and OH^- concentrations. However, biological fluids contain **buffers,** substances that resist changes in pH by accepting H^+ ions when they are in excess and donating H^+ ions when they are depleted. This buffering process, however, is not foolproof. The biological damage from an unfavorable pH is apparent in the toll that acid rain can take on an ecosystem such as a pond or forest (Figure 2.18). It is a daunting reminder that the chemistry of life is linked to the chemistry of the environment.

Investigate acid rain in The Process of Science activity available on the website and CD-ROM.

CheckPoint

1. Why is it unlikely that two neighboring water molecules would be arranged like this?

2. Explain why, if you pour very carefully, you can actually "stack" water slightly above the rim of a cup.

3. Why is it more dangerous to stay neck-deep in a 105°F (41°C) hot tub for an hour than it is to sit for an hour outside when the air temperature is 105°F (41°C)?

4. Explain why ice floats.

5. Why are blood and most other biological fluids classified as aqueous solutions?

6. Compared to a basic solution of pH 8, the same volume of an acidic solution at pH 5 has _____ times more hydrogen ions (H^+).

Answers: 1. The positively charged hydrogen regions would repel each other. **2.** Surface tension due to water's cohesion will hold the water together until it is far enough above the rim for gravity to overcome the cohesion. **3.** Evaporative cooling in the hot tub is limited to the skin of the head and neck. **4.** Ice is less dense than liquid water because the more stable hydrogen bonds "lock" the molecules in a spacious crystal. **5.** Because the solvent is water **6.** 1000

Evolution Link:
Earth Before Life

Chemical reactions and physical processes on the early Earth created an environment that made life possible. And life, once it began, transformed the planet's chemistry. Biological and geological histories are inseparable.

Earth began as a cold world when gravity drew together dust and ice orbiting a young sun about 4.5 billion years ago. The planet eventually melted from the heat produced by compaction, radioactive decay, and the impact of meteorites. Molten material sorted into layers of varying density. Most of the iron and nickel sank to the center and formed a dense core. Less dense material became concentrated in a layer called the mantle, which surrounds the core. And the least dense material solidified to form a thin crust. The present continents, including North America, are attached to plates of crust that float on the flexible mantle.

The first atmosphere, which was probably composed mostly of hot hydrogen gas (H_2), escaped. The gravity of Earth was not strong enough to hold such small molecules. Volcanoes belched gases that formed a new atmosphere (Figure 2.19). Based on analysis of gases vented by modern volcanoes, scientists have speculated that the second early atmosphere consisted mostly of water vapor (H_2O), carbon monoxide (CO), carbon dioxide (CO_2), nitrogen (N_2), methane (CH_4), and ammonia (NH_3). The first seas formed from torrential rains that began when Earth had cooled enough for water in the atmosphere to condense. In addition to an atmosphere very different from the one we know, lightning, volcanic activity, and ultraviolet radiation were much more intense when Earth was young. In such a world, life began about 3.5–4.0 billion years ago. In Chapter 14, we will examine hypotheses and experiments of scientists who investigate the origin of life.

> Discover more about Earth before life by going to the Chapter 2 Evolution Link on the *Essential Biology Place* website.

Figure 2.19 The gaseous exhaust of a volcano. Such vulcanism was probably much more common when Earth was young.

Chapter Review

Summary of Key Concepts

Overview: Tracing Life Down to the Chemical Level

- The levels of biological order, beginning with the smallest, build from atoms → molecules → cells → tissues → organs → organ systems → organisms → populations → communities → ecosystems.
  - Web/CD Activity 2A *The Levels of Life Card Game*

Some Basic Chemistry

- **Matter: Elements and Compounds** Matter consists of elements and compounds, which combine two or more elements. There are 25 elements that are essential for life, with carbon, oxygen, nitrogen, and hydrogen being the most abundant in living matter.

- **Atoms** Protons, neutrons, and electrons are the parts of an atom. All the atoms of a particular element have the same unique proton number. However,

neutron number varies for isotopes of that element. Some isotopes are radioactive. Only electrons participate directly in chemical reactions, and it is the number of electrons in the outermost energy shell that defines the chemical behavior of an atom.
- Web/CD Activity 2B *The Structure of Atoms*
- Web/CD Activity 2C *Electron Arrangement*
- Web/CD Activity 2D *Build an Atom*

- **Chemical Bonding and Molecules** Electron transfers that complete outer electron shells result in charged atoms, or ions. Oppositely charged ions attract one another in what are called ionic bonds. Salts are ionic compounds. In covalent bonds, atoms complete their outer electron shells by sharing electrons. A molecule consists of two or more atoms connected by covalent bonds.
- Web/CD Activity 2E *Ionic Bonds*
- Web/CD Activity 2F *Covalent Bonds*

- **Chemical Reactions** By breaking bonds in reactants and forming new bonds in products, chemical reactions rearrange matter.

Water and Life

- **The Structure of Water** Water is a polar molecule, with its oxygen end negative in charge and its hydrogen end positive. This results in weak electrical attractions called hydrogen bonds between neighboring water molecules.
- Web/CD Activity 2G *The Structure of Water*

- **Water's Life-Supporting Properties** The ability of leaves to pull water up microscopic tubes without the water columns breaking is an example of how water's cohesiveness supports life. Water moderates temperature by absorbing heat in warm environments and releasing heat in cold environments. Evaporative cooling also helps stabilize the temperatures of oceans and organisms. The fact that ice floats because it is less dense than liquid water prevents the oceans from freezing solid. Blood and other biological fluids are aqueous solutions with a diversity of solutes dissolved in water, a most versatile solvent.
- Web/CD Activity 2H *Cohesion of Water*

- **Acids, Bases, and pH** Acidic solutions have more H^+ ions than OH^- ions and a pH less than 7. Basic solutions have fewer H^+ ions than OH^- ions and a pH greater than 7. At pH 7, H^+ and OH^- ions balance (a neutral solution). Each pH unit represents a tenfold difference in H^+ concentration. Most cells are close to neutral (pH 7) and cannot tolerate much deviation from that pH.
- Web/CD Activity 2I *Acids, Bases, and pH*
- Web/CD The Process of Science *Investigating Acid Rain*

Evolution Link: Earth Before Life

- Life began about 3.5–4.0 billion years ago on Earth when the environment was quite different from our present-day world.
- Web Evolution Link *Earth Before Life*

Self-Quiz

1. Which of the following represents the correct sequence of levels in life's hierarchy, proceeding downward from an individual animal?
   - a. organ, organ system, cell, tissue
   - b. organ system, population of cells, tissue, organ
   - c. organism, organ system, tissue, cell, organ
   - d. organ system, organ, tissue, cell
   - e. organ system, tissue, molecule, cell

2. A chemical compound is to a(an) _____ as a body organ is to a tissue.

3. Changing _____ would change it into an atom of a different element.
   - a. the number of electrons surrounding the nucleus of an atom
   - b. the number of bonds formed by an atom
   - c. the number of protons in the nucleus of an atom
   - d. the electrical charge of an atom
   - e. the number of neutrons in the nucleus of an atom

4. Compared to the phosphorus-31 isotope, the radioactive phosphorus-32 isotope has
   - a. a different atomic number.
   - b. one more neutron.
   - c. one more proton.
   - d. one more electron.
   - e. a different electrical charge.

5. A sulfur atom has 6 electrons in its outer shell. As a result, it forms _____ covalent bonds with other atoms.
   - a. 2
   - b. 3
   - c. 4
   - d. 6
   - e. 8

6. This diagram shows the arrangement of electrons in a fluorine atom and a potassium atom. What would happen if a fluorine atom and a potassium atom came into contact? What kind of bond would they form?

 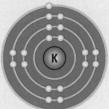

Fluorine atom   Potassium atom

7. Which of the following is *not* a chemical reaction?
   - a. Sugar ($C_6H_{12}O_6$) and oxygen gas (O_2) combine to form carbon dioxide (CO_2) and water (H_2O).
   - b. Sodium metal and chlorine gas unite to form sodium chloride.
   - c. Hydrogen gas combines with oxygen gas to form water.
   - d. Ice melts to form liquid water.
   - e. Methane gas is produced by certain bacteria.

8. Most of the unique properties of water result from the fact that
   - a. water molecules are very small.
   - b. water contains carbon atoms.
   - c. water molecules easily separate from one another.
   - d. water molecules are constantly in motion.
   - e. water molecules tend to stick together.

9. Some people in your study group say they don't understand what a polar molecule is. You explain that a polar molecule
   - a. has an extra electron, giving it a positive charge.
   - b. has an extra electron, giving it a negative charge.
   - c. has covalent bonds.
   - d. has an extra proton, giving it a positive charge.
   - e. is slightly negative at one end and slightly positive at the other end.

10. A can of cola consists mostly of sugar dissolved in water, with some carbon dioxide gas that makes it fizzy and makes the pH less than 7. In chemical terms, you would say that cola is an aqueous solution, where water is the _____, sugar is the _____, and carbon dioxide makes the solution _____ rather than neutral.

- **Go to the website or CD-ROM for more self-quiz questions.**

The Process of Science

1. Plants use a process called photosynthesis to incorporate the carbon from atmospheric carbon dioxide gas into sugar. Outline an experiment using radioactive carbon to test the hypothesis that carbon assimilated by the leaves during photosynthesis can turn up in protein molecules stored in seeds.

2. **Investigate acid rain in The Process of Science activity on the website and CD-ROM.**

Biology and Society

Evaluate this statement: "It's paranoid and ignorant to worry about industry or agriculture contaminating the environment with chemical wastes—this stuff is just made of the same atoms that were already present in our environment."

CHAPTER 3

The Molecules of Life

Americans consume an average of 140 pounds of

sugar per person per year.  Most of the sugar in

soft drinks comes from corn.  Cellulose, found

in plant cell walls, is the most abundant organic

compound on Earth. A typical cell in your body has

about 3 meters of DNA. Cells are

composed of 70–95% water.

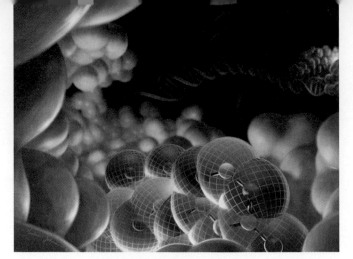

Figure 3.1 A "cellscape" with diverse molecules.

I t is as difficult to imagine life without carbon as it is life without water. Although a cell is mostly water, most of the rest of it consists of carbon-based molecules. Proteins, DNA, carbohydrates, and other molecules that distinguish living matter from inanimate material are all composed of carbon atoms bonded to one another and to atoms of other elements. Carbon is unparalleled in its ability to form the large, complex, diverse molecules that characterize life on Earth (Figure 3.1). The study of carbon compounds is called **organic chemistry.** After an overview of organic molecules, we will be ready to explore the architecture and function of the exquisite chemicals unique to life.

Overview: Organic Molecules

Carbon Skeletons and Functional Groups

Why are carbon atoms so versatile as molecular ingredients? Remember that an atom's bonding ability is related to the number of electrons it must share to complete its outer shell. A carbon atom has 4 outer electrons in an outer shell that holds 8 (see Figure 2.7). Carbon completes its outer shell by sharing electrons with other atoms in four covalent bonds. Each carbon thus acts as an intersection from which an organic molecule can branch off in up to four directions. And because carbon can use one or more of its bonds to attach to other carbon atoms, it is possible to construct an endless diversity of **carbon skeletons** varying in size and branching pattern (Figure 3.2). The carbon atoms of organic molecules can also use one or more of their bonds to partner with other elements, most commonly hydrogen, oxygen, and nitrogen.

In terms of chemical composition, the simplest organic compounds are **hydrocarbons,** organic molecules consisting of carbon skeletons bonded only to hydrogen atoms. And the simplest hydrocarbon is methane, a single carbon atom bonded to four hydrogen atoms (Figure 3.3). Methane is one of the most abundant hydrocarbons in natural gas and is also produced by bacteria that live in swamps and in the digestive tracts of cows. Larger hydrocarbons are the main molecules in the gasoline we burn in cars and other machines. Hydrocarbons are also important fuels in your body; the energy-rich parts of fat molecules are hydrocarbon in structure (Figure 3.4).

> Build a hydrocarbon in Activity 3A on the website or CD-ROM.

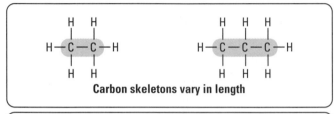

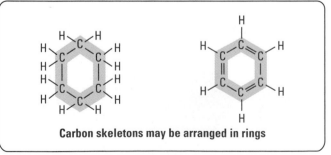

Figure 3.2 **Variations in carbon skeletons.** All of these examples are hydrocarbons, organic molecules consisting only of carbon and hydrogen. Note that each carbon atom forms four bonds, and each hydrogen atom forms one bond.

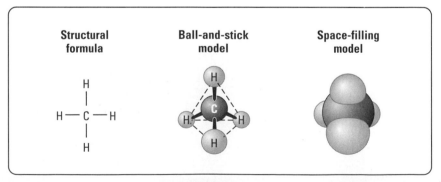

Figure 3.3 **Methane, the simplest hydrocarbon.** In the ball-and-stick model, note that the four single bonds of carbon point to the corners of a tetrahedron.

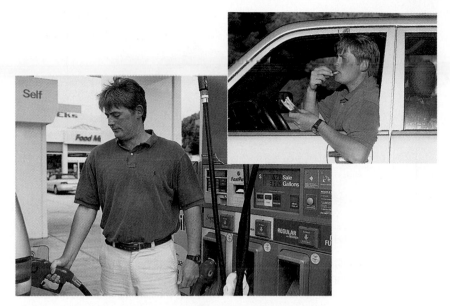

Figure 3.4 **Hydrocarbons as fuel.** Energy-rich hydrocarbons provide fuel for our machines and, in the form of the hydrocarbon content of fats, for our cells.

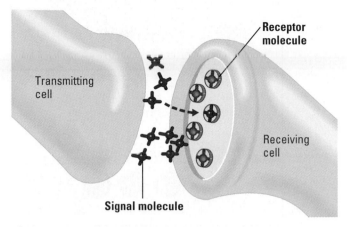

Figure 3.5 **Molecular shape and brain chemistry.** One nerve cell in your brain "talks" to another by releasing signal molecules with a shape that fits receptor molecules located on the surface of the receiving cell. The signal molecules cross the tiny gap between cells and bind to the receptors, stimulating the receiving cell. The actual shapes of the signal and receptor molecules are much more complex than represented here.

Each type of organic molecule has a unique three-dimensional shape. Note in Figure 3.3 that carbon's four bonds point to the corners of an imaginary tetrahedron (an object with four triangular sides). This geometry occurs at each carbon "intersection" where there are four covalent bonds, and thus organic molecules much larger than methane can have very elaborate shapes. We'll encounter many cases of how the molecules of your body recognize one another based on their shapes. Just one example is the chemical signaling your brain cells use to "talk" to each other (Figure 3.5).

The unique properties of an organic compound depend not only on its carbon skeleton but also on the atoms attached to the skeleton. In an organic molecule, the groups of atoms that usually participate in chemical reactions are called **functional groups.** Figure 3.6 shows four of the functional groups important in the chemistry of life. Though the examples in the figure each contain only one functional group, many biological molecules have two or more. For example, compounds called amino acids have carboxyl as well as amino groups. Amino acids are the building blocks of the much larger molecules called proteins. We are now ready to see how your cells make such giant molecules out of smaller organic molecules.

Review functional groups in Web/CD Activity 3B.

Giant Molecules from Smaller Building Blocks

On a molecular scale, many of life's molecules are gigantic; in fact, biologists call them **macromolecules.** Examples are proteins, DNA, and carbohydrates called polysaccharides. Your cells make all these macromolecules by joining smaller organic molecules into chains called **polymers** (from the Greek *polys,* many, and *meros,* part). A polymer consists of many identical or similar molecular units strung together, much as a train consists of many individual cars. The units that serve as the building blocks of polymers are called **monomers.**

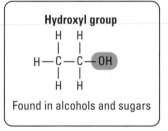

Hydroxyl group

Found in alcohols and sugars

Carbonyl group

Found in sugars

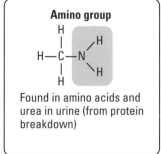

Amino group

Found in amino acids and urea in urine (from protein breakdown)

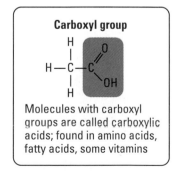

Carboxyl group

Molecules with carboxyl groups are called carboxylic acids; found in amino acids, fatty acids, some vitamins

Figure 3.6 **Some common functional groups.**

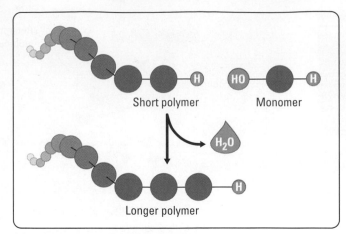

(a) Dehydration synthesis of a polymer

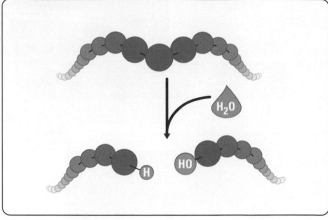

(b) Hydrolysis of a polymer

Figure 3.7 Synthesis and digestion of polymers. **(a)** The only atoms shown in these diagrams are hydrogens and hydroxyl groups (—OH or HO—, depending on orientation) in strategic locations on the monomers. A polymer grows in length when an incoming monomer and the monomer at the end of the existing chain each contribute to the formation of a water molecule. The monomers replace those lost covalent bonds with a bond to each other. **(b)** Hydrolysis reverses the process by digesting (breaking down) the polymer with the addition of water molecules to break the bonds between monomers.

Cells link monomers together to form polymers by a process called **dehydration synthesis** (Figure 3.7a). For each monomer added to a chain, a water molecule (H_2O) is formed by the release of two hydrogen atoms and one oxygen atom from the monomers (hence the term *dehydration* synthesis—the monomers lose water). This same basic process of dehydration synthesis occurs regardless of the specific monomers and the type of polymer the cell is producing.

Organisms not only make macromolecules; they also have to break them down. For example, many of the molecules in your food are macromolecules. You must digest these giant molecules to make their monomers available to your cells for assimilation into your own brand of macromolecules. That digestion occurs by a process called **hydrolysis** (Figure 3.7b). Hydrolysis means to break (*lyse*) with water (*hydro-*). Cells break bonds between monomers by adding water to them, a process essentially the reverse of dehydration synthesis.

You can see an animation of dehydration synthesis and hydrolysis in Activity 3C on the CD-ROM.

CheckPoint

1. Draw a structural formula for C_2H_4. Remember that each carbon has four bonds; each hydrogen has one.

2. Why is the following compound called an amino acid?

3. How many molecules of water are required to hydrolyze a polymer that is 100 monomers long?

Answers: 1.

2. It has both an amino group (—NH_2) and a carboxyl group (—COOH), which also makes it a carboxylic acid. **3.** 99

Biological Molecules

In the remainder of the chapter, we explore the four classes of large molecules in cells: carbohydrates, lipids, proteins, and nucleic acids, the class that includes DNA. We'll see that for each of these classes, cells use dehydration synthesis to connect small molecules to make macromolecules.

Carbohydrates

Athletes know them as "carbs." **Carbohydrates** include the small sugar molecules dissolved in soft drinks as well as the long starch molecules we consume in pasta and potatoes.

Monosaccharides Simple sugars, or **monosaccharides** (from the Greek *mono-*, single, and *sacchar*, sugar), include glucose, found in sports drinks, and fructose, found in fruit. Both of these simple sugars are found in honey (Figure 3.8). Generally, monosaccharides have molecular formulas that are some multiple of CH_2O. For example, the formula for glucose is $C_6H_{12}O_6$. Fructose has the same molecular formula as glucose, $C_6H_{12}O_6$, but its atoms are arranged differently (Figure 3.9). Glucose and fructose are examples of **isomers,** molecules that match in their molecular formulas but have different structures. Seemingly minor differences like this give isomers different properties. In this case, the differences make fructose taste considerably sweeter than glucose.

It is convenient to draw sugars as if their carbon skeletons were linear. However, in aqueous solutions, many monosaccharides form rings, as shown for glucose in Figure 3.10.

See models of glucose in Web/CD Activity 3D.

Monosaccharides, particularly glucose, are the main fuel molecules for cellular work. Analogous to an automobile engine consuming gasoline, your cells break down glucose molecules and extract their stored energy, giving off carbon dioxide as "exhaust." Cells also use the carbon skeletons of monosaccharides as raw material for manufacturing other kinds of organic molecules, including amino acids. Monosaccharides that cells do not use immediately are usually incorporated into larger carbohydrates. Carbohydrates can also be converted to fats. Thus, if we eat more food than we need, we store fat even if our diet is fat-free.

Figure 3.8 Honey, a mixture of two simple sugars. Honey consists mainly of the monosaccharides glucose and fructose, giving honey its sweet taste.

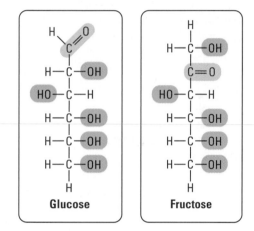

Figure 3.9 Monosaccharides (simple sugars). These molecules have the two trademarks of sugars: several hydroxyl groups (—OH) and a carbonyl group ($>C=O$). Glucose and fructose are isomers of each other: They have identical molecular formulas ($C_6H_{12}O_6$), but their structures differ because the atoms are not arranged the same way. The difference in this case is the location of the carbonyl group.

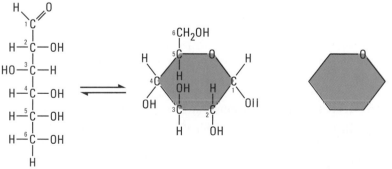

(a) Linear and ring structures

(b) Abbreviated ring structure

Figure 3.10 The ring structure of glucose. (a) Dissolved in water, one part of a glucose molecule can bond to another part to form a ring. The carbon atoms are numbered here so you can relate the linear and ring versions of the molecule. As the double arrows indicate, ring formation is a reversible process, but at any instant in an aqueous solution, most glucose molecules are rings. **(b)** From now on, we'll use this abbreviated ring symbol for glucose.

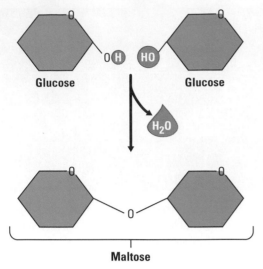

Figure 3.11 Disaccharide (double sugar) formation.
The simple sugars are joined by dehydration synthesis, in this case forming a bond between two glucose units to make the double sugar maltose.

Figure 3.12 A year's supply of sugar. Americans consume an average of 140 pounds of sweetener per person per year—over a third of a pound per day. The 140 pounds of sucrose (table sugar) in this photograph will give you some idea of that consumption. We actually get much of our sugar in the form of high-fructose corn syrup (HFCS) used to sweeten such products as soft drinks and candy. Reading food labels will make you aware of the amount of sugar in processed food.

Disaccharides Cells construct a **disaccharide,** or double sugar, from two monosaccharides by the process of dehydration synthesis. An example of a disaccharide is maltose, also called malt sugar, which consists of two glucose monomers (Figure 3.11). Maltose, which is common in germinating seeds, is used in making beer.

The most common disaccharide is sucrose, which consists of a glucose linked to a fructose. Sucrose is the main carbohydrate in plant sap, and it nourishes all the parts of the plant. We extract sucrose from the stems of sugarcane or the roots of sugar beets to use as table sugar.

We are born preferring sweet-tasting foods, and although this preference lessens after childhood, it continues to influence our diet. Not only may we prefer a peach to a serving of squash, but we also add concentrated sugar to many prepared foods.

The history of sugar refining and consumption by humans is a story with important health and economic consequences. Before the 1980s, refined sugar was processed from cane sugar grown in the tropics and beet sugar grown in temperate regions. Occupying only a small part of the sweetener market was corn syrup, which contains glucose. Because glucose is only half as sweet as sucrose, corn syrup was not a serious rival to sucrose.

The balance among sweeteners was drastically upset in the 1980s, when corn syrup producers developed a commercial method for converting much of the glucose in corn syrup to fructose, an isomer of glucose even sweeter than sucrose. The resulting high-fructose corn syrup (HFCS), which is about half fructose, is inexpensive and has replaced sucrose in many prepared foods. Soft drink manufacturers, once the largest commercial users of sucrose in the world, have almost completely replaced sucrose with HFCS. Because corn syrup is produced mainly in developed countries, the changeover hurt the developing economies of tropical countries where sugarcane is a major crop.

The United States is one of the world's leading markets for sweeteners, with the average American consuming about 140 pounds per year, mainly as sucrose and HFCS (Figure 3.12). This national "sweet tooth" persists in spite of growing awareness about possible health effects. Sugars are a major cause of tooth decay: Bacteria living in the mouth convert sugars to acids, which eat away tooth enamel. However, studies investigating whether sugar consumption causes such diseases as diabetes, cancer, and heart disease have not revealed any direct connections. Similar efforts have failed to show any link between sugar intake and behavioral problems in children.

Besides tooth decay, high sugar consumption seems to be hazardous mainly to the extent that sugars replace other, more varied foods in our diet. The description of sugars as "empty calories" is accurate in the sense that even the less refined sweeteners such as brown sugar and honey contain only negligible amounts of nutrients other than carbohydrates. For good health, we also require proteins, fats, vitamins, and minerals. And we need to include substantial amounts of "complex carbohydrates"—that is, polysaccharides—in our diet.

Polysaccharides Complex carbohydrates, or **polysaccharides,** are long chains of sugar units—polymers of sugar monomers. Foods such as potatoes, rice, and corn are rich in starch, a familiar polysaccharide. **Starch** is a storage polysaccharide that cells break down as needed to obtain sugar. Starch, which is found in roots and other plant organs, consists entirely of glucose monomers (Figure 3.13a). Plant cells often contain starch granules, which serve as sugar stockpiles. Plant cells need sugar for energy and as a raw material for building other molecules. They break starch down into glucose by hydrolyzing the bonds between the glucose monomers. Humans

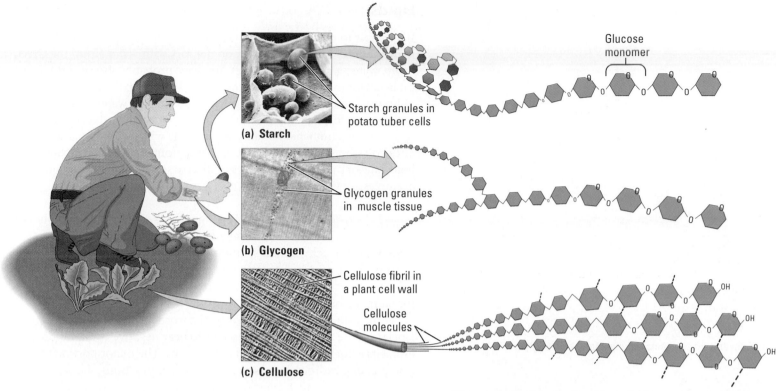

(a) Starch

Starch granules in potato tuber cells

Glucose monomer

(b) Glycogen

Glycogen granules in muscle tissue

(c) Cellulose

Cellulose fibril in a plant cell wall

Cellulose molecules

Figure 3.13 Polysaccharides. (a) Plants store glucose by polymerizing it to form starch. **(b)** Animals also store glucose, but in the form of glycogen, a polysaccharide more extensively branched than plant starch. **(c)** The cellulose of plant cell walls is an example of a structural polysaccharide. The cellulose molecules are assembled into fibrils that make up the main fabric of the walls. Wood, cell wall material consisting of such cellulose fibrils along with other polymers, is strong enough to support trees hundreds of feet high. We take advantage of that structural strength in our use of lumber as a building material.

and most other animals are able to use plant starch as food by hydrolyzing it within their digestive systems. Potatoes and grains, such as wheat, corn, and rice, are the major sources of starch in the human diet.

Animals store excess sugar in the form of a polysaccharide called **glycogen.** Glycogen is similar to starch, but glycogen is more extensively branched (Figure 3.13b). Most of our glycogen is stored as granules in our liver and muscle cells, which hydrolyze the glycogen to release glucose when it is needed for energy.

In addition to storage polysaccharides, there are also polysaccharides that serve as building material for structures that protect cells and support whole organisms. **Cellulose,** the most abundant organic compound on Earth, forms cablelike fibrils in the tough walls that enclose plant cells and is a major component of wood (Figure 3.13c). Cellulose resembles starch and glycogen in being a polymer of glucose, but its glucose monomers are linked together in a different orientation. Unlike the glucose linkages in starch and glycogen, those in cellulose cannot be hydrolyzed by most animals. The cellulose in plant foods, which passes unchanged through our digestive tract, is commonly known as "fiber." It may help keep our digestive system healthy, but it does not serve as a nutrient. Most animals that do derive nutrition from cellulose, such as cows and termites, have bacteria inhabiting their digestive tracts that can break down the cellulose.

Build carbohydrates in Web/CD Activity 3E.

Cellulose and some forms of starch are such large molecules that they do not dissolve in water. If they did, then your cotton bath towels, which are mostly cellulose, would dissolve the first time you put them in a washing machine. In contrast, simple sugars (such as glucose) and double sugars (such as sucrose) do dissolve readily in water, forming sugary solutions, including soft drinks. In spite of this difference, almost all carbohydrates, including cellulose and most other polysaccharides, are **hydrophilic,** which literally means "water-loving." Hydrophilic molecules adhere water to their surface. It is the hydrophilic quality of cellulose that makes a fluffy bath towel so water absorbent.

Lipids

In contrast to carbohydrates and most other biological molecules, **lipids** are **hydrophobic,** which means they do not mix with water (from the Greek *hydro,* water, and *phobos,* fearing). You have probably observed this chemical behavior in an unshaken bottle of salad dressing: the oil, which is a type of lipid, separates from the vinegar, which is mostly water. Shake the bottle and you can force a temporary mixture long enough to douse your salad with dressing, but what remains in the bottle will quickly separate again once you stop shaking. Beyond this distinction of being hydrophobic, the lipid category includes molecules of diverse structure and function. Just two examples are fats and steroids.

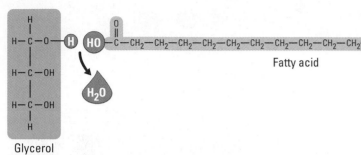

Glycerol

(a) Dehydration synthesis linking a fatty acid to glycerol

(b) A fat molecule

Figure 3.14 Synthesis and structure of a fat, or triglyceride. **(a)** This diagram shows the first of three fatty acids that will attach to glycerol by dehydration synthesis. **(b)** The finished fat has a glycerol "head" and three fatty acid "tails." The fatty acids consist mainly of energy-rich hydrocarbon.

Fats The ingredients of a **fat** are an alcohol called glycerol joined to three molecules called fatty acids. Cells use dehydration synthesis to attach the three fatty acids to the glycerol (Figure 3.14). The fatty acid trio accounts for the term *triglyceride,* a synonym for fat that you may see on food labels or on medical test results for fat in the blood.

Perhaps the food industry prefers the word *triglyceride* on their product labels because the very word *fat* seems to bear negative connotations. However, fats actually perform essential functions. The major portion of a fatty acid is a long hydrocarbon, which, like the hydrocarbons of gasoline, store much energy. In fact, a pound of fat packs more than twice as much energy as a pound of carbohydrate such as starch. This compact energy storage enables a mobile animal such as a human to get around much better than if the animal had to lug its stored energy around in the bulkier form of carbohydrate. The downside to this energy efficiency, of course, is that it is very difficult for a person trying to lose weight to "burn off" excessive body fat. What's important to understand is that a reasonable amount of body fat is both normal and healthy as a fuel reserve. We stock these long-term food stores in specialized reservoirs called adipose cells that swell and shrink when we deposit and withdraw the fat from storage. In addition to storing energy, adipose tissue cushions such vital organs as the kidneys. A layer of fat beneath the skin also insulates us, helping us maintain a warm body temperature even when the outside air is cold.

"Saturated" versus "unsaturated" fats is a comparison you have probably encountered in advertisements for foods. What's *that* all about? Notice in Figure 3.14b that one of the fatty acids bends where there is a double bond in the carbon skeleton. Less hydrogen is attached to the carbon skeleton at that location, and so the fatty acid is said to be **unsaturated** because it has less than the maximum number of hydrogens. The other two fatty acids in the fat molecule of Figure 3.14b lack double bonds in their hydrocarbon portions. Those fatty acids are **saturated,** meaning that they are bonded to the maximum number of hydrogen atoms. A saturated fat is one with all three of its fatty acids saturated. If one or more of the fatty acids is unsaturated, then we have an unsaturated fat, such as the one in Figure 3.14b. And if the fatty acids have several double bonds, further reducing the number of hydrogens, then that's a *polyunsaturated* fat.

Most animal fats, such as lard and butter, are saturated. They are solid at room temperature. In contrast, the fats of plants and fishes are generally unsaturated. These unsaturated fats are referred to as oils because they are liquid at room temperature. Corn oil, olive oil, and other vegetable oils are examples. When you see "hydrogenated vegetable oils" on the label of a product such as margarine, it means that unsaturated fats have been converted to saturated fats by adding hydrogen. This chemical processing gives the lipids the solid consistency of margarine.

Diets rich in saturated fats may contribute to cardiovascular disease by promoting atherosclerosis. In this condition, lipid-containing deposits called plaques build up within the walls of blood vessels, reducing blood flow.

Steroids Classified as lipids because they are hydrophobic, **steroids** are very different from fats in structure and function. The carbon skeleton of a steroid is bent to form four fused rings. Perhaps the best known steroid is cholesterol, which gets a lot of bad press because of its association with cardiovascular disease. But cholesterol is also an essential molecule in your body. It is present in the membranes that surround your cells, and cholesterol is also the "base steroid" from which your body produces other steroids—and that includes estrogen and testosterone, the steroids that function as sex hormones (Figure 3.15).

Check out Web/CD Activity 3F for an interactive review of lipids.

The controversial drugs called synthetic **anabolic steroids** are variants of testosterone, the male hormone. Testosterone causes a general buildup in muscle and bone mass during puberty in males and maintains masculine traits throughout life. Because anabolic steroids structurally resemble testosterone, they also mimic some of its effects. (Anabolism is the building of substances by the body.) Pharmaceutical companies first produced and marketed anabolic steroids in the early 1950s as a treatment for general anemia and for certain diseases that destroy muscle. About a decade later, some athletes began using anabolic steroids to build up their muscles quickly and enhance their performance. Today, anabolic steroids, along with many other drugs, are banned by most athletic organizations. Nonetheless, many professional athletes admit to using them heavily, citing such benefits as increased muscle mass, strength, stamina, and aggressiveness. It is not surprising that some of the heaviest users are weight lifters, football players, and bodybuilders (Figure 3.16).

Using anabolic steroids is indeed a fast way to increase body size. With these drugs, an athlete who is willing to cheat at a sport can increase body mass beyond what hard work alone can produce. But at what cost? Although medical researchers still debate the extent of health risks from steroid abuse, there is evidence that these substances can cause serious physical and mental problems. Overdosing can bloat the face and produce violent mood swings and deep depression. Internally, there may be liver damage leading to cancer. Anabolic steroids can also make blood cholesterol levels rise, perhaps increasing a user's chances of developing serious cardiovascular problems. Heavy users may also experience a reduced sex drive and become infertile, just the opposite of what one might expect from using a mimic of a sex hormone. The reason is that anabolic steroids often make the body reduce its normal output of sex hormones. The many potential health hazards of anabolic steroids coupled with the unfairness of an artificial advantage strongly support the argument for banning their use in athletics.

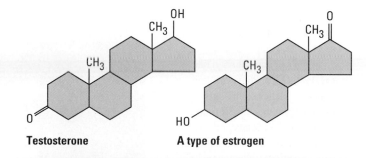

Figure 3.15 Sex hormones, examples of steroids. All steroids have a carbon skeleton consisting of four fused rings, abbreviated here with all the atoms of the rings omitted. Different steroids vary in the functional groups attached to this core set of rings. For example, the subtle contrast between testosterone and estrogen influences the development of the anatomical and physiological differences between male and female mammals, including humans.

Figure 3.16 Worth the risk? Steroids can build muscle mass, but can also cause serious health problems.

Figure 3.17 Some functions of proteins. (a) Structural proteins provide support. Examples include the proteins found in hair, horns, feathers, spiderwebs, and connective tissues such as tendons and ligaments. **(b)** Storage proteins, found in seeds and eggs, provide a source of amino acids for developing plants and animals. **(c)** Muscles are enriched in contractile proteins. **(d)** Transport proteins include hemoglobin, the iron-containing protein in blood that conveys oxygen from our lungs to other parts of the body. The red blood cells in this photograph contain hemoglobin.

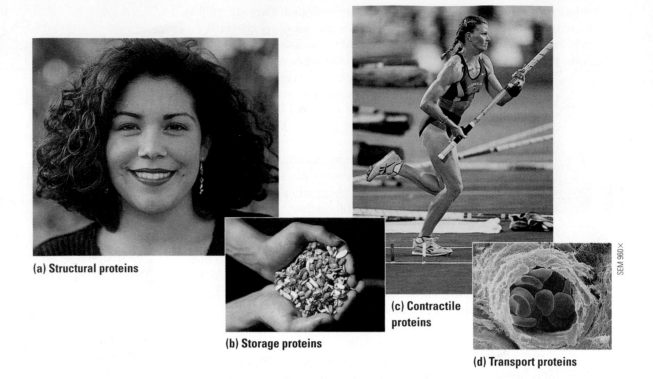

(a) Structural proteins

(b) Storage proteins

(c) Contractile proteins

SEM 960×

(d) Transport proteins

Proteins

The name *protein*, from the Greek word *proteios*, meaning "first place," suggests the importance of this class of macromolecules. A **protein** is a polymer constructed from amino acid monomers. Each of us has tens of thousands of different kinds of proteins, each with a unique, three-dimensional structure that corresponds to a specific function. Diverse proteins provide a cell with a molecular tool kit for almost everything it does. Figure 3.17 surveys the functions of four classes of proteins: structural proteins, storage proteins, contractile proteins, and transport proteins. Other types of proteins include defensive proteins, such as antibodies, and signal proteins that convey messages from one cell to another. Enzymes, another important group of proteins, change the rate of a chemical reaction without being changed in the process. Enzymatic proteins will be the subject of closer inspection later in this chapter. Now let's take a closer look at the architecture of proteins—the most elaborate of life's molecules.

Watch animations of proteins in action in Web/CD Activity 3G.

The Monomers: Amino Acids All proteins are constructed from a common set of just 20 kinds of amino acids. Each **amino acid** consists of a central carbon atom bonded to four covalent partners (carbon, remember, always forms four covalent bonds). Three of those attachments are common to all 20 amino acids: a carboxyl group, an amino group, and a hydrogen atom. The variable component of amino acids, called the side group, is attached to the fourth bond of the central carbon. Each type of amino acid has a unique side group, giving that amino acid its special chemical properties (Figure 3.18).

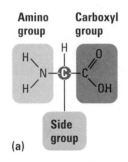

Amino group   **Carboxyl group**

Side group

(a)

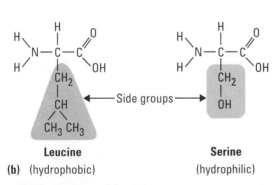

← Side groups →

Leucine   **Serine**
(b) (hydrophobic)   (hydrophilic)

Figure 3.18 Amino acids. (a) The general structure of an amino acid. **(b)** The 20 amino acids vary only in their side groups, which give these monomers their unique properties. For example, the side group of the amino acid leucine is pure hydrocarbon. That region of leucine is hydrophobic, because hydrocarbons don't mix with water (think of gasoline or fats, which are mostly hydrocarbon). In contrast, the side group of the amino acid serine has a hydroxyl group, which is hydrophilic. We'll see later how such differences in amino acids contribute to a protein's three-dimensional shape.

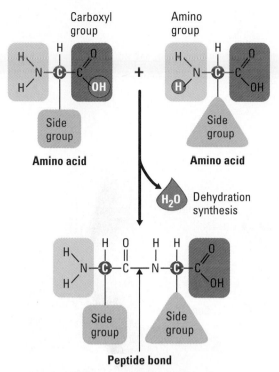

Amino acid + **Amino acid**

Dehydration synthesis

Peptide bond

Figure 3.19 Joining amino acids. Dehydration synthesis links adjacent amino acids by a peptide bond.

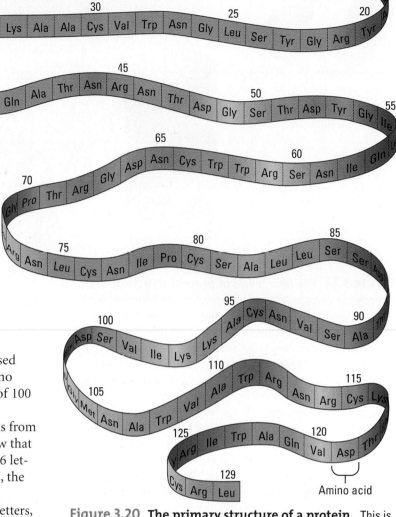

Figure 3.20 The primary structure of a protein. This is the unique amino acid sequence, or primary structure, of a protein called lysozyme. (The chain was drawn in serpentine fashion so that it would fit on the page. The actual shape of lysozyme is much more complex.) The names of the amino acids are given as their three-letter abbreviations, with the positions of lysozyme's 129 amino acids numbered along the chain.

Proteins as Polymers Cells link amino acids together by—you guessed it—dehydration synthesis. The resulting bond between adjacent amino acids is called a **peptide bond** (Figure 3.19). Proteins usually consist of 100 or more amino acids, forming a chain called a **polypeptide**.

How is it possible to make essentially a limitless variety of proteins from just 20 kinds of amino acids? The answer is in arrangement. You know that you can make many different words by varying the sequence of just 26 letters. Though the protein alphabet is slightly smaller (just 20 "letters"), the words are much longer, with polypeptides at least 100 amino acids in length. Just as each word is constructed from a unique succession of letters, each protein has a unique linear sequence of amino acids. This specific amino acid sequence is called the protein's **primary structure** (Figure 3.20).

Even a slight change in primary structure can affect a protein's ability to function. Consider, for example, the substitution of one amino acid for another at a particular position in hemoglobin, the blood protein that carries oxygen. Such an amino acid swap is the basis for sickle-cell disease, an inherited blood malfunction (Figure 3.21).

Figure 3.21 A single amino acid substitution in a protein causes sickle-cell disease. (a) Red blood cells of humans are normally disk-shaped, as seen in this microscopic view. Each cell contains millions of molecules of the protein hemoglobin, which transports oxygen from the lungs to other organs of the body. Next to the photograph, you can see the first 7 of the 146 amino acids in a polypeptide chain of hemoglobin. **(b)** A slight change in the primary structure of hemoglobin—an inherited substitution of one amino acid—causes sickle-cell disease. The substitution—valine in place of the amino acid glutamic acid—occurs in the number 6 position of the polymer. The abnormal hemoglobin molecules tend to crystallize, deforming some of the cells into a sickle shape. The life of someone with the disease is punctuated by dangerous episodes when the angular cells clog tiny blood vessels, impeding blood flow.

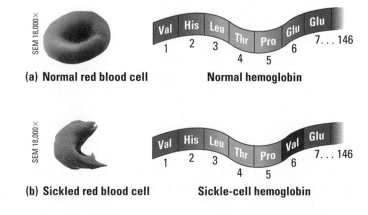

(a) Normal red blood cell   **Normal hemoglobin**

(b) Sickled red blood cell   **Sickle-cell hemoglobin**

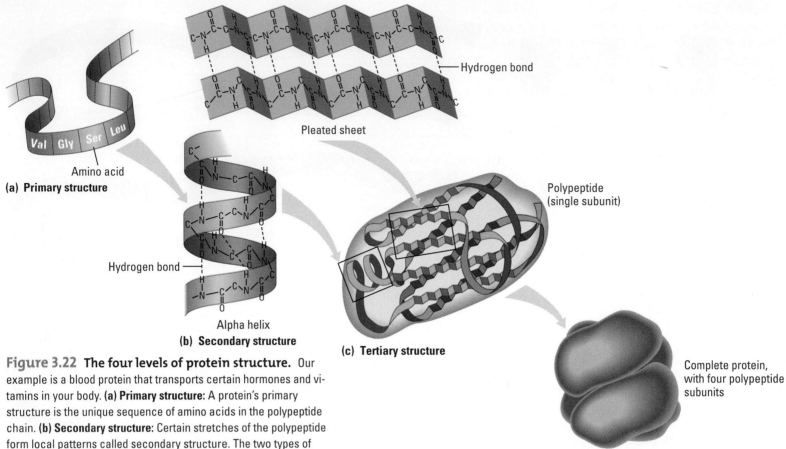

(a) Primary structure

Amino acid

Val Gly Ser Leu

Hydrogen bond

Pleated sheet

Hydrogen bond

Alpha helix

(b) Secondary structure

(c) Tertiary structure

Polypeptide (single subunit)

Complete protein, with four polypeptide subunits

(d) Quaternary structure

Figure 3.22 The four levels of protein structure. Our example is a blood protein that transports certain hormones and vitamins in your body. **(a) Primary structure:** A protein's primary structure is the unique sequence of amino acids in the polypeptide chain. **(b) Secondary structure:** Certain stretches of the polypeptide form local patterns called secondary structure. The two types of secondary structure are named alpha helix and pleated sheet. Secondary structure is reinforced by hydrogen bonds along the polypeptide backbone, similar to the weak hydrogen bonds between water molecules (see Figure 2.11). Dashed lines symbolize the hydrogen bonds in this diagram. The structure illustrated here is simplified, showing only the atoms of the polypeptide backbone, not those of the amino acids' side groups. **(c) Tertiary structure:** The overall three-dimensional shape of the protein is called tertiary structure. It is reinforced by chemical bonds (not shown here) between the side groups of amino acids in different regions of the polypeptide chain. **(d) Quaternary structure:** Some proteins consist of not just one but two or more polypeptide chains. (For example, this blood protein is constructed from four polypeptides.) Such proteins have a quaternary structure, which results from bonding between the polypeptide chains.

Protein Shape By now, you might be thinking that *polypeptide chain* is a synonym for *protein*, but that is not quite true. The distinction is analogous to the relationship between a long strand of yarn and a sweater of particular size and shape that you could knit from the yarn. A functional protein is not just a polypeptide chain, but one or more polypeptides precisely twisted, folded, and coiled into a molecule of unique shape. If we dissect the overall shape of a protein, it is possible to recognize at least three levels of structure: primary, secondary, and tertiary. Proteins having more than one polypeptide chain have a fourth level: quaternary structure. You can examine these levels of protein structure in Figure 3.22.

See animations of the levels of protein structure in Web/CD Activity 3H.

When a cell makes a polypeptide, the chain usually folds spontaneously to form the functional shape for that protein. For example, those regions of the polypeptide enriched with amino acids having hydrophobic ("water-fearing") side groups, such as the leucine in Figure 3.18b, will congregate in the interior of the protein, away from water. Regions rich in amino acids having hydrophilic ("water-loving") side groups will assume locations on the protein surface, immersed in the water of the aqueous solution. As the protein begins to take form, chemical bonding between different parts of the polypeptide reinforces the shape.

It is a protein's three-dimensional shape that enables the molecule to carry out its specific duty in a cell. In almost every case, a protein's function depends on its ability to recognize and bind to some other molecule. For example, the receptors on the brain cell in Figure 3.5 are actually proteins that recognize certain chemical signals from other cells.

A protein's shape is sensitive to the surrounding environment. An unfavorable change in temperature, pH, or some other quality of the aqueous environment can cause a protein to unravel and lose its normal shape. This is called **denaturation** of the protein. If you cook an egg, the transformation of the egg white from clear to opaque is caused by proteins in the egg white denaturing. The denatured proteins become insoluble in water and form a white solid. Though good for you to eat, these storage proteins would be useless to a bird embryo developing in an egg. And that's the main point about denaturation: When a protein denatures and loses its shape, it also loses its ability to work properly, for a protein's function depends on its shape. Let's now see how this relationship of structure to function applies to proteins that perform as enzymes.

Enzymes The many chemical reactions that occur in organisms are collectively called **metabolism.** Few metabolic reactions would occur without the assistance of catalysts, which are agents that speed up chemical reactions. The main catalysts of life are specialized proteins called **enzymes.**

Why are enzymes necessary, and how do they work? The chemistry of life is "cold chemistry" in the sense that cells are not hot enough for most reactions to occur without some help. To initiate a chemical reaction, it is first necessary to break chemical bonds in the reactant molecules. That activation process requires that the molecules absorb energy from their surroundings, usually in the form of heat. This energy is called **activation energy,** because it activates the reactants and triggers a chemical reaction.

One way to speed up a chemical reaction is to increase the supply of energy by heating the mixture, which is why flame-producing devices called Bunsen burners are used in chemistry labs. Boiling your cells, however, is not an option for speeding up your metabolism. Enzymes enable metabolism to occur at cooler temperatures by reducing the amount of activation energy required to break the bonds of reactant molecules. If you think of

Experiment with enzymes in The Process of Science activity available on the website and CD-ROM.

the requirement for activation energy as a barrier to a chemical reaction, then the function of an enzyme is to lower that barrier (Figure 3.23). An enzyme does this by binding to reactant molecules and putting them under some kind of physical or chemical stress, making it easier to break their bonds and start a reaction.

Each enzyme is very selective in the reaction it catalyzes. This specificity is based on the ability of the enzyme to recognize the shape of a certain reactant molecule, which is termed the enzyme's **substrate.** And the ability of the enzyme to recognize and bind to its specific substrate depends on the enzyme's shape. A special region of the enzymatic protein, called the **active site,** has a shape and chemical behavior that fits it to the substrate molecule. When a substrate molecule slips into this docking station, the active site changes shape slightly to embrace the substrate and catalyze the reaction. This interaction is called **induced fit**

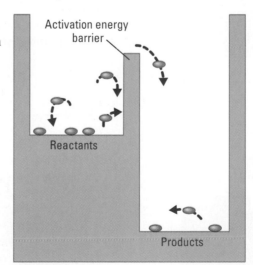

(a) Without enzyme

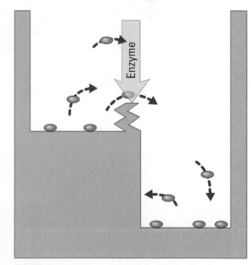

(b) With enzyme

Figure 3.23 Activation energy: A jumping-bean analogy. (a) A chemical reaction requires activation energy to break the bonds of the reactant molecules. The jumping beans in this analogy represent reactant molecules that must overcome the barrier of activation energy before they can reach the "product" side. **(b)** An enzyme speeds the process by lowering the barrier of activation energy.

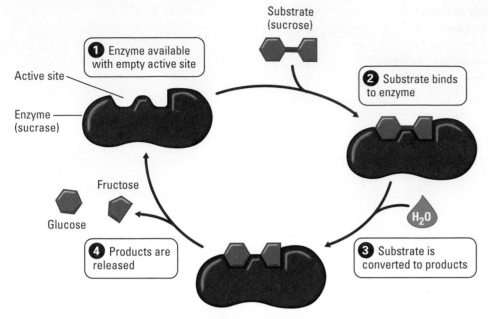

① Enzyme available with empty active site

Active site

Enzyme (sucrase)

Substrate (sucrose)

② Substrate binds to enzyme

H_2O

③ Substrate is converted to products

④ Products are released

Fructose

Glucose

Figure 3.24 How an enzyme works. Our example is the enzyme sucrase, named for its substrate, sucrose. ① With its active site empty, sucrase is receptive to a molecule of its substrate. ② The substrate has a shape that fits the shape of the active site. Sucrose enters the active site, and ③ the enzyme catalyzes the chemical reaction—in this case the hydrolysis of sucrose. ④ The products—glucose and fructose in this example—exit the active site, and ① sucrase is available to receive another molecule of its substrate.

because the entry of the substrate induces the enzyme to change its shape slightly and make the fit between substrate and active site even snugger. The products are then released from the active site, and the enzyme is now available to accept another molecule of its specific substrate. In fact, the ability to function over and over again is a key characteristic of catalysts, including enzymes, the catalysts of life. Figure 3.24 follows the action of a specific enzyme called sucrase, which hydrolyzes the double sugar sucrose (the substrate). Like sucrase, many enzymes are named for their substrates, but with an *-ase* ending.

See enzymes in action in Web/CD Activity 3I.

Certain molecules can inhibit a metabolic reaction by binding to an enzyme and disrupting its function (Figure 3.25). Some of these enzyme inhibitors are actually substrate imposters that plug up the active site. Other inhibitors bind to the enzyme at some site remote to the active site, but the binding changes the shape of the enzyme so that its active site is no longer receptive to the substrate. In some cases, the binding is reversible, enabling certain inhibitors to play a normal role in regulating metabolism. For example, if metabolism is producing more of a certain product than a cell needs, that very product may inhibit an enzyme required for its production. This **feedback regulation** keeps the cell from wasting resources that could be put to better use. However, some enzyme inhibitors act as poisons that block metabolic processes essential to the survival of an organism. For example, an insecticide called malathion inhibits an enzyme required for normal functioning of the insect nervous system. Many antibiotics that kill disease-causing bacteria are also enzyme inhibitors. For instance, penicillin inhibits an enzyme that many bacteria use to make their cell walls. Because humans do not have this enzyme, the antibiotic can help us combat an infection without destroying our own cells.

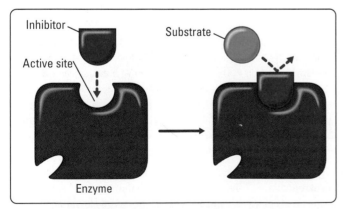

Substrate

Active site

Enzyme

(a) Normal enzyme action

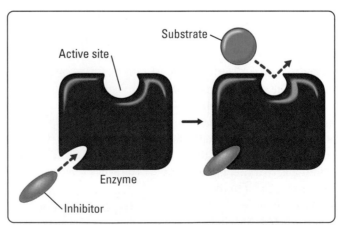

Inhibitor

Active site

Substrate

Enzyme

(b) Enzyme inhibition by a substrate imposter

Figure 3.25 Enzyme inhibitors.

Substrate

Active site

Enzyme

Inhibitor

(c) Enzyme inhibition by a molecule that causes the active site to change shape

What Determines Protein Structure? Our study of enzymes has reinforced the idea that each protein has a unique shape that enables it to do a certain job in a cell. Given an environment suitable for that protein (so that it doesn't denature), the primary structure of a protein causes it to fold into its functional shape. But what determines primary structure, a protein's specific amino acid sequence? If left to chance, it is close to impossible that amino acids would link into a chain of just the right sequence for a particular protein. But cells do not leave their protein building to chance. Each polypeptide chain has a sequence specified by an inherited gene. And that relationship between genes and proteins brings us to this chapter's last category of molecules.

Nucleic Acids

Those most famous of chemical initials, **DNA,** stand for deoxyribonucleic acid. **Nucleic acids** provide the directions for building proteins. There are actually two types of nucleic acids: DNA and **RNA** (for ribonucleic acid). The genetic material that humans and other organisms inherit from their parents consists of DNA. Within the DNA are genes, specific stretches of giant DNA molecules that program the amino acid sequences (primary structure) of proteins. Those directions, however, are written in a kind of chemical code that must be translated from "nucleic acid language" to "protein language." A cell's RNA molecules help translate. You'll learn more about the genetic code and RNA in Chapter 9. We'll focus here on the structure of DNA, the molecule of inheritance.

Nucleic acids are polymers of monomers called **nucleotides** (Figure 3.26a). Each nucleotide is itself a complex organic molecule with three parts: a sugar, a phosphate group (a phosphorus atom bonded to oxygen atoms), and a nitrogen-containing **base** (so called because it behaves as a base, or alkali, in aqueous solutions). The part of the molecule that varies between different types of nucleotides is the base. Each DNA nucleotide has one of the following four bases: adenine (abbreviated A), guanine (G), thymine (T), or cytosine (C). Thus, all genetic information is written in a four-letter alphabet—A, G, C, T—the bases that distinguish the four nucleotides that make up DNA.

The cell uses dehydration synthesis to link nucleotides into long polymers, the DNA strands (Figure 3.26b). One long DNA strand contains many genes, each a specific series of hundreds or thousands of nucleotides. And each of these genes has information written in its unique sequence of nucleotide bases. In fact, it is this information that cells translate into an amino acid sequence to make a specific protein.

How can cells of parents copy their genes to pass along to offspring? Inheritance is based on DNA actually being double-stranded, with the two DNA strands wrapped around each other to form a **double helix** (Figure 3.26c). In the core of the helix, the bases along one DNA strand bond to bases along the other strand. This base pairing is specific: The base A can pair only with T, and G can pair only with C. Thus, if you know the sequence of bases along one DNA strand, you also know the sequence along the complementary strand in the double helix.

Review DNA structure and make copies of a DNA molecule in Web/CD Activity 3J.

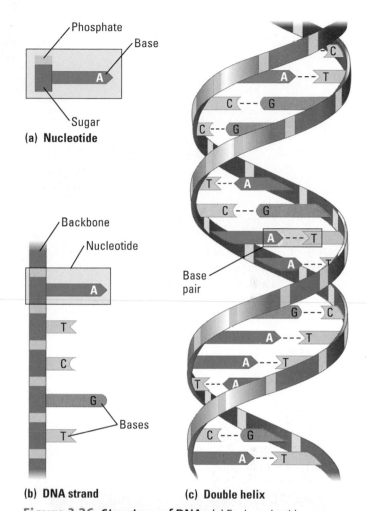

(a) **Nucleotide**

(b) **DNA strand**   (c) **Double helix**

Figure 3.26 Structure of DNA. (a) Each nucleotide monomer consists of three parts: a sugar (deoxyribose in the case of DNA nucleotides); a phosphate (phosphorus atom bonded to oxygen); and a nitrogenous (nitrogen-containing) base. In these simplified diagrams, the three parts of a nucleotide are symbolized with shapes and colors rather than actual chemical structures. **(b)** A DNA strand is a polymer of nucleotides linked into a backbone, with appendages consisting of the bases. A strand has a specific sequence of the four bases, abbreviated A, G, C, and T. **(c)** A double helix consists of two DNA strands held together by bonds between bases. The bonds are individually weak—they are hydrogen bonds, like those between water molecules—but they zip the two strands together with a cumulative strength that gives the double helix its stability. The base pairing is specific: A always pairs with T; G always pairs with C.

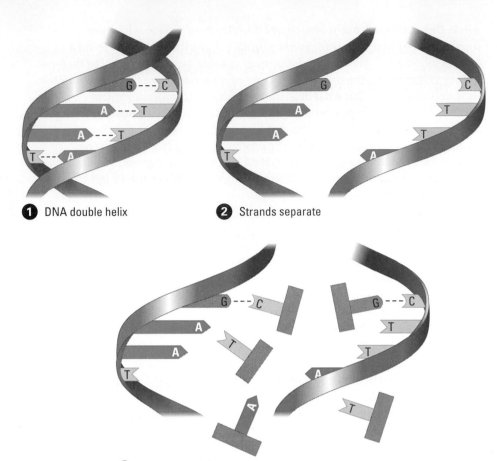

1 DNA double helix

2 Strands separate

3 Each strand serves as template for a new complementary strand

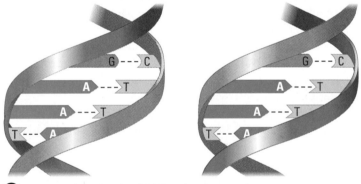

4 Two identical "daughter" DNA molecules

Figure 3.27 DNA replication: The basic concept. DNA replicates when its two strands separate and each old strand serves as a template for ordering nucleotides into a new complementary strand according to the base-pairing rules.

Figure 3.28 The signs of inheritance. The concept that parents pass traits along to their offspring is very visible to us in family resemblance. But it is submicroscopic chemistry, the replication of DNA and its transmission in the father's sperm and mother's eggs, that is behind the adage "Like begets like."

Now imagine separating the two strands and using the base sequence of each to regenerate a new complementary strand (Figure 3.27). Starting with one double helix, you would end up with two exact copies of the "parent" molecule. This DNA replication is the molecular basis of inheritance (Figure 3.28).

1. _____ is to polysaccharide as amino acid is to _____.

2. The formula for glucose is $C_6H_{12}O_6$. What would be the formula for a sugar having just three carbons instead of six?

3. How do manufacturers produce the high-fructose corn syrup listed as an ingredient on a bottle of soda?

4. Why can't you digest wood?

5. On a food package label, what is the meaning of "unsaturated fats"?

6. In classifying the molecules of life, what do fats and human sex hormones have in common?

7. Indicate which of the following is *not* a protein: an enzyme, hemoglobin, a gene.

8. In what way is the production of a polypeptide similar to the production of a polysaccharide?

9. Why does a denatured protein no longer function?

10. How does an enzyme recognize its substrate?

11. How does the antibiotic penicillin work?

12. In a double helix, a region along one DNA strand has this sequence of bases: GAATGC. What is the base sequence along the complementary region of the other strand of the double helix?

Answers: 1. Monosaccharide (or simple sugar); protein (or polypeptide) **2.** $C_3H_6O_3$ **3.** A commercial process converts glucose in the syrup to the much sweeter fructose. **4.** Humans lack an enzyme that digests cellulose, the main ingredient of wood. **5.** Unsaturated fats have less than the maximum hydrogen content. This is because double bonds between carbons in the fatty acids replace some of the bonds between carbon and hydrogen **6.** Both fats and sex hormones, which are steroids, are classified as lipids because they are hydrophobic. **7.** A gene **8.** Both of these polymers consist of monomers linked by dehydration synthesis. **9.** Protein function depends on shape, which is altered by denaturation. **10.** The substrate and the enzyme's active site are complementary in shape. **11.** It inhibits an enzyme that certain bacteria use to make their cell walls. **12.** CTTACG

Evolution Link: DNA and Proteins as Evolutionary Tape Measures

Genes (DNA) and their products (proteins) are historical documents. These information-rich molecules are the records of an organism's hereditary background. The linear sequences of nucleotides in DNA molecules are passed from parents to offspring, and these DNA sequences determine the amino acid sequences of proteins in the offspring. The DNA and proteins of siblings are more similar than the DNA and proteins of unrelated individuals of the same species. This concept of molecular genealogy also extends to relationships between species.

Testable hypotheses are at the heart of science, and analysis of DNA and protein sequences adds a new tool for testing evolutionary hypotheses. For example, fossil evidence and anatomical similarity support the hypothesis that humans and gorillas are closely related animals. This hypothesis is testable— we can use it to make predictions about what to expect if the hypothesis is correct. For example, *if* humans and gorillas are closely related, *then* they should share a greater

Discover more about how DNA and proteins are used as evolutionary tape measures in the Web Evolution Link.

proportion of their inherited DNA and protein sequences than they do with more distantly related species. If molecular analysis did not confirm this prediction, then the test would cast doubt on the hypothesis of a close evolutionary relationship. In fact, however, molecular analysis of DNA and protein sequences in humans and gorillas supports the hypothesis that these two species are very closely related (Figure 3.29). Molecular biology has added a new tape measure to the tool kit biologists use to assess evolutionary relationships.

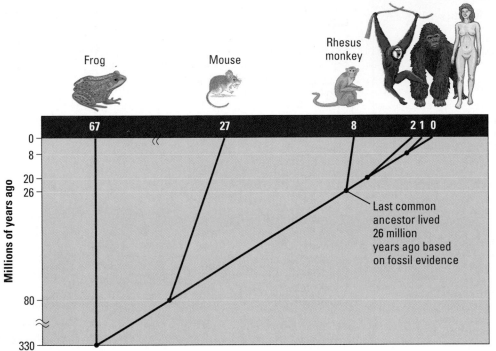

Figure 3.29 Evolutionary relationships among six vertebrates, based on amino acid sequences of a protein. Inherited DNA determines the amino acid sequences of proteins. In this case, the sequences are for one of the polypeptides making up hemoglobin, the oxygen-transporting protein in the blood of vertebrates (animals with backbones). The times marking the evolutionary branch points are based on the fossil history of these animals.

Chapter Review

Summary of Key Concepts

Overview: Organic Molecules

- **Carbon Skeletons and Functional Groups** Carbon atoms can form large, complex, diverse molecules by bonding to four partners, including other carbon atoms. In addition to variations in the size and shape of carbon skeletons, organic molecules also vary in the presence and locations of different functional groups. These groups of atoms are the most chemically reactive parts of the organic molecule.
- Web/CD Activity 3A *Diversity of Carbon-Based Molecules*
- Web/CD Activity 3B *Functional Groups*

- **Giant Molecules from Smaller Building Blocks** Cells build macromolecules called polymers by using dehydration synthesis to join smaller molecules, the monomers. The reverse process, hydrolysis, dismantles polymers to their building blocks (monomers).
- Web/CD Activity 3C *Making and Breaking Giant Molecules*

Biological Molecules

- **Carbohydrates** Simple sugars (monosaccharides) provide cells with energy and carbon skeletons to build other organic molecules. Double sugars

(disaccharides), such as sucrose (table sugar), consist of two monosaccharides joined by dehydration synthesis. Polysaccharides are macromolecules, long polymers of sugar monomers. Starch and glycogen are storage polysaccharides in plants and animals, respectively. The cellulose of plant cell walls is an example of a structural polysaccharide.
- Web/CD Activity 3D *Models of Glucose*
- Web/CD Activity 3E *Carbohydrates*

- **Lipids** Along with other subclasses of lipids, fats are hydrophobic. Fats are the major form of long-term energy storage in animals. A fat, or triglyceride, consists of three fatty acids joined to a glycerol by dehydration synthesis. Most animal fats are saturated, meaning that their fatty acids have the maximum amount of hydrogen. Plant oils are unsaturated fats, having less hydrogen in the fatty acids because of double bonding in the carbon skeletons. Steroids, including cholesterol and sex hormones, are also lipids.
- Web/CD Activity 3F *Lipids*

- **Proteins** The monomers of proteins, amino acids, come in 20 varieties. They are linked by dehydration synthesis to form polymers called polypeptides. A protein consists of one or more polypeptides folded into a specific three-dimensional shape. Contributing to this shape are four levels of structure: the protein's amino acid sequence (primary structure); localized coiling or pleating in certain regions of the protein (secondary structure); other contortions that reinforce the overall shape of the protein (tertiary structure); and association of subunits in proteins with more than one polypeptide (quaternary structure). Shape is sensitive to environment, and if a protein loses its shape because of an unfavorable environment (denaturation), the protein's function is also disrupted. Most processes in cells use proteins. Many of those proteins are enzymes, the biological catalysts that speed up the chemical reactions of metabolism by lowering activation energy. An enzyme has an active site that fits a specific substrate, which the enzyme converts to products.

• **Web/CD Activity 3G** *The Functions of Proteins*
• **Web/CD Activity 3H** *The Structure of Proteins*
• **Web/CD Activity 3I** *How Enzymes Work*
• **Web/CD The Process of Science** *Enzyme Catalysis Lab*

• **Nucleic Acids** The chemical of inheritance is a nucleic acid called DNA. It takes the form of a double helix, two DNA strands (polymers of nucleotides) held together by bonds between nucleotide components called bases. There are four kinds of bases: adenine (A), guanine (G), thymine (T), and cytosine (C). A always pairs with T, and G always pairs with C. It is these base-pairing rules that provide the information for gene copying when two DNA strands separate and regenerate new complementary strands.

• **Web/CD Activity 3J** *Nucleic Acids*

Evolution Link: DNA and Proteins as Evolutionary Tape Measures

• Closely related species share a greater proportion of their inherited DNA and proteins than do more distantly related species.

• **Web Evolution Link** *DNA and Proteins as Evolutionary Tape Measures*

Self-Quiz

1. Draw two isomers of the hydrocarbon C_4H_{10}.

2. Your digestive system is equipped with a diversity of enzymes that break the polymers in your food down to monomers that your cells can assimilate. A generic name for these digestive enzymes is *hydrolase*. What is the chemical basis for that name? hydrolyze

3. Which of the following terms includes all others in the list?
    a. polysaccharide
    b. carbohydrate
    c. monosaccharide
    d. disaccharide
    e. starch

4. Based on your experience with common materials, which of the following is an example of a hydrophobic material?
    a. paper
    b. table salt
    c. wax
    d. sugar
    e. pasta

5. Lipids differ from other large biological molecules in that they
    a. are much larger.
    b. are not truly polymers.
    c. do not have specific shapes.
    d. do not contain carbon.
    e. contain nitrogen atoms.

6. What chemical element is present in proteins that is absent in sugars and fats?

7. The enzyme called pancreatic amylase is a protein whose job is to attach to starch molecules in food and help break them down to disaccharides. Amylase cannot break down cellulose. Why not?
    a. Cellulose is a kind of fat, not a carbohydrate like starch.
    b. Cellulose molecules are much too large.
    c. Starch is made of glucose; cellulose is made of other sugars.
    d. The bonds between sugars in cellulose are much stronger.
    e. The sugars in cellulose bond together differently than in starch, giving cellulose a different shape.

8. Which of the following molecules is *incorrectly* matched with its chemical class?
    a. gene—nucleic acid
    b. enzyme—protein
    c. starch—carbohydrate
    d. sucrose—carbohydrate
    e. sugar—lipid

9. A glucose molecule is to starch as
    a. a steroid is to a lipid.
    b. a protein is to an amino acid.
    c. a nucleic acid is to a polypeptide.
    d. a nucleotide is to a nucleic acid.
    e. an amino acid is to a nucleic acid.

10. A shortage of phosphorus in the soil would make it especially difficult for a plant to manufacture
    a. DNA.   d. fatty acids.
    b. proteins.   e. sucrose.
    c. cellulose.

• **Go to the website or CD-ROM for more self-quiz questions.**

The Process of Science

1. A food manufacturer is advertising a new cake mix as fat-free. Scientists at the U.S. Food and Drug Administration (FDA) are testing the product to see if it truly lacks fat. Hydrolysis of the cake mix yields glucose, fructose, glycerol, a number of amino acids, and several kinds of molecules with long hydrocarbon chains. Further analysis shows that most of the hydrocarbon chains have a carboxyl group at one end. What would you tell the food manufacturer if you were the spokesperson for the FDA?

2. **Experiment with enzymes in The Process of Science activity on the website and CD-ROM.**

Biology and Society

Some amateur and professional athletes take anabolic steroids to help them "bulk up" or build strength. The health risks of this practice are extensively documented. Apart from these health issues, what is your opinion about the ethics of athletes using chemicals to enhance performance? Is this a form of cheating, or is it just part of the preparation required to stay competitive in a sport where anabolic steroids are commonly used? Defend your opinion.

A Tour of the Cell

Your body is a cooperative society of trillions of cells.

If you stacked up 8000 cell membranes

they would only be as thick as a page in this book.

Every second you produce about 2 million red blood

cells. A single nerve cell in the human

body can reach up to a meter in length. The cells of

a whale are about the same size as the cells of a

mouse.

Figure 4.1 Microscopic view of an animal cell.

Overview: The Microscopic World of Cells

Cells are as fundamental to biology as atoms are to chemistry. All organisms are made of cells. Organisms are either single-celled, such as most bacteria and protists, or multicelled, such as plants, animals, and most fungi. Your own body is a cooperative society of trillions of cells of many different specialized types. Just three examples are the muscle cells that move your arms and legs, the nerve cells that control your muscles, and the red blood cells that carry oxygen to your muscles and nerves. Everything you do—every action and every thought—reflects processes occurring at the cellular level. To explore this world of cells, our main tools are microscopes.

Microscopes as Windows to Cells

Human understanding of nature often parallels the invention and refinement of instruments that extend human senses to new limits. The development of microscopes, for example, provided increasingly clear windows to the world of cells.

The type of microscope used by Renaissance scientists, as well as the microscope you will use if your biology course includes a lab, is called a **light microscope (LM).** Visible light passes through the specimen, such as sperm cells (Figure 4.2a). Glass lenses then enlarge the image and project it into a human eye or a camera.

Two important values of microscopes are magnification and resolving power. **Magnification** is how much larger the object appears compared to its actual size. The clarity of that magnified image depends on **resolving power,** which is the ability of an optical instrument to show two objects as separate. For example, what appears to the unaided eye as one star in the sky may be resolved as twin stars with a telescope. Each optical instrument, be it an eye, a telescope, or a microscope, has a limit to its resolving power. For light microscopes, the resolving power is limited to about 0.2 micrometer (μm), the size of a small bacterial cell (1 μm = $\frac{1}{1000}$ millimeter, or 10^{-3} mm). This limits useful magnification to about 1,000×. With greater magnification, the image is blurry.

In 1665, a British scientist named Robert Hooke used a light microscope in first describing cells in a thin slice of cork from the bark of an oak tree. For the next two centuries, scientists found cells in every organism they examined with a microscope. By the mid-1800s, this accumulation of evidence led to the **cell theory,** which includes the induction that all living things are composed of cells.

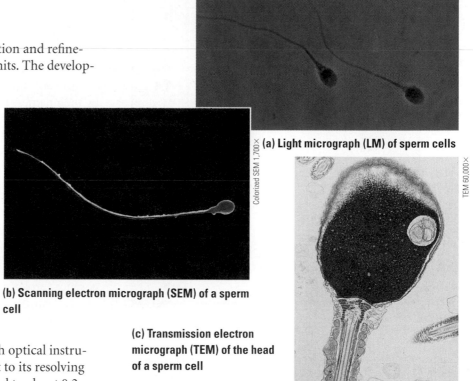

LM 2,090×

(a) Light micrograph (LM) of sperm cells

Colorized SEM 1,700×

(b) Scanning electron micrograph (SEM) of a sperm cell

TEM 60,000×

(c) Transmission electron micrograph (TEM) of the head of a sperm cell

Figure 4.2 Different views of sperm cells. These photographs were taken with different types of microscopes: **(a)** light microscope (LM); **(b)** scanning electron microscope (SEM); and **(c)** transmission electron microscope (TEM). Such photographs taken with microscopes are called micrographs. Throughout this textbook, each micrograph will have a notation along its side. For example, in (a), "LM 2,090×" indicates that the micrograph was taken with a light microscope and the objects are magnified to 2,090× their original size.

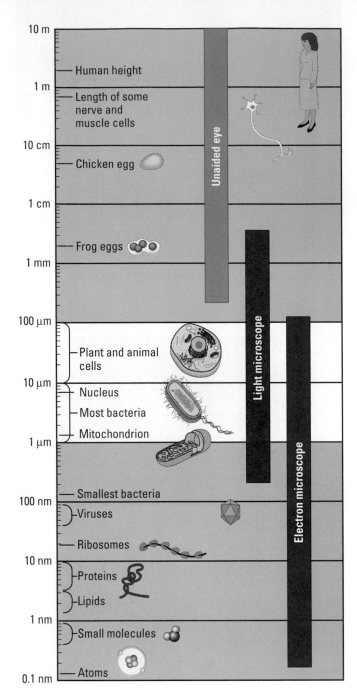

Measurement Equivalents
1 meter (m) = 100 cm = 1000 mm = about 39.4 inches
1 centimeter (cm) = 10^{-2} (1/100) meter (m) = about 0.4 inch
1 millimeter (mm) = 10^{-3} (1/1000) m = 1/10 cm
1 micrometer (μm) = 10^{-6} m = 10^{-3} mm
1 nanometer (nm) = 10^{-9} m = 10^{-3} μm

Our knowledge of cell structure took a giant leap forward as biologists began using electron microscopes in the 1950s. Instead of using light, the **electron microscope (EM)** uses a beam of electrons. The electron microscope has a much better resolving power than the light microscope. In fact, the most powerful modern electron microscopes can distinguish objects as small as 0.2 nanometer (1 nm = $\frac{1}{1000}$ μm, or 10^{-3} μm). This is a thousand-fold improvement over the light microscope. The period at the end of this sentence is about a million times bigger than an object 0.2 nanometer in diameter. The highest-power electron micrographs you will see in this book have magnifications of about 100,000×. Such power reveals the diverse parts, or **organelles** ("little organs"), within a cell (Figure 4.3).

Explore size and scale in The Process of Science activity available on the website and CD-ROM.

Figure 4.2b and c show images taken with two kinds of electron microscopes. Biologists use the **scanning electron microscope (SEM)** to study the detailed architecture of the surface of a cell. The **transmission electron microscope (TEM)** is especially useful for exploring the internal structure of a cell. Preparing specimens for both types of electron microscopes kills and preserves cells before they can be examined. Thus, light microscopes are still very useful, especially as windows to living cells.

The Two Major Classes of Cells

Two basic kinds of cells evolved on Earth: **prokaryotic cells** and **eukaryotic cells** (Figure 4.4). Bacteria and archaea consist of prokaryotic cells (see Chapter 1). All other organisms—protists, plants, fungi, and animals—are composed of eukaryotic cells.

A major difference between these two main classes of cells is indicated by their names. The word *eukaryotic* is from the Greek *eu,* meaning "true," and *karyon,* meaning "kernel." In this case, "kernel" refers to the nucleus, usually the largest organelle within eukaryotic cells. The nucleus, bordered by a membranous envelope, houses most of a eukaryotic cell's genetic material, its DNA (see Figure 3.26). A prokaryotic cell lacks such a nucleus; its DNA is coiled in a nucleoid region, which, unlike a true nucleus, is not partitioned from the rest of the cell by a membrane (Figure 4.5). The word *prokaryotic* (Greek *pro,* before, and *karyon,* kernel) implies that prokaryotic cells appeared on Earth before eukaryotic cells. That evolutionary sequence is evident in the fossil record. Though named for the presence (eukaryotic) or absence (prokaryotic) of a true nucleus, this distinction is just one example of a fundamental difference in complexity between these two major classes of cells. We'll examine prokaryotes in more detail in Chapter 14. Eukaryotic cells are our main focus in this chapter.

Review prokaryotic cell structure in Web/CD Activity 4A.

Figure 4.3 The size range of cells. Starting at the top of this scale with 10 meters and going down, each reference measurement along the left side marks a tenfold decrease in size. Most cells are between 1 and 100 μm in diameter (the yellow area in the diagram), a size range that can be viewed with either a light microscope or an electron microscope, but not the unaided eye.

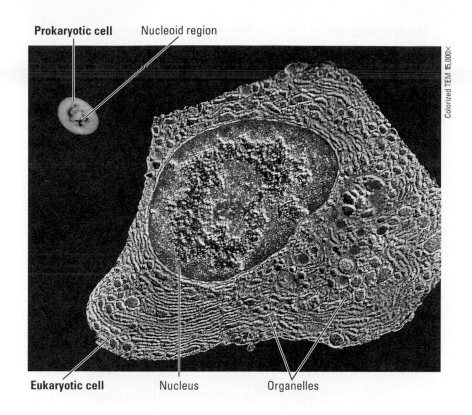

Prokaryotic cell Nucleoid region

Colorized TEM 15,000×

Eukaryotic cell Nucleus Organelles

Figure 4.4 Contrasting the size and complexity of prokaryotic and eukaryotic cells. The smaller cell is a bacterium, an example of a prokaryote. Lacking most of the organelles of eukaryotic cells, prokaryotic cells are much simpler in structure. The prokaryotic cell's DNA is concentrated in a nucleoid region, the denser mass within this cell. No membrane separates the nucleoid region from the rest of the cell. The larger cell here is a eukaryotic cell. Its largest organelle is the nucleus. You can see many other types of organelles outside the nucleus. Note again the difference in size between these two cells. Eukaryotic cells are generally about ten times larger in diameter than prokaryotic cells (see Figure 4.3 for size ranges).

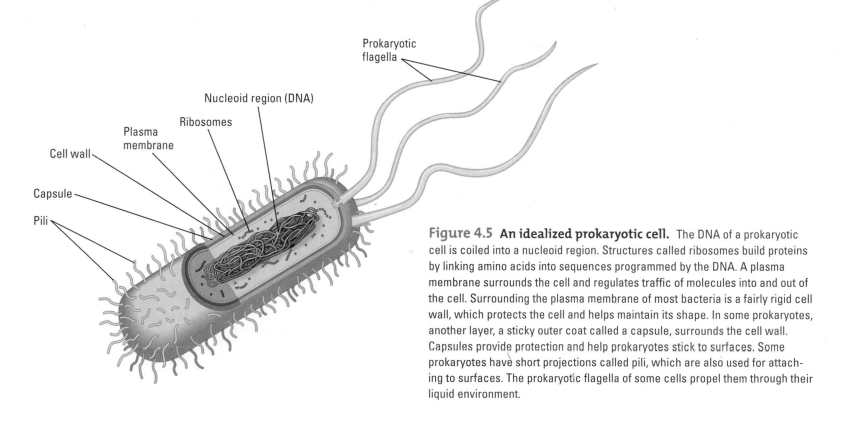

Prokaryotic flagella

Nucleoid region (DNA)

Ribosomes

Plasma membrane

Cell wall

Capsule

Pili

Figure 4.5 An idealized prokaryotic cell. The DNA of a prokaryotic cell is coiled into a nucleoid region. Structures called ribosomes build proteins by linking amino acids into sequences programmed by the DNA. A plasma membrane surrounds the cell and regulates traffic of molecules into and out of the cell. Surrounding the plasma membrane of most bacteria is a fairly rigid cell wall, which protects the cell and helps maintain its shape. In some prokaryotes, another layer, a sticky outer coat called a capsule, surrounds the cell wall. Capsules provide protection and help prokaryotes stick to surfaces. Some prokaryotes have short projections called pili, which are also used for attaching to surfaces. The prokaryotic flagella of some cells propel them through their liquid environment.

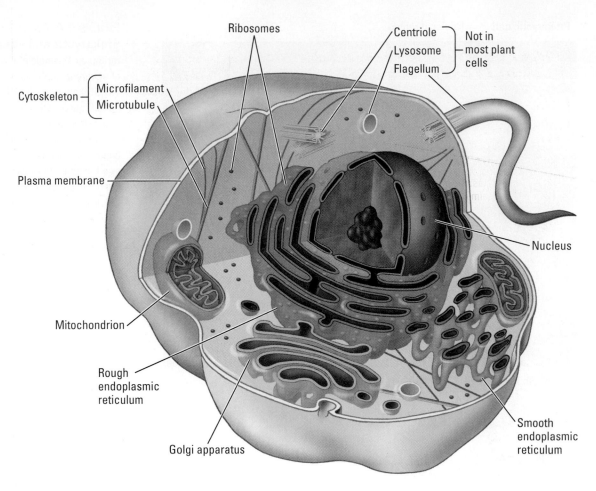

Figure 4.6 **Panoramic view of idealized animal cell and plant cell.** For now, the labels are just words, but these organelles will come to life as we take a closer look at how each of these parts works. To keep from getting lost on our tour of cells, we'll carry miniature versions of these overview diagrams as our road maps, with "you are here" highlighted.

Ribosomes

Centriole
Lysosome
Flagellum
Not in most plant cells

Cytoskeleton — Microfilament
Microtubule

Plasma membrane

Nucleus

Mitochondrion

Rough endoplasmic reticulum

Smooth endoplasmic reticulum

Golgi apparatus

(a) An animal cell

A Panoramic View of Eukaryotic Cells

Figure 4.6 provides a panoramic view of an idealized animal cell and plant cell, two examples of eukaryotic cells. The similarities are more obvious than the differences. Both cells have a very thin outer membrane, the **plasma membrane,** which regulates the traffic of molecules between the cells and their surroundings. (Prokaryotic cells are also bordered by plasma membranes.) Each cell also has a prominent **nucleus,** the gene-containing organelle for which eukaryotic cells are named.

The entire region of the cell between the nucleus and plasma membrane is called the **cytoplasm.** It consists of various organelles suspended in a fluid, the **cytosol.** The structure of each organelle has become adapted during evolution for specific functions. Many of these organelles are bordered by their own membranes, which maintain a unique chemical environment within the organelle. Thus, a eukaryotic cell divides the labor of life among many internal compartments.

As you can see in Figure 4.6, most of the organelles are found in both animal and plant cells. One important difference is the presence of chloroplasts in plant cells but not in animal cells. Chloroplasts are the organelles that convert light energy to the chemical energy of food. Also note that

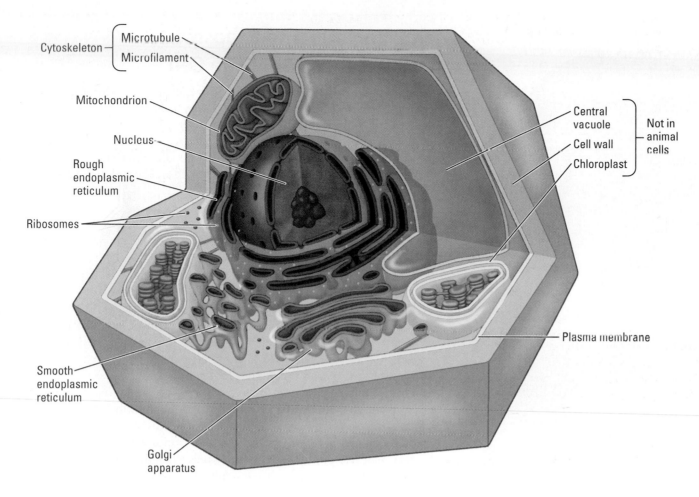

Cytoskeleton — Microtubule
           — Microfilament

Mitochondrion

Nucleus

Rough
endoplasmic
reticulum

Ribosomes

Smooth
endoplasmic
reticulum

Golgi
apparatus

Central
vacuole

Cell wall   } Not in
                   animal
Chloroplast   } cells

Plasma membrane

(b) A plant cell

Build an animal cell
and a plant cell in
Web/CD Activity 4B.

plant cells have walls exterior to their plasma membranes. Animal cells lack cell walls. We'll see other differences and similarities between plant and animal cells as we now take a closer look at the architecture of eukaryotic cells.

CheckPoint

1. Which type of microscope would you use to study **(a)** the changes in shape of a living human white blood cell; **(b)** the finest details of surface texture of a human hair; **(c)** the detailed structure of an organelle in the cytoplasm of a human liver cell?

2. Using a light microscope to examine a thin section of a large spherical cell, you find that it is 0.3 mm in diameter. The nucleus is about one-fourth as wide. What is the diameter of the nucleus in micrometers?

3. How is the nucleoid region of a prokaryotic cell unlike the nucleus of a eukaryotic cell?

Answers: 1. (a) light microscope; **(b)** scanning electron microscope; **(c)** transmission electron microscope **2.** About 75 μm **3.** There is no membrane enclosing the DNA of the nucleoid region.

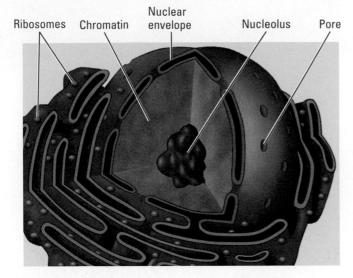

Ribosomes Chromatin Nuclear envelope Nucleolus Pore

Figure 4.7 The nucleus.

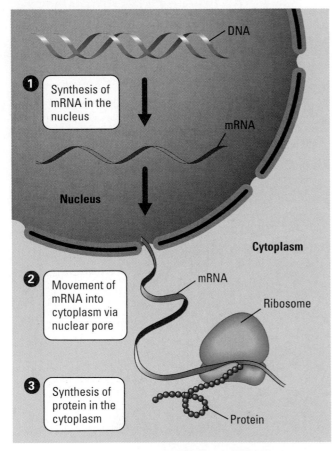

Figure 4.8 DNA → RNA → Protein: How genes in the nucleus control the cell. In a eukaryotic cell, DNA in the nucleus programs protein production in the cytoplasm by dictating the synthesis of messenger RNA (mRNA). The mRNA travels to the cytoplasm and binds to ribosomes. As a ribosome (disproportionately large in this drawing) moves along the mRNA, the genetic message is translated into a protein of specific amino acid sequence. The structure and behavior of a cell reflects the kinds of proteins it has, but it is the inherited DNA that determines a cell's protein composition.

The Nucleus and Ribosomes: Genetic Control of the Cell

If the cell is a factory, then the nucleus is its executive boardroom. The top managers are the genes, the inherited DNA molecules that direct almost all the business of the cell. They do this by specifying the production of enzymes and other proteins. A **gene** is a specific stretch of DNA that contains the code for the structure of a specific protein. The actual sites of protein synthesis are organelles called ribosomes, which are located in the cytoplasm.

Structure and Function of the Nucleus

The nucleus is bordered by a double membrane called the **nuclear envelope** (Figure 4.7). Pores through the envelope allow the passage of material between the nucleus and the cytoplasm. Within the nucleus, the DNA is attached to certain proteins, forming long fibers called **chromatin.** Each fiber constitutes a **chromosome.** The number of chromosomes in a cell depends on the species; most of your cells have 46 chromosomes.

Most of the time, chromatin just looks like a tangled mess to anybody exploring with a microscope. But as a cell starts to divide (reproduce), the fibers coil up to form shorter, thicker structures, and individual chromosomes can be seen with a microscope. Before cell division takes place, the chromosomes replicate (duplicate). The two copies of each chromosome are then allocated to the two "daughter" cells when the parent cell divides. This assures that each new cell gets a full set of the inherited DNA molecules.

In addition to chromosomes, the nucleus also contains a ball-like mass of fibers and granules called the **nucleolus.** It produces the parts that make up ribosomes.

Ribosomes

The small dots in the cells in Figure 4.6 and outside the nucleus in Figure 4.7 are **ribosomes.** Although they are assembled from components made in the nucleus, ribosomes do not begin to work until they move to the cytoplasm. Ribosomal parts travel from the nucleus to the cytoplasm via the pores in the nuclear envelope. It is in the cytoplasm that ribosomes build all the cell's proteins. Some ribosomes are suspended in the cytosol, the fluid of the cytoplasm. They make enzymes and other proteins that will remain dissolved in the cytosol. Other ribosomes are attached to the outside of a membranous organelle called the endoplasmic reticulum. These ribosomes make the proteins of membranes and proteins the cell will secrete.

How DNA Controls the Cell

How does the DNA in the "executive boardroom" (the nucleus) direct the manufacture of proteins in the "hard-hat zone" (the cytoplasm)? DNA does this by transferring its coded information to a portable molecule called messenger RNA (mRNA) (Figure 4.8). These genetic messages exit the nu-

Get an overview of protein synthesis in Web/CD Activity 4C.

cleus through the pores in the envelope. Like a middle manager, an RNA molecule carries the order "Build this type of protein" from the nucleus to the cytoplasm. You'll learn how the message is actually translated into a specific protein in Chapter 9.

The Endomembrane System: Manufacture and Distribution of Cellular Products

Note again in Figure 4.6 that the cytoplasm of a eukaryotic cell is partitioned by membranes. Many of the membranous organelles belong to the **endomembrane system.** This system includes the endoplasmic reticulum, the Golgi apparatus, lysosomes, and vacuoles.

The Endoplasmic Reticulum

To extend our factory metaphor of the cell, the **endoplasmic reticulum (ER)** is one of the main manufacturing facilities. It produces an enormous variety of molecules. The ER is a membranous labyrinth of tubes and sacs running throughout the cytoplasm (*reticulum* is from the Latin word for "network"). The ER membrane separates its internal compartment from the surrounding cytosol (Figure 4.9). There are two distinct types of ER: **rough ER** and **smooth ER.** These two ER components are physically connected, but they differ in structure and function.

Rough ER The "rough" in "rough ER" refers to the appearance of this organelle in electron micrographs (see Figure 4.9). The roughness is due to ribosomes that stud the outside of the ER membrane. These ribosomes produce two main classes of proteins: membrane proteins and secretory proteins. Newly manufactured membrane proteins are inserted right into the ER membrane. Thus, one function of rough ER is the production of new membrane. Secretory proteins are those the cell will actually export (secrete). Cells that secrete a lot of protein, such as the cells of your salivary glands (they secrete an enzyme into your mouth), are especially rich in rough ER. Some of the products manufactured by rough ER are dispatched to other locations via **transport vesicles,** membranous spheres that bud from the ER (Figure 4.10).

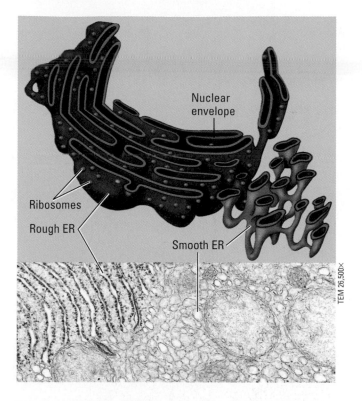

Figure 4.9 Endoplasmic reticulum (ER). The flattened sacs of rough ER and the tubes of smooth ER are continuous, though different in structure and function. Note that the ER is also continuous with the nuclear envelope, which is actually part of the endomembrane system.

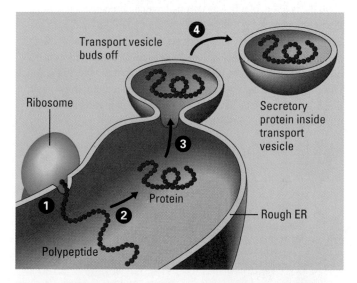

Figure 4.10 How rough ER manufactures and packages secretory proteins. A secretory protein is a protein that a cell exports to the fluid outside the cell. ❶ A ribosome links amino acids to form a polypeptide chain with a unique amino acid sequence (see Chapter 3). The chain threads through the membrane and into the cavity of the ER. ❷ Secretory proteins are often modified in the ER. ❸ Secretory proteins depart in transport vesicles that ❹ bud off from the ER. If the proteins require no further refinement in other organelles, they are secreted from the cell when the vesicles fuse with the plasma membrane, the outer membrane of the cell.

Smooth ER This organelle is so named because it lacks the ribosomes that decorate rough ER (see Figure 4.9). A diversity of enzymes built into the smooth ER membrane enables this organelle to perform many functions. One is the synthesis of lipids, including steroids (see Figure 3.15). For example, the cells in your ovaries or testes that produce sex hormones, which are steroids, are enriched with smooth ER.

In liver cells, the functions of smooth ER include the detoxification of drugs and other poisons that might be present in the bloodstream. For example, there are ER enzymes that detoxify sedatives such as barbiturates, stimulants such as amphetamines, and certain antibiotics (which is why they don't persist in the bloodstream after combating an infection). Undesirable complications may result when liver cells respond to drugs. As the cells are exposed to such chemicals, the amounts of smooth ER and its detoxifying enzymes increase, thereby increasing the body's tolerance to the drugs. This means that higher and higher doses of a drug are required to achieve a particular effect, such as sedation. The growth of smooth ER in response to one drug can also increase tolerance to other drugs, including important medicines. Barbiturate use, for example, may decrease the effectiveness of certain antibiotics by accelerating their breakdown in the liver.

The Golgi Apparatus

The **Golgi apparatus** is a refinery, warehouse, and shipping center. (This organelle is named for its discoverer, Italian scientist Camillo Golgi.) Working in close partnership with the ER, the Golgi apparatus finishes, stores, and distributes chemical products of the cell (Figure 4.11). Products made in the ER reach the Golgi in transport vesicles. Enzymes of the Golgi modify many of these products, and vesicles that bud from the Golgi then distribute the finished molecules to other organelles or to the plasma membrane. Vesicles that bind with the plasma membrane secrete finished chemical products to the outside of the cell.

Figure 4.11 The Golgi apparatus. This component of the endomembrane system consists of flattened sacs arranged something like a stack of pita bread. A cell may contain just a few of these Golgi stacks or hundreds. One function of the Golgi is to receive, refine, and distribute products of the ER. One side of a Golgi stack serves as a receiving dock for vesicles transported from the ER. The ER products are often modified during their stay in the Golgi. The "shipping" side of a Golgi stack serves as a depot from which the finished products can be dispatched in transport vesicles to other organelles or to the plasma membrane.

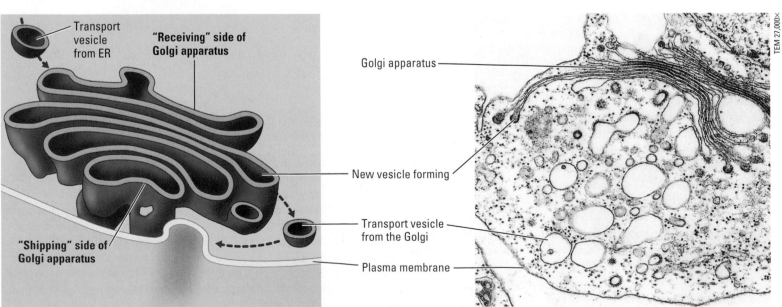

Transport vesicle from ER

"Receiving" side of Golgi apparatus

"Shipping" side of Golgi apparatus

Golgi apparatus

New vesicle forming

Transport vesicle from the Golgi

Plasma membrane

TEM 27,000×

Lysosomes

The name **lysosome** is from the Greek for "breakdown body," a good description of how these organelles function in cells. A lysosome is a membrane-enclosed sac of digestive enzymes (Figure 4.12). These enzymes can break down macromolecules such as proteins, polysaccharides, fats, and nucleic acids (see Chapter 3). The lysosome provides a compartment where the cell can digest macromolecules safely, without committing suicide by unleashing these digestive enzymes on the cell itself.

Lysosomes have several types of digestive functions. Many cells engulf nutrients into tiny cytoplasmic sacs called **food vacuoles.** Lysosomes fuse with the food vacuoles, exposing the food to enzymes that digest it (Figure 4.12a). Small molecular products of digestion, such as amino acids, leave the lysosome and nourish the cell. Lysosomes also help destroy harmful bacteria. Our white blood cells ingest bacteria into vacuoles, and lysosomal enzymes emptied into these vacuoles rupture the bacterial cell walls. Lysosomes also serve as recycling centers for damaged organelles. Without harming the cell, a lysosome can engulf and digest parts of another organelle, making its molecules available for the construction of new organelles (Figure 4.12b). Lysosomes also have sculpturing functions in embryonic development. For example, lysosomal enzymes destroy cells of the webbing that joins the fingers of early human embryos. In this case, the lysosomes really *are* "suicide packs," breaking open and causing the programmed death of whole cells.

Abnormal lysosomes are associated with hereditary disorders called **lysosomal storage diseases.** A person with such a disease is missing one of the digestive enzymes of lysosomes. The abnormal lysosomes become engorged with indigestible substances, and this eventually interferes with other cellular functions. Most of these diseases are fatal in early childhood. An example is Tay-Sachs disease, which ravages the nervous system. In this disorder, lysosomes lack a lipid-digesting enzyme, and nerve cells in the brain are damaged as they accumulate excess lipids. Fortunately, storage diseases are rare in the general population. For Tay-Sachs disease, carriers of the abnormal gene that causes it can be identified, and prospective parents often seek such genetic testing if there is a family history of the disorder.

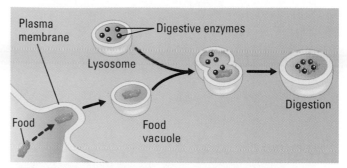

(a) Lysosome digesting food

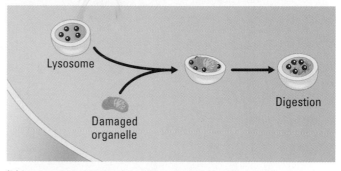

(b) Lysosome breaking down damaged organelle

Figure 4.12 The functions of lysosomes. Lysosomes are part of the endomembrane system because the ER and Golgi contribute to their formation. As sacs of digestive enzymes, lysosomes **(a)** digest food and **(b)** also help recycle the molecules of the cell itself by breaking down damaged organelles.

Vacuoles

Vacuoles are membranous sacs that belong to the endomembrane system because they bud from the ER, Golgi, or plasma membrane. Vacuoles come in different sizes and have a variety of functions. In Figure 4.12a you saw a food vacuole budding from the plasma membrane. Certain freshwater protists have contractile vacuoles (Figure 4.13a). Another type of vacuole is a plant cell's **central vacuole,** which can account for more than half the volume of a mature cell (Figure 4.13b).

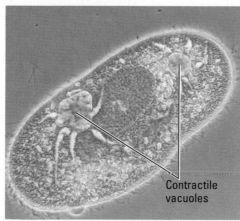

(a) Contractile vacuoles in a protist

Figure 4.13 Two types of vacuoles.
(a) This single-celled organism, a protist named *Paramecium,* has two contractile vacuoles. They function as pumps that expel excess water that

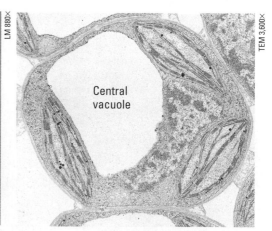

(b) Central vacuole in a plant cell

flows into the cell across its outer membrane from the pond water outside. **(b)** The central vacuole is often the largest organelle in a mature plant cell.

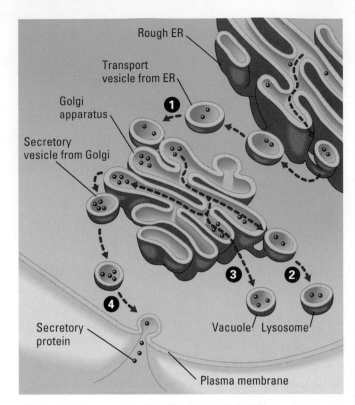

Figure 4.14 **Review of the endomembrane system.** The arrows show some of the pathways of cell product distribution and membrane migration via transport vesicles. For example, ❶ digestive enzymes made by the rough ER are transported via vesicles to the Golgi for processing. ❷ Lysosomes containing the processed digestive enzymes bud off from the Golgi. ❸ Other cell products end up in vacuoles for storage. ❹ Still other cell products are secreted from the cell.

The plant cell vacuole is a versatile compartment. It is a place to store organic nutrients. For example, proteins are stockpiled in the vacuoles of cells in seeds such as beans and peas. Central vacuoles also contribute to plant growth by absorbing water and causing cells to expand. Central vacuoles in flower petals may contain pigments that attract pollinating insects. Central vacuoles can also contain poisons that protect against plant-eating animals.

Figure 4.14 will help you review how all the organelles of the endomembrane system are related. Note that it is possible for a product made in one part of the endomembrane system to eventually exit the cell or become part of another organelle without ever crossing a membrane. Also note that membrane originally fabricated by the ER can eventually turn up as part of the plasma membrane through the fusion of secretory vesicles. In this way, even the plasma membrane is related to the endomembrane system.

> Trace the movement of a protein through the endomembrane system in Web/CD Activity 4D.

Trace the movement of a protein through the endomembrane system in Web/CD Activity 4D.

CheckPoint

1. Which structure includes all others in the list: rough ER, smooth ER, endomembrane system, the Golgi apparatus?
2. What makes rough ER rough?
3. What is the relationship of the Golgi apparatus to the ER in a protein-secreting cell?
4. How can defective lysosomes result in excess accumulation of a particular compound in a cell?

Answers: 1. Endomembrane system **2.** Ribosomes attached to the membranes **3.** The Golgi receives transport vesicles that bud from the ER and that contain proteins synthesized in the ER. The Golgi finishes processing the proteins and then dispatches transport vesicles that secrete the proteins to the outside of the cell. **4.** If the lysosomes lack an enzyme needed to break down the compound, the cell will accumulate an excess of the compound.

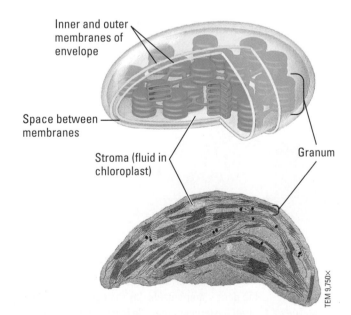

Figure 4.15 **The chloroplast: site of photosynthesis.**

Energy-Converting Organelles

The cellular machinery requires a continuous energy supply to do all the work of life. The two types of cellular power stations are the organelles called chloroplasts and mitochondria.

Chloroplasts

Most of the living world runs on the energy provided by photosynthesis, the conversion of light energy from the sun to the chemical energy of sugar and other organic molecules. **Chloroplasts** are the organelles of plants and protists that perform photosynthesis.

Befitting an organelle that carries out complex, multistep processes, internal membranes partition the chloroplast into three major compartments (Figure 4.15). One compartment is the space between the two membranes that envelop the chloroplast. The stroma, the thick fluid within the chloroplast, is the second compartment. Suspended in that fluid, the interior of a network of membrane-enclosed tubes and disks forms the third compartment. Notice in Figure 4.15 that the disks occur in stacks called **grana** (singular, *granum*). The grana are the chloroplast's solar power packs, the structures that actually trap light energy and convert it to chemical energy. You'll learn in Chapter 6 how the chloroplast works.

Mitochondria

Mitochondria (singular, *mitochondrion*) are the sites of cellular respiration. This process harvests energy from sugars and other food molecules and converts it to another form of chemical energy called ATP. Cells use molecules of ATP as the direct energy source for most of their work. In contrast to chloroplasts, which are unique to the photosynthetic cells of plants and protists, mitochondria are found in almost all eukaryotic cells, including your own.

An envelope of two membranes encloses the mitochondrion, which contains a thick fluid (Figure 4.16). The inner membrane of the envelope has numerous infoldings called **cristae.** Many of the enzymes and other molecules that function in cellular respiration are built into the inner membrane. By increasing the surface area of this membrane, the cristae maximize ATP output. Chapter 5 is the place where you will learn more about how mitochondria convert food energy to ATP energy.

Build a chloroplast and a mitochondrion in Web/CD Activity 4E.

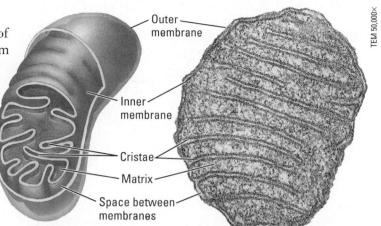

TEM 50,000×

Outer membrane
Inner membrane
Cristae
Matrix
Space between membranes

Figure 4.16 The mitochondrion: site of cellular respiration.

CheckPoint

1. What does photosynthesis accomplish?

2. What is cellular respiration?

Answers: 1. The conversion of light energy to chemical energy stored in food molecules **2.** A process that converts the chemical energy of sugars and other food molecules to chemical energy in the form of ATP

The Cytoskeleton: Movement and Maintenance of Shape

Biologists once thought that the organelles of a cell drifted about freely in an unstructured cytosol. However, improvements in microscopes revealed a **cytoskeleton,** a network of fibers extending throughout the cytoplasm. The cytoskeleton serves as both skeleton and "muscles" for the cell, functioning in both support and movement.

The most obvious function of the cytoskeleton is to give mechanical support to the cell and maintain its shape. This is especially important for animal cells, which lack cell walls. And just as the bony skeleton of your body helps fix the positions of your organs, the cytoskeleton provides anchorage for many organelles in a cell. The cytoskeleton is more dynamic than an animal skeleton, however. It can quickly dismantle in one part of the cell and re-form in a new location, changing the shape of the cell. In other cases, rearrangement of the cytoskeleton can even cause the whole cell or some of its parts to move.

The Cytoskeleton's Three Types of Fibers

Three main kinds of fibers make up the cytoskeleton: microfilaments, the thinnest type of fiber; microtubules, the thickest; and intermediate filaments, in between in thickness.

Microfilaments are solid helical rods composed mainly of a globular protein called actin (Figure 4.17a). Actin microfilaments can help cells change shape and move by assembling actin subunits at one end while disassembling (losing subunits) at the other. This process contributes to the amoeboid (crawling) movements of the protist *Amoeba* and certain of our white blood cells. In addition, actin microfilaments often interact with other kinds of protein filaments to make cells contract. This function of microfilaments is best known from studies of muscle cells.

Intermediate filaments are a varied group. They are made of fibrous proteins rather than globular ones and have a ropelike structure (Figure 4.17b). Intermediate filaments serve mainly as reinforcing rods for bearing tension but also help anchor certain organelles. For instance, the nucleus is often held in place by a cage of intermediate filaments.

Microtubules are straight, hollow tubes composed of globular proteins called tubulins (Figure 4.17c). Microtubules elongate by adding tubulin subunits at one end. They are disassembled in a reverse manner, and the tubulin subunits can then be reused in another microtubule. Microtubules that provide rigidity and shape in one area may disassemble and then reassemble elsewhere in the cell. In addition to reinforcing cell shape, microtubules also act as tracks along which other organelles can move within the cytoplasm. For example, a lysosome might reach a food vacuole by moving along a microtubule. Microtubules also guide the movement of chromosomes when cells divide. As another example, microtubules are responsible for the movement of structures called cilia and flagella.

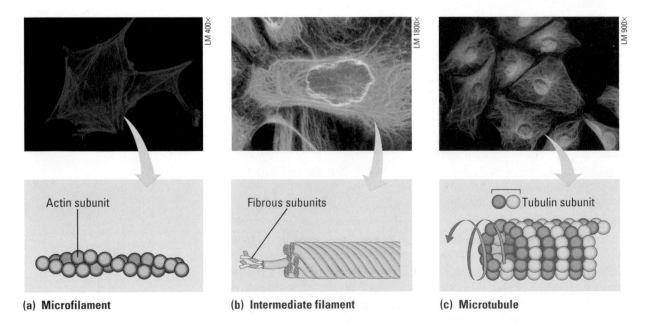

Figure 4.17 Fibers of the cytoskeleton.

(a) **Microfilament**   (b) **Intermediate filament**   (c) **Microtubule**

Cilia and Flagella

In eukaryotic cells, a specialized arrangement of microtubules functions in the beating of **flagella** and **cilia,** which extend from some cells. Cilia are generally shorter and more numerous than flagella. Flagella propel the sperm cells of humans and other animals (Figure 4.18a). And both cilia and flagella propel various protists through water (Figure 4.18b). If cilia or flagella extend from cells that are held in place as part of a tissue layer, they function to move fluid over the surface of the tissue. For example, the ciliated lining of your windpipe helps cleanse your respiratory system by sweeping mucus with trapped debris out of your lungs (Figure 4.18c). Tobacco smoke irritates these ciliated cells, inhibiting or destroying the cilia. This interferes with the normal cleansing mechanisms and allows more toxin-laden smoke particles to reach the lungs. Frequent coughing—common in heavy smokers—then becomes the body's attempt to cleanse its respiratory system.

Though different in length, number per cell, and beating pattern, cilia and flagella have the same basic architecture. Each has a core of microtubules wrapped in an extension of the plasma membrane. A ring of nine microtubule doublets surrounds a central pair of microtubules. This arrangement, found in nearly all eukaryotic flagella and cilia, is called the

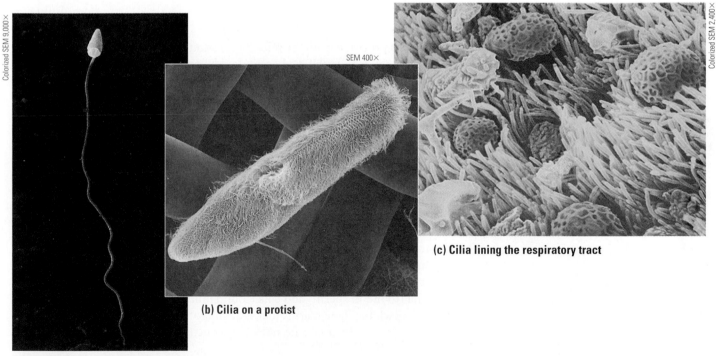

(a) Flagellum of a sperm cell

(b) Cilia on a protist

(c) Cilia lining the respiratory tract

Figure 4.18 Flagella and cilia. (a) A flagellum usually undulates, its snakelike motion driving a cell such as this sperm cell. **(b)** Cilia have a back-and-forth motion, alternating power strokes with recovery strokes, something like the oars of a galley ship. A dense nap of beating cilia covers this *Paramecium,* a freshwater protist that is able to dart rapidly through its watery home. **(c)** The cilia lining your respiratory tract sweep mucus with trapped debris out of your lungs.

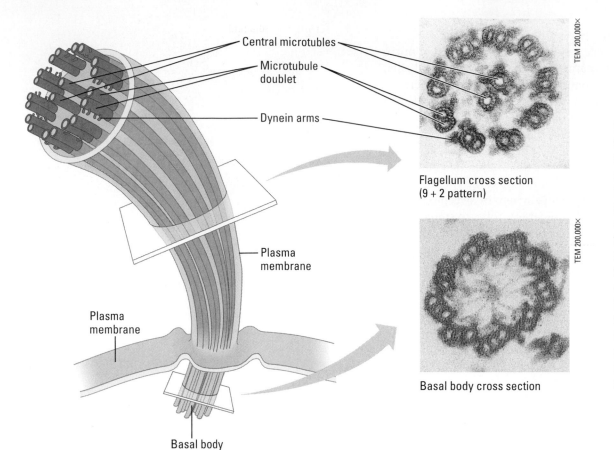

Central microtubules

Microtubule doublet

Dynein arms

Plasma membrane

Plasma membrane

Basal body (structurally identical to centriole)

(a)

Flagellum cross section (9 + 2 pattern)

TEM 200,000×

Basal body cross section

TEM 200,000×

Figure 4.19 Structure and function of a flagellum. (a) The basic structure of cilia and flagella evolved very early in the history of eukaryotic life. Our cilia and those of a protist such as *Paramecium* share the same complex arrangement of microtubules, the so-called 9 + 2 pattern: nine microtubule doublets surrounding two single microtubules. (b) Clawlike structures called dynein arms reach from one doublet to the next. The motor proteins of these dynein arms slide the doublets past one another, causing the bending movements of the flagellum or cilium. Molecules of ATP power the motor proteins.

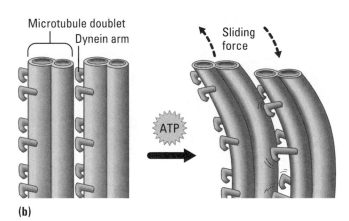

Microtubule doublet
Dynein arm

Sliding force

ATP

(b)

9 + 2 pattern (Figure 4.19). Arms made of special motor proteins called dyneins protrude from the microtubule doublets. Using energy from ATP, the dynein arms grab an adjacent doublet and "walk" along it. This bends the cilium or flagellum.

A basal body, which anchors the flagellum or cilium in the cytoplasm, has a different arrangement of microtubules. When a cilium or flagellum begins to grow, the basal body may act as a foundation for microtubule assembly from tubulin subunits. Basal bodies are identical to structures called **centrioles** (shown in Figure 4.6a). In Chapter 7, you will learn how centrioles function in the reproduction (cell division) of your cells.

See an animation of a flagellum moving in Web/CD Activity 4F.

Because human sperm rely on flagella for movement, it's easy to understand why problems with flagella can lead to male infertility: The sperm are unable to travel up the female reproductive tract to fertilize the ovum (egg). But why do some males with a certain type of hereditary sterility also suffer from respiratory problems? The answer lies in the similarities between flagella (found in sperm) and cilia (found lining the respiratory tract). These males lack the dynein arms that slide microtubules, and so their sperm do not swim and their cilia do not sweep mucus out of their lungs.

CheckPoint

1. Which component of the cytoskeleton is most important in **(a)** holding the nucleus in place within the cell; **(b)** guiding chromosomes during cell division; **(c)** contracting muscle cells?

2. How do cilia and flagella bend?

Answers: 1. (a) intermediate filaments; **(b)** microtubules; **(c)** microfilaments **2.** Dynein arms, powered by ATP, slide neighboring doublets of microtubules past one another.

Cell Surfaces: Protection, Support, and Cell-Cell Interactions

Having crisscrossed the interior of the cell to explore various organelles, our tour now takes us to the surface of this microscopic world. Most cells secrete materials for coats of one kind or another that are external to the plasma membrane. These extracellular coats help support and protect cells and function in certain interactions between cellular neighbors in tissues.

Plant Cell Walls and Cell Junctions

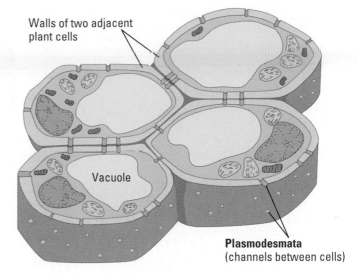

Plant cells are encased by **cell walls.** The wall protects the cell, maintains its shape, and restrains the cell from absorbing so much water that it would pop. Cell walls do not totally isolate plant cells from each other. To function in a coordinated way as part of a tissue, cells must have **cell junctions,** structures that connect them to one another. In plants, channels called **plasmodesmata** pass through the cell walls, connecting the cytoplasm of each cell to its neighbors. These pores allow water and other small molecules to move between cells, integrating the activities of a tissue (Figure 4.20).

Plants do not have bones, but the walls of their cells are strong enough to hold even the tallest trees up against the pull of gravity. Much thicker and stronger than the plasma membrane, the plant cell wall is made from cellulose fibrils embedded in a matrix of other molecules (see Figure 3.13). This combination of materials, strong fibers in a "ground substance" (matrix), is the same basic architectural design found in steel-reinforced concrete and in fiberglass. It is a composite that combines the benefits of strength and flexibility. You've witnessed how this works if you've ever watched a tall pine tree or palm tree flex back and forth in a strong wind without breaking. It is cell wall material in the tree trunk that makes that possible. We exploit this combination of strength and flexibility by using lumber cut from trees as building material.

Animal Cell Surfaces and Cell Junctions

Animal cells lack the cell walls seen in plants, but most animal cells secrete a sticky coat called the **extracellular matrix.** This layer helps hold cells together in tissues and can also have protective and supportive functions.

You can see animations of plant and animal cell junctions in Web/CD Activity 4G.

Adjacent cells in many animal tissues connect by various types of cell junctions (Figure 4.21). **Tight junctions** bind cells together tightly. **Anchoring junctions** attach cells with fibers but still allow materials to pass along the spaces between cells. **Communicating junctions** are channels that allow water and other small molecules to flow between neighboring cells.

Review animal and plant cell structure and function in Web/CD Activities 4H and 4I.

We have seen that both plant and animal cells have coats external to their plasma membranes. The plasma membrane, however, is generally regarded as the boundary of the living cell itself. To complete our exploration of cells, we need to take a closer look at how biological membranes work.

Figure 4.20 Plant cell walls with plasmodesmata.

Walls of two adjacent plant cells

Vacuole

Plasmodesmata (channels between cells)

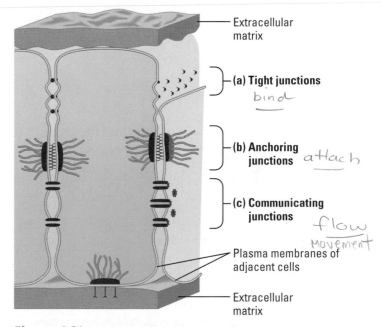

Extracellular matrix

(a) Tight junctions — *bind*

(b) Anchoring junctions — *attach*

(c) Communicating junctions — *flow movement*

Plasma membranes of adjacent cells

Extracellular matrix

Figure 4.21 Animal cell surfaces and junctions. Three types of cell–cell junctions integrate the cells of a tissue. **(a)** Tight junctions bind cells together, forming a leakproof sheet. Such a sheet of tissue lines the digestive tract, preventing the contents of the tract from leaking into surrounding tissues. **(b)** Anchoring junctions attach adjacent cells to each other or to the extracellular matrix. These junctions rivet cells together with cytoskeletal fibers but still allow materials to pass along the spaces between cells. **(c)** Communicating junctions are channels that allow water and other small molecules to flow between neighboring cells. These junctions are especially common in animal embryos, where chemical communication between cells is essential for development.

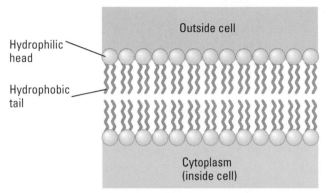

(a) Phospholipid bilayer of membrane

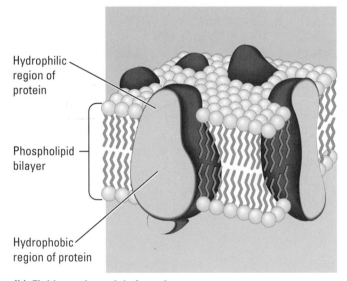

(b) Fluid mosaic model of membrane

Figure 4.22 Membrane structure. (a) At the interface between two aqueous compartments, phospholipids arrange themselves into a bilayer. The symbol for phospholipids that we'll use throughout this book is like a lollipop with two sticks instead of one. The "head" of the lollipop is the end with the phosphate group, which is hydrophilic. The two sticks, the hydrocarbon tails of the phospholipid, are hydrophobic. Notice how the bilayer arrangement keeps the heads exposed to water while keeping the tails in the dry interior of the membrane. **(b)** Membrane proteins, like the phospholipids, have both hydrophilic and hydrophobic regions. The membrane behaves as a fluid mosaic because its phospholipids and many of its diverse proteins drift about in the plane of the membrane.

Membrane Structure and Function

We have encountered diverse membranes throughout our tour of the cell. There is, of course, the plasma membrane, the cell's outer membrane. But we have also examined the different organelles of the endomembrane system and the membranous envelopes of chloroplasts and mitochondria. All these membranes help keep the cell's functions organized. So many chemical reactions occur simultaneously in a cell that chaos would result if membranes did not impose a structural order. As partitions, membranes sequester teams of enzymes within a cell's functional compartments. But membranes are more than mere cellular room dividers. Many of the most complex metabolic processes are catalyzed by enzymes that are built right into the fabric of membranes. And membranes, unlike walls in a building, allow certain substances to pass, but not others. In this way, membranes maintain a specific chemical environment within each compartment they enclose. Each membrane of the cell has its own specific functions, but we will focus on the plasma membrane.

Membrane Structure: A Fluid Mosaic of Lipids and Proteins

The plasma membrane is the edge of life, the boundary that separates the living cell from its nonliving surroundings. The plasma membrane is a remarkable film so thin that you would have to stack 8000 of them to equal the thickness of the page you are reading. Yet the plasma membrane can regulate the traffic of chemicals into and out of the cell. The key to how a membrane works is its structure.

The plasma membrane and other membranes of the cell are composed mostly of proteins and lipids. The lipids belong to a special class called **phospholipids.** They are related to fats, but have only two fatty acids instead of three (see Figure 3.14). A phospholipid has a phosphate group (phosphorus and oxygen) in place of the third fatty acid. The phosphate group is electrically charged, which makes it hydrophilic ("water-loving"). The rest of the phospholipid, however, consisting of the two fatty acids, is hydrophobic ("water-fearing"). Thus, phospholipids have a kind of chemical ambivalence in their interactions with water. The phosphate group "head" mixes with water while the fatty acid "tails" avoid it. This makes phospholipids good membrane material. By forming a two-layered membrane, or *phospholipid bilayer,* the hydrophobic parts of the molecules hide from water, while the hydrophilic portions are wet (Figure 4.22a). A membrane's specific proteins are inserted into this phospholipid bilayer (Figure 4.22b). Figure 4.23 surveys some of the functions of membrane proteins.

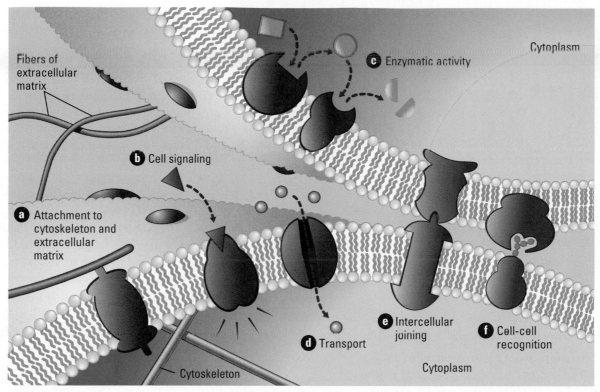

Figure 4.23 Some functions of membrane proteins. (a) Attachment to the cytoskeleton and extracellular matrix. Microfilaments or other elements of the cytoskeleton may be bonded to membrane proteins, a function that helps maintain cell shape and fixes the location of certain membrane proteins. Proteins that adhere to the fibers of the extracellular matrix can coordinate extracellular and intracellular changes. **(b) Cell signaling.** A membrane protein may have a binding site with a specific shape that fits the shape of a chemical messenger, such as a hormone. The external messenger (signal) may cause a change in the protein that relays the message to the inside of the cell. **(c) Enzymatic activity.** A protein built into the membrane may be an enzyme with its active site exposed to substances in the adjacent solution. In some cases, several enzymes in a membrane are organized as a team that carries out sequential steps of a metabolic pathway. **(d) Transport.** A protein that spans the membrane may provide a channel across the membrane that is selective for a particular solute. **(e) Intercellular joining.** Membrane proteins of adjacent cells may be hooked together in various kinds of junctions. **(f) Cell-cell recognition.** Some proteins with short chains of sugars serve as identification tags that are specifically recognized by other cells.

See an animation of the fluid nature of membranes in Web/CD Activity 4J.

Membranes are not static sheets of molecules locked rigidly in place. The phospholipids and most of the proteins are free to drift about in the plane of the membrane. This behavior is captured in the description of a membrane as a **fluid mosaic**—fluid in its molecular wanderings and mosaic in the diversity of proteins that float like icebergs in the sea of phospholipids.

Selective Permeability

The plasma membrane and other membranes of the cell are **selectively permeable.** That means that a membrane allows some substances to cross more easily than others and blocks passage of some substances altogether. For example, the plasma membrane allows the cell to take up such materials as oxygen and nutrients and to dispose of wastes such as carbon dioxide. The traffic of some substances can only occur through avenues called **transport proteins,** which are among the specialized proteins built into membranes. The two main mechanisms of passage across membranes are passive transport and active transport.

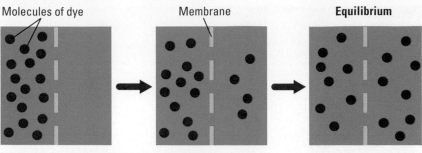

(a) **Passive transport of one type of molecule**

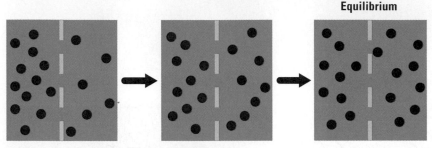

(b) **Passive transport of two types of molecules**

Figure 4.24 Passive transport: diffusion across a membrane. (a) The membrane is permeable to these dye molecules, which diffuse down their concentration gradient. At equilibrium, the molecules are still restless, but the rate of transport is equal in both directions. (b) If solutions have two or more solutes, each will diffuse down its own concentration gradient.

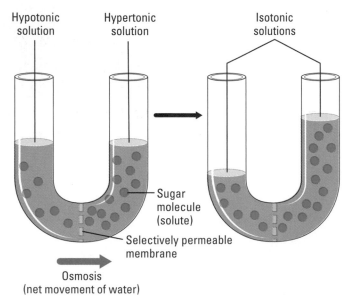

Figure 4.25 Osmosis. A membrane separates two solutions with different sugar concentrations. Water molecules can pass through the membrane, but sugar molecules cannot. Osmosis, the passive transport of water across the membrane, reduces the difference in sugar concentrations and changes the volumes of the two solutions.

Passive Transport: Diffusion Across Membranes

Molecules are restless. They vibrate and wander randomly with an energy called thermal motion, or heat. One result of this motion is **diffusion,** the tendency for molecules of any substance to spread out into the available space. Each molecule moves randomly, and yet diffusion of a population of molecules may be directional. For example, imagine a membrane separating pure water from a solution of a dye dissolved in water (Figure 4.24). Assume that this membrane is permeable to the dye molecules—meaning that the membrane allows the dye molecules to pass through. Although each dye molecule moves randomly, there will be a net migration across the membrane to the side that began as pure water. The spreading of the dye across the membrane will continue until both solutions have equal concentrations of the dye. Once that point is reached, there will be a dynamic equilibrium, with as many dye molecules moving per second across the membrane in one direction as the other.

We can now state a simple rule of diffusion: Each substance will diffuse from where it is more concentrated to where it is less concentrated. Put another way, a substance tends to diffuse down its concentration gradient.

Diffusion across a membrane is called **passive transport**—passive because the cell does not expend any energy for it to happen. The membrane does, however, play a regulatory role by being selectively permeable. For example, small molecules generally pass through more readily than large molecules such as proteins (otherwise, the cell would lose its macromolecules). However, the membrane is relatively impermeable to even some very small substances. Examples are sodium ions and other inorganic ions, which are too hydrophilic to pass through the phospholipid bilayer of the membrane. Specific transport proteins, acting as selective corridors, facilitate the passive transport of certain substances, including some ions.

Passive transport is extremely important to all cells. In our lungs, for example, passive transport along concentration gradients is the sole means by which oxygen (O_2, essential for metabolism) enters red blood cells and carbon dioxide (CO_2, a metabolic waste) passes out of them. Water is another substance that crosses membranes by passive transport.

Watch diffusion in action in Web/CD Activity 4K.

Osmosis and Water Balance in Cells

The passive transport of water across a selectively permeable membrane is called **osmosis.** Imagine the case of a membrane separating two solutions with different concentrations of a solute—say, the sugar sucrose. The membrane is permeable to water but not to the solute (Figure 4.25). The solution with a higher concentration of solute is said to be **hypertonic** (*hyper,* above; *tonos,* tension). The solution with the lower solute concentration is **hypotonic** (*hypo,* below). Note that the hypotonic solution, by having the lower solute concentration, has the higher water concentration. Therefore, water will diffuse across the membrane from the hypotonic solution to the hypertonic solution. Osmosis has occurred, reducing the difference in solute concentrations and changing the volumes of the two solutions.

The example in Figure 4.25 shows only one type of solute molecule, but the same net movement of water would occur no matter how many kinds of solutes were present. The direction of osmosis is determined only by the difference in *total* solute concentration, not by the nature of the solutes. For example, seawater has a great variety of solutes, but it will lose water to a solution containing a high enough concentration of a single solute. Only if the total solute concentrations are the same on both sides of the membrane will water molecules move at the same rate in both directions. Solutions of equal solute concentration are said to be **isotonic** (*isos,* equal).

Water Balance in Animal Cells The survival of a cell depends on its ability to balance water uptake and loss. When an animal cell, such as a red blood cell, is immersed in an isotonic solution, the cell's volume remains constant because the cell gains water at the same rate that it loses it (Figure 4.26a, top). In this situation, the cell and its surroundings are in water balance because the two solutions have the same total concentration of solutes. We describe an organism or a cell in this situation as being isotonic to the surrounding solution. Many marine animals, such as sea stars and crabs, are isotonic to seawater. What happens if an animal cell finds itself in a hypotonic solution, which has a lower solute concentration than the cell? The cell gains water, swells, and may pop (lyse) like an overfilled water balloon (Figure 4.26b, top). A hypertonic environment is also harsh on an animal cell. The cell shrivels and can die from water loss (Figure 4.26c, top).

Perform osmosis experiments in Web/CD Activity 4L.

For an animal to survive if its cells are exposed to a hypotonic or hypertonic environment, the animal must have a way to balance excessive uptake or excessive loss of water. The control of water balance is called **osmoregulation.** For example, a freshwater fish, which lives in a hypotonic environment, has kidneys and gills that work constantly to prevent an excessive buildup of water in the body. And if you return to Figure 4.13a, you'll see *Paramecium*'s contractile vacuole, which bails out the excess water that continuously enters the cell from the hypotonic pond water.

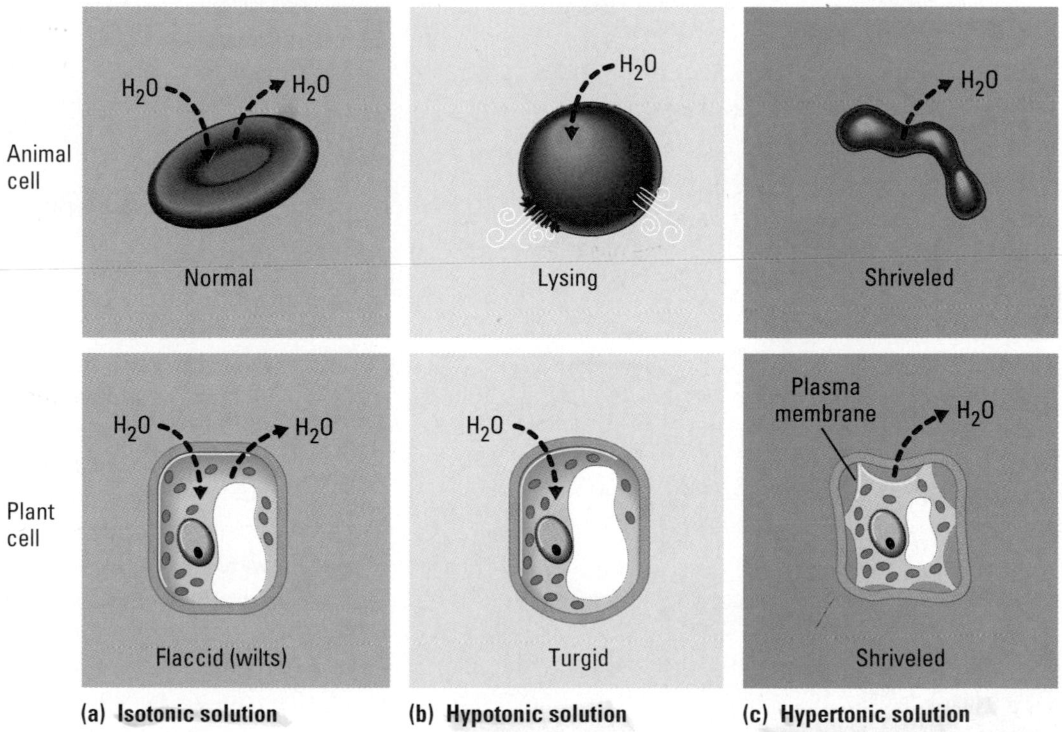

Figure 4.26 The behavior of animal and plant cells in different osmotic environments.

(a) **Isotonic solution**   (b) **Hypotonic solution**   (c) **Hypertonic solution**

Water Balance in Plant Cells Water balance problems are somewhat different for plant cells because of their cell walls. A plant cell immersed in an isotonic solution is flaccid (floppy), and a plant wilts in this situation (see Figure 4.26a, bottom). In contrast, a plant cell is turgid (firm) and healthiest in a hypotonic environment (Figure 4.26b, bottom). A net inflow of water results in a plant cell becoming turgid. Although the elastic cell wall expands a bit, the back pressure it exerts prevents the cell from taking in too much water and bursting, as an animal cell would in this environment. However, in a hypertonic environment, a plant cell is no better off than an animal cell. As a plant cell loses water, it shrivels, and its plasma membrane pulls away from the cell wall (Figure 4.26c, bottom). This process, called **plasmolysis,** usually kills the cell.

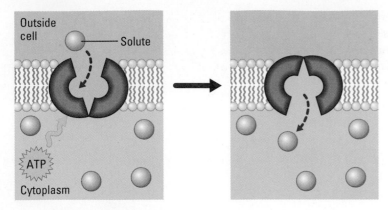

Figure 4.27 Active transport. Like enzymes, transport proteins are specific in their recognition of molecules. This transport protein has a binding site that accepts only a certain solute. Using energy from ATP, the protein pumps the solute against a concentration gradient.

Active Transport: The Pumping of Molecules Across Membranes

In contrast to passive transport, **active transport** requires that a cell expend energy to move molecules across a membrane. In active transport, a specific transport protein actively pumps a solute across a membrane *against* the solute's concentration gradient—that is, away from the side where it is less concentrated (Figure 4.27). Membrane proteins usually use ATP as their energy source for active transport. (ATP, remember, is the energy-carrying molecule that cells produce by extracting energy from food molecules in mitochondria.)

Active transport enables cells to maintain internal concentrations of small molecules that differ from environmental concentrations. For example, compared to its surroundings, an animal cell has a much higher concentration of potassium ions and a much lower concentration of sodium ions. The plasma membrane helps maintain these steep gradients by pumping sodium out of the cell and potassium into the cell. This particular case of active transport is central to how your nerve cells work.

Watch a cell pump sodium and potassium ions in Web/CD Activity 4M.

Exocytosis and Endocytosis: Traffic of Large Molecules

So far, we've focused on how water and small solutes enter and leave cells by moving through the plasma membrane. The story is different for large molecules such as proteins. Their traffic into and out of the cell depends on being packaged in vesicles. You have already seen an example in the packaging and secretion of proteins. A vesicle containing the secretory protein fuses with the plasma membrane, spilling its contents outside the cell (see Figure 4.14). That process is called **exocytosis** (Figure 4.28a). The reverse process, **endocytosis,** takes material into the cell within vesicles that bud inward from the plasma membrane (Figure 4.28b).

See Web/CD Activity 4N for an animation of exocytosis and endocytosis.

There are three types of endocytosis. In **phagocytosis** ("cellular eating"), a cell engulfs a particle and packages it within a food vacuole (Figure 4.29). In **pinocytosis** ("cellular drinking"), the cell "gulps" droplets of fluid

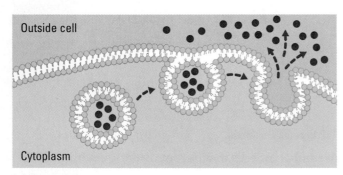

(a) Exocytosis

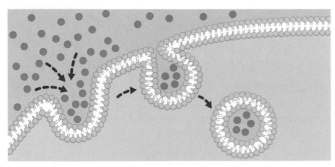

(b) Endocytosis

Figure 4.28 Exocytosis and endocytosis.

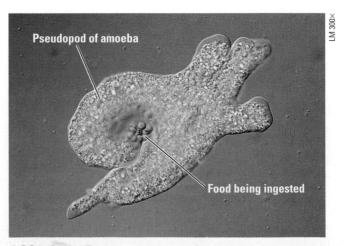

Figure 4.29 Phagocytosis. An amoeba uses a cellular extension called a pseudopod to engulf food and package it in a food vacuole.

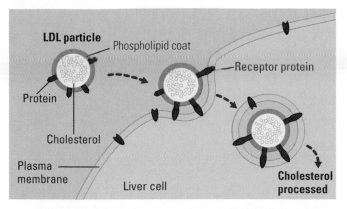

Figure 4.30 Cholesterol uptake by human cells. Normal liver cells remove excess cholesterol from the blood by receptor-mediated endocytosis. This prevents the excessive cholesterol levels in the blood that contribute to cardiovascular disease. Cholesterol circulates in the blood mainly in particles called low-density lipoproteins, or LDLs. An LDL is a cholesterol globule coated by phospholipid. Proteins embedded in the phospholipid coat attach to receptor proteins built into liver cell membranes. This enables liver cells to take up LDLs and process the cholesterol. Faulty liver cell membranes can overload the blood with cholesterol. For example, in one type of hereditary disorder, LDL receptors on liver cell membranes are missing or reduced in number. This results in a heavy load of blood cholesterol, which accumulates on artery walls and causes early onset of cardiovascular disease.

by forming tiny vesicles. Because any and all solutes dissolved in the droplet are taken into the cell, pinocytosis is unspecific in the substances it transports. In contrast, **receptor-mediated endocytosis** is very specific. That is because it is triggered by the binding of certain external molecules to specific receptor proteins built into the plasma membrane. An example of receptor-mediated endocytosis is the mechanism human liver cells use to take up cholesterol particles that circulate in the blood (Figure 4.30).

The Role of Membranes in Cell Signaling

The cells of your body talk to each other by chemical signaling. The three stages of cell signaling are reception, transduction, and response. Reception of an extracellular signal, such as a hormone, depends on a specific receptor protein built into the plasma membrane (Figure 4.31). This triggers changes in one or more molecules that function in transduction. The proteins and other molecules of the **signal transduction pathway** relay the signal and convert it to chemical forms that work within the cell. This leads to chemical responses, such as activation of certain metabolic functions, and structural responses, such as rearrangements of the cytoskeleton. This is just another example of how the plasma membrane serves as a cell's interface with its surroundings.

You can see a signal-transduction pathway in action in Web/CD Activity 4O.

Figure 4.31 An example of cell signaling. When a person gets "psyched up" for an athletic contest, cells in adrenal glands call muscle cells to action. The adrenal cells secrete a hormone called epinephrine (also called adrenaline) into the bloodstream. When that signal reaches muscle cells, it is recognized by receptor proteins built into the plasma membrane. This triggers responses within the muscle cells, without the hormone even entering. One of the responses is the hydrolysis of glycogen (a storage polysaccharide) that makes the sugar glucose available as an energy source for the muscle cells. This chain of events is part of the "fight-or-flight" response that enables us to attack or run when we're in danger—or keep our edge during an intense competition.

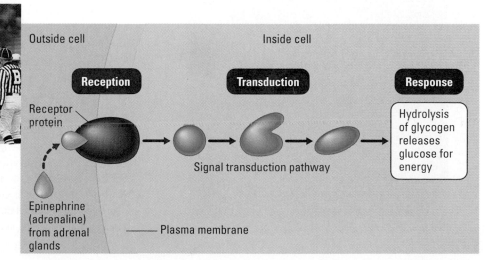

1. Why do phospholipids tend to organize into a bilayer in an aqueous environment?

2. Explain why it is not enough just to say that a solution is "hypertonic."

3. What is the energy source for active transport?

4. Explain how a protein-secreting cell can synthesize and secrete its product without the protein ever having to cross a membrane.

5. Explain how a person can have an excessive cholesterol level in the blood even without producing or eating more cholesterol than other people.

6. The hormone epinephrine can cause a liver cell to break down its stored glycogen and release sugar without the hormone even entering the cell. Explain.

Answers: 1. This structure shields the hydrophobic tails of the phospholipids from water while exposing the hydrophilic heads to water. **2.** Hypertonic and hypotonic are relative terms: A solution that is hypertonic to tap water could be hypotonic to seawater. In using these terms, you must provide a comparison, as in "The solution is hypertonic to the cell." **3.** ATP **4.** From the time the protein is made by rough ER, it is topographically "outside" the cell, first in the ER interior, then within the Golgi and transport vesicles, and finally outside the plasma membrane as the vesicles release their contents by exocytosis. **5.** The person may have an impairment in the ability of cells to remove cholesterol from the blood because of the absence or shortage of LDL receptors on liver cell plasma membranes. **6.** Epinephrine binds to a receptor on the liver cell surface, activating a signal transduction pathway inside the cell that leads to sugar release.

Evolution Link: The Origin of Membranes

One of the earliest episodes in the evolution of life on Earth may have been the formation of membranes. The ability of cells to regulate their chemical exchanges with the environment is a basic requirement for life. A selectively permeable membrane can enclose a solution that is different in composition from the surrounding solution, while still permitting the uptake of nutrients and elimination of waste products. Phospholipids, the key ingredients of membranes, were probably among the organic molecules that formed from chemical reactions on Earth before life. And when mixed with water, phospholipids self-assemble into membranes spontaneously (Figure 4.32). This requires no genes or other information except the unique characteristics of the phospholipids themselves (see Figure 4.22a). The origin of membranes on a primordial Earth may have been an early step in the evolution of the first cells. We will evaluate this hypothesis in more detail in Chapter 14.

Discover more about the origin of membranes in the Web Evolution Link.

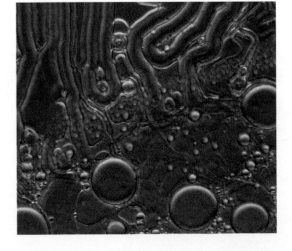

Figure 4.32 The spontaneous formation of membranes: a key step in the origin of life. When phospholipids are mixed with water, they self-assemble to form films. Agitation breaks the films into spheres. Even these primitive membranes have some ability to control the traffic of substances between the contents of a sphere and the aqueous environment outside.

Chapter Review

Summary of Key Concepts

Overview: The Microscopic World of Cells

- **Microscopes as Windows to Cells** Using early microscopes, biologists discovered that all organisms are made of cells. Resolving power limits the useful magnification of microscopes. A light microscope (LM) has useful magnifications of up to about 1,000×. Electron microscopes, both scanning (SEM) and transmission (TEM), are much more powerful.
- Web/CD The Process of Science *Exploring Size and Scale*

- **The Two Major Classes of Cells** Prokaryotic cells, which lack nuclei and most other membranous organelles, characterize bacteria and archaea. All other organisms, including plants and animals, consist of the much more complex eukaryotic cells.
- Web/CD Activity 4A *Prokaryotic Cell Structure and Function*

- **A Panoramic View of Eukaryotic Cells** Many cellular functions are partitioned by membranes in the complex organization of eukaryotic cells. The largest organelle is usually the nucleus. Other organelles are located in the cytoplasm, the region between the nucleus and the plasma membrane.
- Web/CD Activity 4B *Build an Animal Cell and a Plant Cell*

The Nucleus and Ribosomes: Genetic Control of the Cell

- **Structure and Function of the Nucleus** An envelope consisting of two membranes encloses the nucleus. Within the nucleus, DNA and proteins make up chromatin fibers. Each fiber is a chromosome, visible in dividing cells. The nucleus also contains the nucleolus, which produces components of ribosomes.

- **Ribosomes** Ribosomes produce proteins in the cytoplasm.

- **How DNA Controls the Cell** Genetic messages are transmitted to the ribosomes via messenger RNA that travels from the nucleus to the cytoplasm.
- Web/CD Activity 4C *Overview of Protein Synthesis*

The Endomembrane System: Manufacture and Distribution of Cellular Products

- **The Endoplasmic Reticulum** The ER consist of membrane-enclosed tubes and sacs within the cytoplasm. Rough ER, named for the ribosomes attached to its surface, makes membranes and secretory proteins. The functions of smooth ER include lipid synthesis and detoxification.

- **The Golgi Apparatus** The Golgi refines certain ER products and packages them in transport vesicles targeted for other organelles or export from the cell.

- **Lysosomes** Lysosomes, sacs containing hydrolytic enzymes, function in digestion within the cell and chemical recycling.

- **Vacuoles** These membrane-enclosed organelles include the contractile vacuoles that expel water from certain freshwater protists and the large, multifunctional central vacuoles of plant cells.
- Web/CD Activity 4D *The Endomembrane System*

Energy-Converting Organelles

- **Chloroplasts** The sites of photosynthesis in plant cells, chloroplasts convert light energy to the chemical energy of food. Grana, stacks of membranous sacs within the chloroplasts, trap the light energy.

- **Mitochondria** These are the sites of cellular respiration, which converts food energy to ATP energy. ATP drives most cellular work.
- Web/CD Activity 4E *Build a Chloroplast and a Mitochondrion*

The Cytoskeleton: Movement and Maintenance of Shape

- **The Cytoskeleton's Three Types of Fibers** The cytoskeleton consists of actin microfilaments (function in cellular movement and shape changes), intermediate filaments (help maintain shape and hold certain organelles in place), and microtubules (reinforce shape and function in movement).

- **Cilia and Flagella** Microtubules in the 9 + 2 arrangement function in the beating of cilia and flagella.
- Web/CD Activity 4F *Cilia and Flagella*

Cell Surfaces: Protection, Support, and Cell-Cell Interactions

- **Plant Cell Walls and Cell Junctions** The walls that encase plant cells consist of cellulose fibers embedded in other materials. Cell walls support plants against the pull of gravity and also prevent cells from absorbing too much water. Channels called plasmodesmata connect the cytoplasm of adjacent cells.

- **Animal Cell Surfaces and Cell Junctions** Animal cells are coated by a sticky extracellular matrix. Junctions between animal cells help hold them together and also integrate cells of a tissue.
- Web/CD Activity 4G *Cell Junctions*
- Web/CD Activity 4H *Review: Animal Cell Structure and Function*
- Web/CD Activity 4I *Review: Plant Cell Structure and Function*

Membrane Structure and Function

- **Membrane Structure: A Fluid Mosaic of Lipids and Proteins** With diverse proteins floating in a phospholipid bilayer, a membrane can be described as a fluid mosaic. The middle zone of a membrane, with the hydrocarbon tails of the phospholipids, is hydrophobic, while the membrane surfaces are hydrophilic.
- Web/CD Activity 4J *Membrane Structure*

- **Selective Permeability** Membranes are selective in transporting molecules in and out of the cell.

- **Passive Transport: Diffusion Across Membranes** Molecules tend to diffuse down their concentration gradients across membranes. This is called passive transport because it requires no energy expenditure by the cell.
- Web/CD Activity 4K *Diffusion*

- **Osmosis and Water Balance in Cells** Osmosis is the passive transport of water across membranes. The net direction of osmosis is from a hypotonic to a hypertonic solution. Most animal cells require an isotonic environment. Plant cells need a hypotonic environment, which keeps the walled cells turgid.
- Web/CD Activity 4L *Osmosis*

- **Active Transport: The Pumping of Molecules Across Membranes** This is the energy-dependent pumping of solutes against their concentration gradients. Specific membrane proteins are the pumps, and ATP provides the energy.
- Web/CD Activity 4M *Active Transport*

- **Exocytosis and Endocytosis: Traffic of Large Molecules** These processes transport large molecules in bulk quantity. Exocytosis exports secretory proteins and other cell products by the fusion of transport vesicles with the plasma membrane. Endocytosis imports material within vesicles that bud inward from the plasma membrane. The three kinds of endocytosis are phagocytosis ("cellular eating"); pinocytosis ("cellular drinking"); and receptor-mediated endocytosis, which enables the cell to take in specific large molecules.
- Web/CD Activity 4N *Exocytosis and Endocytosis*

- **The Role of Membranes in Cell Signaling** Receptors on the cell surface trigger signal transduction pathways that control processes within the cell.
- **Web/CD Activity 4O** *Cell Signaling*

Evolution Link: The Origin of Membranes

- The spontaneous self-assembly of phospholipids into membranes may have been an early step in the evolution of the first cells.
- **Web Evolution Link** *The Origin of Membranes*

Self-Quiz

1. Which of the following clues would tell you whether a cell is prokaryotic or eukaryotic?
   a. the presence or absence of a rigid cell wall
   b. whether or not the cell has a nucleus
   c. the presence or absence of a plasma membrane
   d. whether or not the cell produces proteins
   e. whether or not the cell contains DNA

2. Emily would like to film the movement of chromosomes during cell division. Her best choice for a microscope would be a
   a. light microscope, because of its magnifying power.
   b. transmission electron microscope, because of its resolving power.
   c. scanning electron microscope, because the chromosomes are on the cell surface.
   d. transmission electron microscope, because it shows cells in color.
   e. light microscope, because the specimen must be kept alive.

3. A type of cell called a lymphocyte makes proteins that are exported from the cell. You can track the path of these proteins within the cell by labeling them with radioactive isotopes. Which of the following might be the path of a protein from the site where its polypeptides are made to the lymphocyte's plasma membrane?
   a. chloroplast → Golgi → plasma membrane
   b. Golgi → rough ER → plasma membrane
   c. rough ER → Golgi → plasma membrane
   d. smooth ER → lysosome → plasma membrane
   e. nucleus → Golgi → rough ER → plasma membrane

4. Which of the following organelles is least closely associated with the endomembrane system?
   a. chloroplast   d. Golgi
   b. plasma membrane   e. lysosome
   c. rough ER

5. Which of the following is an *incorrect* match of organelle and function?
   a. ribosome—protein synthesis   d. Golgi—photosynthesis
   b. microtubules—movement   e. lysosome—digestion
   c. mitochondria—cellular respiration

6. Prokaryotic cells are characteristic of
   a. plants.   d. bacteria.
   b. protists.   e. fungi.
   c. animals.

7. Cellular respiration is to _____ as _____ is to chloroplasts.
   a. nucleus; cytoplasm
   b. mitochondria; photosynthesis
   c. ATP; light
   d. nucleus; protein synthesis
   e. grana; cristae

8. Which best describes the structure of the plasma membrane?
   a. proteins sandwiched between two layers of phospholipid
   b. proteins embedded in two layers of phospholipid
   c. a layer of protein coating a layer of phospholipid
   d. phospholipids sandwiched between two layers of protein
   e. phospholipids embedded in two layers of protein

9. The total solute concentration in a red blood cell is about 2%. Sucrose cannot pass through the membrane, but water and urea can. Osmosis would cause such a cell to shrink the most when the cell is immersed in which of the following?
   a. a hypertonic sucrose solution   d. a hypotonic urea solution
   b. a hypotonic sucrose solution   e. pure water
   c. a hypertonic urea solution

10. What process links reception of cell signals to responses within the cell?
   a. a signal transduction pathway
   b. protein synthesis by ribosomes
   c. budding of transport vesicles from the Golgi
   d. active transport of the signal into the cell
   e. rearrangement of the cytoskeleton

- **Go to the website or CD-ROM for more self-quiz questions.**

The Process of Science

1. The cells of plant seeds store oils in the form of droplets enclosed by membranes. Unlike the membranes you saw in this chapter, the oil droplet membrane consists of a *single* layer of phospholipids rather than a bilayer. Draw a model for a membrane around an oil droplet. Explain why this arrangement is more stable than a bilayer.

2. **Explore size and scale in The Process of Science activity available on the website and CD-ROM.**

Biology and Society

Doctors at a university medical center removed John Moore's spleen, standard treatment for his type of leukemia. The disease did not recur. Researchers kept the spleen cells alive in a nutrient medium. They found that some of the cells produced a blood protein that showed promise as a treatment for cancer and AIDS. The researchers patented the cells. Moore sued, claiming a share in profits from any products derived from his cells. The U.S. Supreme Court ruled against Moore, stating that his suit "threatens to destroy the economic incentive to conduct important medical research." Moore argued that the ruling left patients "vulnerable to exploitation at the hands of the state." Do you think Moore was treated fairly? Is there anything else you would like to know about this case that might help you decide?

CHAPTER 5

Cellular Respiration: Harvesting Chemical Energy

A gram of fat has more than twice as many

calories as a gram of carbohydrate or protein.

Even sitting still, you radiate about as much heat as a

100 watt light bulb. It takes about

10 million ATP molecules per second to power

an active muscle cell. You would have to

walk about 4 miles to burn off the energy in a

candy bar.

Figure 5.1 **Fuel stop.**

T he photo in Figure 5.1 makes a connection between the energy demands of our cells and those of our machines. Food and gasoline are both fuels. They contain chemicals that provide energy to perform work. Without fuel, a car stops running, and so do our cells. We need energy to move. We need energy to think. We even need energy just to stay alive while sleeping—to keep the heart beating and the lungs breathing, for example. The human body has trillions of cells, all hard at work, all demanding fuel continuously. We all know how it feels to be hungry at certain times of the day; that "hunger drive" is an evolutionary adaptation that reminds us that it's time for a fuel stop—for a meal. In this chapter, you'll learn how cells harvest food energy and put it to work.

Overview: Energy Flow and Chemical Cycling in the Biosphere

In an indirect way, the fuel molecules in food represent solar energy. We can trace the energy stored in all our food to the sun. Humans and other animals depend on plants to convert the energy of sunlight to the chemical energy of sugars and other organic molecules.

Producers and Consumers

The star we call the sun is at the center of the solar system, almost 100 million miles away. A giant thermonuclear reactor, the sun converts about 120 million metric tons of matter to energy each minute. The energy that radiates into space from the sun includes visible light. A miniscule portion of the light energy reaches planet Earth. And a tiny fraction of the light that illuminates Earth powers life. The process that makes this possible is called **photosynthesis.** *Photo* means "light," and in photosynthesis, it refers to light energy from the sun. To *synthesize* is to make something. In photosynthesis, it is organic material that is made. Putting this together, photosynthesis uses light energy to power a chemical process that makes organic molecules.

Here on land, photosynthesis occurs mainly in green cells within the leaves of plants. Leaves owe their greenness to chlorophyll, the pigment contained in chloroplasts. Chloroplasts are the organelles that house the equipment for photosynthesis (see Figure 4.15). Chloroplasts trap light energy and use it to produce sugars and other energy-rich organic molecules. (You'll learn more about photosynthesis in Chapter 6.)

Most ecosystems depend entirely on photosynthesis for food. That is why biologists refer to plants and other photosynthetic organisms as **producers.** The organic matter they produce not only nourishes the plants, but also supplies food for other organisms. **Consumers,** including humans and other animals, obtain their food by eating plants (Figure 5.2) or by eating animals that have eaten plants. We animals depend on food not only for fuel, but also for the raw organic materials we need to build our cells and tissues.

Figure 5.2 **Producer and consumer.** Plants and other photosynthetic organisms use light energy to drive production of their own organic material; animals and other consumers depend on this photosynthetic product for energy and building material.

Chemical Cycling Between Photosynthesis and Cellular Respiration

The ingredients for photosynthesis are carbon dioxide (CO_2) and water (H_2O). The carbon dioxide, a gas in the surrounding air, enters the plant through tiny pores in the surface of the leaves. The water is absorbed from the damp soil by the plant's roots. Water moves up the plant's veins from the roots to the leaves (see Figure 2.12). Chloroplasts rearrange the atoms of these inorganic ingredients to produce sugars and other organic molecules. One key product of photosynthesis is the sugar glucose ($C_6H_{12}O_6$). A by-product of this food production is oxygen, O_2 (Figure 5.3).

Just as animals rely on plants for fuel, plants themselves use some of the organic molecules they make for their own fuel. A chemical process called **cellular respiration** harvests energy that is stored in sugars and other organic molecules. Cellular respiration uses oxygen to help convert energy extracted from organic fuel to another form of chemical energy called ATP. Cells spend the ATP for almost all their work. Production of ATP during cellular respiration occurs mainly in the organelles called mitochondria (see Figure 4.16). Animals lack chloroplasts and are not capable of photosynthesis, but animals and plants *both* have mitochondria that perform respiration.

Notice in Figure 5.3 that the waste products of cellular respiration are carbon dioxide (CO_2) and water (H_2O), which are the chemical ingredients for photosynthesis. Plants recycle chemicals as they store chemical energy by photosynthesis and harvest chemical energy by cellular respiration. However, plants usually make more organic molecules than they need for fuel. This photosynthetic surplus provides the organic material for the plant to grow. It is also the source of food for humans and other consumers. Analyze any food chain, and you can trace the energy and raw materials for growth back to photosynthesis. The biosphere is solar powered.

Build your own chemical cycling system in Web/CD Activity 5A.

Build your own chemical cycling system in Web/CD Activity 5A.

CheckPoint

1. What are the chemical ingredients for photosynthesis?
2. Why are plants called producers?
3. What is misleading about the following statement? "Plants do photosynthesis and animals do cellular respiration."

Answers: 1. Carbon dioxide (CO_2) and water (H_2O) **2.** They produce organic molecules by photosynthesis. **3.** It implies that cellular respiration does not occur in plants. It does.

Some Basic Energy Concepts

What exactly *is* energy? And how is it stored in the chemicals of food? Our first step to answering these questions is to learn a few basic concepts about energy.

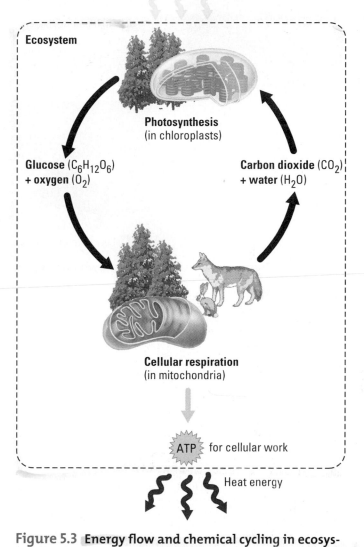

Sunlight energy

Ecosystem

Photosynthesis
(in chloroplasts)

Glucose ($C_6H_{12}O_6$)
+ oxygen (O_2)

Carbon dioxide (CO_2)
+ water (H_2O)

Cellular respiration
(in mitochondria)

ATP for cellular work

Heat energy

Figure 5.3 Energy flow and chemical cycling in ecosystems. Energy enters a forest or other ecosystem as sunlight and exits in the form of heat. Organisms temporarily trap the energy for their work. Photosynthesis in the chloroplasts of plants converts light energy to the chemical energy of sugars, such as glucose, and other organic molecules. Cellular respiration in the mitochondria of plants, animals, and other eukaryotes harvests the food energy to generate ATP. These molecules of ATP directly drive most cellular work. Chemical elements essential for life recycle between cellular respiration and photosynthesis.

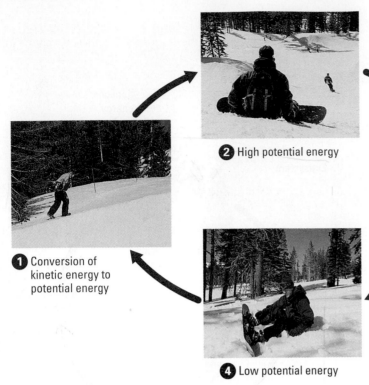

2 High potential energy

1 Conversion of kinetic energy to potential energy

3 Conversion of potential energy to kinetic energy

4 Low potential energy

Figure 5.4 Energy conversions while snowboarding.

Conservation of Energy

Anything that can do work has energy. In fact, **energy** is defined as the capacity to perform work. And work is performed whenever an object is moved against an opposing force. In other words, work moves things in directions in which they would not move if left alone. As an example, imagine a snowboarder climbing up and sliding down a hill that doesn't have a chairlift. You know that the person's legs must do some work to climb to the top of the hill. The boarder expends some food energy to do that work, moving his or her body against the opposing force of gravity. In the process of walking uphill, that food energy is converted to **kinetic energy,** the energy of motion (*kinetic* is from a Greek word meaning "motion"). Anything that moves has kinetic energy.

What happens to the kinetic energy once the snowboarder reaches the top of the hill? Has the energy disappeared? The answer is no. It is not possible to destroy or create energy. This is a principle known as **conservation of energy.** However, energy *can* be converted from one form to another. That is exactly what happened in climbing the hill. The kinetic energy of the snowboarder's climb is now stored in a form called **potential energy.** This is energy that an object has because of its location or arrangement. In our example, the snowboarder's body has potential energy because of its elevated location. That potential energy is converted back to kinetic energy as the person slides down the hill (Figure 5.4).

Entropy

If energy cannot be destroyed, where has it gone once the snowboarder reaches the bottom of the hill again? It has been converted to heat. Most of that heat was generated by friction between the snowboard and the snow. Energy hasn't disappeared, because heat is a type of kinetic energy. Heat is the random motion of atoms and molecules. The snowboarding increased this heat energy in the snowboard, in the surrounding air, and in the snow. In fact, the process has actually melted some snow.

All energy conversions generate some heat. While that does not destroy energy, it does make it less useful. Heat, of all energy forms, is the most difficult to "tame"—the most difficult to harness for useful work. It is energy in its most randomized form, the energy of aimless molecular movement. By melting snow, the snowboard also randomized matter: Liquid water is much less ordered than ice (see Figure 2.15).

Scientists use the term **entropy** as a measure of disorder, or randomness. Everything that happens in nature increases the entropy of the universe. The energy conversions during a climb up and a slide down a hill contribute to this randomizing of matter and energy. All of the energy the snowboarder expended has now been lost to the atmosphere as heat. To climb up the hill for another run down, the snowboarder must use some more stored food energy.

Increase your understanding of energy with Web/CD Activity 5B.

Chemical Energy

How can molecules derived from the food we eat provide energy for our working cells? The molecules of food, gasoline, and other fuels have a special form of potential energy called **chemical energy.** We have already defined potential energy as energy stored because of location or arrangement. In the case of chemical energy, the potential to perform work is due to the arrangement of atoms within the fuel molecules. Carbohydrates, fats, and gasoline have structures that make them especially rich in chemical energy.

Living cells and automobile engines use the same basic process to make the chemical energy stored in their fuel available for work. In both cases, this process breaks the organic fuel into smaller waste molecules that have much less chemical energy than the fuel molecules did. The engine of an automobile mixes oxygen with gasoline in an explosive chemical reaction that breaks down the fuel molecules and pushes pistons. The waste products emitted from the exhaust pipe of the car are mostly carbon dioxide and water (Figure 5.5a). About 25% of the energy an automobile engine extracts from its fuel is converted to the kinetic energy of the car's movement. The rest is converted to heat energy. That is why it is very hot under the hood. The car's radiator and fan disperse enough heat to the atmosphere to prevent an engine meltdown.

Your cells also use oxygen to help harvest chemical energy (Figure 5.5b). And as in a car engine, the "exhaust" is mostly carbon dioxide and water. "Combustion" of fuel in cells, remember, is called cellular respiration. Fortunately, cellular respiration is a slow, gradual "burn" rather than an explosive combustion. This is one reason your cellular "engines" are more efficient than an automobile engine. You convert about 40% of your food energy to useful work, such as contraction of your muscles. Nevertheless, your fuel consumption does generate considerable heat (the other 60% of the energy released by the breakdown of the fuel molecules). This explains why you feel so hot after running, rollerblading, or dancing vigorously. Sweating and other cooling mechanisms enable your body to lose the excess heat, much as a car's radiator keeps the engine from overheating. Even sitting still at room temperature, a human radiates about as much heat as a 100-watt lightbulb. You've probably experienced the discomfort of being in an unairconditioned room crowded with human "lightbulbs."

The heat generated from fuel consumption in cells is not completely unproductive; mammals, including humans, and birds can use some of this heat to keep the body at an almost constant, warm temperature, even when the surrounding air is much colder than body temperature. Humans, for example, use heat generated from "burning" fuel to maintain body temperature very close to 37°C (98.6°F).

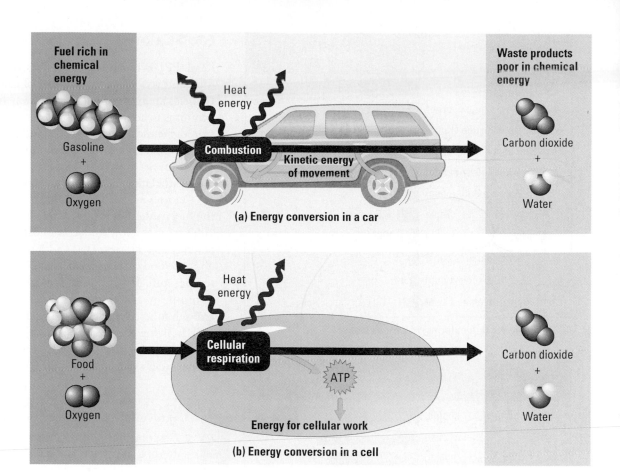

Figure 5.5 Energy conversions in (a) a car and (b) a cell. Like an automobile engine, cellular respiration uses oxygen to harvest the chemical energy of organic fuel molecules. Cellular respiration breaks sugars and other organic molecules down to smaller waste products. This chemical breakdown releases energy that was stored in the fuel molecules. Cells use this energy for their work.

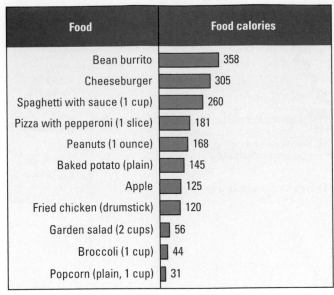

(a) Food calories (kilocalories) in various foods

Food	Food calories
Bean burrito	358
Cheeseburger	305
Spaghetti with sauce (1 cup)	260
Pizza with pepperoni (1 slice)	181
Peanuts (1 ounce)	168
Baked potato (plain)	145
Apple	125
Fried chicken (drumstick)	120
Garden salad (2 cups)	56
Broccoli (1 cup)	44
Popcorn (plain, 1 cup)	31

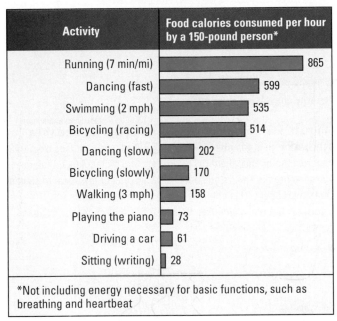

Activity	Food calories consumed per hour by a 150-pound person*
Running (7 min/mi)	865
Dancing (fast)	599
Swimming (2 mph)	535
Bicycling (racing)	514
Dancing (slow)	202
Bicycling (slowly)	170
Walking (3 mph)	158
Playing the piano	73
Driving a car	61
Sitting (writing)	28

*Not including energy necessary for basic functions, such as breathing and heartbeat

(b) Food calories (kilocalories) we burn in various activities

Figure 5.6 Some caloric accounting.

Food Calories

Read the label on a bag of peanuts, a box of cereal, or any other packaged food and you'll find the number of calories in each serving of that food. Calories are units of energy. One way to measure the caloric content of a peanut is to actually burn it under a container of water to convert all of the stored chemical energy to heat and then measure the temperature increase of the water. A **calorie** is the amount of energy that raises the temperature of 1 gram (g) of water by 1 degree Celsius (1°C). Of course, our cells don't "burn" fuel molecules from a peanut in this manner.

Calories are such tiny units of energy that they are not very practical for the fuel content of foods. We can scale up with the **kilocalorie,** which is 1000 calories. Now for a complication: The "calories" on a food package are actually kilocalories. That's a lot of energy. For example, just one peanut has about 5 food calories (kilocalories). That's enough energy to increase the temperature of 1 kilogram (kg; a little more than a quart) of water by 5°C in our peanut-burning experiment. And just a handful of peanuts packs enough food calories, if converted to heat, to bring a quart of water to a boil. In living organisms, food isn't used to boil water, of course, but to fuel the activities of life. Figure 5.6 shows the number of food calories in several foods and how many food calories are burned off by some typical activities.

CheckPoint

1. How can an object at rest have energy?
2. Describe the energy transformations that occur when you climb to the top of a stairway.
3. Which form of energy is the most randomized and the most difficult to put to useful work?
4. Walking at 3 miles per hour (mph), how long would you have to walk to "burn off" the equivalent of three slices of pepperoni pizza? How far would you have to travel? (Consult Figure 5.6.)

Answers: 1. It can have potential energy because of its location. **2.** You convert the chemical energy of food to the kinetic energy of your upward climb. At the top of the stairs, some of the energy has been stored as potential energy because of your higher elevation. The rest has been converted to heat. **3.** Heat energy **4.** You would have to walk about 3.4 hours, traveling 10.3 miles.

ATP and Cellular Work

It's a good thing food doesn't fuel cells by actually burning. In fact, the carbohydrates, fats, and other fuel molecules we obtain from food do not drive the work in our cells in any direct way. Providing the chemical energy for cellular work is the job of one very special substance called ATP. So, what becomes of the chemical energy that is released by the breakdown of organic molecules during cellular respiration? It is used to generate the ATP molecules that actually power cellular work directly.

The Structure of ATP

The initials **ATP** stand for adenosine triphosphate. ATP consists of a complex organic molecule called adenosine plus a tail of three phosphate groups, the "triphosphate" part of ATP (Figure 5.7).

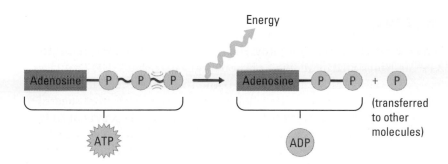

Energy

(transferred to other molecules)

ATP

ADP

Figure 5.7 ATP power. Each (P) in the triphosphate tail of ATP represents a phosphate group, a phosphorus bonded to oxygen atoms. The triphosphate tail is unstable, partly because of repulsion between the phosphate groups, which are negatively charged. There is a tendency for the phosphate group at the end of the triphosphate tail to "break away" from ATP and bond to other molecules instead. This phosphate transfer, catalyzed by enzymes, energizes cellular work. The leftover molecule is ADP (adenosine diphosphate).

The triphosphate tail is the "business" part of ATP, the part that provides energy for cellular work. Each phosphate group, consisting of phosphorus and oxygen, is negatively charged. Negative charges repel each other. The crowding of negative charge in the triphosphate tail contributes to the potential energy of ATP. It's analogous to storing energy by compressing a spring; if you release the spring, it will "relax," and you could use that springiness to do some useful work. For ATP power, it is release of the phosphate at the tip of the triphosphate tail that makes energy available to working cells. What is left behind is now called **ADP**, for adenosine diphosphate (two phosphate groups instead of three; see Figure 5.7).

Make and break ATP in Web/CD Activity 5C.

Phosphate Transfer

When ATP drives work in cells, phosphate groups don't just fly off into space, as it seems in Figure 5.7. ATP energizes other molecules in cells by actually transferring phosphate groups to those molecules (Figure 5.8). As an example, let's imagine another climb up a snowboarding hill. In the muscle cells of the snowboarder's legs, ATP transfers phosphate groups to special motor proteins. The proteins change their shape, causing the muscle cells to contract (see Figure 5.8a). By a similar process, ATP energizes proteins in the cilia that line the human windpipe (see Chapter 4). The beating cilia remove mucus that has trapped dust, pollen, and other inhaled debris.

Your cells perform three main kinds of work: mechanical work, transport work, and chemical work. The movement of muscles is an example of mechanical work; so is the movement of cilia. As an example of transport work, ATP enables brain cells to pump ions across their membranes (see Figure 5.8b). This prepares the brain cells to transmit signals. And ATP drives the chemical work of making a cell's giant molecules. An example is the linking of amino acids to make a protein (see Figure 5.8c). In Chapter 3, we saw that dehydration synthesis links amino acids, but it is ATP that provides the energy for this process. Notice again in Figure 5.8 that all these types of work occur when target molecules accept phosphate from ATP. Special enzymes catalyze these phosphate transfers that energize the working parts of cells.

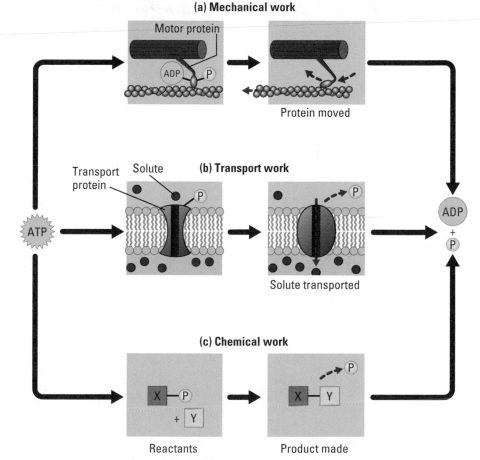

(a) Mechanical work

Motor protein

ADP · P

Protein moved

(b) Transport work

Transport protein · Solute

Solute transported

ATP

ADP + P

(c) Chemical work

X P + Y

X Y

Reactants

Product made

Figure 5.8 How ATP drives cellular work. Each type of work is powered when enzymes transfer phosphate from ATP to a recipient molecule: **(a)** a motor protein (mechanical work), **(b)** a transport protein (transport work), or **(c)** a chemical reactant (chemical work).

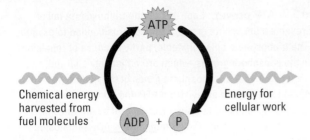

Figure 5.9 The ATP cycle.

Chemical energy harvested from fuel molecules

Energy for cellular work

ATP

ADP + P

The ATP Cycle

You spend your ATP continuously. Fortunately, it is a renewable resource. ATP can be restored by adding a phosphate group back to ADP. That takes energy, like compressing a spring again. And that's where food reenters the story. The chemical energy that cellular respiration harvests from sugars and other organic fuels is put to work regenerating a cell's supply of ATP. Cells have an ATP cycle: Cellular work spends ATP, which is recycled from ADP and phosphate using energy from food (Figure 5.9). Thus, ATP functions in what is called **energy coupling,** the transfer of energy from processes that yield energy, such as breakdown of organic fuels, to processes that consume energy, such as muscle contraction and other types of cellular work.

The ATP cycle runs at an astonishing pace. A working muscle cell recycles all of its ATP about once each minute. That amounts to about 10 million ATP molecules spent and regenerated per second per cell. If ATP could not be recycled and had to be completely fabricated from scratch instead, your body would consume its own weight in ATP each day. Let's now take a closer look at how cellular respiration keeps pace with this ATP demand by harvesting chemical energy from food.

CheckPoint

1. Explain how ATP powers cellular work.
2. What is the source of energy for regenerating ATP from ADP?

Answers: 1. ATP transfers phosphate groups to other molecules, increasing the energy content of those molecules. **2.** Chemical energy harvested from sugars and other organic fuels

Cellular Respiration: Aerobic Harvest of Food Energy

You have already learned that a cell, like a car's engine, consumes oxygen in breaking down its fuel. It is this living version of "internal combustion" that we call cellular respiration. And it is mainly cellular respiration that harvests chemical energy from food and converts it to ATP energy. Cellular respiration is an **aerobic** process, which is just another way of saying that it requires oxygen. We can now define cellular respiration as the aerobic harvesting of chemical energy from organic fuel molecules.

Relationship of Cellular Respiration to Breathing

We sometimes use the word *respiration* to mean breathing. While respiration on that level should not be confused with cellular respiration, the two processes are closely related. If you return to Figure 5.5, you'll see that cellular respiration requires a cell to exchange two gases with its surroundings. The cell takes in oxygen in the form of the gas O_2. It gets rid of waste in the form of the gas carbon dioxide, or CO_2. Your bloodstream keeps your cells supplied with O_2 and carries away the CO_2 "exhaust." Breathing exchanges these same gases between your blood and the outside air. Oxygen present in the air you inhale diffuses across the lining of your lungs and into your bloodstream. And the CO_2 in your bloodstream diffuses across your lungs' lining and exits when you exhale (Figure 5.10).

Overall Equation for Cellular Respiration

A common fuel molecule for cellular respiration is glucose, a six-carbon sugar with the formula $C_6H_{12}O_6$ (see Figure 3.10). Here is the overall equation for what happens to glucose during cellular respiration:

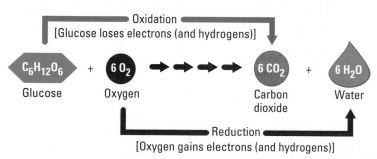

The series of arrows indicates that cellular respiration consists of many chemical steps, not just a single chemical reaction. Remember, the main function of cellular respiration is to generate ATP for cellular work. In fact, the process can produce up to 38 ATPs for each glucose molecule consumed.

Notice that cellular respiration also transfers hydrogen atoms from glucose to oxygen, forming water. That hydrogen transfer turns out to be the key to how oxygen enhances energy harvest during cellular respiration.

The Role of Oxygen in Cellular Respiration

In tracking hydrogen from sugar to oxygen, we are also following the transfer of electrons. The atoms of sugar and other molecules are bonded together by their sharing of electrons (see Figure 2.9). During cellular respiration, hydrogen and its bonding electrons change partners, from sugar to oxygen, forming water as a product.

Redox Reactions Chemical reactions that transfer electrons from one substance to another are called oxidation-reduction reactions, or **redox reactions** for short. The loss of electrons during a redox reaction is called **oxidation.** Glucose is oxidized during cellular respiration, losing electrons to oxygen. The acceptance of electrons during a redox reaction is called **reduction.** Oxygen is reduced during cellular respiration, accepting electrons (and hydrogen) lost from glucose:

When hydrogen and its bonding electrons change partners, from sugar to oxygen, energy is released.

Why does electron transfer to oxygen release energy? In redox reactions, oxygen is an "electron grabber." An oxygen atom pulls on electrons harder than almost any other type of atom. When electrons move (along with hydrogen) from glucose to oxygen, it is as though they were falling. They are

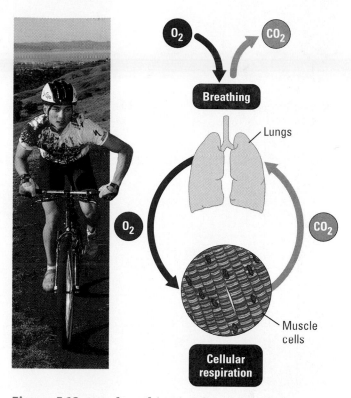

Figure 5.10 How breathing is related to cellular respiration. When you inhale, you breathe in oxygen, which diffuses across the lining of your lungs and into your bloodstream. Oxygen is delivered to your cells, where it is used in cellular respiration. Carbon dioxide, a waste product of cellular respiration, diffuses from your cells to your blood and travels to your lungs. CO_2 diffuses across the lining of your lungs and exits when you exhale.

not really falling in the sense of an apple dropping from a tree. However, in the case of both falling objects and electron transfer to oxygen, potential energy is unlocked. Instead of gravity, it is the pull of oxygen on electrons that causes the "fall" and energy release during cellular respiration.

A very rapid electron "fall" generates an explosive release of energy in the form of heat and light. For example, a spark will trigger a reaction between hydrogen gas (H_2) and oxygen gas (O_2) that produces water (H_2O). The reaction also releases a large amount of energy as the electrons of the hydrogen "fall" into their new bonds with oxygen (Figure 5.11). The result is similar if you burn glucose to measure its caloric content, as we burned the peanut earlier. The hydrogen and electrons from the glucose "fall" to the oxygen in the air. The flame represents energy released as heat and light.

Cellular respiration is a more controlled "fall" of electrons—more like a stepwise cascade of electrons down an energy staircase. Instead of liberating food energy in a burst of flame, cellular respiration unlocks chemical energy in smaller amounts that cells can put to productive use.

NADH and Electron Transport Chains Let's take a closer look at the path that electrons take on their way down from glucose to oxygen (Figure 5.12). First stop is an electron acceptor called NAD^+ (an abbreviation for a very long chemical name: nicotinamide adenine dinucleotide). Transfer of electrons from organic fuel to NAD^+ reduces the NAD^+ to **NADH** (the H represents the transfer of hydrogen along with the electrons). In our staircase analogy, the electrons have taken only one baby step down in their trip

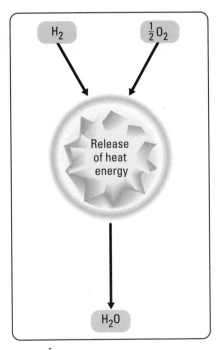

Figure 5.11 A rapid electron "fall." An all-at-once redox reaction between hydrogen and oxygen to form water releases a burst of energy. It is difficult to use such an explosion for productive work.

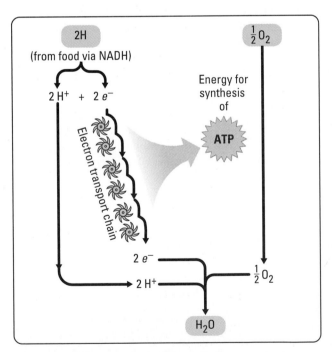

Figure 5.12 The role of oxygen in harvesting food energy. Like the explosive formation of water in Figure 5.11, cellular respiration combines oxygen and hydrogen to produce water. However, respiration breaks up the "fall" of electrons from food to oxygen into steps. NADH transfers electrons (e^-) from food to an electron transport chain. Oxygen pulls the electrons down the chain. Cells use the stepwise release of energy to make ATP. The oxygen combines with the electrons and hydrogen from food to produce water. The oxygen molecules (O_2) we inhale are the source of the oxygen atoms we use during cellular respiration, written here as $\frac{1}{2} O_2$.

from glucose to oxygen. The rest of the staircase consists of an **electron transport chain**. Each link in an electron transport chain is actually a molecule, usually a protein. In a series of redox reactions, each member of the chain can first accept and then donate electrons. The electrons give up a small amount of energy with each transfer. At the "uphill" end, the first molecule of the chain accepts electrons from NADH. Thus, the function of NADH is to carry electrons from glucose and other fuel molecules and deposit them at the top of an electron transport chain. The electrons then cascade down the chain, from molecule to molecule. The molecule at the bottom of the chain finally drops the electrons to oxygen. The oxygen also picks up hydrogen to form water (see Figure 5.12).

The overall effect of all this electron traffic during cellular respiration is a downhill trip for electrons from glucose to oxygen via NADH and electron transport chains. It is the stepwise freeing of chemical energy during electron transport that our cells use to make most of their ATP. Oxygen, the "electron grabber," makes it all possible. By pulling electrons down the transport chain from fuel molecules, oxygen functions somewhat like gravity pulling objects downhill. This is how the oxygen we breathe functions in our cells and why we cannot survive more than a few minutes without it.

The Metabolic Pathway of Cellular Respiration

Cellular respiration is an example of **metabolism**, the general term for *all* the chemical processes that occur in cells. More specifically, cellular respiration is a metabolic pathway. That means that it is not a single chemical reaction, but a series of reactions. A specific enzyme catalyzes each reaction in a metabolic pathway. More than two dozen reactions are involved in cellular respiration. We can group them into three main metabolic stages: glycolysis, the Krebs cycle, and electron transport (which you've already encountered). Let's see how these stages cooperate to harvest food energy.

A Road Map for Cellular Respiration Figure 5.13 is a map that will help you follow glucose through the metabolic pathway of cellular respiration. The map also shows you where the three stages of respiration occur in your cells.

Glycolysis splits glucose into two molecules of a compound called pyruvic acid. The enzymes for glycolysis are located outside the mitochondria, dissolved in the cytosol. The **Krebs cycle** completes the breakdown of sugar all the way to CO_2, the waste product of cellular respiration. The enzymes for the Krebs cycle are dissolved in the fluid within mitochondria. Glycolysis and the Krebs cycle generate a small amount of ATP directly. They generate much more ATP indirectly, via redox reactions that transfer electrons from fuel molecules to NAD^+. The third stage of cellular respiration is **electron transport.** Electrons captured from food by NADH "fall" down electron transport chains to oxygen. The proteins and other molecules that make up electron transport chains, which you learned about in the previous section, reside within mitochondria. It is electron transport from NADH to oxygen that releases the energy your cells use to make most of their ATP.

Get the big picture of cellular respiration with Web/CD Activity 5D.

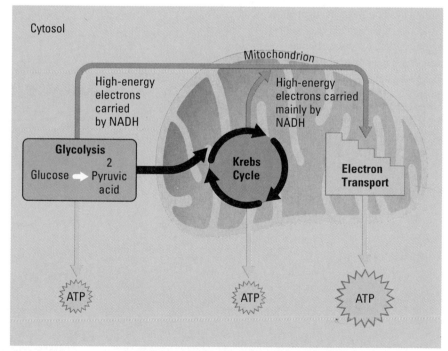

Figure 5.13 A road map for cellular respiration.
The three main stages—glycolysis, Krebs cycle, and electron transport—are color coded in this diagram. We'll carry a smaller version of this map with us so you can keep the overall process of cellular respiration in plain view even as we take a closer look at its three stages.

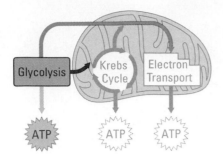

Stage 1: Glycolysis The word *glycolysis* means "splitting of sugar." That is exactly what happens (Figure 5.14). Glycolysis breaks the six-carbon glucose in half, forming two three-carbon molecules. These molecules then donate high-energy electrons to NAD$^+$, the electron carrier. Glycolysis also makes some ATP directly when enzymes transfer phosphate groups from fuel molecules to ADP (Figure 5.15). What remains of the fractured glucose at the end of glycolysis are two molecules of pyruvic acid. The pyruvic acid still holds most of the energy of glucose, and that energy is harvested in the Krebs cycle.

Watch the process of glycolysis in Web/CD Activity 5E.

Figure 5.14 Glycolysis. Each gray ball in this diagram represents a carbon atom. A team of enzymes splits glucose, eventually forming two molecules of pyruvic acid. Along the way, energy is stored as ATP and NADH. Note that the cell actually invests some ATP to get glycolysis started. Enzymes attach phosphate groups (P) to the fuel molecules during this energy investment phase. That investment is paid back with dividends during the energy harvest phase. Glycolysis generates some ATP directly, but it also donates high-energy electrons to NAD$^+$, forming NADH.

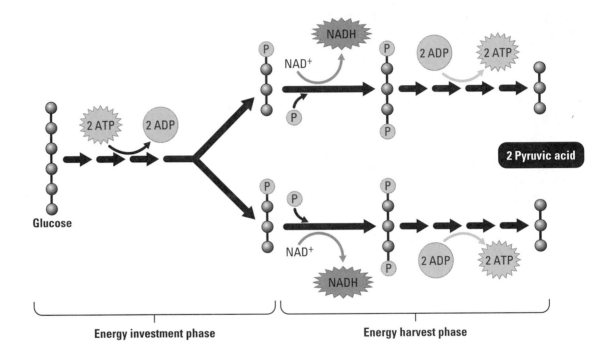

Energy investment phase

Energy harvest phase

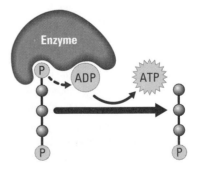

Figure 5.15 ATP synthesis by direct phosphate transfer. Glycolysis generates ATP when enzymes transfer phosphate groups directly from fuel molecules to ADP.

Stage 2: The Krebs Cycle

This stage of cellular respiration is named for Hans Krebs, the pioneer of biochemistry who worked out the steps of the cycle in the 1930s.

Pyruvic acid, the fuel that remains after glycolysis, is not quite ready for the Krebs cycle. First, the fuel must be "prepped"—converted to a form the Krebs cycle can use (Figure 5.16). The actual fuel consumed by the Krebs cycle is a two-carbon compound called acetic acid. It must enter the Krebs cycle in the form of acetyl-CoA, in which the acetic acid is bonded to a carrier molecule called coenzyme A (the CoA in acetyl-CoA).

The Krebs cycle finishes extracting the energy of sugar by breaking the acetic acid molecules (two per glucose) all the way down to CO_2 (Figure 5.17). The cycle uses some of this energy to make ATP by the direct method (see Figure 5.15). However, the Krebs cycle captures much more energy in the form of NADH and a second electron carrier, $FADH_2$. Electron transport then converts NADH and $FADH_2$ energy to ATP energy.

> See the Krebs cycle in action in Web/CD Activity 5F.

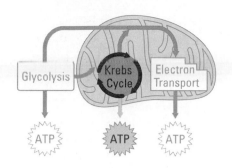

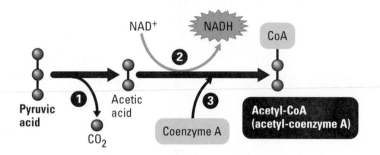

Figure 5.16 The link between glycolysis and the Krebs cycle: conversion of pyruvic acid to acetyl-CoA. ❶ Each pyruvic acid (there are two per starting glucose, remember) loses a carbon as CO_2. This is the first of this waste product we've seen so far in the breakdown of glucose. The remaining fuel molecules, each with only two carbons left, are called acetic acid (the same acid that's in vinegar). ❷ Oxidation of the fuel generates NADH. ❸ And finally, the acetic acid is attached to an adapter molecule called coenzyme A (CoA) to form acetyl-CoA. The CoA escorts the acetic acid into the first reaction of the Krebs cycle.

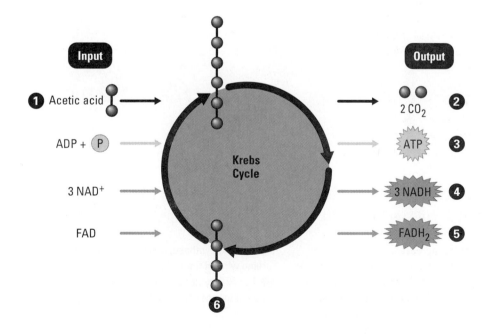

Figure 5.17 The Krebs cycle. This simplified version of the Krebs cycle emphasizes inputs and outputs. ❶ Input includes fuel in the form of acetic acid from acetyl-CoA. It joins a four-carbon acceptor molecule to form a six-carbon product. For every acetic acid molecule that enters the cycle this way as fuel, ❷ two CO_2 molecules eventually exit as exhaust (waste product). Along the way, the Krebs cycle harvests energy from the fuel. ❸ Some of the energy is used to produce ATP directly. ❹ Most of the energy is trapped by NADH. ❺ The $FADH_2$ you see in the diagram is another electron carrier that works something like NADH. ❻ All the carbon atoms that entered the cycle as fuel are accounted for as CO_2 exhaust, and the four-carbon acceptor molecule is recycled. This is what makes the Krebs cycle a cycle. We have tracked only one acetic acid molecule through the Krebs cycle here. But recall that glycolysis splits glucose in two (see Figure 5.14). So the Krebs cycle actually turns twice for each glucose molecule that fuels a cell.

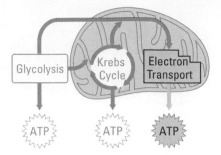

Stage 3: Electron Transport The molecules of electron transport chains are built into the inner membranes of mitochondria (Figure 5.18). An entire electron transport chain functions as a chemical machine that uses the energy released by the "fall" of electrons to pump hydrogen ions (H^+) across the inner mitochondrial membrane. Remember from Chapter 2 that H^+ ions are present in varying amounts in biological fluids. Small as they are, H^+ ions are solutes because they are dissolved in water. By pumping H^+ ions, electron transport chains store potential energy by making the ions more concentrated on one side of the membrane than the other.

The energy stored by electron transport behaves something like the elevated reservoir of water behind a dam. There is a tendency for the H^+ ions to gush back to where they are less concentrated, much as there is a tendency for water to flow downhill. The membrane, analogous to the dam, temporarily restrains the H^+ ions.

The energy of dammed water can be harnessed to do work. Gates in the dam allow the water to rush downhill, turning giant wheels called turbines as it goes. The spinning turbines can do work—generate electricity, for example. Your mitochondria also have turbines. Each of these miniature machines, called an **ATP synthase,** is constructed from several proteins. The ATP synthase complexes are built into the inner mitochondrial membrane,

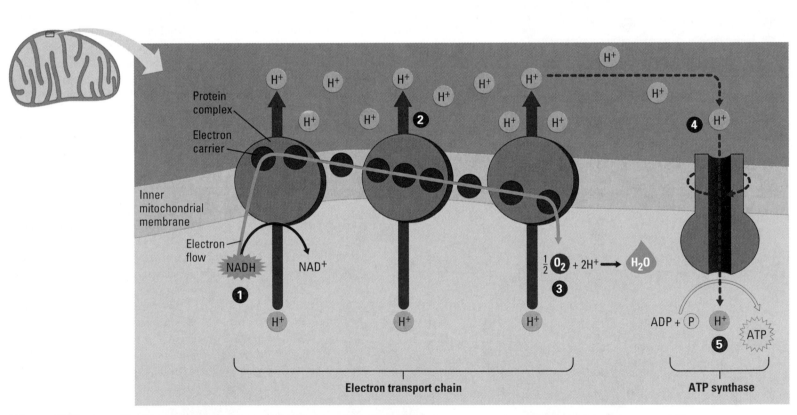

Figure 5.18 How electron transport drives ATP synthase machines. ❶ NADH transfers electrons from food to electron transport chains. **❷** Electron transport chains use this energy supply to pump H^+ ions across the inner membrane of the mitochondrion. The infoldings of the inner membrane increase surface area, maximizing the number of electron transport chains that are built into the membrane. **❸** Again, notice that the function of the oxygen you breathe is to pull electrons down the transport chain. **❹** The H^+ ions flow back through an ATP synthase. This spins a part of the synthase, much like water turns a turbine when it flows through the gates in a dam. **❺** The ATP synthase uses the energy of the H^+ gradient to regenerate ATP from ADP.

the same membrane where electron transport chains are located (see Figure 5.18). Electron transport provides energy to operate the ATP synthase machines, but indirectly. Hydrogen ions pumped by electron transport rush back "downhill" through an ATP synthase. This spins a component of the ATP synthase, just as water turns the turbines in a dam. The rotation activates catalytic sites in the synthase that attach phosphate groups to ADP molecules to regenerate ATP. The turbines of mitochondria are marvelous molecular dynamos that scientists are only beginning to understand.

Produce ATP using electron transport in Web/CD Activity 5G. Learn how to measure the rate of cellular respiration in The Process of Science.

Some of the deadliest poisons do their damage by disrupting electron transport in mitochondria. For example, carbon monoxide and cyanide both kill by blocking the transfer of electrons from electron transport chains to oxygen. With its energy-harvesting mechanism shut down, the mitochondrial membrane can no longer convert food energy to ATP energy. Cells stop working, and the organism dies.

The Versatility of Cellular Respiration We have seen so far that food provides the energy to make the ATP our cells use for all their work. We have concentrated on the sugar glucose as the fuel that is broken down in cellular respiration, but respiration is a versatile metabolic furnace that can "burn" many other kinds of food molecules. Figure 5.19 diagrams some metabolic routes for the use of diverse carbohydrates, fats, and proteins as fuel for cellular respiration.

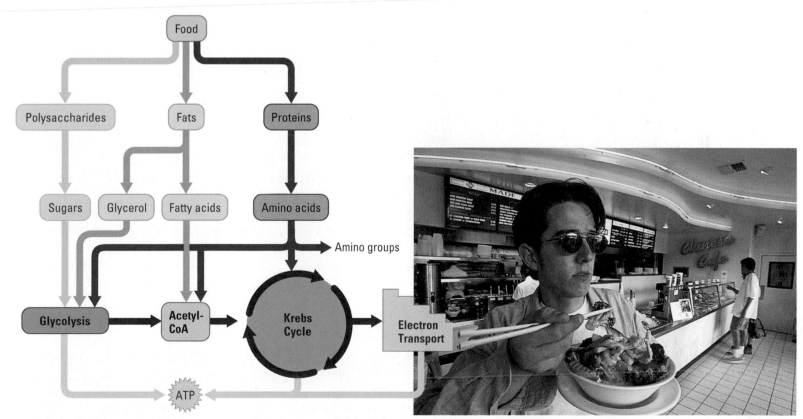

Figure 5.19 Energy from food. Carbohydrates, fats, and proteins can all serve as fuel for cellular respiration. Digestion hydrolyzes the large food molecules to their monomers (see Chapter 3), which are distributed to cells by the circulatory system. The monomers "feed" into glycolysis and the Krebs cycle at various points.

Figure 5.20 Summary of ATP yield during cellular respiration. A cell can convert the energy of each glucose molecule to as many as 38 ATPs. Glycolysis and the Krebs cycle each contribute 2 ATPs by direct synthesis (see Figure 5.15). The other 34 ATPs are produced by the ATP synthase machines. It is the "fall" of electrons from food to oxygen that powers ATP synthase. The electrons are carried from the organic fuel to electron transport chains by NADH and $FADH_2$. Each electron pair "dropped" down a transport chain from NADH can power the synthesis of up to 3 ATPs. Each electron pair transferred to an electron transport chain from $FADH_2$ is worth up to 2 ATPs.

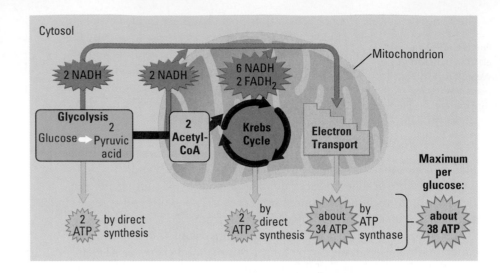

Adding Up the ATP Yield from Cellular Respiration Taking cellular respiration apart to see how all the molecular nuts and bolts of its metabolic machinery work, it's easy to lose sight of the overall function: to generate ATP. Figure 5.20 will help you add up the ATP molecules a cell can make for each glucose molecule it consumes as fuel. Notice that most of that ATP production is powered by electron transport. And electron transport depends on the presence of oxygen. After a CheckPoint, we'll see what happens when cells harvest food energy without the help of oxygen.

CheckPoint

1. How is your breathing related to your cellular respiration?

2. In the following redox reaction, which compound is oxidized and which is reduced?

$$C_4H_6O_5 + NAD^+ \rightarrow C_4H_4O_5 + NADH + H^+$$

3. At the "downhill" end of the electron transport chain, when electrons from NADH are finally passed to oxygen, what waste product of cellular respiration is produced? (*Hint:* Review Figure 5.12)

4. What is the potential energy source that directly drives ATP production by ATP synthase?

5. Of the three main stages of cellular respiration represented in Figure 5.13, which one uses oxygen directly to extract chemical energy from organic compounds?

6. The oxidation of acetic acid by NAD^+ extracts some chemical energy from the acetic acid. How can the cell harness that energy to make ATP?

7. Of the three stages of cellular respiration, which occurs in the cytosol, outside mitochondria?

8. What effect would an absence of oxygen (O_2) have on the process illustrated in Figure 5.18?

Answers: 1. In breathing, your lungs exchange CO_2 and O_2 between your body and the atmosphere. In cellular respiration, your cells consume the O_2 in extracting energy from food and release CO_2 as a waste product. **2.** $C_4H_6O_5$ is oxidized and NAD^+ is reduced. **3.** Water (H_2O) **4.** A concentration gradient of hydrogen ions across the inner membrane of a mitochondrion **5.** Electron transport **6.** The NADH can supply electrons to the electron transport chain, which generates a hydrogen ion gradient that drives ATP synthesis. **7.** Glycolysis **8.** There would be no production of ATP by ATP synthase. Without oxygen to "pull" electrons down the electron transport chain, the energy stored in NADH cannot be extracted and harnessed for ATP synthesis.

Fermentation: Anaerobic Harvest of Food Energy

Although you must breathe to stay alive, some of your cells can actually work for short periods without oxygen. Your muscle cells are a good example. They can produce some ATP under conditions that are **anaerobic,** meaning "without oxygen." This anaerobic harvest of food energy is called **fermentation.**

Fermentation in Human Muscle Cells

When you walk between classes on campus, ATP powers the muscle cells of your legs. Cellular respiration regenerates the ATP supply to keep you going. Blood provides your muscle cells with enough oxygen to keep electrons "falling" down transport chains in your mitochondria. During moderate activity, such as walking, your muscle cells are operating under aerobic conditions (with oxygen) because the oxygen supply to your cells can keep pace with your activity level. But if you run across campus because you're late for class, your muscles are forced to keep working under anaerobic conditions. That is because they are spending ATP at a rate that outpaces the delivery by the bloodstream of oxygen from your lungs to your muscles.

Muscle cells have enough ATP to support such an anaerobic activity for about 5 seconds. A secondary supply of phosphate bond energy, called **creatine phosphate,** can keep you going another 10 seconds or so. But what if you still haven't made it to class? To keep running, your muscles must generate ATP by the anaerobic process of fermentation.

The metabolic pathway that provides ATP during fermentation is glycolysis, the same pathway that functions as the first stage of cellular respiration. Glycolysis does not require oxygen (see Figure 5.14). And glycolysis, remember, does produce a small amount of ATP directly: 2 ATPs for each glucose broken down to pyruvic acid. That doesn't seem very efficient compared to the 38 ATPs each glucose generates during cellular respiration, but it can energize your leg muscles fast enough and long enough for you to make it to class. However, you'll have to consume more fuel per second, since you get so much less ATP per glucose under anaerobic conditions.

There is more to fermentation than just glycolysis. To harvest food energy during glycolysis, NAD^+ must be present as an electron acceptor (see Figure 5.14). This is no problem under aerobic conditions. The cell regenerates its NAD^+ when NADH drops its electron cargo down electron transport chains to oxygen (see Figure 5.12). However, this productive recycling of NAD^+ cannot occur under anaerobic conditions. Instead, NADH disposes of electrons by adding them to the pyruvic acid produced by glycolysis (Figure 5.21a). This restores NAD^+ and keeps glycolysis working as an ATP source. Reduction (additions of electrons) of pyruvic acid produces a waste

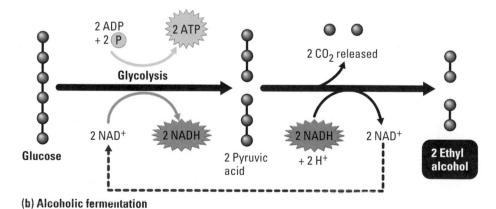

(a) Lactic acid fermentation

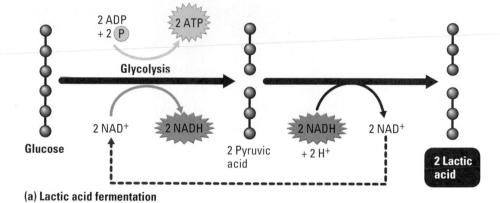

(b) Alcoholic fermentation

Figure 5.21 Fermentation. Glycolysis produces ATP without the help of oxygen. This requires a continuous supply of NAD^+ to accept electrons from glucose. The NAD^+ is regenerated when NADH transfers the electrons it removed from food to pyruvic acid. This produces **(a)** lactic acid, **(b)** ethyl alcohol, or other waste products, depending on the species of organism.

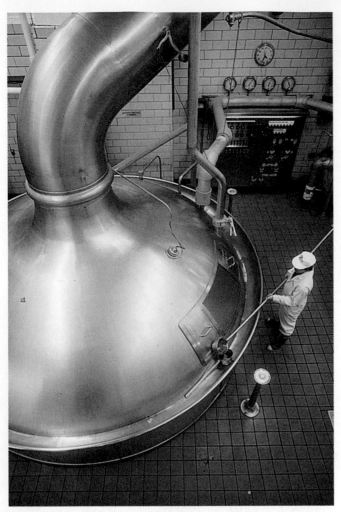

Figure 5.22 Fermentation tank at a brewery.

product called lactic acid. A temporary accumulation of lactic acid in muscle cells probably contributes to the soreness or burning you feel after an exhausting run or other anaerobic spurt of activity. The lactic acid is eventually transported in the blood from muscles to the liver, where liver cells convert the lactic acid back to pyruvic acid. That metabolic process requires oxygen, and thus fermentation in muscles results in an "oxygen debt" that must be paid back by heavy breathing for some time after vigorous activity ceases.

Fermentation in Microorganisms

Like human muscle cells, yeast, a microscopic fungus, is capable of both cellular respiration and fermentation. Keep yeast cells in an anaerobic environment, and they are forced to ferment their sugars and other foods. In contrast to muscle cells, the yeast style of fermentation produces ethyl alcohol as a waste product instead of lactic acid. This alcoholic fermentation also releases CO_2 (Figure 5.21b on previous page). For thousands of years, humans have put yeast to work producing alcoholic beverages such as beer and wine (Figure 5.22). The CO_2 is what makes beer and champagne bubbly. And as every baker knows, the CO_2 bubbles from fermenting yeast also cause bread dough to rise.

Compare lactic acid fermentation and alcoholic fermentation in Web/CD Activity 5H.

There are also fungi and bacteria that produce lactic acid as their waste product during fermentation, just as our muscle cells do. We have domesticated such microbes to transform milk to cheese and yogurt. These foods owe their sharp or sour flavor mainly to lactic acid. The food industry uses the same kind of fermentation to turn soybeans into soy sauce, cucumbers into pickles, and cabbage into sauerkraut.

Yeast is an example of what is called a **facultative anaerobe,** an organism with the metabolic versatility to harvest food energy by either respiration or fermentation. In contrast are **obligate anaerobes,** organisms that are actually poisoned by oxygen. Examples are certain bacteria that live in stagnant ponds or deep in the soil. At the cellular level, our muscle cells, like yeast cells, behave as facultative anaerobes. But as whole organisms, we are best described, in metabolic terms, as **obligate aerobes,** so dependent are we on oxygen and cellular respiration to stay alive.

CheckPoint

1. A glucose-fed yeast cell is moved from an aerobic environment to an anaerobic one. If the cell continues to generate ATP at the same rate, how will its rate of glucose consumption compare with its consumption in the aerobic environment?

2. _____ acid is to human muscle cells as ethyl _____ is to yeast.

Answers: 1. The cell must consume glucose at a rate of about 19 times the consumption rate in the aerobic environment (2 ATPs by fermentation versus 38 ATPs by cellular respiration). **2.** Lactic; alcohol

Evolution Link:
Life on an Anaerobic Earth

The role of glycolysis in both fermentation and respiration has an evolutionary basis. Ancient bacteria probably used glycolysis to make ATP long before oxygen was present in Earth's atmosphere. The oldest known fossils of bacteria date back over 3.5 billion years, but appreciable quantities of oxygen did not begin to accumulate in the atmosphere until about 2.5 billion years ago. For a billion years, bacteria must have generated ATP exclusively from glycolysis, which does not require oxygen. Glycolysis is the most widespread metabolic pathway, which also suggests that it evolved very early in ancestors common to all kingdoms of life. And the location of glycolysis in the cytosol of cells implies great antiquity, too. Glycolysis does not require mitochondria, which first evolved in eukaryotes long after the origin of prokaryotic life (bacteria and archaea). Glycolysis is a metabolic heirloom from the earliest cells that continues to function today in the harvest of food energy.

Discover more about life on an anaerobic Earth in the Web Evolution Link.

Chapter Review

Summary of Key Concepts

Overview: Energy Flow and Chemical Cycling in the Biosphere

- **Producers and Consumers** Plants and other photosynthetic organisms are the producers of organic matter in most ecosystems. Animals and other consumers depend on that photosynthetic product for their nutrition. The food we consume provides both our fuel (energy supply) and building material.

- **Chemical Cycling Between Photosynthesis and Cellular Respiration** Photosynthesis produces $C_6H_{12}O_6$ (sugar) and O_2 from CO_2 and H_2O. Cellular respiration consumes the $C_6H_{12}O_6$ and O_2 and produces CO_2 and H_2O as waste products, which are also the ingredients for photosynthesis.
- **Web/CD Activity 5A** *Build a Chemical Cycling System*

Some Basic Energy Concepts

- **Conservation of Energy** Machines and organisms can transform kinetic energy (energy of motion) to potential energy (stored energy) and vice versa. In all such energy transformations, total energy is conserved. Energy cannot be created or destroyed.

- **Entropy** Every energy conversion releases some randomized energy in the form of heat. This is an example of the tendency for the entropy, or disorder, of the universe to increase.
- **Web/CD Activity 5B** *Energy Concepts*

- **Chemical Energy** Molecules store varying amounts of potential energy in the arrangement of their atoms. Organic molecules are relatively rich in such chemical energy.

- **Food Calories** Actually kilocalories, food calories are units for the amount of energy in our foods and also for the amount of energy we expend in various activities.

ATP and Cellular Work

- **The Structure of ATP** Energy is made available for cellular work when adenosine triphosphate loses a phosphate group. This converts ATP to ADP.
- **Web/CD Activity 5C** *ATP*

- **Phosphate Transfer** ATP power is put to work by the direct transfer of phosphate from ATP to other molecules in the cell.

- **The ATP Cycle** Cells use energy from organic fuel to recycle ATP from ADP.

Cellular Respiration: Aerobic Harvest of Food Energy

- **Relationship of Cellular Respiration to Breathing** The bloodstream distributes O_2 from the lungs to all the cells of the body and transports CO_2 waste from the cells to the lungs for disposal.

- **Overall Equation for Cellular Respiration**

$$C_6H_{12}O_6 \ + \ 6\,O_2 \ \longrightarrow \ 6\,CO_2 \ + 6\,H_2O + \text{ATPs}$$

Glucose   Oxygen   Carbon dioxide   Water   Energy

- **The Role of Oxygen in Cellular Respiration** Redox reactions transfer electrons from food molecules to an electron acceptor called NAD^+, forming NADH. The NADH then passes the high-energy electrons to an electron transport chain that eventually "drops" them to oxygen. The energy released during this electron transport is used to regenerate ATP from ADP. It is the affinity of oxygen for electrons that keeps the redox reactions of cellular respiration working.

- **The Metabolic Pathway of Cellular Respiration** Respiration consists of glycolysis, the Krebs cycle, and electron transport. Glycolysis, which occurs in the cytosol, and the Krebs cycle, located in mitochondria, break sugar down to the waste product CO_2. These metabolic pathways produce a small amount of ATP directly, but mostly their redox reactions harvest food energy by transferring electrons from the fuel to NAD^+, forming NADH. That energy is

converted to ATP energy by the inner membranes of mitochondria, location of both electron transport chains and ATP synthases. The electron transport chains pump H^+ across the membrane as they ease electrons stepwise from NADH to oxygen. Backflow of the H^+ across the membrane powers the ATP synthases, which attach phosphate to ADP to make ATP. In all, cellular respiration can generate up to 38 ATPs for each glucose molecule it consumes.
• Web/CD Activity 5D *Overview of Cellular Respiration*
• Web/CD Activity 5E *Glycolysis*
• Web/CD Activity 5F *The Krebs Cycle*
• Web/CD Activity 5G *Electron Transport*
• Web/CD The Process of Science *Cellular Respiration Lab*

Fermentation: Anaerobic Harvest of Food Energy

• **Fermentation in Human Muscle Cells** When muscle cells consume ATP faster than oxygen can be supplied for cellular respiration, they regenerate ATP by fermentation. The waste product under these anaerobic conditions is lactic acid. The ATP yield per glucose is almost 20 times less during fermentation compared to respiration.

• **Fermentation in Microorganisms** Yeast and other facultative anaerobes can survive with or without oxygen. Wastes from fermentation can be ethyl alcohol, lactic acid, or other compounds, depending on the species. Some microorganisms are obligate anaerobes, which are poisoned by oxygen.
• Web/CD Activity 5H *Fermentation*

Evolution Link: Life on an Anaerobic Earth

• Ancient bacteria probably used glycolysis to make ATP long before oxygen was present in Earth's atmosphere.
• Web Evolution Link *Life on an Anaerobic Earth*

Self-Quiz

1. Which of the following directly produces the most ATP molecules per glucose molecule consumed?
   a. lactic acid fermentation
   b. the Krebs cycle
   c. electron transport and ATP synthase
   d. alcoholic fermentation
   e. glycolysis

2. Which of the following molecules are you least likely to find in a human muscle cell working under anaerobic conditions?
   a. ATP
   b. ADP
   c. acetyl-CoA
   d. NADH
   e. glucose

3. For a muscle cell to produce ATP at the same rate during fermentation as it does during cellular respiration, it must
   a. speed up the Krebs cycle.
   b. consume more sugar as fuel.
   c. speed up its use of oxygen.
   d. consume more lactic acid.
   e. increase the efficiency of electron transport chains.

4. In glycolysis, _____ is oxidized and _____ is reduced.
   a. NAD^+; glucose
   b. glucose; oxygen
   c. ATP; ADP
   d. glucose; NAD^+
   e. ADP; ATP

5. Sports physiologists at an Olympic training center wanted to monitor athletes to determine at what point their muscles were functioning anaerobically. They could do this by checking for a buildup of
   a. ATP.
   b. lactic acid.
   c. carbon dioxide.
   d. ADP.
   e. oxygen.

6. Which metabolic pathway is common to both fermentation and cellular respiration?
   a. glycolysis
   b. Krebs cycle
   c. electron transport
   d. synthesis of acetyl-CoA from pyruvic acid
   e. reduction of pyruvic acid to lactic acid

7. The final electron acceptor of electron transport chains in mitochondria is
   a. ADP.
   b. NAD^+.
   c. oxygen.
   d. glucose.
   e. pyruvic acid.

8. Cellular respiration that consumes oxygen occurs in all of the following *except*
   a. aerobic consumers.
   b. plants.
   c. human muscle cells.
   d. obligate anaerobes.
   e. yeast.

9. Cells can harvest the most total chemical energy from which of the following?
   a. an ATP moleucle
   b. an NADH molecule
   c. a glucose molecule
   d. six carbon dioxide molecules
   e. two pyruvic acid molecules

10. Which of the following best captures the function of photosynthesis?
    a. produces food for animals
    b. converts light energy to chemical energy
    c. replenishes the atmosphere's supply of CO_2
    d. harvests the chemical energy of ATP
    e. creates energy that sustains life

• **Go to the website or CD-ROM for more self-quiz questions.**

The Process of Science

1. Your body makes NAD^+ and FAD from two B vitamins, niacin and riboflavin. You need only tiny amounts of vitamins; the recommended dietary allowance for niacin is 20 mg daily and for riboflavin, 1.7 mg. These amounts are thousands of times less than the amount of glucose your body needs each day to fuel its energy needs. How many NAD^+ and FAD molecules are needed for the breakdown of each glucose molecule? Why do you think your daily requirement for these substances is so small?

2. **Learn how to measure the rate of cellular respiration in The Process of Science activity on the website and CD-ROM.**

Biology and Society

Nearly all human societies use fermentation to produce alcoholic drinks such as beer and wine. The technology dates back to the earliest civilizations. Suggest a hypothesis for how humans first discovered fermentation. In preindustrial cultures, why do you think wine was a more practical beverage than the grape juice from which it was made?

CHAPTER 6

Photosynthesis: Converting Light Energy to Chemical Energy

Life on Earth is solar powered.

Photosynthesis produces 160 billion metric tons of carbohydrates each year. Without the greenhouse effect of the atmosphere, the average air temperature on Earth would be about −18°C (0°F). Each square millimeter of a leaf contains about 500,000 photosynthesis factories called chloroplasts.

Figure 6.1 **A forest's photosynthetic organisms convert light energy to the chemical energy of food.**

On a global scale, the productivity of photosynthesis is astounding. Earth's plants and other photosynthetic organisms make about 160 billion metric tons of organic material per year (a metric ton is 1000 kg, about 1.1 tons). That much organic material is equivalent in weight to about 80 trillion copies of this textbook—25 stacks of books reaching from Earth to the sun! No other chemical process on the planet can match the output of photosynthesis. And no process is more important than photosynthesis to life on Earth (Figure 6.1). In this chapter, you will learn how plants convert solar energy to food energy.

Overview: Our Dependence on Plants

Plants and other photosynthetic organisms convert the energy of sunlight to the chemical energy of sugar and other organic compounds. It is the source of all our food. And we depend on plants for more than our food. You are probably wearing underwear, jeans, or other clothing made of another product of photosynthesis, cotton. Most of our homes are framed with lumber, which is wood produced by photosynthetic trees. Even the text you are now reading is printed on paper, still another material that can be traced to photosynthesis in plants. But mostly, photosynthesis is all about feeding the biosphere.

> Discover more ways we are linked to plants in Web/CD Activity 6A.

Plants are **autotrophs,** which means "self-feeders" in Greek. The term is a bit misleading in its implication that plants do not require nutrients—they do. But those nutrients are entirely inorganic: carbon dioxide from the air, and water and minerals from the soil. From that inorganic diet, plants can make all their own organic molecules, including carbohydrates, lipids, proteins, and nucleic acids. That is the definition of an autotroph: an organism that makes all its own organic matter from inorganic nutrients. In contrast, humans and other animals are **heterotrophs,** meaning "other-feeders." We heterotrophs cannot make organic molecules from inorganic ones. That is why we must eat. Heterotrophs depend on autotrophs for their organic fuel and material for growth and repair.

The most common form of autotrophic nutrition is photosynthesis, which uses light energy to drive the synthesis of organic molecules. Almost all plants are photosynthetic, and so are certain groups of protists and bacteria (Figure 6.2).

(a) Mosses, ferns, and flowering plants

(b) Kelp

Figure 6.2 **Photosynthetic autotrophs: producers for most ecosystems.** **(a)** On land, plants are the predominant producers of food. Three major groups of plants—mosses, ferns, and flowering plants—are represented in this scene. In oceans, lakes, ponds, streams, and other aquatic habitats, photosynthetic organisms include: **(b)** large algae, such as this kelp; **(c)** certain microscopic protists, such as *Euglena;* **(d)** bacteria called cyanobacteria; and **(e)** other photosynthetic bacteria, such as these purple sulfur bacteria.

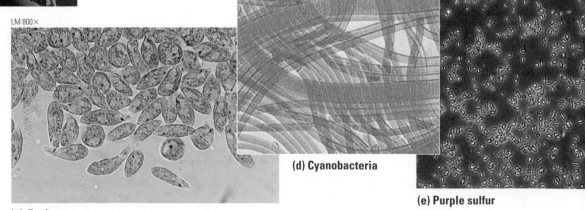

LM 800×

LM 200×

LM 400×

(d) Cyanobacteria

(c) *Euglena*

(e) Purple sulfur bacteria

We can connect the autotroph/heterotroph distinction back to another pair of terms you learned in Chapter 5. Autotrophs are also known as **producers,** in honor of their central role as the ultimate source of all food in any ecosystem. Animals and other heterotrophs are also called **consumers,** because all of their food traces back to producers.

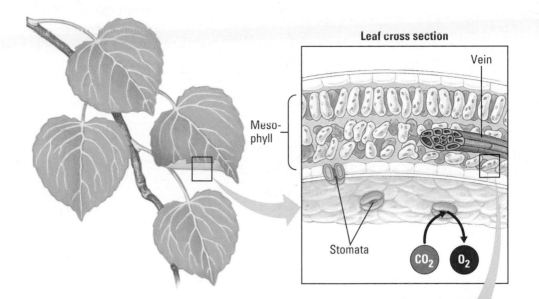

Leaf cross section

Vein

Meso- phyll

Stomata

CO_2 O_2

The Basics of Photosynthesis

Photosynthesis is complex, as you would expect for a process that converts solar energy to food energy. But there are some basic concepts that will keep us oriented as we dissect photosynthesis for a closer look.

Chloroplasts: Sites of Photosynthesis

In Chapter 4, you learned that photosynthesis occurs within **chloroplasts,** organelles in certain plant cells. All green parts of a plant have chloroplasts and can carry out photosynthesis. However, in most plants, the leaves have the most chloroplasts and are the major sites of photosynthesis. The green color in plants is from chlorophyll pigments in the chloroplasts. Chlorophyll absorbs the light energy that the chloroplast puts to work in making food.

Chloroplasts are concentrated in the cells of the **mesophyll,** the green tissue in the interior of the leaf (Figure 6.3). Carbon dioxide enters the leaf, and oxygen exits, by way of tiny pores called **stomata** (singular, *stoma*, which means "mouth"). In addition to carbon dioxide, photosynthesis also requires water as an inorganic ingredient. The water is mainly absorbed by the plant's roots, then travels via veins to the leaves.

Membranes in the chloroplast form the apparatus where many of the reactions of photosynthesis occur. Like the mitochondrion, the chloroplast has a double-membrane envelope (see Figures 4.15 and 4.16). The chloroplast's inner membrane encloses a compartment filled with **stroma,** a thick fluid. The stroma is where sugars are made from carbon dioxide. Suspended in the stroma is an elaborate system of disklike membranous sacs called **thylakoids.** The thylakoids are concentrated in stacks called **grana** (singular, *granum*). The chlorophyll molecules that capture light energy are built into the thylakoid membranes.

Test your knowledge of the sites of photosynthesis in Web/CD Activity 6B.

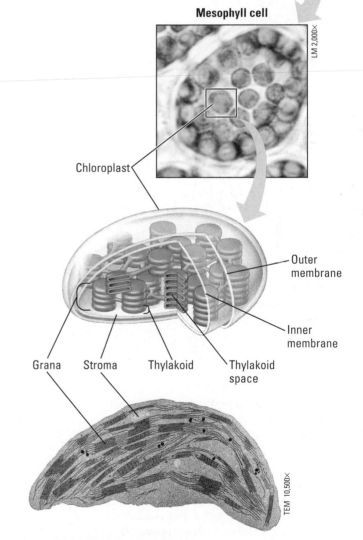

Mesophyll cell

LM 2,000×

Chloroplast

Outer membrane

Inner membrane

Grana   Stroma   Thylakoid   Thylakoid space

TEM 10,500×

Figure 6.3 Journey into a leaf. This series of blowups takes you into the mesophyll of a leaf, then into a cell, and finally into a chloroplast, site of photosynthesis.

Overall Equation for Photosynthesis

The following chemical equation provides an overall look at the reactants and products of photosynthesis:

$$6 CO_2 + 6 H_2O \xrightarrow{\text{Light energy}} C_6H_{12}O_6 + 6 O_2$$

Carbon dioxide + Water → Glucose + Oxygen gas

Photosynthesis

Notice that the reactants of photosynthesis, carbon dioxide (CO_2) and water (H_2O), are also the waste products of cellular respiration (see Figure 5.3). And photosynthesis produces what respiration uses, namely glucose ($C_6H_{12}O_6$) and oxygen (O_2). In other words, photosynthesis takes the "exhaust" of cellular respiration and rearranges its atoms to produce food and oxygen. It's a chemical transformation that requires much energy, and it is sunlight absorbed by chlorophyll that provides the energy.

You learned in Chapter 5 that cellular respiration is a process of electron transfer, or reduction and oxidation (redox). A "fall" of electrons from food molecules to oxygen to form water releases the energy that mitochondria can use to make ATP (see Figure 5.12). The opposite occurs in photosynthesis: Electrons are boosted "uphill" and added to carbon dioxide to produce sugar. Hydrogen is moved along with the electrons, so the redox process takes the form of hydrogen transfer from water to carbon dioxide. This requires the chloroplast to actually split water molecules into hydrogen and oxygen. The chloroplast transfers the hydrogen along with electrons to carbon dioxide to form sugar. The oxygen from the splitting of water escapes through stomata into the atmosphere as O_2, a waste product of photosynthesis. It takes a lot of energy to split water, which is a very stable molecule (that's why Earth has so much of it). Again, it is sunlight that provides the energy (Figure 6.4).

Figure 6.4 Life's ultimate source of energy. Sunlight provides the energy that photosynthetic organisms, such as trees, use to split water and make organic material. Hydrogen and electrons from water are transferred to carbon dioxide to form sugars.

Photosynthesis Road Map

The equation for photosynthesis is a deceptively simple summary of a very complex process. Actually, photosynthesis is not a single process, but two, each with many steps. These two stages of photosynthesis are called the **light reactions** and the **Calvin cycle** (Figure 6.5).

The light reactions convert solar energy to chemical energy. They use the light energy to drive the synthesis of two molecules: ATP and NADPH. We have already met ATP as the molecule that drives most cellular work. **NADPH,** chemical cousin of the NADH that appeared in Chapter 5, is an electron carrier. In cellular respiration, remember, NADH carries electrons from food molecules. In photosynthesis, light drives electrons from water to $NADP^+$ (oxidized form of the carrier) to form NADPH (reduced form of the carrier). Although the light reactions convert light energy to the chemical energy of ATP and NADPH, this stage of photosynthesis does not produce sugar.

It is the Calvin cycle that actually makes sugar from carbon dioxide. The ATP generated by the light reactions provides the energy for sugar synthesis. And the NADPH produced by the light reactions provides the high-energy electrons for the reduction of carbon dioxide (CO_2) to sugar ($C_6H_{12}O_6$). Thus, the Calvin cycle depends on light, but only indirectly—it depends on the supply of ATP and NADPH produced by the light reactions.

Go to Web/CD Activity 6C to see simple animations of the overall process of photosynthesis.

The road map in Figure 6.5 will help you keep oriented in this energy traffic from the light reactions to the Calvin cycle as we now take a closer look at how these two stages of photosynthesis work.

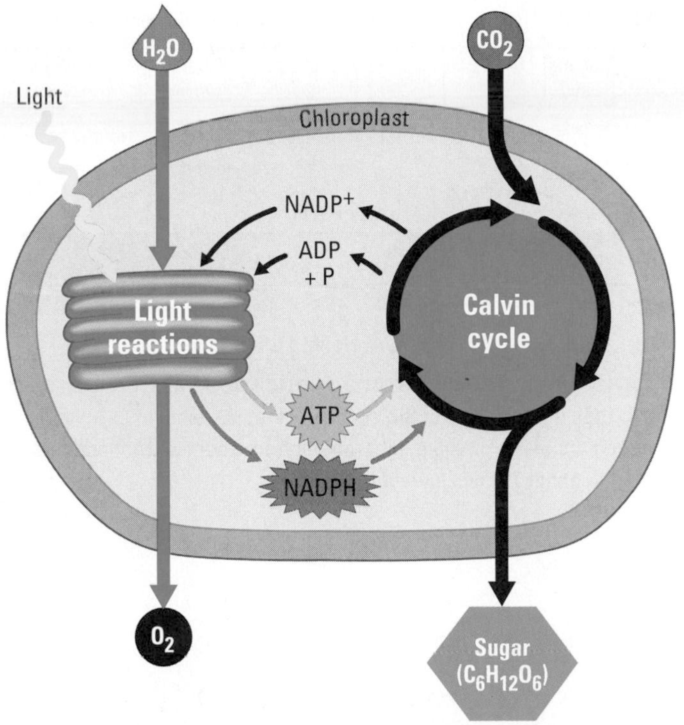

Figure 6.5 Road map for photosynthesis. Thylakoids, the membranous sacs stacked as grana in chloroplasts, are the sites of the light reactions. Chlorophyll built into the thylakoid membranes absorbs light energy. The thylakoids convert the light energy to the chemical energy of ATP and NADPH. The Calvin cycle uses these two products of the light reactions to power production of sugar from carbon dioxide. ATP provides energy and NADPH is a source of high-energy electrons to convert CO_2 to sugar. The enzymes for the Calvin cycle are dissolved in the stroma, the thick fluid within the chloroplast. We'll carry a smaller version of this road map for orientation as we take a closer look at the light reactions and the Calvin cycle.

CheckPoint

1. For chloroplasts to produce sugar from carbon dioxide in the dark, they would require an artificial supply of _____ and _____.

2. For each CO_2 molecule it consumes, how many O_2 molecules does a plant release to the atmosphere?

Answers: 1. ATP and NADPH **2.** One

The Light Reactions: Converting Solar Energy to Chemical Energy

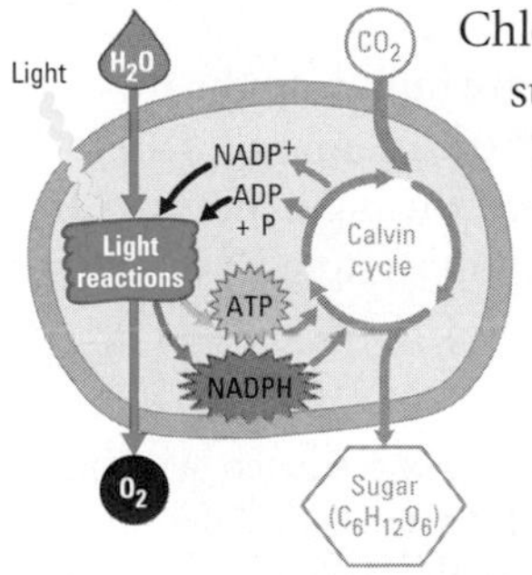

Chloroplasts are chemical factories powered by the sun, an energy source over 90 million miles from Earth. In this section, we'll track sunlight into a chloroplast to see how it is converted to the chemical energy of ATP and NADPH.

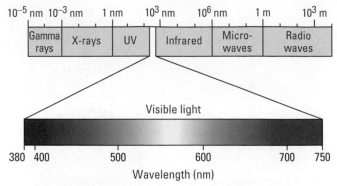

Figure 6.6 **The electromagnetic spectrum.** This diagram
expands the thin slice of the spectrum that is visible to us as different colors of light. Visible light ranges from about 380 nanometers
(nm) to about 750 nm in wavelength.

The Nature of Sunlight

Sunlight is a type of energy called radiation, or electromagnetic energy. Electromagnetic energy travels through space as rhythmic waves analogous to those made by a pebble dropped in a puddle of water. The distance between the crests of two adjacent waves is called a **wavelength.** The full range of radiation, from the very short wavelengths of gamma rays to the very long wavelengths of radio signals, is called the **electromagnetic spectrum.** Visible light forms only a small fraction of the spectrum. It consists of those wavelengths that our eyes see as different colors (Figure 6.6).

When sunlight shines on a pigmented material, certain wavelengths (colors) of the visible light are absorbed and disappear from the light that is reflected by the material. For example, we see a pair of jeans as blue because the pigments in the material absorb the other colors, leaving only light in the blue part of the spectrum to be reflected from the material into our eyes. And the leaves of plants appear green because pigments in their chloroplasts absorb mainly blue-violet and red-orange wavelengths (Figure 6.7). Of course, energy cannot be destroyed. If a pigment has subtracted certain wavelengths of light from the spectrum through absorption, that energy has been converted to other forms. Chloroplasts are able to convert some of the solar energy they absorb into chemical energy.

Chloroplast Pigments

Different pigments absorb light of different wavelengths, and chloroplasts contain several kinds of pigments. One, **chlorophyll a,** absorbs mainly blue-violet and red light. It looks grass-green because it reflects mainly green light. Chlorophyll a is the pigment that participates directly in the light reactions. A very similar molecule, chlorophyll b, absorbs mainly blue and orange light and reflects (appears) yellow-green. Chlorophyll b does not participate directly in the light reactions, but it broadens the range of light that a plant can use by conveying absorbed energy to chlorophyll a, which then puts the energy to work in the light reactions. Chloroplasts also contain a family of yellow-orange pigments called carotenoids, which absorb mainly blue-green light. Some may pass energy to chlorophyll a. Other carotenoids have a protective function: They absorb and dissipate excessive light energy that would otherwise damage chlorophyll. (Similar carotenoids, which we obtain from carrots and certain other plants, may help protect our eyes from very bright light.)

All of these chloroplast pigments are built into the thylakoid membranes, especially where those membranes are stacked as grana (see Figure 6.3). There, the pigments are part of light-harvesting complexes called photosystems.

Experiment with
wavelengths and
pigments in Web/CD
Activity 6D and The
Process of Science.

Light

**Reflected
light**

**Absorbed
light**

**Transmitted
light**

Chloroplast

Figure 6.7 **Why are leaves green?** Chlorophyll and the other pigments built into the membranes of grana mainly absorb light in the
blue-violet and red-orange part of the spectrum. The pigments do not absorb much green light, which is reflected to our eyes.

How Photosystems Harvest Light Energy

The theory of light as waves explains most of light's properties. However, light also behaves as discrete packets of energy called photons. A **photon** is a fixed quantity of light energy. The shorter the wavelength, the greater the energy. For example, a photon of violet light packs nearly twice as much energy as a photon of red light (see Figure 6.6).

When a pigment molecule absorbs a photon, one of the pigment's electrons gains energy—we say that the electron has been raised from a ground state to an excited state. The excited state is very unstable, and generally the electron loses the excess energy and falls back to its ground state almost immediately. Most pigments merely release heat energy as their light-excited electrons fall back to their ground state. (That is why a dark surface, such as a green automobile roof, gets so hot on a sunny day.) Some pigments, including isolated chlorophyll that has been extracted from chloroplasts, emit light as well as heat after absorbing photons (Figure 6.8).

Light-excited chlorophyll behaves very differently in an intact chloroplast than it does in isolation. In its native habitat of the thylakoid membrane, chlorophyll is organized with other molecules into **photosystems.** Each photosystem has a cluster of a few hundred pigment molecules, including chlorophylls *a* and *b* and carotenoids (Figure 6.9). This cluster of pigment molecules functions as a light-gathering antenna. When a photon strikes one pigment molecule, the energy jumps from pigment to pigment until it arrives at what is called the **reaction center** of the photosystem. The reaction center consists of a chlorophyll *a* molecule that sits next to another molecule called a **primary electron acceptor.** This acceptor traps the light-excited electron from the reaction-center chlorophyll. Another team of molecules built into the thylakoid membrane then uses that trapped energy to make ATP and NADPH.

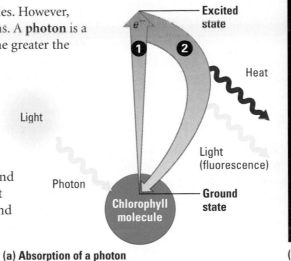

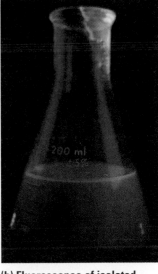

(a) Absorption of a photon

(b) Fluorescence of isolated chlorophyll in solution

Figure 6.8 Behavior of isolated chlorophyll. (a) ❶ Absorption of a photon drives an electron (e^-) from its ground state to an excited state. ❷ In a billionth of a second, the excited electron falls back to its ground state, releasing heat and light. **(b)** The afterglow from chlorophyll is called fluorescence. This flask contains a solution of chlorophyll illuminated with ultraviolet light. The red-orange wavelengths of the fluorescence are less energetic than the ultraviolet light that excited the chlorophyll. Remember that energy cannot be destroyed, however; heat makes up the difference.

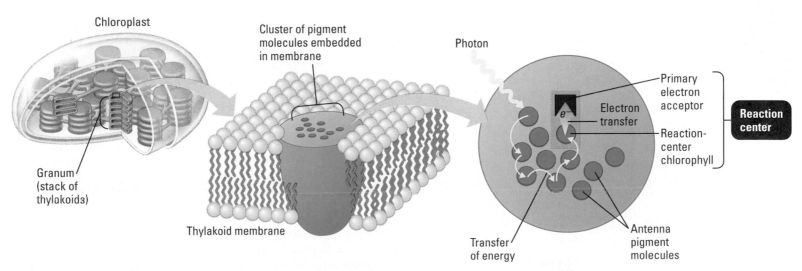

Figure 6.9 A photosystem. The pigment cluster of the photosystem functions as a light-gathering antenna that focuses energy onto the reaction center. At the reaction center of the photosystem, a chlorophyll molecule transfers its light-excited electron (e^-) to a primary electron acceptor.

How the Light Reactions Generate ATP and NADPH

Two types of photosystems cooperate in the light reactions. One is a water-splitting photosystem, which uses light energy to extract electrons from H_2O (Figure 6.10). This is the process that releases O_2 as a waste product of photosynthesis. The second photosystem is an NADPH-producing photosystem. It produces NADPH by transferring light-excited electrons from chlorophyll to $NADP^+$. An electron transport chain connecting the two photosystems releases energy that the chloroplast uses to make ATP. The traffic of electrons through the two photosystems is analogous to the cartoon in Figure 6.11.

Figure 6.12 places the light reactions in the thylakoid membrane. Notice that the mechanism of ATP production during the light reactions is very similar to the ATP production we saw in cellular respiration (see Chapter 5). In both cases, an electron transport chain pumps hydrogen ions (H^+) across a membrane—the inner mitochondrial membrane in the case of respiration and the thylakoid membrane in photosynthesis. And in both cases, ATP synthases use the energy stored by the H^+ gradient to make ATP. The main difference is that food provides the high-energy electrons in cellular respiration, while it is light-excited electrons that flow down the transport chain during photosynthesis.

To help you understand the light reactions, watch the animations in Web/CD Activity 6E.

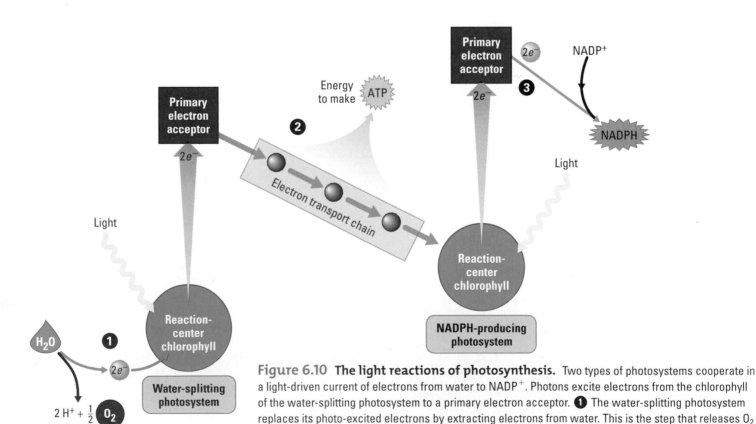

Figure 6.10 The light reactions of photosynthesis. Two types of photosystems cooperate in a light-driven current of electrons from water to $NADP^+$. Photons excite electrons from the chlorophyll of the water-splitting photosystem to a primary electron acceptor. ❶ The water-splitting photosystem replaces its photo-excited electrons by extracting electrons from water. This is the step that releases O_2 during photosynthesis. ❷ Energized electrons from the water-splitting photosystem pass down an electron transport chain to the NADPH-producing photosystem. The chloroplast uses the energy released by this electron "fall" to make ATP. ❸ The NADPH-producing photosystem transfers its photo-excited electrons to $NADP^+$, reducing it to NADPH. The electron transport chain replaces the electrons lost from the photosystem's chlorophyll.

In tracking photons into a chloroplast, we have seen how the light reactions convert solar energy to the chemical energy of ATP and NADPH. Notice again, however, that the light reactions produce no sugar. That is the job of the Calvin cycle, which spends the ATP and NADPH produced by the light reactions.

CheckPoint

1. Compared to a solution of isolated chlorophyll, why do intact chloroplasts release less heat and fluorescence when illuminated?

2. Why is water required as a reactant in photosynthesis?

3. In addition to conveying electrons from the water-splitting photosystem to the NADPH-producing photosystem, the electron transport chains of chloroplasts also provide the energy for synthesis of _____.

Answers: 1. In the chloroplasts, the light-excited electrons are trapped by a primary electron acceptor rather than immediately giving up all their energy as heat and light. **2.** It is the splitting of water that provides electrons for converting CO_2 to sugar (via electron transfer by NADPH). **3.** ATP

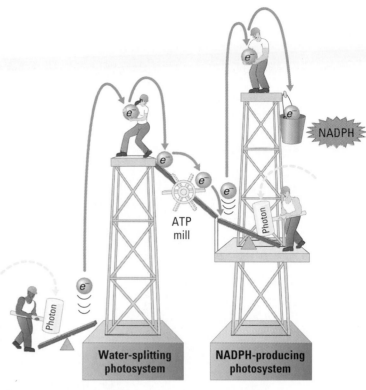

Figure 6.11 A hard-hat analogy for the light reactions.

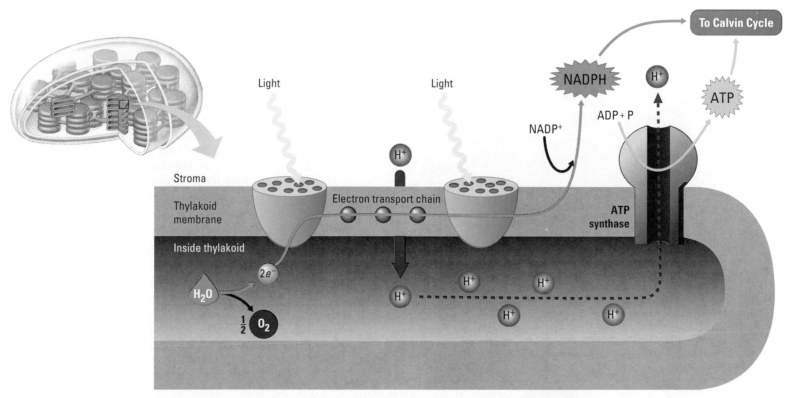

Figure 6.12 How the thylakoid membrane converts light energy to the chemical energy of NADPH and ATP. The two photosystems and the electron transport chain that connects them transfer electrons from water to NADP$^+$. The electron transport chain also functions as a hydrogen ion (H$^+$) pump. ATP synthase molecules, much like the ones in mitochondria, use the energy of the H$^+$ gradient to make ATP.

The Calvin Cycle: Making Sugar from Carbon Dioxide

A Trip Around the Calvin Cycle

The Calvin cycle functions like a sugar factory within a chloroplast. It is called a cycle because, like the Krebs cycle in cellular respiration, the starting material is regenerated with each turn of the cycle (Figure 6.13). And with each turn there are chemical inputs and outputs. The inputs are CO_2 from the air and ATP and NADPH produced by the light reactions. Using carbon from CO_2, energy from ATP, and high-energy electrons from NADPH, the

The animations in Web/CD Activity 6F will help clarify the Calvin cycle.

Calvin cycle constructs an energy-rich sugar molecule. That sugar is not glucose, but a smaller sugar named glyceraldehyde 3-phosphate (G3P). The plant cell can then use this G3P as the raw material to make the glucose and other organic molecules it needs.

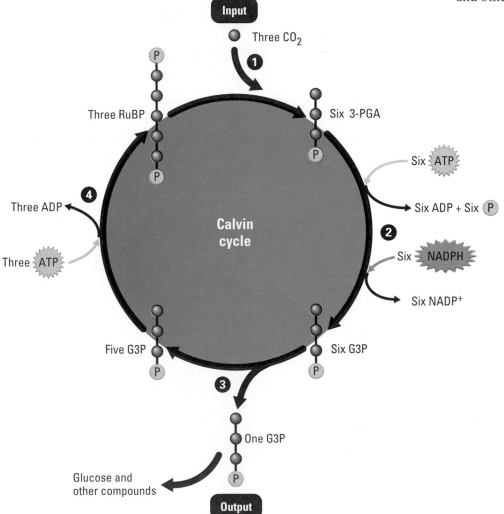

Figure 6.13 The Calvin cycle. The gray balls in this diagram symbolize carbon atoms. The cycle produces one G3P sugar molecule for every three CO_2 molecules that enter the cycle. **①** Carbon enters the cycle as CO_2. An enzyme adds the CO_2 to RuBP (ribulose bisphosphate), a five-carbon sugar already present in the chloroplast. (The ⓟ s in the diagram are phosphate groups.) The product then breaks into a three-carbon compound called 3-PGA (3-phosphoglyceric acid). **②** ATP and NADPH from the light reactions provide energy and electrons. Enzymes use the ATP energy and high-energy electrons from NADPH to convert the 3-PGA to a three-carbon sugar, G3P (glyceraldehyde-3-phosphate). **③** Carbon exits the cycle as sugar. The cycle has converted three CO_2 molecules to one molecule of the sugar G3P. This is the direct product of photosynthesis, but plant cells can use the G3P to make glucose and other organic compounds for growth and fuel. **④** The cycle regenerates its starting material. Note that of the six G3P molecules produced in step 3, only one of them represents net sugar output. That's because we started with a total of 15 sugar carbons in the three RuBP molecules that accepted CO_2 back in step 1. Enzymes now regenerate the RuBP by rearranging the five G3P molecules that are left after one of those sugars exits the cycle.

Water-Saving Adaptations of C₄ and CAM Plants

During the long evolution of plants in diverse environments, natural selection has refined photosynthetic adaptations that enable certain plants to continue producing food even in arid conditions.

Plants in which the Calvin cycle uses CO_2 directly from the air are called **C₃ plants** because the first organic compound produced is the three-carbon compound 3-PGA (see Figure 6.13). C₃ plants are common and widely distributed, and some of them, such as soybeans, oats, wheat, and rice, are important in agriculture. One of the problems that farmers face in growing C₃ plants, however, is that dry weather can reduce the rate of photosynthesis and decrease crop productivity. On a hot, dry day, plants close their stomata, the pores in the leaf surface. Closing stomata is an adaptation that reduces water loss, but it also prevents CO_2 from entering the leaf. As a result, CO_2 levels can get very low in the leaf, and sugar production ceases, at least in C₃ plants.

In contrast to C₃ plants, so-called **C₄ plants** have special adaptations that save water without shutting down photosynthesis. When the weather is hot and dry, a C₄ plant keeps its stomata closed most of the time, thus conserving water. At the same time, it continues making sugars by photosynthesis, using the route shown in Figure 6.14a. A C₄ plant has an enzyme that incorporates carbon from CO_2 into a four-carbon (4-C) compound instead of into 3-PGA. This enzyme has an intense affinity for CO_2 and can continue to mine it from the air spaces of the leaf even when the stomata are closed. The four-carbon compound the enzyme produces acts as a carbon shuttle; it donates the CO_2 to the Calvin cycle in a nearby cell, which therefore keeps on making sugars even though the plant's stomata are closed. Corn, sorghum, and sugarcane are examples of agriculturally important C₄ plants. All three evolved in hot regions of the tropics where there are dry seasons.

Another photosynthetic adaptation that conserves water evolved in pineapples, many cacti, and most of the so-called succulent plants (those with very juicy tissues), such as ice plants and jade plants. Collectively called **CAM plants,** most such species are adapted to very dry climates. CAM stands for crassulacean acid metabolism, after the plant family Crassulaceae (jade plants and others), in which this important water-saving adaptation was first discovered. A CAM plant conserves water by opening its stomata and admitting CO_2 mainly at night (Figure 6.14b). When CO_2 enters the leaves, it is incorporated into a four-carbon compound, as in C₄ plants. The four-carbon compound in a CAM plant banks CO_2 at night and releases it to the Calvin cycle during the day. This keeps photosynthesis operating during the day, even though the leaf admits no more CO_2 because the stomata are closed. Note that in all plants—C₃, C₄, and CAM types—it is the Calvin cycle that is ultimately responsible for sugar synthesis.

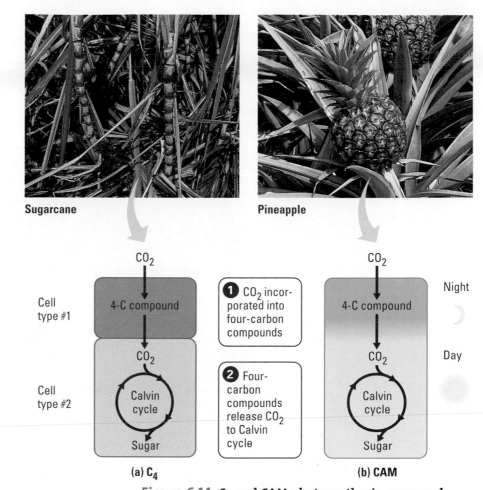

Figure 6.14 C₄ and CAM photosynthesis compared. Both adaptations are characterized by ❶ preliminary incorporation of CO_2 into 4-carbon compounds, followed by ❷ transfer of the CO_2 to the Calvin cycle. **(a)** In C₄ plants, such as sugarcane, these two steps are separated spatially; they are segregated into two cell types. **(b)** In CAM plants, such as pineapple, the two steps are separated temporally; carbon incorporation into 4-carbon compounds occurs at night, and the Calvin cycle operates during the day. The C₄ and CAM pathways are two evolutionary solutions to the problem of maintaining photosynthesis with stomata partially or completely closed on hot, dry days.

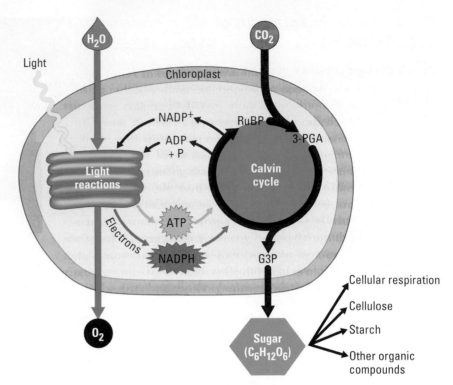

Figure 6.15 Review of photosynthesis.

Review: The Environmental Impact of Photosynthesis

Figure 6.15 reviews how the light reactions and the Calvin cycle cooperate in converting light energy to the chemical energy of food. What the diagram doesn't show is the transfer of this organic material from the producers (plants and other photosynthetic organisms) to consumers (such as the animals that eat plants). Even the energy we acquire when we eat meat was originally captured by photosynthesis. The energy in a hamburger, for instance, came from sunlight that was originally converted to chemical energy in chloroplasts in grasses eaten by cattle.

In addition to producing food, photosynthesis also has an enormous impact on the atmosphere by swapping O_2 for CO_2. This O_2, of course, sustains cellular respiration in all organisms. But gas exchange by plants also helps moderate temperatures on Earth, as you will see in the next section.

CheckPoint

1. In terms of the spatial organization of photosynthesis within the chloroplast, what is the advantage of the light reactions producing NADPH and ATP on the stroma side of the thylakoid membrane?

2. What is the function of NADPH in the Calvin cycle?

3. How do special enzymes enable C_4 and CAM plants to conserve water during photosynthesis?

Answers: 1. The Calvin cycle, which consumes the NADPH and ATP, occurs in the stroma. **2.** It provides the high-energy electrons for the reduction of CO_2 to form sugar. **3.** By allowing photosynthesis to continue even when stomata are closed during dry conditions

How Photosynthesis Moderates the Greenhouse Effect

Old-growth forests are the focus of a long-standing controversy between the timber industry and conservationists (Figure 6.16). The giant trees of these forests contain a lot of marketable lumber, and economic arguments can be made for harvesting them. On the other side of the controversy, conservationists argue that old-growth forests are home to many species of plants and wildlife that can survive nowhere else and that we should save these remnants of our ancient forests for future generations.

The photosynthesis carried out by old-growth forests has direct bearing on the controversy. At center stage is CO_2, the gas that plants use to make sugars in photosynthesis and that all organisms give off as waste from cellular respiration. Carbon dioxide normally makes up about 0.03% of the air we breathe. This amount of CO_2 in the atmosphere provides plants with plenty of carbon. It also helps moderate world climates, because CO_2 retains heat from the sun that would otherwise radiate from Earth back into space. Warming induced by CO_2 is called the **greenhouse effect** because atmospheric CO_2 traps heat and warms the air just as clear glass does in a greenhouse (Figure 6.17). The greenhouse effect keeps the average temperature on Earth some 10°C warmer than it would be otherwise.

Ironically, planet Earth may now be overheating from the greenhouse effect. The amount of atmospheric CO_2 has been on the rise in the past century, mainly because of worldwide industrialization and the increased use of oil, gas, coal, and wood as fuels. When these substances are burned, the carbon in them is released as CO_2. The steady increase in atmospheric CO_2 seems to be contributing to a rise in global temperatures. Photosynthesis consumes CO_2, tending to counteract the greenhouse effect, but the global rate of photosynthesis may decline as we clear huge tracts of forest for logging, farming, and urban expansion. Extensive deforestation continues, especially in the tropics, the northwestern and southeastern United States, Canada, and Siberia.

> Perform experiments in Web/CD Activity 6G to see how various factors affect global temperatures.

The greenhouse effect has come up on both sides of the old-growth forest controversy. Those who favor saving the old-growth trees argue that these large plants remove a lot of potentially harmful CO_2 from the atmosphere. Those favoring harvesting argue that replacing the old trees (which have a lot of nonphotosynthetic wood tissue) with seedlings would actually increase photosynthesis and reduce atmospheric CO_2.

Relative to their size and weight, young, rapidly growing trees do, in fact, take up CO_2 at a faster rate than old ones. However, when old trees are harvested, much less than half their bulk becomes lumber. All the roots and many branches are left behind to decompose. Most of the wood itself is turned into paper, sawdust, or fuel—products that usually decompose or are burned within a few years. Decomposition and burning turn the carbon compounds that were in the trees into CO_2, and this puts much more CO_2 into the atmosphere than young trees can take up.

> Investigate global warming in The Process of Science activity available on the website and CD-ROM.

Figure 6.16 An old-growth forest in the Pacific Northwest. "Old growth" refers to ancient forests that have never been seriously disturbed by humans. Many of the trees—mainly Douglas fir here—are thousands of years old. The forest they dominate is one of a few remaining undisturbed areas that contain harvestable timber in the United States.

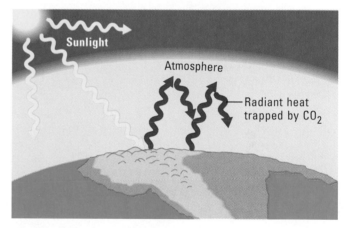

Figure 6.17 The greenhouse effect due to CO_2 in the atmosphere.

Evolution Link:
The Oxygen Revolution

The atmospheric oxygen we breathe and use for cellular respiration is a by-product of the water-splitting step of photosynthesis. The first photosynthetic organisms with the metabolic equipment to split water were prokaryotes called cyanobacteria (see Figure 6.2d). They changed Earth forever by adding O_2 to the atmosphere.

Cyanobacteria evolved between 2.5 and 3.4 billion years ago. The gradual accumulation of O_2 in the atmosphere created a crisis for ancient life, because oxygen attacks the bonds of organic molecules. The corrosive atmosphere probably caused the extinction of many prokaryotic forms unable to cope. Other species survived in habitats that remained anaerobic (such as deep in the soil), where we find their descendants living today. These organisms die if they are exposed to oxygen. Still other species adapted to the changing environment, actually putting the oxygen to use in extracting energy from food, the key process of cellular respiration. The oxygen revolution was a major episode in the history of life on Earth.

Learn more about the oxygen revolution in the Web Evolution Link.

Chapter Review

Summary of Key Concepts

Overview: Our Dependence on Plants

• Photosynthetic organisms are autotrophs, which means they can make all their organic molecules from inorganic nutrients. In contrast, heterotrophs, including animals, must consume organic material as food. In terms of their ecological roles in food chains, autotrophs and heterotrophs are also called producers and consumers, respectively.
• Web/CD Activity 6A *Plants in Our Lives*

The Basics of Photosynthesis

• **Chloroplasts: Sites of Photosynthesis** Chloroplasts are concentrated in the mesophyll cells of leaves. They contain a thick fluid called stroma surrounding a network of membranes called thylakoids.
• Web/CD Activity 6B *The Sites of Photosynthesis*

• **Overall Equation for Photosynthesis**

$$6\ CO_2\ +\ 6\ H_2O\ \rightarrow\ C_6H_{12}O_6\ +\ 6\ O_2$$

Carbon dioxide   Water   Glucose   Oxygen

• **Photosynthesis Road Map** The two stages of photosynthesis are the light reactions, performed by the thylakoids, and the Calvin cycle, which occurs in the stroma. The light reactions convert solar energy to the chemical energy of ATP and NADPH, which the Calvin cycle uses to make sugar from CO_2.
• Web/CD Activity 6C *Overview of Photosynthesis*

The Light Reactions: Converting Solar Energy to Chemical Energy

• **The Nature of Sunlight** Visible light is part of the electromagnetic spectrum of radiation. It travels through space as waves.

- **Chloroplast Pigments** Pigments absorb light energy of certain wavelengths and reflect other wavelengths. We see the reflected wavelengths as the color of the pigment. Several chloroplast pigments absorb light of various wavelengths, but it is the green pigment chlorophyll *a* that participates directly in the light reactions.
- **Web/CD Activity 6D** *Light Energy and Pigments*
- **Web/CD The Process of Science** *Photosynthesis, Pigments, and Autumn Color Changes in Leaves*

- **How Photosystems Harvest Light Energy** A light-harvesting antenna of pigment molecules passes energy to a photosystem's reaction center. At the reaction center, a chlorophyll *a* molecule loses its light-excited electron to a primary electron acceptor.

- **How the Light Reactions Generate ATP and NADPH** Two classes of photosystems cooperate in a flow of electrons from water to $NADP^+$, storing high-energy electrons in NADPH. The electron current also drives ATP synthesis with the help of an electron transport chain.
- **Web/CD Activity 6E** *The Light Reactions*

The Calvin Cycle: Making Sugar from Carbon Dioxide

- **A Trip Around the Calvin Cycle** Using the ATP energy and NADPH electrons from the light reactions, the Calvin cycle converts CO_2 to a three-carbon sugar. This sugar provides the raw material for making other organic molecules.
- **Web/CD Activity 6F** *The Calvin Cycle*

- **Water-Saving Adaptations of C_4 and CAM Plants** Photosynthetic adaptations of C_4 and CAM plants enable sugar production to continue even when stomata are closed, which reduces water loss in arid environments.

- **Review: The Environmental Impact of Photosynthesis** The process of photosynthesis provides organic material and chemical energy for life on Earth. Photosynthesis also swaps O_2 for CO_2 in the atmosphere.

How Photosynthesis Moderates the Greenhouse Effect

- Atmospheric CO_2 traps heat and contributes to our planet's relatively warm temperature. Photosynthesis makes use of some of this CO_2. CO_2 levels are increasing, raising concerns about global warming. Deforestation and burning of fossil fuels may be contributing to these global changes in atmosphere and climate.
- **Web/CD Activity 6G** *The Greenhouse Effect*
- **Web/CD The Process of Science** *Global Warming: Evaluating Alternatives*

Evolution Link: The Oxygen Revolution

- The production of oxygen by cyanobacteria changed Earth forever. Many species probably became extinct, others survived in anaerobic environments, and still others adapted, using oxygen to extract energy from food in cellular reopiration.
- **Web Evolution Link** *The Oxygen Revolution*

Self-Quiz

1. Photosynthesis consumes _____ and produces _____.
   a. sugar; CO_2
   b. CO_2; O_2
   c. chlorophyll *a*; chlorophyll *b*
   d. O_2; CO_2
   e. organic nutrients; inorganic nutrients

2. Autotroph is to _____ as _____ is to consumer.
   a. cellular respiration; photosynthesis
   b. producer; heterotroph
   c. light reactions; Calvin cycle
   d. organic molecules; food
   e. heterotroph; animal

3. Which of the following are produced by reactions that take place in the thylakoids and are consumed by reactions in the stroma?
   a. CO_2 and H_2O
   d. glucose and O_2
   b. $NADP^+$ and ADP
   e. CO_2 and ATP
   c. ATP and NADPH

4. When light strikes chlorophyll molecules, they lose electrons, which are ultimately replaced by
   a. splitting water.
   b. breaking down ATP.
   c. removing them from NADPH.
   d. extracting them from sugar.
   e. removing them from CO_2.

5. The reactions of the Calvin cycle are not directly dependent on light, but they usually do not occur at night. Why?
   a. It is often too cold at night for these reactions to take place.
   b. Carbon dioxide concentrations decrease at night.
   c. The Calvin cycle depends on products of the light reactions.
   d. Plants don't need sugar at night.
   e. At night, plants cannot produce the water needed for the Calvin cycle.

6. Which of the following statements is a correct distinction between autotrophs and heterotrophs?
   a. Only heterotrophs require chemical compounds from the environment.
   b. Cellular respiration is unique to heterotrophs.
   c. Only heterotrophs have mitochondria.
   d. Only autotrophs can live on nutrients that are entirely inorganic.
   e. Only heterotrophs require oxygen.

7. Of the following metabolic processes, which one is common to photosynthesis and cellular respiration?
   a. reactions that convert light energy to chemical energy
   b. reactions that split water molecules and release oxygen
   c. reactions that store energy by pumping hydrogen ions across membranes
   d. reactions that convert CO_2 to sugar
   e. reactions that transfer electrons from chlorophyll to $NADP^+$

8. Combustion of fossil fuels may be contributing to global warming mainly by

    a. raising atmospheric CO_2 concentrations.
    b. lowering atmospheric O_2 concentrations.
    c. releasing hot exhaust to the atmosphere.
    d. producing gases that make the atmosphere more transparent to solar radiation.
    e. decreasing the greenhouse effect.

9. Which of the following colors of light would be least effective in driving photosynthesis?

    a. red   d. violet
    b. orange   e. green
    c. blue

10. The O_2 that a plant releases to the atmosphere is derived from which of the following molecules?

    a. CO_2 d. ATP
    b. $C_6H_{12}O_6$ e. H_2O
    c. NADPH

• **Go to the website or CD-ROM for more self-quiz questions.**

The Process of Science

1. Tropical rain forests cover only about 3% of Earth's surface, but they are estimated to be responsible for more than 20% of global photosynthesis. For this reason, rain forests are often referred to as the "lungs" of the planet, providing O_2 for life all over the planet. However, most experts believe that rain forests make little or no *net* contribution to global oxygen production. From your knowledge of photosynthesis and cellular respiration, can you explain why they might think this? (What happens to the food produced by a rain forest tree when it is eaten by animals or when the tree dies?)

2. **Learn more about pigments in fall foliage and investigate global warming in The Process of Science activities available on the website and CD-ROM.**

Biology and Society

There is evidence that Earth is getting warmer, and the cause could be an intensified greenhouse effect resulting from increased CO_2 from industry, vehicles, and the burning of forests. Global warming could influence agriculture and perhaps even melt polar ice and flood coastal regions. Several European countries, the United States, Australia, and New Zealand have made a commitment to reduce carbon dioxide emissions significantly. However, some countries oppose taking strong action at this time for two main reasons. First, some experts think the apparent warming trend may be just a random fluctuation in temperature. Second, if the temperature increase is real, it is not yet certain that it is caused by increased CO_2. Some people also believe that it would be difficult to cut CO_2 emissions without sacrificing comfort, convenience, and economic growth. Do you think we should have more evidence that greenhouse warming is real before taking action? Or is it better to play it safe and act now to reduce CO_2 emissions? What are the possible costs and benefits of following each of these two strategies? Which do you favor, and why?

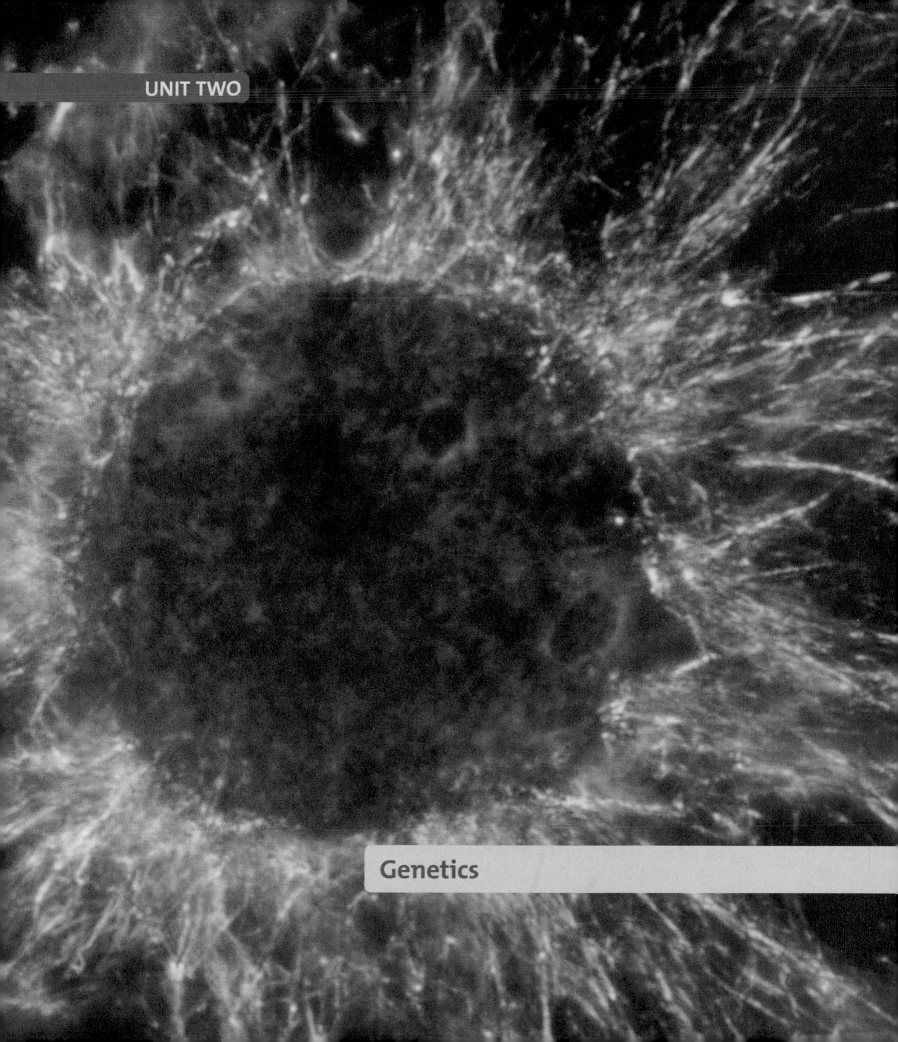

Genetics

CHAPTER 7

The Cellular Basis of Reproduction and Inheritance

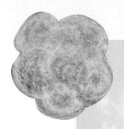

You began life as a single cell, but there are now more cells in your body than stars in the Milky Way.

Just in the past second, millions of your cells have divided in two.  The dance of the chromosomes in a dividing cell is so precise that only one error occurs in 100,000 cell divisions. Each sperm or egg produced in your reproductive organs carries one of over 8 million possible combinations of parental chromosomes.

onsider the skin on your arm. The surface is a protective layer of dead cells, but underneath are layers of living cells busy carrying out the chemical reactions you studied in Unit One. The living cells are also engaged in another vital activity: They are reproducing themselves. The new cells are moving outward toward the skin's surface, replacing dead cells that have rubbed off. This renewal of your skin goes on throughout your life. And when your skin is injured, additional cell reproduction helps heal the wound. In this chapter, you'll learn what actually happens inside reproducing cells and how cell reproduction functions in the perpetuation of all life.

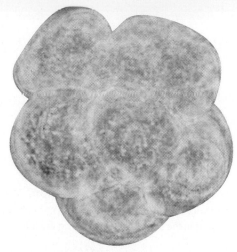

Figure 7.1 Human embryo at the eight-cell stage, the result of three rounds of cell division. The eight cells have identical sets of genes, as will the cells that they give rise to. As development continues, the ball of cells will transform itself into a baby and eventually into an adult.

Overview: What Cell Reproduction Accomplishes

The replacement of lost or damaged cells is just one of the important roles that cell reproduction—or **cell division,** as we commonly call it—has played in your life. Another is growth. All of the trillions of cells in your body result from repeated cell divisions that began in your mother's body with a single fertilized egg cell. (Figure 7.1 shows a human embryo at the eight-cell stage, after three rounds of cell division.)

Passing On the Genes from Cell to Cell

When a cell divides, the two "daughter" cells that result are ordinarily genetically identical to each other and to the original "parent" cell. (Biologists traditionally use the word *daughter* in this context; it does not imply gender.) Before the parent cell splits into two, it duplicates its **chromosomes,** the DNA-containing structures that carry the organism's genes. Then, during the division process, the two sets of chromosomes are distributed to the daughter cells. As a rule, the daughter cells receive identical sets of chromosomes, with identical genes.

The Reproduction of Organisms

Some organisms reproduce by simple cell division. Single-celled organisms such as amoebas reproduce this way, and the offspring are replicas of the parent (Figure 7.2). Because it does not involve fertilization of an egg by a sperm, this type of reproduction is called **asexual reproduction.** Offspring produced by asexual reproduction inherit all their chromosomes from a single parent. In addition to single-celled creatures, many multicellular organisms can also reproduce asexually. For example, some sea stars can divide into two pieces that regrow into two whole new individuals. And if you've ever grown an African violet from a clipping, you've observed asexual reproduction in plants. In asexual reproduction, there is one simple principle of inheritance: The parent and each of its offspring have identical genes.

Learn more about asexual and sexual reproduction in Web/CD Activity 7A.

Sexual reproduction, which requires fertilization of an egg by a sperm, is different. The production of egg and sperm cells involves a special type of cell division, called meiosis, that occurs only in reproductive organs (such as testes and ovaries). As we'll discuss later, a sperm or egg cell has only half as many chromosomes as the parent cell that gave rise to it. So two kinds of

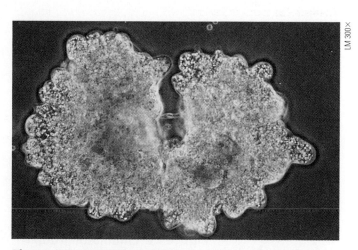

Figure 7.2 Asexual reproduction of an amoeba. This single-celled organism is reproducing by dividing in half. Its chromosomes have been duplicated, and the two identical sets of chromosomes have been allocated to opposite sides of the parent. When division is complete, the two daughter amoebas will be genetically identical to each other and to their parent.

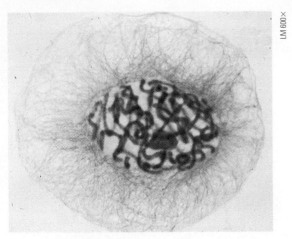

Figure 7.3 **A plant cell just before division.** The colors result from staining. The purple threads are the chromosomes. (The thinner red threads in the surrounding cytoplasm are the cytoskeleton.)

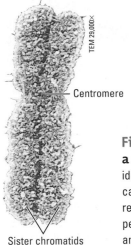

Figure 7.4 **Electron micrograph of a duplicated chromosome.** The two identical copies of the chromosome are called sister chromatids. The constricted region is the centromere. The fuzzy appearance comes from the intricate twists and folds of the chromatin fibers.

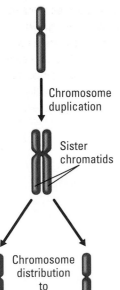

Figure 7.5 **Chromosome duplication and distribution.** In the course of cell reproduction, the cell duplicates each chromosome and distributes the two copies to the daughter cells. This diagram focuses on a single chromosome.

cell division—meiosis and ordinary cell division—are involved in the lives of sexually reproducing organisms.

The remainder of the chapter is divided into two main sections. The first section deals with the cell cycle and mitosis, the type of cell division responsible for asexual reproduction and for the growth and maintenance of multicellular organisms. The second section focuses on meiosis, the special type of cell division that produces sperm and eggs for sexual reproduction.

CheckPoint

Ordinary cell division produces two daughter cells that are genetically identical. Give three functions of this type of cell division.

Answer: Cell replacement, growth of an organism, asexual reproduction of an organism

The Cell Cycle and Mitosis

Almost all the genes of a eukaryotic cell—around 75,000 genes in humans—are located on chromosomes in the cell nucleus. (The main exceptions are genes on small DNA molecules found in mitochondria and chloroplasts.) As leading players in cell division, the chromosomes deserve a little more of our attention before we broaden our focus to the cell as a whole.

Eukaryotic Chromosomes

Chromosomes get their name (Greek *chroma,* colored, and *soma,* body) from their attraction for certain stains used in microscopy. In the micrograph of a plant cell in Figure 7.3, each dark purple thread is an individual chromosome. Chromosomes are clearly visible under the light microscope as structures like these only when a cell is in the process of dividing. The rest of the time, the chromosomes exist as a diffuse mass of very long fibers that are too thin to distinguish in a light micrograph. The chromosomal material, called **chromatin,** is a combination of DNA and protein molecules. As a cell prepares to divide, its chromatin fibers coil up, forming compact, distinct chromosomes.

Each eukaryotic chromosome contains one long DNA molecule, typically bearing thousands of genes. The attached protein molecules help organize the chromatin and help control the activity of its genes. The number of chromosomes in a eukaryotic cell, like the number of genes, depends on the species. For example, human body cells generally have 46 chromosomes.

Well before a cell begins the division process, it duplicates all of its chromosomes. The DNA molecule of each chromosome is replicated, and new protein molecules attach as needed. The result is that each chromosome now consists of two copies called **sister chromatids,** which contain identical genes. Figure 7.4 is an electron micrograph of a human chromosome that has been duplicated. The two chromatids are joined together especially tightly at a region called the **centromere.**

When the cell divides, the sister chromatids of a duplicated chromosome separate from each other, as shown in the simple diagram in Figure 7.5. Once separated from its sister, each chromatid is considered a full-fledged chromosome, and it is identical to the chromosome we started with.

One of the new chromosomes goes to one daughter cell, and the other goes to the other daughter cell. In this way, each daughter cell receives a complete and identical set of chromosomes. A dividing human skin cell, for example, has 46 duplicated chromosomes, and each of the two daughter cells that results from it has 46 single chromosomes.

Let's now summarize how cell division fits into the life of an organism. Cell division is the basis of reproduction for every organism, and it enables a multicellular organism to grow from a single cell. It also replaces worn-out or damaged cells, keeping the total cell number in a mature individual relatively constant. In your own body, for example, millions of cells must divide every second to maintain the total number of about 60 trillion cells.

The Cell Cycle

How do chromosome duplication and cell division fit into the life of a cell? How often a cell divides depends on the role of the cell in the organism's body. Some cells divide once a day, others less often, and highly specialized cells, such as mature muscle cells, not at all. Eukaryotic cells that do divide undergo a **cell cycle,** an orderly sequence of events that extends from the time a cell first arises from cell division until it itself divides.

As Figure 7.6 shows, most of the cell cycle is spent in **interphase.** This is a time when a cell metabolizes and performs its various other functions within the organism. For example, a cell in your stomach lining might be making and releasing enzyme molecules that aid in digesting the food you eat. During interphase, a cell roughly doubles everything in its cytoplasm. It increases its supply of proteins, increases the number of many of its organelles (such as mitochondria and ribosomes), and grows in size. Typically, interphase lasts for at least 90% of the cell cycle.

From the cell reproduction standpoint, the most important event of interphase is chromosome duplication, when the DNA in the nucleus is precisely doubled. This occurs approximately in the middle of interphase, and the period when it is occurring is called the S phase (for DNA *synthesis*). The interphase periods before and after the S phase are called the G_1 and G_2 phases, respectively (G stands for *gap*). During G_2, each chromosome in the cell consists of two identical sister chromatids, and the cell is preparing to divide.

The part of the cell cycle when the cell is actually dividing is called the **mitotic phase** (M phase). It includes two overlapping processes, mitosis and cytokinesis. In **mitosis,** the nucleus and its contents, notably the duplicated chromosomes, divide and are evenly distributed to form two daughter nuclei. In **cytokinesis,** the cytoplasm is divided in two. Cytokinesis usually begins before mitosis is completed. The combination of mitosis and cytokinesis produces two genetically identical daughter cells, each with a single nucleus, surrounding cytoplasm, and plasma membrane.

Mitosis is a remarkably accurate mechanism for allocating identical copies of a large amount of genetic material to two daughter cells. Experiments with yeast cells, for example, indicate that an error in chromosome distribution occurs only once in about 100,000 cell divisions. Mitosis is unique to eukaryotes. Prokaryotes have only a single small chromosome (see Chapter 9) and use a simpler mechanism for allocating DNA to daughter cells.

Test your knowledge of the cell cycle with Web/CD Activity 7B.

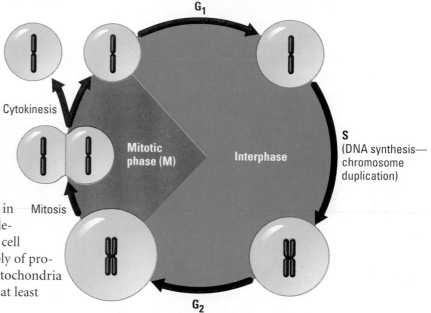

Figure 7.6 The eukaryotic cell cycle. The cell cycle extends from the "birth" of a cell as a result of cell reproduction to the time the cell itself divides in two. The cell spends most of the cycle in interphase. A key event of interphase is the duplication of the chromosomes; the period during which this occurs is called the S phase (for DNA synthesis). Before S, the cell is said to be in the G_1 phase; after S, the cell is in G_2. The cell metabolizes and grows throughout interphase. The actual division process occurs during the mitotic phase (M phase), which includes mitosis (the division of the cell's nucleus) and cytokinesis (the division of the cytoplasm). During interphase, the chromosomes are diffuse masses of thin fibers; they do not actually appear in the rodlike form you see here.

Mitosis and Cytokinesis

The light micrographs in Figure 7.7 show the cell cycle for an animal cell, with most of the figure devoted to the mitotic phase. With the onset of mitosis, striking changes are visible in the nucleus and other cellular structures. The text under the figure describes the events occurring at each stage. Mitosis is a continuum, but biologists distinguish four main stages: **prophase, metaphase, anaphase,** and **telophase.**

The chromosomes are the stars of the mitotic drama, and their movements depend on the **mitotic spindle,** a football-shaped structure of microtubules that guides the separation of the two sets of daughter chromosomes. The spindle microtubules grow from two **centrosomes,** clouds of

Figure 7.7 Cell reproduction: A dance of the chromosomes. After the chromatin doubles during interphase, the elaborately choreographed stages of mitosis—prophase, metaphase, anaphase, and telophase—distribute the duplicate sets of chromosomes to two separate nuclei. Cytokinesis then divides the cytoplasm, yielding two genetically identical daughter cells. The micrographs here show cells from a fish. The drawings include details not visible in the micrographs. For simplicity, only four chromosomes appear in the drawings.

See mitosis animations and videos in Web/CD Activities 7C and 7D.

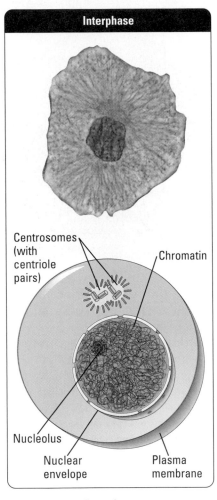

Interphase

Interphase is the period of cell growth, when the cell makes new molecules and organelles. At the point shown here, late interphase (G₂), the cytoplasm contains two centrosomes. Within the nucleus, the chromosomes are duplicated, but they cannot be distinguished individually because they are still in the form of loosely packed chromatin fibers. The prominent nucleolus is an indication that the cell is making ribosomes.

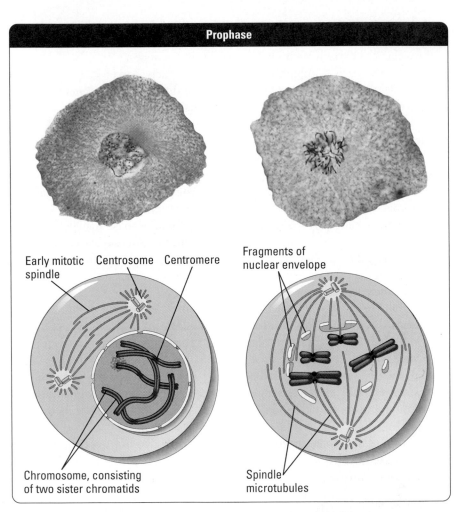

Prophase

During prophase, changes occur in both nucleus and cytoplasm. In the nucleus, the chromatin fibers coil, so that the chromosomes become thick enough to be seen with the light microscope. The nucleoli disappear. Each chromosome appears as two identical sister chromatids joined together, with a narrow "waist" at the centromere. In the cytoplasm, the mitotic spindle begins to form as microtubules grow out from the centrosomes, which are moving away from each other.

Late in prophase, the nuclear envelope breaks up. The spindle microtubules can now reach the chromosomes, which are thick and have a protein structure (black dot) at their centromeres. Some of the spindle microtubules capture chromosomes by attaching to these structures, throwing the chromosomes into agitated motion. Other microtubules make contact with microtubules coming from the opposite spindle pole. The spindle moves the chromosomes toward the center of the cell.

cytoplasmic material that in animal cells contain centrioles. (Centrioles are can-shaped structures made of microtubules; microtubules and centrioles were introduced in Chapter 4.) The role of centrioles in cell division is a mystery; destroying them experimentally does not interfere with normal spindle formation, and plant cells lack them entirely.

Cytokinesis, the actual division of the cytoplasm into two cells, typically occurs during telophase. In animal cells, the cytokinesis process is known as cleavage. The first sign of cleavage is the appearance of a **cleavage furrow,** an indentation at the equator of the cell (Figure 7.8a). A ring of microfilaments in the cytoplasm just under the plasma membrane is responsible for the cleavage furrow. The ring contracts like the pulling of a drawstring, deepening the furrow and pinching the parent cell in two. Microfilaments are made of actin, a protein that also enables muscle cells to contract.

Metaphase	Anaphase	Telophase and Cytokinesis
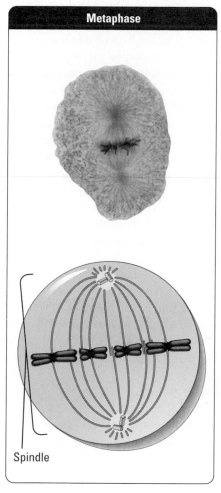	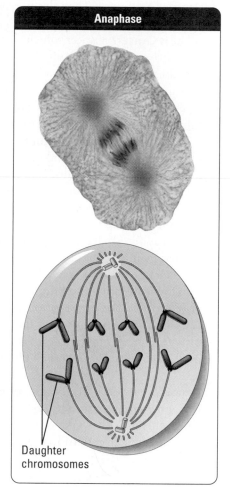	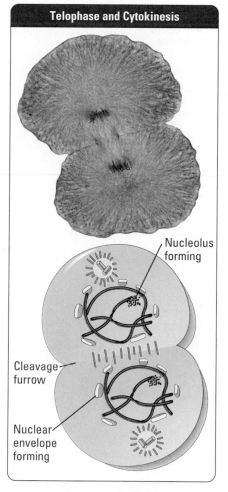

Metaphase

The mitotic spindle is now fully formed. The chromosomes convene on an imaginary plate equidistant from the two poles of the spindle. The centromeres of all the chromosomes are lined up at this plate. For each chromosome, the spindle microtubules attached to the two sister chromatids pull toward opposite poles. This tug of war keeps the chromosomes in the middle of the cell.

homologous

Anaphase

Anaphase begins suddenly, when the sister chromatids of each chromosome separate. Each is now considered a full-fledged (daughter) chromosome. Motor proteins at the centromeres "walk" the daughter chromosomes along their microtubules toward opposite poles of the cell (see motor proteins in Figure 5.8). Meanwhile, these microtubules shorten. However, the microtubules *not* attached to chromosomes lengthen, pushing the poles farther apart and elongating the cell.

Telophase and Cytokinesis

Telophase begins when the two groups of chromosomes have reached the cell poles. Telophase is the reverse of prophase: Nuclear envelopes form, the chromosomes uncoil, nucleoli reappear, and the spindle disappears. Mitosis, the division of one nucleus into two genetically identical daughter nuclei, is now finished.

Cytokinesis, the division of the cytoplasm, usually occurs with telophase. In animals, a cleavage furrow pinches the cell in two, producing two daughter cells.

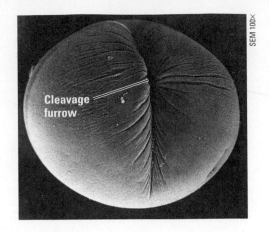

Cleavage furrow

SEM 100×

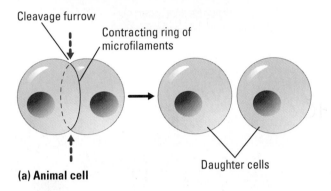

Cleavage furrow
Contracting ring of microfilaments

Daughter cells

(a) Animal cell

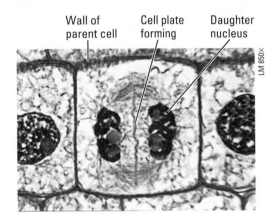

Wall of parent cell

Cell plate forming

Daughter nucleus

LM 850×

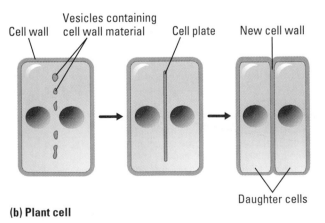

Cell wall

Vesicles containing cell wall material

Cell plate

New cell wall

Daughter cells

(b) Plant cell

Cytokinesis in a plant cell occurs differently (Figure 7.8b). A **cell plate,** a membranous disk containing cell-wall material, forms inside the cell and grows outward. Eventually the new piece of wall divides the cell in two.

Cancer Cells: Growing out of Control

For a plant or animal to grow and develop normally and maintain its tissues once full grown, it must be able to control the timing of cell division in different parts of its body. The sequential events of the cell cycle are directed by a *cell cycle control system* that consists of special proteins within the cell. When this control system malfunctions, cells may reproduce at the wrong time or in the wrong place. This can result in a **benign tumor,** an abnormal mass of essentially normal cells. Benign tumors can cause problems if they grow in certain organs, such as the brain, but usually they can be completely removed by surgery. They always remain at their original site in the body.

What Is Cancer? Cancer, which currently claims the lives of one out of every five people in the United States and other developed nations, is a serious disease of the cell cycle. Unlike normal cells of the body, **cancer cells** have a severely deranged cell cycle control system; not only do they divide excessively, but they also exhibit other kinds of bizarre behavior. A lump resulting from the reproduction of a cancer cell is called a **malignant tumor.**

The most dangerous attribute of cancer cells is their ability to spread into neighboring tissues and often to other parts of the body. Like a benign tumor, a malignant tumor displaces normal tissue as it grows (Figure 7.9). But if a malignant tumor is not killed or removed, it can spread into surrounding tissues. More alarming still, cells may split off from the tumor, invade the circulatory system (lymph vessels and blood vessels), and travel to new locations, where they can form new tumors. The spread of cancer cells beyond their original site is called **metastasis.**

Cancers are named according to where they originate. They are grouped into four categories. **Carcinomas** are cancers that originate in the external or internal coverings of the body, such as the skin or the lining of the intestine. **Sarcomas** arise in tissues that support the body, such as bone and muscle. Cancers of blood-forming tissues, such as bone marrow and lymph nodes, are called **leukemias** and **lymphomas.**

Cancer Treatment In addition to removing malignant tumors with surgery, physicians combat cancers in two other ways: radiation therapy and chemotherapy. Both types of cancer treatment attempt to stop cancer cells from dividing. In **radiation therapy,** parts of the body that have cancerous

Figure 7.8 Cytokinesis in animal and plant cells. (a) In a dividing animal cell, a contracting ring of microfilaments pinches the cell in two. **(b)** In a plant cell, the development of a new piece of cell wall brings about cytokinesis. First, membrane-enclosed vesicles containing cell wall material collect at the middle of the cell. The vesicles gradually fuse, forming a membranous disk called the cell plate. The cell plate grows outward, accumulating more cell wall material as more vesicles join it. Eventually, the membrane of the cell plate fuses with the plasma membrane, and the cell plate's contents join the parental cell wall. The result is two daughter cells, each bounded by its own continuous plasma membrane and a cell wall.

tumors are exposed to high-energy radiation, which disrupts cell division. Because cancer cells divide more often than most normal cells, they are more likely to be dividing at any given time. So radiation can often destroy cancer cells without seriously injuring the normal cells of the body. However, there is sometimes enough damage to normal body cells to produce bad side effects. For example, damage to cells of the ovaries or testes can cause sterility.

Chemotherapy generally uses the same strategy as radiation; in this case, drugs that disrupt cell division are administered to the patient. These drugs work in a variety of ways. Some, called antimitotic drugs, prevent cell division by interfering with the mitotic spindle. One antimitotic drug, vinblastine, prevents the spindle from forming in the first place; another, Taxol, freezes the spindle after it forms, keeping it from functioning.

Vinblastine was first obtained from the periwinkle, a flowering plant native to tropical rain forests in Madagascar. Taxol is made from a chemical found in the bark of the Pacific yew, a tree found mainly in the northwestern United States. Taxol has fewer side effects than many anticancer drugs and seems to be effective against some hard-to-treat cancers of the ovary and breast.

In the laboratory, researchers can grow cancer cells in culture. The cells are placed in a glass container, and nutrients are provided by an artificial liquid medium (Figure 7.10). Normal mammalian cells grow and divide in culture for only about 50 cell generations. But cancer cells are "immortal"—they can go on dividing indefinitely, as long as they have a supply of nutrients. It is by studying cancer cells in culture that researchers are learning about the molecular changes that make a cell cancerous. We will return to the topic of cancer in Chapter 10, after learning more about genes.

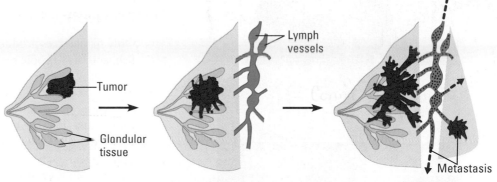

A tumor grows from a single cancer cell.

Cancer cells invade neighboring tissue.

Cancer cells spread through lymph and blood vessels to other parts of the body.

Figure 7.9 Growth and metastasis of a malignant (cancerous) tumor of the breast.

Figure 7.10 Growing cancer cells in the lab. This researcher is working under a fume hood to help prevent contamination of the cells by microbes from the air.

CheckPoint

1. When in the cell cycle does each chromosome consist of two chromatids?

2. An organism called a plasmodial slime mold is one huge cytoplasmic mass with many nuclei. Explain how this "monster cell" could arise.

3. In what sense are the two daughter cells produced by mitosis identical?

4. When a cancer patient is treated with vinblastine, which prevents the mitotic spindle from forming, in what stage of mitosis are dividing cells trapped?

Answers: 1. During G_2, the last part of interphase, and during prophase and metaphase of the next mitosis **2.** Mitosis occurs repeatedly without cytokinesis. **3.** They have identical genes (DNA). **4.** Prophase

Figure 7.11 The varied products of sexual reproduction. Eddie Murphy, his wife Nicole, and their children pose for a family snapshot. Each child has inherited a unique combination of genes from the parents and displays a unique combination of traits.

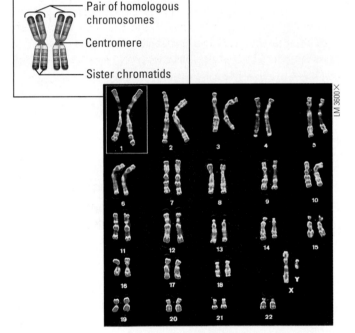

Figure 7.12 Pairs of homologous chromosomes. To make this karyotype (chromosome display) of a man, a scientist broke open a cell in metaphase of mitosis, stained the released chromosomes with special dyes, made a micrograph, and then arranged the chromosome images in matching pairs. The result: 22 well-matched pairs (autosomes) and a twenty-third pair that consists of an X chromosome and a Y chromosome (sex chromosomes). Each chromosome consists of two sister chromatids closely attached all along their lengths. Notice that with the exception of X and Y, the homologous chromosomes of each pair match in size, centromere position, and staining pattern.

Meiosis, the Basis of Sexual Reproduction

Only maple trees produce more maple trees, only goldfish make more goldfish, and only people make more people. These simple facts of life have been recognized for thousands of years and are reflected in the age-old saying "Like begets like." But in a strict sense, "Like begets like" applies only to asexual reproduction, such as the reproduction of the amoeba in Figure 7.2. In that case, because offspring inherit all their DNA from a single parent, they are exact genetic replicas of that one parent and of each other, and their appearances are very similar.

The family photo in Figure 7.11 makes the point that in a sexually reproducing species, like does not exactly beget like. You probably resemble your parents more closely than you resemble a stranger, but you do not look exactly like your parents or your siblings. Each offspring of sexual reproduction inherits a unique combination of genes from its two parents, and this combined set of genes programs a unique combination of traits. As a result, sexual reproduction can produce great variation among offspring. Notice in the photograph that despite the family resemblances, each Murphy child has a unique appearance. You'll find the same sort of similarities and differences in pictures of your own relatives.

Long before anyone knew about genes or chromosomes, people recognized that individuals of sexually reproducing species are highly varied. What's more, they learned to develop domestic breeds of plants and animals by controlling sexual reproduction. A domestic breed displays particular traits from among the great variety of traits found in the species as a whole. All dachshunds, for instance, have sausage-shaped bodies and short legs, whereas Saint Bernards are much taller and bulkier. The ancestry of dog breeds can be traced back for many generations, during which breeders reduced variability in the breed by mating only those dogs with specific traits. In a sense, selective breeding is an attempt to make like beget like more than it does in nature.

Sexual reproduction depends on the cellular processes of meiosis and fertilization. But before discussing these processes, we return to chromosomes and their role in the life cycles of sexually reproducing organisms.

Homologous Chromosomes

If we examine a number of cells from any individual organism, we discover that virtually all of them have the same number and types of chromosomes. Likewise, if we examine cells from different individuals of a single species—sticking to one gender, for now—we find that they have the same number and types of chromosomes. Viewed with a microscope, your chromosomes would look just like those of Queen Elizabeth (if you're a woman) or Michael Jordan (if you're a man).

A typical body cell, called a **somatic cell,** has 46 chromosomes in humans. If we break open a human cell in metaphase of mitosis, make a micrograph of the chromosomes, and arrange the chromosome images in an orderly array, we produce a display called a **karyotype** (Figure 7.12). Every (or almost every) duplicated chromosome has a twin that resembles it in size and shape. The two chromosomes of such a matching pair, called **homologous chromosomes,** carry the same sequence of genes controlling the same inherited characteristics. For example, if a gene influencing eye color is located at a particular place on one chromosome—for example, within the yellow band in the Figure 7.12 inset—then the homologous chromosome has a similar gene for eye color there. However, the two genes may be slightly different versions, unlike the ones on sister chromatids, which are identical. Altogether, we humans have 23 homologous pairs of chromosomes. Other species have different numbers of chromosomes, but these, too, usually match in pairs.

For a human female, the 46 chromosomes fall neatly into 23 homologous pairs, with the members of each pair essentially identical in appearance. For a male, however, one pair of chromosomes do *not* look alike (see Figure 7.12). The nonmatching pair, called the **sex chromosomes,** determines the person's gender. Like all mammals, human males have one X chromosome and one Y chromosome. Only small parts of the X and Y are homologous; most of the genes carried on the X chromosome do not have counterparts on the tiny Y, and the Y has genes lacking on the X. Females have two X chromosomes. The remaining chromosomes, found in both males and females, are called **autosomes.** For both autosomes and sex chromosomes, we inherit one chromosome of each pair from our mother and the other from our father.

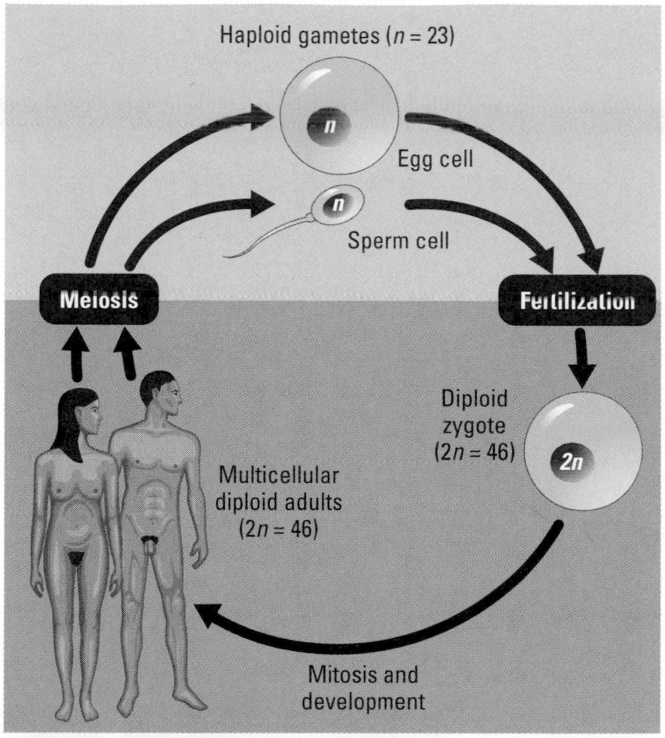

Gametes and the Life Cycle of a Sexual Organism

Having two sets of chromosomes, one inherited from each parent, is a key factor in the human life cycle, outlined in Figure 7.13, and in the life cycles of all other species that reproduce sexually. The **life cycle** of a multicellular organism is the sequence of stages leading from the adults of one generation to the adults of the next. Let's follow the human chromosomes through the human life cycle.

Humans are said to be **diploid** organisms because almost all our cells are diploid: They contain two homologous sets of chromosomes. The total number of chromosomes, 46 in humans, is the diploid number (abbreviated 2*n*). The exceptions are the egg and sperm cells, known as **gametes.** Made by meiosis in an ovary or testis, each gamete has a single set of chromosomes: 22 autosomes plus a single sex chromosome, X or Y. A cell with a single chromosome set is called a **haploid** cell; it has only one member of each homologous pair. For humans, the haploid number (abbreviated *n*) is 23.

In the human life cycle, sexual intercourse allows a haploid sperm cell from the father to reach and fuse with a haploid egg cell of the mother in the process known as **fertilization.** The resulting fertilized egg, called a **zygote,** is diploid. It has two homologous sets of chromosomes, one set from each parent. The life cycle is completed as a sexually mature adult develops from the zygote. Mitotic cell division ensures that all somatic cells of the human body receive copies of all of the zygote's 46 chromosomes.

All sexual life cycles involve an alternation of diploid and haploid stages. Producing haploid gametes by meiosis keeps the chromosome number from doubling in every generation (Figure 7.14).

Review sexual life cycles in Web/CD Activity 7E.

Figure 7.13 The human life cycle. In each generation, the doubling of chromosome number that results from fertilization is offset by the halving of chromosome number that occurs in meiosis. For humans, the number of chromosomes in a haploid cell (sperm or egg) is 23 (that is, *n* = 23). The number of chromosomes in the diploid zygote and all somatic cells arising from it is 46 (2*n* = 46).

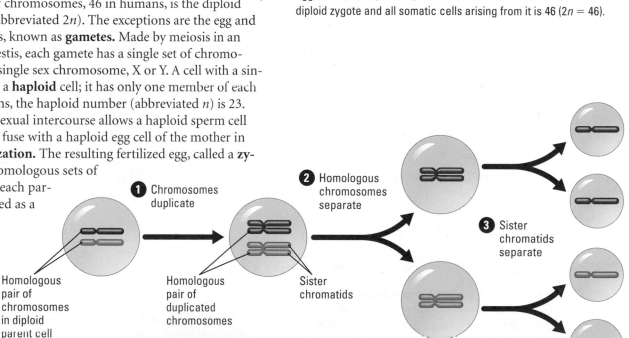

Figure 7.14 How meiosis halves chromosome number. This simplified diagram tracks just one pair of homologous chromosomes. ❶ Each of the chromosomes is duplicated during the preceding interphase. ❷ The first division, meiosis I, segregates the two chromosomes of the homologous pair, packaging them in separate (haploid) daughter cells. But each chromosome is still doubled. ❸ Meiosis II separates the sister chromatids. Each of the four daughter cells is haploid and contains only one single chromosome from the homologous pair.

The Process of Meiosis

Meiosis, the process that produces haploid gametes in diploid organisms, resembles mitosis, but with two special features. The first is the halving of the number of chromosomes. In meiosis, a cell that has duplicated its chromosomes undergoes *two consecutive divisions*, called *meiosis I* and *meiosis II*. Because the two divisions of meiosis are preceded by only one duplication of the chromosomes, each of the four daughter cells resulting from meiosis has only half as many chromosomes as the starting cell—a haploid set of chromosomes.

Watch an animation of meiosis in Web/CD Activity 7F.

The chromosome number is actually haploid by the end of meiosis I, although there are still two sister chromatids per chromosome at that point.

Figure 7.15 The stages of meiosis. The drawings here show the two cell divisions of meiosis, starting with a diploid animal cell containing four chromosomes. Each homologous pair consists of a red chromosome and a blue chromosome of the same size. The colors remind us that the members of a homologous pair were inherited from different parents and carry different versions of some genes.

Meiosis I: Homologous chromosomes separate

Interphase	Prophase I	Metaphase I	Anaphase I

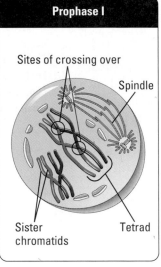
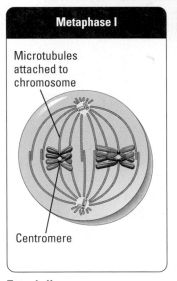
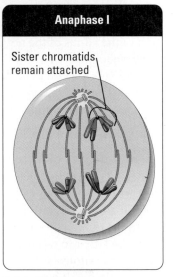

Chromosomes duplicate — **Homologous chromosomes pair and exchange segments** — **Tetrads line up** — **Pairs of homologous chromosomes split up**

Interphase

Like mitosis, meiosis is preceded by an interphase during which the chromosomes duplicate. Each chromosome then consists of two identical sister chromatids.

Meiosis I

Prophase I Prophase I is the most complicated stage of meiosis. As the chromatin condenses, special proteins cause the homologous chromosomes to stick together in pairs. The resulting structure has four chromatids and is called a **tetrad.** Within each tetrad, chromatids of the homologous chromosomes exchange corresponding segments—they "cross over." Because the versions of the genes on a chromosome (or one of its chromatids) may be different from those on its homologue, crossing over rearranges genetic information.

As prophase I continues, the chromosomes condense further, a spindle forms, and the tetrads are moved toward the center of the cell.

Metaphase I At metaphase I, the tetrads are aligned in the middle of the cell. The sister chromatids of each chromosome are still attached at their centromeres, where they are anchored to spindle microtubules. Notice that for each tetrad, the spindle microtubules attached to one homologous chromosome come from one pole of the cell, and the microtubules attached to the other chromosome come from the opposite pole. With this arrangement, the homologous chromosomes of each tetrad are poised to move toward opposite poles of the cell.

Anaphase I As in anaphase of mitosis, chromosomes now migrate toward the poles of the cell. But in contrast to mitosis, the sister chromatids migrate as a pair instead of splitting up. They are separated not from each other, but from their homologous partners. So in the drawing, you see two still-doubled chromosomes moving toward each pole.

The second special feature of meiosis is an exchange of genetic material —pieces of chromosomes—between homologous chromosomes. This exchange, called **crossing over,** occurs during the first prophase of meiosis. We'll look more closely at crossing over later. For now, study Figure 7.15 and the text below it, which describe the stages of meiosis in detail.

As you go through Figure 7.15, keep in mind the difference between homologous chromosomes and sister chromatids: The two chromosomes of a homologous pair are individual chromosomes that were inherited from different parents. Homologues appear alike in the microscope, but they have different versions of some of their genes (for example, a gene for freckles on one chromosome and a gene for the absence of freckles at the same place on the homologue). The homologues in Figure 7.15 (and later figures) are colored red and blue to remind you that they differ in this way. In the interphase just before meiosis, each homologue replicates to form sister chromatids that remain together until anaphase of meiosis II. Before crossing over occurs, sister chromatids are identical and carry the same versions of all their genes.

Telophase I and Cytokinesis	**Meiosis II: Sister chromatids separate**			
	Prophase II	**Metaphase II**	**Anaphase II**	**Telophase II and Cytokinesis**

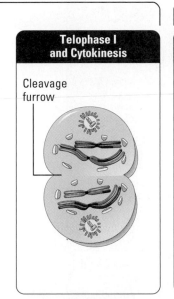

Cleavage furrow

Sister chromatids separate

Haploid daughter cells forming

Two haploid cells form; chromosomes are still double

During another round of cell division, the sister chromatids finally separate; four haploid daughter cells result, containing single chromosomes

Telophase I and Cytokinesis

In telophase I, the chromosomes arrive at the poles of the cell. When they finish their journey, each pole has a haploid chromosome set, although each chromosome is still in duplicate form. Usually, cytokinesis occurs along with telophase I, and two haploid daughter cells are formed.

Depending on the species, the nuclei may or may not return to an interphase state. But in either case, there is no further chromosome duplication.

Meiosis II

Meiosis II is essentially the same as mitosis. The important difference is that meiosis II starts with a haploid cell.

During prophase II, a spindle forms and moves the chromosomes toward the middle of the cell. During metaphase II, the chromosomes are aligned as they are in mitosis, with the microtubules attached to the sister chromatids of each chromosome coming from opposite poles. In anaphase II, the centromeres of sister chromatids finally separate, and the sister chromatids of each pair, now individual daughter chromosomes, move toward opposite poles of the cell. In telophase II, nuclei form at the cell poles, and cytokinesis occurs at the same time. There are now four daughter cells, each with the haploid number of single chromosomes.

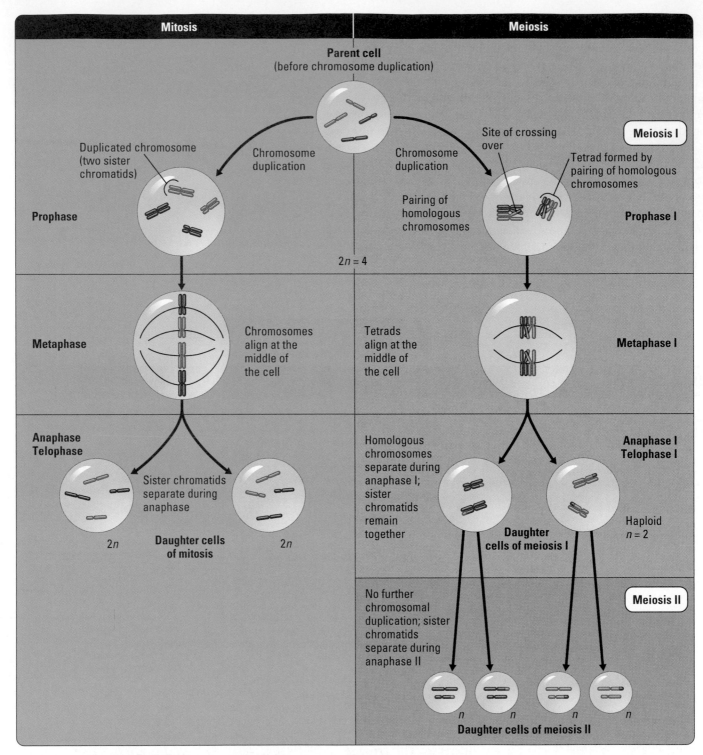

Figure 7.16 Comparing mitosis and meiosis. The events unique to meiosis occur during meiosis I: In prophase I, duplicated homologous chromosomes pair to form tetrads, and crossing over occurs between homologous (nonsister) chromatids. In metaphase I, tetrads (rather than individual chromosomes) are aligned at the center of the cell. During anaphase I, sister chromatids of each chromosome stay together and go to the same pole of the cell as homologous chromosomes separate. At the end of meiosis I, there are two haploid cells, but each chromosome still has two sister chromatids. Meiosis II is virtually identical to mitosis and separates sister chromatids. But unlike mitosis, meiosis II yields daughter cells with a haploid set of chromosomes.

Review: Comparing Mitosis and Meiosis

We have now described the two ways that cells of eukaryotic organisms divide. Mitosis, which provides for growth, tissue repair, and asexual reproduction, produces daughter cells genetically identical to the parent cell. Meiosis, needed for sexual reproduction, yields haploid daughter cells—cells with only one member of each homologous chromosome pair.

For both mitosis and meiosis, the chromosomes duplicate only once, in the preceding interphase. Mitosis involves one division of the nucleus, and it is usually accompanied by cytokinesis, producing two diploid cells. Meiosis entails two nuclear and cytoplasmic divisions, yielding four haploid cells.

Figure 7.16 (at the left) compares mitosis and meiosis, tracing these two processes for a diploid parent cell with four chromosomes. As before, homologous chromosomes are those matching in size. Notice that all the events unique to meiosis occur during meiosis I.

The Origins of Genetic Variation

As we discussed earlier, offspring that result from sexual reproduction are genetically different from their parents and from one another. When we discuss natural selection and evolution in Unit Three, we will see that this genetic variety in offspring is the raw material for natural selection. For now, let's take another look at meiosis and fertilization to see how genetic variety arises.

Independent Assortment of Chromosomes Figure 7.17 illustrates one way in which meiosis contributes to genetic variety. The figure shows how the arrangement of homologous chromosome pairs at metaphase of meiosis I affects the resulting gametes. Once again, our example is from an organism with a diploid chromosome number of 4, with red and blue used to differentiate homologous chromosomes. These colors highlight the fact that homologous chromosomes differ genetically, although the two look alike under a microscope. (For example, the chromosome 3 you inherited from your mother undoubtedly carries many genes that are slightly different from those on the homologous chromosome 3 you received from your father.)

The orientation of the homologous pairs of chromosomes (tetrads) at metaphase I is a matter of chance, like the flip of a coin. In this example, there are two possible ways that the two tetrads can align during metaphase I. In possibility 1, the tetrads are oriented with both red chromosomes on the same side. In this case, each of the gametes produced at the end of meiosis II has only red or only blue chromosomes (combinations 1 and 2). In possibility 2, the tetrads are oriented differently. This arrangement produces gametes that each have one red and one blue chromosome. Furthermore, half the gametes have a big blue chromosome and a small red one (combination 3), and half have a big red one and a small blue one (combination 4).

So we see that for this example, a total of four chromosome combinations is possible in the gametes, and the organism will in fact produce gametes of all four types. This variety in gametes arises because each homologous pair

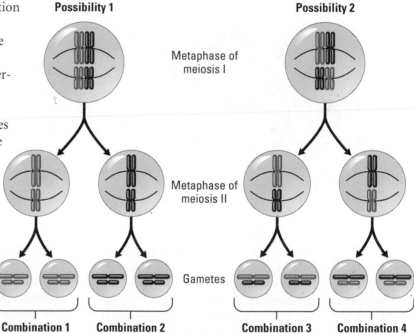

Figure 7.17 Results of alternative arrangements of chromosomes at metaphase of meiosis I. In this figure, we consider the consequences of meiosis in a diploid organism with four chromosomes (two homologous pairs). The positioning of each homologous pair of chromosomes (tetrad) at metaphase of meiosis I is random; the two red chromosomes can be on the same side (possibility 1) or on opposite sides (possibility 2). The arrangement of chromosomes at metaphase I determines which chromosomes will be packaged together in the haploid gametes. Because possibilities 1 and 2 are equally likely, the four possible types of gametes will be made in approximately equal numbers.

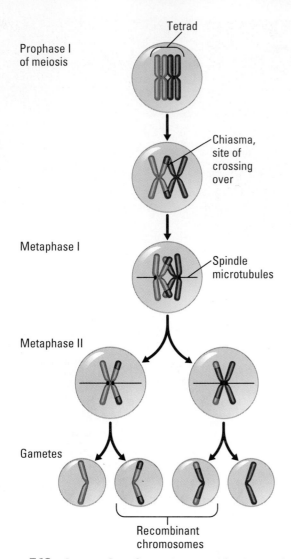

Tetrad

Prophase I
of meiosis

Chiasma,
site of
crossing
over

Metaphase I

Spindle
microtubules

Metaphase II

Gametes

Recombinant
chromosomes

Figure 7.18 The results of crossing over during meiosis. This diagram focuses on a single pair of homologous chromosomes (a tetrad). Early in prophase I of meiosis, homologous (*non*sister) chromatids exchange corresponding segments, remaining attached at the crossover points. Sister chromatids are joined at their centromeres. Following these chromosomes through the rest of meiosis, we see that crossing over gives rise to *recombinant* chromosomes—individual chromosomes that combine genetic information originally derived from different parents. With multiple pairs of homologous chromosomes, the result is a huge variety of gametes.

of chromosomes orients itself at metaphase I independently of the other pair. For a species with more than two pairs of chromosomes, such as the human, every chromosome pair orients independently of all the others at metaphase I. (Chromosomes X and Y behave as a homologous pair in meiosis.)

For any species, the total number of chromosome combinations that can appear in gametes is 2^n, where n is the haploid number. For the organism in this figure, $n = 2$, so the number of chromosome combinations is 2^2, or 4. For a human ($n = 23$), there are 2^{23}, or about 8 million, possible chromosome combinations! This means that every gamete a human produces contains one of about 8 million possible combinations of maternal and paternal chromosomes.

Random Fertilization How many possibilities are there when a gamete from one individual unites with a gamete from another individual during fertilization? A human egg cell, representing one of about 8 million possibilities, is fertilized at random by one sperm cell, representing one of about 8 million other possibilities. By multiplying 8 million by 8 million, we find that a man and a woman can produce a diploid zygote with any of 64 trillion combinations of chromosomes! So we see that the random nature of fertilization adds a huge amount of potential variability to the offspring of sexual reproduction.

These large numbers suggest that independent orientation of chromosomes at metaphase I and random fertilization could account for all the variety we see among people. Actually, these two events are only part of the picture, as we see next.

Crossing Over So far, we have focused on genetic variability in gametes and zygotes at the whole-chromosome level. We have ignored crossing over, the exchange of corresponding segments between two homologous chromosomes, which occurs during prophase I of meiosis. Figure 7.18 shows crossing over between two homologous chromosomes and the results in the gametes. At the time that crossing over begins, homologous chromosomes are closely paired all along their lengths, with a precise gene-by-gene alignment. The sites of crossing over appear as X-shaped regions; each is called a **chiasma** (Greek for cross; plural, *chiasmata*). The homologous chromatids remain attached to each other at chiasmata until anaphase I.

The exchange of segments by homologous chromatids adds to the genetic variety that results from the independent orientation of chromosome pairs at metaphase I. In Figure 7.18, if there were no crossing over, meiosis could produce only two types of gametes. These would be the ones ending up with the "parental" types of chromosomes, either all blue or all red (as in Figure 7.17). With crossing over, gametes arise that have chromosomes that are part red and part blue. These chromosomes are called "recombinant" because they result from **genetic recombination,** the production of gene combinations different from those carried by the parental chromosomes.

Because most chromosomes contain thousands of genes, a single crossover event can affect many genes. When we also consider that multiple crossovers can occur in each tetrad, it's no wonder that gametes and the offspring that result from them can be so varied. In fact, it's surprising that even siblings resemble one another as much as they do.

We have now examined three sources of genetic variability in sexually reproducing organisms: crossing over during prophase I of meiosis, independent orientation of chromosome pairs at metaphase I, and random fer-

To help you understand the origins of genetic variability, go to Web/CD Activity 7G.

tilization. When we take up molecular genetics in Chapter 9, we will see yet another source of variation—mutations, which are rare changes in the DNA of genes. The different versions of genes found on homologous chromosomes originally arose from mutations, and it is mutations that are ultimately responsible for the genetic diversity in living organisms.

When Meiosis Goes Amok

So far, our discussion of meiosis has focused on the process as it normally and correctly occurs. But what happens when an error occurs in the process?

Down Syndrome: An Extra Chromosome 21 Figure 7.12 showed a normal human complement of 23 pairs of chromosomes. Figure 7.19 is different; besides having two X chromosomes (because it's from a female), it has *three* number 21 chromosomes. This condition is called **trisomy 21.**

In most cases, a human embryo with an abnormal number of chromosomes is spontaneously aborted (miscarried) long before birth. However, some aberrations in chromosome number, including trisomy 21, seem to upset the genetic balance less drastically, and individuals carrying them survive. These people usually have a characteristic set of symptoms, called a syndrome. A person with trisomy 21, for instance, is said to have **Down syndrome** (named after John Langdon Down, who described it in 1866).

Trisomy 21 is the most common chromosome number abnormality. Affecting about one out of every 700 children born, it is the most common serious birth defect in the United States. Chromosome 21 is one of our smallest chromosomes, but an extra copy produces a number of effects. Down syndrome includes characteristic facial features—frequently a fold of skin at the inner corner of the eye (epicanthic fold), a round face, flattened nose bridge, and small, irregular teeth—as well as short stature, heart defects, and susceptibility to respiratory infection, leukemia, and Alzheimer's disease.

People with Down syndrome usually have a life span shorter than normal. They also exhibit varying degrees of mental retardation. However, individuals with the syndrome may live to middle age or beyond, and many are socially adept and able to hold a job. A few women with Down syndrome have had children, though most people with the syndrome are sexually underdeveloped and sterile. Half the eggs produced by a woman with Down syndrome will have an extra chromosome 21, so there is a 50% chance that she will transmit the syndrome to her child.

As indicated in Figure 7.20, the incidence of Down syndrome in the offspring of normal parents increases markedly with the age of the mother. Down syndrome strikes less than 0.05% of children (fewer than one in 2000) born to women under age 30. The risk climbs to 1% for mothers in their late 30s and is even higher for older mothers. Because of this relatively high risk, pregnant women over 35 are candidates for fetal testing for trisomy 21 and other chromosomal abnormalities (see Chapter 8).

What causes trisomy 21? We address that question next.

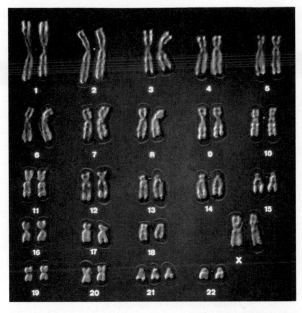

Figure 7.19 Trisomy 21 and Down syndrome. The karyotype (left) shows trisomy 21; notice the three copies of chromosome 21. The child displays the characteristic facial features of Down syndrome.

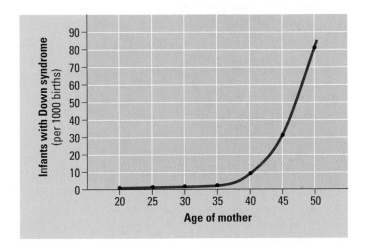

Figure 7.20 Maternal age and Down syndrome. The chance of having a baby with Down syndrome rises with the age of the mother.

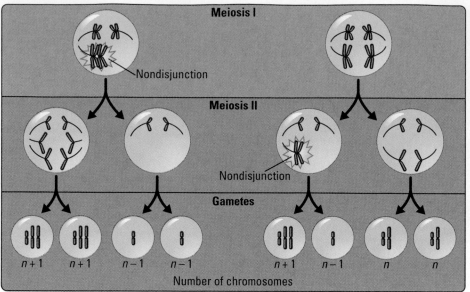

Meiosis I

Nondisjunction

Meiosis II

Nondisjunction

Gametes

$n+1$ $n+1$ $n-1$ $n-1$ $n+1$ $n-1$ n n

Number of chromosomes

(a) Nondisjunction in meiosis I   **(b) Nondisjunction in meiosis II**

Figure 7.21 Two types of nondisjunction. In both parts of the figure, the cell at the top is diploid (2n), with two pairs of homologous chromosomes. **(a)** A pair of homologous chromosomes fails to separate during anaphase of meiosis I, even though the rest of meiosis occurs normally. In this case, all the resulting gametes end up with abnormal numbers of chromosomes. **(b)** Meiosis I is normal, but a pair of sister chromatids fail to move apart in one of the cells during anaphase of meiosis II. In this case, two gametes have the normal complement of two chromosomes each, but the other two gametes are abnormal.

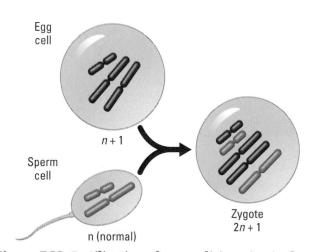

Egg cell

$n+1$

Sperm cell

n (normal)

Zygote
$2n+1$

Figure 7.22 Fertilization after nondisjunction in the mother. Assuming that the organism has a diploid number of 4 (2n = 4), the sperm is a normal haploid cell (n = 2). The egg cell, however, contains an extra copy of the larger chromosome as a result of nondisjunction in meiosis; it has a total of $n + 1 = 3$ chromosomes. When the sperm and egg fuse during fertilization, the result is an abnormal zygote with an extra chromosome; it has $2n + 1 = 5$ chromosomes.

How Accidents During Meiosis Can Alter Chromosome Number Meiosis occurs repeatedly in our lifetime as our testes or ovaries produce gametes. Almost always, the meiotic spindle distributes chromosomes to daughter cells without error. But occasionally there is an accident, called a **nondisjunction,** in which the members of a chromosome pair fail to separate at anaphase. Nondisjunction can occur in meiosis I or II (Figure 7.21). In either case, gametes with abnormal numbers of chromosomes result.

Figure 7.22 shows what can happen when an abnormal gamete produced by nondisjunction unites with a normal gamete in fertilization. When a normal sperm fertilizes an egg cell with an extra chromosome, the result is a zygote with a total of $2n + 1$ chromosomes. Mitosis then transmits the abnormality to all embryonic cells. If the organism survives, it will have an abnormal karyotype and probably a syndrome of disorders caused by the abnormal number of genes.

Nondisjunction can lead to an abnormal chromosome number in either sex of any sexually reproducing, diploid organism, including humans. If, for example, there is nondisjunction affecting human chromosome 21 during meiosis I, half the resulting gametes will carry an extra chromosome 21. Then if one of these gametes unites with a normal gamete in fertilization, trisomy 21 will result.

Nondisjunction explains how abnormal chromosome numbers come about, but what causes nondisjunction in the first place? We do not know the answer, nor do we fully understand why offspring with trisomy 21 are more likely to be born as a woman ages. We do know, however, that meiosis begins in a woman's ovaries before she is born but is not completed until years later, at the time of an ovulation. Because only one egg cell usually matures each month, a cell might remain arrested in the middle of meiosis for decades. Perhaps damage to the cell during this time leads to meiotic errors. It seems that the longer the time lag, the greater the chance that there will be errors such as nondisjunction when meiosis is completed.

Abnormal Numbers of Sex Chromosomes Nondisjunction in meiosis does not affect just autosomes, such as chromosome 21. It can also lead to abnormal numbers of sex chromosomes (X and Y). Unusual numbers of sex chromosomes seem to upset the genetic balance less than unusual numbers of autosomes. This may be because the Y chromosome is very small and carries fewer genes than other chromosomes. Also, most of the genes on the Y chromosome affect maleness but not functions that are essential to the person's survival. A peculiarity of X chromosomes in humans and other mammals also helps an individual tolerate unusual numbers of X chromosomes: In mammals, the cells usually operate with only one functioning X chromosome because extra copies of the chromosome become inactivated in each cell (see Chapter 10).

Table 7.1 lists the most common sex chromosome abnormalities. An extra X chromosome in a male, making him XXY, occurs approximately once

Table 7.1		Abnormalities of Sex Chromosome Number in Humans		

Sex Chromosomes	Syndrome	Origins of Nondisjunction	Frequency in Population
XXY	Klinefelter syndrome (male)	Meiosis in egg or sperm formation	$\frac{1}{2000}$
XYY	Normal male	Meiosis in sperm formation	$\frac{1}{2000}$
XXX	Metafemale	Meiosis in egg or sperm formation	$\frac{1}{1000}$
XO	Turner syndrome (female)	Meiosis in egg or sperm formation	$\frac{1}{5000}$

in every 2000 live births (once in every 1000 male births). This disorder is called *Klinefelter syndrome.* Men with this disorder have male sex organs, but the testes are abnormally small and the individual is sterile. The syndrome often includes breast enlargement and other feminine body contours (Figure 7.23, right). The person is usually of normal intelligence. Klinefelter syndrome is also found in individuals with more than one additional sex chromosome, such as XXYY, XXXY, or XXXXY. These abnormal numbers of sex chromosomes probably result from multiple nondisjunctions. Such men are more likely to be mentally retarded than XY or XXY individuals.

Human males with a single extra Y chromosome (XYY) do not have any well-defined syndrome, although they tend to be taller than average. Females with an extra X chromosome (XXX) are called *metafemales.* They have limited fertility but are otherwise apparently normal.

Females who are lacking an X chromosome are designated XO; the O simply indicates the absence of a second sex chromosome. These women have *Turner syndrome.* They have a characteristic appearance, including short stature and often a web of skin extending between the neck and shoulders (Figure 7.23, right). Women with Turner syndrome are sterile because their sex organs do not fully mature at adolescence, and they have poor development of breasts and other secondary sex characteristics. However, they are usually of normal intelligence. The XO condition occurs in about one in 5000 babies born (about one in 2500 female births).

The sex chromosome abnormalities described here illustrate the crucial role of the Y chromosome in determining a person's sex. In general, a single Y chromosome is enough to produce "maleness," even when it is combined with several X chromosomes. The absence of a Y chromosome results in "femaleness."

Alterations of Chromosome Structure Even if all chromosomes are present in normal numbers, abnormalities in chromosome *structure* may cause disorders. Breakage of a chromosome can lead to a variety of rearrangements

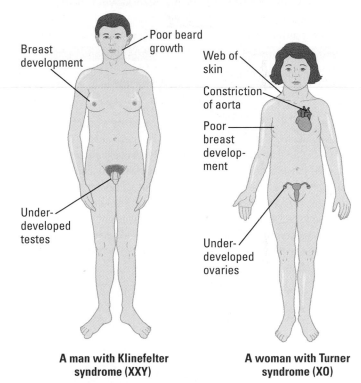

A man with Klinefelter syndrome (XXY) — Breast development, Poor beard growth, Underdeveloped testes

A woman with Turner syndrome (XO) — Web of skin, Constriction of aorta, Poor breast development, Underdeveloped ovaries

Figure 7.23 Syndromes associated with unusual numbers of sex chromosomes.

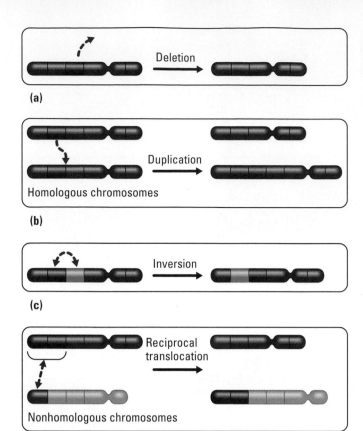

Figure 7.24 Alterations in chromosome structure.
(a) A deletion is the removal of a chromosome segment. **(b)** A duplication is the repetition of a segment. When a duplication results from the movement of a chromosomal segment to a homologous chromosome, as shown here, the "duplicates" may be slightly different. **(c)** An inversion is the reversal of a segment within a chromosome. **(d)** A translocation is the movement of a segment to a *nonhomologous* chromosome. In the case shown here, the two nonhomologous chromosomes trade segments, producing a *reciprocal* translocation.

affecting the genes of that chromosome. Figure 7.24 shows four types of rearrangement. If a fragment of a chromosome is lost, the remaining chromosome has a **deletion.** If a fragment from one chromosome is inserted into a homologous chromosome, it produces a **duplication** there. If a fragment reattaches to the original chromosome but in the reverse direction, an **inversion** results.

Inversions are less likely than deletions or duplications to produce harmful effects, because in inversions, all genes are still present in their normal number. Deletions, especially large ones, tend to have the most serious effects. One example in humans is a specific deletion in chromosome 5 that causes the *cri du chat ("cat-cry") syndrome*. A child born with this syndrome is mentally retarded and has a small head and a cry like the mewing of a cat. Death usually occurs in infancy or early childhood.

Another type of chromosome change is chromosomal **translocation,** the attachment of a chromosome fragment to a nonhomologous chromosome. Figure 7.24d shows a translocation that is reciprocal; that is, two nonhomologous chromosomes exchange segments. Like inversions, translocations may or may not be harmful. Some people with Down syndrome have only part of a third chromosome 21; as the result of a translocation, it is attached to another (nonhomologous) chromosome.

Whereas chromosome alterations in sperm or egg can cause congenital disorders, such changes in a somatic cell may contribute to the development of cancer. For example, a chromosomal translocation in somatic cells in the bone marrow is associated with *chronic myelogenous leukemia (CML)*. Leukemias are the cancers affecting cells that give rise to white blood cells (leukocytes). In the cancerous cells of most CML patients, a part of chromosome 22 has switched places with a small fragment from chromosome 9 (Figure 7.25). This reciprocal translocation activates a gene that leads to leukemia. The chromosome ending up with the activated cancer-causing gene is called the "Philadelphia chromosome," after the city where it was discovered.

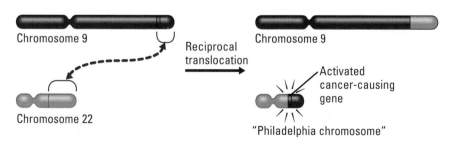

Figure 7.25 The translocation associated with a type of leukemia. This is a reciprocal translocation in which one end of chromosome 9 trades places with about half of chromosome 22. One of the genes moved from chromosome 9 is activated to become a cancer-causing gene in its new home, called a "Philadelphia chromosome." When a Philadelphia chromosome arises in a bone marrow cell that is destined to develop into a white blood cell (leukocyte), the result is a type of leukemia called chronic myelogenous leukemia (CML).

CheckPoint

1. _____ is to somatic cells as haploid is to _____.

2. If a diploid cell with 18 chromosomes undergoes meiosis, the resulting gametes will each have _____ chromosomes.

3. Explain how mitosis conserves chromosome number while meiosis reduces the number in half.

4. In what important way is anaphase of meiosis II like anaphase of mitosis?

5. Name two events during meiosis that contribute to genetic variety among gametes.

6. How does the karyotype of a human female differ from that of a male?

7. What is the chromosomal basis of Down syndrome?

8. Explain how nondisjunction in meiosis could result in a diploid gamete.

9. How is reciprocal translocation different from normal crossing over?

Answers: 1. Diploid; gametes. **2.** 9 **3.** In mitosis, a single replication of the chromosomes is followed by one division of the cell. In meiosis, a single replication of the chromosomes is followed by two cell divisions. **4.** Sister chromatids separate. **5.** Crossing over between homologous chromosomes during prophase I and independent orientation of tetrads at metaphase I **6.** A female has two X chromosomes; a male has an X and a Y. **7.** Three copies of chromosome 21 (trisomy 21) **8.** A diploid gamete would result if there were nondisjunction of *all* the chromosomes during meiosis I or II. **9.** Normal crossing over is an exchange of segments between homologous chromosomes; reciprocal translocation occurs between nonhomologous chromosomes.

Evolution Link: New Species from Errors in Cell Division

Errors in meiosis or mitosis do not always lead to problems. In fact, biologists believe that such errors have been instrumental in the evolution of many species. Numerous plant species, in particular, seem to have originated from accidents during cell division that resulted in extra sets of chromosomes. The new species is **polyploid,** meaning that it has more than two sets of homologous chromosomes in each somatic cell. At least half of all species of flowering plants are polyploid, including such useful ones as wheat, potatoes, apples, and cotton.

Let's consider one scenario by which a diploid (2*n*) plant species might generate a tetraploid (4*n*) plant. Imagine that, like many plants, our diploid plant produces both sperm and egg cells and can self-fertilize. If meiosis fails to occur in the plant's reproductive organs and gametes are instead produced by mitosis, the gametes will be diploid. The union of a diploid (2*n*) sperm with a diploid (2*n*) egg in self-fertilization will produce a tetraploid (4*n*) zygote, which may develop into a mature tetraploid plant that can itself reproduce by self-fertilization. The tetraploid plants will constitute a new species, one that has evolved in just one generation.

Learn more about new species resulting from errors in cell division in the Web/CD Process of Science and the Web Evolution Link.

Although polyploid animal species are less common than polyploid plants, they are known to occur among the fishes and amphibians. Recently, researchers in Chile have identified the first candidate for polyploidy among

Figure 7.26 Chock full of chromosomes—a tetraploid mammal? The somatic cells of this red viscacha rat from Argentina have about twice as many chromosomes as those of closely related species. (Interestingly, the heads of its sperm are unusually large, presumably a necessity for holding all that genetic material.) Scientists think that this rat is a tetraploid species that arose when an ancestor somehow doubled its chromosome number, probably by errors in mitosis or meiosis within the animal's reproductive organs. Researchers are studying the rat's chromosomes to verify that it actually has four homologous sets.

the mammals, a rat whose cells seem to be tetraploid (Figure 7.26). Tetraploid organisms are sometimes strikingly different from their recent diploid ancestors—larger, for example. Scientists don't yet understand exactly how polyploidy brings about such differences.

You'll learn more about the evolution of polyploid species in Chapter 13. In Chapter 8, we continue our study of genetic principles by looking at the rules governing the inheritance of biological traits and the connection between these traits and the organism's chromosomes.

Chapter Review

Summary of Key Concepts

Overview: What Cell Reproduction Accomplishes

Cell reproduction, usually called cell division, enables a multicellular organism to grow and develop and to replace damaged or lost cells.

- **Passing On the Genes from Cell to Cell** Most cell division involves a duplication of all the chromosomes, followed by the distribution of the two identical sets of chromosomes to two "daughter" cells when the cell divides in two. The daughter cells are genetically identical.

- **The Reproduction of Organisms** Some organisms use ordinary cell division to reproduce. Their offspring are therefore genetically identical to the one parent and to each other. Organisms that reproduce sexually, by the union of a sperm with an egg cell, carry out another type of cell division in their reproductive organs. This process, meiosis, yields sperm and egg cells with only half as many chromosomes as ordinary body cells.
- Web/CD Activity 7A *Asexual and Sexual Reproduction*

The Cell Cycle and Mitosis

- **Eukaryotic Chromosomes** The many genes of a eukaryotic cell are grouped into multiple chromosomes in the nucleus. Each chromosome contains a very long DNA molecule with thousands of genes. Individual chromosomes are visible only when the cell is in the process of dividing; otherwise, they are in the form of thin, loosely packed chromatin fibers. Before a cell starts dividing, the chromosomes duplicate, producing sister chromatids (containing identical DNA) joined together at the centromere. Cell division involves the separation of sister chromatids and results in two daughter cells, each containing a complete and identical set of chromosomes.

- **The Cell Cycle** Cell division is only one phase, called the mitotic phase, of the eukaryotic cell cycle. Most of the cycle is spent in interphase, when meta-

bolic activity is high, chromosomes duplicate, many cell parts are made, and the cell grows in size. Eukaryotic cell division consists of two processes: mitosis and cytokinesis. Mitosis is the process that distributes the duplicated chromosomes to daughter nuclei; cytokinesis is the division of the cytoplasm to create two daughter cells.
- Web/CD Activity 7B *The Cell Cycle*

- **Mitosis and Cytokinesis** At the start of mitosis, the chromosomes coil up, becoming thick enough to be visible with a light microscope. The nuclear envelope breaks down, and a mitotic spindle made of microtubules moves the chromosomes to the middle of the cell. The sister chromatids then separate and are moved to opposite poles of the cell, where two new nuclei form. Cytokinesis overlaps the end of mitosis. Mitosis and cytokinesis produce genetically identical cells. In animals, cytokinesis occurs by cleavage, which pinches the cell apart. In plants, a membranous cell plate splits the cell in two.
- Web/CD Activity 7C *Mitosis and Cytokinesis Animation*
- Web/CD Activity 7D *Mitosis and Cytokinesis Video*

- **Cancer Cells: Growing out of Control** When the cell cycle control system malfunctions, a cell may divide excessively and form a tumor. Cancer cells have highly abnormal cell cycles. They can grow to form malignant tumors, invade other tissues (metastasize), and even kill the organism. Radiation and chemotherapy are effective as treatments because they interfere with cell division.

Meiosis, the Basis of Sexual Reproduction

- **Homologous Chromosomes** The somatic cells (body cells) of each species contain a specific number of chromosomes; for example, human cells have 46, making up 23 pairs (two sets) of homologous chromosomes. The chromosomes of a homologous pair carry genes for the same characteristics at the same places. In mammalian males, one pair of chromosomes are only partially homologous: the sex chromosomes X and Y. Females have two X chromosomes.

• **Gametes and the Life Cycle of a Sexual Organism** Cells with two sets of homologous chromosomes are said to be diploid. Gametes—eggs and sperm—are haploid cells. Each gamete contains a single set of chromosomes. At fertilization, a sperm fuses with an egg, forming a diploid zygote. Repeated mitotic cell divisions lead to a multicellular adult made of diploid cells. The diploid adult produces haploid gametes by meiosis, the kind of cell division that reduces the chromosome number by half. Although sexual life cycles differ with the species, they all involve the alternation of haploid and diploid stages.
• **Web/CD Activity 7E** *Sexual Life Cycles*

• **The Process of Meiosis** Meiosis, like mitosis, is preceded by chromosome duplication. But in meiosis, the cell divides twice to form four daughter cells. The first division, meiosis I, starts with the pairing of homologous chromosomes. In crossing over, homologous chromosomes exchange corresponding segments. Meiosis I separates the members of the homologous pairs and produces two daughter cells, each with one set of (duplicated) chromosomes. Meiosis II is essentially the same as mitosis; in each of the cells, the sister chromatids of each chromosome separate.
• **Web/CD Activity 7F** *Meiosis Animation*

• **Review: Comparing Mitosis and Meiosis** See Figure 7.16 for a review and comparison of the two processes.

• **The Origins of Genetic Variation** Because the chromosomes of a homologous pair come from different parents, they carry different versions of many of their genes. The large number of possible arrangements of chromosome pairs at metaphase of meiosis I leads to many different combinations of chromosomes in eggs and sperm. This is one source of the variation in offspring that results from sexual reproduction. Random fertilization of eggs by sperm greatly increases the variation. Crossing over during prophase of meiosis I increases variation still further.
• **Web/CD Activity 7G** *The Origins of Genetic Variation*

• **When Meiosis Goes Amok** Sometimes a person has an abnormal number of chromosomes, which causes problems. Down syndrome is caused by an extra copy of chromosome 21. The abnormal chromosome count is a product of nondisjunction, the failure of a homologous pair of chromosomes to separate during meiosis I or of sister chromatids to separate during meiosis II . Nondisjunction can also produce gametes with extra or missing sex chromosomes, which lead to varying degrees of malfunction in humans but do not usually affect survival. Chromosome breakage can lead to rearrangements—deletions, duplications, inversions, and translocations—that can produce genetic disorders or, if the changes occur in somatic cells, cancer.

Evolution Link: New Species from Errors in Cell Division

New species can arise very quickly when errors in meiosis or mitosis create polyploid cells, which have more than two sets of homologous chromosomes. Many plants are polyploid, as well as certain animals.
• **Web Evolution Link** *New Species from Errors in Cell Division*
• **Web/CD The Process of Science** *Polyploid Plants*

Self-Quiz

1. If an intestinal cell in a grasshopper contains 24 chromosomes, a grasshopper sperm cell would contain _____ chromosomes.
   a. 3
   b. 6
   c. 12
   d. 24
   e. 48

2. Which of the following phases of mitosis is essentially the opposite of prophase in terms of nuclear changes?
   a. telophase
   b. metaphase
   c. S phase
   d. interphase
   e. anaphase

3. A biochemist measures the amount of DNA in cells growing in the laboratory and finds that the quantity of DNA in a cell has doubled
   a. between prophase and anaphase of mitosis.
   b. between the G_1 and G_2 phases of the cell cycle.
   c. during the M phase of the cell cycle.
   d. between prophase I and prophase II of meiosis.
   e. between anaphase and telophase of mitosis.

4. Which of the following is not a function of mitosis in humans?
   a. repair of wounds
   b. growth
   c. production of gametes from diploid cells
   d. replacement of lost or damaged cells
   e. multiplication of somatic cells

5. A micrograph of a dividing cell from a mouse shows 19 chromosomes, each consisting of two sister chromatids. During which of the following stages of cell division could this picture have been taken? (*Explain your answer.*)
   a. prophase of mitosis
   b. telophase II of meiosis
   c. prophase I of meiosis
   d. anaphase of mitosis
   e. prophase II of meiosis

6. Cytochalasin B is a chemical that disrupts microfilament formation. This chemical would interfere with
   a. DNA replication.
   b. formation of the mitotic spindle.
   c. cleavage.
   d. formation of the cell plate.
   e. crossing over.

7. It is difficult to observe individual chromosomes during interphase because
   a. the DNA has not been replicated yet.
   b. they are in the form of very long, thin strands.
   c. they leave the nucleus and are dispersed to other parts of the cell.
   d. homologous chromosomes do not pair up until division starts.
   e. the spindle must move them to the center of the cell before they become visible.

8. A fruit fly somatic cell contains eight chromosomes. This means that _____ different combinations of chromosomes are possible in its gametes.
   a. 4
   b. 8
   c. 16
   d. 32
   e. 64

9. If a fragment of a chromosome breaks off and then reattaches to the original chromosome but in the reverse direction, the resulting chromosome abnormality is called
   a. a deletion.
   b. an inversion.
   c. a translocation.
   d. a nondisjunction.
   e. a reciprocal translocation.

10. Why are individuals with an extra chromosome 21, which causes Down syndrome, more numerous than individuals with an extra chromosome 3 or chromosome 16?
    a. There are probably more genes on chromosome 21 than on the others.
    b. Chromosome 21 is a sex chromosome and 3 and 16 are not.
    c. Down syndrome is not more common, just more serious.
    d. Extra copies of chromosomes 3 or 16 are probably fatal.
    e. Nondisjunction of chromosome 21 probably occurs much more frequently.

11. The following light micrograph shows dividing cells near the tip of an onion root. Identify the stage of mitosis for each of the outlined cells, a–d.

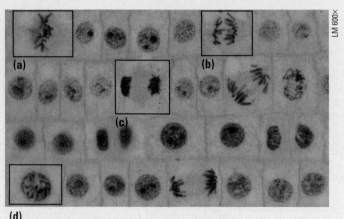

• **Go to the website or CD-ROM for more self-quiz questions.**

The Process of Science

1. A mule is the offspring of a horse and a donkey. A donkey sperm contains 31 chromosomes and a horse egg 32 chromosomes, so the zygote contains a total of 63 chromosomes. The zygote develops normally. The combined set of chromosomes is not a problem in mitosis, and the mule combines some of the best characteristics of horses and donkeys. However, a mule is sterile; meiosis cannot occur normally in its testes or ovaries. Explain why mitosis is normal in cells containing both horse and donkey chromosomes but the mixed set of chromosomes interferes with meiosis.

2. **Explore how new species can result from errors in cell division in The Process of Science activity available on the website and CD-ROM.**

Biology and Society

Every year about a million Americans are diagnosed as having cancer. This means that about 75 million Americans now living will eventually have cancer, and one in five will die of the disease. There are many kinds of cancers and many causes of the disease. For example, smoking causes most lung cancers. Overexposure to ultraviolet rays in sunlight causes most skin cancers. There is evidence that a high-fat, low-fiber diet is a factor in breast, colon, and prostate cancers. And agents in the workplace such as asbestos and vinyl chloride are also implicated as causes of cancer. Hundreds of millions of dollars are spent each year in the search for effective treatments for cancer; far less money is spent on preventing cancer. Why might this be the case? What kinds of lifestyle changes could we make to help prevent cancer? What kinds of prevention programs could be initiated or strengthened to encourage these changes? What factors might impede such changes and programs? Should we devote more of our resources to treating cancer or preventing it? Why?

Patterns of Inheritance

The first geneticist was a monk named Gregor

Mendel, the first genetics research site the garden of his

abbey. The variant gene that causes sickle-cell disease

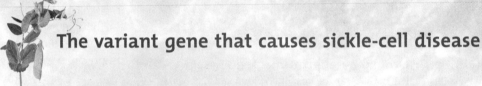

can help protect you against malaria. If

you're male and color-blind, it's probably your mother's

"fault." Hemophilia has plagued the royal

families of Europe, apparently originating in one of the sons

of Queen Victoria.

yes of brown, blue, green, or gray; hair of black, brown, blond, or red; bodies compact or gangly—these are just a few examples of heritable variations that we see in the human population. What genetic principles account for the transmission of such traits from parents to offspring? A little over a century ago, a monk named Gregor Mendel discovered that the inheritance of many genetic traits could be explained by a few simple principles. Just as a game with many possible sequences of moves may be based on just a few rules—think of checkers—so is inheritance. In this chapter, you will learn the basic rules of the genetic "game" and how the behavior of the chromosomes accounts for these rules.

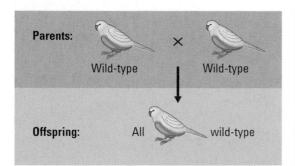

Figure 8.1 Heritable variation in budgies. In nature, budgies (also called parakeets) are most often green and yellow like the second from the right in this photograph. This is the "wild-type" coloration. Birds of other colors are rarer in nature, but a wide variety have been bred in captivity.

Overview: Heritable Variation and Patterns of Inheritance

Budgies, also known as parakeets, are small, long-tailed parrots native to Australia and kept as pets throughout the world. Wild budgies typically have green underparts and yellow upperparts barred with black, like the bird second from the right in Figure 8.1. Geneticists call such traits—the ones found most commonly in nature—the **wild type.** The white and blue birds in the photo are uncommon in nature, but large numbers of these and other color variants are raised in captivity and are available in pet stores. Blue budgies called sky-blues are especially popular.

Feather color in budgies is an inherited characteristic, and pet breeders can predict which color variants will result from particular parents. For instance, if two wild-type (green and yellow) budgies are mated (Figure 8.2a), it's likely that all their offspring will be green and yellow. The same is true when a wild-type bird and a sky-blue are mated (Figure 8.2b). But, if two offspring of a wild-type budgie and a sky-blue are mated, their offspring, the second generation, will most likely include both green and yellow birds and sky-blues. Knowing how these feather color traits are inherited—their *patterns of inheritance*—a breeder would predict that about one in four second-generation birds would be sky-blue. The predictions are based on the hereditary principles discovered by Mendel. Although Mendel himself undoubtedly never even saw a budgie, his principles turn out to apply to budgies and all other sexually reproducing organisms.

Learn more about heritable variation in Web/CD Activity 8A.

(a) Offspring from the mating of two wild-type birds

Figure 8.2 Patterns of inheritance in budgies. (a) In nature, when two wild-type (green and yellow) budgies mate, all the offspring usually look like the parents. (The cross symbol indicates a mating.) **(b)** When such a wild-type bird mates with a sky-blue bird, all their offspring will also be wild-type in appearance. But when these first-generation offspring mate with each other, about one-quarter of their offspring, the second generation, will be sky-blue.

(b) Two generations of offspring from the mating of a wild-type with a sky-blue bird

CheckPoint

What is a wild-type trait?

Answer: A trait that is the prevailing one in nature

Mendel's Principles

Gregor Mendel was the first to analyze patterns of inheritance in a systematic, scientific way. During the 1860s, using the garden of an abbey in Brunn, Austria (now Brno, in the Czech Republic), Mendel deduced the fundamental principles of genetics by breeding garden peas (Figure 8.3). His work is a classic in the history of biology. Strongly influenced by his study of physics, mathematics, and chemistry at the University of Vienna, his research was both experimental and mathematically rigorous, and these qualities were largely responsible for his success.

In a paper published in 1866, Mendel correctly argued that parents pass on to their offspring discrete heritable factors that are responsible for inherited traits. He stressed that the heritable factors (today called genes) retain their individuality generation after generation. In other words, genes are like marbles of different colors: Just as marbles retain their colors permanently, no matter how they are mixed up or temporarily covered up, genes permanently retain their identities.

Figure 8.3 Gregor Mendel in his garden. This painting shows Mendel examining some garden peas, the plant he chose for his experiments.

In an Abbey Garden

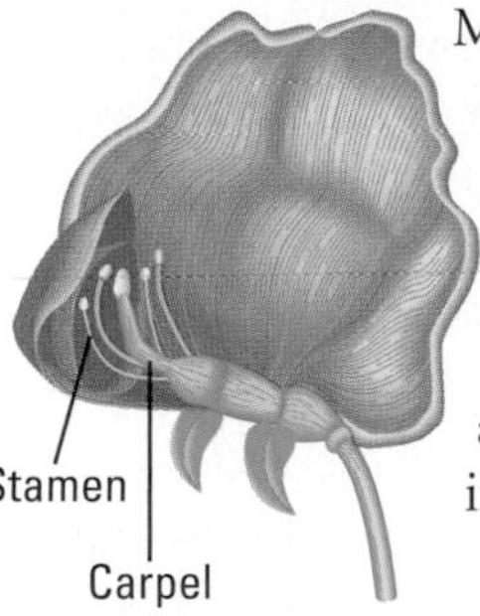

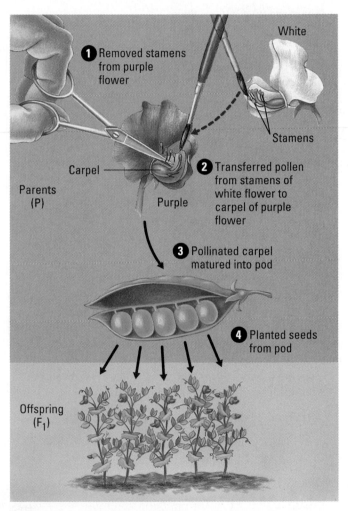

Mendel probably chose to study garden peas because they were easy to grow and available in many readily distinguishable varieties. Also, with pea plants, Mendel was able to exercise strict control over plant matings. The petals of the pea flower almost completely enclose the female and male parts—carpel and stamens, respectively (Figure 8.4). Consequently, in nature, pea plants usually **self-fertilize,** when sperm-carrying pollen grains released from the stamens land on the tip of the egg-containing carpel of the same flower. Mendel could ensure self-fertilization by covering a flower with a small bag so that no pollen from another plant could reach the carpel. When he wanted **cross-fertilization** (fertilization of one plant by pollen from a different plant), he pollinated the plants by hand, as shown in Figure 8.5. So whether Mendel let a pea plant self-fertilize or cross-fertilized it with a known source of pollen, he could always be sure of the parentage of new plants.

Figure 8.4 The structure of a pea flower. To reveal the reproductive organs—stamens and carpel—one of the petals has been removed. The carpel produces egg cells; the stamens make pollen, which carries sperm.

Mendel's success was due not only to his experimental approach and choice of organism, but also to his selection of characteristics to study. Each of the characteristics he chose to study, such as flower color, occurs in two distinct forms. Mendel worked with his plants until he was sure he had *true-breeding* varieties—that is, varieties for which self-fertilization produced offspring all identical to the parent. For instance, he identified a purple-flowered variety that, when self-fertilized, produced offspring plants that all had purple flowers.

Now Mendel was ready to ask what would happen when he crossed his different true-breeding varieties with each other. For example, what offspring would result if plants with purple flowers and plants with white flowers were cross-fertilized as shown in Figure 8.5? In the language of breeders and geneticists, the offspring of two different true-breeding varieties are

Figure 8.5 Mendel's technique for cross-fertilizing pea plants. ❶ To prevent self-fertilization he cut off the stamens from an immature flower before they produced pollen. This stamenless plant would be the female parent in the experiment. ❷ To cross-fertilize this female, he dusted its carpel with pollen from another plant. After pollination, ❸ the carpel developed into a pod, containing seeds (peas) that ❹ he planted. The seeds grew into offspring plants.

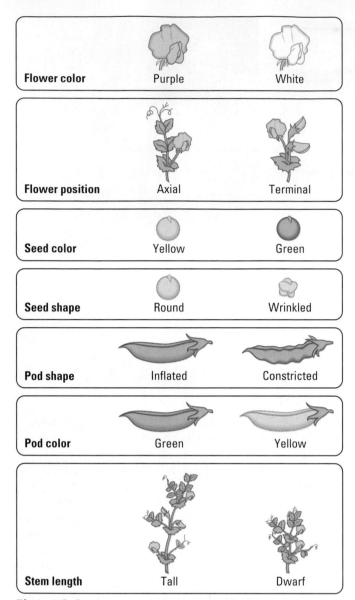

Flower color	Purple	White
Flower position	Axial	Terminal
Seed color	Yellow	Green
Seed shape	Round	Wrinkled
Pod shape	Inflated	Constricted
Pod color	Green	Yellow
Stem length	Tall	Dwarf

Figure 8.6 **The seven characteristics of pea plants studied by Mendel.** Each characteristic comes in the two alternatives shown here. The alternatives in the left column (such as purple flower color) are dominant; the ones in the right column (such as white flower color) are recessive. You will learn about dominant and recessive genes shortly.

called **hybrids,** and the cross-fertilization itself is referred to as a hybridization or simply a **cross.** The parental plants are called the **P generation** (P for parental), and their hybrid offspring are the **F₁ generation** (F for filial, from the Latin word for "son"). When F₁ plants self-fertilize or fertilize each other, their offspring are the **F₂ generation.**

Mendel's Principle of Segregation

Mendel performed many experiments in which he tracked the inheritance of characteristics, such as flower color, which come in two alternatives (Figure 8.6). The results led him to formulate several hypotheses about inheritance. Let's look at some of his experiments and follow the reasoning that led to his hypotheses.

Monohybrid Crosses Figure 8.7 a starts with a cross between a pea plant with purple flowers and one with white flowers. This is called a **monohybrid cross** because the parent plants differ in only one characteristic. Mendel saw that the F₁ plants (monohybrids) all had truly purple flowers (and not a blend of purple and white). Was the heritable factor for white flowers now lost as a result of the hybridization? By mating the F₁ plants, Mendel found the answer to be no. Of the F₂ plants, one-fourth had white flowers. Mendel concluded that the heritable factor for white flowers did not disappear in the F₁ plants, but that only the purple-flower factor was affecting F₁ flower color. He also deduced that the F₁ plants must have carried two factors for the flower color characteristic, one for purple and one for white. From these results and others, Mendel developed four hypotheses. Using modern terminology (including "gene" instead of "heritable factor"), his hypotheses are:

1. There are alternative forms of genes, the units that determine heritable traits. For example, the gene for flower color in pea plants exists in one form for purple and another for white. We now call alternative forms of genes **alleles.**

2. For each inherited characteristic, an organism has two genes, one from each parent. These genes may be the same allele or different alleles.

3. A sperm or egg carries only one allele for each inherited characteristic, because allele pairs segregate (separate) during the production of gametes. Then when sperm and egg unite at fertilization, each contributes its allele, restoring the paired condition in the offspring.

4. When the two genes of a pair are different alleles and one is fully expressed while the other has no noticeable effect on the organism, the alleles are called the **dominant allele** and the **recessive allele,** respectively.

Do Mendel's hypotheses account for the 3:1 ratio he observed in the F₂ generation? His hypotheses predict that when alleles segregate during gamete formation in the F₁ plants, half the gametes will receive a purple-flower allele and the other half a white-flower allele. During pollination among the F₁ plants, the gametes unite randomly. An egg with a purple-flower allele has an equal chance of being fertilized by a sperm with a purple-flower allele or one with a white-flower allele. Since the same is true for an egg with a white-flower allele, there are a total of four equally likely combinations of sperm and egg. Figure 8.7b illustrates these combinations using a diagram called a **Punnett square,** a handy device for predicting the results of a genetic cross. Notice

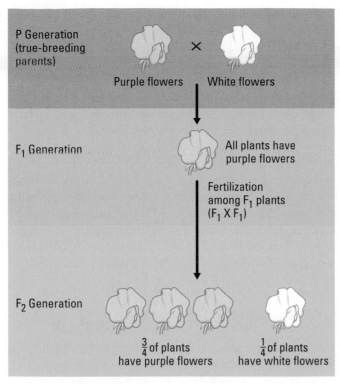

(a) Mendel's crosses tracking one characteristic (flower color)

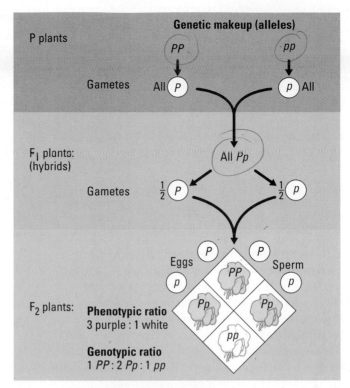

(b) Explanation of the results in part (a)

Figure 8.7 The principle of segregation. (a) Crossing true-breeding purple-flowered plants with true-breeding white-flowered plants produced F_1 plants with purple flowers. When Mendel cross-pollinated F_1 plants with other F_1 plants (or allowed F_1 plants to self-pollinate), he found that about three-quarters of the F_2 plants had purple flowers and about one-quarter had white flowers. In other words, there was a 3:1 ratio of the two varieties in the F_2 generation. Mendel looked at a total of 929 F_2 plants. **(b)** This diagram illustrates Mendel's explanation for the inheritance pattern shown in part (a) using modern terminology. Uppercase and lowercase letters distinguish dominant from recessive alleles, with P representing the dominant allele, for purple flowers, and p standing for the recessive allele, for white flowers. At the top of the diagram, you see the alleles carried by the parental plants. Both plants were true-breeding, so one parental variety must have had two alleles for purple flowers (PP), while the other variety had two alleles for white flowers (pp). Gametes, symbolized with circles, each contained only one allele for the flower color gene. Union of the parental gametes produced F_1 hybrids having a Pp combination. Because the purple-flower allele is dominant, all these hybrids had purple flowers. When the hybrid plants produced gametes, the two alleles segregated, half the gametes receiving the P allele and the other half receiving the p allele. Random combination of these gametes resulted in the 3:1 ratio that Mendel observed in the F_2 generation. The box at the bottom of the figure is a Punnett square, a useful tool for showing all possible combinations of alleles in offspring. Each square represents an equally probable product of fertilization. For example, the box at the right corner of the Punnett square shows the genetic combination resulting from a p sperm fertilizing a P egg.

that we use a capital letter to symbolize the dominant allele (P, the purple-flower allele) and a lowercase letter to represent the recessive allele (p, the white-flower allele).

What will be the physical appearance of these F_2 offspring? One-fourth of the plants have two alleles specifying purple flowers; clearly, these plants will have purple flowers. One-half (two-fourths) of the F_2 offspring have inherited one allele for purple flowers (P) and one allele for white flowers (p); like the F_1 plants, these plants will also have purple flowers, the dominant trait. Finally, one-fourth of the F_2 plants have inherited two alleles specifying white flowers (pp) and will express this recessive trait. Thus, Mendel's model accounts for the 3:1 ratio that he observed in the F_2 generation.

Because an organism's appearance does not always reveal its genetic composition, geneticists distinguish between an organism's expressed, or physical, traits, called its **phenotype** (such as purple or white flowers), and its genetic makeup, its **genotype** (in this example, PP, Pp, or pp). So now we see that Figure 8.7a shows the phenotypes and Figure 8.7b the genotypes in our sample crosses. For the F_2 plants, the ratio of plants with purple flowers to those with white flowers (3:1) is called the phenotypic ratio. The genotypic ratio is $1(PP) : 2(Pp) : 1(pp)$.

Mendel found that each of the seven characteristics he studied had the same inheritance pattern. One parental trait disappeared in the F_1 generation, only to reappear in one-fourth of the F_2 offspring. The underlying mechanism is stated by **Mendel's principle of segregation:** *Pairs of alleles segregate (separate) during gamete formation; the fusion of gametes at fertilization creates allele pairs again.* Research since Mendel's day has established that the principle of segregation applies to all sexually reproducing organisms.

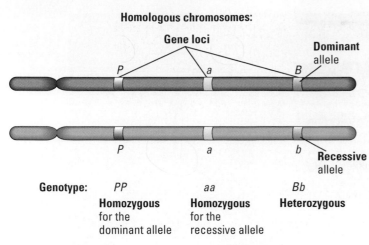

Homologous chromosomes:

Gene loci

Dominant allele

Recessive allele

Genotype: *PP* *aa* *Bb*

Homozygous for the dominant allele | Homozygous for the recessive allele | Heterozygous

Figure 8.8 The relationship between alleles and homologous chromosomes. The two chromosomes shown here are a homologous pair. The labeled bands on the chromosomes represent three gene loci, specific locations of genes along the chromosome. The matching colors of corresponding loci on the two homologues highlight the fact that homologous chromosomes carry genes for the same characteristics at the same positions along their lengths. However, as the uppercase and lowercase gene labels indicate, the two chromosomes may bear either identical alleles or different ones at any one locus. In other words, the organisms may be homozygous or heterozygous for the gene at that locus. We use uppercase for dominant alleles, lowercase for recessive alleles.

Genetic Alleles and Homologous Chromosomes Before continuing with Mendel's experiments, let's see how some of the concepts we discussed in Chapter 7 fit with what we've said about genetics so far. The diagram in Figure 8.8 shows a pair of homologous chromosomes. Recall from Chapter 7 that every diploid individual, whether pea plant or human, has two sets of homologous chromosomes. One set comes from the organism's female parent, the other from the male parent. On the chromosomes in the figure are marked three gene **loci** (singular, *locus*), the specific locations of three genes on the chromosomes. You can see the connection between Mendel's principle of segregation and homologous chromosomes: Alleles (alternative forms) of a gene reside at the same locus on homologous chromosomes.

Notice that the homologous chromosomes may bear either the same alleles or different ones at a given locus. When an organism has a pair of identical alleles for a gene, it is said to be **homozygous** for that gene. (True-breeding organisms are homozygous.) When the two alleles are different, it is **heterozygous.** We will return to the chromosomal basis of Mendel's principles later in the chapter.

Mendel's Principle of Independent Assortment

Two of the seven characteristics Mendel studied were seed shape and seed color. Mendel's seeds were either round or wrinkled in shape, and they were either yellow or green in color. From tracking these characteristics one at a time in monohybrid crosses, Mendel knew that the allele for round shape (designated R) was dominant to the allele for wrinkled shape (r) and that the allele for yellow seed color (Y) was dominant to the allele for green seed color (y). What would result from a mating of parental varieties differing in two characteristics—a **dihybrid cross**? Mendel crossed homozygous plants having round yellow seeds (genotype *RRYY*) with plants having wrinkled green seeds (*rryy*). As shown in Figure 8.9, the union of *RY* and *ry* gametes yielded hybrids heterozygous for both characteristics (*RrYy*)—that is, dihybrids. As we would expect, all of these offspring, the F_1 generation, had round yellow seeds. But were the two characteristics transmitted from parents to offspring as a package, or was each characteristic inherited independently of the other?

The question was answered when Mendel allowed fertilization to occur among the F_1 plants. If the genes for the two characteristics were inherited together (Figure 8.9a), then the F_1 hybrids would produce only the same two kinds (genotypes) of gametes that they received from their parents. In that case, the F_2 generation would show a 3:1 phenotypic ratio (three plants with round yellow seeds for every one with wrinkled green seeds), as in the left Punnett square. If, however, the two seed characteristics segregated independently, then the F_1 generation would produce four gamete genotypes— *RY, rY, Ry,* and *ry*—in equal quantities. The Punnett square in Figure 8.9b shows all possible combinations of alleles that can result in the F_2 generation from the union of four kinds of sperm with four kinds of eggs. If you study the Punnett square, you will see that it predicts nine different genotypes in the F_2 generation. These will produce four different phenotypes in a ratio of 9:3:3:1.

The right-hand Punnett square also reveals that a dihybrid cross is equivalent to two monohybrid crosses occurring simultaneously. From the 9:3:3:1 ratio, we can see that there are 12 plants with round seeds to 4 with wrinkled seeds, and 12 yellow-seeded plants to 4 green-seeded ones. These 12:4 ratios each reduce to 3:1, which is the F_2 ratio for a

Conduct crosses with pea plants in Web/CD Activity 8B and The Process of Science.

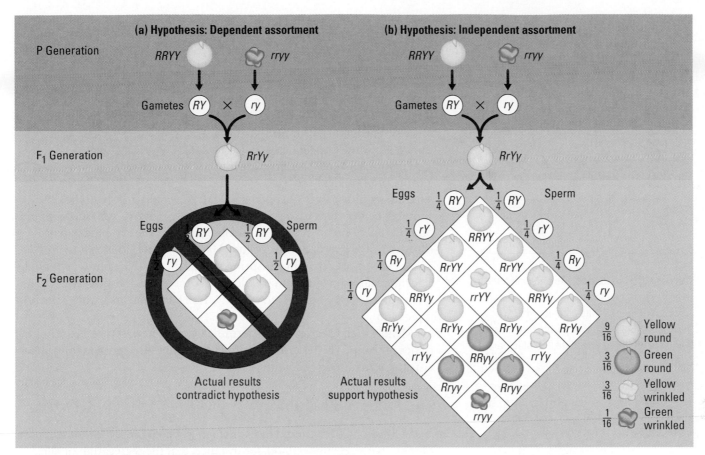

(a) Hypothesis: Dependent assortment

P Generation   RRYY ⊗   🌰 rryy

Gametes (RY) ✕ (ry)

F₁ Generation   RrYy

F₂ Generation

Eggs   Sperm
½ (RY)   ½ (RY)
½ (ry)   ½ (ry)

Actual results
contradict hypothesis

(b) Hypothesis: Independent assortment

P Generation   RRYY ⊗   🌰 rryy

Gametes (RY) ✕ (ry)

F₁ Generation   RrYy

Eggs ¼(RY) ¼(RY) Sperm

¼(rY)   RRYY   ¼(rY)
RrYY   RrYY
¼(Ry)   ¼(Ry)
RRYy   rrYY   RRYy
¼(ry)   ¼(ry)
RrYy   RrYy   RrYy   RrYy
rrYy   RRyy   rrYy
Rryy   Rryy
rryy

$\frac{9}{16}$ Yellow round
$\frac{3}{16}$ Green round
$\frac{3}{16}$ Yellow wrinkled
$\frac{1}{16}$ Green wrinkled

Actual results
support hypothesis

Figure 8.9 Testing two hypotheses for gene assortment in a dihybrid cross. **(a)** One hypothesis leads to the prediction that every plant in the F₂ generation will have seeds exactly like one of the parents, either round and yellow or wrinkled and green. **(b)** The other hypothesis, independent assortment, predicts F₂ plants with four different seed phenotypes. The results of experiments support this hypothesis. See the text for further explanation.

monohybrid cross. Mendel tried his seven pea characteristics in various dihybrid combinations and always observed a 9:3:3:1 ratio (or two simultaneous 3:1 ratios) of phenotypes in the F₂ generation. These results supported hypothesis (b) in Figure 8.9—that *each pair of alleles segregates independently of the other pairs during gamete formation.* This is **Mendel's principle of independent assortment.** For another application of this principle, examine Figure 8.10, showing the inheritance of feather color in budgies.

Figure 8.10 Independent assortment in budgies. **(a)** Feather color is controlled by two genes, each with two alleles (*Y* and *y*, *M* and *m*). Birds with *Y* have yellow pigment; those with the *M* allele have a pigment called melanin, which gives feathers a blue or black color. The alleles *m* and *y* produce no pigments. (Blanks in the genotypes indicate alleles that can be dominant or recessive.) Green feathers result from the combination of yellow pigment and melanin. **(b)** When we mate two doubly heterozygous (*YyMm*) budgies, the phenotypic ratio of the offspring is 9:3:3:1. These results resemble the F₂ results in Figure 8.9 and demonstrate that the *Y/y* and *M/m* genes assort independently.

Phenotypes			
Green and yellow	Blue ("sky-blue")	Yellow ("black-eyed yellow")	White
Genotypes Y_M_	yyM_	Y_mm	yymm

(a)

Mating of heterozygotes   (Green and yellow) *YyMm* ✕ *YyMm* (Green and yellow)

Phenotypic ratio of offspring: 9 Green and yellow | 3 Blue | 3 Yellow | 1 White

(b)

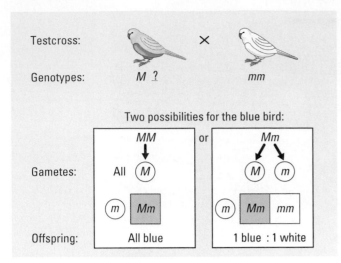

Testcross:

Genotypes: M ?   ×   mm

Two possibilities for the blue bird:

MM	or	Mm

Gametes: All M | M m

	m	Mm

	m	Mm	mm

Offspring: All blue | 1 blue : 1 white

Figure 8.11 A testcross with budgies. To determine the genotype of a sky-blue bird with respect to the gene for melanin production (*M*), it is crossed with a bird that is homozygous recessive (*mm*). If all the offspring are blue, the blue parent bird must have genotype *MM*. If half the offspring are white, the blue parent must be heterozygous (*Mm*).

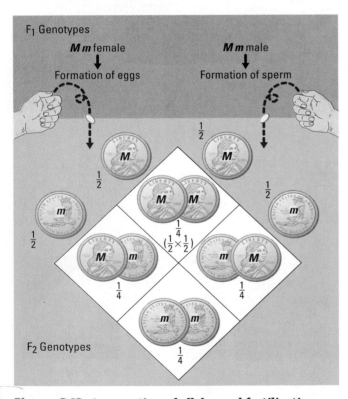

F₁ Genotypes

Mm female   *Mm* male

Formation of eggs   Formation of sperm

F₂ Genotypes

Figure 8.12 Segregation of alleles and fertilization as chance events. When a heterozygote (*Mm*) forms gametes, segregation of alleles is like the toss of a coin. An egg cell has a 50% chance of receiving the dominant allele and a 50% chance of receiving the recessive allele. The same odds apply to a sperm cell. Like two separately tossed coins, segregation during sperm and egg formation occurs as two independent events. To determine the probability that an individual offspring will inherit the dominant allele from both parents, we multiply the probabilities of each required event: $\frac{1}{2} \times \frac{1}{2} = \frac{1}{4}$.

Using a Testcross to Determine an Unknown Genotype

Suppose you have a sky-blue budgie and you want to determine its genotype. You know that it carries two copies of the recessive allele *y* because it shows no evidence of yellow pigment. But is it homozygous (*MM*) or heterozygous (*Mm*) for the melanin gene? To answer this question, you need to perform what geneticists call a **testcross,** a mating between an individual of unknown genotype (your bird) and a homozygous recessive individual—in this case, a white budgie.

Figure 8.11 shows the offspring that could result from a mating between a white (*mm*) budgie and a sky-blue one (*MM* or *Mm*). If, as shown on the left, the sky-blue parent's genotype is *MM*, we would expect all the offspring to be sky-blue, because a cross between genotypes *MM* and *mm* can produce only *Mm* offspring. On the other hand, if the sky-blue parent is *Mm*, we would expect both sky-blue (*Mm*) and white (*mm*) offspring. Thus, the appearance of the offspring reveals the sky-blue bird's genotype. The figure also shows that we would expect the white and sky-blue offspring of a *Mm* × *mm* cross to exhibit a 1:1 (1 sky-blue to 1 white) phenotypic ratio.

Mendel used testcrosses to determine whether he had true-breeding varieties of plants. The testcross continues to be an important tool of geneticists for determining genotypes.

The Rules of Probability

Mendel's strong background in mathematics served him well in his studies of inheritance. He understood, for instance, that the segregation of allele pairs during gamete formation and the re-forming of pairs at fertilization obey the rules of probability—the same rules that apply to the tossing of coins, the rolling of dice, and the drawing of cards. Mendel also appreciated the statistical nature of inheritance. He knew that he needed to obtain large samples—count many offspring from his crosses—before he could begin to interpret inheritance patterns.

Let's see how the rules of probability apply to inheritance. The probability scale ranges from 0 to 1. An event that is certain to occur has a probability of 1, while an event that is certain not to occur has a probability of 0. The probabilities of all possible outcomes for an event must add up to 1. With a coin, the chance of tossing heads is $\frac{1}{2}$, and the chance of tossing tails is $\frac{1}{2}$. In a standard deck of 52 playing cards, the chance of drawing a jack of diamonds is $\frac{1}{52}$, and the chance of drawing any card other than the jack of diamonds is $\frac{51}{52}$.

An important lesson we can learn from coin tossing is that for each and every toss of the coin, the probability of heads is $\frac{1}{2}$. In other words, the outcome of any particular toss is unaffected by what has happened on previous attempts. Each toss is an *independent event.*

If two coins are tossed simultaneously, the outcome for each coin is an independent event, unaffected by the other coin. What is the chance that both coins will land heads up? The probability of such a *compound event* is the product of the separate probabilities of the independent events—for the coins, $\frac{1}{2} \times \frac{1}{2} = \frac{1}{4}$. This is called the **rule of multiplication,** and it holds true for genetics as well as coin tosses, as illustrated in Figure 8.12. In our dihybrid cross of budgies, the genotype of the F₁ birds for melanin pigment was *Mm*. What is the probability that a particular F₂ bird will have the *mm* genotype? To produce an *mm* offspring, both egg and sperm must carry the *m* allele. The probability that an egg will have the *m* allele is $\frac{1}{2}$, and the probability that a sperm will have the m allele is also $\frac{1}{2}$. By the rule of multiplication, the probability that two m alleles will come together at fertilization is $\frac{1}{2} \times \frac{1}{2} = \frac{1}{4}$. This is exactly the answer given by the Punnett square. If we know the genotypes of the parents, we can predict the probability for any genotype among the offspring.

Now let's consider the probability that an F_2 budgie will be heterozygous for the melanin gene. As the figure shows, there are two ways in which F_1 gametes can combine to produce a heterozygous offspring. The dominant (M) allele can come from the egg and the recessive (m) allele from the sperm, or vice versa. The probability of an event occurring when there are two or more alternative ways is the *sum* of the separate probabilities of the different ways. This is known as the **rule of addition.** Using this rule, we can calculate the probability of an F_2 heterozygote as $\frac{1}{4} + \frac{1}{4} = \frac{1}{2}$.

By applying the rules of probability to segregation and independent assortment, we can solve some rather complex genetics problems. For instance, we can predict the results of trihybrid crosses, in which three different characteristics are involved. Consider a cross between two organisms that both have the genotype *AaBbCc*. What is the probability that an offspring from this cross will be a recessive homozygote for all three genes (*aabbcc*)? Because each allele pair assorts independently, we can treat this trihybrid cross as three separate monohybrid crosses:

Aa × *Aa:* Probability of *aa* offspring = $\frac{1}{4}$

Bb × *Bb:* Probability of *bb* offspring = $\frac{1}{4}$

Cc × *Cc:* Probability of *cc* offspring = $\frac{1}{4}$

Because the segregation of each allele pair is an independent event, we use the rule of multiplication to calculate the probability that the offspring will be *aabbcc*:

$$\tfrac{1}{4}\,aa \times \tfrac{1}{4}\,bb \times \tfrac{1}{4}\,cc = \tfrac{1}{64}$$

We could reach the same conclusion by constructing a 64-section Punnett square, but that would take a lot of time—and also a lot of space!

Family Pedigrees

Mendel's principles apply to the inheritance of many human traits. Figure 8.13 illustrates alternative forms of three human characteristics that are thought to be determined by simple dominant-recessive inheritance at one gene locus. If we call the dominant allele of any such gene *A,* the dominant phenotype results from either the homozygous genotype *AA* or the heterozygous genotype *Aa.* Recessive phenotypes always result from the homozygous genotype *aa.* In genetics, the word *dominant* does not imply that a phenotype is either normal or more common than a recessive phenotype; wild-type traits (those prevailing in nature) are not necessarily specified by dominant alleles. In genetics, dominance means that a heterozygote (*Aa*), carrying only a single copy of a dominant allele, displays the dominant phenotype. By contrast, the phenotype of the corresponding recessive allele is seen only in a homozygote (*aa*). Recessive traits are often more common in the population than dominant ones. For example, the absence of freckles is more common than their presence.

How do we know how particular human traits are inherited? Unlike researchers working with pea plants or budgies, geneticists who study humans cannot control the mating of their subjects. Instead, they must analyze the

Probability Problem. What is the probability that a throw of the dice will come up "snake eyes" (two dots) as shown?

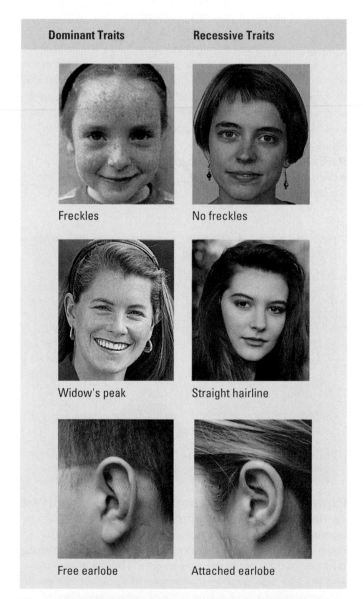

Dominant Traits	Recessive Traits
Freckles	No freckles
Widow's peak	Straight hairline
Free earlobe	Attached earlobe

Figure 8.13 Examples of inherited traits in humans.

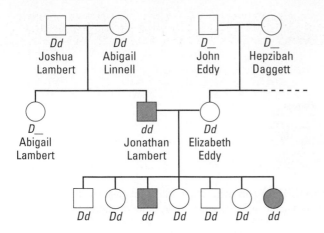

Figure 8.14 Pedigree showing inheritance of deafness. The letter *D* stands for the hearing allele, and *d* symbolizes the recessive allele for deafness. The first deaf individual to appear in this pedigree is Jonathan Lambert. Because only people who are homozygous for the recessive allele are deaf, his genotype must have been *dd*. Therefore, both his parents must have carried a *d* allele along with the *D* allele that gave them normal hearing. Two of Jonathan's children were deaf (*dd*), so his wife, who had normal hearing, must also have carried a *d* allele. Likewise, all of the couple's normal children must have been heterozygous (*Dd*). What are the genotypes of Elizabeth Eddy's parents and Jonathan's sister Abigail? These three people had normal hearing, so they must have carried at least one *D* allele. And at least one of Elizabeth's parents must have had a *d* allele. But more than that we cannot say without additional information.

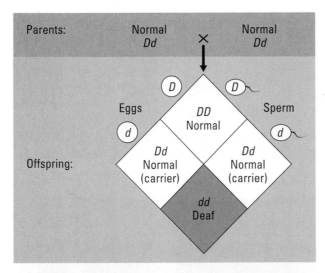

Figure 8.15 Predicted offspring when both parents are carriers for a recessive disorder. Carriers are heterozygotes who have one copy of a potentially harmful recessive allele. Because carriers also have the dominant allele, they do not display the disorder. As in the last figure, *d* stands for a recessive allele associated with deafness and *D* for the dominant allele. As you can see in the Punnett square, one-quarter of the offspring of two carriers would be expected to be homozygous recessive and therefore deaf.

results of matings that have already occurred. Suppose you wanted to study the inheritance of a type of deafness that is inherited as a recessive trait. First you would collect as much information as possible about a family's history for the trait. Then you would assemble this information into a family tree—the family **pedigree.** Finally, to analyze the pedigree, you would use Mendel's concept of dominant and recessive alleles and his principle of segregation.

Let's apply this approach to the example in Figure 8.14, which shows part of a real pedigree. This pedigree is from a family that lived on Martha's Vineyard, an island off the coast of Massachusetts, where a particular kind of inherited deafness was once prevalent. In the pedigree, squares represent males, circles represent females, and colored symbols indicate deafness. The earliest generation studied is at the top of the pedigree. Notice that deafness did not appear in this generation and that it showed up in only two of the seven children in the third generation. By applying Mendel's principles, we can deduce that the deafness allele is recessive. Mendel's principles also enable us to deduce the genotypes that are shown for most of the people in the pedigree, including all the homozygous recessives and many of the heterozygotes. People who have one copy of the allele for a recessive disorder and do not exhibit symptoms are called **carriers** of the disorder.

Human Disorders Controlled by a Single Gene

The hereditary deafness found on Martha's Vineyard is just one of over a thousand human genetic disorders currently known to be inherited as dominant or recessive traits controlled by a single gene locus; ten more examples are listed in Table 8.1. These disorders show simple inheritance patterns like the traits Mendel studied in pea plants. The genes involved are all located on autosomes (chromosomes other than sex chromosomes X and Y).

Recessive Disorders Most human genetic disorders are recessive. They range in severity from relatively harmless conditions, such as albinism (lack of pigmentation), to deadly diseases. The vast majority of people afflicted with recessive disorders are born to normal parents who are both heterozygotes—that is, who are carriers of the recessive allele for the disorder but are phenotypically normal. For example, in Figure 8.14, the hearing parents of Jonathan Lambert were both carriers.

Using Mendel's principles, we can predict the fraction of affected offspring likely to result from a marriage between two carriers. Suppose one of Jonathan Lambert's hearing sons (*Dd*) married a hearing cousin whose pedigree indicated that her genotype was also *Dd*. What is the probability that they would have deaf children? As the Punnett square in Figure 8.15 shows, each child of two carriers has a $\frac{1}{4}$ chance of inheriting two recessive alleles. Thus, we can say that about one-fourth of the children of this marriage are likely to be deaf. We can also say that a hearing ("normal") child from such a marriage has a $\frac{2}{3}$ chance of being a carrier (that is, two out of three of the offspring with the hearing phenotype are likely to be carriers). We can apply this same method of pedigree analysis and prediction to any genetic trait controlled by a single gene locus.

The most common lethal genetic disease in the United States is **cystic fibrosis.** Though the disease affects only about one in 17,000 African Americans and about one in 90,000 Asian Americans, it occurs in approximately one out of every 3,300 Caucasian births (European ancestry). The cystic fibrosis allele is recessive and is carried by about one in every 25 Caucasians. A person with two copies of this allele has cystic fibrosis, which is character-

Table 8.1   Some Autosomal Disorders in Humans

Disorder	Major Symptoms	Incidence	Comments
Recessive disorders			
Albinism	Lack of pigment in skin, hair, and eyes	$\frac{1}{22,000}$	Very easily sunburned
Cystic fibrosis	Excess mucus in lungs, digestive tract, liver; increased susceptibility to infections; death in infancy unless treated	$\frac{1}{1800}$ Caucasians	See pp. 152–153 and Chap. 11
Galactosemia	Accumulation of galactose in tissues, mental retardation; eye and liver damage	$\frac{1}{100,000}$	Treated by eliminating galactose from diet
Phenylketonuria (PKU)	Accumulation of phenylalanine in blood; lack of normal skin pigment; mental retardation	$\frac{1}{10,000}$ in U.S. and Europe	See p. 156
Sickle-cell disease (homozygous)	Sickled red blood cells; damage to many tissues	$\frac{1}{500}$ African Americans	Alleles are codominant; see pp. 159–160
Tay-Sachs disease	Lipid accumulation in brain cells; mental deficiency; blindness; death in childhood	$\frac{1}{3500}$ Jews from central Europe	See Chap. 4
Dominant disorders			
Achondroplasia	Dwarfism	$\frac{1}{25,000}$	See p. 154
Alzheimer's disease (one type)	Mental deterioration; usually strikes late in life	Not known	
Huntington's disease	Mental deterioration and uncontrollable movements; strikes in middle age	$\frac{1}{25,000}$	See p. 154 and Chap. 11
Hypercholesterolemia	Excess cholesterol in blood; heart disease	$\frac{1}{500}$ is heterozygous	Incomplete dominance; see p. 158

ized by an excessive secretion of very thick mucus from the lungs, pancreas, and other organs. This mucus can interfere with breathing, digestion, and liver function and makes the person vulnerable to pneumonia and other infections. Untreated, most children with cystic fibrosis die by the time they are 5 years old. However, a special diet, antibiotics to prevent infection, frequent pounding of the chest and back to clear the lungs, and other treatments can prolong life to adulthood.

Like cystic fibrosis, most genetic disorders are not evenly distributed across all ethnic groups. Such uneven distribution is the result of prolonged geographic isolation of certain populations. For example, the isolated lives of the Martha's Vineyard inhabitants between 1700 and 1900 fostered marriage between close relatives. Consequently, the frequency of deafness remained high, and the deafness allele was not transmitted to outsiders.

With the increased mobility in most societies today, it is relatively unlikely that two carriers of a rare, harmful allele will meet and mate. However, the probability increases greatly if close relatives marry and have children. People with recent common ancestors are more likely to carry the same recessive alleles than are unrelated people. Therefore, a mating of close relatives is more likely to produce offspring homozygous for a harmful recessive trait.

Most societies have taboos and laws forbidding marriage between very close relatives. These rules may have arisen out of the observation that stillbirths and birth defects are more common when parents are closely related. Such effects can also be observed in many types of inbred animals. For example, dogs that have been inbred for appearance may have serious genetic disorders, such as weak hip joints and undesirable behaviors. The detrimental

Figure 8.16 Achondroplasia, a dominant trait. The late David Rappaport, an actor, had this form of dwarfism, in which the head and torso of the body develop normally, but the arms and legs are short.

effects of inbreeding are also seen in some endangered species, such as cheetahs (see Figure 12.21).

Although inbreeding is clearly dangerous, geneticists debate the extent to which human inbreeding increases the risk of inherited diseases. Many harmful mutations have such severe effects that a homozygous embryo spontaneously aborts long before birth—perhaps so early that the miscarriage goes undetected. Furthermore, as some geneticists argue, marriage between relatives is just as likely to concentrate favorable alleles as harmful ones. There are, in fact, some human populations, such as the Tamils of India, in which marriage between first cousins has long been common and has not produced ill effects.

Dominant Disorders Although most harmful alleles are recessive, a number of human disorders are due to dominant alleles. Some are nonlethal conditions, such as extra fingers and toes, or fingers and toes that are webbed.

One serious disorder caused by a dominant allele is **achondroplasia,** a form of dwarfism. The head and torso of the body develop normally, but the arms and legs are short (Figure 8.16). About one out of 25,000 people have achondroplasia. Only heterozygotes, individuals with a single copy of the defective allele, have this disorder. The homozygous dominant genotype causes death of the embryo. Therefore, all those who do not have achondroplasia, more than 99.99% of the population, are homozygous for the normal, recessive allele. This example makes it clear that a dominant allele is not necessarily better than the corresponding recessive allele or likely to be more plentiful in a population.

Dominant alleles that are lethal are, in fact, much less common than lethal recessives. One reason for this difference is that the dominant lethal allele cannot be carried by heterozygotes without affecting them. Many lethal dominant alleles result from mutations in a sperm or egg that subsequently kill the embryo. And if the afflicted individual is born but does not survive long enough to reproduce, he or she will not pass on the lethal allele. This is in contrast to lethal recessive mutations, which are perpetuated from generation to generation by the reproduction of heterozygous carriers.

A lethal dominant allele can escape elimination, however, if it does not cause death until a relatively advanced age. By the time the symptoms become evident, the afflicted individual may have already transmitted the lethal gene to his or her children. **Huntington's disease,** a degeneration of the nervous system that usually does not begin until middle age, is one example. As the disease progresses, it causes uncontrollable movements in all parts of the body. Loss of brain cells leads to loss of memory and judgment and contributes to depression. Loss of motor skills eventually prevents swallowing and speaking. Death usually ensues 10 to 20 years after the onset of symptoms.

Fetal Testing: Spotting Inherited Disorders Early in Pregnancy

Many genetic disorders can be detected before birth. Tests done in conjunction with **amniocentesis** (Figure 8.17a) can determine, between the fourteenth and sixteenth weeks of pregnancy, whether the developing fetus has the condition. By this time, the fetus is about 15 cm (6 inches) long and is surrounded by a pool of liquid called amniotic fluid. A sample of this fluid is withdrawn and tested as described in Figure 8.17a. Some genetic disorders can be detected by immediate biochemical tests because of the pres-

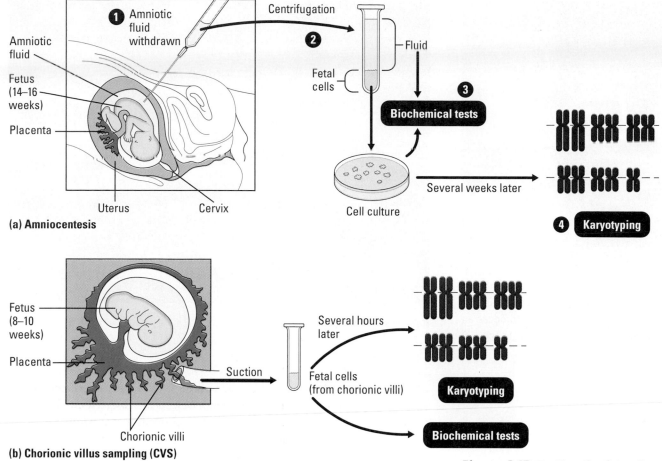

(a) Amniocentesis

① Amniotic fluid withdrawn

Centrifugation

② Fluid

Fetal cells

③ **Biochemical tests**

Cell culture

Several weeks later

④ **Karyotyping**

Amniotic fluid

Fetus (14–16 weeks)

Placenta

Uterus

Cervix

(b) Chorionic villus sampling (CVS)

Fetus (8–10 weeks)

Placenta

Chorionic villi

Suction

Fetal cells (from chorionic villi)

Several hours later

Karyotyping

Biochemical tests

Figure 8.17 Testing the fetus for genetic disorders.
(a) To perform amniocentesis, a physician carefully inserts a needle through the mother's abdomen into her uterus, avoiding the fetus. The physician extracts about 10 milliliters (2 teaspoonsful) of the amniotic fluid. The sample is then centrifuged to separate the fluid from fetal cells suspended in it. (The fetal cells had been sloughed off from the fetus's skin and mouth cavity.) The fluid can be tested immediately for the presence of chemicals that indicate certain disorders. The cells must usually be cultured for several weeks before testing in order to obtain a sufficient number of cells. Dividing cells are then karyotyped to detect chromosomal abnormalities and may be used for biochemical tests. **(b)** In chorionic villus sampling (CVS), the physician inserts a narrow, flexible tube through the mother's vagina and cervix into the uterus and suctions off a small amount of fetal tissue (chorionic villi) from the placenta. These cells can be used for immediate karyotyping and certain biochemical tests.

ence of certain telltale chemicals in the amniotic fluid. Tests for other disorders are performed on fetal cells that are present in the fluid. The cells are usually grown in culture before testing. Biochemical tests can be performed on the cultured cells to reveal conditions such as Tay-Sachs disease (see Table 8.1), and karyotyping is used to detect chromosomal abnormalities such as Down syndrome (see Figure 7.19).

In a newer technique called **chorionic villus sampling (CVS)** (Figure 8.17b), fetal cells are obtained by taking a small piece of fetal tissue from the placenta, the organ that carries nourishment and wastes between the fetus and the mother. The tissue is taken from *chorionic villi,* small, fingerlike projections on the outside of the placenta. Because the cells of chorionic villi are multiplying rapidly, enough cells are undergoing mitosis to allow karyotyping to be carried out immediately. Results of the karyotyping, along with some biochemical tests, are available within a few hours. The speed of CVS is an advantage over amniocentesis. Another advantage is that CVS can be performed early, between the eighth and tenth week of pregnancy. However, CVS is not suitable for tests requiring amniotic fluid, and it is is less widely available than amniocentesis. Recently, medical scientists have developed methods for isolating fetal cells that have escaped into the mother's blood. Although very few in number, the cells can be cultured and then tested.

Other techniques enable a physician to examine a fetus directly for anatomical deformities. One such technique is **ultrasound imaging,** which

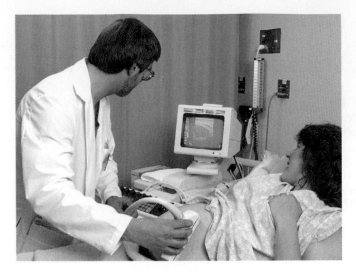

(a) An ultrasound scanner

Figure 8.18 Ultrasound imaging. (a) An ultrasound scanner produces high-frequency sounds, beyond the range of hearing. When the sound waves bounce off the fetus, the echoes produce an image on the monitor. **(b)** This color-enhanced ultrasound image shows a fetus at about 18 weeks.

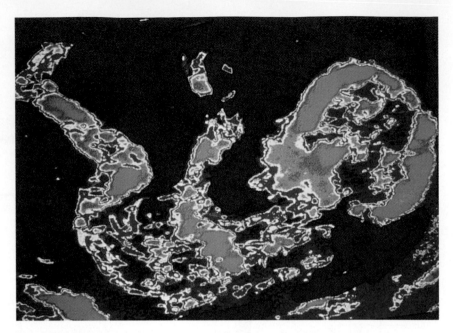

(b) Ultrasound image

uses sound waves to produce an image of the fetus (Figure 8.18). Ultrasound imaging is noninvasive (no foreign objects are inserted into the mother's body) and has no known risk. This procedure is used during amniocentesis and CVS to determine the position of the fetus as well as the position of the needle or tube. With another technique, **fetoscopy,** a needle-thin tube containing a viewing scope is inserted into the uterus, giving the physician a direct view of the fetus.

Fetoscopy, amniocentesis, and chorionic villus sampling all involve some risk of complications, such as maternal bleeding, miscarriage, or premature birth. The complication rate for fetoscopy is the highest (up to about 10%). Complication rates for CVS and amniocentesis are about 2% and 1%, respectively. Because of the risks, all these techniques are usually reserved for situations in which the possibility of a genetic disorder or other type of birth defect is significant. For example, amniocentesis or chorionic villus sampling to test for Down syndrome is usually reserved for pregnant women age 35 or older.

Family histories, blood tests, genetic counseling, and fetal testing offer couples a great deal of information about their unborn children. If fetal tests reveal a genetic disorder that cannot be helped by routine surgery or other therapy, the parents must choose between terminating the pregnancy and preparing themselves for a baby with a serious birth defect. CVS provides a chance to make a decision while the fetus is still quite young.

For some, no matter how undeveloped the fetus, an abortion for any reason is an unthinkable waste of human life. For others, it is a regrettable but permissible way to prevent suffering and avoid a situation that could adversely affect their own lives, the life of the child, and the lives of other family members and close friends. Many genetic disorders, even treatable ones, require constant parental vigilance. Treatment for phenylketonuria (PKU), for instance, demands a strict diet during early childhood and reduced intake of the amino acid phenylalanine throughout life (see Table 8.1). A disorder such as Down syndrome can lead to the agony of childhood leukemia or heart attack.

1. What is meant by "self-fertilization" of a plant?

2. How can two plants that have different genotypes for flower color be identical in phenotype?

3. Which of the following is Mendel's principle of segregation?

   a. Each pair of alleles segregates independently during gamete formation.

   b. Alleles segregate during gamete formation; fertilization creates pairs of alleles once again.

4. An imaginary creature called a glump has blue skin, controlled by the allele B, which is dominant to allele b. Glumps of genotype bb are albino (white). The independently inherited characteristic of eye color is specified by alleles E and e. Glumps of genotype EE have orange eyes; glumps that are homozygous recessives for the eye color gene have pink eyes.

   a. If a homozygous blue-skinned glump mates with an albino glump, what will be the phenotypes of their offspring?

   b. You use a testcross to determine the genotype of an orange-eyed glump. Half of the offspring of the testcross have orange eyes and half have pink eyes. What is the genotype of the orange-eyed parent?

   c. A glump of genotype BbEe is crossed with one that is bbEe. What is the probability of their producing an albino baby glump with pink eyes?

5. A man and a woman who are both carriers of cystic fibrosis allele c, which is recessive to the normal allele C, have had three children without cystic fibrosis. If the couple has a fourth child, what is the probability that the child will have the disorder?

6. Peter is a 28-year-old man whose father died of Huntington's disease, a dominant disorder whose symptoms usually begin to appear between ages 30 and 50. Neither Peter's much older sister, who is 44, nor his 60-year-old mother show any signs of the disease. What is the probability that Peter has inherited Huntington's disease?

7. Recall from Chapter 7 that Down syndrome is caused by an extra copy of chromosome 21. Which technique would allow earliest diagnosis of a fetus: amniocentesis, chorionic villus sampling, or fetoscopy?

Answers: 1. An egg cell in the plant's carpel is fertilized by a sperm-carrying pollen grain from a stamen of the same plant. **2.** One could be homozygous for the dominant allele, while the other is heterozygous. **3. b** (Statement a is the principle of independent assortment.) **4a.** All will have blue skin ($BB \times bb \rightarrow Bb$). **4b.** Ee **4c.** $\frac{1}{8}$ (that is, $\frac{1}{2} \times \frac{1}{4}$) **5.** $\frac{1}{4}$ (The genotypes and phenotypes of their other children are irrelevant.) **6.** $\frac{1}{2}$ **7.** Chorionic villus sampling

Variations on Mendel's Principles

Mendel's two principles explain inheritance in terms of discrete factors—genes—that are passed along from generation to generation according to simple rules of probability. Mendel's principles are valid for all sexually reproducing organisms, including garden peas, birds, and human beings. But just as the basic rules of musical harmony cannot account for all the rich sonorities of a symphony, Mendel's principles stop short of explaining some patterns of genetic inheritance. In particular, they do not explain the many cases of inherited characteristics that exist in more than two clear-cut variants—

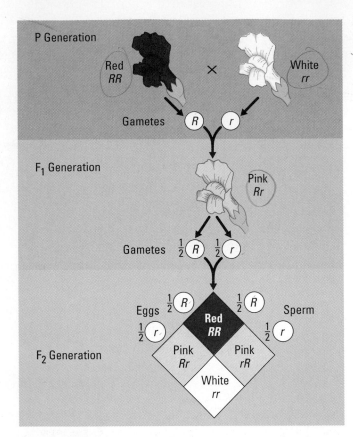

Red
RR

×

White
rr

Gametes

R   r

F₁ Generation

Pink
Rr

Gametes   ½ R   ½ r

Eggs   ½ R   ½ R   Sperm

½ r   ½ r

F₂ Generation

Red
RR

Pink
Rr

Pink
rR

White
rr

Figure 8.19 Incomplete dominance in snapdragons.
In incomplete dominance, a heterozygote has a phenotype in between the phenotypes of the two kinds of homozygotes. The phenotypic ratios in the F₂ generation are the same as the genotypic ratios, 1:2:1. Compare this diagram with Figure 8.7, where one of the alleles displays *complete* dominance.

Aliens demonstrate incomplete dominance in Web/CD Activity 8C.

such as flower color that can be red, white, or pink or human skin color in all its range of shades. We will now add extensions to Mendel's principles that account for these.

Incomplete Dominance in Plants and People

The F₁ offspring of Mendel's crosses always looked like one of the two parent plants because the dominant alleles had the same effect in one or two copies. But for some characteristics, the F₁ hybrids have an appearance in between the phenotypes of the two parents, an effect called **incomplete dominance.** For instance, when red snapdragons are crossed with white snapdragons, all the F₁ hybrids have pink flowers (Figure 8.19). And in the F₂ generation, the genotypic ratio and the phenotypic ratio are the same, 1:2:1.

We also see examples of incomplete dominance in humans. One case involves the recessive allele (*h*) responsible for **hypercholesterolemia,** a condition characterized by dangerously high levels of cholesterol in the blood. Normal individuals are *HH*. Heterozygotes (*Hh;* about one in 500 people) have blood cholesterol levels about twice normal. They are unusually prone to atherosclerosis, the blockage of arteries by cholesterol buildup in artery walls, and they may have heart attacks from blocked heart arteries by their mid-30s. Hypercholesterolemia is even more serious in homozygous individuals (*hh;* about one in a million people). These homozygotes have about five times the normal amount of blood cholesterol and may have heart attacks as early as age 2.

If we look at the molecular basis for hypercholesterolemia, we can understand the intermediate phenotype of heterozygotes (Figure 8.20). The *H* allele specifies a cell surface receptor protein that certain cells use to mop up excess cholesterol from the blood. With only half as many receptors as *HH* individuals, heterozygotes can remove much less of the excess cholesterol.

Multiple Alleles and Blood Type

So far, we have discussed inheritance patterns involving only two alleles per gene. But many genes have multiple alleles. Although each individual carries, at most, two different alleles for a particular gene, in cases of multiple alleles, more than two possible alleles exist in the population. The **ABO blood groups** in humans are examples of multiple alleles. There are three alleles for the characteristic of ABO blood type, which in various combinations produce four phenotypes. A person's blood type may be O, A, B, or AB. These letters refer to two carbohydrates, designated A and B, which are found on the surface of red blood cells. A person's red blood cells may be coated with one substance or the other (type A or B), with both (type AB), or with neither (type O). Matching compatible blood groups is critical for blood transfusions. If a donor's blood cells have a carbohydrate (A or B) that is foreign to the recipient, then proteins called antibodies in the recipient's blood bind specifically to the

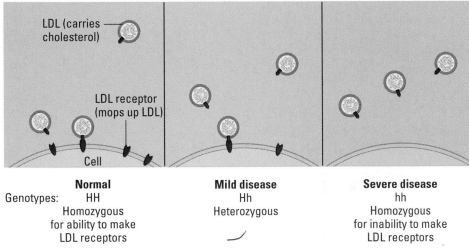

LDL (carries cholesterol)

LDL receptor (mops up LDL)

Cell

Normal	**Mild disease**	**Severe disease**
Genotypes: HH	Hh	hh
Homozygous for ability to make LDL receptors	Heterozygous	Homozygous for inability to make LDL receptors

Figure 8.20 Incomplete dominance in human hypercholesterolemia. The dominant allele, which normal individuals carry in duplicate (*HH*), specifies a cell surface protein called an LDL receptor. LDLs, or low-density lipoproteins, are cholesterol-containing particles in the blood. The LDL receptors pick up LDL particles from the blood and promote their uptake by cells that break down the cholesterol (see Figure 4.30). This process helps prevent the accumulation of cholesterol in the arteries. Heterozygotes (*Hh*) have only half the normal number of LDL receptors, and homozygotes (*hh*) have none, allowing dangerous levels of LDL to build up in the blood.

foreign carbohydrates and cause the donated blood cells to clump together. This clumping can kill the recipient. It is also the basis of blood-typing (Figure 8.21). You'll see blood cell clumping as shown in the figure if you type your own blood in the laboratory.

The four blood types result from various combinations of the three different alleles, symbolized as I^A (for the ability to make substance A), I^B (for B), and i (for neither A nor B). Each person inherits one of these alleles from each parent. Because there are three alleles, there are six possible genotypes, as listed in the figure. Both the I^A and I^B alleles are dominant to the i allele. Thus, $I^A I^A$ and $I^A i$ people have type A blood, and $I^B I^B$ and $I^B i$ people have type B. Recessive homozygotes (ii), have type O blood; they make neither the A nor the B substance. The I^A and I^B alleles exhibit **codominance,** meaning that both alleles are expressed in heterozygous individuals ($I^A I^B$), who have type AB blood.

Blood Group (Phenotype)	Genotypes	Antibodies Present in Blood	Reaction When Blood from Groups Below Is Mixed with Antibodies from Groups at Left			
			O	A	B	AB
O	ii	Anti-A Anti-B				
A	$I^A I^A$ or $I^A i$	Anti-B				
B	$I^B I^B$ or $I^B i$	Anti-A				
AB	$I^A I^B$	—				

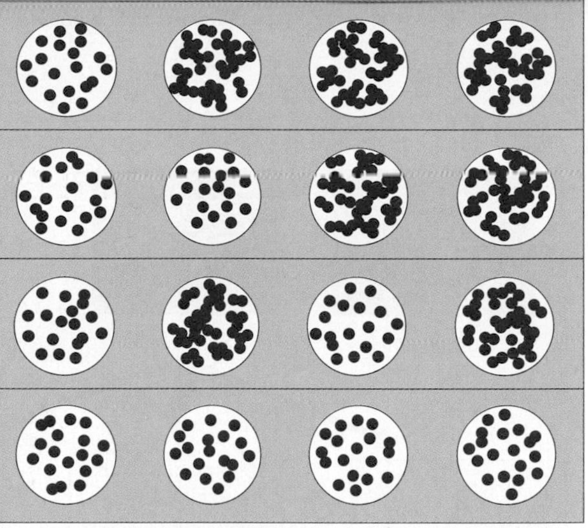

Figure 8.21 Multiple alleles for the ABO blood groups. For the gene that controls a person's ABO blood type, three alleles exist in the human population: I^A, I^B, and i. Because each person carries two alleles, six genotypes are possible. Whenever the I^A or I^B allele is present, the corresponding carbohydrate (A or B, respectively) is present on the surface of red blood cells. Both of these alleles, which are codominant, are dominant to the i allele, which does not specify any surface carbohydrate. A person produces antibodies against foreign blood carbohydrates. When these antibodies meet up with red blood cells carrying the corresponding carbohydrate—anti-A with A, for example—the cells clump. This is the basis of blood-typing and of the adverse reaction that occurs when someone receives a transfusion of incompatible blood.

How Pleiotropy Explains Sickle-Cell Disease

All of our genetic examples to this point have been cases in which each gene specified one hereditary characteristic. But in many cases, one gene influences several characteristics. The impact of a single gene on more than one characteristic is called **pleiotropy** (from the Greek *pleion*, more).

An example of pleiotropy in humans is **sickle-cell disease,** a disorder characterized by the diverse symptoms shown in Figure 8.22 (next page). All of these possible phenotypic effects result from the action of a single kind of allele when it is present on both homologous chromosomes. The direct effect of the sickle-cell allele is to make red blood cells produce abnormal hemoglobin molecules. As mentioned in Chapter 3, these molecules tend to link together and crystallize, especially when the oxygen content of the blood is lower than usual because of high altitude, overexertion, or respiratory ailments. As the hemoglobin crystallizes, the normally disk-shaped red blood cells deform to a sickle shape with jagged edges. Sickling of the cells, in turn, can lead to the cascade of symptoms in Figure 8.22. Blood transfusions and certain drugs may relieve some of the symptoms, but there is no cure, and sickle-cell disease kills about 100,000 people in the world annually.

In most cases, only people who are homozygous for the sickle-cell allele suffer from the disease. Heterozygotes, who have one sickle-cell allele and one nonsickle allele, are usually healthy, although in rare cases they may experience some pleiotropic effects when oxygen in the blood is severely reduced, such as at very high altitudes. These effects may occur because the nonsickle and sickle-cell alleles are actually codominant: Both alleles are expressed in heterozygous individuals, and the red blood cells contain both normal and abnormal hemoglobin. Heterozygotes are said to have "sickle-cell trait." A simple blood test can identify homozygotes and heterozygotes (Figure 8.23).

Sickle-cell disease is by far the most common inherited illness among black people, striking one in 500 African American children born in the

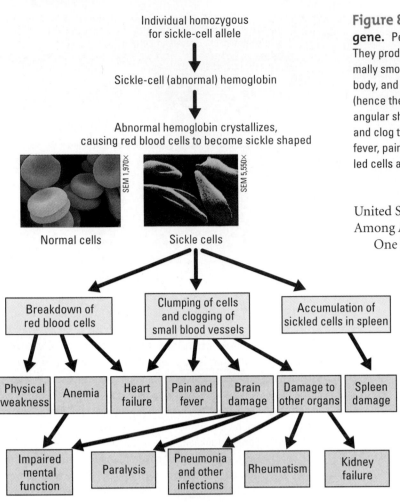

Individual homozygous
for sickle-cell allele

Sickle-cell (abnormal) hemoglobin

Abnormal hemoglobin crystallizes,
causing red blood cells to become sickle shaped

SEM 1,970×

SEM 5,550×

Normal cells

Sickle cells

Breakdown of
red blood cells

Clumping of cells
and clogging of
small blood vessels

Accumulation of
sickled cells in spleen

Physical weakness

Anemia

Heart failure

Pain and fever

Brain damage

Damage to other organs

Spleen damage

Impaired mental function

Paralysis

Pneumonia and other infections

Rheumatism

Kidney failure

Figure 8.22 Sickle-cell disease, multiple effects of a single human gene. People who are homozygous for the sickle-cell allele have sickle-cell disease. They produce abnormal hemoglobin molecules, which cause their red blood cells, normally smooth disks, to become sickle-shaped. Sickled cells are destroyed rapidly by the body, and their destruction may cause anemia and general weakening of the body (hence the alternative name for the disorder, sickle-cell anemia). Also, because of their angular shape, sickled cells do not flow smoothly in the blood and tend to accumulate and clog tiny blood vessels. Blood flow to body parts is reduced, resulting in periodic fever, pain, and damage to various organs, including the heart, brain, and kidneys. Sickled cells accumulate in the spleen, damaging that organ.

United States. About one in ten African Americans is a heterozygote. Among Americans of other ancestry, the sickle-cell allele is very rare.

One in ten is an unusually high frequency of heterozygotes for an allele with such harmful effects in homozygotes. We might expect that the frequency of the sickle-cell allele in the population would be much lower because many homozygotes die before passing their genes to the next generation. The high frequency appears to be a vestige of the roots of African Americans. Sickle-cell disease is most common in tropical Africa, where the deadly disease malaria is also prevalent. The microorganism that causes malaria spends part of its life cycle inside red blood cells. When it enters those of a person with the sickle-cell allele, it triggers sickling. The body destroys most of the sickled cells, and the microbe does not grow well in those that remain. Consequently, sickle-cell carriers (heterozygotes) are resistant to malaria, and in many parts of Africa they live longer and have more offspring than noncarriers who are exposed to malaria. In this way, malaria has kept the frequency of the sickle-cell allele relatively high in much of the African continent. To put it in evolutionary terms, as long as malaria is a danger, individuals with the sickle-cell allele have a selective advantage. (You'll learn more about this topic in Chapters 9 and 12.)

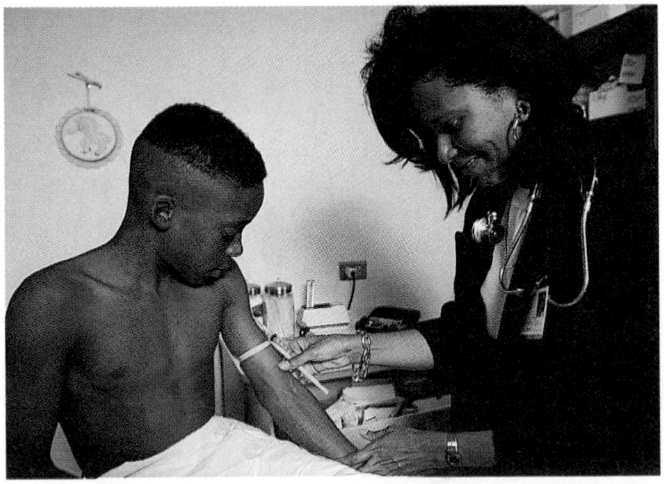

Polygenic Inheritance

Mendel studied genetic characteristics that could be classified on an either-or basis, such as purple or white flower color. However, many characteristics, such as human skin color and height, vary along a continuum in a population. Many such features result from **polygenic inheritance,** the additive effects of two or more genes on a single phenotypic characteristic. (This is the converse of pleiotropy, in which a single gene affects several characteristics.)

There is evidence, for instance, that skin pigmentation in humans is controlled by several genes. Let's consider how three hypothetical genes could produce some of the continuous variation we observe in human skin pigmentation. Assume that the three genes are inherited separately, like Mendel's pea genes. The "dark-skin" allele for each gene (*A*, *B*, and *C*) contributes one "unit" of darkness to the phenotype and is incompletely dominant to the other alleles (*a*, *b*, and *c*). A person who is *AABBCC* would be very dark, while an *aabbcc* individual would be very light. An *AaBbCc* person (resulting, for example, from a marriage between an *AABBCC* person and an *aabbcc* person) would have skin of an intermediate shade. Because the alleles have an additive effect, the genotype *AaBbCc* would produce the same skin color as any other genotype with just three dark-skin alleles, such

Figure 8.23 Collection of a blood sample for sickle-cell testing. In combination with community education and family counseling, early diagnosis and treatment can help people with sickle-cell disease avoid some of the potential effects listed in Figure 8.22.

Figure 8.24 A model for polygenic inheritance of skin color. According to this model, three separately inherited genes affect the darkness of skin. For each gene, there is an allele for dark skin (*A*, *B*, or *C*) that is incompletely dominant to an allele for light skin (*a*, *b*, or *c*). Each dominant allele brings one "unit" of skin pigmentation. An individual who is heterozygous for all three genes (*AaBbCc*) or has some other combination of three dominant alleles and three recessive alleles has three units of darkness, indicated in the figure by three black dots in the diamonds. Thus, diamonds representing the triply heterozygous F$_1$ individuals have three black dots. The Punnett square shows all possible genotypes of F$_2$ offspring. The row of diamonds below the Punnett square shows the seven skin pigmentation phenotypes that would theoretically result. The seven bars in the graph at the bottom of the figure depict the relative numbers of each of the phenotypes in the F$_2$ generation. The bell-shaped curve indicates the distribution of an even greater variety of skin shades in the population that might result from the combination of heredity and environmental effects such as sun-tanning.

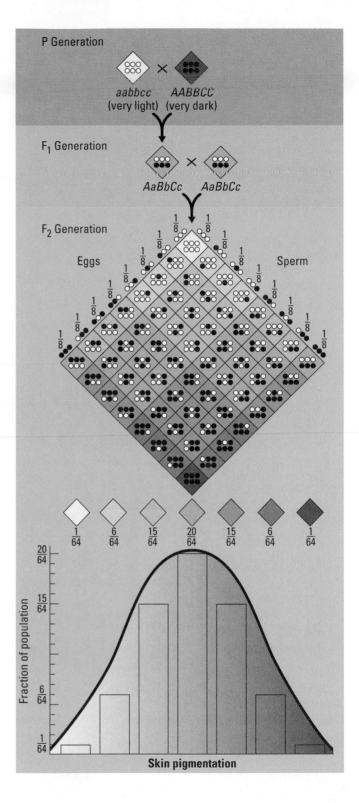

as *AABbcc*. Figure 8.24 shows how inheritance of these three genes could lead to a range of skin color in a population. Seven levels of pigmentation would arise at the frequencies indicated by the bars in the graph.

If this were a real human population, the skin colors would form even more of a continuum, perhaps similar to the entire spectrum of color under the bell-shaped curve in the graph. Shades intermediate between the seven variants shown in the row of diamonds would result from the effects of environmental factors, such as sun-tanning. Of course, any effect of environment would not be passed on to the next generation.

The Chromosomal Basis of Inheritance

Mendel published his results in 1866, but not until long after he died did biologists understand the significance of his work. Cell biologists worked out the processes of mitosis and meiosis in the late 1800s (see Chapter 7 to review these processes). Then, around 1900, researchers began to notice parallels between the behavior of chromosomes and the behavior of Mendel's heritable factors. One of biology's most important concepts—the chromosome theory of inheritance—was emerging.

The **chromosome theory of inheritance** states that genes are located on chromosomes and that the behavior of chromosomes during meiosis and

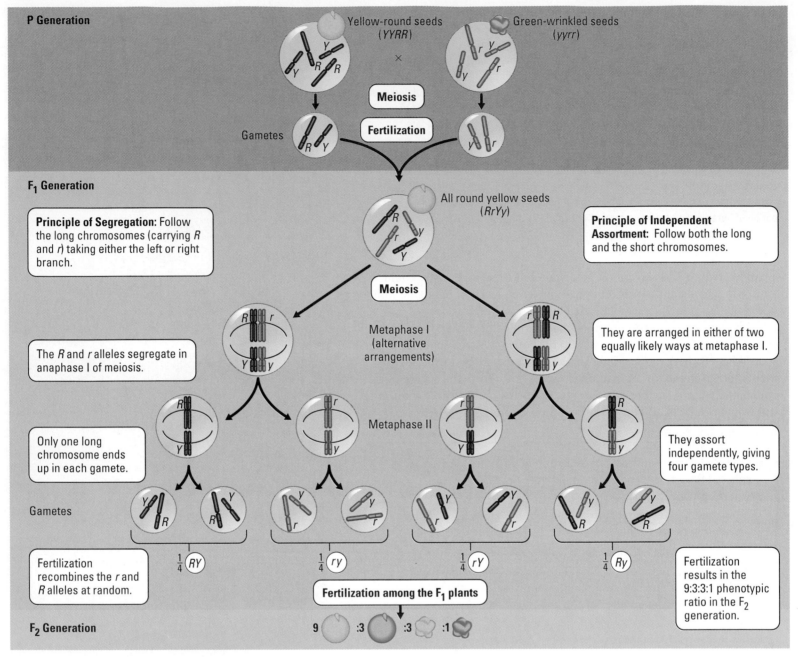

Figure 8.25 The chromosomal basis of Mendel's principles. Here we correlate the results of one of Mendel's dihybrid crosses (see Figure 8.9) with the behavior of chromosomes. Starting with two true-breeding parental plants, the diagram follows two genes through the F₁ and F₂ generations. The two genes specify seed shape (alleles *R* and *r*) and seed color (alleles *Y* and *y*) and are on different chromosomes.

fertilization accounts for inheritance patterns. Indeed, it is chromosomes that undergo segregation and independent assortment during meiosis and thus account for Mendel's principles. We can see the chromosomal basis of Mendel's principles by following the fates of two genes during meiosis and fertilization in pea plants (Figure 8.25).

Beyond Mendel: Gene Linkage

In 1908, British biologists William Bateson and Reginald Punnett (originator of the Punnett square) discovered an inheritance pattern that seemed

(a) Experiment:

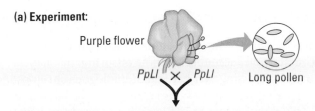

Purple flower

$PpLl$ × $PpLl$

Long pollen

Phenotypes	Observed offspring	Prediction (9:3:3:1)
Purple long	284	215
Purple round	21	71
Red long	21	71
Red round	55	24

Figure 8.26 Linked genes and crossing over. **(a)** In a famous experiment, Bateson and Punnett looked at the progeny of sweet-pea plants heterozygous for flower color (*Pp*) and pollen shape (*Ll*). The parent plants had purple flowers (expression of the dominant *P* allele) and long pollen grains (expression of the *L* allele). The corresponding recessive traits are red flowers (in *pp* plants) and round pollen (in *ll* plants). When they tabulated the phenotypes of their 381 new plants, the results were consistent with Mendel's principle of segregation but not with Mendel's principle of assortment. They did not see the 9:3:3:1 ratio predicted for a dihybrid cross. Instead, as shown in the table, they found a disproportionately large number of plants with either purple flowers and long pollen (284 of 381) or red flowers and round pollen (55 of 381). **(b)** The disproportionately large numbers of two of the phenotypes among the offspring is explained by gene linkage, the fact that the gene loci for the two characteristics are on the same chromosome and tend to remain together during meiosis and fertilization. In the starting plants in this experiment, alleles *P* and *L* were together on one chromosome and *p* and *l* on its homologue. **(c)** The small numbers of the other two offspring phenotypes is explained by crossing over during gamete formation. Some of the gametes end up with recombinant chromosomes, carrying new combinations of alleles, either *P* and *l* or *p* and *L*. When the recombinant gametes participate in fertilization, recombinant offspring can result.

totally inconsistent with Mendelian principles. Bateson and Punnett were working with two characteristics in sweet-pea plants, flower color and pollen shape. In an experiment that at first seemed just like one of Mendel's dihybrid crosses, they bred plants that were heterozygous for both characteristics. These plants exhibited the dominant traits of purple flowers and long pollen grains, as expected. But when the biologists crossed these (F_1) heterozygotes, the offspring (F_2) plants gave them a surprise. The plants did not show the 9:3:3:1 ratio predicted for a dihybrid cross. Instead, there were too many plants in two of the phenotypic categories and too few plants in the other two categories (Figure 8.26a).

These results were explained several years later, when other studies revealed that the sweet-pea genes for flower color and pollen shape are on the same chromosome (Figure 8.26b). Alleles that start out together on the same chromosome—such as *P* with *L* and *p* with *l*—would be expected to travel together during meiosis and fertilization. In general, genes that are located close together on a chromosome, called **linked genes,** do tend to be inherited together. They do not follow Mendel's principle of independent assortment. In the Bateson-Punnett experiment, meiosis in the heterozygous sweet-pea plant yielded mostly two genotypes of gametes (*PL* and *pl*), rather than equal numbers of the four types of gametes that would result if the flower-color and pollen-shape genes were not linked. The large numbers of plants with purple long and red round traits in the experiment resulted from fertilization among these two types of gametes.

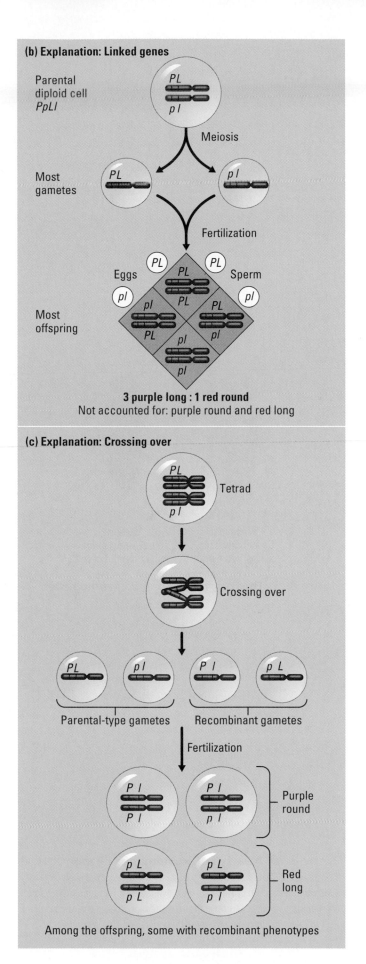

(b) Explanation: Linked genes

Parental diploid cell *PpLl*

PL
p l

Meiosis

Most gametes

PL

p l

Fertilization

Eggs

PL Sperm
p l

Most offspring

3 purple long : 1 red round
Not accounted for: purple round and red long

(c) Explanation: Crossing over

PL
p l
Tetrad

Crossing over

PL   *p l*   *P l*   *p L*

Parental-type gametes   Recombinant gametes

Fertilization

P l
P l

P l
p l

} Purple round

p L
p L

p L
p l

} Red long

Among the offspring, some with recombinant phenotypes

Genetic Recombination: Consequences of Crossing Over

But what about the smaller numbers of plants with purple round and red long traits? Crossing over accounts for these offspring. In Chapter 7, we saw that during meiosis, crossing over between homologous chromosomes shuffles chromosome segments in the haploid daughter cells (see Figure 7.18) and hence produces new combinations of alleles. Figure 8.26c reviews crossing over, showing how two linked genes can give rise to four different gamete genotypes. Two of the gamete genotypes reflect the presence of parental-type chromosomes, which have not been altered by crossing over. In contrast, the other two gamete genotypes are recombinant. The chromosomes of these gametes carry new combinations of alleles that result from the exchange of chromosome segments in crossing over. So in the Bateson-Punnett experiment, the small fraction of offspring plants with recombinant (purple round and red long) phenotypes must have resulted from fertilization involving recombinant gametes. The percentage of recombinant offspring among the total is called the **recombination frequency**. In this case, it is the sum of the recombinant plants, 21 + 21, divided by 381, or 11%.

> They're back! See aliens with linked genes in Web/CD Activity 8D.

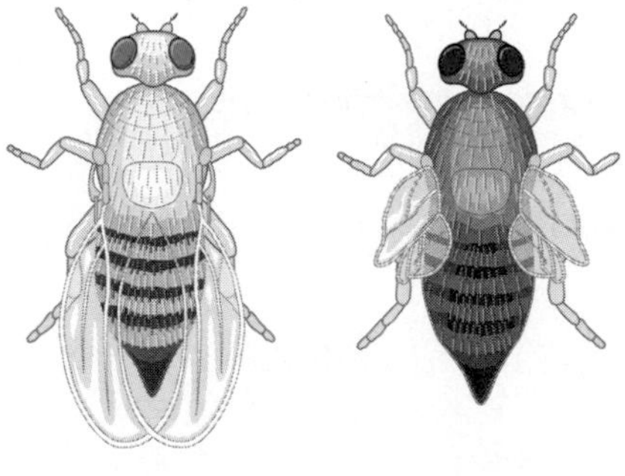

T. H. Morgan and Fruit Fly Genetics

The discovery of how crossing over adds to gamete diversity confirmed the relationship between chromosome behavior and inheritance. Some of the most important early studies of crossing over were performed in the laboratory of American embryologist Thomas Hunt Morgan in the early 1900s. Morgan and his colleagues used the fruit fly *Drosophila melanogaster* in many of their experiments. Often seen flying around overripe fruit, *Drosophila* is a good research animal for studies of inheritance (Figure 8.27). It can be grown in small containers on a mixture of cornmeal and molasses and will produce hundreds of offspring in a few weeks. Using fruit flies, geneticists can trace the inheritance of a trait through several generations in a matter of months.

Working mostly with *Drosophila*, T. H. Morgan and his students (Figure 8.28) produced a virtual explosion in genetics. Among their achievements was an approach for using crossover data to map gene loci, the invention of

Wild type	Variant
(gray body, long wings, red eyes)	(black body, short wings, cinnabar eyes)

Figure 8.27 The fruit fly *Drosophila melanogaster*.

Figure 8.28 A party in T. H. Morgan's "fly room." In this photo, Morgan (back row, far right), several students, and a skeleton are celebrating the return of Alfred H. Sturtevant (left foreground) from World War I military service. Sturtevant later made a number of major contributions to genetics, including a way to map genes on chromosomes.

Alfred H. Sturtevant. Sturtevant started by assuming that the chance of crossing over is approximately equal at all points on a chromosome. He then hypothesized that the farther apart two genes are on a chromosome, the higher the probability that a crossover would occur between them. His reasoning was elegantly simple: The greater the distance between two genes, the more points there are between them where crossing over can occur. With this principle in mind, Sturtevant began using recombination data from fruit fly crosses to assign to genes relative positions on chromosomes —that is, to map genes. Figure 8.29 shows the reasoning Sturtevant used to map three genes that reside on one of the *Drosophila* chromsomes. The result is called a **linkage map.**

Years later, it was learned that Sturtevant's assumption that crossovers are equally likely at all points on a chromosome was not exactly correct. Still, his method of mapping genes worked, and it proved extremely valuable in establishing the relative positions of many other fruit fly genes. Eventually, enough data were accumulated to reveal that *Drosophila* has four groups of genes, corresponding to its four pairs of chromosomes.

Today, with DNA technology, geneticists can determine the physical distances in nucleotides between linked genes. These newer gene maps generally confirm the relative positions established by Sturtevant's mapping method.

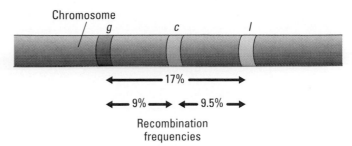

Figure 8.29 Using crossover data to map genes. This diagram represents a part of the *Drosophila* chromosome that carries three linked genes. Their recessive alleles specify black body (*g*), cinnabar eyes (*c*), and short wings (*l*). The corresponding dominant alleles specify the wild-type traits of gray body, red eyes, and long wings. Under the chromosome are the actual recombination frequencies (crossover frequencies) between these genes, taken two at a time: 17% between *g* and *l*, 9% between *g* and *c*, and 9.5% between *l* and *c*. Sturtevant reasoned that these values represent the relative distances between the genes. Because the recombination frequencies between *g* and *c* and between *l* and *c* are approximately half that between *g* and *l*, gene *c* must lie roughly midway between *g* and *l*. So the sequence of these genes on their chromosome must be *g-c-l* (or the equivalent *l-c-g*).

CheckPoint

1. Which of Mendel's principles have their physical basis in the following phases of meiosis?

   a. the orientation of homologous chromosome pairs in metaphase I

   b. the separation of homologues in anaphase I

2. What are linked genes?

3. If the order of three genes on a chromosome is *A-B-C,* between which two genes will the recombination frequency be highest?

Answers: 1a. The principle of independent assortment **1b.** The principle of segregation **2.** Genes on the same chromosome that tend to be inherited together **3.** Between *A* and *C*

Sex Chromosomes and Sex-Linked Genes

The patterns of inheritance we've discussed so far have involved only genes located on autosomes, not on the sex chromsomes. We're now ready to look at the role of sex chromosomes in humans and fruit flies and the inheritance patterns exhibited by the traits they control.

Sex Determination in Humans and Fruit Flies

A number of organisms, including fruit flies and all mammals, have a pair of **sex chromosomes,** designated X and Y, that determine an individual's sex. Figure 8.30 reviews what you learned in Chapter 7 about sex determination in humans. Individuals with one X chromosome and one Y chromosome are males; XX individuals are females. Human males and females both have 44 autosomes (nonsex chromosomes). As a result of chromosome segregation during meiosis, each gamete contains one sex chromosome and a haploid set of autosomes (22 in humans). All eggs contain a single X chromosome.

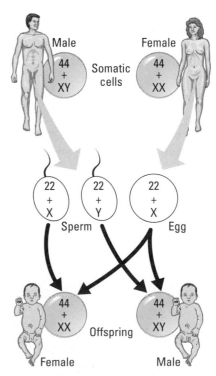

Figure 8.30 Sex determination in humans: A review. See the text at the left for explanation.

Of the sperm cells, half contain an X chromosome and half contain a Y chromosome. An offspring's sex depends on whether the sperm cell that fertilizes the egg bears an X or a Y. The same is true for *Drosophila*.

The genetic basis of sex determination in humans is not yet completely understood, but one gene on the Y chromosome plays a crucial role. This gene, discovered by a British research team in 1990, is called *SRY* and triggers testis development. In the absence of a functioning version of *SRY*, an individual develops ovaries rather than testes. Other genes on the Y chromosome are also necessary for normal sperm production. The X–Y system in other mammals is similar to that in humans. In the fruit fly's X–Y system, some genetic details are different, although a Y chromosome is still essential for sperm formation.

Sex-Linked Genes

Search for alien sex-linked genes in Web/CD Activity 8E.

Besides bearing genes that determine sex, the so-called sex chromosomes also contain genes for characteristics unrelated to maleness or femaleness. Any gene located on a sex chromosome is called a **sex-linked gene.** (Note that the use of "linked" here is somewhat different from its use in the term *linked genes.*) Most sex-linked genes unrelated to sex determination are found on the X chromosome.

Sex linkage was discovered by T. H. Morgan while he was studying the inheritance of white eye color in fruit flies. (In fact, this was the first association of a specific gene with a specific chromosome.) White eye color turned out to be a recessive trait whose gene is on the fly's X chromosome; the gene and trait are said to be X-linked (or sex-linked). Wild-type fruit flies have red eyes; white eyes are very rare (Figure 8.31).

Figure 8.32 shows the eye color inheritance patterns that Morgan observed, along with the genotypes that he eventually deduced. When he mated a white-eyed male fly with a red-eyed female, all the offspring had red eyes, suggesting that the wild type was dominant (Figure 8.32a). And when he bred those offspring to each other, he got the classical 3:1 phenotypic ratio of wild type to white eyed among the offspring (Figure 8.32b). However, there was a surprising twist: The white-eye trait showed up only in males. All the females had red eyes, while half the males had red eyes and half had white eyes. Somehow, Morgan realized, a fly's eye color was tied to its sex.

From this and other evidence, Morgan deduced that the gene involved in this inheritance pattern is located exclusively on the X chromosome; there is no corresponding eye color locus on the Y. Thus, females (XX) carry two copies of the gene for this characteristic, while males (XY) carry only one. Because the white-eye allele is recessive, a female will have white eyes only if she receives that allele on both X chromosomes. For a male, on the other hand, a single copy of the white-eye allele confers white eyes. Since a male has only one X chromosome, there can be no wild-type allele present to offset the recessive allele.

Sex-Linked Disorders in Humans

Fruit fly genetics has taught us much about human inheritance. A number of human conditions, including red-green color blindness, hemophilia, and a type of muscular dystrophy, result from sex-linked (X-linked) recessive alleles that are inherited in the same way as the white-eye allele in fruit flies.

(a) **(b)**

Figure 8.31 Fruit-fly eye color. (a) The wild-type color for fruit fly eyes is red. **(b)** White eye color is a rare variant. Study of the inheritance patterns of the white-eye trait led T. H. Morgan to the conclusion that the gene controlling it was carried on the fly's X chromosome. Such a gene and trait are said to be sex-linked.

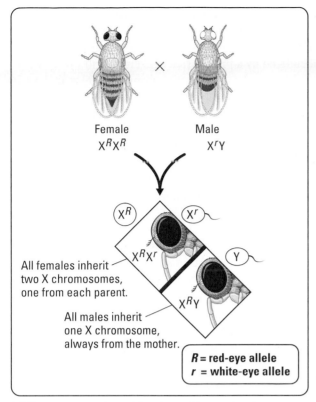

All females inherit two X chromosomes, one from each parent.

All males inherit one X chromosome, always from the mother.

R = red-eye allele
r = white-eye allele

(a) Homozygous red-eyed female × white-eyed male

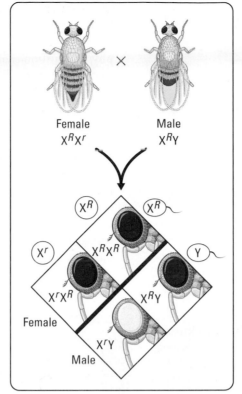

Female

Male

(b) Heterozygous female × red-eyed male

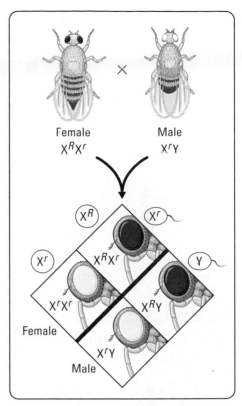

Female

Male

(c) Heterozygous female × white-eyed male

Figure 8.32 Inheritance of white eye color, a sex-linked trait. We use the uppercase letter *R* for the dominant, wild-type, red-eye allele and *r* for the recessive, white-eye allele. To indicate that these alleles are on the X chromosome, we show them as superscripts to the letter X. Thus, red-eyed male fruit flies have the genotype $X^R Y$; white-eyed males are $X^r Y$. The Y chromosome does not have a gene locus for eye color; therefore, the male's phenotype results entirely from his single X-linked gene. In the female, $X^R X^R$ and $X^R X^r$ flies have red eyes, and $X^r X^r$ flies have white eyes.

The fruit fly model also shows us why recessive sex-linked traits are expressed much more frequently in men than in women. Like a male fruit fly, if a man inherits only one sex-linked recessive allele—from his mother—the allele will be expressed. In contrast, a woman has to inherit two such alleles—one from each parent—to exhibit the trait.

Red-green color blindness is a common sex-linked disorder characterized by a malfunction of light-sensitive cells in the eyes. It is actually a class of disorders involving several X-linked genes. A person with normal color vision can see more than 150 colors. In contrast, someone with red-green color blindness can see fewer than 25. For some affected people, red hues appear gray; others see gray instead of green; still others are green-weak or red-weak, tending to confuse shades of these colors. Mostly males are affected, but heterozygous females have some defects. (If you have red-green color blindness, you probably cannot see the numeral in Figure 8.33.)

Hemophilia is a sex-linked recessive trait with a long, well-documented history. Hemophiliacs bleed excessively when injured because they have inherited an abnormal allele for a factor involved in blood clotting. The most seriously affected individuals may bleed to death after relatively minor bruises or cuts.

A high incidence of hemophilia has plagued the royal families of Europe. The first royal hemophiliac seems to have been a son of Queen Victoria (1819–1901) of England. It is likely that the hemophilia allele arose through a mutation in one of the gametes of Victoria's mother or father, making Victoria a carrier of the deadly allele. Hemophilia was eventually introduced into the royal families of Prussia, Russia, and Spain through the marriages of two of Victoria's daughters who were carriers. In this way, the age-old practice of strengthening international alliances by marriage effectively

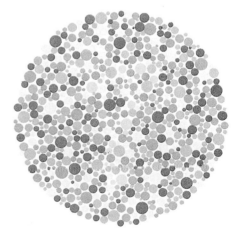

Figure 8.33 A test for red-green color blindness. Can you see a green numeral 7 against the reddish background? If not, you probably have some form of red-green color blindness, an X-linked trait.

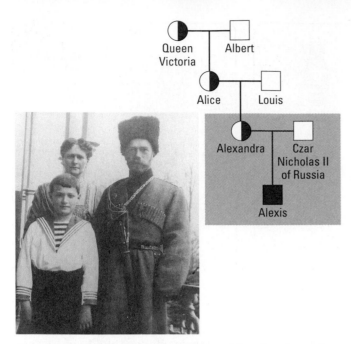

Figure 8.34 Hemophilia in the royal family of Russia. The photograph shows Queen Victoria's granddaughter Alexandra, her husband Nicholas, who was the last czar of Russia, and their son Alexis. The pedigree uses half-colored symbols to represent heterozygous carriers of the hemophilia allele. As you can see in the pedigree, Alexandra, like her mother and grandmother, was a carrier, and Alexis had the disease.

spread hemophilia through the royal families of several nations (Figure 8.34).

Another sex-linked recessive disorder is **Duchenne muscular dystrophy,** a condition characterized by a progressive weakening and loss of muscle tissue. Almost all cases are males, and the first symptoms appear in early childhood, when the child begins to have difficulty standing up. He is inevitably wheelchair-bound by age 12. Eventually, he becomes severely wasted, and normal breathing becomes difficult. Death usually occurs by age 20. For such a severe disease, Duchenne muscular dystrophy is relatively common. In the general U.S. population, about one in 3500 male babies are affected, and the disease is even more common in some inbred populations. In one Amish community in Indiana, for instance, one out of every 100 males are born with the disease.

With the help of DNA technology (discussed in Chapter 11), the gene whose defectiveness causes Duchenne muscular dystrophy has been mapped at a particular point on the X chromosome. The gene's wild-type allele codes for a protein that is present in normal muscle but missing in Duchenne patients.

Evolution Link:
The Telltale Y Chromosome

The Y chromosome of human males is only about one-third the size of the X chromosome and carries only $\frac{1}{100}$ as many genes. As mentioned earlier, most of the Y genes seem to function in maleness and male fertility and are not present on the X. In prophase I of meiosis, only two tiny regions of the X and Y chromosomes can cross over (recombine). Crossing over requires that the DNA in the recombining regions line up and match very closely, and for the human X and Y chromosomes, this can only happen at their tips.

Nevertheless, biologists believe that X and Y were once a fully homologous pair, having evolved from a pair of autosomes about 300 million years ago. Since that time, four major episodes of change, the most recent about 40 million years ago, have rearranged pieces of the Y chromosome in a way that prevents the matching required for recombination with the X. Much of the reshuffling seems to have resulted from inversions (see Figure 7.24c).

Over millions of years, many Y genes have disappeared, shrinking the chromosome. Meanwhile, the lack of substantial exchange between X and Y chromosomes has prevented male-determining genes from migrating to the X chromosome—which could have had the disastrous consequence of making XX individuals male.

Because most of the DNA of the Y chromosome passes more or less intact from father to son—the main changes are rare mutations—the Y chromosome provides a window to the ancestry of the male lines of humanity.

To learn more about the Y chromosome, go to the Web Evolution Link.

Recently, researchers used comparisons of Y DNA to confirm the claim by an African tribe, the Lemba of southern Africa, to descent from ancient Jews (Figure 8.35). Certain Y DNA sequences had previously been shown to be distinctive of Jews and in particular of the priestly caste called Cohanim (descendants of Moses' brother Aaron, according to the Bible). These same sequences were found at equally high frequencies among the Lemba. And it was a study of Y DNA that confirmed the oral tradition among the descendants of the slave Sally Hemmings that Thomas Jefferson was their ancestor. On a grander scale, Y chromosome studies are providing evidence bearing on the question of when and where fully modern humans first evolved.

Figure 8.35 A Lemba tribesman. DNA sequences from the Y chromosomes of Lemba men suggest that the Lemba are descended from ancient Jews.

Chapter Review

Summary of Key Concepts

Overview: Heritable Variation and Patterns of Inheritance

- Breeders of budgies can predict the colors of offspring birds by using the genetic principles discovered by Gregor Mendel.
- **Web/CD Activity 8A *Heritable Variation***

Mendel's Principles

- **In an Abbey Garden** Mendel started with true-breeding varieties of pea plants representing two alternative variants of each of seven hereditary characteristics, such as flower color and seed shape. He then crossed the different varieties and traced the inheritance of traits from generation to generation.

- **Mendel's Principle of Segregation** From the results of experiments tracking one characteristic at a time, Mendel hypothesized that there are alternative forms of genes. He deduced that an organism has two genes for each inherited characteristic, one originating from each parent. Alternative forms of a gene are called alleles. A sperm or egg carries only one allele of each pair, because allele pairs separate when gametes form—Mendel's principle of segregation. If an individual's genotype (genetic makeup) has two different alleles for a gene and only one influences the organism's phenotype (appearance), that allele is said to be dominant and the other allele recessive. We now know that alleles of a gene reside at the same locus, or position, on homologous chromosomes. Where the allele pair match, the organism is homozygous; where they're different, the organism is heterozygous.

- **Mendel's Principle of Independent Assortment** By following two characteristics at once, Mendel found that the alleles of a pair segregate independently of other allele pairs during gamete formation—Mendel's principle of independent assortment.
- **Web/CD Activity 8B/The Process of Science *Gregor's Garden***

- **Using a Testcross to Determine an Unknown Genotype** The offspring of a testcross, a mating between an individual of unknown genotype and a homozygous recessive individual, can reveal the unknown genotype.

- **The Rules of Probability** Inheritance follows the rules of probability. The chance of inheriting a recessive allele from a heterozygous parent is $\frac{1}{2}$. The chance of inheriting it from both of two heterozygous parents is $\frac{1}{2} \times \frac{1}{2} = \frac{1}{4}$, illustrating the rule of multiplication for calculating the probability of two independent events. There are two ways a dominant and recessive allele from heterozygous parents can combine in offspring: the dominant from the father and the recessive from the mother, or the recessive from the father and the dominant from the mother. Because of these two possibilities, the probability of heterozygous parents producing a heterozygous offspring is $\frac{1}{4} + \frac{1}{4} = \frac{1}{2}$, illustrating the rule of addition for calculating the probability of an event that can occur in alternative ways.

- **Family Pedigrees** The inheritance of many human traits, from freckles to genetic diseases, follows Mendel's principles and the rules of probability. In human genetics, we use family pedigrees to determine patterns of inheritance and individual genotypes.

- **Human Disorders Controlled by a Single Gene** Many inherited disorders in humans are controlled by a single gene (represented by two alleles). Most such disorders, such as cystic fibrosis, are caused by autosomal recessive alleles. A few, such as Huntington's disease, are caused by dominant alleles.

- **Fetal Testing: Spotting Inherited Disorders Early in Pregnancy** Testing of fetal cells obtained by amniocentesis or chorionic villus sampling and examination of the fetus with ultrasound or fetoscopy can detect disorders in a fetus.

Variations on Mendel's Principles

- **Incomplete Dominance in Plants and People** When an offspring's phenotype—flower color, for example—is in between the phenotypes of its parents, we say that it displays incomplete dominance.
- **Web/CD Activity 8C** *Incomplete Dominance in Aliens*

- **Multiple Alleles and Blood Type** In a population, there often exist multiple kinds of alleles for a characteristic, such as the three alleles for the ABO blood group. The alleles determining the A and B blood factors are codominant; that is, both are expressed in a heterozygote.

- **How Pleiotropy Explains Sickle-Cell Disease** A single gene may affect phenotype in multiple ways, a phenomenon called pleiotropy. For example, the presence of two copies of the sickle-cell allele at a single gene locus brings about the many symptoms of sickle-cell disease.

- **Polygenic Inheritance** A single characteristic, such as skin color, may be influenced by multiple genes, creating a continuum of phenotypes.

The Chromosomal Basis of Inheritance

- Genes are located on chromosomes, whose behavior during meiosis and fertilization accounts for inheritance patterns (see Figure 8.25 to review).

- **Beyond Mendel: Gene Linkage** Certain genes are linked: They tend to be inherited together because they lie close together on the same chromosome.

- **Genetic Recombination: Consequences of Crossing Over** Crossing over can separate linked alleles, producing gametes with recombinant chromosomes and offspring with recombinant phenotypes.
- **Web/CD Activity 8D** *Linked Genes and Crossing Over in Aliens*

- **T. H. Morgan and Fruit Fly Genetics** In the first decades of the twentieth century, researchers in T. H. Morgan's group used the fruit fly *Drosophila melanogaster* to make important advances in the science of genetics. Among their discoveries was a way of mapping the relative positions of genes on chromosomes, taking advantage of the fact that crossing over between linked genes is more likely to occur between genes that are farther apart.

Sex Chromosomes and Sex-Linked Genes

- **Sex Determination in Humans and Fruit Flies** In humans and many other animals, including fruit flies, a male has one X and one Y sex chromosome, and a female has two X chromosomes. Whether a sperm cell contains an X or a Y determines the sex of the offspring. The Y chromosome has genes for the development of testes, whereas an absence of the Y allows ovaries to develop.

- **Sex-Linked Genes** Genes on the sex chromosomes are said to be sex-linked. However, in both fruit flies and humans, the X chromosome carries many genes unrelated to sex. Their inheritance pattern reflects the fact that females have two homologous X chromosomes, but males have only one.
- **Web/CD Activity 8E** *Sex-Linked Genes in Aliens*

- **Sex-Linked Disorders in Humans** Most sex-linked human disorders, such as red-green color blindness and hemophilia, are due to recessive alleles and are seen mostly in males. A male receiving a single X-linked recessive allele from his mother will have the disorder; a female has to receive the allele from both parents to be affected.

Evolution Link: The Telltale Y Chromosome

- The human Y chromosome seems to have evolved from the X chromosome through rearrangement and loss of genes. Because there is almost no crossing over (genetic recombination) between X and Y during meiosis, the DNA of the Y chromosome almost solely reflects the evolutionary history of males.
- **Web Evolution Link** *The Telltale Y Chromosome*

Self-Quiz

1. Edward was found to be heterozygous (*Ss*) for sickle-cell trait. The alleles represented by the letters *S* and *s* are
    a. on the X and Y chromosomes.
    b. linked.
    c. on homologous chromosomes.
    d. both present in each of Edward's sperm cells.
    e. on the same chromosome but far apart.

2. Whether an allele is dominant or recessive depends on
    a. how common the allele is, relative to other alleles.
    b. whether it is inherited from the mother or the father.
    c. which chromosome it is on.
    d. whether it or another allele determines the phenotype when both are present.
    e. whether or not it is linked to other genes.

3. Two fruit flies with eyes of the usual red color are crossed and their offspring are as follows: 77 red-eyed males, 71 ruby-eyed males, 152 red-eyed females. The gene that controls whether eyes are red or ruby is
    a. autosomal (carried on an autosome) and dominant.
    b. autosomal and recessive.
    c. sex-linked and dominant.
    d. sex-linked and recessive.
    e. impossible to determine without more information.

4. All the offspring of a white hen and a black rooster are gray. The simplest explanation for this pattern of inheritance is
    a. pleiotropy.   d. independent assortment.
    b. sex linkage.   e. incomplete dominance.
    c. linkage.

5. A man who has type B blood and a woman who has type A blood could have children of which of the following phenotypes?
    a. A or B only   d. A, B, or O
    b. AB only   e. A, B, AB, or O
    c. AB or O

• **Go to the website or CD-ROM for more self-quiz questions.**

More Genetics Problems

1. In fruit flies, the genes for wing shape and body stripes are linked. In a fly whose genotype is *WwSs, W* is linked to *S,* and *w* is linked to *s.* Show how this fly can produce gametes containing four different combinations of alleles. Which are parental-type gametes? Which are recombinant gametes? What process produces recombinant gametes?

2. Adult height in humans is at least partially hereditary; tall parents tend to have tall children. But humans come in a range of sizes, not just tall or short. What extension of Mendel's model could produce this variation in height?

3. A brown mouse is repeatedly mated with a white mouse, and all their offspring are brown. If two of these brown offspring are mated, what fraction of the F_2 mice will be brown?

4. How could you determine the genotype of one of the brown F_2 mice in problem 3? How would you know whether a brown mouse is homozygous? Heterozygous?

5. Tim and Jan both have freckles (a dominant trait), but their son Michael does not. Show with a Punnett square how this is possible. If Tim and Jan have two more children, what is the probability that both of them will have freckles?

6. Incomplete dominance is seen in the inheritance of hypercholesterolemia. Mack and Toni are both heterozygous for this characteristic, and both have elevated levels of cholesterol. Their daughter Zoe has a cholesterol level six times normal; she is apparently homozygous, *hh.* What fraction of Mack and Toni's children are likely to have elevated but not extreme levels of cholesterol, like their parents? If Mack and Toni have one more child, what is the probability that the child will suffer from the more serious form of hypercholesterolemia seen in Zoe?

7. A female fruit fly with forked bristles on her body is mated with a male fly with normal bristles. Their offspring are 121 females with normal bristles and 138 males with forked bristles. Explain the inheritance pattern for this trait.

8. A couple are both phenotypically normal, but their son suffers from hemophilia, a sex-linked recessive disorder. What fraction of their children are likely to suffer from hemophilia? What fraction are likely to be carriers?

9. Heather was surprised to discover that she suffered from red-green color blindness. She told her biology professor, who said, "Your father is color-blind too, right?" How did her professor know this? Why did her professor not say the same thing to the color-blind males in the class?

The Process of Science

1. In 1981, a stray cat with unusual curled-back ears was adopted by a family in Lakewood, California. Hundreds of descendants of this cat have since been born, and cat fanciers hope to develop the "curl" cat into a show breed. The curl allele is apparently dominant and carried on an autosome. Suppose you owned the first curl cat and wanted to develop a true-breeding variety. Describe tests that would determine whether the curl gene is dominant or recessive and whether it is autosomal or sex-linked.

2. **Experiment with Mendel's peas in The Process of Science activity available on the website and CD-ROM.**

Biology and Society

Gregor Mendel never saw a gene, yet he concluded that "heritable factors" were responsible for the patterns of inheritance he observed in peas. Similarly, Morgan and Sturtevant never actually observed the linkage of genes on chromosomes. Their maps of *Drosophila* chromosomes (and the very idea that genes are carried on chromosomes) were conceived by observing the patterns of inheritance of linked genes, not by observing the genes directly. Is it legitimate for biologists to claim the existence of objects and processes they cannot actually see? How do scientists know whether an explanation is correct?

Molecular Biology of the Gene

In the replication of our DNA, the new DNA strands grow at a rate of about 50 nucleotides per second; in bacteria, the rate is 500 per second. Because all organisms use the same genetic code, scientists can make a plant glow like a 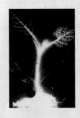 firefly. The loss of a single nucleotide from a 1000-nucleotide gene can completely destroy the gene's function. Many viruses have genes that are not made of DNA.

When you read about sickle-cell disease in Chapter 8, you learned about the variant hemoglobin gene that is ultimately responsible for the many symptoms of the disease. But how does this variant gene—the sickle-cell allele—differ from the normal allele? In this chapter, you will learn the amazingly simple answer to that question as we begin to explore heredity at the molecular level, the subject of *molecular biology*. Here you will learn in detail how the DNA of genes exerts its effects on the cell and on the whole organism. You'll learn, too, how DNA replicates—the molecular basis for the similarities between parents and offspring—and how it can mutate. And because viruses have played a key role in the history of molecular biology and continue to be important, these tiny entities are also a major topic of the chapter.

Overview: The Discovery That DNA Is the Genetic Material

DNA was known as a chemical in cells 100 years ago, but Mendel, Morgan, and other early geneticists did all their work without any knowledge of DNA's role in heredity. So we begin our account of molecular biology with the story of how biologists learned that DNA was the genetic material.

Griffith's Transforming Factor

Our story opens in 1928. That year, English bacteriologist Frederick Griffith reported results of experiments on a species of bacterium that causes pneumonia. Griffith studied two varieties of the bacterium, a pathogenic (disease-causing) one and a variant that was harmless. He found that when he killed pathogenic bacteria with heat and then mixed the cell remains with living bacteria of the harmless variety, some of the living cells were converted to the pathogenic form. Furthermore, this new trait of pathogenicity was inherited by all the descendants of the transformed bacteria (Figure 9.1). Clearly, some substance in the dead pathogenic cells—some "transforming factor"—caused this heritable change, but the identity of the factor was not known.

By the late 1930s, other experimental studies had convinced most biologists that a specific kind of molecule, rather than some complex chemical mixture, was the basis of inheritance. Attention focused on chromosomes, which were already known to carry genes. By the 1940s, scientists knew that chromosomes consisted of two types of chemicals: DNA and protein. Most researchers thought the protein was the material of genes. Proteins were known to be made of 20 kinds of building blocks (amino acids) and to have elaborate and varied structures and functions. DNA, on the other hand, had only four types of building blocks (nucleotides), and its properties seemed too monotonous to account for the multitude of traits inherited by every organism. These arguments were so persuasive that even after researchers established in 1944 that Griffith's transforming factor was DNA, the scientific community remained skeptical.

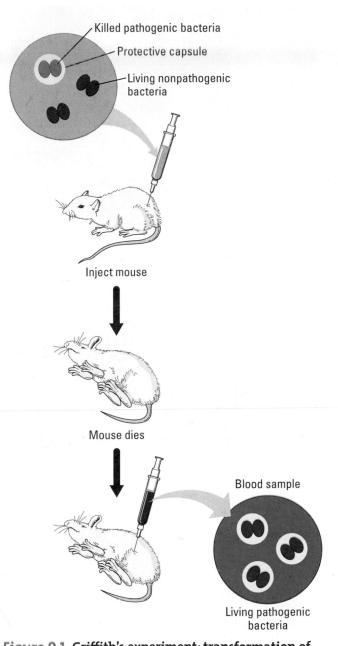

Figure 9.1 Griffith's experiment: transformation of bacteria. In his famous experiment in 1928, Frederick Griffith worked with two strains of the bacterium *Streptococcus pneumoniae,* a pathogenic strain that caused fatal pneumonia when injected into mice and a nonpathogenic strain, which was harmless. He discovered that if he killed a sample of the pathogenic strain with heat and then mixed it with living cells of the nonpathogenic strain, the combination produced pneumonia. He was then able to find living pathogenic bacteria in the blood of the dead mice. Griffith concluded that a substance from the heat-killed pathogenic cells had genetically transformed some of the living nonpathogenic bacteria into pathogens. The pathogenic bacteria differed from the nonpathogenic ones by having an outer capsule that protected them from destruction by the mouse's immune system.

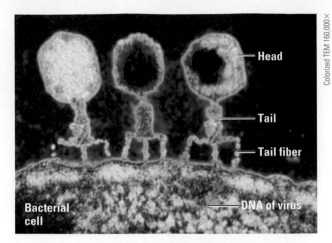

Figure 9.2 Bacteriophages infecting a bacterial cell.
The "landing crafts" settling on the surface of the "planet" in this electron micrograph are actually viruses in the process of infecting a bacterium. The viruses are bacteriophage T4, a close relative of the T2 used by Hershey and Chase, and the bacterium is *E. coli*. The phage consists of a molecule of DNA enclosed within an elaborate structure made of proteins. The "legs" of the phage (called tail fibers) bend when they touch the cell surface. The tail is a hollow rod enclosed in a springlike sheath. As the legs bend, the spring compresses, the bottom of the rod punctures the cell membrane, and the viral DNA passes from inside the head of the virus into the cell. Once inside, the viral DNA takes over the molecular machinery of the cell and directs the manufacture of hundreds of new viruses, which are released when the cell bursts open. Each phage is only about 200 nm tall (two ten-thousandths of a millimeter).

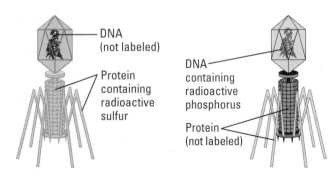

Figure 9.3 Radioactively labeled phages. In preparation for the experiment described in Figure 9.4, Hershey and Chase grew two batches of phage T2, each labeled in a different way. First, they incubated T2 with *E. coli* in a growth medium containing radioactive sulfur. Protein contains sulfur but DNA does not, so as the phages multiplied within the cells, the radioactive sulfur atoms were incorporated only into their proteins (the head exterior, tail, and tail fibers). Next, the researchers grew a separate batch of phages in medium containing radioactive phosphorus. Because nearly all the phages' phosphorus is in their DNA, this procedure labeled the phage DNA.

The Genetic Material of a Bacterial Virus

Gradually, evidence built up in support of DNA as the genetic material, and in 1952, American biologists Alfred Hershey and Martha Chase performed one of the most convincing experiments. They were working with the bacterium *Escherichia coli (E. coli)*, an inhabitant of the mammalian large intestine, and a virus that infects it. Viruses that attack bacteria are called **bacteriophages** (meaning "bacteria-eaters"), or **phages** for short.

Phages and other viruses sit on the fence between life and nonlife. A virus is lifelike in having genes and a highly organized structure; but it differs from a living organism in not being cellular or able to reproduce on its own. Viruses are essentially genetic material enclosed in a protein coat. A virus can survive only by infecting a living cell with genetic material that directs the cell's molecular machinery to make more virus. Because they are much simpler than cellular organisms, viruses are far easier to study on the molecular level than Mendel's peas or Morgan's fruit flies.

Hershey and Chase demonstrated that DNA, and not protein, was the genetic material of a phage called T2 (Figure 9.2). At the time of this work, the biologists already knew that T2 was composed solely of DNA and protein and that it could somehow reprogram its host cell to produce new phages. But they did not know which phage component—DNA or protein—was responsible. To find out, they devised an experiment to determine which phage component enters *E. coli* during infection. Their experiment used only a few, relatively simple tools: chemicals containing radioactive isotopes (see Chapter 2); a device for detecting radioactivity; a kitchen blender; and a centrifuge, an extremely fast merry-go-round for test tubes that is used to separate particles of different weights. (These are still basic tools of molecular biology.) Hershey and Chase used different radioactive isotopes to label the DNA and protein in two separate batches of T2. One batch of T2 had DNA containing atoms of radioactive phosphorus; the other batch had protein containing radioactive sulfur atoms (Figure 9.3).

Armed with the radioactively labeled batches of T2, Hershey and Chase were ready to perform the experiment outlined in Figure 9.4. They found that when they infected *E. coli* with the T2 containing labeled proteins, very little of the radioactivity got into the bacteria. But when they used phages with radioactive DNA, most of the radioactivity entered the bacteria and ended up in progeny phages.

Hershey and Chase concluded that the phage DNA entered the cell but that most of the phage proteins remained outside. More important, they showed that it is the injected DNA molecules that cause the cells to produce additional phage DNA and new phage proteins—indeed, new, complete phages.

See how phage T2 reproduces in Web/CD Activity 9A.

The Hershey-Chase results, added to earlier evidence, convinced the scientific world that DNA was the hereditary material. What came next was one of the most celebrated quests in the history of science—the effort to figure out the structure of DNA.

CheckPoint

How did researchers Hershey and Chase use radioactively labeled viruses to show that DNA, not protein, is the genetic material?

Answer: They prepared two batches of phage T2, one with labeled DNA and one with labeled proteins. When they used these viruses to infect bacterial cells, they found that radioactively labeled viral DNA, but not labeled proteins, entered the host cell and directed the synthesis of a new crop of phages.

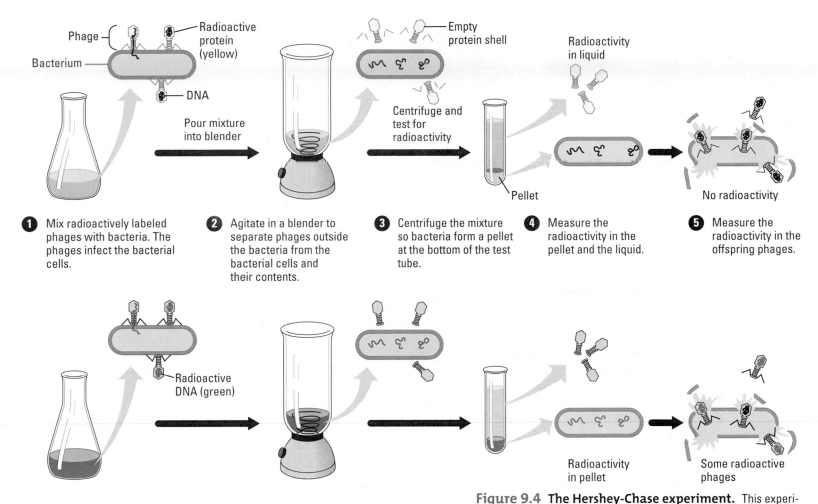

① Mix radioactively labeled phages with bacteria. The phages infect the bacterial cells.

② Agitate in a blender to separate phages outside the bacteria from the bacterial cells and their contents.

③ Centrifuge the mixture so bacteria form a pellet at the bottom of the test tube.

④ Measure the radioactivity in the pellet and the liquid.

⑤ Measure the radioactivity in the offspring phages.

Figure 9.4 The Hershey-Chase experiment. This experiment demonstrated that DNA, not protein, is the genetic material of the phage T2. After preparing the radioactively labeled phages shown in Figure 9.3, ① Hershey and Chase allowed the protein-labeled and DNA-labeled batches of T2 to infect separate samples of nonradioactive bacteria. ② A few minutes later, they agitated the cultures in a blender to shake loose any parts of the phages that remained outside the bacterial cells. ③ They then spun the mixtures in a centrifuge. The bacterial cells, being heavier, were deposited as pellets at the bottom of the centrifuge tubes, but free phages and phage parts remained suspended in the liquid. ④ The researchers measured and compared the radioactivity in the pellets and the liquid. ⑤ They then returned the infected bacteria to liquid medium and allowed the production of progeny phages to proceed. When the progeny phages were released, only the ones from the batch with labeled DNA contained some radioactivity. These results showed that the only part of the T2 phage that enters the cell during infection is its DNA. Therefore, it must be the phage DNA that directs the manufacture of new phages.

The Structure and Replication of DNA

By the time Hershey and Chase performed their experiments, a good deal was already known about DNA. Scientists had identified all its atoms and knew how they were covalently bonded to one another. What was not understood was the specific arrangement that gave DNA its unique properties —the capacity to store genetic information, copy it, and pass it from generation to generation. However, only one year after Hershey and Chase published their results scientists figured out DNA's three-dimensional structure and the basic strategy of how it works. We will describe that momentous discovery shortly. First, let's look at the underlying chemical structure of DNA and its chemical cousin RNA.

DNA and RNA: Polymers of Nucleotides

Recall from Chapter 3 that both DNA and RNA are nucleic acids, which consist of long chains (polymers) of chemical units (monomers) called **nucleotides.** A very simple diagram of such a polymer, or **polynucleotide,** is shown at the left in Figure 9.5 (next page). This sample polynucleotide chain shows only one possible arrangement of the four different types of nucleotides (abbreviated A, C, T, and G) that make up DNA. The nucleotides

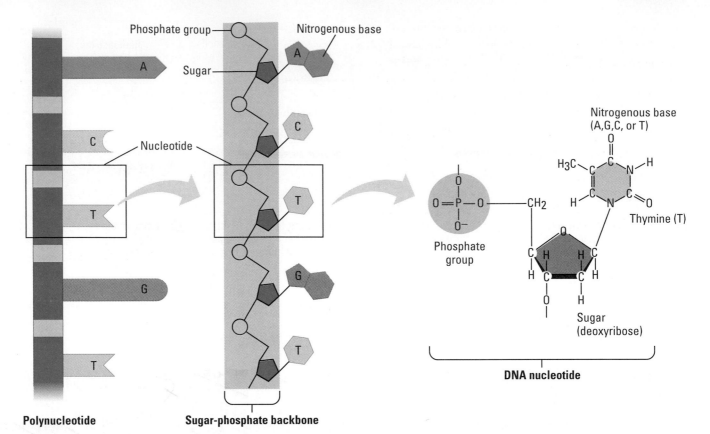

Figure 9.5 The structure of a DNA polynucleotide. The polynucleotide is a chain of nucleotides. Each nucleotide consists of a nitrogenous base, a sugar (blue), and a phosphate group (gold). The nucleotides are linked, the sugar of one connected to the phosphate of the next, forming a sugar-phosphate backbone, with the bases protruding from the sugars. The chemical structure at the right shows the details of a DNA nucleotide. The sugar has five carbon atoms (shown in red for emphasis) and is called deoxyribose. The phosphate group has given up an H⁺ ion, acting as an acid. Hence, the full name for DNA is deoxyribonucleic acid.

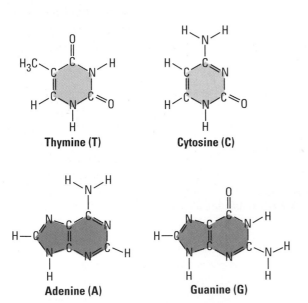

Figure 9.6 The nitrogenous bases of DNA. Thymine and cytosine have single-ring structures. Adenine and guanine have double-ring structures.

of a polynucleotide can occur in any sequence, and polynucleotides vary in length, from long to very long. So the number of possible polynucleotide sequences is very great.

Zooming in on our polynucleotide in Figure 9.5, we see that each nucleotide consists of three components: a nitrogenous base, a sugar (blue), and a phosphate group (gold). The nucleotides are joined to one another by covalent bonds between the sugar of one nucleotide and the phosphate of the next. This results in a **sugar-phosphate backbone,** a repeating pattern of sugar-phosphate-sugar-phosphate. The nitrogenous bases are arranged like appendages all along this backbone.

Examining a single nucleotide even more closely (Figure 9.5, right), we see the chemical structure of its three components. The phosphate group, with a phosphorus atom (P) at its center, is the source of the *acid* in nucleic acid. (The phosphate has already given up a hydrogen ion, H⁺, leaving a negative charge on one of its oxygen atoms.) The sugar has five carbon atoms: four in its ring and one extending above the ring. The ring also includes an oxygen atom. The sugar is called *deoxyribose* because, compared to the sugar ribose, it is missing an oxygen atom. The full name for DNA is *deoxyribonucleic acid,* with the "nucleic" part coming from DNA's location in the nuclei of eukaryotic cells. The nitrogenous base (thymine, in our example) has a ring of nitrogen and carbon atoms with various functional groups attached. In contrast to the acidic phosphate group, nitrogenous bases are basic; hence their name.

The four nucleotides found in DNA differ only in their nitrogenous bases. Figure 9.6 (facing page) shows the structures of DNA's four nitrogenous bases. At this point, the structural details are not as important as the fact that the bases are of two types. **Thymine (T)** and **cytosine (C)** are single-ring structures. **Adenine (A)** and **guanine (G)** are larger, double-ring structures. The one-letter abbreviations can be used for either the bases alone or for the nucleotides containing them.

What about RNA? As its name—ribonucleic acid—implies, its sugar is ribose rather than deoxyribose. Notice the ribose in the RNA nucleotide in Figure 9.7; the sugar ring has an —OH group attached to the C atom at its lower right corner. Another difference between RNA and DNA is that instead of thymine, RNA has a nitrogenous base called **uracil (U).** (Uracil is very similar to thymine.) Except for the presence of ribose and uracil, an RNA polynucleotide chain is identical to a DNA polynucleotide chain.

Watson and Crick's Discovery of the Double Helix

Once biologists were convinced that DNA was the genetic material, knowing the molecule's building blocks was not enough. A race was on to determine how the structure of this molecule could account for its role in heredity. The three-dimensional structures of proteins were starting to yield fascinating clues about the functions of those macromolecules, and biologists hoped that the three-dimensional structure of DNA might do the same. Among the scientists working on the problem were several who had already made discoveries in deciphering protein structure: Linus Pauling in California and Maurice Wilkins and Rosalind Franklin (Figure 9.8a) in London. First to the finish line with DNA, however, were two scientists who were relatively unknown at the time—American James D. Watson and Englishman Francis Crick (Figure 9.8b). In 1962, Watson, Crick, and Wilkins received the Nobel Prize for their work. (Franklin probably would have received the prize as well but for her death from cancer in 1958.)

The celebrated partnership that determined the structure of DNA began soon after the 23-year-old Watson journeyed to Cambridge University, where Crick was studying protein structure with a technique called X-ray crystallography. While visiting the laboratory of Maurice Wilkins at King's College in London, Watson saw an X-ray crystallographic photograph of DNA, produced by Wilkins's colleague Rosalind Franklin. The photograph clearly revealed the basic shape of DNA to be a helix (spiral). On the basis of Watson's later recollection of the photo, he and Crick deduced that the diameter of the helix was uniform. The thickness of the helix suggested that it was made up of two polynucleotide strands—in other words, a **double helix.**

Using wire models of the nucleotides, Watson and Crick began trying to construct a double helix that would conform both to Franklin's data and to what was then known about the chemistry of DNA. After failing to make a satisfactory model that placed the sugar-phosphate backbones inside the double helix, Watson tried putting the backbones on the outside and forcing the nitrogenous bases to swivel to the interior of the molecule. It occurred to him that the four kinds of bases might pair in a specific way. This idea of specific base pairing was a flash of inspiration that enabled Watson and Crick to solve the DNA puzzle.

At first, Watson imagined that the bases paired like with like—for example, A with A, C with C. But that kind of pairing did not fit with the fact that the DNA molecule has a *uniform* diameter. An AA pair (made of double-ringed bases) would be almost twice as wide as a CC pair, causing

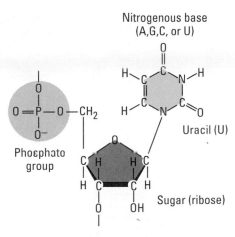

Figure 9.7 An RNA nucleotide. Notice that this RNA nucleotide differs from the DNA nucleotide in Figure 9.5 in two ways. The RNA sugar is ribose rather than deoxyribose, and the base is uracil (U) instead of thymine (T). The other three kinds of RNA nucleotide have the bases A, C, and G, as in DNA.

(a) Rosalind Franklin

(b) James Watson and Francis Crick

Figure 9.8 Discoverers of the double helix. (a) Rosalind Franklin, who performed the X-ray crystallography that revealed the helical nature of the DNA molecule. **(b)** James Watson (left) and Francis Crick, who used Franklin's data to deduce that the DNA helix consisted of two intertwined polynucleotide chains, with hydrogen-bonded base pairs on the inside of the molecule. Watson and Crick are shown in 1953 with their model of DNA.

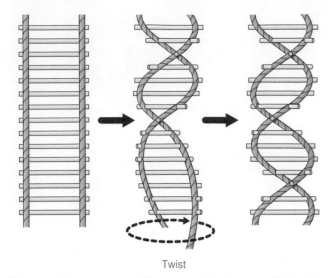

Twist

Figure 9.9 A rope-ladder model for the double helix.
The ropes at the sides represent the sugar-phosphate backbones. Each wooden rung stands for a pair of bases connected by hydrogen bonds.

bulges in the molecule. It soon became apparent that a double-ringed base must always be paired with a single-ringed base on the opposite strand. Moreover, Watson and Crick realized that the individual structures of the bases dictated the pairings even more specifically. Each base has chemical side groups that can best form hydrogen bonds with one appropriate partner (to review the hydrogen bond, see Figure 2.11). Adenine can best form hydrogen bonds with thymine, and guanine with cytosine. In the biologist's shorthand, A pairs with T, and G pairs with C. A is also said to be "complementary" to T, and G to C.

Watson and Crick's pairing scheme not only fit what was known about the physical attributes and chemical bonding of DNA, but also explained some data obtained several years earlier by American biochemist Erwin Chargaff. Chargaff and his co-workers had discovered that the amount of adenine in the DNA of any one species was equal to the amount of thymine and that the amount of guanine was equal to that of cytosine. Chargaff's rules, as they are called, are explained by the fact that A on one of DNA's polynucleotide chains always pairs with T on the other polynucleotide chain, and G on one chain pairs only with C on the other chain.

You can picture the model of the DNA double helix proposed by Watson and Crick as a rope ladder having rigid, wooden rungs, with the ladder twisted into a spiral (Figure 9.9). Figure 9.10 shows three more detailed representations of the double helix. The ribbonlike diagram in part (a) sym-

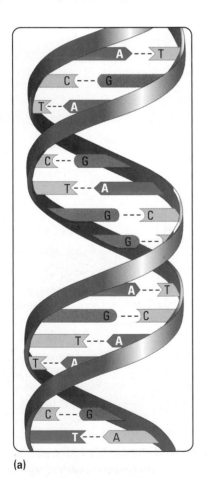

(a)

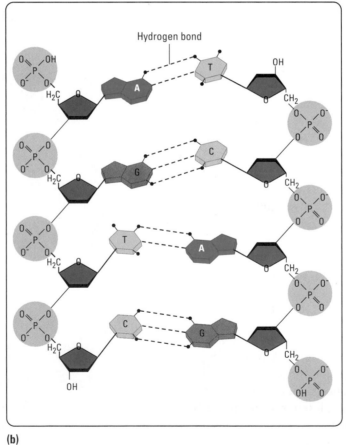

Hydrogen bond

(b)

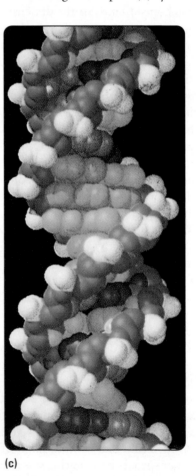

(c)

Figure 9.10 Three representations of DNA. **(a)** In this model, the sugar-phosphate backbones are blue ribbons and the bases are complementary shapes in shades of green and orange. **(b)** In this more chemically detailed structure, you can see the individual hydrogen bonds (dashed lines). You can also see that the strands run in opposite directions; notice that the sugars on the two strands are upside down with respect to each other. **(c)** In this computer graphic, the atoms of the deoxyribose sugars are bright blue, phosphate groups are yellow, and nitrogenous bases are shades of green, light blue, and orange.

Review DNA and RNA structure in Web/CD Activity 9B.

bolizes the bases with shapes that emphasize their complementarity. Part (b) is a more chemical version showing only four base pairs, with the helix untwisted and the individual hydrogen bonds specified by dotted lines; you can see that the two sugar-phosphate backbones of the double helix are oriented in opposite directions. Part (c) is a computer graphic showing every atom of part of a double helix.

Although the Watson-Crick base-pairing rules dictate the side-by-side combinations of nitrogenous bases that form the rungs of the double helix, they place no restrictions on the *sequence* of nucleotides along the length of a DNA strand. In fact, the sequence of bases can vary in countless ways. Consequently, it is not surprising that the DNA of different species, which have different genes, have different proportions of the bases in their DNA.

In April 1953, Watson and Crick shook the scientific world with a succinct, two-page announcement of their molecular model for DNA in the journal *Nature*. Few milestones in the history of biology have had as broad an impact as their double helix, with its AT and CG base pairing.

The Watson-Crick model gave new meaning to the words *gene* and *chromosome*—and to the chromosome theory of inheritance. With a complete picture of DNA, we can see that the genetic information in a chromosome must be encoded in the nucleotide sequence of its DNA molecule. The structure of DNA also suggests a molecular explanation for life's unique properties of reproduction and inheritance, as we see next.

DNA Replication

When a cell or a whole organism reproduces, a complete set of genetic instructions must pass from one generation to the next. For this to occur, there must be a means of copying the instructions. Long before DNA was identified as the genetic material, some people argued that gene replication must be based on a concept called complementary surfaces. According to this idea, when a gene replicates, a "negative image" is created along the original ("positive") surface, just as clay or plaster forms a negative shape when it is packed around an object. The gene's negative image, like the plaster, could serve as a template (mold) for making copies of the original positive image. Photography provides another example of the template principle: A print can be used to make a negative, which can then be used to make copies of the original print. Watson and Crick's model for DNA structure immediately suggested to them that DNA replicated by a template mechanism.

The logic behind the Watson-Crick proposal for how DNA is copied—by specific pairing of complementary bases—is quite simple, as you saw in Chapter 3 (see Figure 3.27). If you know the sequence of bases in one strand of the double helix, you can very simply determine the sequence of bases in the other strand by applying the base-pairing rules: A pairs with T, G with C. Watson and Crick predicted that a cell applies the same rules when copying its genes. Figure 9.11 shows the template hypothesis for a piece of DNA. The two strands of parental DNA separate, and each becomes a template for the assembly of a complementary strand from a supply of free nucleotides. The nucleotides line up one at a time along the template strand in accordance with the base-pairing rules. Enzymes link the nucleotides to form the new DNA strands. The completed new molecules,

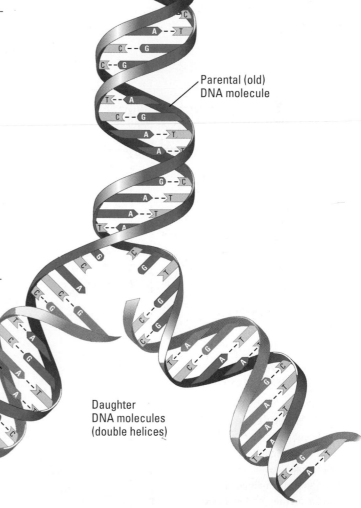

Parental (old) DNA molecule

Daughter (new) strand

Daughter DNA molecules (double helices)

Figure 9.11 A template model for DNA replication. The two strands of the original ("parental") DNA molecule, shown in blue, serve as templates for making new ("daughter") strands (orange). Replication results in two daughter DNA molecules, each consisting of one old strand and one new strand. The parental DNA untwists as its strands separate, and the daughter DNA rewinds as it forms.

Figure 9.12 Damage to DNA by ultraviolet light. The ultraviolet (UV) radiation in sunlight can damage the DNA in skin cells. Fortunately, the cells can repair some of the damage, using enzymes that include some of those that catalyze DNA replication.

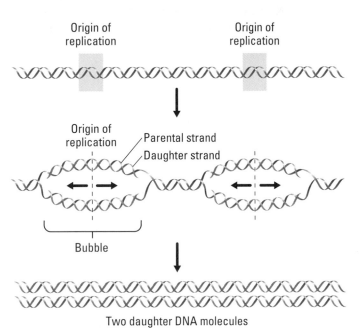

Figure 9.13 Multiple "bubbles" in replicating DNA. DNA replication begins at sites called origins of replication and proceeds in both directions, producing the "bubbles" shown here. The long DNA molecules of eukaryotic chromosomes each have many origins, leading to multiple bubbles as they replicate. Eventually, all the bubbles merge, producing two completed daughter DNA molecules.

identical to the parental molecule, are known as daughter DNA. This hypothesis was confirmed by experiments performed in the 1950s.

Although the general mechanism of DNA replication is conceptually simple, the actual process is complex and requires the cooperation of more than a dozen enzymes and other proteins. The enzymes that actually make the covalent bonds between the nucleotides of a new DNA strand are called **DNA polymerases.** As an incoming nucleotide base-pairs with its complement on the template strand, a DNA polymerase adds it to the end of the growing daughter strand (polymer). The process is both fast and amazingly accurate; typically, only about one in a billion nucleotides in DNA is incorrectly paired. In addition to their roles in DNA replication, DNA polymerases and some of the associated proteins are also involved in repairing damaged DNA. DNA can be harmed by toxic chemicals in the environment or by high-energy radiation, such as X-rays and ultraviolet light (Figure 9.12).

DNA replication begins at specific sites on a double helix, called origins of replication. It then proceeds in both directions, creating what are called replication "bubbles" (Figure 9.13). The parental DNA strands open up as daughter strands elongate on both sides of each bubble. The DNA molecule of a eukaryotic chromosome has many origins where replication can start simultaneously, shortening the total time needed for the process. Eventually, all the bubbles merge, yielding two completed double-stranded daughter DNA molecules.

> Build your own DNA molecule in Web/CD Activity 9C.

DNA replication ensures that all the somatic cells in a multicellular organism carry the same genetic information. It is also the means by which genetic instructions are copied for the next generation of the organism.

CheckPoint

1. Compare and contrast DNA with RNA.

2. Along one strand of a DNA double helix is the nucleotide sequence GGCATAGGT. What is the sequence for the other DNA strand?

3. How does complementary base pairing make possible the replication of DNA?

4. What is the function of DNA polymerase in DNA replication?

Answers: 1. Both are polymers of nucleotides. A nucleotide consists of a sugar + a nitrogenous base + a phosphate group. In RNA, the sugar is ribose; in DNA, it is deoxyribose. Both RNA and DNA have the bases A, G, and C; for a fourth base, DNA has T and RNA has U. **2.** CCGTATCCA **3.** When the two strands of the double helix separate, each serves as a template on which nucleotides can be arranged by specific base pairing into new complementary strands. **4.** This enzyme covalently connects nucleotides one at a time to one end of a growing daughter strand, as the nucleotides line up along a template strand according to the base-pairing rules.

The Flow of Genetic Information from DNA to RNA to Protein

We are now ready to address the question of how DNA functions as the inherited directions for a cell and for the organism as a whole. What exactly are the instructions carried by the DNA, and how are these instructions carried out? You learned the general answers to these questions in Chapter 4, but here we will explore them in more detail.

How an Organism's DNA Genotype Produces Its Phenotype

Knowing the structure of DNA, we can now define genotype and phenotype more precisely than we did in Chapter 8. An organism's *genotype,* its genetic makeup, is the sequence of nucleotide bases in its DNA. The *phenotype* is the organism's specific traits. The molecular basis of the phenotype lies in proteins with a variety of functions. For example, structural proteins help make up the body of an organism, and enzymes catalyze its metabolic activities.

What is the connection between the genotype and the protein molecules that more directly determine the phenotype? As you read in Chapter 4, DNA specifies the synthesis of proteins. A gene does not build a protein directly, but rather dispatches instructions in the form of RNA, which in turn programs protein synthesis. This central concept in biology is summarized in Figure 9.14. The chain of command is from DNA in the nucleus of the cell to RNA to protein synthesis in the cytoplasm. The two main stages are **transcription,** the transfer of genetic information from DNA into an RNA molecule, and **translation,** the transfer of the information in the RNA into a protein.

Get an overview of protein synthesis in Web/CD Activity 9D.

The relationship between genes and proteins was first proposed in 1909, when English physician Archibald Garrod suggested that genes dictate phenotypes through enzymes, the proteins that catalyze chemical processes. Garrod's idea came from his observations of inherited diseases. He hypothesized that an inherited disease reflects a person's inability to make a particular enzyme, and he referred to such diseases as "inborn errors of metabolism." He gave as one example the hereditary condition called alkaptonuria, in which the urine appears dark red because it contains a chemical called alkapton. Garrod reasoned that normal individuals have an enzyme that breaks down alkapton, whereas alkaptonuric individuals lack the enzyme. Garrod's hypothesis was ahead of its time, but research conducted decades later proved him right for alkaptonuria and many other genetic diseases. In the intervening years, biochemists accumulated evidence that cells make and break down biologically important molecules via metabolic pathways, as in the synthesis of an amino acid or the breakdown of a sugar. As you learned in Unit One, each step in a metabolic pathway is catalyzed by a specific enzyme. If a person lacks one of the enzymes, the pathway cannot be completed.

The major breakthrough in demonstrating the relationship between genes and enzymes came in the 1940s from the work of American geneticists George Beadle and Edward Tatum with the orange bread mold *Neurospora crassa* (Figure 9.15). Beadle and Tatum studied strains of the mold that were unable to grow on the usual simple growth medium. Each of these strains turned out to lack an enzyme in a metabolic pathway that produced some molecule the mold needed, such as an amino acid. Beadle and Tatum also showed that each mutant was defective in a single gene. Accordingly, they formulated the one gene–one enzyme hypothesis, which states that the function of an individual gene is to dictate the production of a specific enzyme.

The one gene–one enzyme hypothesis has been amply confirmed, but with some important modifications. First it was extended beyond enzymes to include *all* types of proteins. For example, alpha-keratin, the structural protein of your hair, and the structural proteins that make up the outside of phage T2 are as much the products of genes as enzymes are. So biologists

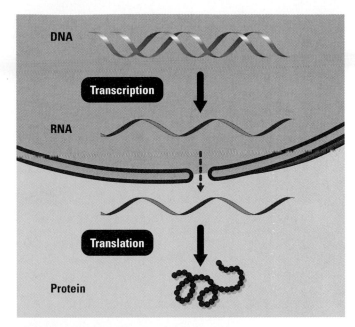

Figure 9.14 The flow of genetic information in a eukaryotic cell: a review. A sequence of nucleotides in the DNA is transcribed into a molecule of RNA in the cell's nucleus (purple area). The RNA travels to the cytoplasm (blue-green area), where it is translated into the specific amino acid sequence of a protein.

Figure 9.15 *Neurospora crassa* growing in a culture dish. Studies of this mold led Beadle and Tatum to their one gene–one enzyme hypothesis, the idea that a gene is the entity that directs the manufacture of a particular enzyme.

soon began to think in terms of one gene–one *protein*. Then it was discovered that many proteins have two or more different polypeptide chains (see Figure 3.22), and each polypeptide is specified by its own gene. Thus, Beadle and Tatum's hypothesis has come to be restated as one gene–one *polypeptide*.

From Nucleotide Sequence to Amino Acid Sequence: An Overview

Stating that genetic information in DNA is transcribed into RNA and then translated into polypeptides does not explain *how* these processes occur. Transcription and translation are linguistic terms, and it is useful to think of nucleic acids and polypeptides as having languages, too. To understand how genetic information passes from genotype to phenotype, we need to see how the chemical language of DNA is translated into the different chemical language of polypeptides.

What exactly is the language of nucleic acids? Both DNA and RNA are polymers made of monomers in specific sequences that carry information, much as specific sequences of letters carry information in English or Russian, for example. In DNA, the monomers are the four types of nucleotides, which differ in their nitrogenous bases (A, T, C, and G). The same is true for RNA, although it has the base U instead of T.

DNA's language is written as a linear sequence of nucleotide bases, a sequence such as the one you see on the enlarged DNA strand in Figure 9.16. Specific sequences of bases, each with a beginning and an end, make up the genes on a DNA strand. A typical gene consists of thousands of nucleotides, and a DNA molecule may contain thousands of genes.

When DNA is transcribed, the result is an RNA molecule. The process is called transcription because the nucleic acid language of DNA has simply been rewritten as a sequence of bases of RNA; the language is still that of nucleic acids. When a reporter transcribes a political speech, the language remains the same, although it changes its form from spoken to written. The nucleotide bases of the RNA molecule are complementary to those on the DNA strand. As you will soon see, this is because the RNA was synthesized using the DNA as a template.

Translation is the conversion of the nucleic acid language into the polypeptide language. Like nucleic acids, polypeptides are polymers, but the monomers that make them up—the letters of the polypeptide alphabet—are the 20 amino acids common to all organisms. Again, the language is written in a linear sequence, and the sequence of nucleotides of the RNA molecule dictates the sequence of amino acids of the polypeptide. But remember, RNA is only a messenger; the genetic information that dictates the amino acid sequence is based in DNA.

What are the rules for translating the RNA message into a polypeptide? What is the correspondence between the nucleotides of an RNA molecule and the amino acids of a polypeptide? Keep in mind that there are only four different kinds of nucleotides in DNA (A, G, C, T) and RNA (A, G, C, U). In translation, these four must somehow specify 20 amino acids. If each nucleotide base coded for one amino acid, only 4 of the 20 amino acids could be accounted for. What if the language consisted of two-letter code words? If we read the bases of a gene two at a time, AG, for example, could specify one amino acid, while AT could designate a different amino acid. However, when the four bases are taken two by two, there are only 16 (that is, 4^2) possible arrangements—still not enough to specify all 20 amino acids.

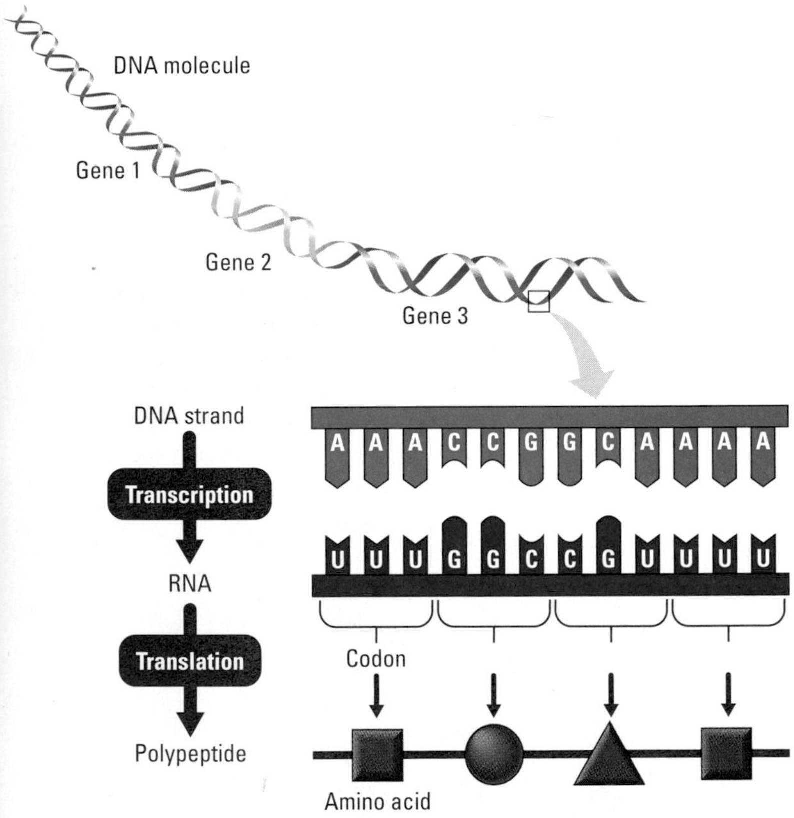

Figure 9.16 Transcription and translation of codons. This figure focuses on a small region of one of the genes carried by a DNA molecule. The enlarged segment from one strand of gene 3 shows its specific sequence of bases. The red strand underneath represents the results of transcription: an RNA molecule. Its base sequence is complementary to that of the DNA. The purple chain represents the results of translation: a polypeptide. The brackets indicate that *three* RNA nucleotides code for each amino acid.

Triplets of bases are the smallest "words" of uniform length that can specify all the amino acids. There can be 64 (that is, 4^3) possible code words of this type—more than enough to specify the 20 amino acids. Indeed, there are enough triplets to allow more than one coding for each amino acid. For example, the base triplets AAT and AAC could both code for the same amino acid—and, in fact, they do.

Experiments have verified that the flow of information from gene to protein is based on a *triplet* code. The genetic instructions for the amino acid sequence of a polypeptide chain are written in DNA and RNA as a series of three-base words called **codons.** Three-base codons in the DNA are transcribed into complementary three-base codons in the RNA, and then the RNA codons are translated into amino acids that form a polypeptide. Next we turn to the codons themselves.

The Genetic Code

In 1799, a large stone tablet was found in Rosetta, Egypt, carrying the same lengthy inscription in three ancient scripts: Greek, Egyptian hieroglyphics, and Egyptian written in a simplified script. This stone provided the key that enabled scholars to crack the previously indecipherable hieroglyphic code.

In cracking the **genetic code,** the set of rules relating nucleotide sequence to amino acid sequence, scientists wrote their own Rosetta stone. It was based on a series of elegant experiments that revealed the amino acid translations of each of the nucleotide-triplet code words. The first codon was deciphered in 1961 by American biochemist Marshall Nirenberg. He synthesized an artificial RNA molecule by linking together identical RNA nucleotides having uracil as their base. No matter where this message started or stopped, it could contain only one type of triplet codon: UUU. Nirenberg added this "poly U" to a test-tube mixture containing ribosomes and the other ingredients required for polypeptide synthesis. This mixture translated the poly U into a polypeptide containing a single kind of amino acid, phenylalanine. In this way, Nirenberg learned that the RNA codon UUU specifies the amino acid phenylalanine (Phe). By variations on this method, the amino acids specified by all the codons were determined.

As Figure 9.17 shows, 61 of the 64 triplets code for amino acids. The triplet AUG has a dual function: It not only codes for the amino acid methionine (Met) but also can provide a signal for the start of a polypeptide chain. Three of the other codons do not designate amino acids. They are the stop codons that instruct the ribosomes to end the polypeptide.

Notice in Figure 9.17 that there is redundancy in the code but no ambiguity. For example, although codons UUU and UUC both specify phenylalanine (redundancy), neither of them ever represents any other amino acid (no ambiguity). The codons in the figure are the triplets found in RNA. They have a straightforward, complementary relationship to the codons in DNA. The nucleotides making up the codons occur in a linear order along the DNA and RNA, with no gaps or "punctuation" separating the codons.

Almost all of the genetic code is shared by all organisms, from the simplest bacteria to the most complex plants and animals. In experiments, genes can be transcribed and translated after transfer from one species to another, even when the organisms are as different as bacterium and human, or firefly and tobacco plant (Figure 9.18). The universality of the genetic vocabulary suggests that it arose very early in evolution and was passed on over the eons to all the organisms living on Earth today.

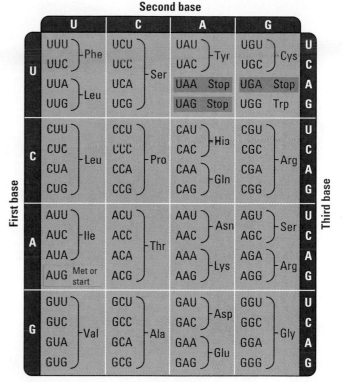

Figure 9.17 The dictionary of the genetic code (RNA codons). The three bases of an RNA codon are designated here as the first, second, and third bases. Practice using this dictionary by finding the codon UGG. This is the only codon for the amino acid tryptophan, but most amino acids are specified by two or more codons. For example, both UUU and UUC stand for the amino acid phenylalanine (Phe). Notice that the codon AUG not only stands for the amino acid methionine (Met) but also functions as a signal to "start" translating the RNA at that place. Three of the 64 codons function as "stop" signals. Any one of these termination codons marks the end of a genetic message.

Figure 9.18 A tobacco plant expressing a firefly gene. Because diverse organisms share a common genetic code, it is possible to program one species to produce a protein characteristic of another species by transplanting DNA. This photo shows the results of an experiment in which researchers incorporated a gene from a firefly into the DNA of a tobacco plant. The gene codes for the firefly enzyme that causes a glow.

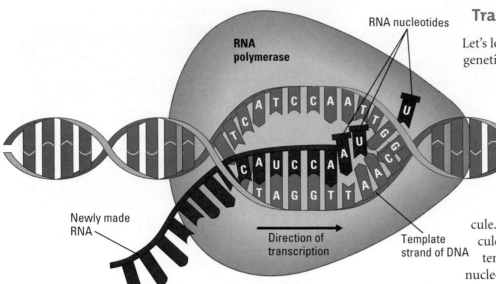

RNA nucleotides

RNA polymerase

U

Newly made RNA

Direction of transcription

Template strand of DNA

(a) A close-up view of transcription

Transcription: From DNA to RNA

Let's look more closely at transcription, the transfer of genetic information from DNA to RNA. An RNA molecule is transcribed from a DNA template by a process that resembles the synthesis of a DNA strand during DNA replication. Figure 9.19a is a close-up view of this process. As with replication, the two DNA strands must first separate at the place where the process will start. In transcription, however, only one of the DNA strands serves as a template for the newly forming molecule. The nucleotides that make up the new RNA molecule take their places one at a time along the DNA template strand by forming hydrogen bonds with the nucleotide bases there. Notice that the RNA nucleotides follow the same base-pairing rules that govern DNA replication, except that U, rather than T, pairs with A. The RNA nucleotides are linked by the transcription enzyme **RNA polymerase.**

Figure 9.19b is an overview of the transcription of an entire gene. Special sequences of DNA nucleotides tell the RNA polymerase where to start and where to stop the transcribing process.

Initiation of Transcription The "start transcribing" signal is a nucleotide sequence called a **promoter,** which is located in the DNA at the beginning of the gene. A promoter is a specific place where RNA polymerase attaches. The first phase of transcription, called initiation, is the attachment of RNA polymerase to the promoter and the start of RNA synthesis. For any gene, the promoter dictates which of the two DNA strands is to be transcribed (the particular strand varying from gene to gene).

RNA Elongation During the second phase of transcription, elongation, the RNA grows longer. As RNA synthesis continues, the RNA strand peels away from its DNA template, allowing the two separated DNA strands to come back together in the region already transcribed.

Termination of Transcription In the third phase, termination, the RNA polymerase reaches a special sequence of bases in the DNA template called a **terminator.** This sequence signals the end of the gene. At that point, the polymerase molecule detaches from the RNA molecule and the gene.

Watch an animation of transcription in Web/CD Activity 9E.

In addition to producing RNA that encodes amino acid sequences, transcription makes two other kinds of RNA that are involved in building polypeptides. We discuss these kinds of RNA a little later.

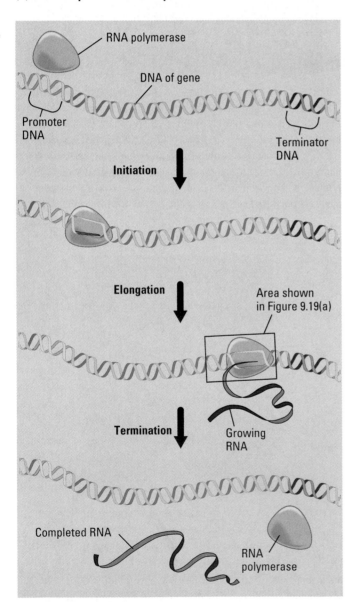

RNA polymerase

DNA of gene

Promoter DNA

Terminator DNA

Initiation

Elongation

Area shown in Figure 9.19(a)

Termination

Growing RNA

Completed RNA

RNA polymerase

(b) Transcription of a gene

Figure 9.19 Transcription. **(a)** As RNA nucleotides base-pair one by one with DNA bases on one DNA strand (called the template strand), the enzyme RNA polymerase links the RNA nucleotides into an RNA chain. The orange shape in the background is the RNA polymerase. **(b)** The transcription of an entire gene occurs in three phases: initiation, elongation, and termination of the RNA. The section of DNA where the RNA polymerase starts is called the promoter; the place where it stops is called the terminator.

The Processing of Eukaryotic RNA

In prokaryotic cells, the RNA transcribed from a gene immediately functions as the messenger molecule that is translated, called **messenger RNA (mRNA).** But this is not the case in eukaryotic cells. The eukaryotic cell not only localizes transcription in the nucleus but also modifies, or *processes,* the RNA transcripts there before they move to the cytoplasm for translation by the ribosomes.

One kind of RNA processing is the addition of extra nucleotides to the ends of the RNA transcript. These additions, called the **cap** and **tail,** protect the RNA from attack by cellular enzymes and help ribosomes recognize the RNA as mRNA.

Another type of RNA processing is made necessary in eukaryotes by noncoding stretches of nucleotides that interrupt the nucleotides that actually code for amino acids. It is as if unintelligible sequences of letters were randomly interspersed in an otherwise intelligible document. Most genes of plants and animals, it turns out, include such internal noncoding regions, which are called **introns.** (The functions of introns, if any, and how introns evolved remain a mystery.) The coding regions—the parts of a gene that are expressed—are called **exons.** As Figure 9.20 illustrates, both exons and introns are transcribed from DNA into RNA. However, before the RNA leaves the nucleus, the introns are removed, and the exons are joined to produce an mRNA molecule with a continuous coding sequence. This process is called **RNA splicing.**

With capping, tailing, and splicing completed, the "final draft" of eukaryotic mRNA is ready for translation.

Translation: The Players

As already discussed, translation is a conversion between different languages —from the nucleic acid language to the protein language—and it involves more elaborate machinery than transcription. Like the molecular machinery used for many cellular processes, the machinery used to translate mRNA requires enzymes and sources of chemical energy, such as ATP. But the heavy-duty components of the machine are ribosomes and a kind of RNA called transfer RNA.

Transfer RNA (tRNA) Translation of any language into another language requires an interpreter, a person or device that can recognize the words of one language and convert them into the other. Translation of the genetic message carried in mRNA into the amino acid language of proteins also requires an interpreter. To convert the three-letter words (codons) of nucleic acids to the one-letter, amino acid words of proteins, a cell employs a molecular interpreter, a type of RNA called **transfer RNA,** abbreviated **tRNA** (Figure 9.21).

A cell that is ready to have some of its genetic information translated into polypeptides has a supply of amino acids in its cytoplasm. It has either made them from other chemicals or obtained them from food. The amino acids themselves cannot recognize the codons arranged in sequence along messenger RNA. The amino acid tryptophan, for example, is no more attracted by UGG, the codon for tryptophan, than by any other codon. It is up to the cell's molecular interpreters, tRNA molecules, to match amino acids to the appropriate codons to form the new polypeptide. To perform this task, tRNA molecules must carry out two distinct functions: (1) picking up the appropriate amino acids and (2) recognizing the appropriate codons

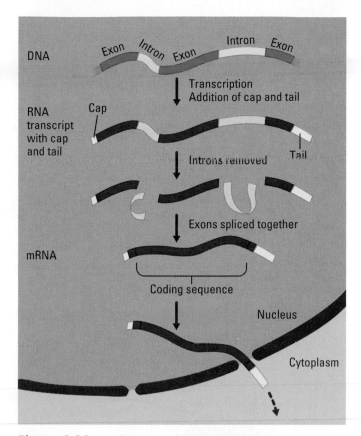

Figure 9.20 Production of messenger RNA in a eukaryotic cell. Both exons and introns are transcribed from the DNA. Additional nucleotides, making up the cap and tail, are attached at the ends of the RNA transcript. The exons are spliced together. The product, a molecule of messenger RNA (mRNA), then travels to the cytoplasm of the cell. There the coding sequence will be translated.

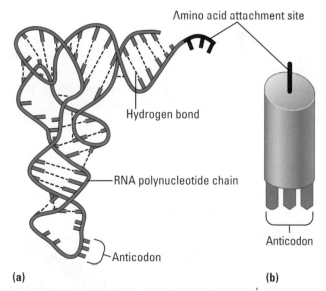

Figure 9.21 The structure of tRNA. (a) The RNA polynucleotide is a "rope" whose appendages are the nitrogenous bases. Dashed lines are hydrogen bonds, which connect some of the bases. The site where an amino acid will attach is a three-nucleotide segment at one end (purple). Note the three-base anticodon at the bottom of the molecule. The overall shape of a tRNA molecule is like the letter "L." **(b)** This is the representation of tRNA that we use in later diagrams.

in the mRNA. The unique structure of tRNA molecules enables them to perform both tasks.

As shown in Figure 9.21 (preceding page), a tRNA molecule is made of a single strand of RNA—one polynucleotide chain—consisting of only about 80 nucleotides. The chain twists and folds upon itself, forming several double-stranded regions in which short stretches of RNA base-pair with other stretches. At one end of the folded molecule there is a special triplet of bases called an **anticodon.** The anticodon triplet is complementary to a codon triplet on mRNA. During translation, the anticodon on tRNA recognizes a particular codon on mRNA by using base-pairing rules. At the other end of the tRNA molecule is a site where an amino acid can attach. Although all tRNA molecules are similar, there is a slightly different version of tRNA for each amino acid. An enzyme that recognizes both a tRNA and its amino acid partner links the two together, using energy from ATP. There is a whole family of such enzymes, at least one for each amino acid.

Ribosomes We have now discussed most of the things a cell needs to carry out translation: instructions in the form of mRNA molecules, tRNA to interpret the instructions, a supply of amino acids, enzymes for attaching amino acids to tRNA, and ATP for energy. Still needed are the organelles that coordinate the functioning of the mRNA and tRNA and actually make polypeptides: ribosomes.

As you can see in Figure 9.22a, a ribosome consists of two subunits. Each subunit is made up of proteins and a considerable amount of yet another kind of RNA, **ribosomal RNA (rRNA).** A fully assembled ribosome has a binding site for mRNA on its small subunit and binding sites for tRNA on its large subunit. Figure 9.22b shows how two tRNA molecules get together with an mRNA molecule on a ribosome. One of the tRNA-binding sites, the P site, holds the tRNA carrying the growing polypeptide chain, while another, the A site, holds a tRNA carrying the next amino acid to be added to the chain. The anticodon on each tRNA base-pairs with a codon on mRNA. The subunits of the ribosome act like a vise, holding the tRNA and mRNA molecules close together. The ribosome can then connect the amino acid from the A site tRNA to the growing polypeptide.

Now let's examine translation in more detail, starting at the beginning.

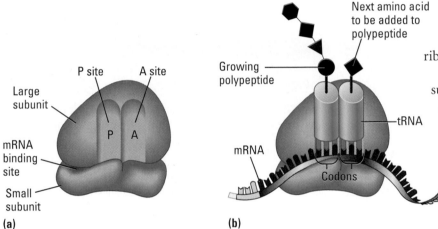

Figure 9.22 The ribosome. (a) A simplified diagram of a ribosome, showing its two subunits and sites where mRNA and tRNA molecules bind. **(b)** When functioning in polypeptide synthesis, a ribosome holds one molecule of mRNA and two molecules of tRNA. The growing polypeptide is attached to one of the tRNAs.

Translation: The Process

Translation can be divided into the same three phases as transcription: initiation, elongation, and termination.

Initiation This first phase brings together the mRNA, the first amino acid with its attached tRNA, and the two subunits of a ribosome. An mRNA molecule, even after splicing, is longer than the genetic message it carries (Figure 9.23). Nucleotide sequences at either end of the molecule are not part of the message but, along with the cap and tail in eukaryotes, help the mRNA bind to the ribosome. The initiation process determines exactly where translation will begin so that the mRNA codons will be translated into the correct sequence of amino acids. Initiation occurs in two steps, as shown in Figure 9.24.

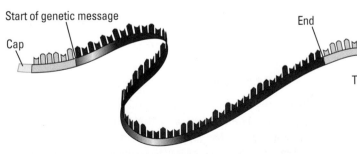

Figure 9.23 A molecule of mRNA. The pink ends are nucleotides that are not part of the message; that is, they are not translated. These nucleotides, along with the cap and tail (yellow), help the mRNA attach to the ribosome.

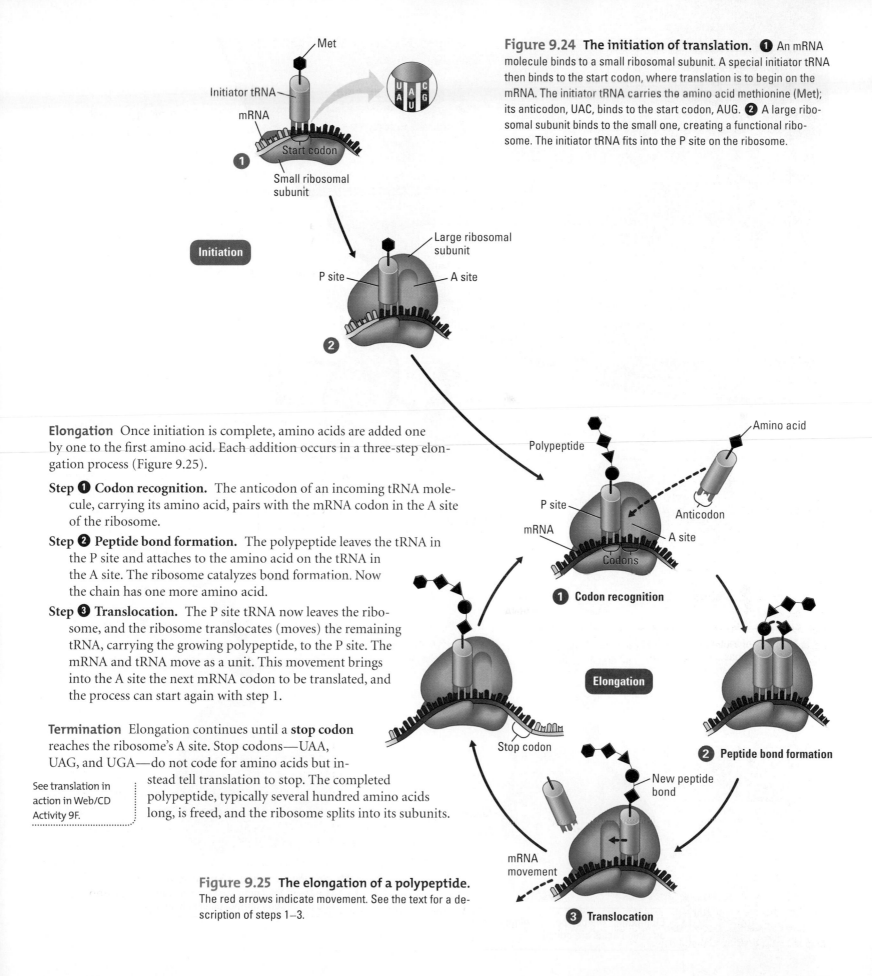

Figure 9.24 The initiation of translation. ❶ An mRNA molecule binds to a small ribosomal subunit. A special initiator tRNA then binds to the start codon, where translation is to begin on the mRNA. The initiator tRNA carries the amino acid methionine (Met); its anticodon, UAC, binds to the start codon, AUG. ❷ A large ribosomal subunit binds to the small one, creating a functional ribosome. The initiator tRNA fits into the P site on the ribosome.

Met
Initiator tRNA
mRNA
Start codon
❶
Small ribosomal subunit

Initiation

Large ribosomal subunit
P site
A site
❷

Elongation Once initiation is complete, amino acids are added one by one to the first amino acid. Each addition occurs in a three-step elongation process (Figure 9.25).

Step ❶ Codon recognition. The anticodon of an incoming tRNA molecule, carrying its amino acid, pairs with the mRNA codon in the A site of the ribosome.

Step ❷ Peptide bond formation. The polypeptide leaves the tRNA in the P site and attaches to the amino acid on the tRNA in the A site. The ribosome catalyzes bond formation. Now the chain has one more amino acid.

Step ❸ Translocation. The P site tRNA now leaves the ribosome, and the ribosome translocates (moves) the remaining tRNA, carrying the growing polypeptide, to the P site. The mRNA and tRNA move as a unit. This movement brings into the A site the next mRNA codon to be translated, and the process can start again with step 1.

Termination Elongation continues until a **stop codon** reaches the ribosome's A site. Stop codons—UAA, UAG, and UGA—do not code for amino acids but instead tell translation to stop. The completed polypeptide, typically several hundred amino acids long, is freed, and the ribosome splits into its subunits.

See translation in action in Web/CD Activity 9F.

Amino acid
Polypeptide
P site
Anticodon
mRNA
A site
Codons
❶ Codon recognition

Elongation

New peptide bond
❷ Peptide bond formation

Stop codon

mRNA movement
❸ Translocation

Figure 9.25 The elongation of a polypeptide. The red arrows indicate movement. See the text for a description of steps 1–3.

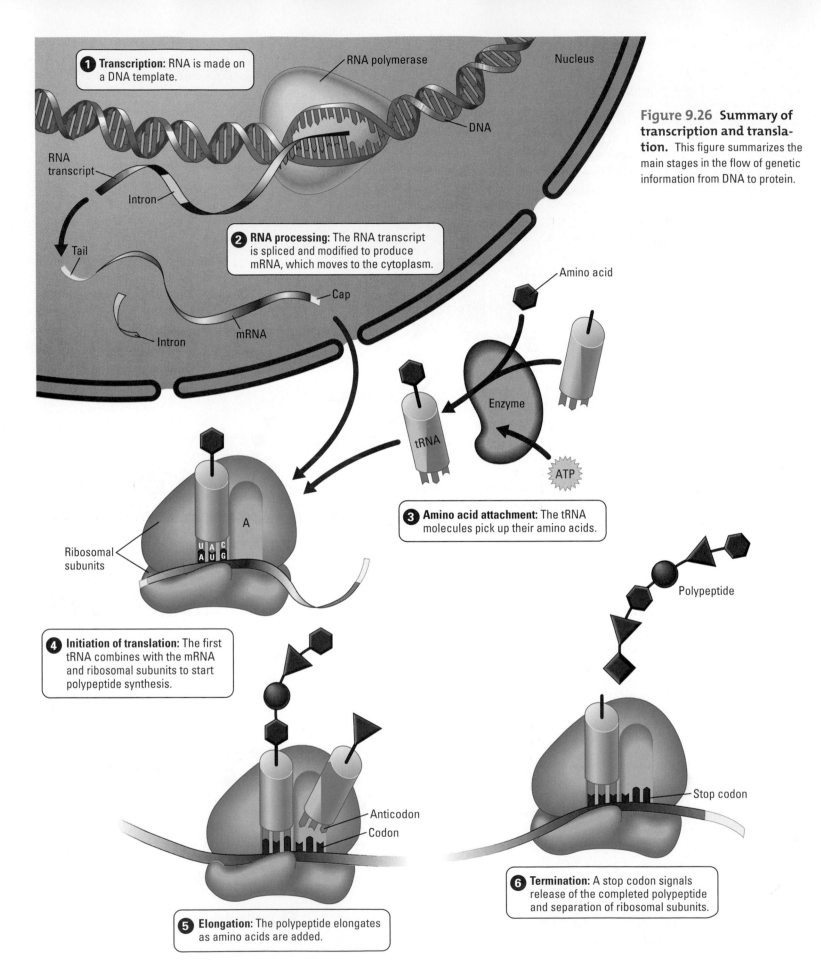

① **Transcription:** RNA is made on a DNA template.

RNA polymerase

Nucleus

DNA

Figure 9.26 Summary of transcription and translation. This figure summarizes the main stages in the flow of genetic information from DNA to protein.

RNA transcript

Intron

Tail

② **RNA processing:** The RNA transcript is spliced and modified to produce mRNA, which moves to the cytoplasm.

Cap

Intron

mRNA

Amino acid

Enzyme

tRNA

ATP

③ **Amino acid attachment:** The tRNA molecules pick up their amino acids.

A

Ribosomal subunits

U A C
A U G

Polypeptide

④ **Initiation of translation:** The first tRNA combines with the mRNA and ribosomal subunits to start polypeptide synthesis.

Anticodon

Codon

Stop codon

⑥ **Termination:** A stop codon signals release of the completed polypeptide and separation of ribosomal subunits.

⑤ **Elongation:** The polypeptide elongates as amino acids are added.

Review: DNA → RNA → Protein

Figure 9.26 (at left) reviews the flow of genetic information in the cell, from DNA to RNA to protein. In eukaryotic cells, transcription—the stage from DNA to RNA—occurs in the nucleus, and the RNA is processed before it enters the cytoplasm. Translation is rapid; a single ribosome can make an average-sized polypeptide in less than a minute. As it is made, a polypeptide coils and folds, assuming a three-dimensional shape, its tertiary structure. Several polypeptides may come together, forming a protein with quaternary structure (see Figure 3.22).

What is the overall significance of transcription and translation? These are the processes whereby genes control the structures and activities of cells—or, more broadly, the way the genotype produces the phenotype. The chain of command originates with the information in a gene, a specific linear sequence of nucleotides in DNA. The gene serves as a template, dictating transcription of a complementary sequence of nucleotides in mRNA. In turn, mRNA specifies the linear sequence in which amino acids appear in a polypeptide. Finally, the proteins that form from the polypeptides determine the appearance and capabilities of the cell and organism.

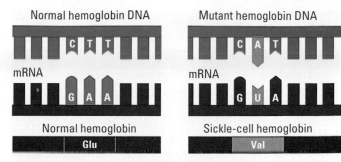

Figure 9.27 The molecular basis of sickle-cell disease. The sickle-cell allele differs from its normal counterpart, a gene for hemoglobin, by only one nucleotide. This difference changes the mRNA codon from one that codes for the amino acid glutamic acid (Glu) to one that codes for valine (Val).

Mutations

Since discovering how genes are translated into proteins, scientists have been able to describe many heritable differences in molecular terms. For instance, when a child is born with sickle-cell disease (see Figure 8.22), the condition can be traced back through a difference in a protein to one tiny change in a gene. In one of the polypeptides in the hemoglobin protein, the sickle-cell child has a single different amino acid, as you may recall from Chapter 3. This difference is caused by a single nucleotide difference in the coding strand of DNA (Figure 9.27). In the double helix, a *base pair* is changed.

The sickle-cell allele is not a unique case. We now know that the various alleles of many genes result from changes in single base pairs in DNA. Any change in the nucleotide sequence of DNA is called a **mutation.** Mutations can involve large regions of a chromosome or just a single nucleotide pair, as in the sickle-cell allele. Let's consider how mutations involving only one or a few nucleotide pairs can affect gene translation.

Mutations within a gene can be divided into two general categories: base substitutions and base insertions or deletions (Figure 9.28). A base substitution is the replacement of one base or nucleotide with another. Depending on how a base substitution is translated, it can result in no change in the protein, in an insignificant change, or in a change that might be crucial to the life of the organism. It is because of the redundancy of the genetic code that some substitution mutations have no effect. For example, if a mutation causes an mRNA codon to change from GAA to GAG, no change in the protein product would result, because GAA and GAG both code for the same amino acid (Glu). Other changes of a single nucleotide may alter an amino acid but have little effect on the function of the protein.

Some base substitutions, as we saw in the sickle-cell case, cause changes in the protein that prevent it from performing normally. Occasionally, a base substitution leads to an improved protein or one with new capabilities that enhance the success of the mutant organism and its descendants. Much more often, though, mutations are harmful.

Mutations involving the insertion or deletion of one or more nucleotides in a gene often have disastrous effects. Because mRNA is read as a

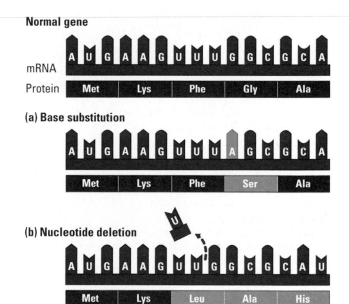

Figure 9.28 Two types of mutations and their effects. Mutations are changes in DNA, but they are represented here as reflected in mRNA and its polypeptide product. **(a)** In the base substitution shown here, an A replaces a G in the fourth codon of the mRNA. The result in the polypeptide is a Ser (serine) instead of a Gly (glycine). This amino acid substitution may or may not affect the protein's function. **(b)** When a nucleotide is deleted (or inserted), the reading frame is altered so that all the codons from that point on are misread. The resulting polypeptide is likely to be completely nonfunctional.

series of nucleotide triplets during translation, adding or subtracting nucleotides may alter the **reading frame** (triplet grouping) of the genetic message. All the nucleotides that are "downstream" of the insertion or deletion will be regrouped into different codons. The result will most likely be a nonfunctional polypeptide.

The creation of mutations, the process of **mutagenesis,** can occur in a number of ways. Mutations resulting from errors during DNA replication or recombination are called spontaneous mutations, as are other mutations of unknown cause. Another source of mutation is a physical or chemical agent, called a **mutagen.** The most common physical mutagen is high-energy radiation, such as X-rays and ultraviolet light. Chemical mutagens are of various types. One type, for example, consists of chemicals that are similar to normal DNA bases but that base-pair incorrectly when incorporated into DNA.

Although mutations are often harmful, they are also extremely useful, both in nature and in the laboratory. It is because of mutations that there is such a rich diversity of genes in the living world, a diversity that makes evolution by natural selection possible. Mutations are also essential tools for geneticists. Whether naturally occurring (as in Mendel's peas) or created in the laboratory (Morgan used X-rays to make most of his fruit fly mutants), mutations create the different alleles needed for genetic research.

Mutations are the ultimate source of the diversity of life in this underwater scene near the Fiji Islands.

CheckPoint

1. What are transcription and translation?

2. How many nucleotides are necessary to code for a polypeptide that is 100 amino acids long?

3. An mRNA molecule contains the nucleotide sequence CCAUUUACG. Using Figure 9.17, translate this sequence into the corresponding amino acid sequence.

4. How does RNA polymerase "know" where to start transcribing a gene?

5. What is an anticodon?

6. What is the function of the ribosome in protein synthesis?

7. What would happen if a mutation changed a start codon to some other codon?

8. Once polypeptide synthesis starts, what are the three main steps by which it grows (elongate)?

9. Which of the following does not participate directly in translation: ribosomes, transfer RNA, messenger RNA, DNA, enzymes, ATP?

10. What happens when one nucleotide is lost from the middle of a gene?

Answers: 1. Transcription is the transfer of genetic information from DNA to RNA—the synthesis of RNA on a DNA template. Translation is the use of the information in an mRNA molecule for the synthesis of a polypeptide. **2.** 300 **3.** Pro-Phe-Thr **4.** It recognizes the gene's promoter, a specific nucleotide sequence. **5.** An anticodon is the base triplet of a tRNA molecule that couples the tRNA to a complementary codon in the mRNA. The base pairing of anticodon to codon is a key step in translating mRNA to polypeptide. **6.** The ribosome holds mRNA and tRNAs together and connects amino acids from the tRNAs to the growing polypeptide chain. **7.** Messenger RNA (if any) transcribed from the mutated gene would be nonfunctional because ribosomes would not initiate translation at the correct point. **8.** Codon recognition by an incoming tRNA, peptide bond formation by the ribosome, and translocation of the tRNA and mRNA through the ribosome **9.** DNA **10.** In the mRNA, the reading frame downstream from the deletion is shifted, leading to a long string of incorrect amino acids in the polypeptide. The polypeptide will be nonfunctional.

Viruses: Genes in Packages

As you saw at the beginning of the chapter, viruses—in particular, phages—provided some of the first glimpses into the molecular details of heredity. Now let's take a closer look at viruses, focusing on the relationship between viral structure and the processes of nucleic acid replication, transcription, and translation.

In many cases, a virus is nothing more than packaged genes, a bit of nucleic acid wrapped in a protein coat (Figure 9.29). For research in molecular biology, viruses may seem like ideal tools—everything you need to study the flow of genetic information and nothing more. It's really not that simple, however, because viruses must use the resources of cells to reproduce. We'll start our exploration of the world of viruses with a return to phages.

Bacteriophages

In reading about the Hershey-Chase experiment, you learned what happens after the DNA of phage T2 enters an *E. coli* cell. The way T2 reproduces is called a **lytic cycle** because it always leads to the lysis (breaking open) of the bacterial cell. The study of another phage of *E. coli*, called lambda, led to the discovery that some viruses can also reproduce by an alternative route called a lysogenic cycle. During a **lysogenic cycle,** viral DNA replication occurs without phage production or the death of the cell.

Figure 9.30 illustrates the two kinds of cycles for phage lambda. Like T2, lambda has a head (containing DNA) and a tail, but no long tail fibers. Before embarking on one of the two cycles, the injected DNA forms a circle. In the lytic cycle, this DNA immediately turns the cell into a virus-producing

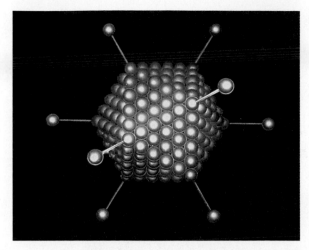

Figure 9.29 Adenovirus. A virus that infects the human respiratory system, adenovirus consists of DNA enclosed in a protein shell shaped like a 20-sided polyhedron, shown here in a computer-generated model. At each vertex of the polyhedron is a protein spike, which helps the virus attach to a susceptible cell.

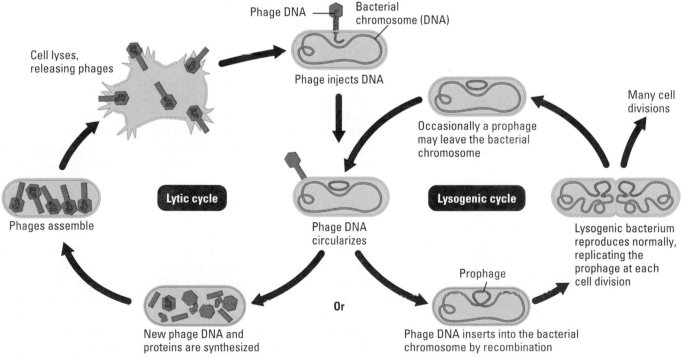

Figure 9.30 Alternative types of phage reproductive cycles. Certain phages, like lambda, can undergo alternative reproductive cycles. After entering the bacterial cell, the phage DNA can either integrate into the bacterial chromosome (lysogenic cycle) or immediately start the production of progeny phages (lytic cycle), destroying the cell. In most cases, the phage follows the lytic pathway, but once it enters a lysogenic cycle, the prophage (the phage's DNA) may be carried in the host cell's chromosome for many generations. The bacterial chromosome is a single DNA molecule in the form of a closed loop.

factory. In the lysogenic cycle, the DNA inserts by genetic recombination into the bacterium's DNA. Called a **bacterial chromosome,** this DNA is a single, circular molecule. Once inserted into the bacterial chromosome, the

Check out the reproductive cycles of phage lambda in Web/CD Activity 9G.

phage DNA is referred to as a **prophage,** and most of its genes are inactive. Survival of the prophage depends on reproduction of the cell where it resides. The host cell replicates the prophage DNA along with its cellular DNA and then, upon dividing, passes on both the prophage and the cellular DNA to its two daughter cells. A single infected bacterium can quickly give rise to a large population of bacteria carrying prophages. The prophages may remain in the bacterial cells indefinitely. Occasionally, however, one leaves its chromosome; this event may be triggered by environmental conditions. Once separate, the lambda DNA usually switches to the lytic cycle.

Sometimes the few prophage genes active in a bacterial cell can cause medical problems. For example, the bacteria that cause diphtheria, botulism, and scarlet fever would be harmless to humans if it were not for the prophage genes they carry. Certain of these genes direct the bacteria to produce toxins that are responsible for making people ill.

Plant Viruses

Viruses that infect plant cells can stunt plant growth and diminish crop yields. Most plant viruses discovered to date have RNA rather than DNA as their genetic material. Many of them, like the tobacco mosaic virus in Figure 9.31, are rod-shaped with a spiral arrangement of proteins surrounding the nucleic acid.

To infect a plant, a virus must first get past the plant's outer protective layer of cells (the epidermis). For this reason, a plant damaged by wind, chilling, injury, or insects is more susceptible to infection than a healthy plant. Besides injuring plants, some insects also carry and transmit plant viruses. Farmers and gardeners, too, may spread plant viruses through the use of pruning shears and other tools. And infected plants may pass viruses to their offspring.

Once a virus enters a plant cell and begins reproducing, it can spread throughout the entire plant through plasmodesmata, the channels that connect the cytoplasms of adjacent plant cells (see Figure 4.20). As with animal viruses, there is no cure for most viral diseases of plants, and agricultural scientists focus on reducing the number of plants that are infected and on breeding genetic varieties of crop plants that resist viral infection.

Animal Viruses

Viruses that infect animal cells are common causes of disease. We have all suffered from viral infections. Figure 9.32 shows the structure of an influenza (flu) virus. Like many animal viruses, this one has an outer envelope made of phospholipid membrane, with projecting spikes of protein. The envelope helps the virus enter and leave a cell. Also, like many other animal viruses, flu viruses have RNA rather than DNA as their genetic material. Other RNA viruses include those that cause the common cold, measles, and mumps, as well as ones that cause more serious human diseases, such as AIDS and polio. Diseases caused by DNA viruses include hepatitis, chicken pox, and herpes infections.

Figure 9.33 shows the reproductive cycle of an enveloped RNA virus (the mumps virus). When the virus contacts a susceptible cell, the protein

Protein   RNA

Figure 9.31 Tobacco mosaic virus. The photo shows the mottling of leaves in tobacco mosaic disease. The rod-shaped virus causing the disease has RNA as its genetic material.

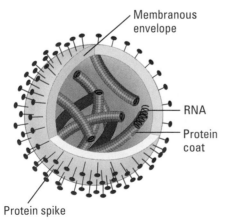

Membranous envelope

RNA

Protein coat

Protein spike

Figure 9.32 An influenza virus. The genetic material of this virus consists of eight separate molecules of RNA, each wrapped in a protein coat. Around the outside of the virus is an envelope made of membrane, studded with protein spikes.

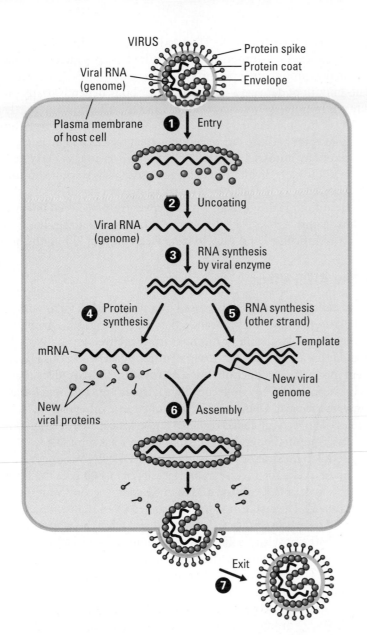

VIRUS

Protein spike
Protein coat
Envelope

Viral RNA
(genome)

Plasma membrane
of host cell

1 Entry

2 Uncoating

Viral RNA
(genome)

3 RNA synthesis
by viral enzyme

4 Protein
synthesis

5 RNA synthesis
(other strand)

mRNA

Template

New viral
genome

New
viral proteins

6 Assembly

Exit

7

Figure 9.33 The reproductive cycle of an enveloped virus. The virus here is the one that causes mumps. Like the flu virus, it has a membranous envelope with protein spikes, but its genome is a single molecule of RNA. See the discussion of the numbered steps in the text.

spikes attach to receptor proteins on the cell's plasma membrane. The viral envelope fuses with the cell's membrane, allowing the protein-coated RNA to ❶ enter the cytoplasm. ❷ Enzymes then remove the protein coat. ❸ An enzyme that entered the cell as part of the virus uses the virus's RNA genome as a template for making complementary strands of RNA (purple strand). The new strands have two functions: ❹ They serve as mRNA for the synthesis of new viral proteins, and ❺ they serve as templates for synthesizing new viral genome RNA. ❻ The new coat proteins assemble around the new viral RNA. ❼ Finally, the viruses leave the cell by cloaking themselves in plasma membrane. In other words, the virus obtains its envelope from the cell, leaving the cell without necessarily lysing it.

Not all animal viruses reproduce in the cytoplasm. For example, the viruses called herpesviruses—which cause chicken pox, shingles, cold sores, and genital herpes—are enveloped DNA viruses that reproduce in a cell's nucleus, and they get their envelopes from the cell's nuclear membranes. While inside the nucleus, herpesvirus DNA may insert itself into the cell's DNA as a **provirus** (like a prophage) and remain dormant within the body. From time to time, physical stress, such as a cold or sunburn, or emotional

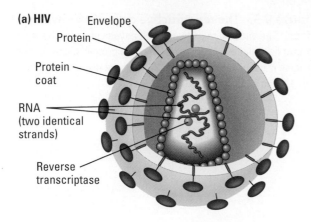

(a) HIV

Envelope

Protein

Protein coat

RNA (two identical strands)

Reverse transcriptase

(b) The behavior of HIV nucleic acid in an infected cell

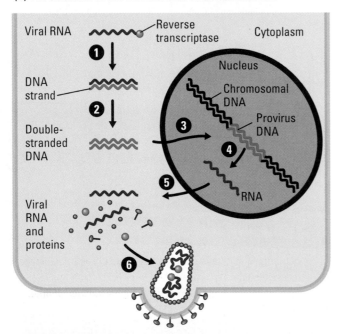

Viral RNA — Reverse transcriptase — Cytoplasm

1

DNA strand

2

Double-stranded DNA

Nucleus

Chromosomal DNA

Provirus DNA

3

4

5

RNA

Viral RNA and proteins

6

Colorized SEM 11,000×

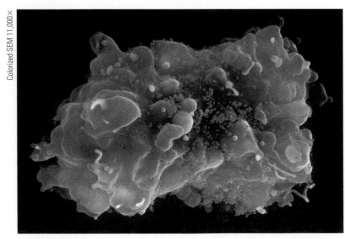

(c) HIV infecting a white blood cell

Figure 9.34 HIV, the AIDS virus. (a) Structure of HIV, the human immunodeficiency virus. **(b)** Once the virus enters a cell and its RNA is unwrapped, it can produce new viruses as shown here. The text describes the numbered steps. **(c)** HIV (blue spots) in the process of infecting a white blood cell.

stress may cause the herpesvirus DNA to begin production of the virus, resulting in unpleasant symptoms. Once acquired, herpes infections may flare up repeatedly throughout a person's life.

The amount of damage a virus causes our body depends partly on how quickly our immune system responds to fight the infection and partly on the ability of the infected tissue to repair itself. We usually recover completely from colds because our respiratory tract tissue can efficiently replace damaged cells by mitosis. In contrast, the poliovirus attacks nerve cells, which do not usually divide. The damage to such cells by polio, unfortunately, is permanent. In such cases, we try to prevent the disease with vaccines. The antibiotic drugs that help us recover from bacterial infections are powerless against viruses. The development of antiviral drugs has been slow because it is difficult to find ways to kill a virus without killing the host cells.

HIV, the AIDS Virus

The devastating disease AIDS is caused by a type of RNA virus with some special twists. In outward appearance, the AIDS virus (HIV, Figure 9.34a) somewhat resembles the flu or mumps virus. Its envelope enables HIV to enter and leave a cell much the way the mumps virus does. But HIV has a different mode of reproduction. It is a **retrovirus,** an RNA virus that reproduces by means of a DNA molecule. Retroviruses are so named because they reverse the usual DNA → RNA flow of genetic information. They carry molecules of an enzyme called **reverse transcriptase,** which catalyzes reverse transcription, the synthesis of DNA on an RNA template.

Figure 9.34b illustrates what happens after HIV RNA is uncoated in the cytoplasm of a cell. The reverse transcriptase (green) **1** uses the RNA as a template to make a DNA strand and then **2** adds a second, complementary DNA strand. **3** The resulting double-stranded DNA then enters the cell nucleus and inserts itself into the chromosomal DNA, becoming a provirus. Occasionally the provirus is **4** transcribed into RNA and **5** translated into viral proteins. **6** New viruses assembled from these components eventually leave the cell. They can then infect other cells. This is the standard reproductive cycle for retroviruses.

Watch an animation of HIV reproduction in Web/CD Activity 9H. Learn more about AIDS and HIV in The Process of Science on the website and CD.

AIDS stands for acquired immune deficiency syndrome, and **HIV** for human immunodeficiency virus; these terms describe the main effect of the virus on the body. HIV infects and eventually kills several kinds of white blood cells that are important in our immune system (Figure 9.34c).

Evolution Link: Emerging Viruses

Acknowledging the persistent threat that viruses pose to human health, geneticist and Nobel Prize winner Joshua Lederberg once warned, "We live in evolutionary competition with microbes. There is no guarantee that we will be the survivors." Lederberg cited the AIDS epidemic and recurrent flu epidemics as examples of the human population's vulnerability to viral attacks.

The AIDS virus (HIV), which seemed to arise abruptly in the early 1980s, and the new varieties of flu virus that frequently appear, are not the only examples of newly dangerous viruses. A deadly virus called the Ebola virus (Figure 9.35a) menaces central African nations periodically, and many biologists fear its emergence as a global threat. Another newly dangerous virus was responsible for killing dozens of people in the southwestern United States in 1993. It was first thought to be a new virus but turned out to have been known to Western medicine for about 45 years, since the Korean War. Named for a river in Korea, it is called hantavirus (Figure 9.35b). Hantavirus is a potential problem worldwide, although the number of cases reported recently in the United States is small (about 170 since 1994).

Although the viruses mentioned here may not have originated as recently as once thought, they still raise intriguing questions: How do new viruses arise, and what causes certain viruses to emerge as major threats? Most biologists favor the hypothesis that the very first viruses arose from fragments of cellular nucleic acid that had some means of moving from one cell to another. Consistent with this idea, viral nucleic acid usually has more in common with its host cell's DNA than with the nucleic acid of viruses that infect other hosts. Indeed, some viral genes are virtually identical to genes of their host. The earliest viruses may have been naked bits of nucleic acid that traveled from one cell to another via injured cell surfaces. Genes coding for viral coat proteins may have evolved later.

Today, the mutation of existing viruses is a major source of new viral diseases. The RNA viruses tend to have an unusually high rate of mutation because the replication of their nucleic acid does not involve proofreading steps as does DNA replication. Some mutations may enable existing viruses to evolve into new genetic varieties that can cause disease in individuals who had developed immunity to the ancestral virus. Flu epidemics are caused by viruses that are genetically different enough from earlier years' viruses that people have little immunity to them.

Another source of new viral diseases is the spread of existing viruses to a new host species. For example, hantavirus is common in rodents, especially deer mice. The population of deer mice in the southwestern United States exploded in 1993 after unusually wet weather increased the rodents' food supply. Humans acquired hantavirus when they inhaled dust containing traces of urine and feces from infected mice.

> Discover more about emerging viruses in the Web Evolution Link.

Finally, a viral disease may start out in a small, isolated population and then rather suddenly become widespread. AIDS, for example, went unnamed and virtually unnoticed for decades before starting to spread around the world. In this case, technological and social factors, including affordable international travel, blood transfusion technology, sexual promiscuity, and the abuse of intravenous drugs, allowed a previously rare human disease to become a global epidemic. It is likely that when we do find the means to control HIV and other deadly viruses , genetic research —in particular, molecular biology—will be responsible for the discovery.

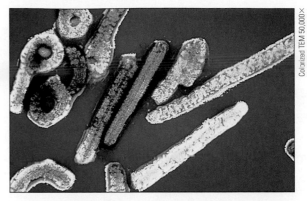

Colorized TEM 50,000×

(a) Ebola virus

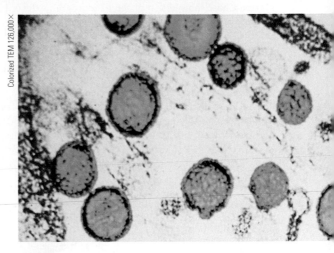

Colorized TEM 126,000×

(b) Hantavirus

Figure 9.35 Emerging viruses. The term *emerging viruses* is used for viruses that have recently appeared or recently come to the attention of medical scientists. **(a)** Ebola virus. Each one is an enveloped thread of protein-coated RNA. **(b)** Hantavirus, another enveloped RNA virus.

Chapter Review

Summary of Key Concepts

Overview: The Discovery That DNA Is the Genetic Material

- **Griffith's Transforming Factor** Frederick Griffith showed in 1928 that a "factor" from pathogenic bacteria could transform harmless bacteria to pathogens. Years later, the transforming factor was shown to be DNA.

- **The Genetic Material of a Bacterial Virus** In 1952, Alfred Hershey and Martha Chase showed that phage T2 reprograms the cell it infects to produce progeny phages by injecting its DNA.
- **Web/CD Activity 9A** *Phage T2 Reproductive Cycle*

The Structure and Replication of DNA

- **DNA and RNA: Polymers of Nucleotides** DNA is a nucleic acid made of long chains of nucleotide monomers. Each nucleotide consists of a sugar connected to a phosphate group and a nitrogenous base. Alternating sugars and phosphates form a backbone. DNA has four kinds of bases, abbreviated A, T, C, and G. RNA is also a nucleic acid, with a slightly different sugar and U instead of T.

- **Watson and Crick's Discovery of the Double Helix** James Watson and Francis Crick worked out the three-dimensional structure of DNA, which is two polynucleotide strands wrapped around each other in a double helix. Hydrogen bonds between bases hold the strands together. Each base pairs with a complementary partner: A with T, and G with C.
- **Web/CD Activity 9B** *DNA and RNA Structure*

- **DNA Replication** In DNA replication, the DNA strands separate, and DNA polymerases use each strand as a template to assemble new nucleotides into a new complementary strand. The two daughter DNA molecules are identical to the parent molecule.
- **Web/CD Activity 9C** *DNA Replication*

The Flow of Genetic Information from DNA to RNA to Protein

- **How an Organism's DNA Genotype Produces Its Phenotype** The information constituting an organism's genotype is carried in the sequence of its DNA bases. Studies of inherited metabolic defects first suggested that phenotype is expressed through proteins. A particular gene—a linear sequence of many nucleotides—specifies a polypeptide. The DNA of the gene is transcribed into RNA, which is translated into the polypeptide.
- **Web/CD Activity 9D** *Overview of Protein Synthesis*

- **From Nucleotide Sequence to Amino Acid Sequence: An Overview** The DNA of a gene is transcribed into RNA using the usual base-pairing rules, except that an A in DNA pairs with U in RNA. In the translation of a genetic message, each triplet of nucleotide bases in the RNA, called a codon, specifies one amino acid in the polypeptide.

- **The Genetic Code** In addition to codons that specify amino acids, the genetic code has one that is a start signal and three that are stop signals for translation. The genetic code is redundant: There is more than one codon for each amino acid.

- **Transcription: From DNA to RNA** In transcription, RNA polymerase binds to the promoter of a gene, opens the DNA double helix there, and catalyzes the synthesis of an RNA molecule using one DNA strand as template. As the single-stranded RNA transcript peels away from the gene, the DNA strands rejoin.
- **Web/CD Activity 9E** *Transcription*

- **The Processing of Eukaryotic RNA** The RNA transcribed from a eukaryotic gene is processed before leaving the nucleus to serve as messenger RNA (mRNA). Introns are spliced out, and a cap and tail are added.

- **Translation: The Players** In addition to mRNA, translation requires ribosomes, transfer RNA, enzymes, and ATP. Transfer RNAs (tRNAs) act as interpreters; each L-shaped molecule has a base triplet called an anticodon at one end and carries an amino acid at the other end. A ribosome is constructed of two subunits and has binding sites for mRNA and tRNAs.

- **Translation: The Process** In initiation, a ribosome assembles with the mRNA and the initiator tRNA bearing the first amino acid. One by one, the codons of the mRNA are recognized by tRNAs bearing succeeding amino acids. The ribosome bonds the amino acids together. With each addition, the mRNA translocates by one codon through the ribosome. When a stop codon is reached, the completed polypeptide is released.
- **Web/CD Activity 9F** *Translation*

- **Review: DNA → RNA → Protein** Figure 9.26 summarizes transcription, RNA processing, and translation. The sequence of codons in DNA, via the sequence of codons in mRNA, spells out the primary structure of a polypeptide.

- **Mutations** Mutations are changes in the DNA base sequence, caused by errors in DNA replication or by mutagens. Substituting, inserting, or deleting nucleotides in a gene has varying effects on the polypeptide and organism.

Viruses: Genes in Packages

- **Bacteriophages** Viruses can be regarded simply as genes packaged in protein. When phage DNA enters a lytic cycle inside a bacterium, it is replicated, transcribed, and translated. The new viral DNA and protein molecules then assemble into new phages, which burst from the cell. In the lysogenic cycle, phage DNA inserts into the cell's chromosome and is passed on to generations of daughter cells. Much later, it may initiate phage production .
- **Web/CD Activity 9G** *Phage Lambda Reproductive Cycles*

- **Plant Viruses** Viruses that infect plants can be a serious agricultural problem. Most have RNA genomes. They enter plants via breaks in the plant's outer layers.

- **Animal Viruses** Many animal viruses, such as flu viruses, have RNA genomes; others, such as hepatitis viruses, have DNA. Some animal viruses "steal" a bit of cell membrane as a protective envelope. Some insert their DNA into a cellular chromosome and remain latent for long periods.

- **HIV, the AIDS Virus** HIV is a retrovirus. Inside a cell it uses its RNA as a template for making DNA, which is then inserted into a chromosome.
- **Web/CD Activity 9H** *HIV Reproductive Cycle*
- **Web/CD The Process of Science** *Investigating AIDS and HIV: The Discovery and Epidemiology of AIDS*

Evolution Link: Emerging Viruses

- Emerging viral diseases, caused by newly dangerous viruses, pose a threat to human health. Viruses probably first arose from fragments of cellular nucleic acid that had some means of moving fom one cell to another.
- **Web Evolution Link** *Emerging Viruses*

Self-Quiz

1. Scientists have discovered how to put together a bacteriophage with the protein coat of phage T2 and the DNA of phage T4. If this composite phage were allowed to infect a bacterium, the phages produced in the cell would have
   a. the protein of T2 and the DNA of T4.
   b. the protein of T4 and the DNA of T2.
   c. a mixture of the DNA and proteins of both phages.
   d. the protein and DNA of T2.
   e. the protein and DNA of T4.

2. A geneticist found that a particular mutation had no effect on the polypeptide coded by the gene. This mutation probably involved
   a. deletion of one nucleotide.
   b. alteration of the start codon.
   c. insertion of one nucleotide.
   d. deletion of the entire gene.
   e. substitution of one nucleotide.

3. Which of the following correctly ranks the structures in order of size, from largest to smallest?
   a. gene, chromosome, nucleotide, codon
   b. chromosome, gene, codon, nucleotide
   c. nucleotide, chromosome, gene, codon
   d. chromosome, nucleotide, gene, codon
   e. gene, chromosome, codon, nucleotide

4. The nucleotide sequence of a DNA codon is GTA. A messenger RNA molecule with a complementary codon is transcribed from the DNA. In the process of protein synthesis, a transfer RNA pairs with the mRNA codon. What is the nucleotide sequence of the tRNA anticodon?
   a. CAT   d. CAU
   b. CUT   e. GT
   c. GUA

5. Describe the process by which the information in a gene is transcribed and translated into a protein. Correctly use these words in your description: tRNA, amino acid, start codon, transcription, mRNA, gene, codon, RNA polymerase, ribosome, translation, anticodon, peptide bond, stop codon.

• Go to the website or CD-ROM for more self-quiz questions.

The Process of Science

1. A cell containing a single chromosome is placed in a medium containing radioactive phosphate so that any new DNA strands formed by DNA replication will be radioactive. The cell replicates its DNA and divides. Then the daughter cells (still in the radioactive medium) replicate their DNA and divide, so that a total of four cells are present. Sketch the DNA molecules in all four cells, showing a normal (nonradioactive) DNA strand as a solid line and a radioactive DNA strand as a dashed line.

2. **Learn more about AIDS and HIV in The Process of Science activity on the website and CD-ROM.**

Biology and Society

Researchers on the Human Genome Project are determining the nucleotide sequences of human genes. In some cases, they are identifying the proteins encoded by the genes. Scientists at the National Institutes of Health (NIH) have worked out thousands of sequences, and similar analysis is being carried out at universities and private companies. Knowledge of the nucleotide sequences of genes might be used to treat genetic defects or produce lifesaving medicines. The NIH and some U.S. biotechnology companies have applied for patents on their discoveries. In Britain, the courts have ruled that a naturally occurring gene cannot be patented. Do you think individuals and companies should be able to patent genes and gene products? Before answering, consider the following: What are the purposes of a patent? How might the discoverer of a gene benefit from a patent? How might the public benefit? What negative effects might result from patenting genes?

CHAPTER 10

The Control of Gene Expression

Shortly after you drink a milk shake, bacteria living

in your large intestine turn on certain genes.

If a salamander loses a leg, it can grow a new one.

One type of mutant fruit fly has legs growing out of its

head. Cancer-causing genes were first

discovered in a chicken virus. Each of your 46 chromo-

somes has a single DNA molecule that is thousands of

times longer than the diameter of a cell nucleus.

Although the upper fruit fly in Figure 10.1 is in most ways similar to the ones you saw in Chapter 8, there's something very strange about it. Compare it with the lower one, which is a normal wild-type fly. The upper fly has four wings instead of two—an anatomical peculiarity as bizarre as if you had four arms instead of two! The four-winged fly is a mutant. Although wing formation involves many genes, the underlying cause of this abnormality turns out to be surprisingly simple: mutation of a single gene. This gene controls the activity of many other genes as an embryonic fruit fly develops from a zygote, so its mutation can lead to a large-scale abnormality in the adult fly. How genes are controlled and the roles of gene regulation in the lives of flies and other organisms are the subject of this chapter.

Mutant fruit fly

Normal fruit fly

Figure 10.1 A fruit fly mutation resulting in extra wings.

Overview: Why and How Genes Are Regulated

What kind of organism will develop from a particular zygote (fertilized egg)? We know that the DNA in the zygote nucleus contains all the genes that dictate what the organism will look like. But this is not the whole story. Every cell that develops from a zygote by mitotic division (that is, every cell except the organism's gametes) has DNA that is virtually identical to the DNA in the zygote. If all that DNA in every cell in a developing embryo were constantly active, every cell would look like every other cell. The only way that cells with the same genetic information can develop into cells with different structures and functions is if gene activity is regulated. In some way, control mechanisms must determine that certain genes will be turned on while others remain turned off in a particular cell. During the repeated cell divisions that lead from a unicellular zygote to a multicellular organism, individual cells must undergo **cellular differentiation**—that is, become specialized in structure and function. It is the regulation of genes that leads to this specialization. For instance, for cells in the eye of a developing animal to create a lens, genes for lens proteins must be turned on, and other genes must be turned off.

The micrographs in Figure 10.2 show four of the many different types of human cells. Each type of differentiated cell has a unique structure appropriate for carrying out its function. For example, a sperm cell has a powerful flagellum that can propel it through a woman's reproductive tract to meet an egg; and a nerve cell has long extensions for conducting nerve signals between widely separated parts of the body. Together, the various types of human cells enable a person to function as a whole.

What do we actually mean when we say that genes are active or inactive, turned on or off? As we discussed in Chapter 9, genes determine the nucleotide sequences of specific mRNA molecules, and mRNA in turn determines the sequences of amino acids in protein molecules. A gene that is turned on is being transcribed into RNA, and that message is being translated into specific protein molecules. The overall process by which genetic information flows from genes to proteins—that is, from genotype to phenotype—is called **gene expression.**

The turning on and off of transcription is the main way that gene expression is regulated. However, especially in eukaryotic cells, the pathway from gene to functioning protein is a long one, and it provides a number of

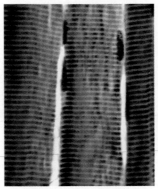

(a) Three muscle cells (partial)

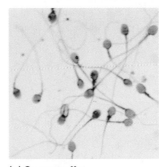

(c) Sperm cells

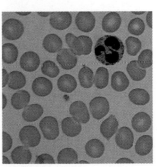

(b) A nerve cell (partial)

(d) Blood cells

Figure 10.2 Four types of human cells. All four micrographs are shown at the same magnification (750×). **(a)** This micrograph shows short segments of three muscle cells, which are long fibers with multiple nuclei (dark vertical rods here). The horizontal stripes result from the arrangement of the proteins that enable the cells to contract. **(b)** The large orange shape is part of a nerve cell, with its nucleus appearing as a lighter orange disk. (The dark spot in the nucleus is a nucleolus.) You can see only a little of this cell's three long extensions, which carry nerve signals from one part of the body to another. **(c)** Sperm cells have long flagella that can propel them to meet an egg cell. **(d)** A single white blood cell (with a three-part darkly stained nucleus) is surrounded by red blood cells. The small size and disklike shape of red blood cells help them efficiently transport oxygen molecules through the narrowest of blood vessels.

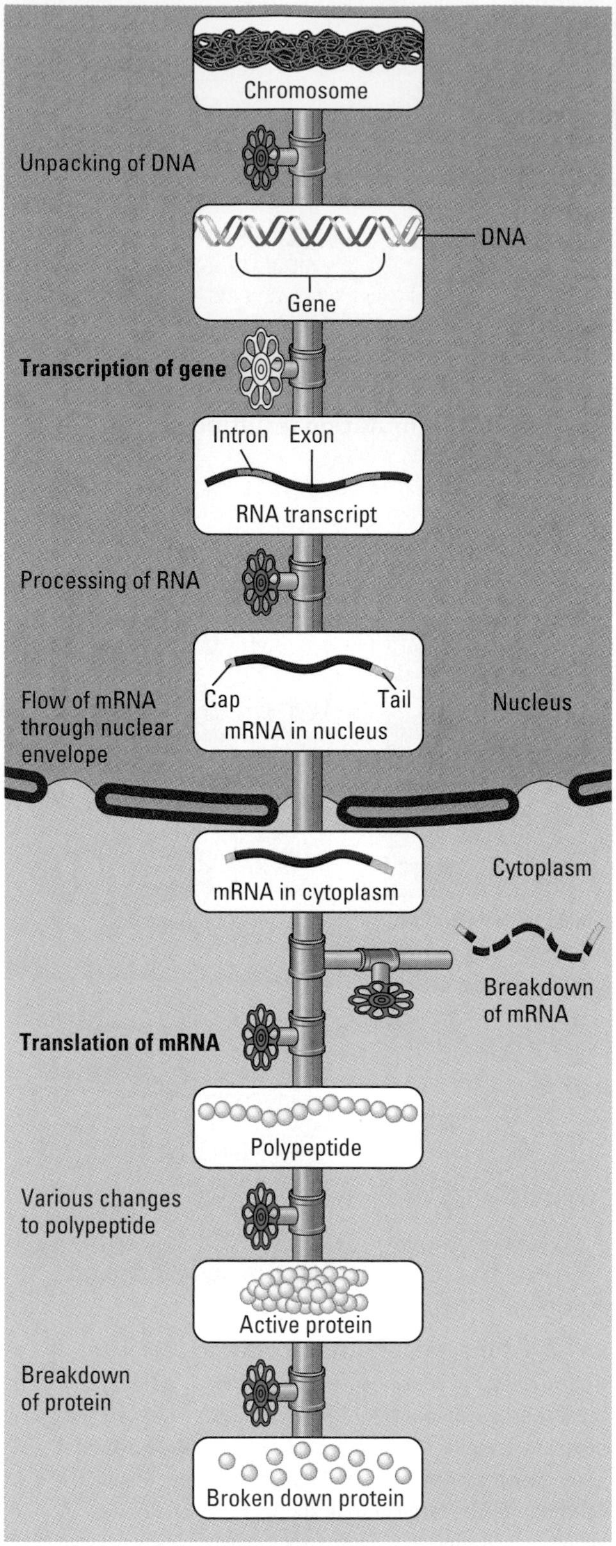
Chromosome
Unpacking of DNA
DNA
Gene
Transcription of gene
Intron
Exon
RNA transcript
Processing of RNA
Cap
Tail
Flow of mRNA through nuclear envelope
mRNA in nucleus
Nucleus
mRNA in cytoplasm
Cytoplasm
Breakdown of mRNA
Translation of mRNA
Polypeptide
Various changes to polypeptide
Active protein
Breakdown of protein
Broken down protein

Figure 10.3 The gene expression "pipeline" in a eukaryotic cell. Each valve in the pipeline represents a stage at which the pathway from chromosome to functioning protein can be regulated. Most regulation of gene expression, in both eukaryotes and prokaryotes, occurs at the transcription stage (yellow knob). By regulating the pathway at this early stage (high in the pipeline), the cell avoids wasting valuable resources.

other points where the process can be regulated: turned on or off, speeded up, or slowed down. Picture the series of pipes that carry water from your local water supply, perhaps a reservoir, to a faucet in your home. At various points, valves control the flow of water. We use this model in Figure 10.3 to illustrate the flow of genetic information from a chromosome—a reservoir of genetic information—to an active protein that has been made in the cell's cytoplasm. The multiple mechanisms that control gene expression are analogous to the control valves in water pipes. In the figure, each gene expression "valve" is indicated by a control knob. These knobs represent *possible* control points; for a typical protein, only one or a few control points are likely to be important. In the figure, the yellow color of the transcription control knob highlights its crucial role.

Get an overview of the control of gene expression in Web/CD Activity 10A.

We take a closer look at how eukaryotic cells control gene expression a little later in the chapter. In the next section we look at how simpler organisms do it.

CheckPoint

What is cellular differentiation?

Answer: The structural and functional specialization of cell types during development of an organism

Gene Regulation in Bacteria

Our earliest understanding of gene control came from the bacterium *Escherichia coli.* As a single cell, this organism does not require the elaborate regulation of gene expression that leads to cell specialization in fruit flies and humans. Nevertheless, *E. coli* changes its activities from time to time in response to changes in its environment. Let's see how the regulation of gene transcription helps *E. coli* accomplish these changes.

The *lac* Operon

Picture an *E. coli* cell living in your intestine. If you eat only a sweet roll for breakfast, the bacterium will be bathed in the sugars glucose and fructose and the digestion products of fats. Later on, if you have a fat-free milk shake and a salad for lunch, *E. coli*'s environment will change drastically.

Let's focus on your milk shake for a moment. One of the main nutrients in milk is the sugar lactose. When lactose is plentiful in the intestine, *E. coli* can make the enzymes necessary to absorb the sugar and use it as an energy source. When lactose is *not* plentiful, *E. coli* does not waste its energy producing these enzymes.

Remember that enzymes are proteins; their production is an example of gene expression. *E. coli* can make lactose-utilization enzymes because it has genes that code for these enzymes. In 1961, French biologists François Jacob and Jacques Monod proposed a hypothesis describing how an *E. coli* cell can adjust its enzyme production in response to changes in its environment. The Jacob and Monod model explains how genes coding for lactose-utilization enzymes are turned off or on, depending on whether lactose is available. Figure 10.4 shows the Jacob-Monod model. (The names are pronounced *zha-KO* and *mo-NO.*)

E. coli uses three enzymes to take up lactose from its environment and start metabolizing it, and the genes coding for these enzymes are adjacent in the DNA and regulated as a unit. At one end of the three enzyme genes are two *control sequences,* short sections of DNA that help control them. One such stretch of nucleotides is a **promoter,** a site where the transcription enzyme, RNA polymerase, attaches and initiates transcription—in this case, transcription of all three lactose enzyme genes. Between the promoter and the enzyme genes, a DNA segment called an **operator** acts as a switch. The operator determines whether RNA polymerase can attach to the promoter and start transcribing the genes.

Such a cluster of genes with related functions, along with a promoter and an operator, is called an **operon;** operons exist only in prokaryotes. The key advantage to the grouping of genes into operons is that it helps coordinate the expression of the genes. The operon discussed here is called the *lac* operon, short for lactose operon. When an *E. coli* cell encounters lactose, all the enzymes needed for its use are made at once, because the operon's genes are all controlled by a single switch, the operator. But what determines whether the operator switch is on or off?

Figure 10.4a shows the *lac* operon in "off" mode, its status when there is no lactose around. Transcription is turned off by a molecule called a **repressor** (shown in red), a protein that functions by binding to the operator and blocking the attachment of RNA polymerase to the promoter. Where does the repressor protein come from? A gene called a **regulatory gene,** located outside the

Drinking a milk shake provides a large supply of lactose to the *E. coli* bacteria in your large intestine. The lactose causes certain of the bacteria's genes to turn on.

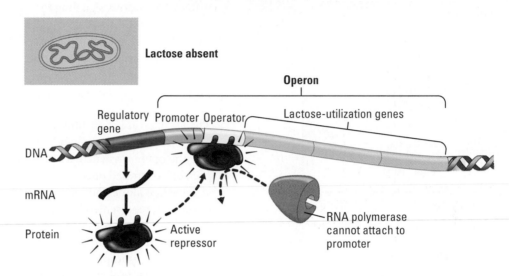

(a) Operon turned off (default state when no lactose is present)

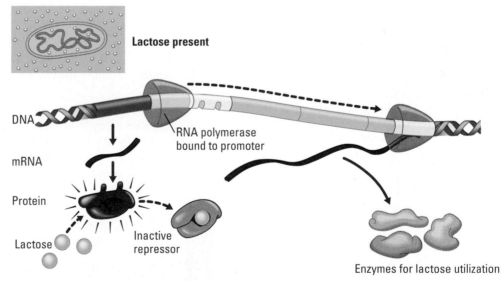

(b) Operon turned on (repressor inactivated by lactose)

Figure 10.4 The *lac* operon of *E. coli.* In the DNA represented in the figure, a small segment of the bacterium's chromosome, the three genes that code for the lactose-utilization enzymes are next to each other (light blue). These genes and two control sequences, a promoter and an operator, make up the *lac* operon. The gene for a repressor is nearby. **(a)** In the absence of lactose, the repressor binds to the operator, preventing transcription of the operon from starting. **(b)** Lactose *de*represses the operon by inactivating the repressor. The result is the production of the three lactose-utilization enzymes.

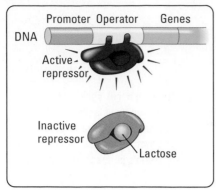

(a) *lac* operon

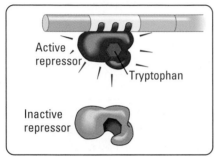

(b) *trp* operon

Figure 10.5 Two types of operons. Both are controlled by repressors, but they differ in whether the repressor is active or inactive in blocking transcription when alone. **(a)** The *lac* repressor is active in the absence of lactose; the binding of lactose causes a shape change that inactivates it. **(b)** The *trp* repressor, on the other hand, assumes its active form only when tryptophan is present. Keep in mind that for both *lac* and *trp,* an active repressor leads to an inactive operon.

operon, codes for the repressor. The regulatory gene is expressed continually, so the cell always has a supply of repressor molecules.

How can an operon be turned on if its repressor is always present? As Figure 10.4b shows, lactose interferes with the attachment of the *lac* repressor to the promoter by binding to the repressor and changing its shape. In its new shape, the repressor is inactive. It cannot bind to the operator, and the operator switch remains on. RNA polymerase can now bind to the promoter and from there transcribe the genes of the operon. The resulting mRNA carries coding sequences for all three enzymes needed for lactose use. The cell can translate this message into separate polypeptides because the mRNA has codons signaling the start and stop of translation.

Interact with the *lac* operon in Web/CD Activity 10B.

The *lac* mRNA and protein molecules remain intact for only a short while before cellular enzymes break them down. When their synthesis stops because lactose is no longer present, they quickly disappear.

Other Kinds of Operons

The *lac* operon is only one type of operon in bacteria. Other types also have a promoter, an operator, and several adjacent genes, but they differ in the way the operator switch is controlled. Figure 10.5 compares two types of repressor-controlled operons. As we already said, the *lac* operon's repressor is active when alone and inactive when combined with lactose. A second type of operon, represented here by the *trp* operon (pronounced "trip"), is controlled by a repressor that is *inactive* alone. To be active, this type of repressor must combine with a specific small molecule. In our example, the small molecule is tryptophan, an amino acid essential for protein synthesis. *E. coli* can make tryptophan from scratch if it has to, using enzymes encoded in the *trp* operon. But it will stop making tryptophan and simply absorb it from its surroundings whenever possible. When *E. coli* is swimming in tryptophan, which occurs in large amounts in such foods as milk and poultry, tryptophan binds to the repressor of the *trp* operon. This activates the *trp* repressor, enabling it to switch off the operon. Thus, this type of operon allows bacteria to stop making certain essential molecules when the molecules are already present in the environment, saving materials and energy for the cells.

A third type of operon (not illustrated) uses **activators,** proteins that turn operons *on* by binding to DNA. These proteins act by somehow making it easier for RNA polymerase to bind to the promoter. Armed with a variety of operons, regulated by repressors and activators, *E. coli* and other prokaryotes can thrive in frequently changing environments.

CheckPoint

A certain mutation in *E. coli* makes the *lac* operator unable to bind the active repressor. How would this affect the cell?

Answer: The cell would wastefully produce the enzymes for lactose metabolism continuously, even in the absence of lactose.

Regulation of Eukaryotic Gene Expression

Eukaryotic organisms are not only larger and more complex than prokaryotes; they also have much more DNA and many more genes. The human genome, for example, has about 600 times as much DNA as the *E. coli* genome and about 40 times as many genes. Not surprisingly, eukaryotic cells have more elaborate mechanisms than bacteria for regulating the expression of their genes.

The Genetic Potential of Cells

Before we discuss eukaryotic control of gene expression, let's examine the evidence that the differentiated body cells of a multicellular eukaryote actually have complete genomes. For example, does a cell in your intestine still have the genetic instructions for making an eye, or have the genes needed for eye formation been permanently inactivated or lost? One way to approach this question is to see if it is possible to reverse differentiation so that a differentiated cell expresses all its genes and generates a complete organism. If you have ever grown a plant from a small cutting, you've seen evidence that plant cells, at least, may have that capability. In fact, in plants the ability of even a single differentiated cell to develop into a whole new individual is common. Figure 10.6 shows this process using a carrot plant. A single cell removed from the root (the carrot) and placed in culture medium may begin dividing and eventually grow into an adult plant. This technique can be used to produce hundreds or thousands of genetically identical organisms —**clones**—from the somatic (body) cells of a single plant. In this way, it is possible to propagate large numbers of plants that have desirable traits such as high fruit yield or resistance to disease. The fact that a mature plant cell can dedifferentiate and then give rise to all the specialized cells of a new plant shows that differentiation does not necessarily involve irreversible changes in the DNA.

But is cloning possible in animals? An indication that differentiation need not impair an animal cell's genetic potential is the natural process of **regeneration,** the regrowth of lost body parts. When a salamander loses a leg, for example, certain cells in the leg stump reverse their differentiated state, divide, and then redifferentiate to give rise to a new leg. Many animals can regenerate lost parts, and in a few animals, isolated differentiated cells can be made to dedifferentiate and then develop into an organism.

However, in many other animals, including humans, we do not see such dramatic examples of natural dedifferentiation. In these animals, are genes lost or permanently turned off in the differentiated cells? Scientists have looked for an answer by replacing the nucleus of an egg or zygote with the nucleus from a differentiated cell. If genes are lost or irreversibly turned off during differentiation, the transplanted nucleus will not allow the development of a normal embryo. The pioneering experiments in nuclear transplantation were performed by American embryologists Robert Briggs and Thomas King in the 1950s (Figure 10.7). Replacing the nuclei of frog egg

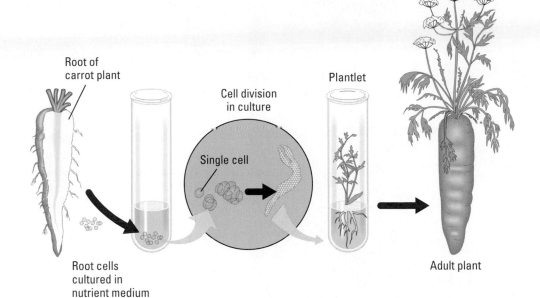

Figure 10.6 Test-tube cloning of a carrot plant. An entire carrot plant can be grown from a differentiated root cell. The new plant is a genetic duplicate of the parent plant.

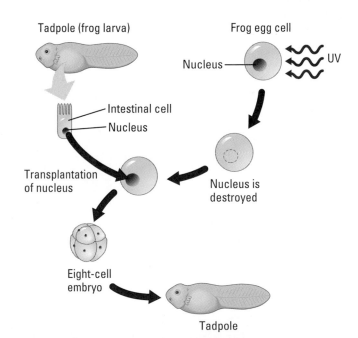

Figure 10.7 A nuclear transplantation experiment using the frog. Researchers Briggs and King destroyed the nuclei of frog egg cells with ultraviolet (UV) light and then transplanted nuclei from tadpole intestinal cells into the eggs. (The tadpole is the juvenile form of the frog; its intestinal cells are already differentiated.) Many of the eggs containing transplanted nuclei started to develop, and a few even developed into normal tadpoles. This result showed that at least some differentiated animal cells retain the full genetic potential of a zygote.

Figure 10.8 Cloning of a mammal. The young sheep at the left is the famous Dolly; the sheep at the right is her surrogate mother. The nucleus of the egg that gave rise to Dolly came from a differentiated sheep mammary cell, making Dolly genetically identical to the mammary cell donor. She is not related to the surrogate mother who carried her to term.

cells with nuclei from tadpole intestinal cells, these researchers demonstrated that at least some of the nuclei from the differentiated tadpole cells could drive the development of a normal frog. In other words, Briggs and King were able to clone frogs using nuclei from differentiated cells. Their work showed that nuclei from differentiated animal cells can retain all of their genetic potential.

A breakthrough in animal cloning came in 1997 with the first cloning of a mammal using a differentiated nucleus. In this case, researchers used an electric shock to fuse a specially treated sheep mammary cell with an egg from which they had removed the nucleus. The egg began to divide, was implanted in the uterus of an unrelated sheep (the surrogate mother), and developed into the celebrated Dolly. As predicted, Dolly resembles her genetic parent, the mammary cell donor, not the egg donor or the surrogate mother (Figure 10.8). Within two years of Dolly's birth, researchers in Hawaii cloned mice by transplanting ovary cell nuclei into enucleated eggs, and a Japanese team cloned a cow.

Patterns of Gene Expression in Differentiated Cells

If all the differentiated cells in an individual organism contain the same genes, and all the genes have the potential of being expressed, the great differences among cells in an organism must result from the selective expression of genes. As a developing embryo undergoes successive cell divisions, specific genes must be activated in different cells at different times. Groups of cells follow diverging developmental pathways, and each group develops into a particular kind of tissue. Finally, in the mature organism, each cell type—nerve or pancreas, for instance—has a different pattern of turned-on genes.

Figure 10.9a illustrates patterns of gene expression for four genes in three different specialized cells of a human. The genes for the enzymes of the metabolic pathway of glycolysis are active in all metabolizing cells, including the pancreas cell, embryonic lens cell, and nerve cell shown here. However, the genes for specialized proteins are expressed only by particular cells.

The specialized proteins we use as examples are the transparent protein crystallin, which forms the lens of the eye; insulin, a hormone made in the pancreas; and the oxygen transport protein hemoglobin. Notice that the hemoglobin genes are not active in any of the cell types shown here. They are turned on only in cells that are developing into red blood cells. Insulin genes are activated only in the cells of the pancreas that produce that hormone. Nerve cells express genes for other specialized proteins not shown here. Mature lens cells, like mature red blood cells, achieve the ultimate in differentiation. After a frenzy of expressing crystallin genes and accumulating the protein products, these cells lose their nuclei and thus all their genes. Using current DNA technology, researchers can now look at patterns of activation of hundreds or even thousands of genes at a time (Figure 10.9b).

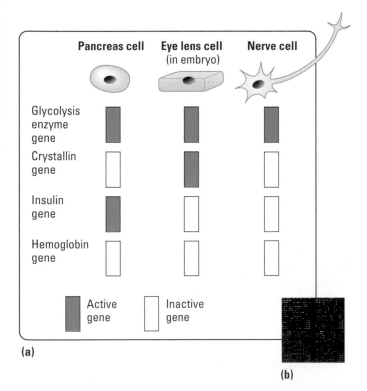

(a)

(b)

Figure 10.9 Patterns of gene expression. (a) Shown here are the activity states of four genes in three types of human cells. Different types of cells express different combinations of genes. **(b)** Shown actual size, this "microarray" contains tiny samples of hundreds of human genes arranged in an orderly pattern. Microarrays are used for testing the activity of many genes at one time.

So we see that eukaryotic cells become specialized because they express only certain genes. Thus, cellular differentiation in multicellular organisms results from selective gene expression, as does the ability of bacteria to produce different enzymes as needed. Now it's time to examine the control of gene expression in eukaryotes more closely. We begin by looking at how DNA is packaged in a eukaryotic cell, because this influences transcription.

DNA Packing in Eukaryotic Chromosomes

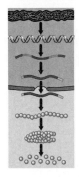

The nucleus in any one of your cells would be the envy of a computer manufacturer. Computer chips can store large amounts of information, but they are still a long way from the extraordinary storage capacity of a eukaryotic cell. Each eukaryotic chromosome contains a single DNA molecule that is thousands of times longer than the 5-μm diameter of a typical nucleus. The total DNA in a human cell's 46 chromosomes would stretch for 3 meters! All this DNA can fit into the nucleus because of an elaborate, multilevel system of coiling and folding, or *packing*, of the DNA in each chromosome. A crucial aspect of DNA packing is the association of the DNA with small proteins called **histones,** found only in eukaryotes. (Bacteria have analogous proteins, but they lack the degree of DNA packing found in eukaryotes.)

Figure 10.10 outlines a simplified model for the main levels of DNA packing. As shown at the top, the double helix has a diameter of 2 nm, which is not altered by packing. At the first level of packing, histones attach to the DNA. In electron micrographs, the combination of DNA and histones has the appearance of beads on a string. Each "bead," called a **nucleosome,** consists of DNA wound around a protein core of eight histone molecules. At the next level of packing, the beaded string is wrapped into a tight helical fiber. Then this fiber coils further into a thick supercoil with a diameter of about 200 nm. Looping and folding can further compact the DNA, as you can see in the metaphase chromosome at the bottom of the figure. (DNA packing during mitosis is so specific and precise that individual genes always end up at the same positions in the metaphase chromosome.) Viewed as a whole, Figure 10.10 gives a sense of how successive levels of coiling and folding enable a huge amount of DNA to fit into a cell nucleus.

DNA packing tends to prevent gene expression, presumably by preventing RNA polymerase and other transcription proteins from contacting the DNA. At the nucleosome level, histones participate in the short-term switching on and off of genes: For a gene to be transcribed, the histones must loosen their grip on the DNA. Nonhistone chromosomal proteins (not shown in Figure 10.10) control how tightly the histones bind to the DNA.

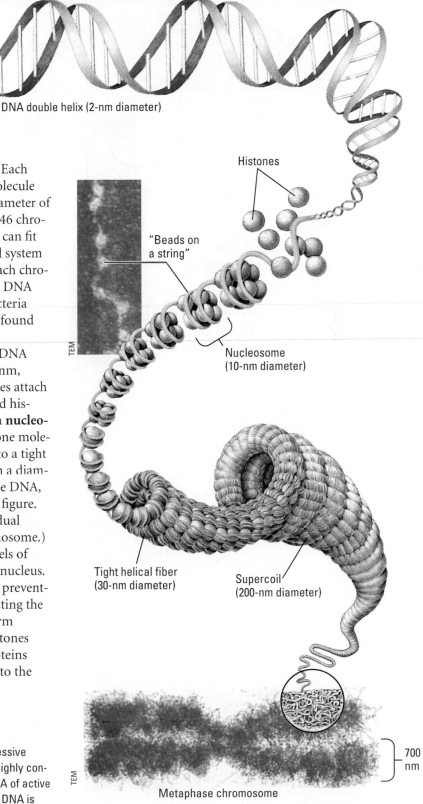

DNA double helix (2-nm diameter)

Histones

"Beads on a string"

Nucleosome (10-nm diameter)

Tight helical fiber (30-nm diameter)

Supercoil (200-nm diameter)

700 nm

Metaphase chromosome

Figure 10.10 DNA packing in a eukaryotic chromosome. Successive levels of coiling of the DNA and associated proteins ultimately results in the highly condensed chromosomes seen in mitosis and meiosis. During interphase, the DNA of active genes is only lightly packed, in the "beads on a string" arrangement. Inactive DNA is packed more tightly, evidence that the packing plays a role in keeping genes turned off.

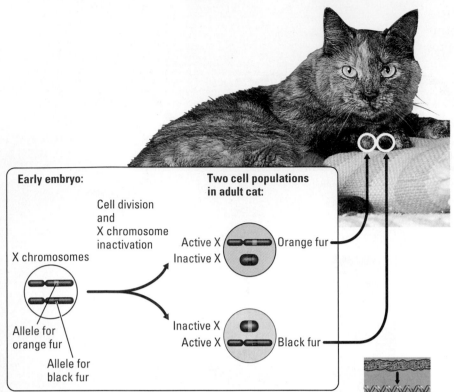

Figure 10.11 Tortoiseshell pattern on a cat, a result of X chromosome inactivation. The tortoiseshell gene is on the X chromosome, and the tortoiseshell phenotype requires the presence of two different alleles, one for orange fur and one for nonorange (black) fur. Normally, only females can have both alleles, because only they have two X chromosomes. If a female is heterozygous for the tortoiseshell gene, she is tortoiseshell. Orange patches are formed by populations of cells in which the X chromosome with the orange allele is active; black patches have cells in which the X chromosome with the nonorange allele is active. (Calico cats also have white areas, which are determined by another gene.)

Cells may use the higher levels of packing for long-term inactivation of genes. Highly compacted chromatin, which is found not only in mitotic chromosomes but also in some regions of interphase chromosomes, is generally not expressed at all. One intriguing case is seen in female mammals, where one entire X chromosome in each somatic cell is highly compacted and almost entirely inactive, during interphase. This **X chromosome inactivation** first takes place early in embryonic development, when one of the two X chromosomes in each cell is inactivated at random. The inactivation is inherited by a cell's descendants. Consequently, a female heterozygous for genes on the X chromosome has populations of cells that express different X-linked alleles. A striking effect of X chromosome inactivation is the tortoiseshell cat, which has orange and black patches of fur (Figure 10.11).

Regulation of Transcription and RNA Processing

The packing and unpacking of chromosomal DNA provide a coarse adjustment for eukaryotic gene expression by making a region of DNA either more available or less available for transcription. The fine-tuning begins with the initiation of transcription. In both prokaryotes and eukaryotes, this is the most important stage for regulating gene expression.

Initiation of Transcription Like prokaryotes, eukaryotes employ regulatory proteins that attach to DNA and turn the transcription of genes on and off. The eukaryotic control mechanisms involve proteins that, like prokaryotic repressors and activators, bind to specific segments of DNA. However, eukaryotic cells have more regulatory proteins and more control sequences in their DNA. A model for the eukaryotic regulation of transcription is shown in Figure 10.12. This model features an intricate array of regulatory proteins that interact with DNA and with one another to turn genes on or off.

In contrast to the genes of bacterial operons, each eukaryotic gene usually has its own promoter and other control sequences. Moreover, activator proteins seem to be more important than repressors. A typical animal or plant cell needs to turn on (transcribe) only a small percentage of its genes, those required for the cell's specialized structure and function. In multicellular organisms, the "default" state for most genes seems to be "off." However, housekeeping genes, those continually active in virtually all cells for routine activities such as glycolysis, may be in an "on" state by default.

Turning on a eukaryotic gene involves regulatory proteins called **transcription factors** in addition to RNA polymerase. Activators are one type of transcription factor. In the model depicted in Figure 10.12, the first step in initiating gene transcription is the binding of activator proteins to DNA sequences called **enhancers.** In contrast to the operators of prokaryotic operons, enhancers are usually far away from the gene they help regulate and may be on either side of the gene. How does the binding of an activator protein to an enhancer sequence in the DNA influence a distant gene? Apparently, the DNA bends, and the bound activators interact with other transcription factors, which then bind together at the gene's promoter. This large assembly of proteins somehow facilitates the correct attachment of RNA polymerase to the promoter and the initiation of transcription. As shown in the figure, several enhancers and activators may be involved. Not shown are *repressor* proteins that may bind to DNA sequences called **silencers** and function analogously to *inhibit* the start of transcription.

In summary, both eukaryotes and prokaryotes regulate transcription by using regulatory proteins that bind to DNA. However, many more regulatory proteins are involved in eukaryotes, and the interactions among these proteins are more complicated.

If eukaryotic genomes do not have operons, how does the eukaryotic cell deal with genes of related function that need to be turned on or off at the same time? Genes coding for the enzymes of a metabolic pathway, for example, are often scattered over different chromosomes. However, if the same specific enhancer (or collection of enhancers) is associated with every gene of a dispersed group, the genes will be activated at the same time. The transcription factors that recognize these DNA sequences will promote simultaneous transcription of the genes.

RNA Processing Both prokaryotes and eukaryotes regulate gene expression at transcription, but the eukaryotic cell also has additional opportunities for control at later stages (see Figure 10.3). The eukaryotic cell not only localizes transcription in the nucleus but (as you learned in Chapter 9) also modifies, or *processes,* the RNA transcripts there, before they move to the cytoplasm for translation by the ribosomes. This processing includes the addition of a cap and a tail to the RNA, as well as the splicing out of any introns, the noncoding DNA segments that interrupt the genetic message.

The removal of introns and the splicing together of the remaining exons provide several possible ways for regulating gene expression. Some scientists think that the splicing process itself may help control the flow of mRNA from nucleus to cytoplasm, because until splicing is completed, the RNA is attached to the molecules of the splicing machinery and cannot pass through the nuclear pores. Moreover, in some cases, the cell can carry out splicing in more than one way, generating different mRNA molecules from the same RNA transcript. Notice in Figure 10.13, for example, that one mRNA molecule ends up with the green exon and the other with the brown exon. With this sort of alternative RNA splicing, an organism can get more than one type of polypeptide from a single gene. The fruit fly provides an interesting example of two-way mRNA splicing. In this animal, the differences between males and females are controlled by two sets of regulatory proteins, one set specific to the male and one to the female. The two sets of proteins, however, are coded by the same set of genes; the sex differences are largely due to different patterns of RNA splicing.

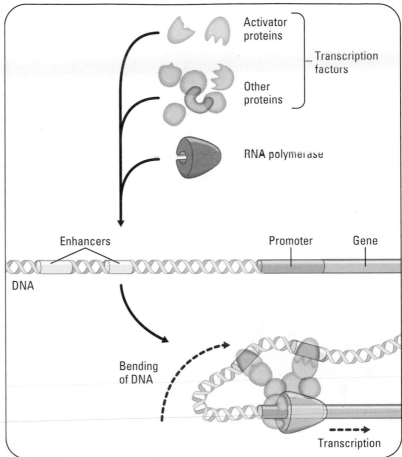

Figure 10.12 A model for the turning on of a eukaryotic gene. A large assembly of proteins and several control sequences in the DNA are involved in initiating the transcription of a eukaryotic gene. See the detailed description in the text.

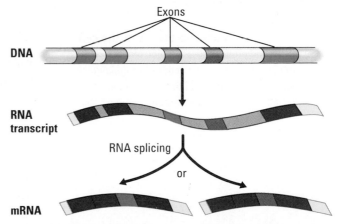

Figure 10.13 Production of two different mRNAs from the same gene. By recognizing different parts of a gene as exons, two different cells can use a given DNA gene to synthesize somewhat different mRNAs and proteins. The cells might be cells of different tissues or the same type of cell at different times in the organism's life. Alternative RNA splicing is a means by which the cell can exert differential control over the expression of certain genes.

Another potential control spot in the eukaryotic gene expression pathway is the transport of mRNA from nucleus to cytoplasm. By facilitating or blocking this transport, the cell can regulate the timing and amount of protein synthesis from a molecule of mRNA.

Regulation in the Cytoplasm

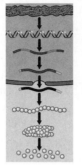

After eukaryotic mRNA is transported to the cytoplasm, there are additional opportunities for regulation. These include mRNA breakdown, translation, protein modifications, and protein breakdown.

Breakdown of mRNA The lifetime of mRNA molecules is one important factor regulating the amounts of various proteins the cell produces. Long-lived mRNAs get translated into many more protein molecules than do short-lived ones. Prokaryotic mRNAs have very short lifetimes; they are degraded by enzymes after only a few minutes. This is one reason bacteria can change their proteins relatively quickly in response to environmental changes. In contrast, the mRNA of eukaryotes can have lifetimes of hours or even weeks. But eventually, they are broken down and their parts recycled (Figure 10.14).

A striking example of long-lived mRNA is found in vertebrate red blood cells, which act like factories for manufacturing the protein hemoglobin. In most species of vertebrates, the mRNAs for hemoglobin are unusually stable. They probably last as long as the red blood cells that contain them—about a month in birds and perhaps longer in reptiles, amphibians, and fishes—and are translated repeatedly. Mammals are an exception. When their red blood cells mature, they lose their ribosomes (along with their other organelles) and thus cease to make new hemoglobin. However, mammalian hemoglobin itself lasts about as long as the red blood cells last, around four months.

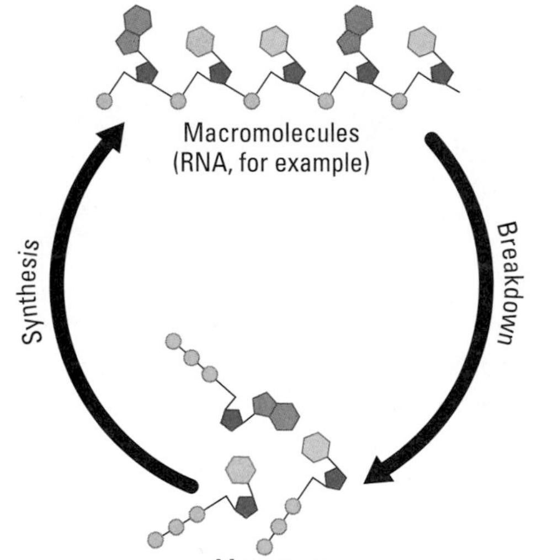

Regulation of Translation The process of translation also offers opportunities for regulation. Among the molecules involved in translation are a great many proteins that have regulatory functions. Red blood cells, for instance, have a protein that prevents translation of hemoglobin mRNA unless the cell has a supply of heme, the iron-containing chemical group essential for hemoglobin function.

Protein Alterations The final opportunities for the control of gene expression occur after translation. Post-translational control mechanisms in eukaryotes often involve cutting eukaryotic polypeptides into smaller, active final products. The hormone insulin, for example, is synthesized in cells of the pancreas as one long polypeptide, which is not active as a hormone. Then, as illustrated in Figure 10.15, a large center portion is cut away, leaving two shorter chains. This combination of two shorter polypeptides is the active form of insulin.

Macromolecules (RNA, for example)

Synthesis

Breakdown

Monomers (nucleotides, for example)

Figure 10.14 Recycling of macromolecules in the cell. The breakdown of mRNA and proteins after varying lengths of time—mRNA and protein "turnover"—are modes of regulation that occur in the cytoplasm. The components of these macromolecules can be recycled and used for making new macromolecules.

Protein Breakdown Another control mechanism operating after translation is the selective breakdown of proteins. Though mammalian hemoglobin may last as long as the red blood cell housing it, the lifetimes of many other proteins are closely regulated. Some of the proteins that trigger metabolic changes in cells are broken down within a few minutes or hours. This regulation allows a cell to adjust the kinds and amounts of its proteins in response to changes in its environment. It also enables the cell to maintain its proteins in prime working order. When proteins are damaged, they are usually broken down right away and replaced by new ones that function properly.

You can review the various control points in the eukaryotic pathway of gene expression by turning back to Figure 10.3. However, that figure omits

> Review the mechanisms that control gene expression in eukaryotes in Web/CD Activity 10C.

the web of control that connects different genes, often through their products. We have seen examples in both prokaryotes and eukaryotes of the actions of gene products (proteins) on other genes or on other gene products within the same cell. The genes of bacterial operons are controlled by repressor or activator proteins encoded by regulatory genes on the same DNA molecule. In eukaryotes, many genes are controlled by proteins encoded by regulatory genes on different chromosomes.

In a multicellular organism, the web of regulation crosses cell boundaries. This starts in the early embryo. A cell of the embryo can produce and secrete chemicals such as hormones that induce a neighboring or distant cell to develop in a certain way. In the next section, we will look at some examples of this cell-to-cell regulation in the developing fruit fly embryo.

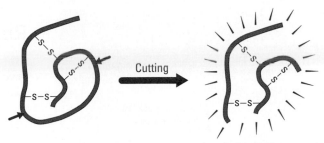

Initial polypeptide   Insulin (active hormone)

Figure 10.15 Formation of an active insulin molecule. One type of post-translational control is the cutting of a polypeptide to convert it to its active form. The hormone insulin, for example, is synthesized as one polypeptide. A special enzyme then removes an interior section. The remaining polypeptide chains are held together by covalent bonds between the sulfur atoms of sulfur-containing amino acids. Only in this form does the protein act as a hormone.

CheckPoint

1. If your nerve cells and skin cells have exactly the same genes, then how can they be so different?

2. What evidence supports the view that differentiation is based on the control of gene expression rather than on irreversible changes in the genome?

3. How does dense packing of DNA in chromosomes prevent gene expression?

4. In stimulating transcription of a specific eukaryotic gene, an enhancer does not act directly on the gene's promoter, but has its effect via DNA-binding proteins called _____ _____.

5. Explain why many eukaryotic genes are longer than the forms of their mRNA that leave the nucleus.

6 Once mRNA encoding a particular protein reaches the cytoplasm, what are four mechanisms that can regulate the amount of the active protein in the cell?

Answers: 1. Each cell type must be expressing certain genes that are present in, but not expressed in, the other cell type. **2.** The nuclear transplantation experiments in animals and the cloning of plants and animals from differentiated cells described in Figures 10.6, 10.7, and 10.8 **3.** RNA polymerase and other proteins required for transcription do not have access to the DNA in tightly packed regions of a chromosome. **4.** transcription factors **5.** Many eukaryotic genes have introns; RNA splicing removes these regions from the RNA transcripts before the RNA leaves the nucleus. **6.** Breakdown of the mRNA; regulation of translation; activation of the protein (by polypeptide cutting, for example); and breakdown of the protein

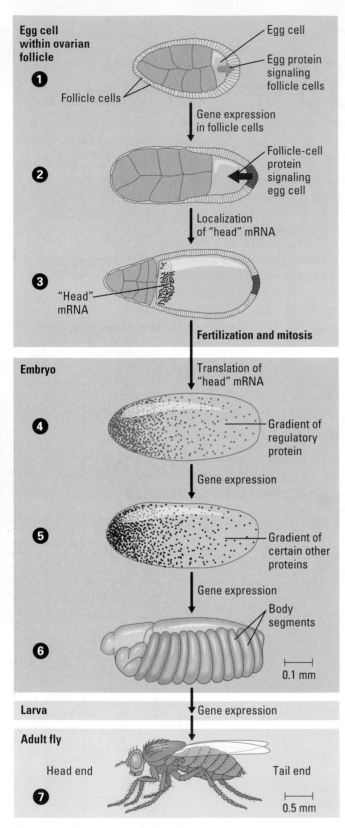

Egg cell within ovarian follicle

1

Follicle cells

Egg cell

Egg protein signaling follicle cells

Gene expression in follicle cells

2

Follicle-cell protein signaling egg cell

Localization of "head" mRNA

3

"Head" mRNA

Fertilization and mitosis

Embryo

Translation of "head" mRNA

4

Gradient of regulatory protein

Gene expression

5

Gradient of certain other proteins

Gene expression

6

Body segments

0.1 mm

Larva

Gene expression

Adult fly

Head end

Tail end

7

0.5 mm

Figure 10.16 The early development of a fruit fly. This figure shows how head-tail orientation is determined in *Drosophila melanogaster*. Important events begin inside the mother fly while the egg cell is still maturing (steps 1–3). Once the egg is fertilized and laid, successive activations turn on the batteries of genes that direct development of the larva and eventually the adult fly. Most of the gene products that regulate other genes are transcription factors.

Regulation of Embryonic Development

Mutant fruit flies, such as the one with four wings that opened this chapter, gave researchers some of their earliest glimpses into the relationship between gene expression and the development of an organism. Such mutants have led to the identification of many of the genes that program development in the normal fly. The use of this genetic approach has revolutionized developmental biology, the study of how multicellular organisms arise from a fertilized egg. It turns out that the regulation of gene expression within cells and across cell boundaries plays a central role in the development of fruit flies and all other multicellular organisms.

Cascades of Gene Activation and Cell Signaling in *Drosophila*

Among the earliest events in fruit fly development are ones that determine which end of the egg cell will become the head and which end will become the tail. These events occur in the ovaries of the mother fly and involve communication between an unfertilized egg cell and cells adjacent to it in its follicle (egg chamber). As Figure 10.16 indicates, **1** one of the first genes activated in the egg cell codes for a protein that leaves the egg cell and signals adjacent follicle cells. **2** These follicle cells are stimulated to turn on genes for other proteins, which signal back to the egg cell. **3** One of the egg cell's responses is to localize a specific type of mRNA at one end of the cell. This mRNA marks the end of the egg where the fly's head will develop and thus defines the animal's head-tail axis. (One piece of evidence from mutant studies is that a defective gene for the "head" mRNA causes the embryo to develop tails at both ends!) Similar changes in the unfertilized egg directed by the mother fly establish the other axes (top to bottom and side to side) and thus lay out a plan for the positioning of body parts as they develop. As one developmental biologist has put it, "Mom tells junior which way is up."

Experiment on fruit fly embryos in Web/CD Activity 10D/The Process of Science.

After the egg is fertilized and laid, repeated mitoses transform the zygote into an embryo. **4** Translation of the "head" mRNA in the early embryo produces a regulatory protein (green dots) that diffuses through the embryo but remains most concentrated at the head end. In turn, this protein gradient triggers a corresponding gradient of transcription in the embryo's nuclei. **5** The proteins (purple dots) resulting from translation of this RNA initiate more rounds of gene expression. Cell signaling—now among the cells of the embryo—helps drive the process. **6** The result is the subdivision of the embryo's body into segments.

Now finer details of the fly can take shape. Protein products of some of the axis-specifying genes and segment-forming genes activate yet another set of genes. These genes, called homeotic genes, determine what body parts will develop from each segment. A **homeotic gene** is a master control gene that regulates batteries of other genes that actually create the anatomical identity of parts of the body. For example, one set of homeotic genes in fruit flies instructs cells in the segments of the head and thorax (midbody) to form antennae and legs, respectively. Elsewhere, these homeotic genes remain turned off, while others are turned on. The eventual outcome, shown in step **7** of the figure, is an adult fly. Notice that the adult's body segments correspond to those of the embryo in step 6.

Cascades of gene expression, with the protein products of one set of genes activating another set of genes, and so on, are a common theme in development. Many of the proteins, such as the one in step 1 of Figure 10.16, do not act on genes directly. Instead, they act as molecular signals that indirectly trigger expression of a gene in a neighboring cell. Now let's take a closer look at how this happens.

Cell Signaling

Cell-to-cell signaling, with proteins or other kinds of molecules carrying messages from signaling cells to receiving (target) cells, is a key mechanism in development, as well as in the coordination of cellular activities in the mature organism. As you saw in Figure 4.31, a signal molecule usually acts by binding to a receptor protein in the plasma membrane of the target cell and initiating a signal transduction pathway in the cell. This is a series of molecular changes that converts a signal received on a target cell's surface into a specific response inside the cell. Figure 10.17 shows the main elements of a signal transduction pathway in which the target cell's response is the transcription (turning on) of a gene.

Watch an animation of a signal transduction pathway in Web/CD Activity 10E.

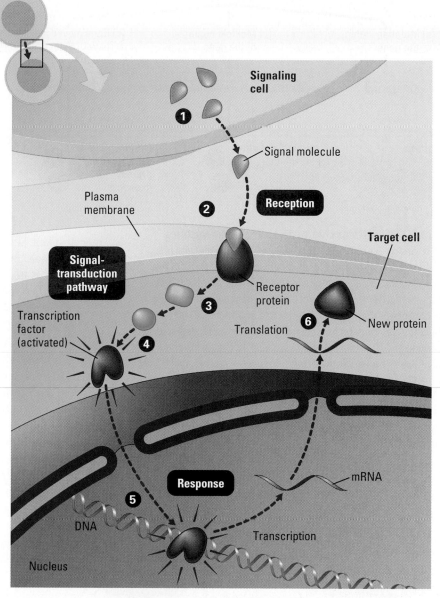

Figure 10.17 **A cell-signaling pathway that turns on a gene.** ❶ First, the signaling cell secretes the signal molecule. ❷ This molecule binds to a specific receptor protein embedded in the target cell's plasma membrane—the reception step. ❸ The binding activates the first in a series of relay proteins (green) within the target cell. Each relay molecule activates another. This is a signal transduction pathway. ❹ The last relay molecule in the series activates a transcription factor that ❺ triggers transcription of a specific gene, the response to the signal. ❻ Translation of the mRNA produces a protein.

CheckPoint

1. What determines which end of a developing fruit fly becomes the head?
2. How can a signal molecule from one cell alter gene expression in a target cell without even entering the target cell?

Answers: 1. A specific kind of mRNA localizes at the end of the unfertilized egg that will become the head. After fertilization of the egg, this mRNA is translated into a regulatory protein that forms a gradient starting at the head end. **2.** By binding to a receptor protein in the membrane of the target cell and triggering within the cell a signal transduction pathway that activates transcription factors

The Genetic Basis of Cancer

In Chapter 7, we introduced cancer as a variety of diseases in which cells escape from the control mechanisms that normally limit their growth and division. In recent years, scientists have learned that this escape from normal controls is due to changes in some of the cells' genes or to changes in the way certain genes are expressed.

Genes That Cause Cancer

The abnormal behavior of cancer cells was observed years before anything was known about the cell cycle, its control, or the role genes play in making cells cancerous. One of the earliest clues to the cancer puzzle was the discovery, in 1911, of a virus that causes cancer in chickens. Recall that viruses are simply molecules of DNA or RNA coated with protein and in some cases a membranous envelope. (We referred to viruses as "packaged genes"

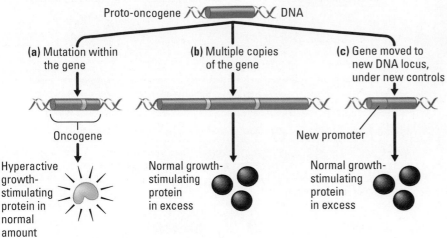

Proto-oncogene DNA

(a) Mutation within the gene

Oncogene

Hyperactive growth-stimulating protein in normal amount

(b) Multiple copies of the gene

Normal growth-stimulating protein in excess

(c) Gene moved to new DNA locus, under new controls

New promoter

Normal growth-stimulating protein in excess

Figure 10.18 Alternative ways to make oncogenes from a proto-oncogene. Let's assume that the starting proto-oncogene codes for a protein that stimulates cell division. **(a)** A mutation (green) in the proto-oncogene itself may create an onco-gene that codes for a hyperactive protein, one whose stimulating effect is stronger than normal. **(b)** An error in DNA replication or re-combination may generate multiple copies of the gene, which are all transcribed and translated. The result would be an excess of the nor-mal stimulatory protein. **(c)** The proto-oncogene may be moved from its normal location in the cell's DNA to another location. At its new site, the gene may be under the control of a different promoter and other transcriptional control sequences, which cause it to be tran-scribed more often than normal. In that case, the normal protein would be made in excess.

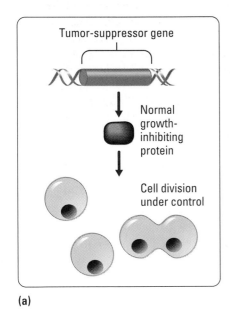

Tumor-suppressor gene

Normal growth-inhibiting protein

Cell division under control

(a)

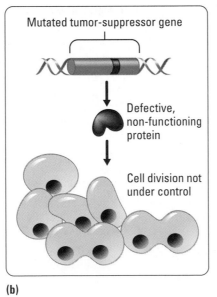

Mutated tumor-suppressor gene

Defective, non-functioning protein

Cell division not under control

(b)

Figure 10.19 Tumor-suppressor genes. **(a)** A tumor-suppressor gene normally codes for a protein that inhibits cell growth and division. In this way, the gene helps pre-vent cancerous tumors from arising. **(b)** When a mutation in a tumor-suppressor gene makes its protein defective, cells that are usually under the control of the normal protein may divide excessively, eventually forming a tumor.

in Chapter 9.) Viruses that cause cancer can become per-manent residents in host cells by inserting their nucleic acid into the DNA of host chromosomes. It is now known that a number of viruses that can produce cancer carry specific cancer-causing genes in their nucleic acid. When inserted into a host cell, these genes can make the cell can-cerous. A gene that causes cancer is called an **oncogene** (from the Greek *onkos*, tumor).

Oncogenes In 1976, American molecular biologists J. Michael Bishop, Harold Varmus, and their colleagues made a startling discovery. They found that a virus that causes cancer in chickens contains an oncogene that is an altered version of a gene found in normal chicken cells. Apparently, the virus had picked up the gene from a for-mer host cell. Subsequent research has shown that the chromosomes of many animals, including humans, contain genes that can be converted to oncogenes. A normal gene with the potential to become an oncogene is called a **proto-oncogene.** A cell can acquire an oncogene from a virus or from the conversion of one of its own genes.

The work by Bishop and Varmus focused cancer research on proto-oncogenes. Searching for the normal role of these genes in the cell, re-searchers found that many proto-oncogenes code for **growth factors**—pro-teins that stimulate cell division—or for other proteins that somehow affect growth factor function or some other aspect of the cell cycle. When all these proteins are functioning normally, in the right amounts at the right times, they help keep the rate of cell division at an appropriate level.

For a proto-oncogene to become an oncogene, a mutation must occur in the cell's DNA. Mutations that produce most types of cancer occur in so-matic cells, those not involved in gamete formation. Figure 10.18 illustrates three kinds of changes in somatic cell DNA that can produce active onco-genes, starting with a proto-oncogene that encodes a pro-tein that stimulates cell division. In all three cases, normal gene expression is changed, and the cell is stimulated to di-vide excessively.

Tumor-Suppressor Genes Changes in genes whose prod-ucts *inhibit* cell division are also involved in cancer. These genes are called **tumor-suppressor genes** because the pro-teins they encode normally help prevent uncontrolled cell growth (Figure 10.19). Any mutation that keeps the nor-mal tumor-suppressor protein from being made or functioning may contribute to uncontrolled cell division. So we have now seen several ways that genetic change may be connected to a malfunction of the cell cycle.

Effects of Cancer Genes on Cell-Signaling Pathways Re-searchers are now working out the details of how onco-genes and defective tumor-suppressor genes contribute to uncontrolled cell growth. Normal proto-oncogenes and tumor-suppressor genes often code for proteins involved in signal transduction pathways leading to gene expres-sion, pathways similar to the one described in Figure 10.17. Let's consider the case of a proto-oncogene named *ras*, which in mutant form is an oncogene involved in many kinds of cancer. The protein encoded by *ras* is a component of signal transduction pathways leading to growth-stimulating responses. So, for example, if an oncogene-creating mutation causes the *ras* protein (called Ras) to be hyperactive, the pathway may be overstimulated, and excessive cell division may result. To produce cancer, though, further abnormalities are usually needed, as we discuss next.

The Progression of a Cancer Including baseball player Daryl Strawberry and perhaps some of your own friends or relatives over 130,000 Americans were stricken by cancer of the colon (large intestine) or rectum in 1998 and similar numbers in 1999 and 2000. One of the best-understood types of human cancer, colon cancer illustrates an important principle about how cancer develops: *More than one somatic mutation is needed to produce a full-fledged cancer cell.* As in many cancers, the development of a colon cancer that metastasizes is gradual. (See Figure 7.9 for the example of breast cancer.)

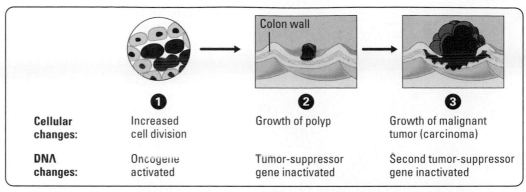

(a) Stepwise development of a typical colon cancer

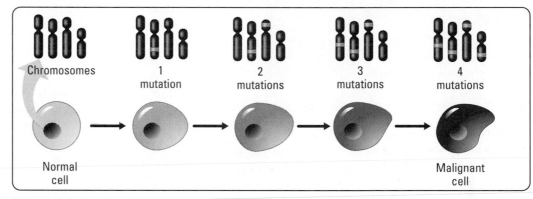

(b) Accumulation of mutations in the development of a cancer cell

As shown in Figure 10.20a, the first sign of a colon cancer is the unusually frequent division of apparently normal cells in the colon lining. This and later cellular changes parallel changes at the DNA level, including the activation of a cellular oncogene (*ras*) and the inactivation of two tumor-suppressor genes. These genetic changes (mutations) result in altered signal transduction pathways. The requirement for several mutations—the total number is usually four or more—explains why cancers can take a long time to develop. Figure 10.20b shows how mutations leading to cancer may accumulate in a lineage of somatic cells. By the time a fourth mutation has occurred, the structure of this cell and its descendants is grossly altered.

Our understanding of the genetic basis of cancer has grown by leaps and bounds in recent years. Among the recent discoveries is a new category of tumor-suppressor genes. Rather than being components of pathways that directly inhibit cell growth, the normal proteins encoded by these tumor-suppressor genes function in the repair of damaged DNA. When they are mutated, other cancer-causing mutations are more likely to accumulate.

"Inherited" Cancer Cancer is always a genetic disease in the sense that it always results from changes in DNA. Most mutations that lead to cancer arise in the organ where the cancer starts—the colon, for example. Because these mutations do not affect the cells that give rise to eggs or sperm, they are not passed from parent to child. In some families, however, there are inherited mutations in one or more of these same genes. Such a mutation *is* passed on and predisposes the recipient to cancer. We say that this cancer is

Figure 10.20 Development of a colon cancer. Like most cancers, colon cancer usually results from a series of mutations. **(a) ❶** The first sign of a colon cancer is the unusually frequent division of apparently normal cells in the colon lining. **❷** Later, a benign tumor (polyp) appears in the colon wall and **❸** eventually the malignant tumor (a carcinoma). These cellular changes parallel changes at the DNA level, including the activation of a cellular oncogene and the inactivation of two tumor-suppressor genes. **(b)** Mutations leading to cancer accumulate in a lineage of somatic cells. Colors distinguish the normal cell from cells with one or more mutations leading to increased cell division and cancer. Once a cancer-promoting mutation occurs (orange band on chromosome), it is passed to all the descendants of the cell carrying it. In our example, the first two mutations make the cells divide more rapidly; otherwise, the cells appear normal. The third mutation increases the rate of cell division even more and also causes some changes in the cells' appearance. Finally, a cell accumulates a fourth cancer-promoting mutation and begins dividing uncontrollably. The structure of this cell and its descendants is grossly altered.

Figure 10.21 Mary-Claire King. Dr. King, now at the University of Washington, is a leader in breast cancer research.

familial or "inherited," even though it doesn't appear unless the person acquires additional somatic mutations.

Some of the earliest studies of "inherited" cancer were carried out by American geneticist Mary-Claire King (Figure 10.21). She has spent over 25 years exploring the genetic basis of breast cancer, a disease that strikes one out of every ten American women. The vast majority of breast cancer cases seem to have nothing to do with inherited mutations. But there are many accounts, going back to the ancient Greeks, of families in which breast cancer appears frequently, suggesting that an inherited trait might be involved. After almost two decades of work, Dr. King and her colleagues succeeded in identifying a gene on chromosome 17 that is mutated in many families with familial breast cancer. Mutations in gene *BRCA1* put a woman at high risk of breast cancer (and ovarian cancer as well)—a more than 80% risk of developing cancer during her lifetime. Research from other laboratories suggests that the protein encoded by the normal version of *BRCA1* acts as a tumor suppressor by helping repair damaged DNA in the cell.

In recent years, clinical tests have been developed for the presence of mutations affecting *BRCA1* and *BRCA2*, another breast cancer gene. Should healthy women be tested, especially those who may be predisposed to breast cancer? King believes that genetic testing is useful only when an individual chooses it with a full understanding of its limitations. As King says, "For a woman today to know that she is predisposed to breast and ovarian cancer offers her a problem—but no solution except preventive surgical removal of her breasts and/or ovaries." In addition, King believes that testing should be carried out only under the condition that the results remain confidential and do not affect the woman's access to jobs or health insurance.

The Effects of Lifestyle on Cancer Risk

Cancer is now one of the leading causes of death in most developed countries, including the United States. Death rates due to certain forms of cancer (including stomach, cervical, and uterine cancers) have decreased, but the overall cancer death rate is still on the rise, currently increasing at about 1% per decade.

Cancer-causing agents are called **carcinogens.** Most mutagens are carcinogens, agents capable of bringing about cancer-causing DNA changes like the ones in Figure 10.20. Two of the most potent carcinogens (and mutagens) are X-rays and ultraviolet (UV) radiation. X-rays are a significant cause of leukemia and brain cancer. And exposure to UV radiation from the sun causes skin cancer, including a deadly type called melanoma.

Learn more about cancer and carcinogens in Web/CD Activity 10F.

The largest group of carcinogens are mutagenic chemical compounds and substances containing them. Among the most important carcinogens, the one substance known to cause more cases and types of cancer than any other single agent is tobacco. In 1900, lung cancer was a rare disease. Since then, largely because of an increase in cigarette smoking, lung cancer has steadily increased, and today more people die of lung cancer than any other form of cancer. Most tobacco-related cancers come from actually smoking, but the passive inhalation of secondhand smoke also poses a risk. As Table 10.1 indicates, tobacco use, sometimes in combination with alcohol consumption, causes a number of other types of cancer in addition to lung cancer. In nearly all cases, cigarettes are the main culprit, but smokeless tobacco products (snuff and chewing tobacco) are linked to cancer of the mouth and throat.

As we have seen, most cancers result from multiple genetic changes, including the activation of oncogenes and inactivation of tumor-suppressor genes. In many cases, these changes result from decades of exposure to the mutagenic effects of carcinogens. Carcinogens can also produce their effect by promoting cell division. Generally, the higher the rate of cell division, the greater the chance for mutations resulting from errors in DNA replication or recombination. Some carcinogens seem to have both effects. For instance, the hormones that contribute to breast and uterine cancers promote cell division and may also cause genetic changes that lead to cancer. In other cases, several different agents, such as viruses and one or more carcinogens, may together produce cancer.

We are still a long way from knowing all the factors that contribute to cancer. The effects of many environmental pollutants have yet to be evaluated, although some substances, such as asbestos, are definitely known to be carcinogenic. Exposure to some of the most important known carcinogens is often a matter of individual choice. Tobacco use, consumption of animal fat and alcohol, and time spent in the sun, for example, are all behavioral factors that affect our cancer risk.

Avoiding carcinogens is not the whole story, for there is growing evidence that some food choices significantly reduce cancer risk. For instance, eating 20–30 g of plant fiber daily (about twice the amount the average American consumes) and at the same time reducing animal fat intake may help prevent colon cancer. There is also evidence that other substances in fruits and vegetables, including vitamins C and E and certain compounds related to vitamin A, may offer protection against a variety of cancers. Cabbage and its relatives, such as broccoli and cauliflower, are thought to be especially rich in substances that help prevent cancer, although the identities of these substances are not yet established. Determining how diet influences cancer has become a major research goal.

The battle against cancer is being waged on many fronts, and there is reason for optimism in the progress being made. It is especially encouraging that we can help reduce our risk of acquiring some of the most common forms of cancer by the choices we make in daily life.

Table 10.1	Cancer in the United States	
Cancer	**Examples of Known or Likely Carcinogens**	**Estimated Number of Cases in 2000**
Breast	Estrogen; possibly dietary fat	184,200
Prostate	Testosterone; possibly dietary fat	180,400
Lung	Cigarette smoke	164,100
Colon and rectum	High dietary fat; low dietary fiber	130,200
Lymphomas	Viruses (for some types)	62,300
Bladder	Cigarette smoke	53,200
Melanoma of skin	Ultraviolet light	47,700
Uterus	Estrogen	36,100
Kidney	Cigarette smoke	31,200
Leukemias	X-rays, benzene; virus (for some types)	30,800
Mouth and throat	Tobacco in various forms; alcohol	30,200
Pancreas	Cigarette smoke	28,300
Ovary	(Large number of ovulation cycles)	23,100
Stomach	Table salt; cigarette smoke	21,500
Brain and nerve	Trauma; X-rays	16,500
Liver	Alcohol; hepatitis viruses	15,300
Cervix	Viruses; cigarette smoke	12,800
All others		152,200

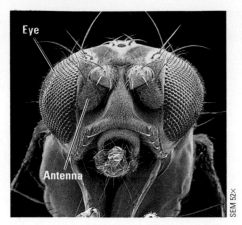

Eye

Antenna

SEM 52×

Head of a wild-type fruit fly

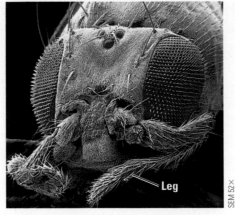

Leg

SEM 52×

Head of a homeotic mutant

Figure 10.22 **Comparison of a wild-type fruit fly with a homeotic mutant.** Due to a mutation in a homeotic (master control) gene, the fly on the right has legs growing out of its head.

Discover more about homeotic genes in the Web Evolution Link.

Evolution Link: Homeotic Genes

Among the most exciting biological discoveries in recent years is that a class of similar genes—homeotic genes (see p. 210)—help direct embryonic development in a wide variety of organisms. Mutations in these master control genes can produce bizarre effects. In fruit flies, for example, homeotic mutants may have extra sets of wings (see Figure 10.1) or legs growing out of their head (Figure 10.22). Researchers studying homeotic genes in fruit flies found a common structural feature: Every homeotic gene they looked at contained a common sequence of 180 nucleotides. Very similar sequences have since been found in virtually every eukaryotic organism examined so far, including yeasts, plants, earthworms, frogs, chickens, mice, and humans. These nucleotide sequences are called **homeoboxes,** and each is translated into a segment (60 amino acids long) of the protein product of the homeotic gene. The polypeptide segment encoded by the homeobox binds to specific sequences in DNA, enabling homeotic proteins that contain it to turn groups of genes on or off during development.

Figure 10.23 highlights some striking similarities in the chromosomal locations and developmental roles of homeobox-containing homeotic genes in two quite different animals. Notice that the order of genes on the fly chromosome is the same as on the four mouse chromosomes and that the gene order on the chromosomes corresponds to analogous body regions in both animals. These similarities suggest that the original version of these homeotic genes arose very early in the history of life and that the genes have remained remarkably unchanged for eons of animal evolution.

By their presence in such diverse creatures, homeotic genes illustrate one of the central themes of biology: unity in diversity. The fact that these key genes are *control* genes underscores the importance of regulation in the lives of organisms.

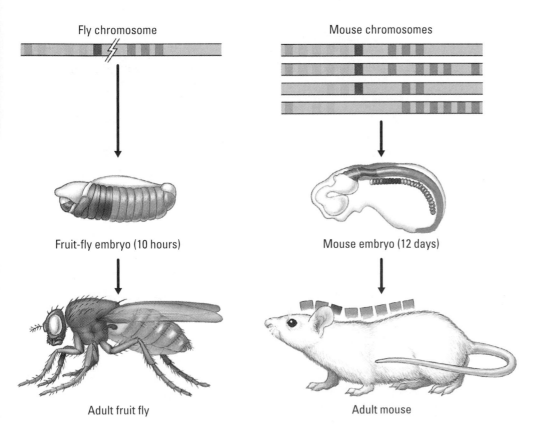

Fly chromosome

Mouse chromosomes

Fruit-fly embryo (10 hours)

Mouse embryo (12 days)

Adult fruit fly

Adult mouse

Figure 10.23 **Comparison of fruit fly and mouse homeotic genes.** At the top of the figure are portions of chromosomes that carry homeotic genes in the fruit fly and the mouse. The colored boxes represent homeotic genes that are very similar (have especially close DNA sequences) in flies and mice. The same color coding identifies the parts of the animals that are affected by these genes.

Summary of Key Concepts

Overview: Why and How Genes Are Regulated

- Gene expression—the flow of information from genes to proteins—is subject to control, mainly by the turning on and off of genes. Cells become specialized in structure and function because only certain genes of the genome are expressed. In eukaryotic cells, there are multiple possible control points in the pathway of gene expression. However, the most important control point in both eukaryotes and prokaryotes is at gene transcription.
- Web/CD Activity 10A *Control of Gene Expression Overview*

Gene Regulation in Bacteria

- **The *lac* Operon** In prokaryotes, genes for enzymes with related functions are often controlled together by being grouped into regulatory units called operons. Regulatory proteins bind to control sequences in the DNA and turn operons on or off in response to environmental changes. The *lac* operon, for example, produces enzymes that break down lactose only when lactose is present.
- Web/CD Activity 10B *The* lac *Operon in* E. Coli

- **Other Kinds of Operons** While the repressor for the *lac* operon is innately active, some operons, such as those that affect the synthesis of certain amino acids, have repressors that are innately *inactive*. The activation of still other operons depends on activator proteins.

Regulation of Eukaryotic Gene Expression

- **The Genetic Potential of Cells** Most differentiated cells retain a complete set of genes, so a carrot plant, for example, can be made to grow from a single carrot cell. Under special conditions, animals can also be cloned.

- **Patterns of Gene Expression in Differentiated Cells** The various cell types of a multicellular organism are different because different combinations of genes are expressed.

- **DNA Packing in Eukaryotic Chromosomes** A chromosome contains a DNA double helix wound around clusters of histone proteins, forming a string of beadlike nucleosomes. This beaded fiber is further wound and folded, packing about 3 m of DNA into the nucleus of each human cell (for example). DNA packing tends to block gene expression, presumably by preventing access of transcription proteins to the DNA. An extreme example is X chromosome inactivation in the cells of female mammals.

- **Regulation of Transcription and RNA Processing** A variety of regulatory proteins interact with DNA and with each other to turn the transcription of eukaryotic genes on or off. There are also opportunities for the control of eukaryotic gene expression after transcription, when introns are cut out of the RNA and a cap and tail are added.

- **Regulation in the Cytoplasm** The lifetime of an mRNA molecule helps determine how much protein is made, as do protein factors involved in translation. Finally, the cell may activate the finished protein in various ways (for instance, by cutting out portions) and later break it down.
- Web/CD Activity 10C *Gene Control in Eukaryotes*

Regulation of Embryonic Development

- **Cascades of Gene Activation and Cell Signaling in *Drosophila*** Studies of mutant fruit flies show that a cascade of gene expression involving genes for regulatory proteins determines how an animal develops from a fertilized egg. The earliest regulatory molecules are placed in the egg cell by the mother fly. Later the embryonic cells make their own. Homeotic genes in the embryo, for example, control batteries of other genes that shape anatomical parts such as wings.
- Web/CD Activity 10D/The Process of Science *Development of the Head-Tail Axis in Fruit Flies*

- **Cell Signaling** Cell-to-cell signaling is key to development. Signal transduction pathways convert molecular messages to cell responses, often the turning on (transcription) of particular genes.
- Web/CD Activity 10E *Signal Transduction Pathway*

The Genetic Basis of Cancer

- **Genes That Cause Cancer** Cancer cells, which divide uncontrollably, can result from mutations in genes whose protein products regulate the cell cycle. A mutation can change a proto-oncogene (a normal gene that promotes cell division) into an oncogene, which causes cells to divide excessively. Mutations that inactivate tumor-suppressor genes have similar effects. Many proto-oncogenes and tumor-suppressor genes code for proteins active in signal transduction pathways regulating cell division. Mutations of these genes cause malfunction of the pathways. Other cancer-causing mutations seem to impair the cell's ability to repair damaged DNA, allowing other mutations to accumulate. Cancers result from a series of genetic changes in a cell lineage. Researchers have gained insight into the genetic basis of breast cancer by studying families in which a disease-predisposing mutation is inherited.

- **The Effects of Lifestyle on Cancer Risk** Reducing exposure to carcinogens (which induce cancer-causing mutations) and other lifestyle choices can help reduce cancer risk.
- Web/CD Activity 10F *Causes of Cancer*

Evolution Link: Homeotic Genes

- Homeotic genes generally contain nucleotide sequences, called homeoboxes, that are very similar in many kinds of organisms. This suggests that homeobox-containing genes arose very early in the history of life.
- Web Evolution Link *Homeotic Genes*

Self-Quiz

1. The control of gene expression is more complex in multicellular eukaryotes than in prokaryotes because _____ (Explain your answer.)
   a. eukaryotic cells are much smaller.
   b. in a multicellular eukaryote, different cells are specialized for different functions.
   c. prokaryotes are restricted to stable environments.
   d. eukaryotes have fewer genes, so each gene must do several jobs.
   e. the genes of eukaryotes carry information for making proteins.

2. Your bone cells, muscle cells, and skin cells look different because
   a. different kinds of genes are present in each kind of cell.
   b. they are present in different organs.
   c. different genes are active in each kind of cell.
   d. they contain different numbers of genes.
   e. different mutations have occurred in each kind of cell.

3. Which of the following methods of gene regulation do eukaryotes and prokaryotes appear to have in common?

    a. elaborate packing of DNA in chromosomes.
    b. activator and repressor proteins, which attach to DNA.
    c. the addition of a cap and tail to mRNA after transcription.
    d. *lac* and *trp* operons.
    e. the removal of noncoding portions of RNA.

4. A eukaryotic gene was inserted into the DNA of a bacterium. The bacterium then transcribed this gene into mRNA and translated the mRNA into protein. The protein produced was useless; it contained many more amino acids than the protein made by the eukaryotic cell. Why?

    a. The mRNA was not spliced as it is in eukaryotes.
    b. Eukaryotes and prokaryotes use different genetic codes.
    c. Repressor proteins interfered with transcription and translation.
    d. The lifetime of the bacterial mRNA was too short.
    e. Ribosomes were not able to bind to tRNA.

5. A homeotic gene does which of the following?

    a. It serves as the ultimate control for prokaryotic operons.
    b. It regulates the expression of groups of other genes during development.
    c. It represses the histone proteins that package eukaryotic DNA.
    d. It helps splice mRNA after transcription.
    e. It inactivates one of the X chromosomes in a female mammal.

6. All your cells contain proto-oncogenes, which can change into cancer-causing genes. Why do cells possess such potential time bombs?

    a. Viruses infect cells with proto-oncogenes.
    b. Proto-oncogenes are genetic "junk" with no known function.
    c. Proto-oncogenes are unavoidable environmental carcinogens.
    d. Cells produce proto-oncogenes as a by-product of mitosis.
    e. Proto-oncogenes are necessary for normal control of cell division.

7. What kinds of evidence demonstrate that differentiated cells in a plant or animal retain the same full genetic potential?

8. A mutation in a single gene may cause a major change in the body of a fruit fly, such as an extra pair of legs or wings. Yet it probably takes the combined action of hundreds or thousands of genes to produce a wing or leg. How can a change in just one gene cause such a big change in the body?

• **Go to the website or CD-ROM for more self-quiz questions.**

The Process of Science

1. Study the depiction of the *lac* operon in Figure 10.4. Normally, the genes are turned off when lactose is not present. Lactose activates the genes, which code for enzymes that enable the cell to use lactose. Mutations can alter the function of this operon; in fact, it was the effects of various mutations that enabled Jacob and Monod to figure out how the operon works. Predict how the following mutations would affect the function of the operon in the presence and absence of lactose:

    a. Mutation of regulatory gene; repressor will not bind to lactose.
    b. Mutation of operator; repressor will not bind to operator.
    c. Mutation of regulatory gene; repressor will not bind to operator.
    d. Mutation of promoter; RNA polymerase will not attach to promoter

2. **Experiment on fruit fly embryos to learn how the head-tail axis is determined in the Process of Science activity on the website and CD-ROM.**

Biology and Society

The possibility of extensive genetic testing raises questions about how personal genetic information should be used. For example, should employers or potential employers have access to such information? Why or why not? Should the information be available to insurance companies? Why or why not? Is there any reason for the government to keep genetic files? Is there any obligation to warn relatives who might share a potentially harmful gene? Might some people avoid being tested for fear of being labeled genetic outcasts? Or might they be compelled to be tested against their wishes? Can you think of other reasons to proceed with caution?

DNA Technology and the Human Genome

The DNA of the human genome—about 3  billion base pairs—will soon be completely sequenced.

On average, we differ from one another in about

one out of every 1000 base pairs in our DNA.

An enzyme isolated from the AIDS virus can be used to make

artificial genes from human mRNA.  Plants

can be genetically modified to produce human

proteins.

Figure 11.1 Genetically modified corn plants. These plants carry a bacterial gene that enables them to resist infestation by a damaging insect pest called the European corn borer.

Figure 11.2 An industrial apparatus for growing genetically engineered bacteria.

T he photograph in Figure 11.1 shows a field of corn that has been genetically engineered to resist attack by insects called corn borers. This resistant property is due to a bacterial gene for an insecticide, which has been inserted into the corn's genome using DNA technology. The genetically modified (GM) food produced by these plants is only one of many that are available—and controversial—today. This chapter examines the basic concepts of DNA technology, its applications, and some of the social and ethical issues raised by these endeavors.

Overview: The Advent of DNA Technology

In 1946, American geneticists Joshua Lederberg and Edward Tatum performed a series of experiments with *E. coli* that demonstrated that these bacteria can carry out a sort of "mating" that results in the combining of genes from two different individuals—a phenomenon that until then was thought to be limited to sexually reproducing eukaryotic organisms. With this work, they and their colleagues pioneered bacterial genetics, a field that within 20 years made *E. coli* the most thoroughly understood organism at the molecular level. In the 1970s, research on *E. coli* led to the development of **recombinant DNA technology.** This is a set of laboratory techniques for combining genes from different sources—even different species—into a single DNA molecule.

In the past two decades, **DNA technology**—including recombinant DNA technology and other methods for studying and manipulating DNA—has paid high dividends in basic biological research. For example, the noncoding sequences within eukaryotic genes were discovered using DNA technology. This technology has also helped uncover mutations that lead to cancer and has shed light on the course of evolution. But perhaps the most exciting use of DNA technology in basic research is the Human Genome Project, the mapping of all the human DNA. This ambitious project is revealing the genetic basis of what it means to be human. On a more practical level, it is expected to help us understand and cure many diseases.

As we hear in the news, DNA technology is already widely used to engineer the genes of many sorts of cells for practical purposes. Scientists have genetically engineered bacteria to mass-produce many useful chemicals, from cancer drugs to pesticides (Figure 11.2). Other kinds of engineered bacteria are being used to clean up toxic wastes and alter crop plants. These applications of DNA technology are the latest incarnation of **biotechnology,** the use of organisms to perform practical tasks. Biotechnology actually goes back thousands of years to the first uses of yeast to make bread and wine.

Explore more applications of DNA technology in Web/CD Activity 11A.

CheckPoint

What is recombinant DNA technology?

Answer: A set of methods for creating a DNA molecule that carries genes originating from different organisms. Recombinant DNA technology is a subset of DNA technology, which includes techniques for manipulating DNA for other purposes.

Using Bacteria to Clone Genes

Bacteria are the workhorses of modern biotechnology. To manipulate genes in the laboratory, biologists make use of both intact bacterial cells and specialized molecules derived from bacteria. In fact, with those tools, scientists can clone—produce in quantity—particular genes and introduce them into other organisms. As a starting point for our exploration of this subject, let's look at how bacteria transfer genes among themselves in nature.

The Transfer of DNA

Unlike the sexual processes in plants and animals, bacteria do not have meiosis, gametes, or fertilization. But as early researchers discovered, they do not lack for ways to produce new combinations of genes. In fact, in the bacterial world, there are three mechanisms by which genes can move from one cell to another. These mechanisms of gene transfer are called transformation, transduction, and conjugation.

Figure 11.3a illustrates **transformation,** which is the taking up of DNA from the fluid surrounding a cell. This is what happened in the famous experiment with pneumonia-causing bacteria performed by Frederick Griffith in the 1920s (see Figure 9.1). A harmless strain of bacteria took up pieces of DNA left from the dead cells of a disease-causing strain. The DNA from the pathogenic bacteria carried a gene that made the cells resistant to an animal's defenses, and when the previously harmless bacteria acquired this gene, they could then cause pneumonia in infected animals.

Bacteriophages provide a second means of bringing together genes of different bacteria in the same cell. The transfer of bacterial genes by a phage is called **transduction** (Figure 11.3b). In this process, a fragment of DNA belonging to a phage's previous host cell has been accidentally packaged within the phage's coat. When the phage infects a new bacterial cell, the DNA stowaway from the former host cell is injected into the new host.

The type of gene transfer discovered by Lederberg and Tatum is called conjugation (from the Latin *conjugatus,* united). In **conjugation,** a "male" bacterial cell attaches to a "female" cell by means of protein projections called *sex pili,* and a bridge forms between the two cells (Figure 11.3c). Through this bridge, DNA passes from the male to the female cell. (The male cell replicates its DNA as it transfers it, so the cell doesn't end up lacking any genes.) Unlike mating in eukaryotes, bacterial conjugation is not a form of reproduction, for the number of individuals does not increase.

Once new DNA gets into a bacterial cell, by whatever mechanism, part of it may then integrate into the recipient's chromosome, replacing part of the recipient cell's original DNA. As Figure 11.4 indicates, integration

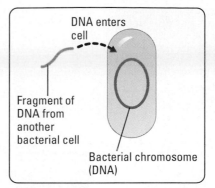

(a) Transformation

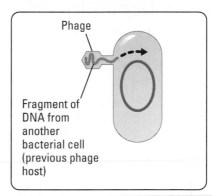

(b) Transduction

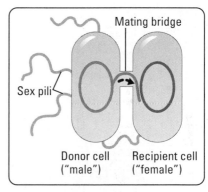

(c) Conjugation

Figure 11.3 DNA transfer in bacteria. Most of a bacterium's DNA is found in a single bacterial chromosome, which is a closed loop of DNA with associated proteins. These diagrams are not to scale: A bacterial chromosome is actually hundreds of times longer than the cell; it fits inside because it is highly folded. **(a)** In transformation, a bacterial cell takes up a piece of naked DNA from its surroundings. **(b)** Transduction is the transfer of a piece of cellular DNA from one cell to another inside a phage. **(c)** In conjugation —a sort of mating—DNA is transferred directly from one bacterium to another via a bridge that forms between them.

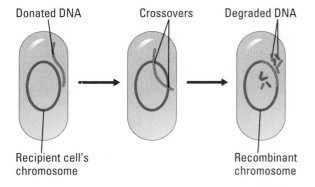

Figure 11.4 Genetic recombination in a bacterium. A piece of DNA from another bacterium that has entered the cell by transformation, transduction, or conjugation can be integrated into the recipient cell's chromosome by crossing over in two places. The extra pieces of DNA are broken down.

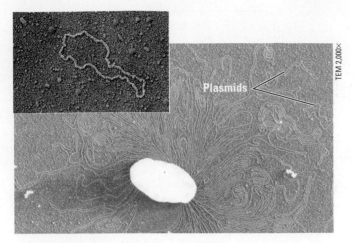

Figure 11.5 **Bacterial plasmids.** The large white shape is the remnant of a bacterium that has ruptured and released all of its DNA. Most of the DNA is the bacterial chromosome, which extends in loops from the cell. But two plasmids are also present. The inset shows an enlarged view of a single plasmid.

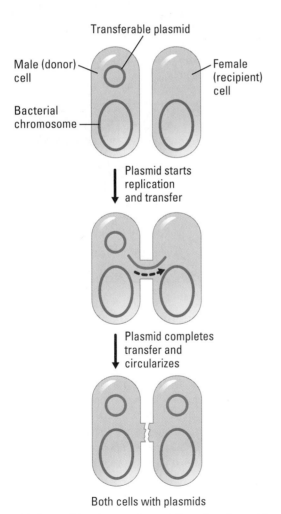

Figure 11.6 **Replication and transfer of a plasmid.** Some plasmids, including some R plasmids, can transfer a copy of themselves to another cell by conjugation. After a short time, the mating bridge breaks spontaneously, separating the cells.

occurs by crossing over between the two DNA molecules, a process similar to crossing over between eukaryotic chromosomes (see Figure 7.18). The leftover pieces of DNA are broken down, leaving the recipient bacterium with a recombinant chromosome.

Bacterial Plasmids

Many bacteria contain plasmids, which are small, circular DNA molecules separate from the much larger bacterial chromosome. You can see several from one cell in Figure 11.5. Every plasmid has an origin of replication, required for its replication within the cell. Some plasmids make their cells "male" and can transfer copies of themselves by conjugation to a "female" cell (Figure 11.6). A plasmid may carry genes other than those needed for replication and conjugation; when a plasmid carries these extra genes to another cell, it is acting as a **vector.**

E. coli and other bacteria have many different kinds of plasmids. Some plasmids carry genes that can affect the survival of the cell. Plasmids of one class, called **R plasmids,** pose serious problems for human medicine. R plasmids carry genes that block the action of antibiotics such as penicillin and tetracycline. Bacteria containing R plasmids are resistant (hence the designation R) to antibiotics that would otherwise kill them. The widespread use of antibiotics in medicine and agriculture has tended to kill off bacteria that lack R plasmids, while those with R plasmids have multiplied. The situation is made worse by the fact that many of these plasmids can spread throughout a bacterial population by conjugation (see Figure 11.6). As a result, an increasing number of bacteria that cause human diseases, such as food poisoning and gonorrhea, are becoming resistant to antibiotics. However, R plasmids can be useful vectors for genetic engineering, in the role we describe next.

Using Plasmids to Customize Bacteria: An Overview

Because plasmids can carry virtually any gene and replicate in bacteria, they are key tools for DNA technology. Figure 11.7 is an overview of how plasmids can be used to give bacteria useful capabilities. Starting at the top left, we see that ❶ a plasmid is first isolated from a bacterium. Meanwhile (top right), ❷ DNA carrying a gene of interest is obtained from another cell—perhaps an animal cell, a plant cell, or another bacterium. The gene of interest could be a human gene encoding a protein of medical value or perhaps a plant gene conferring resistance to pest insects. ❸ A piece of DNA containing the gene is inserted into the plasmid, producing recombinant DNA, and ❹ a bacterial cell takes up the plasmid by transformation. ❺ This genetically engineered, recombinant bacterium is then cloned (allowed to reproduce) to generate many copies of the gene.

At the bottom of Figure 11.7, you can see a few applications of genetically engineered bacteria. In the examples on the left, copies of the gene itself are the desired product. In the examples on the right, the protein product of the gene is harvested from bacteria. Many pharmaceutical companies employ biologists to customize bacteria for making specific products. In fact, genetic engineering has launched a revolution in biotechnology. On the next several pages, we examine in more detail some tools and procedures of this technology.

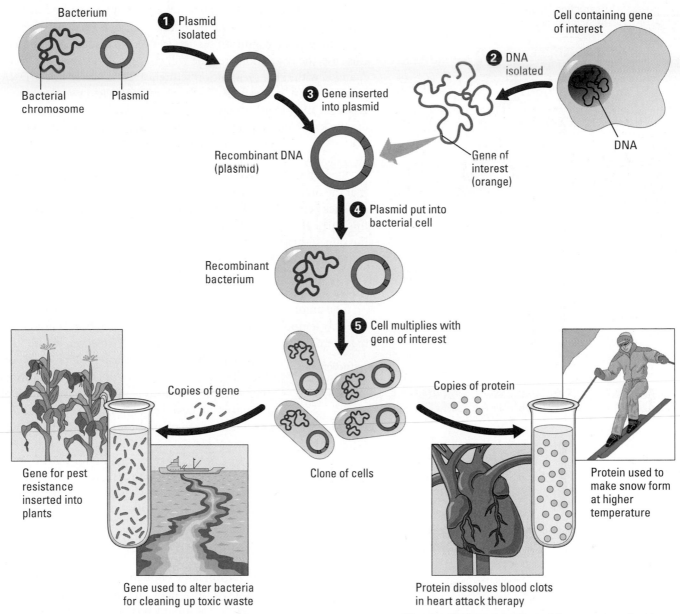

Figure 11.7 Using plasmids to customize bacteria for practical uses. This diagram is described in more detail in the text. The type of plasmid used in such a procedure is not one that can transfer itself by conjugation.

Labels within the figure:

Bacterium
Bacterial chromosome
Plasmid
❶ Plasmid isolated
Cell containing gene of interest
❷ DNA isolated
❸ Gene inserted into plasmid
DNA
Recombinant DNA (plasmid)
Gene of interest (orange)
❹ Plasmid put into bacterial cell
Recombinant bacterium
❺ Cell multiplies with gene of interest
Copies of gene
Clone of cells
Copies of protein
Gene for pest resistance inserted into plants
Gene used to alter bacteria for cleaning up toxic waste
Protein dissolves blood clots in heart attack therapy
Protein used to make snow form at higher temperature

Enzymes That "Cut and Paste" DNA

Extracting a gene from one DNA molecule and inserting it into another requires precise "cutting and pasting." To carry out the procedure outlined in Figure 11.7, a piece of DNA containing the gene of interest must be cut out of a chromosome and "pasted" into a bacterial plasmid.

The cutting tools for making recombinant DNA in a test tube are bacterial enzymes called **restriction enzymes,** which were first discovered in the late 1960s. In nature, these enzymes protect bacteria against intruding DNA from other organisms and phages. They work by chopping up the foreign DNA, a process called restriction because it *restricts* foreign DNA from surviving in the cell. Other enzymes chemically modify the cell's own DNA in a way that protects it from the restriction enzymes.

Most restriction enzymes recognize short nucleotide sequences in DNA molecules and cut at specific points within these recognition sequences.

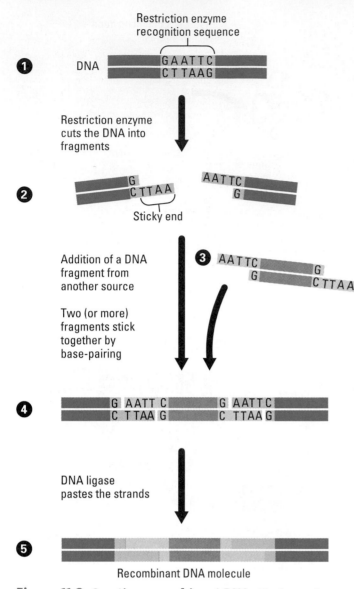

Figure 11.8 Creating recombinant DNA. The key tools are two enzymes: a restriction enzyme, which cuts the original DNA molecules into pieces, and DNA ligase, which connects them back together—in different combinations. The text describes the numbered steps in more detail.

Several hundred restriction enzymes and about a hundred different recognition sequences are known. In Figure 11.8, step ❶ shows a piece of DNA containing one recognition sequence for a particular restriction enzyme. In this case, the restriction enzyme will cut the DNA strands between the bases A and G within the recognition sequence. (The places where DNA is cut are called restriction sites.) ❷ The staggered cuts yield two double-stranded DNA fragments with single-stranded ends, called "sticky ends." Sticky ends are the key to joining DNA restriction fragments originating from different sources. These short extensions can base-pair with complementary single strands of DNA.

Step ❸ shows a piece of DNA (orange) from another source. Notice that the orange DNA has single-stranded ends identical in base sequence to the sticky ends on the blue DNA. The orange, "foreign" DNA has ends with this particular base sequence because it was cut from a larger molecule by the same restriction enzyme used to cut the blue DNA. ❹ The complementary ends on the blue and orange fragments allow them to stick together by base-pairing. (The hydrogen bonds that hold the base pairs together are not shown.)

The union between the blue and orange DNA fragments shown in step 4 is only temporary, because only a few hydrogen bonds hold the fragments together. The union can be made permanent, however, by the "pasting" enzyme **DNA ligase.** This enzyme, which is one of the proteins the cell normally uses in DNA replication, connects the DNA pieces into continuous strands by catalyzing the formation of covalent bonds between adjacent nucleotides. ❺ The final outcome is **recombinant DNA,** a DNA molecule carrying a new combination of genes.

See restriction enzymes in action in Web/CD Activity 11B.

Our molecular toolbox now contains enzymes for cutting and pasting DNA, plasmids for carrying that DNA into cells and providing an origin of replication, and bacterial cells for replicating the DNA. We are ready to make multiple copies of a gene.

Cloning Genes in Plasmids: A Closer Look

Making recombinant DNA in large enough amounts to be useful requires several steps. Consider a typical genetic engineering challenge: A molecular biologist at a pharmaceutical company has identified a human gene that codes for a valuable product—a hypothetical substance called protein V that kills certain human viruses. The biologist wants to set up a system for making large amounts of the gene so that the protein can be manufactured on a large scale. Figure 11.9 illustrates a way to make many copies of the gene using the techniques we have been discussing.

In step ❶, the biologist isolates two kinds of DNA: the bacterial plasmid that will serve as the vector and the human DNA containing the protein-V gene, called gene *V.* In this example, the DNA containing the gene of interest comes from human tissue cells that have been growing in laboratory culture. The plasmid comes from the bacterium *E. coli.*

In step ❷, the researcher treats both the plasmid and the human DNA with the same restriction enzyme. An enzyme is chosen that cleaves the plasmid in only one place. The human DNA, with thousands of restriction sites, is cut into many fragments, one of which carries the protein-V gene. In making the cuts, the restriction enzyme creates sticky ends on both the human DNA fragments and the plasmid. The figure shows the processing of just one human DNA fragment and one plasmid, but actually millions of plasmids and human DNA fragments (most of which do not contain gene *V*) are treated simultaneously.

In step ❸, the human DNA is mixed with the cut plasmid. The sticky ends of the plasmid base-pair with the complementary sticky ends of the human DNA fragment. In step ❹, the enzyme DNA ligase joins the two DNA molecules by covalent bonds, and the result is a recombinant DNA plasmid containing gene *V*.

In step ❺, the recombinant plasmid is added to a bacterium. Under the right conditions, the bacterium takes up the plasmid DNA by transformation (see Figure 11.3a).

The last step here, step ❻, is the actual **gene cloning,** the production of multiple copies of the gene. The bacterium, with its recombinant plasmid, is allowed to reproduce. As the bacterium forms a cell clone (a group of identical cells descended from a single ancestral cell), any genes carried by the recombinant plasmid are also cloned (copied). In our example, the biologist will grow a cell clone large enough to produce protein V in marketable quantities.

> Go to Web/CD Activity 11C to help you understand gene cloning.

At the bottom of Figure 11.9, we see only one bacterial clone, which carries one cloned fragment of foreign DNA. Actually, the procedure described in the figure "hits" a large number of genes in addition to the desired gene *V*, like a shotgun spraying pellets. For this reason, the procedure is called a "shotgun" approach to gene cloning. It yields many different bacterial clones, which carry many different fragments of the human DNA.

Genomic Libraries

The bacterial cells making up each clone produced in the shotgun procedure of Figure 11.9 have identical recombinant plasmids, each carrying the same fragment of human DNA. The entire collection of cloned DNA fragments from such a shotgun experiment, in which the starting material is bulk DNA from whole cells, is called a **genomic library** (Figure 11.10). A typical cloned DNA fragment is big enough to carry one or a few genes, and together the fragments include the entire genome of the organism from which the DNA was derived—in our example, a human. Bacterial plasmids are not the only type of vector that can be used in the shotgun cloning of

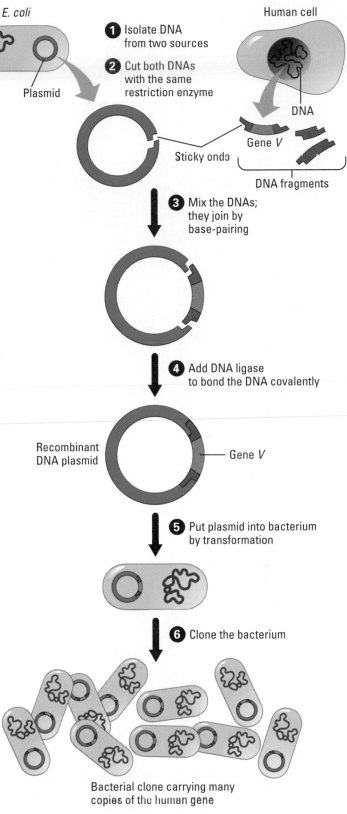

Figure 11.9 Cloning a gene in a bacterial plasmid. The numbered steps are described in more detail in the text.

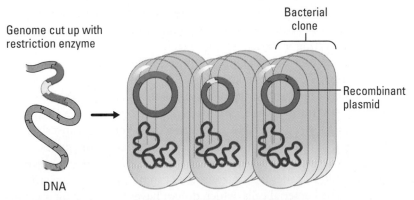

Figure 11.10 A genomic library. In this "library," segments of a genome are stored in bacterial plasmids. The library consists of a large number of bacterial clones. Each cell within a clone carries the same recombinant plasmid—with the same genome segment. The orange, yellow, and green DNA segments represent only three of the thousands of different library "books" shelved in plasmids in bacterial cells.

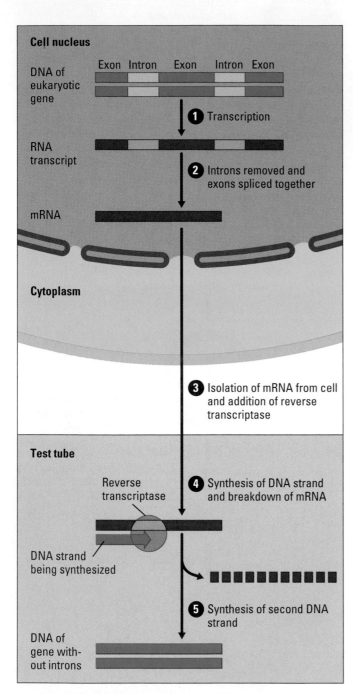

Cell nucleus

DNA of eukaryotic gene

Exon Intron Exon Intron Exon

❶ Transcription

RNA transcript

❷ Introns removed and exons spliced together

mRNA

Cytoplasm

❸ Isolation of mRNA from cell and addition of reverse transcriptase

Test tube

Reverse transcriptase

❹ Synthesis of DNA strand and breakdown of mRNA

DNA strand being synthesized

❺ Synthesis of second DNA strand

DNA of gene without introns

Figure 11.11 Making an intron-lacking gene from eukaryotic mRNA. Using the enzyme reverse transcriptase, a researcher can produce an artificial DNA gene from a molecule of messenger RNA (mRNA). Because the original gene's introns were spliced out during RNA processing, the artificial gene will not include introns. The gene will be shorter than the original gene and can be used for making the gene product in bacteria.

genes. Phages can also serve as vectors. When a phage is used, the DNA fragments are inserted into phage DNA molecules. The fragments are cloned when the phages replicate inside bacteria.

Soon you will learn how to find a particular "book"—a particular sequence of DNA—in a genomic library. But first we'll look at another source of DNA for cloning.

More Tools of DNA Technology

In this section, we discuss additional tools and techniques that scientists use in working with DNA. Their uses range from gene cloning to the analysis of genomes.

Reverse Transcriptase for Synthesizing Genes from mRNA

Not all DNA that is cloned comes directly from cells. Rather than starting with an entire eukaryotic genome, researchers can focus in on the genes *expressed* in a particular kind of cell (for example, muscle cells) by using mRNA as the starting material. As reviewed in steps ❶ and ❷ of Figure 11.11, the cells transcribe the genes and process the transcripts to produce mRNA. The researcher then isolates the mRNA and, with appropriate enzymes, uses it as a template for synthesizing DNA (steps ❸–❺). The key enzyme is reverse transcriptase, obtained from retroviruses (see Chapter 9), because making DNA on an RNA template is the reverse of transcription.

The DNA that results from this procedure represents pure genes—and only the genes actually transcribed in the starting cells. Furthermore, because these artificial genes lack introns, they can be correctly transcribed and translated by bacterial cells, which do not have RNA-splicing machinery. In other words, after the genes are cloned, the "host" bacteria can make the eukaryotic protein product.

Nucleic Acid Probes for Finding Genes of Interest

Often the most difficult task in gene cloning is finding the right shelf in a genomic library—that is, identifying the bacterial or phage clone containing a desired gene. If bacterial clones containing a specific gene actually translate the gene into protein, they can be identified by testing for the protein. This is not always the case, however. Fortunately, researchers can also test directly for the gene itself.

Methods for detecting genes depend on base pairing between the gene and a complementary sequence on another nucleic acid molecule, either DNA or RNA. When at least part of the nucleotide sequence of a gene is already known or can be guessed, this information can be used to advantage. For example, if we know that our hypothetical gene *V* contains the sequence TAGGCT, a biochemist can use nucleotides labeled with a radioactive isotope to synthesize a short single strand of DNA with a complementary sequence (ATCCGA). This sort of labeled nucleic acid molecule is called a **nucleic acid probe** because it is used to find a specific gene or other nucleotide sequence within a mass of DNA. (In practice, a probe molecule would be considerably longer than six nucleotides.)

Figure 11.12 shows how a probe works. The DNA sample to be tested is treated with heat or alkali to separate the DNA strands. When the radioactive DNA probe is added to these strands, it tags the correct molecule—finds the needle in the haystack—by hydrogen-bonding to the complementary sequence in the gene of interest.

Nucleic acid probes are powerful tools that do not require the targeted DNA to be a pure preparation. Figure 11.13 shows how a researcher might actually use such a probe to find a bacterial clone carrying a gene of interest among the thousands of other clones produced by shotgun cloning. In the figure, a collection of bacterial clones, each consisting of millions of identical cells, appear as visible colonies growing on a solid nutrient medium. Once the researcher identifies the desired clones, the cells can be grown further and the gene of interest isolated in large amounts.

Automated Synthesis and Sequencing of DNA

As you have just seen, nucleic acid probes can be chemically synthesized in the laboratory. In some cases, scientists have made entire genes this way, without using mRNA or any other nucleic acid template. If the order of amino acids in a protein is known, it is easy to use the genetic code (see Figure 9.17) to determine a nucleotide sequence that would code for it. More difficult is actually synthesizing gene-length DNA. Among the first genes synthesized artificially were two that encoded the two polypeptide chains making up human insulin. (In nature, the insulin polypeptides are encoded

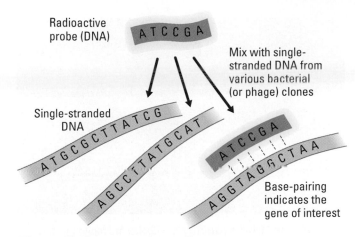

Figure 11.12 How a DNA probe tags a gene. The probe is a short, single-stranded molecule of DNA that is radioactive (it can also be RNA). When it is mixed with single-stranded DNA from a gene with a complementary base sequence, it attaches by hydrogen bonds. In this way, it labels the gene.

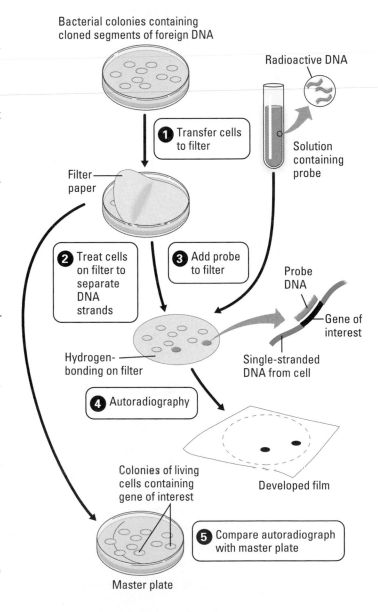

Figure 11.13 Using a DNA probe to identify a bacterial clone carrying a specific gene. The researcher starts with a Petri dish in which a number of bacterial clones (colonies) are growing on solid nutrient medium. The colonies result from a procedure like the one in Figure 11.9. ❶ A piece of filter paper is pressed against the colonies, blotting cells onto the paper. ❷ The paper is treated to break open the cells and separate the strands of their DNA, which stick to the paper. ❸ A solution containing molecules of the probe is poured on the paper. The probe molecules hydrogen-bond to any complementary DNA sequences, and excess probe is rinsed off. ❹ The paper is laid on photographic film, and any radioactive areas expose the film (autoradiography). ❺ The developed film, an autoradiograph, is compared with the master culture plate to determine which colonies carry the desired gene.

Figure 11.14 DNA sequencers. Computers report the data obtained from the DNA-sequencing machines shown here (large gray boxes). These state-of-the-art machines are being used to sequence the human genome for the Human Genome Project. Each machine can sequence DNA at the rate of 350,000 nucleotides per day.

in one long gene; see Figure 10.15.) The task of making these first artificial genes was arduous, feasible only because the two genes together had only about 150 nucleotides. Today, DNA-synthesizing machines are available that can rapidly produce DNA hundreds of nucleotides long.

Automation has also revolutionized DNA sequencing, the process of determining the exact sequence of nucleotides in a stretch of DNA. Sequencing even a short gene used to be very difficult, but thanks to methods developed in the 1970s, the sequences of even large genes can now often be determined in a day or so. DNA sequencing is made possible by restriction enzymes, which are used to cut the very long DNA of cells into discrete fragments that can be analyzed by machine (Figure 11.14).

Thousands of DNA sequences are now being collected in computer data banks, and they are proving of great value for understanding genes and gene control, as well as for biotechnology. With the help of computers, long sequences can be scanned to find particular genes or control sequences, such as promoters. Computers also help researchers compare the DNA of different species.

Both DNA sequencing and virtually every kind of DNA analysis, including DNA sequencing, depends on the technique we describe next.

Gel Electrophoresis for Sorting DNA Molecules

An essential tool of DNA technology, **gel electrophoresis** is a method for sorting macromolecules—proteins or nucleic acids—primarily on the basis of their electric charge and size. Figure 11.15 shows how we would use gel electrophoresis to separate the various DNA molecules in three different mixtures at the same time. A sample of each mixture is placed in a well at one end of a flat, rectangular gel, a thin slab of jellylike material. The gel is supported by two glass plates. A negatively charged electrode is then attached to the DNA-containing end of the gel and a positive electrode to the other end. Because the DNA molecules have a negative charge owing to

> See an animation of gel electrophoresis in Web/CD Activity 11D.

their phosphate groups, they move through the gel toward the positive pole. However, the longer DNA molecules are held back by the molecules of the gel, so they move more slowly, and thus not as far in a given time period, as the shorter DNA molecules. When the current is turned off, we see a series of bands in each "lane" of the gel. Each band consists of DNA molecules of one size.

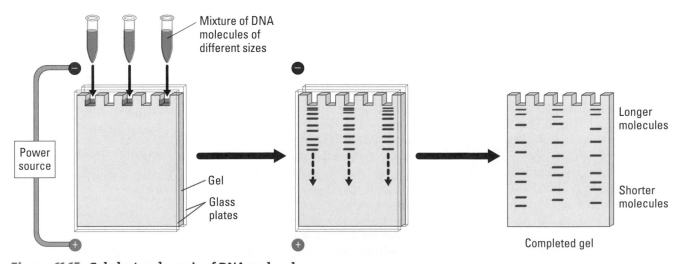

Figure 11.15 Gel electrophoresis of DNA molecules.

Comparing DNA Sequences Indirectly by Restriction Fragment Analysis

Unless you have an identical twin, your DNA is different from everyone else's; its total nucleotide sequence is unique. Some of your DNA consists of your particular sets of alleles for all the human genes, and even more of it is composed of noncoding stretches of DNA. Whether a segment of DNA codes for amino acids or not, it is inherited just like any other part of a chromosome. For this reason, geneticists can use *any* DNA segment that varies from person to person as a **genetic marker,** a chromosomal landmark whose inheritance can be studied. And just like a gene, a noncoding segment of DNA is more likely to be an exact match to the comparable segment in a relative than to the segment in an unrelated individual.

With the help of restriction enzymes and gel electrophoresis, a molecular geneticist could demonstrate the uniqueness of your DNA and also identify particular features of your DNA sequence that are shared by members of your family or other people. To do this, the geneticist would extract DNA from some of your cells and treat it with a restriction enzyme, creating a mixture of DNA pieces called **restriction fragments.** *The number of restriction fragments and their sizes reflect the specific sequence of nucleotides in your DNA.* To understand this statement, we need to examine Figure 11.16.

How Restriction Fragments Reflect DNA Sequence In Figure 11.16a, we see corresponding segments of DNA from two chromosomes—say, two different alleles, 1 and 2, of the same gene. Allele 1 might be from one of your chromosomes; allele 2 might be from your homologous chromosome or from another person's chromosome. Notice that the alleles differ by a single base pair (highlighted in yellow). In this case, the restriction enzyme cuts the DNA between two cytosine (C) bases in the sequence CCGG or in its complement, GGCC. Because allele 1 has two recognition sequences for the restriction enzyme, it is cleaved in two places, yielding three restriction fragments (labeled *w, x,* and *y*). Allele 2, however, has only one recognition sequence and yields only two restriction fragments (*z* and *y*). Notice that fragments *w* and *x* from allele 1 differ in length from fragments *z* and *y*. In other words, the lengths of restriction fragments, as well as their numbers, differ depending on the exact sequence of bases in the DNA.

To detect the differences between the collections of restriction fragments that come from alleles 1 and 2, we need to separate the restriction fragments in the two mixtures and compare their lengths. We can accomplish these things with the method shown in Figure 11.15, gel electrophoresis. As shown in Figure 11.16b, the restriction fragments from allele 1 separate into three bands in the gel, while those from allele 2 separate into only two bands. Notice that the smallest fragment from allele 1 (*y*) produces a band at the same location as the identical small fragment from allele 2. So you can see that electrophoresis allows us to see similarities as well as differences between mixtures of restriction fragments—and similarities as well as differences between the base sequences in DNA from two individuals.

> Use gel electrophoresis to analyse DNA fragments in The Process of Science activity available on the website and CD-ROM.

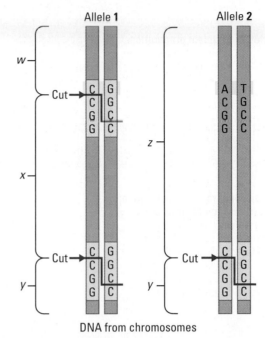

(a) Restriction-site differences between two alleles

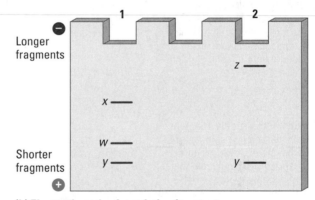

(b) Electrophoresis of restriction fragments

Figure 11.16 How restriction fragment patterns reflect DNA sequence. Different alleles of a gene have slightly different DNA nucleotide sequences. A particular restriction enzyme may therefore cut the alleles at different places. Here we consider two alleles differing by only a single base pair. **(a)** A certain restriction enzyme cuts allele 1 in two places, but it only cuts allele 2 in one place because of the sequence difference (yellow). **(b)** When the mixtures of fragments from alleles 1 and 2 are submitted separately to gel electrophoresis, allele 1 yields three bands on the gel, and allele 2 yields two bands.

1 **Restriction fragment preparation.** DNA is extracted from white blood cells taken from individuals I, II, and III. Individual I is known to carry a disease allele. A restriction enzyme is added to the three samples of DNA to produce restriction fragments.

2 **Gel electrophoresis.** The mixture of restriction fragments from each sample is separated by electrophoresis. Each sample forms a characteristic pattern of bands. (Since all the DNA in the blood cells was used, there would be many more bands than shown here.)

3 **Blotting.** The DNA bands are treated to separate the strands of the molecules, and the single strands are transferred by simple blotting onto special filter paper.

4 **Radioactive probe.** The paper blot is immersed in a solution of a radioactive probe, a single-stranded DNA molecule that is complementary to the DNA sequence of the genetic marker of interest. The probe attaches by base-pairing to restriction fragments arising from the marker DNA.

5 **Detection of radioactivity (autoradiography).** The unattached probe is rinsed off, and a sheet of photographic film is laid over the paper. The radioactivity in the probe exposes the film to form an image corresponding to specific bands—the bands containing DNA that base-pairs with the probe. In this example, the band pattern is the same for individuals I and II but different for individual III. Since we know that individual I carries the disease allele, the test shows that individual II is also a carrier but that individual III is not.

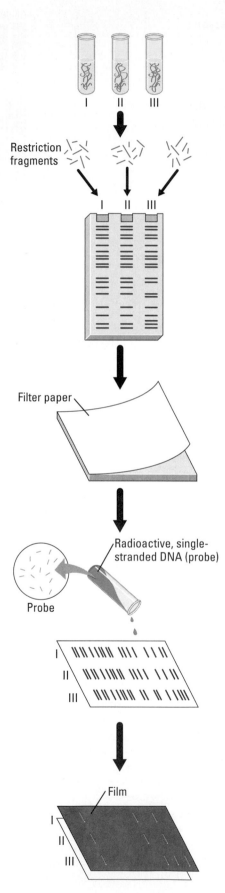

Restriction fragments

Filter paper

Radioactive, single-stranded DNA (probe)

Probe

Film

Using DNA Probes to Tag Specific Restriction Fragments In real life, the samples of DNA used as starting material for preparing restriction fragments would not be pure preparations of single genes, as assumed in Figure 11.16. More likely, the starting material would be bulk DNA from cells, which would yield huge numbers of bands on the gel. Fortunately, we can use a DNA probe to focus in on the bands coming solely from the DNA sequences we are interested in, without having to purify them from the rest of the DNA. Figure 11.17 shows how this is done. Note the crucial blotting step (step 3, which is similar to the first step in Figure 11.13). The paper to which the DNA is transferred (blotted) holds the DNA stationary while it is being exposed to the probe.

An important application of restriction fragment analysis is the detection of potentially harmful alleles in heterozygous individuals who are free of symptoms. The heterozygotes may be carriers of a harmful recessive allele, such as for cystic fibrosis, or a dominant allele that is not expressed until later in life, such as for Huntington's disease. The harmful allele often contains one or more restriction sites that are different from the ones in the normal allele.

Restriction fragment analysis requires only about 1 microgram (μg) of DNA—the amount in a small drop of blood. Still, even this tiny amount of DNA may not be available. Fortunately, another powerful method, the topic of the next section, can help out in those situations.

The Polymerase Chain Reaction (PCR)

The **polymerase chain reaction (PCR)** is a technique by which any segment of DNA can be amplified (cloned) in a test tube without using living cells. In principle, PCR is surprisingly simple. The DNA is mixed with a special version of the DNA replication enzyme DNA polymerase, nucleotide monomers, and short pieces of synthetic single-stranded DNA that serve as primers for DNA synthesis. (The DNA polymerase needs primers because it can only add nucleotides to a preexisting nucleotide chain.) In this mixture, the DNA replicates, producing two daughter DNAs; the daughter molecules, in turn, replicate; and the process continues as long as nucleotides remain. Each time replication occurs, the amount of DNA doubles. Starting with a single DNA molecule, automated PCR can generate 100 billion similar molecules in a few hours. This is actually not a lot of DNA, but it is enough for many purposes. And the time required is much shorter than the days it takes to clone a piece of DNA by attaching it to a plasmid.

Figure 11.17 Identifying particular restriction fragments using a DNA probe. Here the fragments of interest come from a gene associated with a genetic disease. The researcher wants to know if individual II or III carries an allele for the disease that afflicts their relative, individual I.

Figure 11.18 describes the PCR procedure. The key to PCR automation is an unusual DNA polymerase that was first isolated from bacteria living in hot springs. Unlike most proteins, this enzyme can withstand the heat needed to separate the DNA strands at the start of each cycle of replication.

A great advantage of PCR is that it can copy a *specific segment* from within a mass of DNA; the primers target the DNA sequence to be amplified. PCR is so specific and powerful that its starting material does not even have to be purified DNA. Only minute amounts of DNA need to be present in the starting material, and this DNA can be in a partially degraded state. It should be noted, however, that PCR cannot substitute for gene cloning in bacteria or other cells when large amounts of the gene are desired. Occasional errors during PCR replication impose limits on the number of good copies that can be made by this method.

PCR is being applied in a great number of ways. In evolution research, the technique has been used to amplify DNA pieces recovered from an ancient mummified human, from a 40,000-year-old woolly mammoth frozen in a glacier, and from a 30-million-year-old plant fossil. In medicine, PCR has been used to amplify DNA from single embryonic cells for rapid prenatal diagnosis, and it has made possible the detection of viral genes in cells infected with HIV. We'll discuss other applications of PCR later.

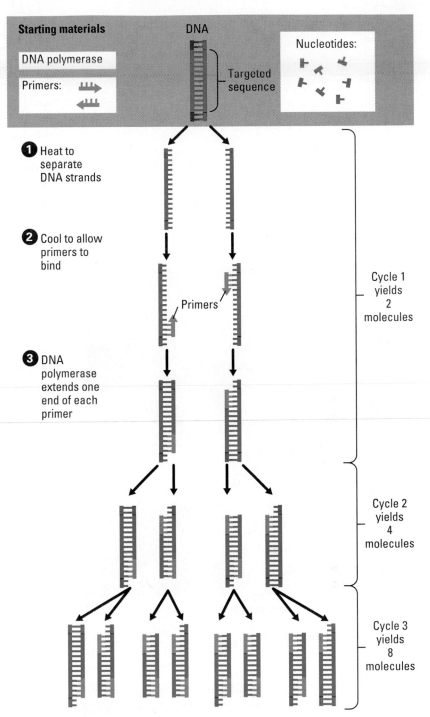

Figure 11.18 **The PCR method.** The polymerase chain reaction (PCR) is a method for making many copies of a specific segment of DNA without using intact cells. The starting material is a solution of DNA containing the nucleotide sequence "targeted" for copying. The scientist adds a heat-resistant DNA polymerase, a supply of all four nucleotides, and primers. The primers are short, synthetic molecules of single-stranded DNA that are complementary to the ends of the targeted DNA. The primers are needed because the DNA polymerase can only add nucleotides to a preexisting DNA chain. ❶ The DNA is briefly heated to separate its strands and then ❷ cooled to allow the primers to bind by hydrogen bonding to the ends of the target sequence, one primer on each strand. ❸ Then the DNA polymerase extends the primers, using the longer DNA strands as templates. Within about 5 minutes, the target DNA sequence has been doubled. The solution is then heated again, starting another cycle of strand separation, primer binding, and DNA synthesis. The cycle runs again and again, duplicating the targeted sequence many times.

CheckPoint

1. Why is an artificial gene made using reverse transcriptase often shorter than the natural form of the gene?

2. How does a probe consisting of radioactive single-stranded DNA enable a researcher to find the bacterial clones carrying a particular gene?

3. Your assignment is to synthesize a gene for a specific human protein and then clone the gene in bacteria. What do you need to know about the protein to make the synthetic DNA?

4. **(a)** What causes DNA molecules to move toward the positive pole during electrophoresis? **(b)** Why do large molecules move more slowly than smaller ones?

5. You use a restriction enzyme to cut a DNA molecule. The base sequence of this DNA is known, and the molecule has a total of three restriction sites clustered close together near one end. When you separate the restriction fragments by electrophoresis, how do you expect the bands to be distributed along the electrophoresis lane?

6. If it takes 3 minutes for each replication cycle of a PCR procedure, how many copies of DNA can be produced in 18 minutes from a single starting DNA molecule?

Answers: 1. Because it does not contain introns **2.** The probe molecules bind to and label DNA only in the specific bacterial clone containing the gene of interest. **3.** Its amino acid sequence **4. (a)** The negatively charged phosphate groups of the DNA are attracted to the positive pole. **(b)** The gel resists their movement. **5.** Three bands near the positive pole (small fragments) and one band near the negative pole (large fragment) **6.** 64

The Challenge of the Human Genome

Within the set of 23 chromosomes making up the haploid human genome, there are approximately 3 billion nucleotide pairs of DNA. To try to get a sense of this quantity of DNA, imagine that its nucleotide sequence is printed in letters (A, T, C, and G) like the letters in this book. At this size, the sequence would fill over 3000 volumes! Perhaps the most exciting application of DNA technology to date has been its use to unravel the mysteries of the human genome. Our genome presents a major challenge not only because of its size, but also because, like that of most complex eukaryotes, most of its DNA does not code for protein.

The Human Genome: Genes and "Junk"

The amount of DNA in a human cell is about 1000 times greater than the amount of DNA in *E. coli.* Does this mean humans have 1000 times as many genes as the 2000 in *E. coli*? The answer is no; the human genome carries between 50,000 and 100,000 genes, which code for various proteins and for tRNA and rRNA, as well. In addition to these genes, humans, like most complex eukaryotes, have a huge amount of noncoding DNA—about 97% of the total human DNA. Some noncoding DNA is made up of gene control sequences such as promoters and enhancers. The remaining DNA has been called "junk DNA," but it is more accurate to say that we do not yet understand its functions. This DNA includes introns (whose total length may be ten times greater than the exons of a gene) and noncoding DNA located between genes.

Much of the DNA between genes consists of **repetitive DNA,** nucleotide sequences present in many copies in the genome. There are two main types of repetitive DNA. In one type, a unit of just a few nucleotide pairs is repeated many times in a row. Stretches of DNA with thousands of such repetitions are prominent at the centromeres and ends of chromosomes, suggesting that this DNA plays a role in chromosome structure: It may help keep the rest of the DNA properly organized during DNA replication and mitosis. Recent research supports the idea that the repetitive DNA at the chromosome ends—called **telomeres** (Figure 11.19)—also has a protective function; a significant loss of telomeric DNA quickly leads to cell death. Furthermore, abnormal lengthening of this DNA may help "immortal" cancer cells evade normal cell aging.

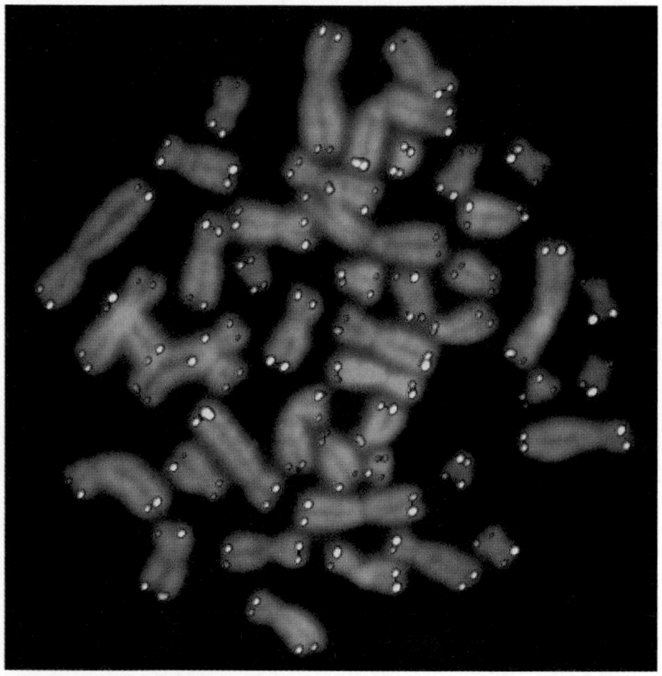

A number of genetic disorders affecting the nervous system are caused by abnormally long stretches of repeated nucleotide triplets. One is Huntington's disease (see Chapter 8), in which a long string of CAG triplets is actually located within a coding region of the gene. The protein produced has a long string of the amino acid glutamine.

In the second main type of repetitive DNA, each repeated unit is hundreds of nucleotide pairs long, and the copies are scattered around the genome. Scientists know little about the functions of this DNA, but they do have an idea how it came to be both abundant and dispersed in the genome. Most of these sequences seem to be associated with transposons, the "jumping genes" discussed next.

(a) The telomeres of human chromosomes

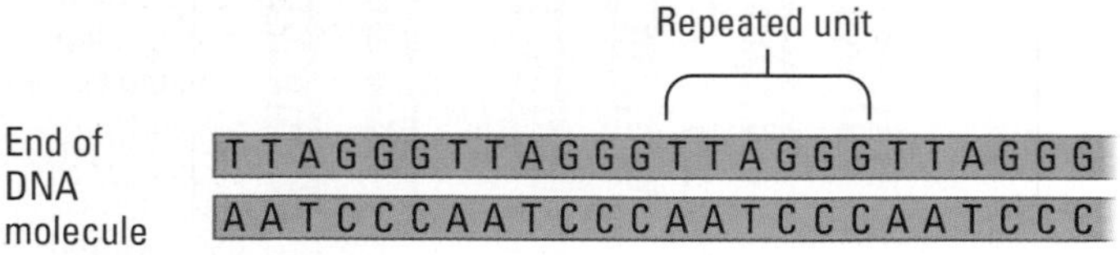

Repeated unit

End of DNA molecule

T T A G G G T T A G G G T T A G G G T T A G G G

A A T C C C A A T C C C A A T C C C A A T C C C

(b) Nucleotide sequence of a human telomere

Figure 11.19 Telomeres. Telomeres are the stretches of repetitive DNA at the ends of eukaryotic chromosomes. **(a)** The fluorescent orange stain marks the telomeres of these human chromosomes. **(b)** The DNA of telomeres consists of repeating units of six nucleotide pairs. Here we see the sequence for human DNA. Each telomere has 100 to 1000 or more repeated units.

Jumping Genes

In the 1940s, while studying inheritance in corn plants, American geneticist Barbara McClintock (Figure 11.20) made a startling discovery. She found that certain genetic elements (now known to be DNA segments) can move from one location to another in a chromosome and even from one chromosome to another. McClintock discovered that these "jumping genes," now called **transposons,** can land in the middle of other genes and disrupt them. In Indian corn, for instance, transposons can disrupt pigment genes in some of the cells, leading to spotted kernels like the ones in Figure 11.21.

McClintock worked largely alone, and few other geneticists appreciated the significance of her discoveries until the 1970s. By that time, transposons had been found in *E. coli,* and the era of recombinant DNA technology was beginning. Finally, in 1983, McClintock received a Nobel Prize for her pioneering work.

Current evidence suggests that all organisms, prokaryotic and eukaryotic alike, have transposons. Figure 11.22 shows the simplest kind of transposon and how its movement can interfere with another gene. ❶ The transposon includes one gene (purple) encoding an enzyme that catalyzes movement of the transposon. Identical noncoding nucleotide sequences (yellow) mark the ends of the transposon. ❷ The enzyme attaches to the ends of the transposon and simultaneously to another site on the DNA. It cuts the DNA and catalyzes insertion of the transposon at the new site. (The DNA of the transposon is never free in the cell.) ❸ Insertion of the transposon at its new site disrupts the nucleotide sequence of the gene shown in green.

In addition to the cut-and-paste type of transposon shown here, molecular biologists have discovered copy-and-paste transposons, which leave the original copy behind when they move. It is this type of transposon that is apparently responsible for the proliferation of dispersed repetitive DNA in the human genome.

Realizing that transposons act as natural mutagens, McClintock suggested that they may help generate genetic diversity and could thus be a significant factor in evolution. In recent years, transposons have been implicated in some cases of cancer.

One of the great biologists of the twentieth century, Barbara McClintock continued working until her death at age 90, in 1992. Like Mendel, she was one of those rare scientists who have profound insights years ahead of their time.

The Human Genome Project

Only 20 years ago, the task of completely deciphering the DNA blueprint for human life was almost inconceivable. But by 1990, advances in DNA technology had led an international consortium of scientists to launch the **Human Genome Project,** with the ultimate goal of determining the nucleotide sequence of all the DNA in the human genome. Private companies, chiefly Celera Genomics in the United States, have joined the efforts of a consortium of researchers funded by the National Institutes of Health (NIH) and the Atomic Energy Commission (AEC) in the United States and comparable government and university researchers in other countries. The public project was originally targeted for completion in the year 2005. But fierce competition between the public consortium and Celera spurred both groups on and led the rivals to announce jointly on June 26, 2000 that the project was largely completed. The results have already yielded exciting insights into the genes of humans and other organisms.

Figure 11.20 Barbara McClintock in 1947. Studying the genetics of corn, she discovered transposons.

Figure 11.21 Effects of transposons. The varying colorations of these corn kernels result from the movement of transposons within the DNA of the corn kernel cells.

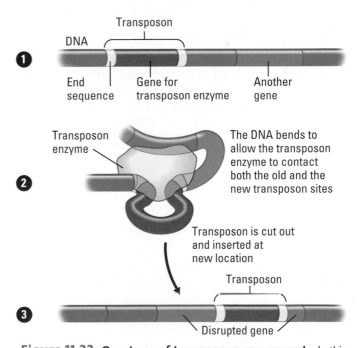

Figure 11.22 One type of transposon movement. In this case, the transposon happens to move to a new site that is within a gene. In its new location, the transposon disrupts the gene, so that a normal mRNA and protein cannot be made from it. The numbered steps are described in the text.

The mapping of the human genome has involved three major stages:

1. *Genetic (linkage) mapping.* In Chapter 8, you learned how geneticists use data from genetic crosses to determine the order of linked genes on a chromosome and the relative distances between the genes. Scientists combined pedigree analysis of large families with DNA technology to map over 5000 genetic markers. These markers include genes, stretches of repetitive DNA, and other noncoding DNA sequences. The resulting low-resolution map has provided a framework that enables researchers to map other markers easily by testing for genetic linkage to the known markers.

2. *Physical mapping.* Researchers used restriction enzymes to break the DNA of each chromosome into a number of identifiable fragments, which they cloned. They then determined the original order of the fragments in the chromosome. Their strategy: Using several different restriction enzymes, they made fragments that overlapped and then matched up the ends. They used probes to relate the fragments to the markers mapped in stage 1.

3. *DNA sequencing.* The most arduous part of the project is determining the nucleotide sequences of a set of DNA fragments covering the entire genome, generally the fragments already mapped in stage 2. Advances in automatic DNA sequencing have been crucial to this endeavor (Figure 11.23). In the "working draft" of the genome announced in June 2000, about 85% of the genome's protein-coding DNA was actually sequenced, including almost all of chromosomes 21 and 22 (the smallest nonsex chromosomes). Because of the presence of repetitive DNA and for other, poorly understood reasons, certain parts of each chromosome resist mapping by the usual methods. These regions are being set aside for later completion.

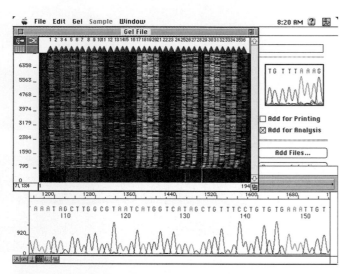

Figure 11.23 The computer screen of a DNA sequencer. Sequencing machines use gel electrophoresis to separate the individual nucleotides of the DNA; notice the band patterns against the black background. At the bottom of the screen is a partial sequence.

While some scientists are mapping the human genome, others are pursuing another important goal of the project, mapping the genomes of certain other important research organisms: *E. coli* and other prokaryotes, a yeast, a nematode, a fruit fly, and a mouse. Work on these genomes has been useful for developing the strategies and techniques needed to map the much larger human genome. Moreover, comparative analysis of the genes of other species helps scientists interpret the human data. For example, when scientists find a nucleotide sequence in the human genome similar to a yeast gene whose function is known, they have a valuable clue to the function of the human sequence. Many genes of disparate organisms are turning out to be astonishingly similar, to the point that one researcher has joked that he now views fruit flies as "little people with wings."

As of July 2000, the genomes of about 40 organisms have been fully sequenced. So far, most are prokaryotes, including *E. coli,* a number of other bacteria (some of medical importance), and several archaea. Yeast was the first eukaryote to have its sequence completed and the nematode *Caenorhabditis elegans,* a simple worm, the first multicellular organism. The fruit fly *Drosophila melanogaster* has been finished, and the plant *Arabidopsis thaliana,* another important research organism, is almost completed.

The potential benefits of having a complete map of the human genome are great. For basic science, the information is already providing insight into such fundamental mysteries as embryonic development and evolution. For human health, the identification of genes will aid in the diagnosis, treatment, and possibly prevention of many of our more common ailments, including heart disease, allergies, diabetes, schizophrenia, alcoholism, Alzheimer's disease, and can-

Learn about some of the genes that have been identified on human chromosome 17 in Web/CD Activity 11E.

cer. Hundreds of disease-associated genes have already been identified as a result of the project. The sequence for chromosome 21 will provide insights into Down syndrome, caused by an extra copy of that chromosome (see Figure 7.19).

The DNA sequences determined by the public consortium are deposited in a database that is available to researchers all over the world via the Internet. Scientists use software to scan and analyze the sequences for genes, control elements, and other features. Then comes the most exciting challenge: figuring out the functions of the genes and other sequences and how they work together to direct the structure and function of a living organism. This challenge and the applications of the new knowledge should keep scientists busy well into the twenty-first century.

Other Applications of DNA Technology

DNA technology will undoubtedly influence human health through the Human Genome Project, but it is already affecting our lives in a number of other ways. Here we address some of these current applications.

DNA Fingerprinting in Courts of Law

"DNA FINGERPRINTING SOUGHT IN MURDER PROSECUTION" blare the headlines. In the last few years, the news media have heralded the use of DNA technology as a new tool for forensic (legal) science. From small-town attorneys' offices to the FBI, lawyers and forensic scientists are having to learn about the new molecular techniques, as are judges and juries (Figure 11.24).

In violent crimes, blood or small fragments of other tissue may be left at the scene of the crime or on the clothes or other possessions of the victim or assailant. If rape is involved, small amounts of semen may be recovered from the victim's body. If enough tissue or semen is available, forensic scientists can determine its blood type or tissue type using older methods that test for proteins. However, such tests require a relatively large amount of fairly fresh tissue. Also, because there are many people in the population with the same blood type or tissue type, this approach can only *exclude* a suspect; it cannot prove guilt.

DNA testing, on the other hand, can theoretically identify the guilty individual with certainty because the DNA sequence of every person is unique (except for identical twins). Restriction fragment analysis (see Figure 11.17) is a powerful method for the forensic detection of similarities and differences in DNA samples and requires only tiny amounts of blood or other tissue (about 1000 cells). For a murder case, for example, restriction

Figure 11.24 DNA data used to solve crimes. In this photograph, the head of the Connecticut State Forensic Laboratory examines DNA data that will be stored in a state database.

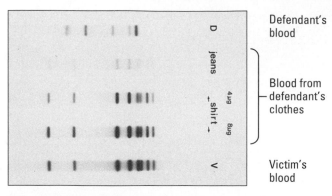

Figure 11.25 DNA fingerprints from a murder case.
DNA from bloodstains on the defendant's clothes matches the DNA fingerprint of the victim but differs from the DNA fingerprint of the defendant. This is evidence that the blood on the defendant's clothes came from the victim, not the defendant, and places the defendant at the scene of the crime.

fragment analysis can be used to compare DNA samples from the defendant, the victim, and bloodstains on the defendant's clothes (Figure 11.25). Radioactive probes mark the electrophoresis bands that contain certain genetic markers. Usually, about five markers are tested, only a few selected portions of the DNA. However, even such a small set of markers from an individual can provide a **DNA fingerprint,** or specific pattern of bands, that is of forensic use, because the probability that two people would have identical sets of markers is very small. The autoradiograph in Figure 11.25 resembles the type of evidence presented (with explanation) to juries in murder trials. As you can see, DNA from blood on the defendant's clothes matches the DNA of the victim but differs from that of the defendant, providing strong evidence of guilt.

The forensic use of DNA fingerprinting extends beyond violent crimes. For instance, comparing the DNA of a mother, her child, and the purported father can conclusively settle a question of paternity. Sometimes paternity is of historical interest: Recently, DNA fingerprinting provided strong evidence that Thomas Jefferson fathered at least one of the children of his slave Sally Hemings.

Today, the markers most often used in DNA fingerprinting are inherited variations in the lengths of repetitive DNA. For example, one person may have the nucleotides AC repeated 65 times at one genome locus, 118 times at a second locus, and so on, whereas another person is likely to have different numbers of repeats at these loci. PCR is often used to amplify particular repetitive-DNA loci before electrophoresis. With PCR, a quantity of DNA sufficient for analysis can be generated from a tissue sample as small as 20 cells!

Just how reliable is DNA fingerprinting? In most legal cases, the probability of two people having identical DNA fingerprints is between one chance in 100,000 and one in a billion. The exact figure depends on how many markers are compared and on how common those markers are in the population. Growing collections of data on the frequencies of various markers—both in the population as a whole and in particular ethnic groups—are enabling forensic scientists to make extremely accurate statistical calculations. Thus, despite problems that can arise from insufficient statistical data, human error, or flawed evidence, DNA fingerprints are now accepted as compelling evidence by legal experts and scientists alike. In fact, many argue that DNA evidence is more reliable than eyewitness testimony in placing a suspect at the scene of a crime.

> Use DNA fingerprinting to analyze evidence from a crime scene in Web/CD Activity 11F.

The Mass Production of Gene Products

DNA technology has been used to create many useful pharmaceutical products. Most of these products are proteins. By transferring the gene for a desired protein product into a bacterium, yeast, or other kind of cell that is easy to grow, one can produce large quantities of proteins that are present naturally in only minute amounts.

Bacteria are often the best organisms for manufacturing a protein product. Major advantages of bacteria include the plasmids and phages available for use as gene-cloning vectors and the fact that bacteria can be grown rapidly and cheaply in large tanks. Furthermore, bacteria can be readily engineered to produce large amounts of particular proteins and in some cases to secrete the protein products into the medium in which they are grown. Secretion into the growth medium simplifies the task of collecting and pu-

Preparing recombinant bacteria. This scientist is purifying plasmid DNA (see Step 1 in Figure 11.7).

Table 11.1	Some Protein Products of Recombinant DNA Technology	

Product	Made in	Use
Human insulin	*E. coli*	Treatment for diabetes
Human growth hormone (GH)	*E. coli*	Treatment for growth defects
Epidermal growth factor (EGF)	*E. coli*	Treatment for burns, ulcers
Tumor necrosis factor	*E. coli*	Killing of certain tumor cells
Interleukin-2 (IL2)	*E. coli*	Possible treatment for cancer
Prourokinase	*E. coli*	Treatment for heart attacks
Porcine growth hormone (PGH)	*E. coli*	Improving weight gain in hogs
Bovine growth hormone (BGH)	*E. coli*	Improving weight gain in cattle
Cellulase	*E. coli*	Breaking down cellulose for animal feeds
Taxol	*E. coli*	Treatment for ovarian cancer
Interferons (alpha and gamma)	*S. cerevisiae; E. coli*	Possible treatment for cancer and virus infections
Hepatitis B vaccine	*S. cerevisiae*	Prevention of viral hepatitis
Colony-stimulating factor (CSF)	Mammalian cells	Treatment for leukemia; boosts resistance to AIDS and other infectious diseases
Erythropoietin (EPO)	Mammalian cells	Treatment for anemia
Factor VIII	Mammalian cells	Treatment for hemophilia
Tissue plasminogen activator (t-PA)	Mammalian cells	Treatment for heart attacks

rifying the products. As Table 11.1 shows, a number of proteins of importance in human medicine and agriculture are being produced in the bacterium *E. coli.*

Despite the advantages of bacteria, it is sometimes desirable or necessary to use eukaryotic cells to produce a protein product. Often the first-choice eukaryotic organism for protein production is the yeast *Saccharomyces cerevisiae,* the single-celled fungus used in making bread, beer, and wine. As bakers and brewers have recognized for centuries, yeast cells are easy to grow. And like *E. coli,* yeast cells can take up foreign DNA and integrate it into their genomes. Yeasts also have plasmids that can be used as gene vectors, and yeasts are often better than bacteria at synthesizing and secreting eukaryotic proteins. The yeast *S. cerevisiae* is currently used to produce a number of proteins. In some cases, the same product (for example, interferons used in cancer research) can be made in either yeast or bacteria. In other cases, such as the hepatitis B vaccine, yeast alone is used.

The cells of choice for making certain gene products come from mammals (see Table 11.1). Genes for these products are often cloned in bacteria as a preliminary step. For example, the genes for colony-stimulating factor (CSF) and for two proteins that affect blood clotting, Factor VIII and t-PA, are cloned in a bacterial plasmid before transfer to mammalian cells for large-scale production. Many proteins that mammalian cells normally secrete are glycoproteins, meaning that they have chains of sugars attached to the protein. Because only mammalian cells can attach the sugars correctly, mammalian cells must be used for making these products.

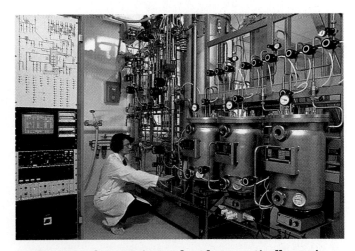

Production of a protein product by genetically engineered animal cells. A scientist monitors "bioreactors" containing cultured animal cells—actually insect cells—that are making an experimental AIDS vaccine. The vaccine is a recombinant form of one of HIV's surface proteins (see discussion of vaccines on next page).

Figure 11.26 Equipment used in the production of a vaccine against hepatitis B. Yeast cells carrying genes from the hepatitis B virus are growing in the stainless steel tank.

The DNA Revolution in the Pharmaceutical Industry and Medicine

DNA technology is already having a major impact on the pharmaceutical industry and on human medicine. Consider the first two products in Table 11.1, human insulin and human growth hormone (GH). These were the first pharmaceutical products made using recombinant DNA technology. In the United States alone, about 2 million people with diabetes depend on insulin treatment. Before 1982, the main sources of this hormone were pig and cattle tissues obtained from slaughterhouses. Insulin extracted from these animals is chemically similar, but not identical, to human insulin, and it causes harmful side effects in some people. Genetic engineering has largely solved this problem by developing bacteria that actually synthesize and secrete human insulin.

GH was harder to produce than insulin because the GH molecule is about twice as big. But because growth hormones from other animals are not effective growth stimulators in humans, GH was urgently needed. Before GH produced in *E. coli* became available in 1985, children with a GH deficiency had to rely on scarce supplies from human cadavers or else face dwarfism.

DNA technology is also helping medical researchers develop vaccines. A **vaccine** is a harmless variant or derivative of a pathogen (usually a bacterium or virus) that is used to prevent an infectious disease. When a person—a potential host for the pathogen—is inoculated, the vaccine stimulates the immune system to develop lasting defenses against the pathogen. Especially for the many viral diseases for which there is no effective drug treatment, prevention by vaccination is virtually the only medical way to fight the disease.

Genetic engineering can be used in several ways to make vaccines. One approach is to use genetically engineered cells to produce large amounts of a protein molecule that is found on the pathogen's outside surface. This method has been used to make the vaccine against hepatitis B virus. Hepatitis is a disabling and sometimes fatal liver disease, and the hepatitis B virus can cause liver cancer. Figure 11.26 shows a tank for growing yeast cells that have been engineered to carry hepatitis B genes.

Another way to use DNA technology in vaccine development is to make a harmless artificial mutant of the pathogen by altering one or more of its genes. When a harmless mutant is used as a vaccine, it multiplies in the body and may trigger a stronger immune response than the protein-molecule type of vaccine. But artificial-mutant vaccines may cause fewer side effects than those that have traditionally been made from natural mutants.

Yet another scheme for making vaccines employs a virus related to the one that causes smallpox. Smallpox was once a dreaded human disease, but it was eradicated worldwide in the 1970s by widespread vaccination with a harmless variant (natural mutant) of the smallpox virus. Using this harmless virus, genetic engineers can replace some of the genes encoding proteins that induce immunity to smallpox with genes that induce immunity to other diseases. In fact, the virus could be engineered to carry the genes needed to vaccinate against several diseases simultaneously. In the future, one inoculation may prevent a dozen diseases.

Among the potential new treatments for viral diseases and some cancers are short, synthetic pieces of single-stranded nucleic acid. Readily taken up by cells, these molecules are designed to base-pair with specific sequences on the DNA or mRNA in cancer cells or cells infected with viruses. One type, called **antisense nucleic acid,** is engineered to bind to, and prevent

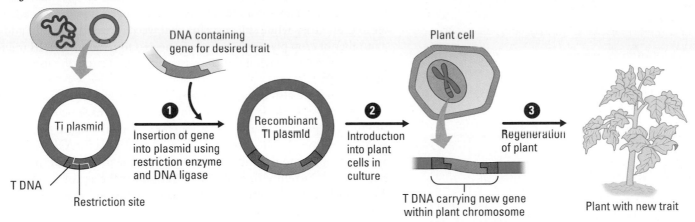

Agrobacterium tumefaciens

DNA containing
gene for desired trait

Plant cell

Ti plasmid

1 Insertion of gene into plasmid using restriction enzyme and DNA ligase

T DNA

Restriction site

Recombinant Ti plasmid

2 Introduction into plant cells in culture

3 Regeneration of plant

T DNA carrying new gene within plant chromosome

Plant with new trait

Figure 11.27 Using the Ti plasmid as a vector. A plasmid called Ti is commonly used to introduce new genes into plants. **1** With the help of a restriction enzyme and DNA ligase, the gene for the desired trait (orange) is inserted into a segment of the plasmid called T DNA (gray). **2** Then the recombinant plasmid is put into a plant cell, where the T DNA carrying the new gene integrates into a plant chromosome. **3** Finally, the recombinant cell is cultured and grows into a whole plant.

translation of, mRNA molecules coding for disease-causing proteins. Trials of antisense nucleic acid are promising, and along with new vaccines, they will probably be part of the future of human medicine.

A completely new development in the use of recombinant DNA technology by the pharmaceutical industry is the genetic modification of *plants* to produce vaccines and rare human proteins of medical importance. We discuss this application, often called "pharming," in the next section.

DNA Technology in Farming and "Pharming"

For a number of years, scientists have been using DNA technology to try to improve the productivity of plants and animals important to agriculture.

Plant Crops Agricultural scientists have already provided a number of crop plants with genes for desirable traits such as delayed ripening and resistance to spoilage and disease. In one striking way, plants are easier to engineer than most animals. For many plant species, a single tissue cell grown in culture can give rise to an adult plant (see Figure 10.6). So genetic manipulations can be performed on a single cell and the cell then used to generate an organism with new traits.

The vector used to introduce new genes into plant cells is most often a plasmid from the soil bacterium *Agrobacterium tumefaciens.* This is the Ti plasmid, so called because in nature it induces tumors in plants infected by the bacterium. For using Ti as a vector, researchers have eliminated its tumor-causing properties while keeping its ability to transfer DNA into plant cells. Figure 11.27 shows how a plant with a new trait can be created using the Ti plasmid. Although the Ti vector does not work with many grain-producing species, researchers can make transgenic varieties of these plants by using a "gene gun" to fire pieces of foreign DNA directly into cultured cells (Figure 11.28). Whatever method is used to introduce the new DNA, the result is a **transgenic** or **genetically modified (GM) organism,** one that has acquired one or more genes by artificial means. The source of the new genetic material may be another organism of the same species or a different species.

Genetic engineering is quickly replacing traditional plant-breeding programs, especially in cases where useful traits are determined by one or only a few genes. In the year 2000, roughly half of the American crops of soybeans and corn were genetically modified in some way. Many of the transgenic plants now being grown have received genes for herbicide resistance.

Figure 11.28 Using a "gene gun" to insert DNA into plant cells. This researcher is preparing to use a modified 22-caliber gun to shoot foreign DNA into plant cells growing in culture.

For example, the development of cotton plants that carry a bacterial gene that makes the plants resistant to weed-killing herbicides have made it easier to grow crops while still ensuring that weeds are destroyed. In addition, a number of crop plants are being engineered to resist infectious microbes and pest insects. Growing insect-resistant plants reduces the use of chemical insecticides.

Scientists are also using gene transfer to improve the nutritional value of crop plants. An exciting recent development is a transgenic rice plant that produces yellow rice grains containing beta-carotene, which our body uses to make vitamin A. This rice could help prevent vitamin A deficiency among people who depend on rice as their staple food—half the world's population. Currently 70% of all children under the age of 5 in Southeast Asia suffer from this condition, which leads to vision impairment and increased susceptibility to disease.

An important potential use of DNA technology in improving the nutrition of the world's population involves nitrogen fixation. Nitrogen fixation is the conversion of atmospheric nitrogen gas, which plants cannot use, into nitrogen compounds that plants can convert to essential nitrogen-containing molecules such as amino acids. In nature, nitrogen fixation is performed by certain bacteria that live in the soil or in plant roots. Even so, the level of nitrogen compounds in the soil is often so low that fertilizers must be applied for crops to grow. Nitrogen-providing fertilizers are costly and contribute to water pollution. DNA technology offers ways to increase bacterial nitrogen fixation and eventually, perhaps, to engineer crop plants to fix nitrogen themselves.

Finally, we arrive at the new and surprising alliance between the pharmaceutical industry and agriculture mentioned earlier. Plants have long been a source of drugs for the pharmaceutical industry. Now, however, DNA technology has made it possible to create plants that produce human proteins for medical use and viral proteins for use as vaccines. Several such "pharm" products are now being tested in clinical trials, including vaccines for hepatitis B and an antibody produced in GM tobacco plants that blocks the bacteria that cause tooth decay. Large amounts of these proteins might be made more economically by plants than by cultured cells. The main competition for the plants comes from animal "pharming."

Farm Animals—and "Pharm" Animals DNA technology is now routinely used to make vaccines and growth hormones for treating farm animals and, on a still largely experimental basis, to make transgenic animals. The goals of creating a transgenic animal are often the same as the goals of traditional breeding—for instance, to make a sheep with better quality wool, a pig with leaner meat, or a cow that will mature in a shorter time. Scientists might, for example, identify and clone a gene that causes the development of larger muscles (which make up most of the meat we eat) in one variety of cattle and transfer it to other cattle or even to sheep.

Another type of transgenic animal is one engineered to be a pharmaceutical "factory"—a producer of a large amount of an otherwise rare biological substance for medical use (Figure 11.29). In most cases to date, a gene for a desired human protein, such as a hormone or blood-clotting factor, has been added to the genome of a farm mammal in such a way that the gene's product is secreted in the animal's milk. It can then be purified, usually more easily than from a cell culture or a transgenic plant.

Human proteins produced by farm animals may or may not be structurally identical to the natural human proteins, so they have to be tested very carefully to make sure they will not cause allergic reactions or other

Figure 11.29 "Pharm" animals. These transgenic sheep carry a gene for a human blood protein, which they secrete in their milk. This protein inhibits an enzyme that contributes to lung damage in patients with cystic fibrosis and some other chronic respiratory diseases. Easily purified from the sheep's milk, the protein is currently being tested as a treatment for cystic fibrosis.

adverse effects in patients receiving them. Also, the health and welfare of farm animals carrying genes from humans and other foreign species are important issues; problems such as low fertility or increased susceptibilty to disease are not uncommon.

How is a transgenic animal created? Scientists first remove egg cells from a female and fertilize them in vitro. Meanwhile, they clone the desired gene from another organism. They then inject the DNA directly into the nuclei of the eggs. Some of the cells integrate the foreign DNA into their genomes and are able to express the foreign gene. The engineered eggs are then transferred to the uterus of a surrogate mother. If an embryo develops successfully, the result is a transgenic animal, containing a gene from a third "parent" that may even be of another species.

Gene Therapy in Humans

Techniques for manipulating DNA hold great potential for treating a variety of diseases by gene therapy—alteration of an afflicted individual's genes. In people with disorders traceable to a single defective gene, it should theoretically be possible to replace or supplement the defective gene with a normal allele.

Ideally, the normal allele would be put into cells that multiply throughout a person's life. Bone marrow cells, which include the stem cells that give rise to all the cells of the blood and immune system, are prime candidates. Figure 11.30 outlines one procedure for correcting a situation in which bone marrow cells are failing to produce a vital protein because of a defective gene. A retrovirus is used as a vector for implanting a normal version of the gene in the bone marrow cells. If the procedure succeeds, the cells will multiply throughout the patient's life and express the normal gene. The engineered cells will supply the missing protein, and the patient will be cured.

Most human gene therapy experiments to date have been preliminary, designed to test the safety and effectiveness of a procedure rather than to attempt a cure. Despite repeated hype in the news media over the past decade, it was not until April 2000 that the first scientifically strong evidence of effective gene therapy was reported. This landmark case involved the treatment of two infants suffering from a form of severe combined immune deficiency (SCID), which prevents the development of the immune system. Unless treated with a bone marrow transplant (effective just 60% of the time), SCID patients quickly die from infections by ever-present microbes that most of us easily fend off. Working at a Paris hospital, the researchers used a procedure similar to the one in Figure 11.30 to provide the infants with functional copies of their defective gene. As of June 2000, 15 months after treatment, the children were still healthy.

Some of the other promising gene therapy trials now going on are not aimed at correcting genetic defects. Instead, researchers are engineering cells from bone marrow in attempts to enhance the ability of immune cells to fight off cancer or resist infection by HIV. This approach may lead to effective treatments for many nonhereditary diseases.

Human gene therapy raises certain technical questions. For example, how can researchers build in gene control mechanisms to ensure that cells with the transferred gene make appropriate amounts of the gene product at the right time and in the right parts of the body? And how can they be sure that the gene's insertion does not harm some other necessary cell function?

There are ethical questions, too. For example, who will have access to gene therapy? The procedures now being tested are expensive and require expertise and equipment found only in major medical centers. A related

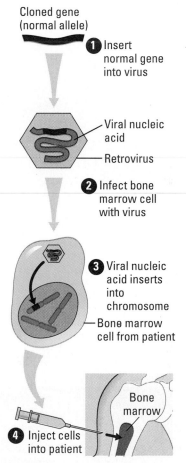

Cloned gene
(normal allele)

❶ Insert normal gene into virus

Viral nucleic acid

Retrovirus

❷ Infect bone marrow cell with virus

❸ Viral nucleic acid inserts into chromosome

Bone marrow cell from patient

Bone marrow

❹ Inject cells into patient

Figure 11.30 One type of gene therapy procedure.
❶ The normal gene is cloned by recombinant DNA techniques. It is then converted into RNA and inserted into the RNA genome of a retrovirus vector that has been rendered harmless. ❷ Bone marrow cells are taken from the patient and infected with the virus. ❸ The virus inserts a DNA copy of its genome, including the human gene, into the cells' DNA. ❹ The engineered cells are then injected back into the patient, where they colonize the bone marrow.

question is, Should gene therapy be reserved for treating serious diseases? And what about its potential use for enhancing athletic ability, physical appearance, or even intelligence?

Technically easier than modifying genes in the somatic cells of children or adults is the genetic engineering of germ cells or zygotes—already accomplished in lab animals. But this possibility raises the most difficult ethical question of all: whether we should try to eliminate genetic defects in our children and their descendants. Should we interfere with evolution in this way? From a biological perspective, the elimination of unwanted alleles from the gene pool could backfire. Genetic variety is a necessary ingredient for the survival of a species as environmental conditions change with time. Genes that are damaging under some conditions may be advantageous under others (one example is the sickle-cell allele, discussed in Chapter 8). Are we willing to risk making genetic changes that could be detrimental to our species in the future? We may have to face this question soon.

Risks and Ethical Questions

As soon as scientists realized the power of DNA technology, they began to worry about potential dangers. Early concerns focused on the possibility that recombinant DNA technology might create hazardous new pathogens. What might happen, for instance, if cancer cell genes were transferred into bacteria or viruses? Scientists developed a set of guidelines that in the United States and some other countries have become formal government regulations.

One type of safety measure is a set of strict laboratory procedures designed to protect researchers from infection by engineered microbes and to prevent the microbes from accidentally leaving the laboratory. In addition, strains of microorganisms to be used in recombinant DNA experiments are genetically crippled to ensure that they cannot survive outside the laboratory. Finally, certain obviously dangerous experiments have been banned. Today, most public concern about possible hazards centers not on recombinant microbes but on transgenic animals and plants.

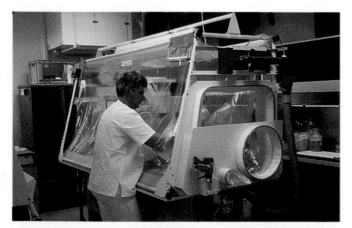

Maximum security laboratory. A scientist in a high-containment laboratory at the Pasteur Institute in Paris uses a "glove box" for working with dangerous microorganisms. To manipulate the microbial cultures inside the box, he inserts his hands into gloves that are sealed to the edges of round holes in the box. In this way, he avoids all contact with the microbes.

The Controversy About GM Foods

Animals that have been genetically modified by artificial means are not yet part of our food supply, but GM crop plants are. In 1999, controversy about the safety of these foods exploded in the United Kingdom (where one of the more extreme headlines warned of "the mad forces of genetic darkness") and soon spread through Europe. In response to these concerns, the European Union suspended the introduction of new GM crops pending new legislation and started considering the possibility of banning the import of all GM foodstuffs. In the United States and other countries where the GM revolution had been proceeding more quietly, the labeling of GM foods as such is now being debated.

Advocates of a cautious approach fear that crops carrying genes from other species might somehow be hazardous to human health or harm the environment. A major concern is that transgenic plants might pass their new genes to close relatives in nearby wild areas. We know that lawn and crop grasses, for example, commonly exchange genes with wild relatives via pollen transfer (Figure 11.31). If domestic plants carrying genes for resistance to herbicides, diseases, or insect pests pollinated wild plants, the offspring might become "superweeds" that would be very difficult to control. However, researchers may be able to prevent the escape of such plant genes by engineering plants so that they cannot hybridize. In April 2000, the U.S. National Academy of Sciences released a study finding no scientific evidence that crops genetically modified to resist pests pose any special health or environmental risks, but the authors of the study also recommended more stringent regulations than now exist. To date, there is little good data on either side; more study is needed.

In late January 2000, negotiators from 130 countries (including the United States) agreed on a Biosafety Protocol that requires exporters to identify GM organisms present in bulk food shipments and allows importing countries to decide whether the shipments pose environmental or health risks. This agreement has been hailed as a breakthrough by environmentalists.

Today, governments and regulatory agencies throughout the world are grappling with how to facilitate the use of biotechnology in agriculture, industry, and medicine while ensuring that new products and procedures are safe. In the United States, all projects are evaluated for potential risks by a number of regulatory agencies, including the Food and Drug Administration, the Environmental Protection Agency, the National Institutes of Health, and the Department of Agriculture. These agencies are under increasing pressure from some consumer groups. Meanwhile, these same agencies also consider some of the ethical questions we discuss next.

Ethical Questions Raised by DNA Technology

In addition to the ethical issues discussed earlier in connection with gene therapy, DNA technology raises other kinds of questions. Consider, for example, the child in Figure 11.32. She is growing at a normal rate, thanks to regular injections of human growth hormone (GH) that was made by genetically engineered *E. coli*. Like any new drug, this GH was subjected to exhaustive laboratory tests before it was released for human use. However, because it is a powerful hormone that affects the body in a number of ways, this drug may be more likely than most ordinary drugs to produce unanticipated side effects in years to come. The increased availability of GH has

Figure 11.31 Pollen transfer. Pollen might transfer genes from genetically engineered crop plants to wild relatives nearby.

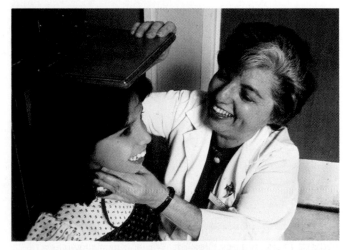

Figure 11.32 Treatment with human growth hormone. This child has been treated with human growth hormone made by bacteria.

Figure 11.33 James D. Watson. Watson, famous as the co-discoverer of the double-helical structure of DNA, was the first head of the Human Genome Project at the U.S. National Institutes of Health.

also raised ethical questions by leading some parents of short but hormonally normal children to seek GH treatment.

A much broader ethical question is, How do we really feel about wielding one of nature's singular powers—the ability to make new microorganisms, plants, and even animals? Some might ask, Do we have any right to alter an organism's genes—or to add our new creations to an already beleaguered environment?

Such questions must be weighed against the apparent benefits to humans and the environment that can be brought about by DNA technology. For example, bacteria are being engineered to clean up mining wastes and a number of industrial and domestic pollutants that threaten our soil, water, and air. These organisms may be the only feasible solutions to some of our most pressing environmental problems.

Ethical issues relating to the human genome are perhaps the most difficult for us. As discussed earlier, human gene therapy is rife with ethical dilemmas, whether it involves changing the DNA in germ cells or simply in somatic cells. And what of the information being obtained in the Human Genome Project? The potential benefits to human health provide strong ethical support for the project. But there is a danger that information about disease-associated genes—especially data about individual genomes that are collected outside the official project—could be abused. In the words of James D. Watson of DNA fame (Figure 11.33):

> *We must work to ensure that society learns to use [genetic] information only in beneficial ways and . . . pass laws . . . to prevent invasion of privacy of an individual's genetic background by either employers, insurers, or government agencies and to prevent discrimination on genetic grounds. . . . We have only to look at how the Nazis used leading members of the German human genetics and psychiatry communities to justify their genocide programs, first against the mentally ill and then the Jews and the Gypsies. We need no more vivid reminders that science in the wrong hands can do incalculable harm.**

Largely because of the events in Nazi Germany, our society today rejects the notion of eugenics—the effort to control the genetic makeup of human populations. The possibility of gene therapy on germ cells raises the greatest fears in this regard. However, many people argue that providing genetically engineered somatic cells to individuals with life-threatening diseases is no different from other medical interventions that save lives. Some compare the transplantation of genes in this case to the transplantation of organs.

With the establishment of databases holding information about individual genomes, the potential for genetic discrimination is already becoming a thorny issue. Insurance companies and employers are increasingly interested in gene testing. Breast cancer researcher Mary-Claire King mentions this issue in Chapter 10. Nancy Wexler (Figure 11.34), a leading Huntington's disease researcher, has put it bluntly:

> *A very big question is the problem of possible discrimination and stigmatization. An insurance company could refuse coverage or an employer could refuse a job if a person was found to carry the gene for a particular disease. People might also be coerced into taking a test that they wouldn't want in order to be considered for a job or an insurance policy.*

Figure 11.34 Nancy Wexler. Wexler is best known as a researcher on Huntington's disease, a disease that killed her mother.

---

*(Excerpted with permission from James Watson, *Science*, April 6, 1990. Copyright © 1990 American Association for the Advancement of Science.)

To what extent should we allow genetic information to be used this way? If we allow its use at all, how do we prevent the information from being used in a discriminatory manner? These are complex issues. As Nancy Wexler has said,

The question could be asked, Do you want people who have genetic suscep-tibility in positions in which they could have a major impact on other indi-viduals? For example, do you want an airline pilot with a genetic predispo-sition toward heart attack? My own feeling is that all of us have genes for something [harmful] that may be quiescent for a major part of our lives. If you start kicking out everybody who has some genetic susceptibility, then you're going to be in tough shape because there won't be anyone left. It's better to provide excellent medical care and preventive measures or early treatment for problems as they arise.

Some people have suggested that the dangers of abusing genetic informa-tion are so great that we should cease research in certain areas. But most scientists would probably tend to agree with the following statement by molecular biologist Leroy Hood (Figure 11.35): "What science does is give society opportunities. What we have to do is look at these oppor-tunities and then set up the constraints and the rules that will allow society to benefit in appropriate ways."

Evaluate scenarios and make decisions about using DNA technology in Web/ CD Activity 11G.

As citizens in the twenty-first century, we must all participate in making the decisions called for by Watson, King, Wexler, and Hood.

Figure 11.35 Leroy Hood. An eminent molecular biologist and immunologist, Hood played a leading role in developing faster DNA-sequencing machines, which have been essential for the Human Genome Project.

CheckPoint

1. What is the main concern about adding genes for herbicide resistance to crop plants?

2. The quotes from James Watson and Leroy Hood highlight the distinc-tion between _____, the basic pursuit of knowledge about the nat-ural world, and _____, the application of scientific knowledge. (*Hint:* Review Chapter 1.)

Answers: 1. The possibility that the genes could escape, via cross-pollination, to weeds that are closely related to the crop species **2.** science; technology

Evolution Link: Genomes Hold Clues to Evolution

The DNA sequences determined to date confirm the evolutionary connec-tions between even distantly related organisms and the relevance of research on simpler organisms to understanding human biology. Similarities between genes of disparate organisms can be surprising. Yeast, for example, has a

Learn more about clues to evolution in genome sequences in the Web Evolution Link.

number of genes close enough to the human versions that they can substitute for them in a human cell. In fact, re-searchers can sometimes work out what a human disease gene does by studying its normal counterpart in yeast. In humans and mice, similarities in gene order as well as DNA sequence remind us of our close evolutionary con-nection to these humble mammals. On a grander scale, comparisons of the completed genome sequences of bacteria, archaea, and eukaryotes strongly support the theory that these are the three fundamental domains of life—a topic we discuss further in the next unit, "Evolution and Biological Diversity."

Chapter Review

Summary of Key Concepts

Overview: The Advent of DNA Technology

- Research on the bacterium *E. coli* led to the development of recombinant DNA technology, a set of techniques for combining genes from different sources in the laboratory. These and other DNA techniques are making possible the Human Genome Project and other important basic research, the manufacture of useful protein products, new ways to improve agriculture, and new treatments for disease.
- Web/CD Activity 11A *Applications of DNA Technology*

Using Bacteria to Clone Genes

- **The Transfer of DNA** Bacterial genes can transfer from cell to cell by one of three processes: transformation, transduction, or conjugation.

- **Bacterial Plasmids** Plasmids, small circular DNA molecules separate from the bacterial chromosome, can serve as carriers for the transfer of genes. R plasmids, which carry genes that make bacteria resistant to various antibiotics, pose medical problems.

- **Using Plasmids to Customize Bacteria: An Overview** A piece of DNA from a eukaryotic organism can be inserted into a bacterial plasmid that will replicate it when returned to a bacterium. When the bacterium reproduces to form a clone of identical cells, the recombinant plasmid and the foreign gene it carries are also cloned. Copies of the gene or its protein product can then be harvested and used for practical purposes.

- **Enzymes That "Cut and Paste" DNA** Restriction enzymes, which cut DNA at specific points, and DNA ligase, which "pastes" DNA fragments together, can together create recombinant DNA.
- Web/CD Activity 11B *Restriction Enzymes*

- **Cloning Genes in Recombinant Plasmids: A Closer Look** Researchers use restriction enzymes and DNA ligase to insert into plasmids the genes they want to clone. Bacteria take up the recombinant plasmids from their surroundings by transformation.
- Web/CD Activity 11C *Cloning a Gene in Bacteria*

- **Genomic Libraries** Recombinant DNA techniques allow the construction of genomic libraries, sets of DNA fragments that together contain all of an organism's genes. Multiple copies of each fragment are stored in a cloned bacterial plasmid (or phage).

More Tools of DNA Technology

- **Reverse Transcriptase** Researchers use the enzyme reverse transcriptase to make DNA on mRNA templates. In this way they can make DNA libraries containing only the genes that are transcribed by a particular type of cell.

- **Nucleic Acid Probes** A nucleic acid probe can tag a desired gene in a library. The probe is a short, single-stranded molecule of radioactively labeled DNA or RNA whose nucleotide sequence is complementary to part of the gene (or other DNA of interest).

- **Automated Synthesis and Sequencing of DNA** Automation speeds up the sequencing and synthesis of genes in the laboratory.

- **Gel Electrophoresis** This method is used to sort DNA molecules, such as restriction fragments, by size.
- Web/CD Activity 11D *Gel Electrophoresis of DNA*

- **Restriction Fragment Analysis** Scientists can compare the DNA sequences of different individuals by cutting up their DNA with a restriction enzyme and then comparing the band patterns formed by the two sets of restriction fragments upon gel electrophoresis. A radioactive probe reveals the bands of interest.
- Web/CD The Process of Science *Analyzing DNA Fragments Using Gel Electrophoresis*

- **The Polymerase Chain Reaction (PCR)** When a DNA sample is very small, specific sequences within it can be quickly replicated in a test tube. Short molecules of single-stranded DNA used to prime the synthesis of new DNA determine the specificity. A heat-resistant DNA polymerase allows automation.

The Challenge of the Human Genome

- **The Human Genome: Genes and "Junk"** The 23 chromosomes of the haploid human genome contain about 3 billion nucleotide pairs. This DNA includes between 40,000 and 100,000 genes and a huge amount of noncoding DNA, much of which consists of repetitive nucleotide sequences. The telomeres protecting the ends of chromosomes are repetitive DNA.

- **Jumping Genes** Barbara McClintock discovered that certain segments of DNA, now called transposons, can move about within a cell's genome, changing the organism's phenotype in the process.

- **The Human Genome Project** The Human Genome Project involves genetic and physical mapping of chromosomes, DNA sequencing, and comparison of human genes to those of other species. The data will give insight into development and evolution and will aid in the diagnosis and treatment of disease.
- Web/CD Activity 11E *Genes on the Human Chromosome 17*

Other Applications of DNA Technology

- **DNA Fingerprinting in Courts of Law** DNA fingerprinting, which depends on small differences in the DNA sequences of different individuals, can be used to solve crimes and determine family relationships.
- Web/CD Activity 11F *DNA Fingerprinting Case Study*

- **The Mass Production of Gene Products** Microbes and cultured cells engineered with recombinant DNA technology are used to mass produce a variety of gene products, mostly proteins.

- **The Pharmaceutical Industry and Medicine** The medically useful or potentially useful pharmaceutical products made using DNA technology include hormones, vaccines, and antisense nucleic acid.

- **DNA Technology in Farming and "Pharming"** DNA technology is being used to produce transgenic plants and animals with economically and/or nutritionally valuable traits. Genetically modified (GM) foods are controversial. Plants and animals are also being engineered to make human substances of medical value.

- **Gene Therapy in Humans** Viruses can be used as vectors to transfer copies of normal human genes into cells with defective versions, potentially curing genetic diseases. A major ethical question is whether we should treat reproductive cells, such as gametes, and thus alter the genotypes of future generations.

Risks and Ethical Questions

- **The Controversy About GM Foods** The debate about genetically modified crops centers on whether they could harm human health or damage the environment.

- **Ethical Questions** We as a society and as individuals must address the ethical questions raised by the use of DNA technology.
• **Web/CD Activity 11G** *Making Decisions About DNA Technology*

Evolution Link: Genomes Hold Clues to Evolution

- Studies of the DNA sequences of genomes confirm evolutionary relationships among living things.
• **Web Evolution Link** *Genomes Hold Clues to Evolution*

Self-Quiz

1. Which of the following would be considered a transgenic organism?
   a. a bacterium that has received genes via conjugation
   b. a human given a corrected version of his own blood-clotting gene
   c. a fern grown in cell culture from a single fern root cell
   d. a rat with rabbit hemoglobin genes
   e. a human treated with insulin produced by *E. coli* bacteria

2. A microbiologist found that some bacteria infected by phages had developed the ability to make a particular amino acid that they could not make before. This new ability was probably a result of
   a. transformation.   d. mutation.
   b. natural selection.   e. transduction.
   c. conjugation.

3. When a typical restriction enzyme cuts a DNA molecule, the cuts are uneven, so that the DNA fragments have single-stranded ends. These ends are useful in recombinant DNA work because
   a. they enable a cell to recognize fragments produced by the enzyme.
   b. they serve as starting points for DNA replication.
   c. the fragments will bond to other fragments with complementary ends.
   d. they enable researchers to use the fragments as molecular probes.
   e. only single-stranded DNA segments can code for proteins.

4. DNA fingerprints used as evidence in a murder trial look something like supermarket bar codes. The pattern of bars in a DNA fingerprint shows
   a. the order of bases in a particular gene.
   b. the presence of various-sized segments of DNA.
   c. the presence of dominant or recessive alleles for particular traits.
   d. the order of genes along particular chromosomes.
   e. the exact location of a specific gene in a genomic library.

5. A biologist isolated a gene from a human cell, attached it to a plasmid, and inserted the plasmid into a bacterium. The bacterium made a new protein, but it was nothing like the protein normally produced in a human cell. Why?
   a. The bacterium had undergone transformation.
   b. The gene did not have sticky ends.
   c. The gene contained introns.
   d. The gene did not come from a genomic library.
   e. The biologist should have cloned the gene first.

6. A paleontologist has recovered a bit of organic material from the 400-year-old preserved skin of an extinct dodo. She would like to compare DNA from the sample with DNA from living birds. Which of the following would be most useful for increasing the amount of DNA available for testing?
   a. restriction fragment analysis   d. electrophoresis
   b. polymerase chain reaction   e. Ti plasmid technology
   c. molecular probe analysis

7. How many genes are there in a human sperm cell?
   a. 23   d. 40,000–100,000
   b. 46   e. about 3 billion
   c. 5,000–10,000

• **Go to the website or CD-ROM for more self-quiz questions.**

The Process of Science

1. A biochemist hopes to find a gene in human liver cells that codes for an important blood-clotting protein. She knows that the nucleotide sequence of a small part of the gene is CTGGACTGACA. Briefly explain how to obtain the desired gene.

2. **Use gel electrophoresis to analyze DNA fragments in The Process of Science activity available on the website and CD-ROM.**

Biology and Society

In the not-too-distant future, gene therapy may be an option for the treatment and cure of many inherited disorders. What do you think are the most serious ethical issues that must be dealt with before human gene therapy is used on a large scale? Why do you think these issues are important?

Evolution

How Populations Evolve

All humans are connected by descent from

African ancestors.  The same type of bones

make up the forelimbs of humans, cats, whales,

and bats.  Abuse of antibiotics has

hastened the evolution of antibiotic-resistant

bacteria.  The A & E channel's *Biography* ranked

Charles Darwin as the fourth most influential

person of the past 1000 years.

There's an expression that the only constant is change. It certainly applies to life, over time scales from milliseconds to millennia. Your brain responds to stimuli on a moment-to-moment basis. Over a longer time frame, your body, now that it is mature, will eventually begin to age. A forest changes with the seasons. Over a longer period of time, the forest may expand or contract its boundaries in response to global changes in climate. And everywhere, all the time, populations of organisms are fine-tuning adaptations to local environments through the evolutionary process of natural selection. It is happening, for example, in the intestines of humans, where resistance to antibiotics is evolving in populations of bacteria, including those that cause disease. Given the dynamics of Earth and its life, it is not surprising that even the kinds of organisms on the planet—the species—have changed over time. It is a historical fact of life documented in the fossil record (Figure 12.1).

Figure 12.1 Fossils of trilobites, animals that lived in the seas hundreds of millions of years ago.

But even as change characterizes life, so does continuity. All members of your family are connected by shared ancestry. In fact, *all* humans are connected by descent from our African ancestors. And all of life, with its dazzling diversity of millions of species, is united by descent from the first microbes that populated the primordial planet. It is this duality of life's unity and diversity that defines modern biology.

Biology came of age on November 24, 1859, the day Charles Darwin published *On the Origin of Species by Means of Natural Selection.* His topic was evolution. Darwin's book drew a cohesive picture of life by connecting the dots of what had once seemed a bewildering array of unrelated facts. *The Origin of Species* focused biologists' attention on the great diversity of organisms—their origins and relationships, their similarities and differences, their geographic distribution, and their adaptations to surrounding environments. We placed the Darwinian revolution in its historical context in Chapter 1. An understanding of evolution continues to inform every field of biology, from exploring life's molecules to analyzing ecosystems. And applications of evolutionary biology are transforming medicine, agriculture, biotechnology, and conservation biology. Because evolution integrates all of biology, it is the thematic thread woven throughout this book. This unit of chapters features mechanisms of evolution and traces the history of life on Earth.

Overview: Descent with Modification

Darwin made two main points in *The Origin of Species.* First, he argued from evidence that the species of organisms inhabiting Earth today descended from ancestral species. In the first edition of his book, Darwin did not actually use the word *evolution.* He referred instead to "descent with modification." Darwin postulated that as the descendants of the earliest organisms spread into various habitats over millions of years, they accumulated different modifications,

Join Charles Darwin as he visits the Galápagos Islands in Web/CD Activity 12A, available on the website and CD-ROM.

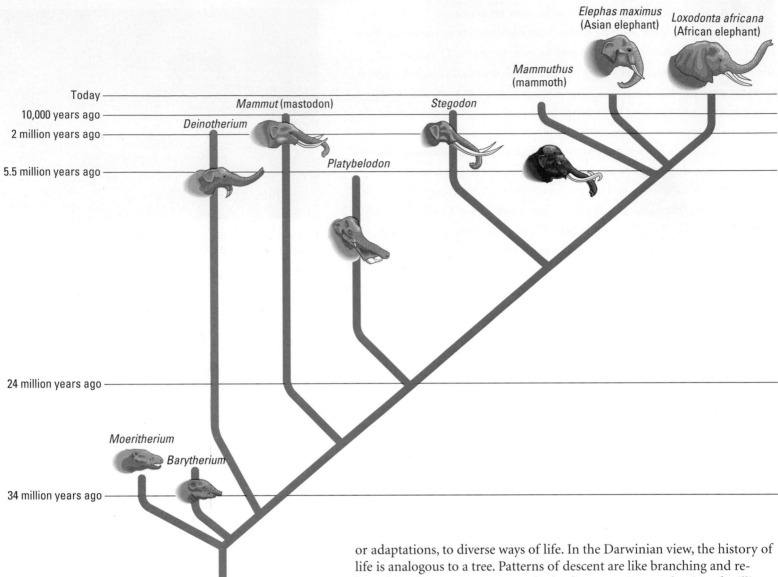

Figure 12.2 Descent with modification. This evolutionary tree of the elephant family is based mainly on evidence from fossils—their anatomy, their order of appearance in geological time, and their geographic distribution.

or adaptations, to diverse ways of life. In the Darwinian view, the history of life is analogous to a tree. Patterns of descent are like branching and re-branching from a common trunk, the first organism, to the tips of millions of twigs representing the species alive today. At each fork of the evolutionary tree is an ancestor common to all evolutionary branches extending from that fork. Closely related species, such as Asian and African elephants, share many characteristics because their lineage of common descent traces to the smallest branches of the tree of life (Figure 12.2).

Darwin's second main point in *The Origin of Species* was his argument for **natural selection** as the mechanism for descent with modification. When biologists speak of "Darwin's theory of evolution," they mean natural selection as a cause of evolution, not the phenomenon of evolution itself.

The basic idea of natural selection is that a population of organisms can change over the generations if individuals having certain heritable traits leave more offspring than other individuals. The result of natural selection is **evolutionary adaptation,** a population's increase in the frequency of traits that are suited to the environment (Figure 12.3). In modern terms, we would say that the genetic composition of the population had changed over time, and that is one way of defining **evolution.** But we can also use the term *evolution* on a much grander scale to mean all of biological history, from the earliest microbes to the enormous diversity of modern organisms.

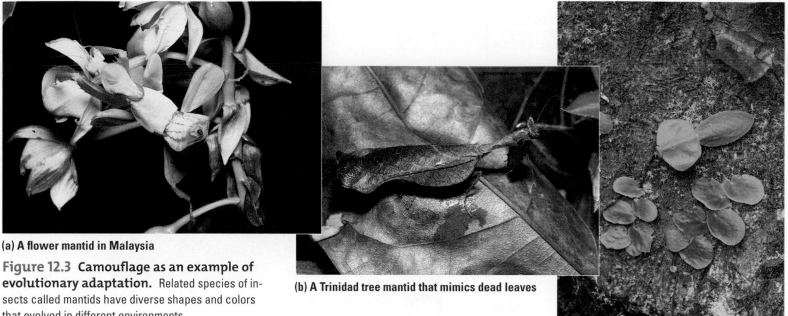

(a) A flower mantid in Malaysia

Figure 12.3 Camouflage as an example of evolutionary adaptation. Related species of insects called mantids have diverse shapes and colors that evolved in different environments.

(b) A Trinidad tree mantid that mimics dead leaves

(c) A Central American mantid that resembles a green leaf

Evidence of Evolution

Evolution leaves observable signs. Such clues to the past are essential to any historical science. Historians of human civilization can study written records from earlier times. But they can also piece together the evolution of societies by recognizing vestiges of the past in modern cultures. Even if we did not know from written documents that Spaniards colonized the Americas, we would deduce this from the Hispanic stamp on Latin American culture. Similarly, biological evolution has left marks—in the fossil record and in the historical vestiges evident in modern life.

The Fossil Record

Fossils are the preserved remnants or impressions left by organisms that lived in the past. Most fossils are found in sedimentary rocks. Sand and silt eroded from the land are carried by rivers to seas and swamps, where the particles settle to the bottom. Over millions of years, deposits pile up and compress the older sediments below into rock. Rock strata, or layers, form when the rates of sedimentation and the types of particles that settle vary over time. When aquatic organisms die, they settle along with the sediments and may leave imprints in the rocks. Some organisms living on land may be

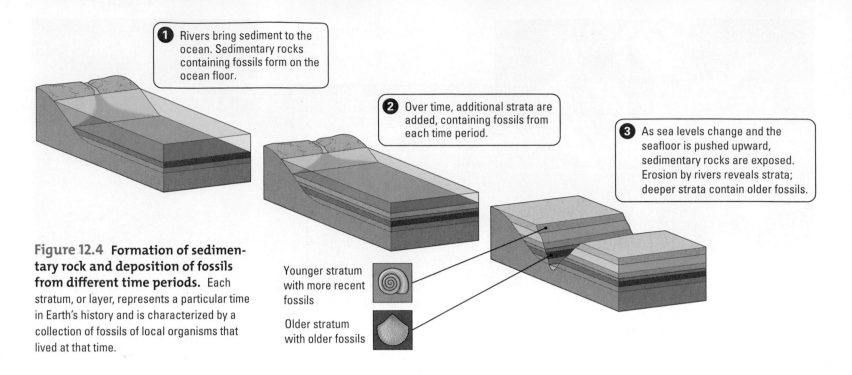

① Rivers bring sediment to the ocean. Sedimentary rocks containing fossils form on the ocean floor.

② Over time, additional strata are added, containing fossils from each time period.

③ As sea levels change and the seafloor is pushed upward, sedimentary rocks are exposed. Erosion by rivers reveals strata; deeper strata contain older fossils.

Younger stratum with more recent fossils

Older stratum with older fossils

Figure 12.4 Formation of sedimentary rock and deposition of fossils from different time periods. Each stratum, or layer, represents a particular time in Earth's history and is characterized by a collection of fossils of local organisms that lived at that time.

swept into swamps and seas. Land organisms that remain in place when they die may first be covered by windblown silt and then buried in water-borne sediments when sea levels rise over them. Thus, each of the rock strata bears a unique set of fossils representing a local sampling of the organisms that lived when the sediment was deposited. Younger strata are on top of older ones. Thus, the positions of fossils in the strata reveal their relative age (Figure 12.4). The **fossil record** is this chronology of fossil appearances in the rock layers, marking the passing of geological time (Figure 12.5).

The fossil record testifies that organisms have appeared in a historical sequence. The oldest known fossils, dating from about 3.5 billion years ago,

Figure 12.5 Strata of sedimentary rock at the Grand Canyon. The Colorado River has cut through over 2000 meters of rock, exposing sedimentary layers that are like huge pages from the book of life. Scan the canyon wall from rim to floor, and you look back through hundreds of millions of years. Each layer entombs fossils that represent some of the organisms from that period of Earth's history.

are prokaryotes (bacteria and archaea). This fits with the molecular and cellular evidence that prokaryotes are the ancestors of all life. Fossils in younger layers of rock chronicle the evolution of various groups of eukaryotic organisms. One example is the successive appearance of the different classes of vertebrates (animals with backbones). Fishlike fossils are the oldest vertebrates in the fossil record. Amphibians are next, followed by reptiles, then mammals and birds.

Paleontologists (scientists who study fossils) have discovered many transitional forms that link past and present. For example, a series of fossils documents the changes in skull shape and size that occurred as mammals evolved from reptiles. Another example is a series of fossilized whales that connect these aquatic mammals to their land-dwelling ancestors (Figure 12.6).

Biogeography

Study of the geographic distribution of species is called **biogeography.** It was biogeography that first suggested to Darwin that today's organisms evolved from ancestral forms. Consider, for example, Darwin's visit to the Galápagos Islands, which are located about 900 km off the west coast of Ecuador (see Figure 1.13). Darwin noted that the Galápagos animals resembled species of the South American mainland more than they resembled animals on similar but distant islands. This is what we should expect if the Galápagos species evolved from South American immigrants.

There are many other examples from biogeography that seem baffling without an evolutionary perspective. Why are the tropical animals of South America more closely related to species in the South American deserts than to species in the African tropics? Why is Australia home to such an impressive diversity of pouched mammals (marsupials) but relatively few placental mammals (those in which embryonic development is completed in the uterus)? It is *not* because Australia is inhospitable to placental mammals. Humans have introduced rabbits, foxes, and many other placental mammals to Australia, where these introduced species have thrived to the point of becoming ecological and economic nuisances. The prevailing hypothesis is that the unique Australian wildlife evolved on that island continent in isolation from regions where early placental mammals diversified (Figure 12.7).

Biogeography makes little sense if we imagine that species were individually placed in suitable environments. In the Darwinian view, we find species where they are because they evolved from ancestors that inhabited those regions.

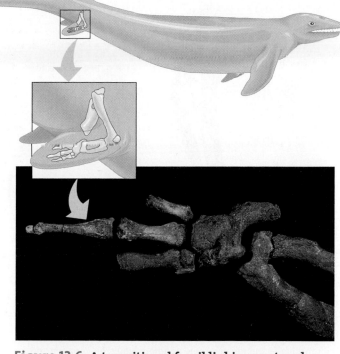

Figure 12.6 A transitional fossil linking past and present. The hypothesis that whales evolved from terrestrial (land-dwelling) ancestors predicts a four-limbed beginning for whales. Paleontologists digging in Egypt and Pakistan have identified extinct whales that had hind limbs. Shown here are the fossilized leg bones of *Basilosaurus,* one of those ancient whales. These whales were already aquatic animals that no longer used their legs to support their weight. The leg bones of an even older fossilized whale named *Ambulocetus* are heftier. *Ambulocetus* may have been amphibious, living on land and in water.

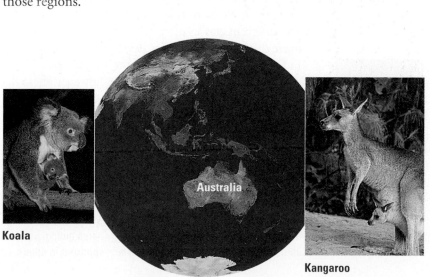

Koala

Australia

Kangaroo

Figure 12.7 Evidence of evolution from biogeography. The continent of Australia is home to many unique plants and animals, such as these marsupials, that evolved in relative isolation from other continents where placental mammals diversified.

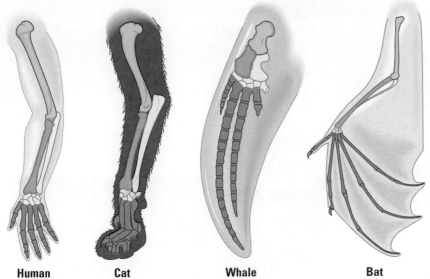

Human   Cat   Whale   Bat

Figure 12.8 Homologous structures: anatomical signs of descent with modification. The forelimbs of all mammals are constructed from the same skeletal elements. The hypothesis that all mammals descended from a common ancestor predicts that their forelimbs, though diversely adapted, would be variations on a common anatomical theme.

Comparative Anatomy

The comparison of body structures between different species is called **comparative anatomy.** Certain anatomical similarities among species bear witness to evolutionary history. For example, the same skeletal elements make up the forelimbs of humans, cats, whales, and bats, all of which are mammals (Figure 12.8). The functions of these forelimbs differ. A whale's flipper does not do the same job as a bat's wing. If these limbs had completely separate origins, we would expect that their basic designs would be very different. However, structural similarity would not be surprising if all mammals descended from a common ancestor with the prototype forelimb. Arms, forelegs, flippers, and wings of different mammals are variations on a common anatomical theme that has become adapted to different functions. Such similarity due to common ancestry is called **homology.** The forelimbs of diverse mammals are homologous structures.

Reconstruct homologous forelimbs in Web/CD Activity 12B.

Comparative anatomy confirms that evolution is a remodeling process. Ancestral structures that originally functioned in one capacity become modified as they take on new functions—descent with modification. The historical constraints of this retrofitting are evident in anatomical imperfections. For example, the human spine and knee joint were derived from ancestral structures that supported four-legged mammals. Almost none of us will reach old age without experiencing knee or back problems. If these structures had first taken form specifically to support our bipedal posture, we would expect them to be less subject to sprains, spasms, and other common injuries. The anatomical remodeling that stood us up was apparently constrained by our evolutionary history.

Comparative Embryology

The comparison of structures that appear during the development of different organisms is called **comparative embryology.** Closely related organisms often have similar stages in their embryonic development. One sign that vertebrates evolved from a common ancestor is that all of them have an embryonic stage in which structures called gill pouches appear on the sides of the throat. At this stage, the embryos of fishes, frogs, snakes, birds, apes—indeed, all vertebrates—look more alike than different (Figure 12.9).

The different classes of vertebrates take on more and more distinctive features as development progresses. In fishes, for example, most of the gill pouches develop into gills. In land vertebrates, however, these embryonic features develop into other kinds of structures, such as bones of the skull, bones supporting the tongue, and the voice box of mammals.

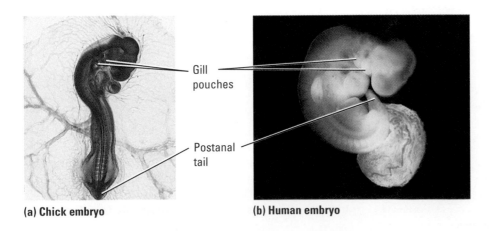

Gill pouches

Postanal tail

(a) Chick embryo   **(b) Human embryo**

Figure 12.9 Evolutionary signs from comparative embryology. At this early stage of development, the kinship of vertebrates is unmistakable. Notice, for example, the gill pouches and tails in both **(a)** the bird embryo and **(b)** the human embryo. Comparative embryology helps biologists identify anatomical homology that is less apparent in adults.

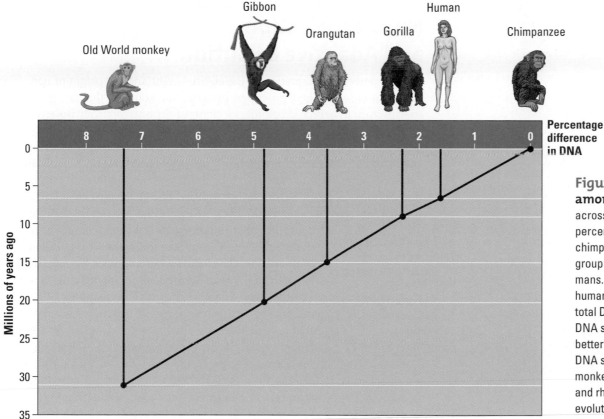

Old World monkey   Gibbon   Orangutan   Gorilla   Human   Chimpanzee

Percentage difference in DNA

Millions of years ago

Figure 12.10 Genetic relationships among some primates. The scale across the top of this diagram gives you the percentage difference in DNA between a chimpanzee and other primates, the animal group that includes monkeys, apes, and humans. For example, note that chimps and humans are less than 2% different in their total DNA sequences. Put another way, the DNA sequences of chimps and humans are better than a 98% match. In contrast, the DNA sequences of chimps and Old World monkeys (macaques, mandrills, baboons, and rhesus monkeys) differ by over 7%. The evolutionary relationships documented by these genetic comparisons are compatible with other types of evidence, including fossils. For example, the fossil record suggests that chimps and humans diverged from a common ancestor only about 6–7 million years ago (left scale on this diagram). The evolutionary split between Old World monkeys and the branch that includes humans and chimps was much earlier, over 30 million years ago.

Molecular Biology

Evolutionary relationships among species leave signs in DNA and proteins — in genes and gene products (see the Evolution Link section in Chapter 3). If two species have libraries of genes and proteins with sequences of monomers that match closely, the sequences must have been copied from a common ancestor (Figure 12.10). By analogy, if two long paragraphs are identical except for the substitution of a letter here and there, we can be quite sure that they came from a common source.

Darwin's boldest hypothesis was that *all* forms of life are related to some extent through branching evolution from the earliest organisms. A century after Darwin's death, molecular biology began confirming the fossil record and other evidence supporting the Darwinian view of kinship among all life. The molecular signs include a common genetic code shared by all species (see Chapter 9). Evidently, the genetic language has been passed along through all branches of evolution ever since its beginnings in an early form of life. Molecular biology has added the latest chapter to the volumes of evidence validating evolution as a natural explanation for the unity and diversity of life.

CheckPoint

1. Why are older fossils generally in deeper rock layers than younger fossils?

2. What is homology?

Answers: 1. Sedimentation adds younger rock layers on top of older ones. **2.** Similarity between species that is due to shared ancestry

Natural Selection and Adaptive Evolution

Darwin perceived adaptation to the environment and the origin of new species as closely related processes. Imagine, for example, that an animal species from a mainland colonizes a chain of distant, relatively isolated islands. In the Darwinian view, populations on the different islands may diverge more and more in appearance as each population adapts to its local environment. Over many generations, the populations on different islands could become dissimilar enough to be designated as separate species. Evolution on the Galápagos Islands is an example. The Galápagos are relatively young islands of volcanic origin. Most of the animal species on the islands

Explore the Galápagos Islands and other places Darwin visited in The Process of Science activity *The Voyage of the Beagle,* available on the website and CD-ROM.

live nowhere else in the world, but they resemble species living on the South American mainland. It is a reasonable hypothesis that the islands were colonized by animals that strayed from the South American mainland and then diversified on the different islands. The example of "Darwin's finches" was introduced in Chapter 1. Among the differences between the finches are their beaks, which are adapted to the specific foods available on each species' home island (Figure 12.11). Darwin anticipated that explaining how such adaptations arise is the key to understanding evolution. And his theory of natural selection remains our best explanation for adaptive evolution.

(a) Large ground finch

(b) Small tree finch

(c) Woodpecker finch

Figure 12.11 Beak adaptations of Galápagos finches. The most striking difference between these closely related species is their beaks, which are adapted for specific diets. **(a)** The large ground finch has a large beak specialized for cracking seeds that fall from plants to the ground. **(b)** The small tree finch uses its beak to grasp insects. **(c)** The woodpecker finch uses tools such as cactus spines to probe for termites and other wood-boring insects.

Figure 12.12 Overproduction of offspring. A cloud of millions of spores is exploding from this puffball, a type of fungus. The wind will disperse the spores far and wide. Each spore, if it lands in a suitable environment, has the potential to grow and develop into a new fungus. Only a tiny fraction of the spores will actually give rise to offspring that survive and reproduce. Otherwise, it wouldn't take too many generations before puffballs filled the universe!

Natural Selection Basics

We traced Darwin's synthesis of natural selection in Chapter 1. Here we review the basics.

Darwin based his theory of natural selection on two key observations. First, Darwin recognized that all species tend to produce excessive numbers of offspring (Figure 12.12). Darwin deduced that because natural resources are limited, the production of more individuals than the environment can support leads to a struggle for existence among the individuals of a population. In most cases, only a small percentage of offspring will survive in each generation. Many eggs are laid, young born, and seeds spread, but only a tiny fraction complete their development and leave offspring of their own. The rest are starved, eaten, frozen, diseased, unmated, or unable to reproduce for other reasons.

The second key observation that led Darwin to natural selection was his awareness of variation among individuals of a population. Just as no two people in a human population are alike, individual variation abounds in almost all species (Figure 12.13). Much of this variation is heritable. Siblings share more traits with each other and with their parents than they do with less closely related members of the population.

From these two observations, Darwin arrived at the conclusion that defines natural selection: Individuals whose inherited traits are best suited to the local environment are more likely than less fit individuals to survive and reproduce. In other words, the individuals that function best tend to leave the most offspring. Darwin's genius was in connecting two observations that anyone could make and drawing an inference that could explain how adaptations evolve.

Figure 12.13 A few of the color variations in a population of Asian lady beetles.

Charles Darwin in 1859.

- *Observation #1: Overproduction.* Populations of many species have the potential to produce far more offspring than the environment can possibly support with food, space, and other resources. This overproduction makes a struggle for existence among individuals inevitable.
- *Observation #2: Individual variation.* Individuals in a population vary in many heritable traits.

→ *Inference: Differential reproductive success (natural selection).* Those individuals with traits best suited to the local environment generally leave a disproportionately large share of surviving, fertile offspring.

Darwin's insight was both simple and profound. The environment screens a population's inherent variability. Differential success in reproduction causes the favored traits to accumulate in the population over the generations. Natural selection causes adaptive evolution.

Natural Selection in Action: The Evolution of Pesticide-Resistant Insects

Natural selection and the adaptive evolution it causes are observable phenomena. A classic and unsettling example is the evolution of pesticide resistance in hundreds of insect species.

Pesticides are poisons used to kill insects that are pests in crops, swamps, backyards, and homes. Examples are DDT, now banned in many countries, and malathion. These chemical weapons against insects have proved to be double-edged swords. We have used pesticides to control insects that eat our crops, transmit diseases such as malaria, or just annoy us around the house or campground. But widespread use of these poisons, which are not specific for the intended targets, has also produced some colossal environmental problems, which we'll examine in Chapter 17. Our focus here is the evolutionary outcome of introducing these chemicals into the environments of insects.

Whenever a new type of pesticide is used to control agricultural pests, the story is usually the same. Early results are encouraging. A relatively small amount of the poison dusted onto a crop may kill 99% of the insects. But subsequent sprayings are less and less effective. One option is to keep increasing the amount of the poison, but that brings high monetary (not to mention environmental) costs. Another strategy is for the farmer to switch to a different pesticide until it, too, becomes ineffective.

Just what is happening on that farm? It's all about natural selection. The relatively few survivors of the first pesticide wave are insects with genes that somehow enable them to resist the chemical attack. In some cases, the lucky few carry genes coding for enzymes that destroy the pesticide. The poison kills most members of the insect population, leaving only the resistant individuals to reproduce. And their offspring inherit the genes for pesticide resistance. In each generation, the proportion of pesticide-resistant individuals in the insect population increases. The population has adapted to a change in its environment.

Make changes in a virtual environment and observe the effects on a population of leafhoppers in Web/CD Activity 12C/ The Process of Science.

This example of insect adaptation to pesticides highlights two key points about natural selection. First, notice that natural selection is more a process of editing than it is a creative mechanism. A pesticide does not create resistant individuals, but selects for resistant insects that were already present in the population (Figure 12.14). Second, note that natural selection is timely and regional. It favors those characteristics in a varying population that fit the current, local environment. Environmental factors vary from place to place and from time to time. An adaptation in one situation may be useless or even detrimental in different circumstances. For example, some genetic mutations that happen to endow houseflies with resistance to the pesticide DDT also reduce a fly's growth rate. Before DDT was introduced to environments, the gene for resistance was a handicap. But the appearance of DDT changed the rules in the environmental arena and favored pesticide-resistant individuals in the reproduction sweepstakes. Such are the dynamics of adaptive evolution by natural selection.

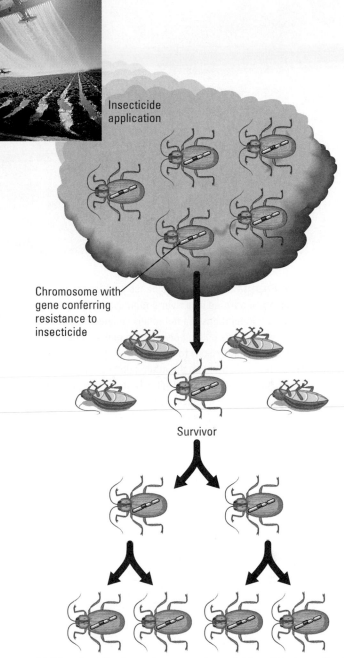

Insecticide application

Chromosome with gene conferring resistance to insecticide

Survivor

Additional applications of the same insecticide will be less effective, and the frequency of resistant insects in the population will grow

Figure 12.14 Evolution of pesticide resistance in insect populations. By spraying crops with poisons to kill insect pests, humans have unwittingly favored the reproductive success of insects with inherent resistance to the poisons.

CheckPoint

1. Define natural selection.
2. Explain why the following statement is incorrect: "Pesticides have created pesticide resistance in insects."

Answers: 1. Natural selection is the differential reproductive success among a population's varying individuals. **2.** An environmental factor does not create new traits such as pesticide resistance, but rather selects among the traits that are already represented in the population.

The Modern Synthesis: Darwinism Meets Genetics

Natural selection requires hereditary processes that Darwin could not explain. How do the variations that are the raw material for natural selection arise in a population? And how are these variations passed along from parents to offspring? Ironically, Darwin and Gregor Mendel lived and worked at the same time. However, Mendel's discoveries went unnoticed or unappreciated by most of the scientific community (see Chapter 8). In fact, by breeding peas in his abbey garden, Mendel illuminated the very hereditary processes required for natural selection to work. Mendelism and Darwinism finally came together in the mid-1900s, decades after both scientists were dead. This fusion of genetics with evolutionary biology came to be known as the **modern synthesis.** One of its key elements is an emphasis on the biology of populations.

Populations as the Units of Evolution

We have already used the term *population* several times in this chapter. Now it's time for a biological definition: A **population** is a group of individuals of the same species living in the same area at the same time. One population may be isolated from others of the same species. If individuals of an isolated population interbreed only rarely with those in another population, there will be little exchange of genes between the populations. Such isolation is common for populations confined to widely separated islands,

(a)

(b)

Figure 12.15 Populations. **(a)** Two dense populations of Douglas fir trees are separated by a river bottom where firs are uncommon. The two populations are not totally isolated. Interbreeding occurs when wind blows pollen between the populations. Nevertheless, trees are more likely to breed with members of the same population than with trees on the other side of the river. **(b)** This nighttime satellite view of the United States shows the lights of human population centers, or cities. People move around the country, of course, and there are suburban and rural communities between cities, but people are more likely to choose mates locally.

unconnected lakes, or mountain ranges separated by lowlands. However, populations are not usually so isolated, and they rarely have sharp boundaries. One population center may blur into another in a region of overlap, where members of both populations are present but less numerous (Figure 12.15). However, individuals are more concentrated in the population centers, and they are likely to breed with other locals. Therefore, organisms of a population are generally more closely related to one another than to members of other populations.

A population is the smallest biological unit that can evolve. A common misconception is that individual organisms evolve (in the Darwinian sense) during their lifetimes. It is true that natural selection acts on individuals. Inherited characteristics affect their survival and reproductive success. However, the evolutionary impact of this natural selection is only apparent in tracking how a population changes over time. In our example of pesticide resistance in insects, we measured evolution by the change in the relative numbers of resistant individuals over a span of generations.

A focus on populations as the evolutionary units led to a new field in science called **population genetics.** Population genetics emphasizes the extensive genetic variation within populations and tracks the genetic makeup of populations over time.

Genetic Variation in Populations

You have no trouble recognizing your friends in a crowd. Each person has a unique genome, reflected in individual variations in appearance and temperament. Individual variation abounds in populations of all species that reproduce sexually. In addition to differences we can see, most populations have a great deal of variation that can be detected only by biochemical means. For example, you cannot tell a person's ABO blood group (A, B, AB, or O) just by looking at her or him.

Not all variation in a population is heritable. Phenotype results from a combination of the genotype, which is inherited, and the many environmental influences. For instance, a strength-training program can build up your muscle mass beyond what would naturally occur from your genetic makeup. However, you would not pass this environmentally induced physique on to your offspring. Only the genetic component of variation is relevant to natural selection.

Many of the variable traits in a population result from the combined effect of several genes (see Chapter 8). This polygenic inheritance produces traits that vary more or less continuously—in human height, for instance, from very short individuals to very tall ones. By contrast, other features, such as human ABO blood group, are determined by a single genetic locus, with different alleles producing only distinct phenotypes; there are no in-between types. In such cases, when a population includes two or more forms of a phenotypic characteristic, the contrasting forms are called morphs. A population is said to be **polymorphic** for a characteristic if two or more morphs are present in readily noticeable numbers—that is, if neither morph is extremely rare (Figure 12.16).

Sources of Genetic Variation Mutations and sexual recombination, which are both random processes, produce genetic variation.

Mutations, random changes in the genetic material, can actually create new alleles. For example, a mutation in a gene may substitute one nucleotide for another. This type of change will be harmless if it does not affect the function of the protein the DNA encodes. However, if it does affect the protein's function, the mutation will probably be harmful. An organism is a refined product of thousands of generations of past selection. A random mutation is like a shot in the dark; it is not likely to improve a genome any more than shooting a bullet through the hood of a car is likely to improve engine performance.

On rare occasions, however, a mutant allele may actually enhance reproductive success. This kind of effect is more likely when the environment is changing in such a way that alleles that were once disadvantageous are favorable under the new conditions. We already considered one example, the changing fortunes of the DDT resistance allele in housefly populations.

Organisms with very short generation spans, such as bacteria, can evolve rapidly with mutations as the only source of genetic variation. Bacteria multiply so quickly that natural selection can increase a population's frequency of a beneficial mutation in just hours or days. For most animals and plants, however, their long generation times prevent new mutations from significantly affecting overall genetic variation from one generation to the next, at least in large populations.

Animals and plants depend mainly on sexual recombination for the genetic variation that makes adaptation possible. The two sexual processes of meiosis and random fertilization shuffle alleles and deal them out to offspring in fresh combinations (see Chapter 7).

Review how sexual recombination produces genetic variation in Web/CD Activity 12D.

While the processes that generate genetic variation—mutations and sexual recombination—are random, natural selection (and hence evolution) is not. The environment selectively promotes the propagation of those genetic combinations that enhance survival and reproductive success.

Analyzing Gene Pools

A key concept of population genetics is the gene pool. The **gene pool** consists of all alleles (alternative forms of genes) in all the individuals making up a population. The gene pool is the reservoir from which the next generation draws its genes.

Figure 12.16 Polymorphism in a garter snake population. These four garter snakes, which belong to the same species, were all captured in one Oregon field. The behavior of each morph (form) is correlated with its coloration. When approached, spotted snakes, which blend in with their background, generally freeze. In contrast, snakes with stripes, which make it difficult to judge the speed of motion, usually flee rapidly when approached.

Imagine a wildflower population with two varieties (morphs) contrasting in flower color. An allele for red flowers, which we will symbolize by R, is dominant to an allele for white flowers, symbolized by r. These are the only two alleles for flower color in the gene pool of this plant population. Now, let's say that 80%, or 0.8, of all flower-color loci in the gene pool have the R allele. We'll use the letter p to represent the relative frequency of the dominant allele in the population. Thus, $p = 0.8$. Because there are only two alleles in this example, the r allele must be present at the other 20% (0.2) of the gene pool's flower-color loci. Let's use the letter q for the frequency of the recessive allele in the population. For the wildflower population, $q = 0.2$. And since there are only two alleles for flower color, we know that

$$p \; + \; q \; = \; 1$$

Frequency of the dominant allele   Frequency of the recesssive allele

Notice that if we know the frequency of either allele in the gene pool, we can subtract it from 1 to calculate the frequency of the other allele.

From the frequencies of alleles, we can also calculate the frequencies of different genotypes in the population if the gene pool is completely stable (not evolving). In the wildflower population, what is the probability of producing an RR individual by "drawing" two R alleles from the pool of gametes? Here we apply the rule of multiplication that you learned in Chapter 8. The probability of drawing an R sperm multiplied by the probability of drawing an R egg is $p \times p = p^2$, or $0.8 \times 0.8 = 0.64$. In other words, 64% of the plants in the population will have the RR genotype. Applying the same math, we also know the frequency of rr individuals in the population: $q^2 = 0.2 \times 0.2 = 0.04$. Thus, 4% of the plants are rr, giving them white flowers. Calculating the frequency of heterozygous individuals, Rr, is trickier. That's because the heterozygous genotype can form in *two* ways, depending on whether the sperm or egg supplies the dominant allele. So the frequency of the Rr genotype is $2pq$, which is $2 \times 0.8 \times 0.2 = 0.32$. In our imaginary wildflower population, 32% of the plants are Rr, with red flowers. Figure 12.17 reviews these calculations.

Now we can write a general formula for calculating the frequencies of genotypes in a gene pool from the frequencies of alleles, and vice versa:

$$p^2 \; + \; 2pq \; + \; q^2 \; = \; 1$$

Frequency of homozygous dominants   Frequency of heterozygotes   Frequency of homozygous recesssives

Notice that the frequencies of all genotypes in the gene pool must add up to 1. This formula is called the **Hardy-Weinberg formula,** named for the two scientists who derived it in 1908.

Population Genetics and Health Science

We can use the Hardy-Weinberg formula to calculate the percentage of a human population that carries the allele for a particular inherited disease. Consider phenylketonuria (PKU), which is an inherited inability to break down the amino acid phenylalanine. If untreated, the disorder causes severe mental retardation. PKU occurs in about one out of 10,000 babies born in

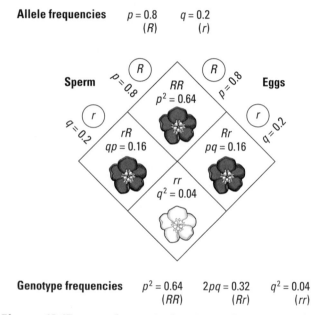

Allele frequencies $p = 0.8$ $q = 0.2$
                (R) (r)

Genotype frequencies $p^2 = 0.64$ $2pq = 0.32$ $q^2 = 0.04$
                     (RR) (Rr) (rr)

Figure 12.17 A mathematical swim in the gene pool. Each of the four boxes in the grid corresponds to an equally probable "draw" of alleles from the gene pool.

the United States. Newborn babies are now routinely tested for PKU, and symptoms can be prevented by following a strict diet (Figure 12.18).

PKU is due to a recessive allele. Thus, the frequency of individuals in the U.S. population born with PKU corresponds to the q^2 term in the Hardy-Weinberg formula. For one PKU occurrence per 10,000 births, $q^2 = 0.0001$. Therefore, q, the frequency of the recessive allele in the population, equals the square root of 0.0001, or 0.01. And p, the frequency of the dominant allele, equals $1 - q$, or 0.99.

Now let's calculate the frequency of heterozygous individuals, who carry the PKU allele in a single dosage. These carriers are free of the disorder but may pass the PKU allele on to offspring. Carriers are represented in the Hardy-Weinberg formula by $2pq$. And that's $2 \times 0.99 \times 0.01$, or 0.0198. Thus, the formula tells us that about 2% (actually 1.98%) of the U.S. population carries the PKU allele. Estimating the frequency of a harmful allele is essential for any public health program dealing with genetic diseases.

Microevolution as Change in a Gene Pool

How can we tell if a population is evolving? It helps, as a basis of comparison, to know what to expect if a population is *not* evolving. A nonevolving population is in genetic equilibrium, also called **Hardy-Weinberg equilibrium.** The population's gene pool remains constant over time. From generation to generation, the frequencies of alleles and genotypes are unchanged. Sexual shuffling of genes cannot by itself change a gene pool. But natural selection can. As an example, let's return to our wildflower population. Imagine the arrival of an insect species that is a vigorous pollinator of plants. If this insect is attracted to white flowers, its presence could enhance the reproductive success of white-flowered plants. Over the generations, this selection factor would increase the frequency of the r allele at the expense of the R allele. In contrast to the Hardy-Weinberg equilibrium of a nonevolving population, we now have the changing gene pool of an evolving population.

One of the products of the modern synthesis was a definition of evolution that is based on population genetics: *Evolution is a generation-to-generation change in a population's frequencies of alleles.* Because this describes evolution on the smallest scale, it is sometimes referred to more specifically as **microevolution.**

Figure 12.18 A warning to individuals with PKU. People with PKU (phenylketonurics) must avoid all foods with the amino acid phenylalanine. In addition to natural sources, phenylalanine is found in aspartame, a common artificial sweetener. The frequency of the PKU allele is high enough to warrant a public health program of including warnings on foods that contain phenylalanine.

CheckPoint

1. What is the smallest biological unit that can evolve?

2. Define microevolution.

3. Which term in the Hardy-Weinberg formula ($p^2 + 2pq + q^2 = 1$) corresponds to the frequency of individuals who have no alleles for the disease PKU?

4. Which of the following variations in a human population is the best example of polymorphism: height, ABO blood group, number of fingers, or math proficiency?

5. Which process, mutation or sexual recombination, results in most of the generation-to-generation variability in human populations?

Answers: 1. A population **2.** Microevolution is a change in a population's frequencies of alleles.
3. p^2 **4.** Blood group **5.** Sexual recombination

Mechanisms of Microevolution

What mechanisms can change a gene pool? Four causes of microevolution are genetic drift, gene flow, mutations, and, of course, natural selection. We'll see that each of these evolutionary mechanisms represents a departure from conditions required for Hardy-Weinberg equilibrium.

Genetic Drift

Flip a coin a thousand times, and a result of 700 heads and 300 tails would make you very suspicious about that coin. But flip a coin ten times, and an outcome of seven heads and three tails would seem within reason. The smaller the sample, the greater the chance of deviation from an idealized result—an equal number of heads and tails, in the case of a sample of coin tosses.

Let's apply coin-toss logic to a population's gene pool. If a new generation draws its alleles at random from the previous generation, then the larger the population (the sample size), the better the new generation will represent the gene pool of the previous generation. Thus, one requirement for a gene pool to maintain the status quo—Hardy-Weinberg equilibrium—is a large population size. The gene pool of a small population may not be accurately represented in the next generation because of sampling error. It is analogous to the erratic outcome from a small sample of coin tosses.

Figure 12.19 applies this concept of sampling error to a small population of wildflowers. Chance causes the frequencies of the alleles for red (R) and white (r) flowers to change over the generations. And that fits our definition of microevolution. This evolutionary mechanism, a change in the gene pool of a small population due to chance, is called **genetic drift.** Two situations that can shrink populations down to a size where genetic drift occurs are known as the bottleneck effect and the founder effect.

Figure 12.19 Genetic drift. This small wildflower population has a stable size of only about ten plants. For generation 1, only the five boxed plants produce fertile offspring. Only two plants of generation 2 manage to leave fertile offspring. Over the generations, genetic drift can completely eliminate some alleles, as is the case for the r allele in generation 3 of this imaginary population.

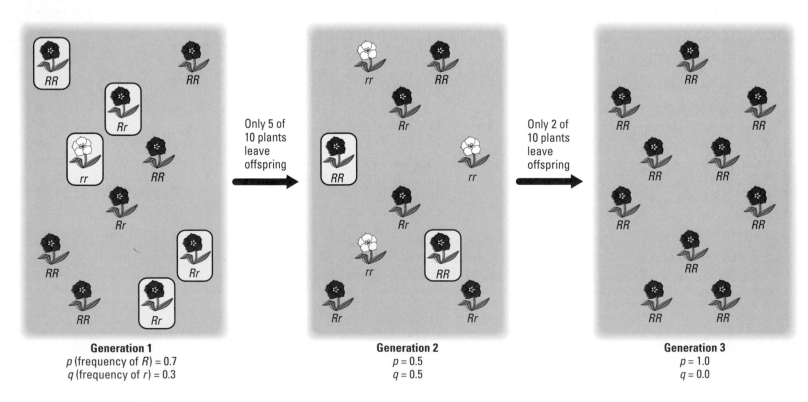

Only 5 of 10 plants leave offspring

Only 2 of 10 plants leave offspring

Generation 1
p (frequency of R) = 0.7
q (frequency of r) = 0.3

Generation 2
$p = 0.5$
$q = 0.5$

Generation 3
$p = 1.0$
$q = 0.0$

The Bottleneck Effect Disasters such as earthquakes, floods, droughts, and fires may reduce the size of a population drastically. The small surviving population may not be representative of the original population's gene pool. By chance, certain alleles will be overrepresented among the survivors. Other alleles will be underrepresented. And some alleles may be eliminated altogether. Chance may continue to change the gene pool for many generations until the population is again large enough for sampling errors to be insignificant. The analogy in Figure 12.20 illustrates why genetic drift due to a drastic reduction in population size is called the **bottleneck effect**.

Bottlenecking usually reduces the overall genetic variability in a population because at least some alleles are likely to be lost from the gene pool. An important application of this concept is the potential loss of individual variation, and hence adaptability, in bottlenecked populations of endangered species, such as the cheetah (Figure 12.21). The fastest of all running animals, cheetahs are magnificent cats that were once widespread in Africa and Asia. Like many African mammals, the number of cheetahs fell drastically during the last ice age some 10,000 years ago. At that time, the species may have suffered a severe bottleneck, possibly as a result of disease, human hunting, and periodic droughts. Some researchers think that the South African cheetah population suffered a second bottleneck during the nineteenth century when South African farmers hunted the animals to near extinction. Today, only three small populations of cheetahs exist in the wild. Genetic variability in these populations is very low compared to populations of other mammals. In fact, genetic uniformity in cheetahs rivals that of highly inbred varieties of laboratory mice! This lack of variability, coupled with an increasing loss of habitat, makes the cheetah's future precarious. The cheetahs remaining in Africa are being crowded into nature preserves and parks as human demands on the land increase. Along with crowding comes an increased potential for the spread of disease. With so little variability, the cheetah may have a reduced capacity to adapt to such environmental challenges. Captive breeding programs are already under way and may be required for the cheetah's long-term survival.

The Founder Effect Genetic drift is also likely when a few individuals colonize an isolated island, lake, or some other new habitat. The smaller the colony, the less its genetic makeup will represent the gene pool of the larger population from which the colonists emigrated. If the colony succeeds, random drift will continue to affect the frequency of alleles until the population is large enough for sampling error to be minimal.

Genetic drift in a new colony is known as the **founder effect**. The effect undoubtedly contributed to the evolutionary divergence of the finches and other South American organisms that arrived as strays on the remote Galápagos Islands that Darwin visited.

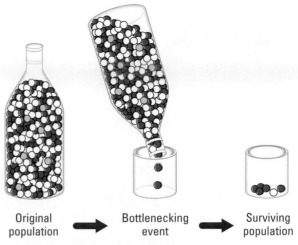

Original population → Bottlenecking event → Surviving population

Figure 12.20 The bottleneck effect. The colored marbles in this analogy represent three morphs in an imaginary population. Shaking just a few of the marbles through the bottleneck is like drastically reducing the size of a population due to some environmental disaster. By chance, blue marbles are overrepresented in the new population, and gold marbles are absent. Similarly, bottlenecking a population of organisms tends to reduce variability.

Figure 12.21 Implications of the bottleneck effect in conservation biology. Endangered species, such as the cheetah, typically have low genetic variability. As a result, they may be less adaptable to environmental changes, such as new diseases, than are species with a greater resource of genetic variation.

Figure 12.22 Residents of Tristan da Cunha in the early 1900s.

Figure 12.23 Gene flow and human evolution. The migration of people throughout the world is transferring alleles between populations that were once isolated. This magazine cover celebrates our changing gene pools and culture with a computer-generated image blending facial features from several races.

Genetic Drift and Hereditary Disorders in Human Populations The founder effect explains the relatively high frequency of certain inherited disorders among some human populations established by small numbers of colonists. In 1814, 15 people founded a British colony on Tristan da Cunha, a group of small islands in the middle of the Atlantic Ocean (Figure 12.22). Apparently, one of the colonists carried a recessive allele for retinitis pigmentosa, a progressive form of blindness. Of the 240 descendants who still lived on the islands in the 1960s, 4 had retinitis pigmentosa. At least 9 others were known to be carriers (heterozygous, with one copy of the recessive allele). That frequency of the retinitis pigmentosa allele is much higher than in Great Britain, the source of the colonists.

Gene Flow

Hardy-Weinberg equilibrium—the stagnant gene pool of a nonevolving population—requires genetic isolation from other populations. Most populations, however, are *not* completely isolated. A population may gain or lose alleles by **gene flow,** which is genetic exchange with another population. Gene flow occurs when fertile individuals or gametes migrate between populations. Perhaps, for example, a population neighboring our hypothetical wildflowers consists entirely of white-flowered individuals. A windstorm may blow pollen to our wildflowers from the neighboring population. Interbreeding would increase the frequency of the white-flower allele in our imaginary population. That qualifies as microevolution.

Gene flow tends to reduce genetic differences between populations. If it is extensive enough, gene flow can eventually amalgamate neighboring populations into a single population with a common gene pool. As humans began to move about the world more freely, gene flow became an important agent of microevolution in populations that were previously isolated (Figure 12.23).

Mutations

A **mutation,** remember, is a change in an organism's DNA. Hardy-Weinberg equilibrium requires that the gene pool is unaffected by mutations. A new mutation that is transmitted in gametes can immediately change the gene pool of a population by substituting one allele for another. For example, a mutation that causes a white-flowered plant *(rr)* in our hypothetical wildflower population to produce gametes bearing the dominant allele for red flowers *(R)* would decrease the frequency of the *r* allele in the population and increase the frequency of the *R* allele.

For any one gene locus, however, mutation alone does not have much quantitative effect on a large population in a single generation. This is because a mutation at any given gene locus is a very rare event. If some new allele produced by mutation increases its frequency by a significant amount in a population, it is not because mutation is generating the allele in abundance, but because individuals carrying the mutant allele are producing a disproportionate number of offspring as a result of natural selection or genetic drift.

Although mutations at a particular gene locus are rare, the cumulative impact of mutations at *all* loci can be significant. This is because each individual has thousands of genes, and many populations have thousands or millions of individuals. Certainly over the long term, mutation is, in itself, very important to evolution because it is the original source of the genetic variation that serves as raw material for natural selection.

Natural Selection: A Closer Look

Genetic drift, gene flow, and mutation can cause microevolution, but they do not necessarily lead to adaptation. In fact, only blind luck could result in random drift, migrant alleles, or shot-in-the-dark mutations improving a population's fit to its environment. Of all causes of microevolution, only natural selection is generally adaptive. And this adaptive evolution, remember, is a blend of chance and sorting—chance in the random generation of genetic variability, and sorting in the differential reproductive success among the varying individuals. Darwin explained the basics of natural selection. But it took the population genetics of the modern synthesis to fill in the details.

> Observe microevolution in a hypothetical bug population in Web/CD Activity 12E.

The Hardy-Weinberg equilibrium, which defines a nonevolving population, demands that all individuals in a population be equal in their ability to survive and reproduce. This condition is probably never completely met. On average, those individuals that work best in the environment leave the most offspring and therefore have a disproportionate impact on the gene pool. When farmers began spraying their fields with pesticides, resistant pests started outreproducing other members of the insect populations. This increased the frequency of alleles for pesticide resistance in gene pools. Microevolution occurred. And it was adaptive. It was evolution by natural selection.

Darwinian Fitness The phrases "struggle for existence" and "survival of the fittest" are misleading if we take them to mean direct competitive contests between individuals. There *are* animal species in which individuals, usually the males, lock horns or otherwise do combat to determine mating privilege. But reproductive success is generally more subtle and passive. In a varying population of moths, certain individuals may average more offspring than others because their wing colors hide them from predators better. Plants in a wildflower population may differ in reproductive success because some are better able to attract pollinators, owing to slight differences in flower color, shape, or fragrance (Figure 12.24). A frog may produce more eggs than her neighbors because she is better at catching insects for food. These examples point to a biological definition of fitness: **Darwinian fitness** *is the contribution an individual makes to the gene pool of the next generation relative to the contributions of other individuals.*

Survival to sexual maturity, of course, is prerequisite to reproductive success. But the biggest, fastest, toughest frog in the pond has a Darwinian fitness of zero if it is sterile. Production of fertile offspring is the only score that counts in natural selection.

Three General Outcomes of Natural Selection Imagine a population of deer mice. Individuals range in fur color from very light to very dark brown. If we graph the number of mice in each color category, we get a bell-shaped curve, like the grading curve some of your instructors draw after an exam. If natural selection favors certain fur color phenotypes over others, the population of mice will change over the generations. Three general outcomes are possible, depending on which phenotypes are favored. These three modes of natural selection are called directional selection, diversifying selection, and stabilizing selection.

Figure 12.24 Darwinian fitness of some flowering plants depends in part on competition in attracting pollinators.

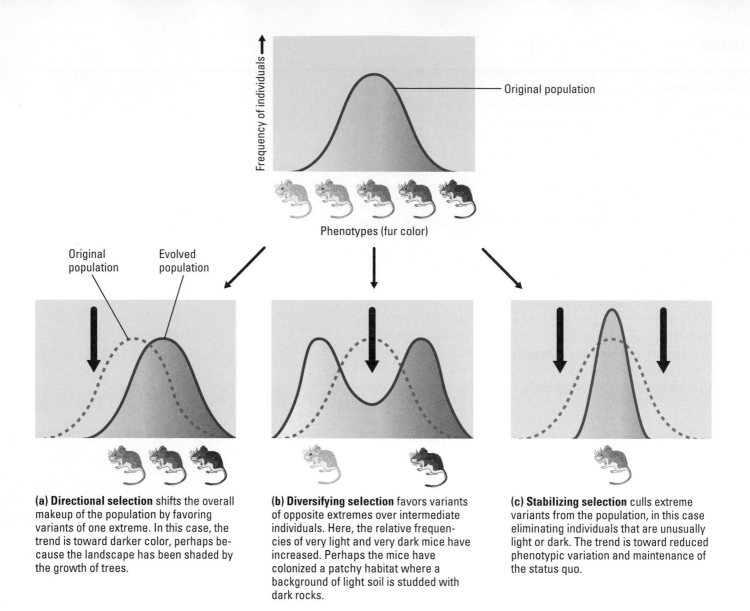

(a) Directional selection shifts the overall makeup of the population by favoring variants of one extreme. In this case, the trend is toward darker color, perhaps because the landscape has been shaded by the growth of trees.

(b) Diversifying selection favors variants of opposite extremes over intermediate individuals. Here, the relative frequencies of very light and very dark mice have increased. Perhaps the mice have colonized a patchy habitat where a background of light soil is studded with dark rocks.

(c) Stabilizing selection culls extreme variants from the population, in this case eliminating individuals that are unusually light or dark. The trend is toward reduced phenotypic variation and maintenance of the status quo.

Figure 12.25 Three general effects of natural selection on a phenotypic character. Here are three possible outcomes for selection working on fur color in imaginary populations of deer mice. The large downward arrows symbolize the pressure of natural selection working against certain phenotypes.

Directional selection (Figure 12.25a) shifts the phenotypic "curve" of a population by selecting in favor of some extreme phenotype—the darkest mice, for example. Directional selection is most common when the local environment changes or when organisms migrate to a new environment. An example is the shift of insect populations toward a greater frequency of pesticide-resistant individuals.

Diversifying selection (Figure 12.25b) can lead to a balance between two or more contrasting morphs (phenotypic forms) in a population. A patchy environment, which favors different phenotypes in different patches, is one situation associated with diversifying selection. The polymorphic snake population in Figure 12.16 is an example of diversifying selection.

Stabilizing selection maintains variation for a particular trait within a narrow range (Figure 12.25c). It typically occurs in a relatively stable environment to which populations are already well adapted. This evolutionary conservatism works by selecting against the more extreme phenotypes. For example, stabilizing selection keeps the majority of human birth weights between 3 and 4 kg (approximately 6.5–9 pounds). For babies much smaller or larger than this, infant mortality is greater.

Of the three selection modes, stabilizing selection probably prevails most of the time, resisting change in well-adapted populations. Evolutionary spurts occur when a population is stressed by a change in the environment or migration to a new place. When challenged with a new set of environmental problems, a population either adapts through natural selection or becomes extinct in that locale. The fossil record tells us that extinction is the most common result. Those populations that do survive crises often change enough to be designated new species, as we will see in Chapter 13.

CheckPoint

1. Compare and contrast the bottleneck effect and the founder effect as causes of genetic drift.

2. Why might new diseases pose a greater threat to cheetah populations than to mammalian populations having more genetic variation?

3. Which mechanism of microevolution has been most affected by the ease of human travel resulting from new modes of transportation?

4. What is the best measure of Darwinian fitness?

5. The thickness of fur in a bear population increases over several generations as the climate in the region becomes colder. This is an example of which type of selection: directional, diversifying, or stabilizing?

Answers: 1. Both processes result in populations being small enough for significant sampling error in the gene pool. A bottleneck reduces the size of an existing population. The founder effect is a new, small population consisting of individuals from a larger population. **2.** Because cheetah populations have so little variation, there is the potential for some new disease against which no individuals are resistant. **3.** Gene flow **4.** The number of fertile offspring an individual leaves **5.** Directional

Evolution Link: Darwinian Medicine

As the capstone of every chapter in this textbook, the Evolution Link section reinforces biology's unifying theme. But *this* unit of chapters is all about evolution itself. Here we'll use this section to relate evolutionary biology to society. And for this chapter, our connection is **Darwinian medicine,** the study of health problems in an evolutionary context. We'll examine just three examples: the population genetics of a hereditary disease; an evolutionary perspective of cardiovascular disease; and the evolution of antibiotic-resistant bacteria.

Population Genetics of the Sickle-Cell Allele

About one out of every 500 African Americans has sickle-cell disease. The disease is named for the abnormal shape of red blood cells in individuals who inherit the disorder (Figure 12.26). Sickle-cell disease is caused by a recessive allele. Only homozygous individuals, who inherit the recessive alleles from both parents, have the disorder. About one in 11 African Americans has a single copy of the sickle-cell allele. These heterozygous individuals do not have sickle-cell disease, but can pass the allele for the disorder on to their children. (You can review the biology of sickle-cell disease in Chapters 3 and 8.)

Why is the sickle-cell allele so much more common in African Americans than in the general U.S. population? And how can we explain such a

Figure 12.26 Sickled red blood cells.

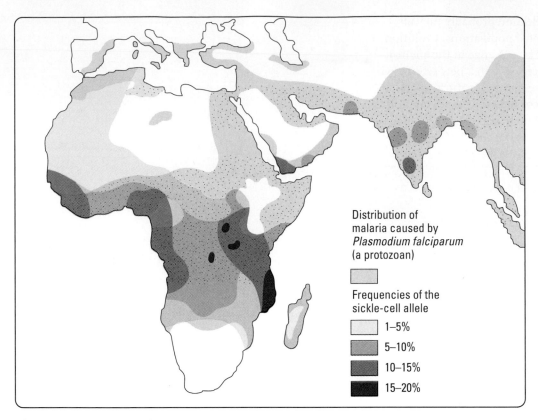

Distribution of
malaria caused by
Plasmodium falciparum
(a protozoan)

Frequencies of the
sickle-cell allele

1–5%

5–10%

10–15%

15–20%

Figure 12.27 Mapping malaria and the sickle-cell allele.

high frequency among African Americans for an allele with the potential to shorten life (and hence reproductive success)? Evolutionary biology holds the answers.

In the African tropics, the sickle-cell allele is both boon and bane. It is true that when inherited in double dosage, the allele causes sickle-cell disease. But heterozygous individuals, who have just one copy of the sickle-cell allele, are relatively resistant to malaria. This is an important advantage in tropical regions where malaria, caused by a parasitic microorganism, is a major cause of death. The frequency of the sickle-cell allele in Africa is generally highest in areas where the malaria parasite is most common (Figure 12.27).

How do we weigh the health impact of an allele that is both beneficial (in heterozygotes) and harmful (in homozygotes)? The Hardy-Weinberg formula is the tool we need. In some African populations, the sickle-cell allele has a frequency of 0.2, or 20%. This is q, the frequency of the recessive allele in the gene pool. And q^2, the frequency of homozygous recessives, is 0.2×0.2, or 0.04. So about 4% of individuals in these populations have sickle-cell disease. Now let's calculate the frequency of heterozygous individuals who have enhanced resistance to malaria. That would be $2pq$ in the Hardy-Weinberg formula: $2 \times 0.8 \times 0.2$, or 0.32. So in this case, the sickle-cell allele benefits about 32% of the population, while it causes sickle-cell disease in only about 4% of the population. That explains the high frequency of the sickle-cell allele compared to other alleles that cause debilitating diseases. And the representation of the allele among American blacks is a vestige of African roots.

An Evolutionary View of Cardiovascular Disease

Cardiovascular disease, which leads to heart attacks and strokes, is the leading cause of death in the United States and most other developed countries. Diet, as you learned in Chapter 3, is one of the factors associated with cardiovascular health. Cardiovascular disease is relatively rare among isolated populations of humans whose diets resemble the diets of early humans.

According to one hypothesis, the high incidence of cardiovascular disease in developed countries is partly due to a modern diet at odds with our ancestry. We are descended from humans who obtained food by gathering and hunting—gathering seeds and other edible plant products and hunting game or scavenging meat from animals killed by other predators. Gathering probably prevailed because it was safer and more reliable than hunting. On the African savanna, where humans first evolved, foods that were fatty or sweet were probably hard to come by. In a feast-or-famine existence, natural selection may have favored individuals who gorged on fatty, rich foods on those rare occasions when such treats were available. Such gluttonous individuals, with their ample reserves, were more likely than thinner friends to survive famines. Perhaps our taste for fats and sugars reflects this evolutionary history. Of course, today most of us only hunt and gather in grocery

Figure 12.28 Stockpiling fats and sugars at the Penn State University Creamery, a tradition since 1896.

stores, fast-food restaurants, and college cafeterias. Although we know it is unhealthful, many of us may find it difficult to overcome the ancient survival behavior of stockpiling for the next famine (Figure 12.28).

Evolution of Antibiotic Resistance in Bacteria

Antibiotics are drugs that help cure certain infections by impairing bacteria. (Don't confuse antibiotics with antibodies, proteins that your body produces on its own as defense against infections.) Antibiotics have saved millions of humans, including one of the authors of this book, who was rescued at age 13 from bacterial meningitis by injections of the antibiotic penicillin. But there's a dark side to the widespread use of antibiotics. It has caused the evolution of antibiotic-resistant populations of the very bacteria the drugs are meant to kill.

Antibiotic resistance evolves by natural selection, much as pesticide resistance evolves in insects. An antibiotic selects among the varying bacteria of a population for those mutant individuals that can survive the drug. In some cases, the mechanism of resistance depends on a gene for an enzyme that destroys the antibiotic—or even uses the drug as a nutrient! While the drug kills most of the bacteria, the resistant bacteria multiply and quickly become the norm in the population rather than the exceptions.

The evolution of antibiotic-resistant bacteria is a huge problem in public health. For example, in New York City, there are now some strains of the tuberculosis-causing bacteria that are resistant to all three of the antibiotics that are used to treat the disease. Unfortunate people infected with one of these resistant strains have no better chance of surviving than did tuberculosis patients a century ago.

Investigate case studies of antibiotic resistance in Web/CD Activity 12F.

Darwinian medicine is beginning to influence the prescription of antibiotics by many physicians. Doctors who understand that the abuse of these drugs is speeding the evolution of resistant bacteria are less likely to prescribe the antibiotics needlessly—for instance, for a patient complaining of a common cold or flu, which are viral diseases against which the antibacterial drugs are powerless.

Discover more about Darwinian medicine in the Web Evolution Link.

These examples of Darwinian medicine are a reminder that biology is the foundation of all medicine. And evolution is the foundation of all biology.

Summary of Key Concepts

Overview: Descent with Modification

• Darwin made two proposals in *The Origin of Species:* (1) modern species descended from ancestral species; (2) natural selection is the mechanism of evolution.
• **Web/CD Activity 12A** *Darwin's Voyage to the Galápagos*

Evidence of Evolution

• **The Fossil Record** The fossil record shows that organisms have appeared in a historical sequence, and many fossils link ancestral species with those living today.

• **Biogeography** Biogeography, the study of the geographic distribution of species, suggests that species evolved from ancestors that inhabited the same region.

• **Comparative Anatomy** Homologous structures among species bear witness to evolutionary history.
• **Web/CD Activity 12B** *Reconstructing Forelimbs*

• **Comparative Embryology** Closely related species often have similar stages in their embryonic development.

• **Molecular Biology** All species share a common genetic code, suggesting that all forms of life are related through branching evolution from the earliest organisms. Also, species that seem to be closely related have similar DNA and protein sequences.

Natural Selection and Adaptive Evolution

• **Natural Selection Basics** Darwin observed that populations produce more offspring than the environment can support and that individuals vary in heritable characteristics. From these observations he arrived at the conclusion that defines natural selection: Individuals best suited for a particular environment are more likely to survive and reproduce than less fit individuals.
• **Web/CD The Process of Science** *The Voyage of the Beagle*

• **Natural Selection in Action: The Evolution of Pesticide-Resistant Insects** Pesticides kill most members of an insect population but leave resistant individuals to reproduce and pass on their genes for pesticide resistance.
• **Web/CD Activity 12C/The Process of Science** *Effects of Environmental Changes on a Population*

The Modern Synthesis: Darwinism Meets Genetics

• The modern synthesis fused genetics (Mendelism) and evolutionary biology (Darwinism) in the mid-1900s.

• **Populations as the Units of Evolution** A population is the smallest biological unit that can evolve. Population genetics emphasizes the extensive genetic variation within populations and tracks the genetic makeup of populations over time.

• **Genetic Variation in Populations** Polygenic ("many gene") inheritance produces traits that vary continuously, whereas traits that are determined by one genetic locus may be polymorphic, producing two or more distinct forms. Mutations and sexual recombination produce genetic variation.
• **Web/CD Activity 12D** *Genetic Variation from Sexual Recombination*

• **Analyzing Gene Pools** The gene pool consists of all alleles in all the individuals making up a population. The Hardy-Weinberg formula can be used to calculate the frequencies of genotypes in a gene pool from the frequencies of alleles, and vice versa.

• **Population Genetics and Health Science** The Hardy-Weinberg formula can be used to estimate the frequency of a harmful allele, which is useful for public health programs dealing with genetic diseases.

• **Microevolution as Change in a Gene Pool** Microevolution is a generation-to-generation change in a population's frequencies of alleles.

Mechanisms of Microevolution

• **Genetic Drift** Genetic drift is a change in the gene pool of a small population due to chance. Bottlenecking and the founder effect are two situations leading to genetic drift.

• **Gene Flow** A population may gain or lose alleles by gene flow, which is genetic exchange with another population.

• **Mutation** Individual mutations have relatively little short-term effect on a large gene pool. Longer-term, mutation is the cumulative source of genetic variation.

• **Natural Selection: A Closer Look** Of all causes of microevolution, only natural selection is generally adaptive. Darwinian fitness is the contribution an individual makes to the gene pool of the next generation relative to the contributions of other individuals. The outcome of natural selection may be directional, diversifying, or stabilizing.
• **Web/CD Activity 12E** *Causes of Microevolution*

Evolution Link: Darwinian Medicine

• **Population Genetics of the Sickle-Cell Allele** Individuals with two recessive alleles for sickle cell develop the sickle-cell disease. However, individuals with just one copy of the sickle-cell allele are relatively resistant to malaria.

• **An Evolutionary View of Cardiovascular Disease** Our tendency to eat fats and sugars may be an evolutionary adaptation to stockpile food when it is available. In developed countries today, where food is always available, stockpiling fats may contribute to cardiovascular disease.

• **Evolution of Antibiotic Resistance in Bacteria** The widespread use of antibiotics has caused the evolution of antibiotic-resistant populations of the bacteria the drugs are meant to kill.
• **Web/CD Activity 12F** *Case Studies of Antibiotic Resistance*
• **Web Evolution Link** *Darwinian Medicine*

Self-Quiz

1. Which of the following is *not* an observation or inference on which Darwin's theory of natural selection is based?
   a. There is heritable variation among individuals.
   b. Poorly adapted individuals never produce offspring.
   c. Because only a fraction of offspring survive, there is a struggle for limited resources.
   d. Individuals whose inherited characteristics best fit them to the environment will generally produce more offspring.
   e. Unequal reproductive success can lead to adaptations.

2. Which of the following pairs of structures is *least* likely to represent homology?
   a. the skull of a cheetah and the skull of a house cat
   b. the gill pouches of a mouse embryo and the gill pouches of a horse embryo
   c. the mitochondria of a plant and those of an animal
   d. the bark of a tree and the protective covering of a lobster
   e. the genetic code of a bacterium and the genetic code of a human

3. In a population with two alleles for a particular genetic locus, *B* and *b*, the allele frequency of *B* is 0.7. What would be the frequency of heterozygotes if the population is in Hardy-Weinberg equilibrium?
   a. 0.7   d. 0.42
   b. 0.49   e. 0.09
   c. 0.21

4. In a population that is in Hardy-Weinberg equilibrium, 16% of the individuals show the recessive trait. What is the frequency of the dominant allele in the population?
   a. 0.84   d. 0.4
   b. 0.36   e. 0.48
   c. 0.6

5. Which of the following is an example of a polymorphism in humans?
   a. variation in height
   b. variation in intelligence
   c. the presence or absence of a widow's peak (see Figure 8.13)
   d. variation in the number of fingers
   e. variation in fingerprints

6. As a mechanism of microevolution, natural selection can be most closely equated with
   a. random mating.
   b. genetic drift.
   c. differential reproductive success.
   d. mutation.
   e. gene flow.

7. A founder event favors microevolution in the founding population mainly because
   a. mutations are more common in a new environment.
   b. a small founding population is subject to extensive sampling error in the composition of its gene pool.
   c. the new environment is likely to be patchy, favoring diversifying selection.
   d. gene flow increases.
   e. members of a small population tend to migrate.

8. In a particular bird species, individuals with average-sized wings survive severe storms more successfully than other birds in the same population with longer or shorter wings. This illustrates
   a. the founder effect.
   b. stabilizing selection.
   c. artificial selection.
   d. gene flow.
   e. diversifying selection.

9. What do we know about a population that is in Hardy-Weinberg equilibrium?
   a. The population is quite small.
   b. The population is not evolving.
   c. Gene flow between the population and surrounding populations does not occur.
   d. Natural selection is occurring.
   e. Genetic drift is occurring.

10. What environmental factor accounts for the relatively high frequency of the sickle-cell allele in tropical Africa?
   a. high temperature
   b. high humidity
   c. an abundance of flu viruses
   d. malaria
   e. intense sunlight

• **Go to the website or CD-ROM for more self-quiz questions.**

The Process of Science

1. A population of snails has recently become established in a new region. The snails are preyed on by birds that break the snails open on rocks, eat the soft bodies, and leave the shells. The snails occur in both striped and unstriped forms. In one area, researchers counted both live snails and broken shells. Their data are summarized here:

	Striped	Unstriped
Living	264	296
Broken	486	377

Based on these data, which snail form is more subject to predation by birds? Predict how the frequencies of striped and unstriped individuals might change over the generations.

2. **Explore the Galápagos Islands and other places Darwin visited in *The Voyage of the Beagle*. Conduct virtual experiments on an evolving population of leafhoppers in *Effects of Environmental Changes on a Population*. Both activities are available in The Process of Science section of Chapter 12 on the website and CD-ROM.**

Biology and Society

To what extent are humans in a technological society exempt from natural selection? Explain your answer.

How Biological Diversity Evolves

Biologists estimate that there are 5–30

million species of organisms currently on Earth.

North America and Europe are drifting apart at a rate

of about 2 cm per year.  Our own scientific name,

Homo sapiens, means "wise man."  Globally,

the rate of species loss may be 50 times higher now

than at any time in the past 100,000 years.

When Darwin traveled to the Galápagos Islands, he realized that he was visiting a place of genesis. Though the volcanic islands are geologically young, they are already home to many plants and animals known nowhere else in the world. Among the islands' unique inhabitants are its giant tortoises, for which the Galápagos are named (*galápago* is the Spanish word for "tortoise") (Figure 13.1). After visiting the Galápagos, Darwin wrote in his diary: "Both in space and time, we seem to be brought somewhat near to that great fact—that mystery of mysteries—the first appearance of new beings on this Earth." In this chapter, we study the birth of species and the methods biologists use to trace the evolution of biological diversity.

Figure 13.1 A unique inhabitant of the Galápagos Islands.

Overview: Macroevolution and the Diversity of Life

To understand the "appearance of new beings," as Darwin put it, it is not enough to explain microevolution and the adaptation of populations, which you learned about in Chapter 12. If that were *all* that happened, then Earth would be populated only by a highly adapted version of the first form of life. Evolutionary theory must also explain **macroevolution,** which encompasses the major biological changes evident in the fossil record. Macroevolution includes the multiplication of species, which has generated biological diversity; the origin of evolutionary novelty, such as the wings and feathers of birds and the big brains of humans; the explosive diversification that follows some evolutionary breakthrough, such as the origin of thousands of plant species after the flower evolved; and mass extinctions, which cleared the way for new adaptive explosions, such as the diversification of mammals that followed the disappearance of dinosaurs.

Explore macroevolution in Web/CD Activity 13A.

The beginning of new forms of life—the origin of species—is at the focal point of our study of macroevolution, for it is in new species that biological diversity arises. Figure 13.2 contrasts two patterns of **speciation,** the origin of new species. In the case of non-branching evolution, a population may change so much through adaptation to a changing environment that it becomes a new species (Figure 13.2a). Such linear evolution cannot multiply species. The second major pattern is branching evolution (Figure 13.2b). In this case, one or more new species branches from a parent species that may continue to exist. Branching evolution is the more important pattern in the history of life; it is probably more common than the non-branching pattern, and only branching evolution can generate biological diversity by increasing the number of species. How did evolution produce over 250,000 species of flowering plants? Or about 35,000 species of fishes? Or over a million insect species? To explain such diversification, we must understand how the origin of new species makes branching evolution possible.

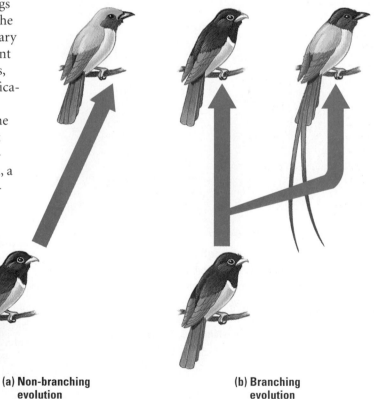

(a) Non-branching evolution

(b) Branching evolution

Figure 13.2 Two patterns of speciation. **(a)** Non-branching evolution can transform a population enough for it to be designated a new species, but this linear pattern of evolution cannot multiply species. **(b)** Branching evolution splits a lineage into two or more species.

The Origin of Species

Species is a Latin wording meaning "kind" or "appearance." Indeed, we learn to distinguish between the kinds of plants and animals—between dogs and cats, for example—from differences in their appearance. Although the basic idea of species as distinct life-forms seems intuitive, devising a more formal definition is not so easy.

What Is a Species?

In 1927, a young biologist named Ernst Mayr led an expedition into the remote Arafak Mountains of New Guinea to study the wildlife (Figure 13.3). He found a great diversity of birds, identifying 138 species on the basis of differences in their appearance. Mayr was surprised to learn that the local tribe of Papuan natives had given names of their own to 137 birds (two birds assigned by Mayr to separate species are very similar, and the Papuans did not distinguish between them). Mayr and the indigenous people agreed almost exactly in their inventory of the local birdlife. The experience sharpened Mayr's view that species are demarcated from one another as distinct forms of life. Put another way, biological diversity is not a continuum of form, but is instead divided into the separate forms that we identify as species. Mayr distilled this observation into the biological species concept.

The **biological species concept** defines species as "groups of interbreeding natural populations that are reproductively isolated from other such groups" (Mayr's words). Put another way, a species is a population or group of populations whose members have the potential to interbreed with one another in nature to produce fertile offspring, but who cannot successfully interbreed with members of other species (Figure 13.4). Geography and culture may conspire to keep a Manhattan businesswoman and a Mongolian dairyman apart. But if the two *did* get together, they could have viable babies that develop into fertile adults. All humans belong to the same species. In contrast, humans and chimpanzees, similar and closely related in their evolution as they are, remain distinct species even where they share territory. The two species do not interbreed. Such reproductive isolation blocks exchange of genes between species and keeps their gene pools separate.

We cannot apply the biological species concept to all situations. For example, basing the definition of species on reproductive compatibility excludes organisms that reproduce exclusively asexually (production of offspring from a single parent rather than from a mating pair). Fossils aren't doing much sexual reproduction either, so they cannot be evaluated by the biolog-

Figure 13.3 Ernst Mayr (on right) in New Guinea, 1927. Mayr was one of the architects of the modern synthesis, which you learned about in Chapter 12. One of his many contributions to the synthesis was the biological species concept. Still an active scientist 75 years after his first New Guinea expedition, Mayr continues to influence modern biology by publishing new books and articles.

(a) Similarity between different species

(b) Diversity within one species

Figure 13.4 The biological species concept is based on reproductive compatibility rather than physical similarity. (a) The eastern meadowlark (*Sturnella magna,* left) and the western meadowlark (*Sturnella neglecta*) are very similar in appearance, but they are separate species that cannot interbreed. Distinct songs help these birds choose mates of the same species. **(b)** In contrast, humans, as diverse in appearance as we are, belong to a single species (*Homo sapiens*), defined by our capacity to interbreed.

ical species concept. Biologists distinguish fossil species mainly by their differences in appearance. But in spite of its limitations, and even though there are alternative species concepts that some biologists now favor, the biological species concept still proves useful in its premise of reproductive barriers that keep species separate.

Reproductive Barriers Between Species

Clearly, a fly will not mate with a frog or a fern. But what prevents biological species that are very similar—that is, closely related—from interbreeding? What, for example, maintains the species boundary between the western spotted skunk and the eastern spotted skunk? Their geographic ranges overlap in the Great Plains region, and they are so similar that only expert zoologists can tell them apart. And yet, these two skunk species do not interbreed. Let's examine the reproductive barriers that isolate the gene pools of species. We can classify reproductive barriers as either pre-zygotic or post-zygotic, depending on whether they block interbreeding before or after the formation of zygotes, or fertilized eggs.

Figure 13.6 Courtship ritual as a behavioral barrier between species. These blue-footed boobies, inhabitants of the Galápagos Islands, will mate only after a specific ritual of courtship displays. Part of the "script" calls for the male to high-step, a dance that advertises the bright blue feet characteristic of the species.

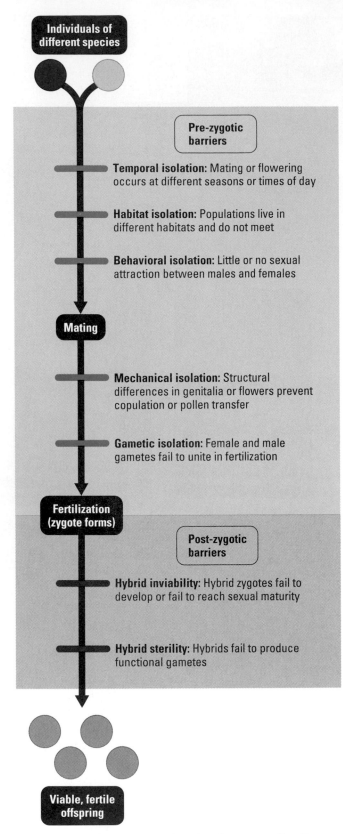

Individuals of different species

Pre-zygotic barriers

Temporal isolation: Mating or flowering occurs at different seasons or times of day

Habitat isolation: Populations live in different habitats and do not meet

Behavioral isolation: Little or no sexual attraction between males and females

Mating

Mechanical isolation: Structural differences in genitalia or flowers prevent copulation or pollen transfer

Gametic isolation: Female and male gametes fail to unite in fertilization

Fertilization (zygote forms)

Post-zygotic barriers

Hybrid inviability: Hybrid zygotes fail to develop or fail to reach sexual maturity

Hybrid sterility: Hybrids fail to produce functional gametes

Viable, fertile offspring

Figure 13.5 Reproductive barriers between closely related species.

Pre-zygotic barriers impede mating between species or hinder fertilization of eggs if members of different species should attempt to mate (Figure 13.5). The barrier may be time based (*temporal isolation*). For example, western spotted skunks breed in the fall, but the eastern species breeds in late winter. Temporal isolation keeps the species separate even where they coexist on the Great Plains. In other cases, species living in the same region are spatially segregated (*habitat isolation*). For example, one species of garter snake lives mainly in water, while a closely related species lives on land. Courtship rituals or other identification requirements for mates can also function as reproductive barriers (*behavioral isolation*). In many bird species, for example, courtship behavior is so elaborate that individuals are unlikely to mistake a bird of a different species as one of their kind (Figure 13.6). In still other cases, the male and female sex organs of different species are anatomically incompatible (*mechanical isolation*). For example, even if insects of closely related species attempt to mate, the male and female copulatory organs may not fit together correctly and no sperm is transferred. In still other cases, individuals of different species may actually copulate, but their gametes are incompatible and fertilization does not occur (*gametic isolation*). In many mammals, for example, sperm may not survive in females of a different species.

Post-zygotic barriers are backup mechanisms that operate should interspecies mating actually occur and form hybrid zygotes (hybrid here meaning that the egg comes from one species and the sperm from another species). In some cases, hybrid offspring die before reaching reproductive maturity (*hybrid inviability*). For example, although certain closely related frog species will hybridize, the offspring fail to develop normally because of genetic incompatibilities between the two species. In other cases of hybridization, offspring may become vigorous adults, but are infertile (*hybrid sterility*). A mule, for example, is the hybrid offspring of a female horse and a male donkey (Figure 13.7). Because mules are sterile, there is no avenue for gene transfer between the two parental species. The horse and donkey remain separate species because their gene pools are isolated, even though they are less than perfect in mate choice.

Horse

Mule (hybrid)

Donkey

Figure 13.7 Hybrid sterility, a post-zygotic barrier. Horses and donkeys remain separate species because their hybrid offspring, mules, are sterile.

In most cases, it is not a single reproductive barrier but some combination of two or more that reinforces boundaries between species. If it is reproductive isolation that keeps species separate, then the evolution of these barriers is the key to the origin of new species.

Mechanisms of Speciation

A key event in the potential origin of a species occurs when the gene pool of a population is somehow severed from other populations of the parent species. With its gene pool isolated, the splinter population can follow its own evolutionary course. Changes in its allele frequencies caused by genetic drift and natural selection are undiluted by gene flow from other populations. Such reproductive isolation can result from two general scenarios: allopatric speciation and sympatric speciation. In **allopatric speciation,** the initial block to gene flow is a geographic barrier that physically isolates the splinter population. The term *allopatric* is derived from the Greek meaning "other country." In contrast, the word *sympatric* means "together country." Thus, **sympatric speciation** is the origin of a new species without geographic isolation. The splinter population becomes reproductively isolated right in the midst of the parent population (Figure 13.8).

Allopatric Speciation Several geological processes can fragment a population into two or more isolated populations. A mountain range may emerge and gradually split a population of organisms that can inhabit only lowlands. A creeping glacier may gradually divide a population. A land bridge, such as the Isthmus of Panama, may form and separate the marine life on either side. A large lake may subside until there are several smaller lakes with their populations now isolated. Even without such geological remodeling, geographic isolation and allopatric speciation can occur if individuals colonize a new, geographically remote area and become isolated from the parent population. An example is the speciation that occurred on the Galápagos Islands following colonization by mainland organisms.

How formidable must a geographic barrier be to keep allopatric populations apart? The answer depends on the ability of the organisms to move about. Birds, mountain

(a) Allopatric speciation　　　　**(b) Sympatric speciation**

Figure 13.8 Two modes of speciation. These cartoons simplify the geographic relationships of new species to their parent species. **(a)** Allopatric speciation. A population forms a new species while geographically isolated from its parent population. **(b)** Sympatric speciation. A small population becomes a new species in the midst of its parent population.

A. harrisi

A. leucurus

Figure 13.9 Allopatric speciation of squirrels in the Grand Canyon. Two species of antelope squirrels inhabit opposite rims of the Grand Canyon. On the south rim is Harris's antelope squirrel (*Ammospermophilus harrisi*). Just a few miles away on the north rim is the closely related white-tailed antelope squirrel (*Ammospermophilis leucurus*). Birds and other organisms that can disperse easily across the canyon have not diverged into different species on opposite rims.

lions, and coyotes can cross mountain ranges, rivers, and canyons. Nor do such barriers hinder the windblown pollen of pine trees, and the seeds of many plants may be carried back and forth on animals. In contrast, small rodents may find a deep canyon or a wide river a formidable barrier (Figure 13.9).

The likelihood of allopatric speciation increases when a population is both small and isolated. A small, isolated population is more likely than a large population to have its gene pool changed substantially by factors such as genetic drift and natural selection. For example, in less than 2 million years, small populations of stray animals and plants from the South American mainland that managed to colonize the Galápagos Islands gave rise to all the new species that now inhabit the islands. But for each small, isolated population that becomes a new species, many more simply perish in their new environment. Life on the frontier is harsh, and most pioneer populations probably become extinct.

Even if a small, isolated population survives, it does not necessarily evolve into a new species. The population may adapt to its local environment and begin to look very different from the ancestral population, but that doesn't make it a new species. Speciation occurs only with the evolution of reproductive barriers between the isolated population and its parent population. In other words, if speciation occurs during geographic separation, the new species will not breed with its ancestral population, even if the two populations should come back into contact (Figure 13.10).

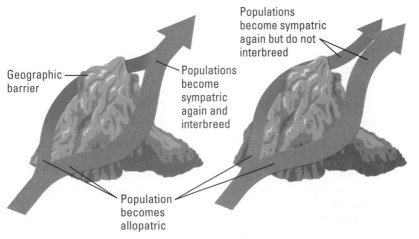

Geographic barrier

Populations become sympatric again and interbreed

Populations become sympatric again but do not interbreed

Population becomes allopatric

(a) Speciation has not occurred

(b) Speciation has occurred

Figure 13.10 Has speciation occurred during geographic isolation? In this analogy, the arrows track populations over time. The mountain symbolizes a period of geographic isolation. What happens if the two populations come back into contact after a long period of separation? **(a)** If the two populations interbreed freely when they become sympatric again, their gene pools merge. Speciation has not occurred. **(b)** If the evolutionary divergence of the two populations results in reproductive isolation, then they will not interbreed, even if they come back into contact. Speciation has occurred.

Sympatric Speciation How can a subpopulation become reproductively isolated while in the midst of its parent population? This can occur in a single generation if a genetic change produces a reproductive barrier between mutants and the parent population. Sympatric speciation does not seem to be widespread among animals, but it may account for over 25% of all plant species.

Many plant species have originated from accidents during cell division that resulted in extra sets of chromosomes (see the Evolution Link section in Chapter 7). A new species that evolves this way has polyploid cells, meaning that each cell has more than two complete sets of chromosomes. The new species cannot produce fertile hybrids with its parent species; reproductive isolation and speciation have occurred in a single generation without geographic isolation. This mechanism of sympatric speciation was first discovered in the early 1900s by Dutch botanist Hugo de Vries. His experiments produced a new species of primrose (Figure 13.11).

Polyploids do not always come from a single parent species. In fact, most polyploid species arise from the hybridization of two parent species (Figure 13.12). This mechanism of sympatric speciation accounts for many of the plant species we grow for food, including oats, potatoes, bananas, peanuts, barley, plums, apples, sugarcane, coffee, and wheat.

To understand sympatric speciation better, go to Web/CD Activity 13B.

O. lamarckiana

O. gigas

Figure 13.11 Botanist Hugo de Vries and his new primrose species. Working in the early 1900s, de Vries studied variation in evening primroses. *Oenothera lamarckiana* (upper left), is a diploid species of primrose with 14 chromosomes. During his breeding experiments, de Vries noted a new primrose variety with 28 chromosomes. It was a tetraploid, which could not interbreed with its parent species. De Vries named the new primrose species *Oenothera gigas,* for its large size (lower right).

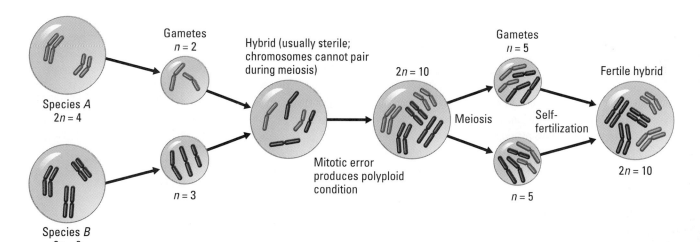

Figure 13.12 Polyploid formation by hybridization of two parent species. A hybrid between two plant species is normally sterile because its chromosomes are not homologous and cannot pair during meiosis (see Chapter 7). However, the hybrids may be able to reproduce asexually. This diagram shows one mechanism that can produce new species from such hybrids. At some instant in the history of the hybrid clone, a mitotic error affecting the reproductive tissue of an individual may double chromosome number. The hybrid will then be able to make gametes because each chromosome has a homologue with which to pair during meiosis. Union of gametes from this hybrid may give rise to a new species of interbreeding plants, reproductively isolated from both parent species (because chromosome sets don't match). The new species has a chromosome number equal to the sum of the chromosomes in the two parent species. Such combinations of hybridization and cell-division accidents make it possible for a new species to arise without geographical isolation from parent species—that is, for sympatric speciation to occur.

Wheat makes a good case study. The most widely cultivated plant in the world, what we call wheat is actually represented by 20 different species. Humans began domesticating wheat from wild grasses at least 11,000 years ago in the Middle East. Figure 13.13 traces the evolutionary path from that first cultivated wheat species to bread wheat, the most important wheat species today.

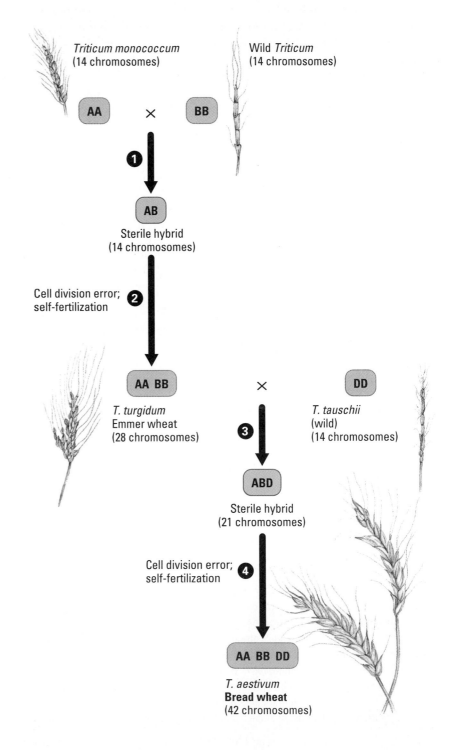

Figure 13.13 The evolution of wheat. The uppercase letters in this diagram represent not alleles but sets of chromosomes that we are tracing in the evolution of wheat, or *Triticum.* ❶ The evolutionary descent of wheat began with hybridization between two diploid wheats. One was the first domesticated species, *T. monococcum* (its diploid chromosome set is symbolized by AA in the diagram). The other species was a wild relative of wheat that probably grew as a weed at the edges of cultivated fields (we're using BB for its diploid chromosome set). Both of these species had a diploid number of 14 chromosomes, producing a hybrid with 14 chromosomes. ❷ Chromosome sets A and B of the two species would not have been able to pair at meiosis, making the AB hybrid sterile. However, an error in cell division in this sterile hybrid and self-fertilization among the resulting gametes produced a new species (AABB) with 28 chromosomes. Today, this species, known as emmer wheat (*T. turgidum*) is still grown widely in Eurasia and western North America. It is used mainly for making macaroni and other noodle products, because its proteins have a tendency to hold their shape better than bread wheat proteins. ❸ The next step in the evolution of bread wheat is believed to have occurred in early farming villages on the shores of European lakes over 8000 years ago. Cultivated emmer wheat, with its 28 chromosomes, hybridized spontaneously with a closely related wild species that has 14 chromosomes. ❹ The hybrid (ABD), with 21 chromosomes, was sterile. A cell-division error in this hybrid and self-fertilization doubled the chromosome number to 42. The result was bread wheat (*T. aestivum*), with two each of the three ancestral sets of chromosomes (AABBDD).

What Is the Tempo of Speciation?

Traditional evolutionary trees that diagram the descent of species sprout branches that diverge gradually (Figure 13.14a). Such trees are based on the idea that the changes big enough to distinguish a new species are an accumulation of many smaller changes occurring over vast spans of time. This concept applies the processes of microevolution you learned about in Chapter 12 to the multiplication of species. To Darwin, the origin of species was an extension of adaptation by natural selection, with isolated populations from common ancestral stock evolving differences gradually as they adapted to their local environments.

On the time scale of the fossil record, new species often appear more abruptly than the traditional model would predict. Paleontologists do find gradual transitions of fossil forms in some cases. However, the more common observation is of species appearing as new forms rather suddenly (in geological terms) in a layer of rocks, persisting essentially unchanged for their tenure on Earth, and then disappearing from the fossil record as suddenly as they appeared. Darwin himself was bewildered by the dearth of connecting fossils and wrote: "Although each species must have passed through numerous transitional stages, it is probable that the periods during which each underwent modification, though many and long as measured by years, have been short in comparison with the periods during which each remained in an unchanged condition."

Over the past 30 years, some evolutionary biologists have addressed the nongradual appearance of species in the fossil record by developing a model now known as **punctuated equilibrium.** According to this model, species most often diverge in spurts of relatively rapid change, instead of slowly and gradually (Figure 13.14b). In other words, species undergo most of their modification in appearance as they first bud from parent species, and then they change little. The term *punctuated equilibrium* is derived from the idea of long periods of stasis (equilibrium) punctuated by episodes of speciation.

Is it likely that most species evolve abruptly and then remain essentially unchanged for most of their existence? One cause of sudden speciation is the polyploidy you learned about as a mechanism of sympatric speciation in plants. Proponents of punctuated equilibrium point out that allopatric speciation can also be quite rapid. In just a few hundred to a few thousand generations, genetic drift and natural selection can cause significant change in the gene pool of a small population cloistered in a challenging new environment.

How can speciation in a few thousand generations, which may require several thousand years, be called an abrupt episode? The fossil record indicates that successful species last for a few million years, on average. Suppose that a particular species survives for 5 million years, but most of its evolutionary changes in anatomy occur during the first 50,000 years of its existence. In this case, the evolution of the species-defining characteristics is compressed into just 1% of the lifetime of the species. In the fossil record, the species will appear suddenly in rocks of a certain age and then linger with little or no change before becoming extinct. During its formative millennia, the species may accumulate its modifications gradually, but relative

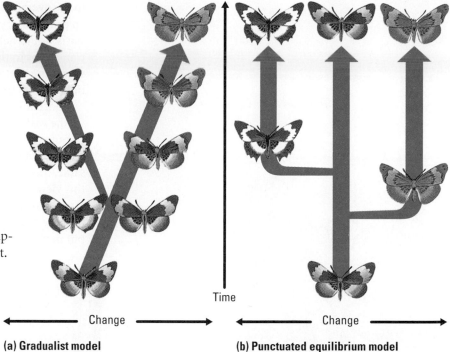

(a) Gradualist model

(b) Punctuated equilibrium model

Figure 13.14 Two models for the tempo of evolution.
(a) In the gradualist model, species descended from a common ancestor diverge more and more in form as they acquire unique adaptations. **(b)** According to the punctuated equilibrium model, a new species changes most as it buds from a parent species. After this speciation episode, there is little change for the rest of the species' existence.

to the overall history of the species, its inception is abrupt. This scenario of an evolutionary spurt preceding a much longer period of morphological stasis would help explain why paleontologists find relatively few smooth transitions in the fossil record of species. Also, since the best candidates for speciation are *small* populations, we are less likely to have fossils of transitional stages than if they occurred in a large population.

Once it is acknowledged that "sudden" may be many thousands of years on the vast scale of geological time, the debate over the tempo of speciation is muted somewhat. The degree to which a species changes after its origin is another issue. If the species is adapted to an environment that stays the same, then natural selection would not favor changes in the gene pool. In this view, stabilizing selection tends to hold a population in a long period of stasis. But some evolutionary biologists argue that stasis is an illusion. They propose that species may continue to change after they come into existence, but in non-anatomical ways that cannot be detected from fossils.

Figure 13.15 An artist's reconstruction of an extinct bird. Called *Archaeopteryx* (Greek *archaios,* ancient, and *pteryx,* wing), this animal lived near tropical lagoons in central Europe about 150 million years ago. Like modern birds, it had flight feathers, but otherwise it was more like some small bipedal dinosaurs of its era; for instance, like those dinosaurs, *Archaeopteryx* had teeth, wing claws, and a tail with many vertebrae. *Archaeopteryx's* flight feathers could have enabled it to fly by flapping its wings. However, it had a fairly heavy body and probably relied mainly on gliding from trees. Despite its feathers, *Archaeopteryx* is not considered an ancestor of modern birds. Instead, it probably represents an extinct side branch of the bird lineage.

Wing claw (like reptile)

Teeth (like reptile)

Feathers

Long tail with many vertebrae (like reptile)

The Evolution of Biological Novelty

The two squirrels in Figure 13.9 are different species, but they are, after all, very similar animals that live very much the same way. When most people think of evolution, they envision much more dramatic transformation. How can we account for such evolutionary products as flight in birds and braininess in humans?

Adaptation of Old Structures for New Functions

Birds are derived from a lineage of earthbound reptiles (Figure 13.15). How could flying vertebrates evolve from flightless ancestors? More generally, how do major novelties of biological structure and function evolve? One mechanism is the gradual refinement of existing structures that take on new functions.

Most biological structures have an evolutionary plasticity that makes alternative functions possible. Biologists use the term **exaptation** for a structure that evolves in one context and becomes adapted for other functions. This concept does not imply that a structure somehow evolves in anticipation of future use. Natural selection cannot predict the future and can only refine a structure in the context of its current utility. Birds have lightweight skeletons with honeycombed bones, which are homologous to the bones of the earthbound reptilian ancestors of birds. However, honeycombed bones could not have evolved in the ancestors as an adaptation for upcoming flights. If light bones predated flight, as is clearly indicated by the fossil record, then they must have had some function on the ground. The probable ancestors of birds were agile, bipedal reptiles that also would have benefited from a light frame. Winglike forelimbs and feathers would have increased the surface area of the forelimbs. These enlarged forelimbs could have been adapted for flight after functioning in some other capacity, such as courtship displays. The first flights may have only been glides or extended hops in an effort to pursue prey or escape from a predator. Once flight itself became an advantage, natural selection would have remodeled feathers and wings to better fit their additional function.

Exaptation is a mechanism for novel features to arise gradually through a series of intermediate stages, each of which has some function in the organism's current context. Harvard zoologist Karel Liem puts it this way: "Evolution is like modifying a machine while it's running." The concept that biological novelties can evolve by the remodeling of old structures for new functions is in the Darwinian tradition of large changes being an accumulation of many small changes crafted by natural selection.

"Evo-Devo": Development and Evolutionary Novelty

Gradual evolutionary remodeling, such as the accumulation of flight adaptations in birds, probably involves a large number of genetic changes in populations. In other cases, relatively few genetic changes can cause major structural modifications. How can slight genetic divergence become magnified into major differences between organisms? Scientists working at the interface of evolutionary biology and developmental biology—the research field abbreviated "evo-devo"—are finding some answers to this question.

Genes that program development control the rate, timing, and spatial pattern of changes in an organism's form as it is transfigured from a zygote into an adult. (Chapter 10 provides a closer look at how genes control development.) A subtle change in a species' developmental program can have profound effects. The animal in Figure 13.16 is a salamander called an axolotl. It illustrates a phenomenon called **paedomorphosis,** which is the retention of juvenile body features in the adult (the term comes from the Greek *paedos,* child, and *morphosis,* shaping). The axolotl grows to full size and reproduces without losing its external gills, a juvenile feature in most species of salamanders.

Figure 13.16 Paedomorphosis. Some species retain as adults features that were juvenile in ancestors. The axolotl, a salamander, becomes an adult and reproduces while retaining certain tadpole characteristics, including gills. It is an example of how changes in the genes controlling development can produce a very different organism.

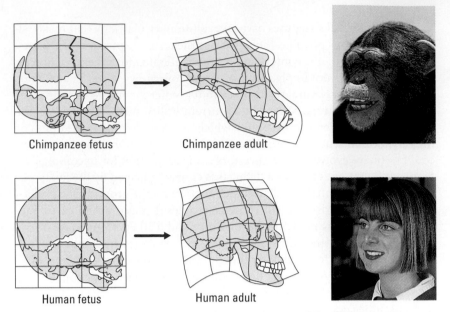

Chimpanzee fetus → **Chimpanzee adult**

Human fetus → **Human adult**

Figure 13.17 Comparison of human and chimpanzee skull development. Starting with fetal skulls that are very similar (left), the differential growth rates of different bones making up the skulls produce adult heads with very different proportions. The grid lines will help you relate the fetal skulls to the adult skulls.

Paedomorphosis has also been important in human evolution. Humans and chimpanzees are even more alike in body form as fetuses than they are as adults. In the fetuses of both species, the skulls are rounded and the jaws are small, making the face rather flat (Figure 13.17). As development proceeds, uneven bone growth makes the chimpanzee skull sharply angular, with heavy browridges and massive jaws. The adult chimpanzee has much greater jaw strength than we have, and its teeth are proportionately larger. In contrast, the adult human has a skull with decidedly rounded, more fetus-like contours. Put another way, our skull is paedomorphic; it retains fetal features even after we are mature. Our large, paedomorphic skull is one of our most distinctive features. Our large, complex brain, which fills that bulbous skull, is another. The human brain is proportionately larger than the chimpanzee brain because growth of the organ is switched off much later in human development. Compared to the brain of chimpanzees, our brain continues to grow for several more years, which can be interpreted as the prolonging of a juvenile process. We owe our culture to this evolutionary novelty, coupled with an extended childhood during which parents and teachers can influence what the young, growing brain stores.

Morph baby chimps and humans into adults in Web/CD Activity 13C.

CheckPoint

1. Explain why the concept of exaptation does not imply that a structure evolves *in anticipation of* some future environmental change.
2. How is the developmental cause of our large, roundish skulls similar to the developmental cause of the gills of axolotls?

Answers: 1. Although an exaptation is adapted for new or additional functions in a new environment, it existed because it worked as an adaptation to the old environment. **2.** Both structures are paedomorphic (juvenile anatomies retained into adulthood).

Earth History and Macroevolution

Having examined mechanisms of speciation and the origin of evolutionary novelty, we are ready to turn our attention to the history of biological diversity. Macroevolution is closely tied to the history of Earth.

Geological Time and the Fossil Record

The fossil record is an archive of macroevolution. Figure 13.18 surveys the diverse ways that organisms can fossilize. Sedimentary rocks are the richest sources of fossils and provide a record of life on Earth in their layers, or strata (see Chapter 12). The trapping of dead organisms in sediments freezes fossils in time. Thus, the fossils in each stratum of sedimentary rock are a local sample of the organisms that existed at the time the sediment was deposited. Because younger sediments are superimposed on older ones, the layers of sediment tell the relative ages of fossils.

Figure 13.18 A gallery of fossils. (a) Sedimentary rocks are the richest hunting grounds for paleontologists, scientists who study the fossil record. This researcher is excavating a fossilized dinosaur skeleton from sandstone in Dinosaur National Monument, located in Utah and Colorado. (b) The hard parts of organisms are the most common fossils. This is a skull of *Australopithecus africanus*, an ancestor of humans that lived about 2.5 million years ago. (c) Fossils may become even harder if minerals replace their organic matter. These petrified (stone) trees in Arizona are about 190 million years old. (d) Some sedimentary fossils, such as this 40-million-year-old leaf, retain organic material, including DNA, which scientists can analyze. (e) Buried organisms that decay may leave molds that are filled by minerals dissolved in water. The casts that form when the minerals harden are replicas of the organisms, as in the case of these casts of animals called brachiopods. (f) Trace fossils are footprints, burrows, and other remnants of an ancient organism's behavior. This boy is standing in a 150-million-year-old dinosaur track in Colorado. (g) This 30-million-year-old scorpion is embedded in amber (hardened resin from a tree). (h) These tusks belong to a whole 23,000-year-old mammoth, which scientists discovered in Siberian ice in 1999.

Table 13.1 The Geological Time Scale

Relative time span of eras

| Cenozoic |
| Mesozoic |
| Paleozoic |
| Precambrian |

Era	Period	Epoch	Age (millions of years ago)	Some important events in the history of life
Cenozoic	Quaternary	Recent		Historical time
			0.01	
		Pleistocene		Ice ages; humans appear
			1.8	
	Tertiary	Pliocene		Apelike ancestors of humans appear
			5	
		Miocene		Continued radiation of mammals and angiosperms
			23	
		Oligocene		Origins of many primate groups, including apes
			35	
		Eocene		Angiosperm dominance increases; origins of most modern mammalian orders
			57	
		Paleocene		Major radiation of mammals, birds, and pollinating insects
			65	
Mesozoic	Cretaceous			Flowering plants (angiosperms) appear; many groups of organisms, including most dinosaur lineages, become extinct at end of period (Cretaceous extinctions)
			145	
	Jurassic			Gymnosperms continue as dominant plants; dinosaurs dominant
			208	
	Triassic			Cone-bearing plants (gymnosperms) dominate landscape; radiation of dinosaurs, early mammals, and birds
			245	
Paleozoic	Permian			Extinction of many marine and terrestrial organisms (Permian extinctions); radiation of reptiles; origins of mammal-like reptiles and most modern orders of insects
			290	
	Carboniferous			Extensive forests of vascular plants; first seed plants; origin of reptiles; amphibians dominant
			363	
	Devonian			Diversification of bony fishes; first amphibians and insects
			409	
	Silurian			Diversity of jawless fishes; first jawed fishes; colonization of land by vascular plants and arthropods
			439	
	Ordovician			Origin of plants; marine algae abundant
			510	
	Cambrian			Origin of most modern animal phyla (Cambrian explosion)
			570	
Precambrian			610	Diverse soft-bodied invertebrate animals; diverse algae
			700	Oldest animal fossils
			1700	Oldest eukaryotic fossils
			2500	Oxygen begins accumulating in atmosphere
			3500	Oldest fossils known (prokaryotes)
			4600	Approximate time of origin of Earth

By studying many different sites, geologists have established a **geological time scale** with a consistent sequence of geological periods (Table 13.1). These periods are grouped into four eras: the Precambrian, Paleozoic, Mesozoic, and Cenozoic eras. Each era represents a distinct age in the history of Earth and its life. The boundaries are marked in the fossil record by explosive diversification of many new forms of life. Mass extinctions also mark many of the boundaries between periods and between eras. For example, the beginning of the Cambrian period is delineated by a great diversity of fossilized animals that are absent in rocks of the late Precambrian. And most of the animals that lived during the late Precambrian became extinct at the end of that era.

Fossils are reliable historical documents only if we can determine their ages. The record of the rocks chronicles the *relative* ages of fossils. It tells us the order in which groups of species evolved. However, the series of sedimentary rocks does not tell the absolute ages of the embedded fossils. The difference is analogous to peeling the layers of wallpaper from the walls of a very old house that has been inhabited by many owners. You could determine the sequence in which the wallpapers had been applied, but not the year that each layer was added. Paleontologists use a variety of methods to determine the ages of fossils in years. The most common method is **radiometric dating,** which is based on the decay of radioactive isotopes (Figure 13.19). The dates you see on the geological time scale in Table 13.1 were established by the radiometric dating of rocks and the fossils they contain.

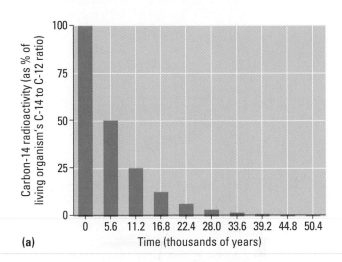

(a)

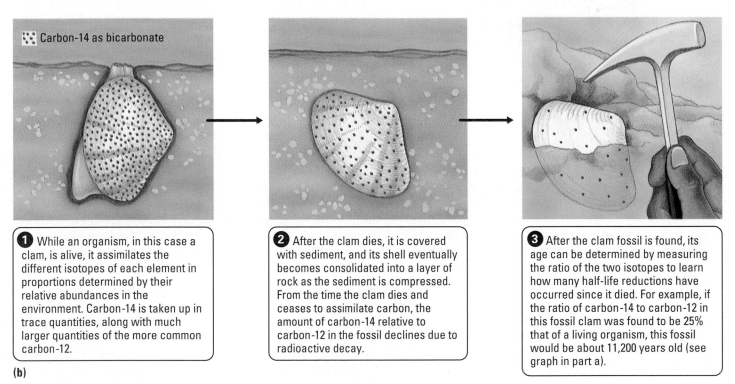

❶ While an organism, in this case a clam, is alive, it assimilates the different isotopes of each element in proportions determined by their relative abundances in the environment. Carbon-14 is taken up in trace quantities, along with much larger quantities of the more common carbon-12.

❷ After the clam dies, it is covered with sediment, and its shell eventually becomes consolidated into a layer of rock as the sediment is compressed. From the time the clam dies and ceases to assimilate carbon, the amount of carbon-14 relative to carbon-12 in the fossil declines due to radioactive decay.

❸ After the clam fossil is found, its age can be determined by measuring the ratio of the two isotopes to learn how many half-life reductions have occurred since it died. For example, if the ratio of carbon-14 to carbon-12 in this fossil clam was found to be 25% that of a living organism, this fossil would be about 11,200 years old (see graph in part a).

(b)

Figure 13.19 Radiometric dating. Amounts of radioactive isotopes can be measured by the radiation they emit as they decompose to more stable atoms (see Chapter 2). The decay rate is unaffected by temperature, pressure, and other environmental variables. Paleontologists use this clock-like decay to date fossils. **(a)** Carbon-14 is a radioactive isotope with a half-life of 5600 years. From the time an organism dies, it takes 5600 years for half of the radioactive carbon-14 to decay; half of the remainder is present in another 5600 years; and so on. Because the half-life of carbon-14 is relatively short, this isotope is only useful for dating fossils less than about 50,000 years old. To date older fossils, paleontologists use radioactive isotopes with longer half-lives. Radiometric dating has an error factor of less than 10%. Other methods for dating fossils generally confirm the ages determined by radiometric dating. **(b)** Here we use carbon-14 dating to determine the vintage of a fossilized clam shell.

Continental Drift and Macroevolution

The continents are not locked in place. They drift about Earth's surface like passengers on great plates of crust floating on the hot, underlying mantle (see the Evolution Link section at the end of Chapter 2). Unless two landmasses are embedded in the same plate, their positions relative to each other change. For example, North America and Europe are presently drifting apart at a rate of about 2 cm per year. Many important geological processes, including mountain building, volcanic activity, and earthquakes, occur at plate boundaries. California's infamous San Andreas Fault is at the border where two plates slide past each other.

Plate movements rearrange geography constantly, but two chapters in the continuing saga of continental drift had an especially strong influence on life. About 250 million years ago, near the end of the Paleozoic era, plate movements brought all the landmasses together into a supercontinent that has been named Pangaea, meaning "all land" (Figure 13.20). Imagine some of the possible effects on life. Species that had been evolving in isolation came together and competed. When the landmasses coalesced, the total amount of shoreline was reduced. There is also evidence that the ocean basins increased in depth, which lowered sea level and drained the shallow coastal seas. Then, as now, most marine species inhabited shallow waters, and the formation of Pangaea destroyed a considerable amount of that habitat. It was probably a long, traumatic period for terrestrial life as well. The continental interior, which has a drier and more erratic climate than coastal regions, increased in area substantially when the land came together. Changing ocean currents also would have affected land life as well as sea life. The formation of Pangaea had a tremendous environmental impact that reshaped biological diversity by causing extinctions and providing new opportunities for the survivors, which diversified through branching evolution.

The second dramatic chapter in the history of continental drift was written about 180 million years ago, during the Mesozoic era. Pangaea began to break up, causing geographic isolation of colossal proportions. As the continents drifted apart, each became a separate evolutionary arena, and the organisms of the different biogeographic realms diverged.

The pattern of continental separations is the solution to many biogeographic puzzles. For example, paleontologists have discovered matching fossils of Mesozoic reptiles in Ghana (West Africa) and Brazil. These two parts of the world, now separated by 3000 km of ocean, were contiguous during the early Mesozoic era. Continental drift also explains much about the current distribution of organisms, such as why the Australian fauna and flora contrast so sharply with the rest of the world (see Figure 12.7).

Mass Extinctions and Explosive Diversifications of Life

The evolutionary road from ancient to modern life has not been smooth. The fossil record reveals an episodic history, with long, relatively stable periods punctuated by briefer intervals when the turnover in species composition was much more extensive. These biological makeovers include mass extinctions as well as explosive diversifications of certain forms of life.

At the end of the Cretaceous period, about 65 million years ago, the world lost an enormous number of species—more than half of its marine animals and many groups of terrestrial plants and animals. For some 150 million years before, dinosaurs had dominated the land and air. Then, in less than 10 million years—a brief period in geological time—all dinosaurs were gone. Scientists have been debating what happened for decades. What do we know about this unusual time? The fossil record shows us that the

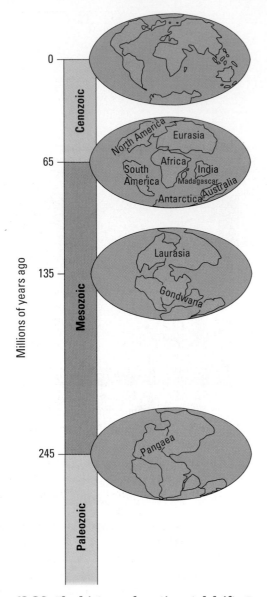

Figure 13.20 The history of continental drift. Pangaea formed about 250 million years ago. About 180 million years ago, Pangaea began to split into northern (Laurasia) and southern (Gondwana) landmasses, which later separated into the modern continents. India collided with Eurasia just 10 million years ago, forming the Himalayas, the tallest and youngest of Earth's mountain ranges. The continents continue to drift, though not at a rate that's likely to cause any motion sickness for their passengers.

climate cooled late in the Cretaceous and that shallow seas were receding from continental lowlands. We also know that many plants required by plant-eating dinosaurs died out first. Perhaps most telling of all, the sediments deposited at the time contain a thin layer of clay rich in iridium, an element very rare on Earth but common in meteorites. Many paleontologists conclude that the iridium layer is the result of fallout from a huge cloud of dust that billowed into the atmosphere when a large meteorite or asteroid hit Earth. The cloud would have blocked light and disturbed climate severely for months, perhaps killing off many plant species.

The asteroid hypothesis predicts that we should find a huge impact crater of the right age. In fact, a large asteroid crater that dates from the late Cretaceous has been found under the Caribbean Sea (Figure 13.21). Many scientists believe that the impact that produced that crater could have caused global climatic changes and mass extinctions. Other researchers propose that climatic changes due to continental drift could have caused the extinctions, whether an asteroid collided with Earth or not. Still others point to evidence in the fossil record in India indicating that during the late Cretaceous, massive volcanic activity released particles into the atmosphere, blocking sunlight and thereby contributing to climatic cooling. The various hypotheses are not mutually exclusive, and researchers continue to debate the extent to which each environmental change contributed to the extinctions.

Extinction is inevitable in a changing world. A species may become extinct because its habitat has been destroyed or because of unfavorable climatic changes. Extinctions occur all the time, but extinction rates have not been steady. There have been six distinct periods of mass extinction over the last 600 million years. During these times, losses escalated to nearly six times the average rate. Of all the mass extinctions, the ones marking the ends of the Cretaceous and the Permian periods have been the most intensively studied. The Permian extinctions, at about the time the continents merged to form Pangaea, claimed over 90% of the species of marine animals and took a tremendous toll on terrestrial life as well. Whatever their causes, mass extinctions affect biological diversity profoundly.

But there is a creative side to the destruction. Each massive dip in species diversity has been followed by explosive diversification of certain survivors. Extinctions seem to have provided the surviving organisms with new environmental opportunities. For example, mammals existed for at least 75 million years before undergoing an explosive increase in diversity just after the Cretaceous. Their rise to prominence was undoubtedly associated with the void left by the extinction of dinosaurs. The world would be a very different place today if many of the dinosaur lineages had escaped the Cretaceous extinctions or if none of the mammals that lived in the Cretaceous had survived. In addition to new diversifications in the wake of mass extinctions, explosive diversification can also result from some major new evolutionary innovation, such as flight.

Scroll through the history of life on Earth in Web/CD Activity 13D.

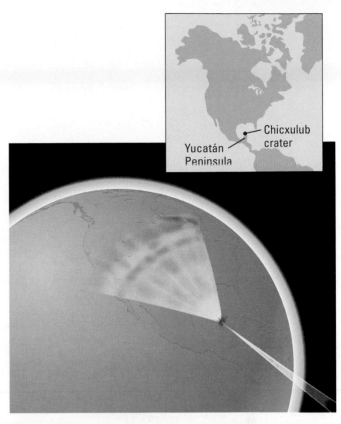

Figure 13.21 Trauma for planet Earth and its Cretaceous life. The 65-million-year-old Chicxulub impact crater is located in the Caribbean Sea near the Yucatán Peninsula of Mexico. The horseshoe shape of the crater and the pattern of debris in sedimentary rocks indicate that an asteroid or comet struck at a low angle from the southeast. This artist's interpretation represents the impact and its immediate effect—a cloud of hot vapor and debris that could have killed most of the plants and animals in North America within minutes. That would explain the higher extinction rates of land animals and plants in North America than elsewhere around the globe. This proposed impact scenario may have been only one of a series of events, including a global cooling trend, that contributed to the Cretaceous crisis.

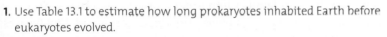

CheckPoint

1. Use Table 13.1 to estimate how long prokaryotes inhabited Earth before eukaryotes evolved.

2. A fossilized skull you unearth has a carbon-14 to carbon-12 ratio about one-sixteenth that of a living organism. What is the approximate age of the skull? (Refer to Figure 13.19a.)

3. How many continents did Earth have at the time of Pangaea?

Answers: 1. About 1800 million years, or 1.8 billion years 2. 22,400 years (four half-life reductions) 3. One

Classifying the Diversity of Life

Reconstructing evolutionary history is part of the science of **systematics,** the study of biological diversity, past and present. Systematics includes **taxonomy,** which is the identification, naming, and classification of species.

Some Basics of Taxonomy

Assigning scientific names to species is an essential part of systematics. Common names, such as monkey, fruit fly, crayfish, and garden pea, may work well in everyday communication, but they can be ambiguous because there are many species of each of these organisms. Biology's more formal taxonomic system dates back to Carolus Linnaeus (1707–1778), a Swedish physician and botanist (plant specialist). The system has two main characteristics: a two-part name for each species and a hierarchical classification of species into broader and broader groups of organisms.

Naming Species Linnaeus's system assigns to each species a two-part latinized name, or **binomial.** The first part of a binomial is the **genus** (plural, *genera*) to which the species belongs. The second part of a binomial refers to one **species** within the genus. An example of a binomial is *Panthera pardus,* the scientific name of the large cat we commonly call the leopard. Notice that the first letter of the genus is capitalized and that the whole binomial is italicized and latinized (you can name a bug you discover after a friend, but you must add the appropriate Latin ending). Our own scientific name, *Homo sapiens,* which Linnaeus made up in a show of optimism, means "wise man."

Hierarchical Classification In addition to defining and naming species, a major objective of systematics is to group species into broader taxonomic categories. The first step of such a hierarchical classification is built into the binomial for a species. We group species that are closely related into the same genus. For example the leopard, *Panthera pardus,* belongs to a genus that also includes the African lion (*Panthera leo*) and the tiger (*Panthera tigris*). Grouping species is natural for us, at least in concept. We lump together several trees we know as oaks and distinguish them from several other species of trees we call maples. Indeed, oaks and maples belong to separate genera. Biology's taxonomic scheme formalizes our tendency to group related objects as a way of structuring our view of the world.

Beyond the grouping of species within genera, taxonomy extends to progressively broader categories of classification. It places similar genera in the same **family,** puts families into **orders,** orders into **classes,** classes into **phyla** (singular, *phylum*), and phyla into **kingdoms.** Figure 13.22 places the leopard in this taxonomic scheme of groups within groups. Classifying a species by kingdom, phylum, and so on, is analogous to sorting mail, first by zip code and then by street, house number, and specific member of the household.

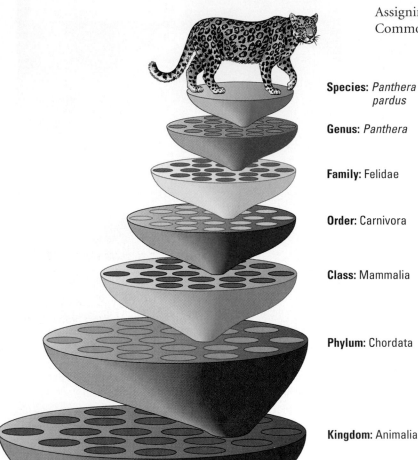

Species: *Panthera pardus*

Genus: *Panthera*

Family: Felidae

Order: Carnivora

Class: Mammalia

Phylum: Chordata

Kingdom: Animalia

Figure 13.22 Hierarchical classification. The taxonomic scheme classifies species into groups belonging to more comprehensive groups. The leopard (*Panthera pardus*) belongs to the genus *Panthera,* which also includes the African lion and tiger. These wild felines belong to the cat family (Felidae), along with the genus *Felis,* which includes the domestic cat and several closely related wild cats, such as the lynx. Family Felidae belongs to the order Carnivora, which also includes the dog family, Canidae, and several other families. Order Carnivora, the carnivores, is grouped with many other orders in the class Mammalia, the mammals. Mammalia is one of several classes belonging to the phylum Chordata in the kingdom Animalia.

Classification and Phylogeny

Ever since Darwin, systematics has had a goal beyond simple organization: to have classification reflect **phylogeny,** which is the evolutionary history of a species. As a systematist classifies species into groups subordinate to other groups in the taxonomic hierarchy, the final product takes on the branching pattern of a **phylogenetic tree.** The principle of common descent is reflected in the branches of the tree (Figure 13.23).

Sorting Homology from Analogy Homologous structures are one of the best sources of information about phylogenetic relationships. Recall from Chapter 12 that homologous structures may look different and function very differently in different species, but they exhibit fundamental similarities because they evolved from the same structure that existed in a common ancestor. Among the vertebrates, for instance, the whale limb is adapted for steering in the water; the bat wing is adapted for flight. Nonetheless, there is a basic similarity in the bones supporting these two structures (see Figure 12.8). The greater the number of homologous structures between two species, the more closely the species are related.

There are pitfalls in the search for homology: Not all likeness is inherited from a common ancestor. Species from different evolutionary branches may have certain structures that are superficially similar if natural selection has shaped analogous adaptations. This is called **convergent evolution.** Similarity due to convergence is termed **analogy,** not homology. For example, the wings of insects and those of birds are analogous flight equipment: They evolved independently and are built from entirely different structures.

To develop phylogenetic trees and classify organisms according to evolutionary history, we must use only homologous similarities. This guideline is simpler in principle than it is in practice. Adaptation can obscure homologies, and convergence can create misleading analogies. As we saw in Chapter 12, comparing the embryonic development of the species in question can often expose homology that is not apparent in the mature structures.

There is another clue to identifying homology and sorting it from analogy: The more complex two similar structures are, the less likely it is they evolved independently. Compare the skulls of a human and a chimpanzee, for example (see Figure 13.17). The skulls are not single bones, but a fusion of many, and the two skulls match almost perfectly, bone for bone. It is highly improbable that such complex structures matching in so many details could have separate origins. Most likely, the genes required to build these skulls were inherited from a common ancestor.

Molecular Biology as a Tool in Systematics If homology is about common ancestry, then comparing the genes and gene products (proteins) of organisms gets right to the heart of their evolutionary relationships. Sequences of nucleotides in DNA are inherited, and they program corresponding sequences of amino acids in proteins (see the Evolution Link section in Chapter 3). At the molecular level, the evolutionary divergence of species parallels the accumulation of differences in their genomes. The more recently two species have branched from a common ancestor, the more similar their DNA and amino acid sequences should be.

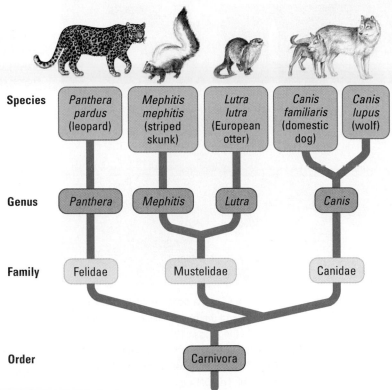

Figure 13.23 The relationship of classification and phylogeny for some members of the order Carnivora. The hierarchical classification is reflected in the finer and finer branching of the phylogenetic tree. Each branch point in the tree represents an ancestor common to species above that branch point.

Figure 13.24 Tyrolean Ice Man, a source of prehistoric human DNA. Hikers discovered the frozen 5000-year-old Stone Age man in a melting glacier in the Alps in 1991. Some skepticism that the man was a fraudulently transplanted South American mummy persisted until 1994. Then researchers analyzed DNA fragments recovered from the fossil. The DNA closely matches that of northern Europeans, not Native Americans. As researchers continue to analyze the Stone Age man's DNA, they may uncover more clues about his place in human evolution.

Analyze proteins to build phylogenetic trees in Web/CD Activity 13E/The Process of Science.

Today, both the amino acid sequences for many proteins and the nucleotide sequences for a rapidly increasing number of genomes are in databases that are available via the Internet (see Chapter 11). This has catalyzed a boom in systematics, as researchers use the databases to compare the hereditary information of different species in search of homology at its most basic level.

Molecular systematics provides a new way to test hypotheses about the phylogeny of species. The strongest support for any such hypothesis is agreement between molecular data and other means of tracing phylogeny, such as evaluating anatomical homology and analyzing the fossil record. And speaking of fossils, some are preserved in such a way that it is possible to extract DNA fragments for comparison with modern organisms (Figure 13.24).

The Cladistic Revolution Systematics entered a vigorous new era in the 1960s. Just as molecular methods became readily available for comparing species, computer technology helped usher in a new approach called **cladistic analysis.**

Cladistic analysis is the scientific search for clades (from the Greek word for "branch"). A clade consists of an ancestor and all its descendants—a distinctive branch in the tree of life. The items in the clade may be species or higher taxonomic groups, such as classes or phyla. Identifying clades centers on the analysis of homologies unique to each group (Figure 13.25). In other words, cladistic analysis focuses on the evolutionary innovations that

Figure 13.25 A simplified example of cladistic analysis. Cladistic analysis requires a comparison between a so-called in-group and an out-group. In our example, the three mammals make up the in-group, while the turtle, a reptile, is the out-group. This provides a reference point for distinguishing primitive characters from derived characters. A primitive character is a homology present in all the organisms being compared and so must also have been present in the common ancestor. All four animals in our example have vertebral columns (backbones), which is a primitive character. The derived characters, such as hair and gestation, are the evolutionary innovations that define the sequence of branch points in the phylogeny of the in-group. ❶ Hair and mammary glands are among the homologies that distinguish mammals from reptiles. ❷ Among the derived characters defining the next branch point is gestation, the carrying of offspring in the womb of the female parent. The duck-billed platypus, though a mammal, lays eggs, as do turtles and other reptiles. We can infer that gestation evolved in an ancestor common to the kangaroo and beaver that is more recent than the ancestor shared by all mammals, including the platypus. ❸ The last branch point in our cladistic analysis is defined by the much longer gestation of the beaver compared to the kangaroo (in which the embryo emerges from the womb very early and completes its development within the mother's pouch).

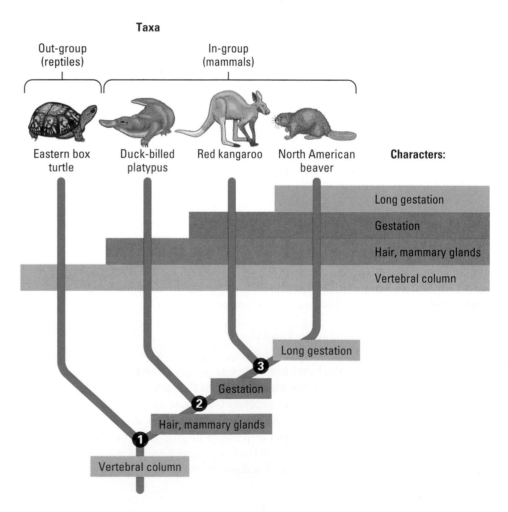

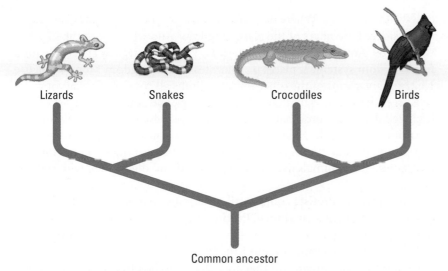

Lizards   Snakes   Crocodiles   Birds

Common ancestor

Figure 13.26 How cladistic analysis is shaking phylogenetic trees. Strict application of cladistic analysis sometimes produces phylogenetic trees that conflict with classical taxonomy. Most biologists (and this textbook) will probably continue to place reptiles and birds in separate vertebrate classes, but the tree you see here is more consistent with cladistic analysis. Based on derived characters, crocodiles are more closely related to birds than they are to lizards and snakes. Birds and crocodiles make up one clade, and lizards and snakes make up another. Or if we go back as far as the ancestor that crocodiles share with lizards and snakes to make up a clade, then the class Reptilia must also include birds. The counterargument is that the numerous flight adaptations of birds justify separating them taxonomically from the reptiles. In other words, classical taxonomy factors in the degree of evolutionary divergence between birds and the organisms we commonly think of as reptiles.

define the branch points in evolution. Identifying clades makes it possible to construct classification schemes that reflect phylogeny.

Cladistic analysis has become the most widely used method in systematics. Its strict application underscores some taxonomic dilemmas. For instance, biologists traditionally place birds and reptiles in separate classes of vertebrate animals (class Aves and class Reptilia, respectively). This classification, however, is inconsistent with a strict application of cladistic analysis. An inventory of homologies suggests that crocodiles are more closely related to birds than they are to lizards and snakes (Figure 13.26).

Arranging Life into Kingdoms: A Work in Progress

Phylogenetic trees are hypotheses about evolutionary history. Like all hypotheses, they are revised, or in some cases completely rejected, in accordance with new evidence. Molecular systematics and cladistic analysis are combining to remodel phylogenetic trees and challenge conventional classifications, even at the kingdom level.

Biologists have traditionally considered the kingdom to be the highest —the most inclusive—taxonomic category. Many of us grew up with the notion that there are only two kingdoms of life—plants and animals—because we live in a macroscopic, terrestrial realm where we are rarely aware of organisms that do not fit neatly into a plant-animal dichotomy. The two-kingdom system also had a long tradition in formal taxonomy; Linnaeus divided all known forms of life between the plant and animal kingdoms. The two-kingdom system prevailed in biology for over 200 years, but it was beset with problems. Where do prokaryotes fit in such a system? Can they be considered members of the plant kingdom? And what about fungi?

In 1969, American ecologist Robert H. Whittaker argued effectively for a **five-kingdom system.** It places prokaryotes in the kingdom Monera (Figure 13.27a). Organisms of the other four kingdoms all consist of eukaryotic cells (see Chapter 4). Kingdoms Plantae, Fungi, and Animalia consist of multicellular eukaryotes that differ in structure, development, and modes of nutrition. Plants make their own food by photosynthesis. Fungi live by decomposing the remains of other organisms and absorbing small organic molecules. Most animals live by ingesting food and digesting it within their bodies.

The kingdom Protista contains all eukaryotes that do not fit the definitions of plant, fungus, or animal. Protista is a taxonomic grab bag in the five-kingdom system. Most protists are unicellular. Amoebas and other so-called protozoa are examples. But Kingdom Protista also includes certain large, multicellular organisms that are believed to be direct descendants of unicellular protists. For example, many biologists now classify the seaweeds as protists because they are more closely related to certain single-celled algae than they are to true plants.

In the last decade, molecular studies and cladistic analysis have led to some reevaluation of the five-kingdom system. Figure 13.27b shows a

Review the history of classification schemes in Web/CD Activity 13F.

three-domain system as one alternative to the five-kingdom system. This newer scheme recognizes three basic groups: two domains of prokaryotes, Bacteria and Archaea, and one domain of eukaryotes, called Eukarya. The Bacteria and the Archaea differ in a number of important structural, biochemical, and functional features, which we will discuss in Chapter 14.

What is most important to understand here is that classifying Earth's diverse species of life will always be a work in progress as we learn more about organisms and their evolution. Charles Darwin envisioned the goals of modern systematics when he wrote in *The Origin of Species,* "Our classifications will come to be, as far as they can be so made, genealogies."

CheckPoint

1. How much of the classification in Figure 13.22 do we share with the leopard?

2. Our forearms and the wings of a bat are derived from the same ancestral prototype, and thus they are _____. In contrast, the wings of a bat and the wings of a bee are derived from totally unrelated structures, and thus they are _____.

3. To distinguish a particular clade of mammals within the larger clade that corresponds to class Mammalia, why is hair not a useful characteristic?

4. In comparing the five-kingdom system with the three-domain system, how many of the kingdoms correspond to domain Eukarya?

Answers: 1. We are classified the same down to the class level: both the cat and human are mammals. We do not belong to the same order. **2.** homologous; analogous **3.** Hair is a primitive character common to *all* mammals and cannot be helpful in distinguishing different mammalian subgroups. **4.** Four

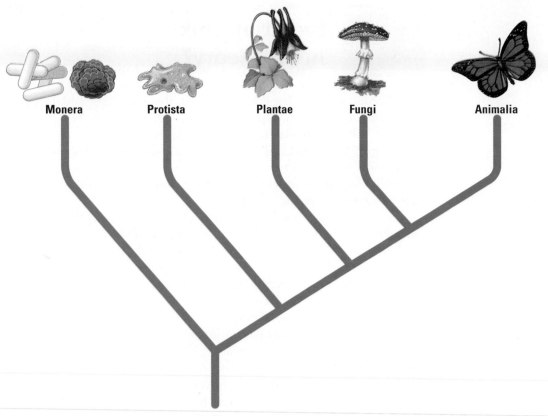

(a) The five-kingdom classification scheme

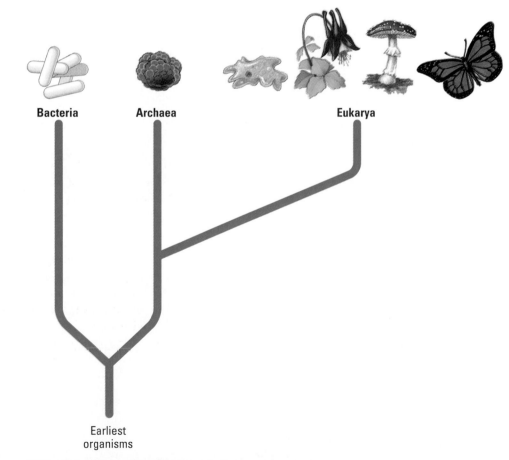

Figure 13.27 Comparison of five-kingdom and three-domain classification schemes.
(a) The traditional five-kingdom system. (b) The three-domain system. Molecular and cellular evidence supports the phylogenetic hypothesis that two lineages of prokaryotes, Bacteria and Archaea, diverged very early in the history of life. Molecular evidence also suggests that the Archaea are more closely related to eukaryotes than to the Bacteria.

(b) The three-domain classification scheme

Evolution Link:
Just a Theory?

So far in our study of evolution, we have come across several key debates. Classification schemes (five kingdoms versus three domains) and the tempo of evolution (gradualism versus punctuated equilibrium) are two examples. These and other controversies stimulate research and lead to refinements in evolutionary theory. But such debates among evolutionary biologists also seem to fuel the "just a theory" argument that certain groups use to pressure politicians to discredit or eliminate the study of evolution in the science curriculum in public schools.

The "just a theory" tactic for nullifying biology's unifying theme has two flaws. First, it fails to separate Darwin's two main points: the evidence that modern species evolved from ancestral forms and the theory that natural selection is the main mechanism for this evolution. To biologists, Darwin's theory of evolution is natural selection—the mechanism Darwin proposed to explain the historical record of evolution documented by fossils, biogeography, and other types of evidence.

This brings us to the second flaw in the "just a theory" argument. The term *theory* has a very different meaning in science than in general use. The colloquial use of "theory" comes close to what scientists mean by "hypothesis." In science, a theory is more comprehensive than a hypothesis (see Chapter 1). A theory, such as Newton's theory of gravity or Darwin's theory of natural selection, accounts for many facts and attempts to explain a great variety of phenomena. Such a unifying theory does not become widely accepted in science unless its predictions stand up to thorough and continual testing by experiments and observations. In other words, a comprehensive theory, to be accepted, is held to a higher standard of evidence than a hypothesis. Even then, science does not allow theories to become dogma. For example, many evolutionary biologists now question whether natural selection alone accounts for the evolutionary history observed in the fossil record. Unpredictable events, such as asteroids crashing into Earth, may be as important as adaptive evolution in the patterns of biological diversity we observe.

The study of evolution is more robust than ever as a branch of science, and questions about how life evolves in no way imply that most biologists consider evolution itself to be "just a theory." Debates about evolutionary theory are like arguments over competing theories about gravity; we know that objects keep right on falling while we debate the cause.

Discover more about evolution in the Web Evolution Link.

By attributing the diversity of life to natural causes rather than to supernatural forces, Darwin gave biology a sound, scientific basis. Nevertheless, the diverse products of evolution are elegant and inspiring. As Darwin said in the closing paragraph of *The Origin of Species,* "There is grandeur in this view of life."

Chapter Review

Summary of Key Concepts

Overview: Macroevolution and the Diversity of Life

- Microevolution is a change in the gene pool of a population; macroevolution is change in life-forms noticeable enough to be evident in the fossil record. Macroevolution includes the multiplication of species, the origin of evolutionary novelty, the explosive diversification that follows some evolutionary breakthroughs, and mass extinctions.
- Web/CD Activity 13A *Overview of Macroevolution*

The Origin of Species

- **What Is a Species?** According to the biological species concept, a species is a population or group of populations whose members have the potential to interbreed with one another in nature to produce fertile offspring, but who cannot successfully interbreed with members of other species.

- **Reproductive Barriers Between Species** Pre-zygotic barriers impede mating between species or hinder fertilization of eggs. Pre-zygotic barriers include temporal isolation, habitat isolation, behavioral isolation, mechanical isolation, and gametic isolation. Post-zygotic barriers include hybrid inviability and hybrid sterility.

- **Mechanisms of Speciation** When the gene pool of a population is severed from other populations of the parent species, the splinter population can follow its own evolutionary course. In allopatric speciation, the initial block to gene flow is a geographic barrier. Sympatric speciation is the origin of a new species without geographic isolation. Hybridization leading to polyploids is a common mechanism of sympatric speciation in plants.
- Web/CD Activity 13B *Sympatric Speciation*

- **What Is the Tempo of Speciation?** According to the punctuated equilibrium model, the time required for speciation in most cases is relatively short compared to the overall duration of the species' existence. This accounts for the relative rarity of transitional fossils linking newer species to older ones.

The Evolution of Biological Novelty

- **Adaptation of Old Structures for New Functions** In exaptation, a structure that evolved in one context gradually becomes adapted for other functions.

- **"Evo-Devo": Development and Evolutionary Novelty** A subtle change in the genes that control a species' development can have profound effects. In paedomorphosis, for example, the adult retains juvenile body features.
- Web/CD Activity 13C *Paedomorphosis: Morphing Chimps and Humans*

Earth History and Macroevolution

- **Geological Time and the Fossil Record** Geologists have established a geological time scale divided into four eras: Precambrian, Paleozoic, Mesozoic, and Cenozoic. The most common method for determining the ages of fossils is radiometric dating.

- **Continental Drift and Macroevolution** About 250 million years ago, plate movements brought all the landmasses together into the supercontinent Pangaea, causing extinctions and providing new opportunities for the survivors to diversify. About 180 million years ago, Pangaea began to break up, causing geographic isolation.

- **Mass Extinctions and Explosive Diversifications of Life** The fossil record reveals long, relatively stable periods punctuated by mass extinctions followed by explosive diversification of certain survivors. For example, during the Cretaceous extinctions, about 65 million years ago, the world lost an enormous number of species, including dinosaurs. Mammals greatly increased in diversity after the Cretaceous.
- Web/CD Activity 13D *A Scrolling Geological Time Scale*

Classifying the Diversity of Life

- Systematics, the study of biological diversity, includes taxonomy, which is the identification, naming, and classification of species.

- **Some Basics of Taxonomy** Each species is assigned a two-part name consisting of the genus and the species. In the taxonomic hierarchy, kingdom > phylum > class > order > family > genus > species.

- **Classification and Phylogeny** The goal of classification is to reflect phylogeny, which is the evolutionary history of a species. Classification is based on the fossil record, homologous structures, comparisons of DNA and amino acid sequences, and cladistic analysis.
- Web/CD Activity 13E/The Process of Science *Classifying Life by Comparing Proteins*

- **Arranging Life into Kingdoms: A Work in Progress** Classification systems have changed from a two-kingdom system (plants and animals) to a five-kingdom system (Monera, Protista, Plantae, Fungi, Animalia) and perhaps now to a three-domain system (Bacteria, Archaea, Eukarya).
- Web/CD Activity 13F *Classification Schemes*

Evolution Link: Just a Theory?

- In science, the word *theory* refers to a unifying principle that explains a variety of phenomena and has withstood repeated experimentation and observations.
- Web Evolution Link *Just a Theory?*

Self-Quiz

1. Bird guides once listed the myrtle warbler and Audubon's warbler as distinct species that lived side by side in parts of their ranges. However, recent books show them as eastern and western forms of a single species, the yellow-rumped warbler. Apparently, it has been found that the two kinds of warblers
    a. live in the same areas.
    b. successfully interbreed.
    c. are almost identical in appearance.
    d. are merging to form a single species.
    e. live in different places.

2. Which of the following is an example of a post-zygotic reproductive barrier?
    a. One lilac species lives on acid soil, another on basic soil.
    b. Mallard and pintail ducks mate at different times of year.
    c. Two species of leopard frogs have different mating calls.
    d. Hybrid offspring of two species of jimsonweed always die before reproducing.
    e. Pollen of one kind of tobacco cannot fertilize another kind.

3. A small, isolated population is more likely to undergo speciation than a large one because a small population
   a. contains a greater amount of genetic diversity.
   b. is more susceptible to gene flow.
   c. is more affected by genetic drift and natural selection.
   d. is more subject to errors during meiosis.
   e. is more likely to survive in a new environment.

4. Many species of plants and animals adapted to desert conditions probably did not arise there. Their success in living in deserts could be due to
   a. paedomorphosis.   d. phylogeny.
   b. convergent evolution.   e. mass extinction.
   c. exaptation.

5. Mass extinctions that occurred in the past
   a. cut the number of species to the few survivors left today.
   b. resulted mainly from the separation of the continents.
   c. occurred regularly, about every million years.
   d. were followed by diversification of the survivors.
   e. wiped out land animals but had little effect on marine life.

6. The animals and plants of India are almost completely different from the species in nearby Southeast Asia. Why might this be true?
   a. They have become separated by convergent evolution.
   b. The climates of the two regions are completely different.
   c. India is in the process of separating from the rest of Asia.
   d. Life in India was wiped out by ancient volcanic eruptions.
   e. India was a separate continent until relatively recently.

7. Two worms in the same class must also be grouped in the same
   a. order.   d. family.
   b. phylum.   e. species.
   c. genus.

8. A paleontologist estimates that when a particular rock formed, it contained 12 mg of the radioactive isotope potassium-40. The rock now contains 3 mg of potassium-40. The half-life of potassium-40 is 1.3 billion years. About how old is the rock?
   a. 0.4 billion years   d. 2.6 billion years
   b. 0.3 billion years   e. 5.2 billion years
   c. 1.3 billion years

9. If you were using cladistic analysis to build a phylogenetic tree of cats, which of the following would be the best choice for an out-group?
   a. lion   d. leopard
   b. domestic cat   e. tiger
   c. wolf

10. In the five-kingdom system, which kingdom consists of prokaryotic organisms?
    a. Monera   d. Fungi
    b. Protista   e. Animalia
    c. Plantae

• Go to the website or CD-ROM for more self-quiz questions.

The Process of Science

1. Distinguish "hypothesis" from "theory" in the vocabulary of science.

2. **Analyze proteins to build phylogenetic trees in The Process of Science activity on the website and CD-ROM.**

Biology and Society

Experts estimate that human activities cause the extinction of hundreds of species every year. The natural rate of extinction is thought to be a few species per year. As we continue to alter the global environment, especially by cutting down tropical rain forests, the resulting extinction will probably rival that at the end of the Cretaceous period. Most biologists are alarmed at this prospect. What are some reasons for their concern? Consider that life has endured numerous mass extinctions and has always bounced back. How is the present mass extinction different from previous extinctions? What might be some of the consequences for the surviving species?

The Evolution of Microbial Life

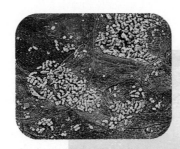

The number of **bacteria in one human's**

mouth is greater than the total number of people who ever

lived. **Each year more than 200 million**

people become **infected with malaria.**

Bacterial fermentation is used to produce **cheese,**

yogurt, buttermilk and many types of sausage.

More than half of our antibiotics come from **soil bacteria**

of the genus *Streptomyces.*

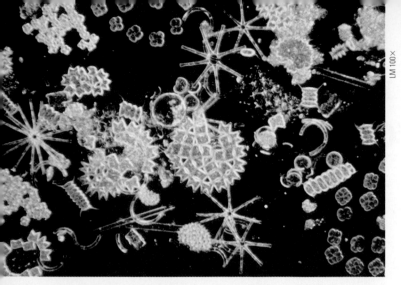

Figure 14.1 A sample of microbes in a droplet of pond water.

LM 100×

You and I and every other creature are relatives. Life is a continuum extending from the earliest organisms through various evolutionary branches to the kaleidoscope of biological diversity we see today. The organisms most familiar to us are macroscopic and multicellular—mainly plants and animals. However, for the first 75% of life's history, Earth was populated only by organisms that were microscopic and mostly unicellular. In this chapter, we trace the origin and diversification of these microbes (Figure 14.1).

Overview: Major Episodes in the History of Life

Life began when Earth was young. The planet was born 4.5 billion years ago, and its crust began to solidify about 4.0 billion years ago (see the Evolution Link section at the end of Chapter 2). A few hundred million years later, by 3.5 billion years ago, Earth was already inhabited by a diversity of organisms. Those earliest organisms were all prokaryotes, their cells lacking true nuclei (see Chapter 4). Within the next billion years, two distinct groups of prokaryotes—bacteria and archaea—diverged (Figure 14.2).

An oxygen revolution began about 2.5 billion years ago. Photosynthetic prokaryotes that split water molecules released oxygen gas, changing Earth's atmosphere profoundly (see the Evolution Link section at the end of Chapter 6). The corrosive O_2 doomed many prokaryotic groups. Among the survivors, a diversity of metabolic modes evolved, including cellular respiration, which uses oxygen to extract energy from food (see Chapter 5). All of this metabolic evolution occurred during the almost 2 billion years that prokaryotes had Earth to themselves.

The oldest eukaryotic fossils are about 1.7 billion years old. Eukaryotic cells, remember, contain nuclei and many other organelles that are absent in prokaryotic cells (see Chapter 4). The eukaryotic cell evolved from a prokaryotic community, a host cell containing even smaller prokaryotes. The mitochondria of our cells and those of every other eukaryote are descendants of those smaller prokaryotes. And so are the chloroplasts of plants and algae.

The origin of more complex cells launched an explosive diversification of eukaryotic forms. They were the protists. Represented today by a great diversity of organisms, protists are mostly microscopic and unicellular. The organisms we call protozoans are protists. So are a great variety of single-celled algae, including diatoms.

The next great evolutionary "experiment" was multicellularity. The first multicellular eukaryotes evolved, perhaps a billion years ago, as colonies of single-celled ancestors. Their modern descendants include multicellular protists, such as seaweeds. Other evolutionary branches stemming from the ancient protists gave rise to animals, fungi, and plants.

The greatest diversification of animals was the so-called Cambrian explosion. The Cambrian was the first period of the Paleozoic era, which began about 570 million years ago. The earliest animals lived in late Precambrian seas, but they diversified extensively over a span of just 10 million years during the early Cambrian. In fact, all the major body plans (phyla) of animals had evolved by the end of that evolutionary eruption.

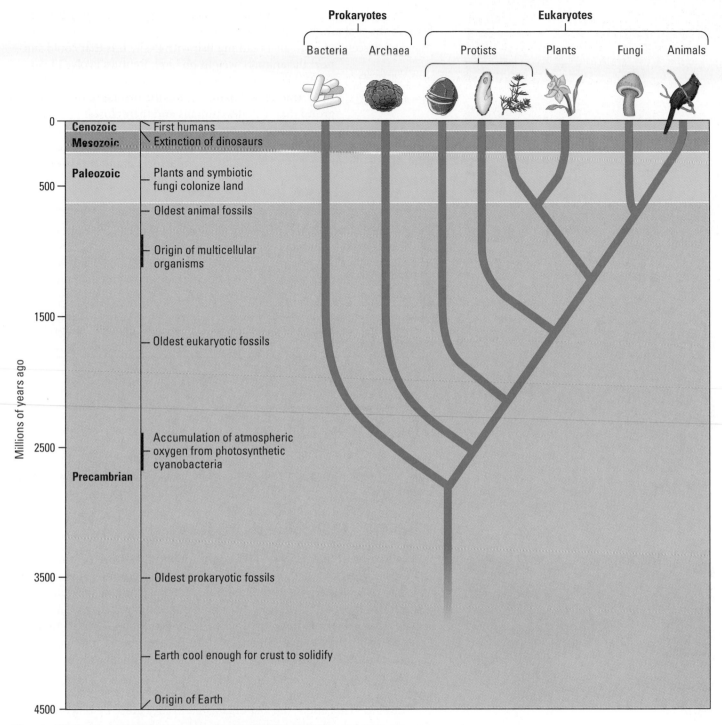

Prokaryotes
Bacteria | Archaea

Eukaryotes
Protists | Plants | Fungi | Animals

0 — Cenozoic — First humans
Mesozoic — Extinction of dinosaurs

Paleozoic — Plants and symbiotic fungi colonize land

500 —

— Oldest animal fossils

— Origin of multicellular organisms

1500 —

— Oldest eukaryotic fossils

2500 —

— Accumulation of atmospheric oxygen from photosynthetic cyanobacteria

Precambrian

3500 — — Oldest prokaryotic fossils

— Earth cool enough for crust to solidify

4500 — — Origin of Earth

Millions of years ago

Figure 14.2 Some major episodes in the history of life. The timing of events is based on fossil evidence and molecular analysis.

For over 85% of biological history—life's first 3 billion years—life was mostly confined to aquatic habitats. The colonization of land was a major milestone in the history of life. Plants, in the company of fungi, led the way about 475 million years ago. Even today, the roots of most plants are associated with fungi that aid in the absorption of water and minerals from soil. Plants transformed the landscape, creating new opportunities for all life-forms, especially herbivorous (plant-eating) animals and their predators.

The evolutionary venture onto land included vertebrate animals in the form of the first amphibians. These prototypes of today's frogs and salamanders descended from air-breathing fish with fleshy fins that could support the animals' weight on land. Reptiles evolved from amphibians, and birds and mammals evolved from reptiles. Among the mammals are the primates, the animal group that includes humans and their closest relatives, apes and monkeys. But trace our genealogy back far enough, and we count certain protists, and before them certain prokaryotes, as our ancestors. To understand life on Earth, we must go back to the origin and diversification of microbes.

Test your knowledge of the history of life in Web/CD Activity 14A.

CheckPoint

Put the following events in order, from the earliest to the most recent: diversification of animals (Cambrian explosion), evolution of eukaryotic cells, first humans, colonization of land by plants and fungi, origin of prokaryotes, evolution of land animals, evolution of multicellular organisms.

Answer: origin of prokaryotes, evolution of eukaryotic cells, evolution of multicellular organisms, diversification of animals (Cambrian explosion), colonization of land by plants and fungi, evolution of land animals, first humans

The Origin of Life

We may never know for sure, of course, how life on Earth began. But we must start with the assumption that science seeks *natural* causes for natural phenomena.

Resolving the Biogenesis Paradox

From the time of the ancient Greeks until well into the nineteenth century, it was common "knowledge" that life could arise from nonliving matter. This idea of life emerging from inanimate material is called **spontaneous generation.** It persisted well into the nineteenth century as an explanation for the rapid growth of microorganisms in spoiled foods. Then in 1862, Louis Pasteur's famous experiments with milk confirmed what many others had suspected: All life today, including microbes, arises only by the reproduction of preexisting life. This "life-from-life" principle is called **biogenesis.**

But wait! What about the *first* organisms? If *they* arose by biogenesis, then they wouldn't be the first organisms. Although there is no evidence that spontaneous generation occurs today, conditions on the early Earth were very different. For instance, there was little atmospheric oxygen to tear apart complex molecules. And such energy sources as lightning, volcanic activity, and ultraviolet sunlight were all more intense than what we experience today. (See the Evolution Link section at the end of Chapter 2.) The resolution to the biogenesis paradox is that life did not begin on a planet anything like the modern Earth, but on a young Earth that was a very different world. Most biologists now think that it is at least a credible hypothesis that chemical and physical processes in Earth's primordial environments could eventually have produced very simple cells through a sequence of stages. Debate abounds about the nature of those steps.

An artist's rendition of Earth about 3 billion years ago.
The pad-like objects in the scene represent colonies of prokaryotes known from the fossil record.

A Four-Stage Hypothesis for the Origin of Life

According to one hypothetical scenario, the first organisms were products of chemical evolution in four stages: (1) the abiotic (nonliving) synthesis of small organic molecules, such as amino acids and nucleotides; (2) the joining of these small molecules (monomers) into polymers, including proteins and nucleic acids; (3) the origin of self-replicating molecules that eventually made inheritance possible; and (4) the packaging of all these molecules into pre-cells, droplets with membranes that maintained an internal chemistry different from the surroundings. This is all speculative, of course, but what makes it science is that the hypothesis leads to predictions that can be tested in the laboratory.

Stage 1: Abiotic Synthesis of Organic Monomers In 1953, Stanley Miller, a 23-year-old graduate student at the University of Chicago, made front-page news. Miller had built a contraption that simulated key conditions on the early Earth. His apparatus produced a variety of small organic molecules that are essential for life, including amino acids, the monomers of proteins (Figure 14.3). In the half century since Miller's seminal experiments, many scientists have repeated and extended the research, varying such conditions as the composition of the ancient "atmosphere" and "sea." These laboratory analogues of the primeval Earth have produced all 20 amino acids, several sugars, lipids, the nucleotides that are the monomers of DNA and RNA, and even ATP, the molecule that powers most biological work. The abiotic synthesis of organic molecules on the early Earth is certainly a plausible scenario.

Conduct experiments on the origin of life in Web/CD Activity 14B.

It is also plausible that some organic compounds reached Earth from space. In 2000, Indian scientists reported that their computer models showed how molecules such as adenine, an ingredient of DNA, could form by reactions of cyanide in the clouds of gas between stars. These simulations would explain why some meteorites that have crashed to Earth contain organic molecules. But whether the primordial Earth was stocked with organic monomers made here or elsewhere, the key point is that the molecular ingredients of life were probably present very early.

Stage 2: Abiotic Synthesis of Polymers The hypothesis of an abiotic origin of life makes another prediction that can be tested in the laboratory. If the hypothesis is correct, then it should be possible to link organic monomers to form polymers such as proteins and nucleic acids without the help of enzymes and other cellular equipment. Researchers have observed such polymerization by dripping solutions of organic monomers onto hot sand, clay, or rock. The heat vaporizes the water in the solutions and concentrates the monomers on the underlying substance. Some of the monomers then

Figure 14.3 Making organic molecules in a laboratory simulation of early-Earth chemistry. A few decades after his famous experiments, Stanley Miller posed with a version of the contraption he built to simulate chemical dynamics on the early Earth. His hand cups a warmed flask containing the primordial "sea." Electrical discharge simulates lightning in an ancient "atmosphere" consisting of ammonia, hydrogen, methane, and water vapor. Glass tubes circulate the water between the "atmosphere" and the "sea." Products of chemical reactions in the "atmosphere" dissolve in the water and accumulate in the "sea" flask. And those products include amino acids and many other organic molecules that eventually color the "sea" a murky brown.

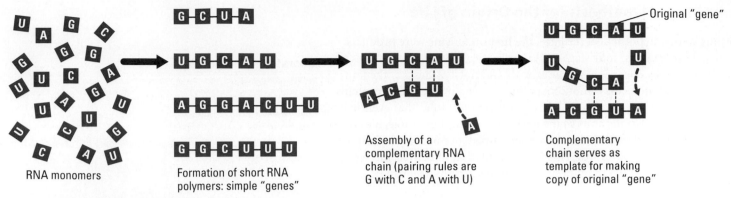

Figure 14.4 Abiotic replication of RNA "genes."

RNA monomers

Formation of short RNA polymers: simple "genes"

Assembly of a complementary RNA chain (pairing rules are G with C and A with U)

Complementary chain serves as template for making copy of original "gene"

Original "gene"

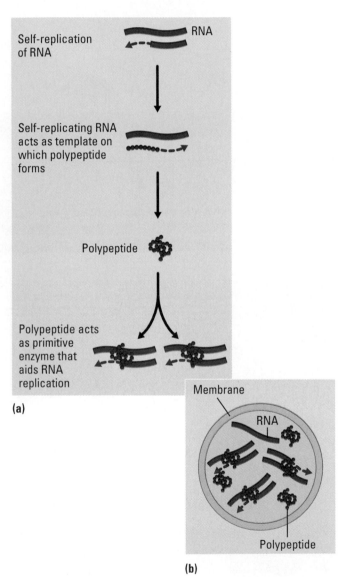

Self-replication of RNA

RNA

Self-replicating RNA acts as template on which polypeptide forms

Polypeptide

Polypeptide acts as primitive enzyme that aids RNA replication

(a)

Membrane

RNA

Polypeptide

(b)

Figure 14.5 The origin of molecular cooperation in the prebiotic world. (a) Possible origin of cooperation between nucleic acids and polypeptides. **(b)** Molecular cooperation within a membrane-enclosed compartment. In such a pre-cell, RNA genes would benefit exclusively from their protein products rather than sharing them with other molecular complexes.

spontaneously bond together to form polymers, including polypeptides, the chains of amino acids that make up proteins. On the early Earth, raindrops or waves may have splashed dilute solutions of organic monomers onto fresh lava or other hot rocks and then rinsed polypeptides and other polymers back into the sea. Alternatively, deep sea vents, where gases and superheated water with dissolved minerals escape from Earth's interior, may have been locales for the abiotic synthesis of organic monomers and polymers. (You'll learn about these unique ecosystems in Chapter 17.)

Stage 3: Origin of Self-Replicating Molecules Life is defined partly by the process of inheritance, which is based on self-replicating molecules. Today's cells store their genetic information as DNA. They transcribe the information into RNA and then translate RNA messages into specific enzymes and other proteins (see Chapter 9). This mechanism of information flow probably emerged gradually through a series of refinements to much simpler processes. What were the first genes like?

One hypothesis is that the first genes were short strands of RNA that replicated without the assistance of enzymes (proteins). In laboratory experiments, short RNA molecules can assemble spontaneously from nucleotide monomers without the presence of cells or enzymes (Figure 14.4). The result is a population of RNA molecules, each with some random sequences of monomers. Some of the molecules self-replicate, but they vary in their success at this reproduction. What happens can only be described as molecular evolution in a test tube. The RNA varieties that replicate fastest increase their frequency in the population.

In addition to the experimental evidence, there is another reason the idea of RNA genes in the primordial world is attractive. Cells actually have catalytic RNAs, which are called **ribozymes.** Perhaps early ribozymes catalyzed their own replication. That would help with the "chicken and egg" paradox of which came first, enzymes or genes. Maybe the "chicken and egg" came together in the same RNA molecules. The molecular biology of today may have been preceded by an "RNA world."

Stage 4: Formation of Pre-Cells The properties of life emerge from an interaction of molecules organized into higher levels of order. For example, Figure 14.5a diagrams one way that RNA and polypeptides could have cooperated in the prebiotic world. But such molecular teams would be much more efficient if they were packaged within membranes that kept the molecules close together in a solution that could be different from the surrounding environment. We'll call these molecular aggregates pre-cells—not really cells, but molecular packages with some of the properties of life (Figure 14.5b).

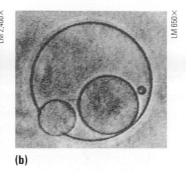

LM 2,400×
LM 650×

(a)　**(b)**

Figure 14.6 Laboratory versions of pre-cells. (a) These tiny spheres are made by cooling solutions of polypeptides that are produced abiotically. The spheres grow by absorbing more polypeptides until they reach an unstable size, when they split to form "daughter" spheres. Of course, this division lacks the precision of cellular reproduction. **(b)** Lipids mixed with water self-assemble into these membrane-enclosed droplets (see the Evolution Link at the end of Chapter 4). In some cases, large droplets bud to give rise to smaller "offspring."

Laboratory experiments demonstrate that pre-cells could have formed spontaneously from abiotically produced organic compounds (Figure 14.6). Some of the pre-cells are coated by membranes that are selectively permeable; osmosis causes the droplets to swell or shrink when placed in solutions of different salt concentrations. Certain types of pre-cells store energy in the form of a voltage across their membranes. They can even discharge the voltage much as nerve cells do. If specific enzymes are included among the ingredients that assemble into pre-cells, the aggregates display a rudimentary metabolism; they absorb substrates from their surroundings and release the products of the reactions.

From Chemical Evolution to Darwinian Evolution

If pre-cells with self-replicating RNA, and later DNA, did form on the young Earth, they would be refined by natural selection—Darwinian evolution. Mutations, errors in the copying of the "genes," would result in variation among the droplets. And the most successful of these droplets would grow, divide (reproduce), and continue to evolve. Of course, the gap between such pre-cells and even the simplest of true cells is enormous. But with millions of years of incremental refinement through natural selection, these molecular cooperatives could have become more and more cell-like. The point at which we stop calling membrane-enclosed units that metabolize and replicate genetic programs pre-cells and start calling them living cells is as fuzzy as our definition of life. But we do know that prokaryotes were already flourishing at least 3.5 billion years ago and that all branches of life arose from those ancient prokaryotes.

CheckPoint

1. According to the four-stage hypothesis we've explored, the first cells were preceded by a chemical evolution that first produced small _____ molecules, which subsequently joined to form larger molecules called _____.

2. What was the hypothesis Stanley Miller was testing with his experiments?

3. What is a ribozyme?

4. Why was the origin of membranes enclosing protein–nucleic acid cooperatives a key step in the onset of Darwinian evolution (natural selection)?

Answers: 1. monomer (or organic); polymers **2.** The hypothesis that conditions on the early Earth favored synthesis of organic molecules from inorganic ingredients **3.** An RNA molecule that functions as a catalyst **4.** In contrast to random mingling of molecules in a nonsubdivided solution, segregation of molecular systems by membranes resulted in selection for the most successful self-replicating aggregates.

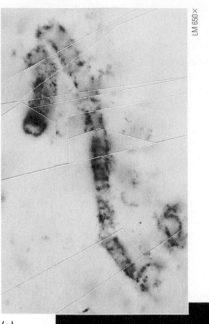

LM 650×

(a)

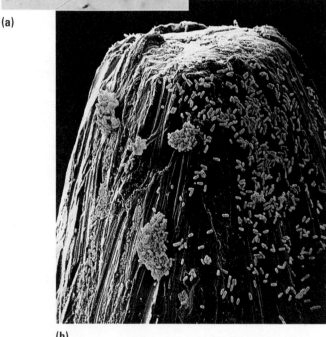

Colorized SEM 550×

(b)

Figure 14.7 Over 3 billion years of prokaryotes. (a) An early prokaryote. This microscopic fossil is a filamentous species consisting of a chain of prokaryotic cells. It is one of a diversity of prokaryotes found in western Australian rocks that are about 3.5 billion years old. **(b)** Modern prokaryotes on the head of a pin. The orange rods are individual bacteria, each about 5 micrometers (μm) long. Most prokaryotic cells have diameters in the range of 1–10 μm, much smaller than most eukaryotic cells (typically 10–100 μm). This micrograph will help you understand why a pin prick can cause infection. And it will help you remember to flame the tip of a needle before using it to remove a splinter. The heat kills the bacteria.

Prokaryotes

The history of prokaryotic life is a success story spanning billions of years (Figure 14.7). Prokaryotes lived and evolved all alone on Earth for 2 billion years. They have continued to adapt and flourish on an evolving Earth and in turn have helped to change the Earth. In this section, you will become more familiar with prokaryotes by studying their diversity and ecological significance.

They're Everywhere!

Today, prokaryotes are found wherever there is life, and they outnumber all eukaryotes combined. More prokaryotes inhabit a handful of fertile soil or the mouth or skin of a human than the total number of people who have ever lived. Prokaryotes also thrive in habitats too cold, too hot, too salty, too acidic, or too alkaline for any eukaryote. In 1999, biologists even discovered prokaryotes growing on the walls of a gold mine 2 miles below Earth's surface.

Though individual prokaryotes are relatively small organisms, they are giants in their collective impact on Earth and its life. We hear most about a few species that cause serious illness. During the fourteenth century, Black Death—bubonic plague, a bacterial disease—spread across Europe, killing an estimated 25% of the human population. Tuberculosis, cholera, many sexually transmissible diseases, and certain types of food poisoning are some other human diseases caused by bacteria.

However, prokaryotic life is no rogues' gallery. Far more common than harmful bacteria are those that are benign or beneficial. Bacteria in our intestines provide us with important vitamins, and others living in our mouth prevent harmful fungi from growing there. Prokaryotes also recycle carbon and other vital chemical elements back and forth between organic matter and the soil and atmosphere. For example, there are prokaryotes that decompose dead organisms. Found in soil and at the bottom of lakes, rivers, and oceans, these decomposers return chemical elements to the environment in the form of inorganic compounds that can be used by plants, which in turn feed animals. If prokaryotic decomposers were to disappear, the chemical cycles that sustain life would come to a halt. All forms of eukaryotic life would also be doomed. In contrast, prokaryotic life would undoubtedly persist in the absence of eukaryotes, as it once did for 2 billion years.

The Two Main Branches of Prokaryotic Evolution: Bacteria and Archaea

Prokaryotes have a cellular organization fundamentally different from that of eukaryotes. Whereas eukaryotic cells have a membrane-enclosed nucleus and numerous other membrane-enclosed organelles, prokaryotic cells lack these structural features (see Chapter 4). The traditional five-kingdom classification scheme emphasizes this fundamental difference in cellular organization. Prokaryotes make up the kingdom Monera, separate from the four eukaryotic kingdoms: Protista, Plantae, Fungi, and Animalia. In the past two decades, however, researchers have learned that a single prokaryotic kingdom may not fit evolutionary history. By comparing genomes among diverse prokaryotes, biologists have identified two major branches of prokaryotic evolution: the **bacteria** and the **archaea.** Though they have prokaryotic cell organization in common, bacteria and archaea differ in many structural, biochemical, and physiological characteristics. And there is also evidence that archaea are more closely related to eukaryotes than they are to bacteria. It was these discoveries that prompted the three-domain classification—domains Bacteria, Archaea, and Eukarya—which you can review in Figure 13.27b. The majority of prokaryotes are bacteria, but the archaea are worth studying for their evolutionary and ecological significance.

The term *archaea* (from the Greek *archaios,* ancient) refers to the antiquity of the group's origin from the earliest cells. Even today, most species of archaea inhabit extreme environments, such as hot springs and salt ponds. Few, if any, other modern organisms can survive in these environments, which may resemble habitats on the early Earth.

Biologists refer to some archaea as "extremophiles," meaning "lovers of the extreme." There are extreme halophiles ("salt lovers"), which are archaea that thrive in such environments as Utah's Great Salt Lake and seawater-evaporating ponds used to produce salt (Figure 14.8). There are also extreme thermophiles ("heat lovers") that live in very hot water; some even populate the deep-ocean vents that gush superheated water hotter than 100°C, the boiling point of water at sea level. And then there are the methanogens, archaea that live in anaerobic environments and give off methane as a waste product. They are abundant in the anaerobic mud at the bottom of lakes and swamps. You may have seen methane, also called marsh gas, bubbling up from a swamp. Great numbers of methanogens also inhabit the digestive tracts of animals. In humans, intestinal gas is largely the result of their metabolism. More importantly, methanogens aid digestion in cattle, deer, and other animals that depend heavily on cellulose for their nutrition. Normally, bloating does not occur, because these animals regularly belch out large volumes of gas produced by the methanogens and other microorganisms that enable them to utilize cellulose. And that may be more than you wanted to know about these gas-producing microbes.

Learn more about the archaea in The Process of Science activities available on the website and CD-ROM.

The Structure, Function, and Reproduction of Prokaryotes

You can use Figure 4.5 to review the general structure of prokaryotic cells. Note again the absence of a true nucleus and the other membrane-enclosed organelles characteristic of the much more complex eukaryotic cells. Another feature to note in prokaryotes is that nearly all species have cell walls exterior to their plasma membranes. These walls are chemically different from the cellulose walls of plant cells. Some antibiotics, including the penicillins, kill certain bacteria by incapacitating an enzyme the microbes use to make their walls.

Review prokaryotic cell structure and function in Web/CD Activity 14C.

Determining cell shape by microscopic examination is an important step in identifying prokaryotes (Figure 14.9). Spherical species are called **cocci** (singular, *coccus*), from the Greek word for "berries." Cocci that occur in clusters are called staphylococci (from the Greek *staphyle,* cluster of grapes), or staph for short (as in "staph infections"). Other cocci occur in chains; they are called streptococci (from the Greek *streptos,* twisted). The bacterium that causes strep throat in humans is a streptococcus. Rod-shaped prokaryotes are called **bacilli** (singular, *bacillus*). A third group of prokaryotes are curved or spiral-shaped. The largest spiral-shaped prokaryotes are

Figure 14.8 Extreme halophiles ("salt-loving" archaea). These are seawater-evaporating ponds at the edge of San Francisco Bay. The colors of the ponds result from dense growth of the prokaryotes that thrive when the salinity of the water reaches 15–20% (before evaporation, seawater has a salt concentration of about 3%). The ponds are used for commercial salt production; the halophilic archaea are harmless.

Colorized SEM 8,000×

(a) Cocci (spheres)

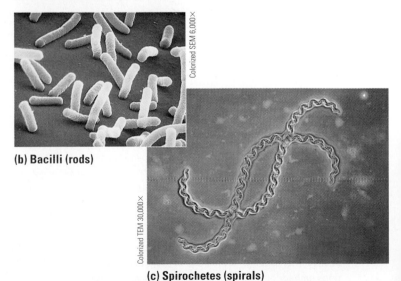

Colorized SEM 6,000×

(b) Bacilli (rods)

Colorized TEM 30,000×

(c) Spirochetes (spirals)

Figure 14.9 Three common shapes of prokaryotes.

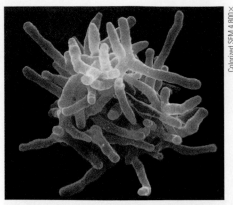

(a) Actinomycete

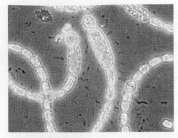

(b) Cyanobacteria

(c) Giant bacterium (and fruit fly for comparison)

Figure 14.10 **Some examples of bacterial diversity.** **(a)** This prokaryotic organism, called an actinomycete, is a mass of branching chains of rod-shaped cells. These bacteria are very common in soil, where they break down organic substances. The filaments enable the organism to bridge dry gaps between soil particles. Most species secrete antibiotics, which inhibit the growth of competing bacteria. Pharmaceutical companies use various species of actinomycetes to produce antibiotic drugs, including streptomycin. **(b)** This filamentous prokaryote belongs to a photosynthetic group called the cyanobacteria. Many species are truly multicellular in having a division of labor among specialized cells. The box on this micrograph highlights a cell that converts atmospheric nitrogen to ammonia, which can then be incorporated into amino acids and other organic compounds. **(c)** There are actually some prokaryotic cells that are gigantic, even by eukaryotic standards. The bright ball in this photo is a marine bacterium discovered in 1997 (the two smaller spheres above it are dead cells). This prokaryotic cell is over half a millimeter in diameter, about the size of a fruit fly's head.

called **spirochetes.** The bacterium that causes syphilis, for example, is a spirochete. So is the culprit that causes Lyme disease.

Most prokaryotes are unicellular and very small, but there are exceptions to both of these generalizations. Some species tend to aggregate transiently into groups of two or more cells, such as the streptococci already mentioned. Others form true colonies, which are permanent aggregates of identical cells (Figure 14.10a). And some species even exhibit a simple multicellular organization, with a division of labor between specialized types of cells (Figure 14.10b). Among unicellular species, there are some giants that actually dwarf most eukaryotic cells (Figure 14.10c).

About half of all prokaryotic species are motile. Many of those travelers have one or more flagella that propel the cells away from unfavorable places or toward more favorable places, such as nutrient-rich locales (Figure 14.11).

Although few bacteria can thrive in the extreme environments favored by many archaea, some bacteria can survive extended periods of very harsh conditions by forming specialized "resting" cells, or **endospores**

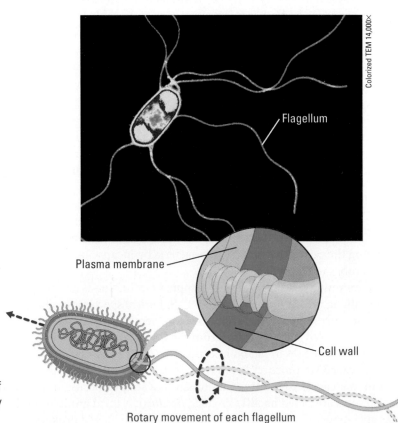

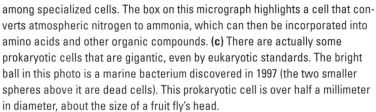

Figure 14.11 **Prokaryotic flagella.** These locomotor appendages are entirely different in structure and mechanics from the eukaryotic flagella discussed in Chapter 4. At the base of the prokaryotic version is a motor and set of rings embedded in the plasma membrane and cell wall. This machinery actually spins like a wheel, rotating the filament of the flagellum.

(Figure 14.12). Some endospores can remain dormant for centuries. Not even boiling water kills most of these resistant cells. To sterilize laboratory equipment, microbiologists use an appliance called an autoclave, a pressure cooker that kills endospores by heating to a temperature of 121°C (250°F) with high-pressure steam. The food-canning industry uses similar methods to kill endospores of dangerous soil bacteria such as *Clostridium botulinum,* which produces a toxin that causes the potentially fatal disease botulism.

Most prokaryotes can reproduce at a phenomenal rate if conditions are favorable. The cells copy their DNA almost continuously and divide again and again by the process called **binary fission.** To understand how this makes explosive population growth possible, flash back to that childhood numbers game: "Would you rather have a million dollars or start out with just a penny and have it doubled every day for a month?" If you opt for the penny, you feel like a loser at mid-month, when you only have a few hundred dollars. But by the end of the month, you've bagged about 10 million bucks. This is the exponential growth that repeated doublings make possible. Now apply that concept to bacterial reproduction, except double the number every 20 minutes. That's the rate at which some bacteria can divide if there is plenty of food and space. In just 24 hours, a single tiny cell could give rise to a bacterial colony equivalent in mass to about 15,000 humans! Fortunately, few bacterial populations can sustain exponential growth for long. Environments are usually limiting in resources such as food and space. The bacteria also produce metabolic waste products that may eventually pollute the colony's environment. Still, you can understand why certain bacteria can make you sick so soon after just a few cells infect you—or why food can spoil so rapidly. Refrigeration retards food spoilage because most microorganisms reproduce only very slowly at such low temperatures.

The Nutritional Diversity of Prokaryotes

Prokaryotic evolution "invented" every type of nutrition we observe throughout life, plus some nutritional modes unique to prokaryotes. Nutrition refers here to how an organism obtains two resources for synthesizing organic compounds: energy and a source of carbon. Species that use light energy are termed phototrophs. Chemotrophs obtain their energy from chemicals taken from the environment. If an organism needs only the inorganic compound CO_2 as a carbon source, it is called an autotroph. Heterotrophs require at least one organic nutrient—the sugar glucose, for instance—as a source of carbon for making other organic compounds (see Chapter 6). We can combine the phototroph-versus-chemotroph (energy source) and autotroph-versus-heterotroph (carbon source) criteria to group organisms according to four major modes of nutrition:

See if you can provide the ideal energy source and carbon source for various prokaryotes in Web/CD Activity 14D.

1. **Photoautotrophs** are photosynthetic organisms that harness light energy to drive the synthesis of organic compounds from CO_2. Among the diverse groups of photosynthetic prokaryotes are the cyanobacteria, such as the species in Figure 14.10b. All photosynthetic eukaryotes—plants and algae—also fit into this nutritional category.

2. **Chemoautotrophs** need only CO_2 as a carbon source. However, instead of using light for energy, these prokaryotes extract energy from certain inorganic substances, such as hydrogen sulfide (H_2S) or ammonia (NH_3). This mode of nutrition is unique to certain prokaryotes. For

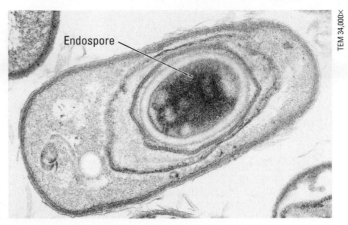

TEM 34,000×

Endospore

Figure 14.12 Endospores. This prokaryote is *Bacillus anthracis,* the bacterium that produces the deadly disease called anthrax in cattle, sheep, and humans. There are actually two cells here, one inside the other. The outer cell produced the specialized inner cell, called an endospore. The endospore has a thick, protective coat. Its cytoplasm is dehydrated, and the cell does not metabolize. Under harsh conditions, the outer cell may disintegrate, but the endospore survives all sorts of trauma, including lack of water and nutrients, extreme heat or cold, and most poisons. When the environment becomes more hospitable, the endospore absorbs water and resumes growth.

Table 14.1	Nutritional Classification of Organisms		
Nutritional Type	**Energy Source**	**Carbon Source**
Photoautotroph (photosynthesizers)	Sunlight	CO_2
Chemoautotroph	Inorganic chemicals	CO_2
Photoheterotroph	Sunlight	Organic compounds
Chemoheterotroph	Organic compounds	Organic compounds

Colorized SEM 4,000×

Figure 14.13 Really bad bacteria. The yellow rods are *Haemophilus influenzae* bacteria on skin cells lining the interior of a human nose. These pathogens are transmitted through the air. *H. influenzae,* not to be confused with influenza (flu) viruses, causes pneumonia and other lung infections that kill about 4 million people worldwide per year. Most victims are children in developing countries, where malnutrition lowers resistance to all pathogens.

example, prokaryotic species living around the hot-water vents deep in the seas are the main food producers in those bizarre ecosystems (see Chapter 17).

3. **Photoheterotrophs** can use light to generate ATP but must obtain their carbon in organic form. This mode of nutrition is restricted to certain prokaryotes.

4. **Chemoheterotrophs** must consume organic molecules for both energy and carbon. This nutritional mode is found widely among prokaryotes, certain protists, and even some plants. And all fungi and animals are chemoheterotrophs.

Table 14.1 reviews the four major modes of nutrition.

The Ecological Impact of Prokaryotes

Organisms as pervasive, abundant, and diverse as prokaryotes are guaranteed to have tremendous impact on Earth and all its inhabitants. Here we survey just a few examples of prokaryotic clout.

Bacteria That Cause Disease We are constantly exposed to bacteria, some of which are potentially harmful (Figure 14.13). Bacteria and other microorganisms that cause disease are called **pathogens.** Most of us are healthy most of the time only because our body defenses check the growth of pathogens. Occasionally, the balance shifts in favor of a pathogen, and we become ill. Even some of the bacteria that are normal residents of the human body can make us sick when our defenses have been weakened by poor nutrition or by a viral infection.

Most pathogenic bacteria cause disease by producing poisons. There are two classes of these poisons: exotoxins and endotoxins. **Exotoxins** are poisonous proteins secreted by bacterial cells. A single gram of the exotoxin that causes botulism could kill a million people. Another exotoxin producer is *Staphylococcus aureus* (abbreviated *S. aureus*). It is a common, usually harmless resident of our skin surface. However, if *S. aureus* enters the body through a cut or other wound or is swallowed in contaminated food, it can cause serious diseases. One type of *S. aureus* produces exotoxins that cause layers of skin to slough off; another can cause vomiting and severe diarrhea; yet another can produce the potentially deadly toxic shock syndrome.

In contrast to exotoxins, **endotoxins** are not cell secretions but are chemical components of the cell walls of certain bacteria. All endotoxins induce the same general symptoms: fever, aches, and sometimes a dangerous drop in blood pressure (shock). The severity of symptoms varies with the host's condition and with the bacterium. Different species of *Salmonella,* for example, produce endotoxins that cause food poisoning and typhoid fever.

During the last 100 years, following the nineteenth-century discovery that "germs" cause disease, the incidence of bacterial infections has declined, particularly in developed nations. Sanitation is generally the most effective way to prevent bacterial disease. The installation of water treatment and sewage systems continues to be a public health priority throughout the world. Antibiotics have been discovered that can cure most bacterial diseases. However, resistance to widely used antibiotics has evolved in many of these pathogens (see the Evolution Link in Chapter 12).

In addition to sanitation and antibiotics, a third defense against bacterial disease is education. A case in point is Lyme disease, currently the most widespread pest-carried disease in the United States. The disease is caused by a spirochete bacterium carried by ticks that live on deer and field mice (Figure 14.14). Lyme disease usually starts as a red rash shaped like a bull's-eye around a tick bite. Antibiotics can cure the disease if administered within about a month after exposure. If untreated, Lyme disease can cause debilitating arthritis, heart disease, and nervous disorders. A vaccine is now available, but it does not give full protection. The best defense is public education about avoiding tick bites and the importance of seeking treatment if a rash develops. When walking through brush, using insect repellent and wearing light-colored clothing can reduce contact with ticks.

Pathogenic bacteria are in the minority among prokaryotes. Far more common are species that are essential to our well-being, either directly or indirectly. Let's turn our attention now to the vital role that prokaryotes play in sustaining the biosphere.

Prokaryotes and Chemical Recycling Not too long ago, in geological terms, the atoms of the organic molecules in your body were parts of the inorganic compounds of soil, air, and water, as they will be again. Life depends on the recycling of chemical elements between the biological and physical components of ecosystems. Prokaryotes play essential roles in these chemical cycles. For example, cyanobacteria not only restore oxygen to the atmosphere; some of them also convert nitrogen gas (N_2) in the atmosphere to nitrogen compounds that plants can absorb from soil and water

> Explore the diversity of prokaryotes in Web/CD Activity 14E.

(see Figure 14.10b). Other prokaryotes, including bacteria living within the roots of bean plants and other legumes, also contribute large amounts of nitrogen compounds to soil. In fact, all the nitrogen that plants use to make proteins and nucleic acids comes from prokaryotic metabolism in the soil. In turn, animals get their nitrogen compounds from plants.

Another vital function of prokaryotes is the breakdown of organic wastes and dead organisms. Prokaryotes decompose organic matter and, in the process, return elements to the environment in inorganic forms that can be used by other organisms. If it were not for such decomposers, carbon, nitrogen, and other elements essential to life would become locked in the organic molecules of corpses and waste products. You'll learn more about the roles that prokaryotes play in chemical cycling in Chapter 20.

Prokaryotes and Bioremediation Humans have put the metabolically diverse prokaryotes to work in cleaning up the environment. The use of organisms to remove pollutants from water, air, and soil is called **bioremediation.** The most familiar example of bioremediation is the use of prokaryotic decomposers to treat our sewage. Raw sewage is first passed through a series of screens and shredders, and solid matter is allowed to settle out from the liquid waste. This solid matter, called sludge, is then gradually added to a culture of anaerobic prokaryotes, including both bacteria and archaea. The microbes decompose the organic matter in the sludge, converting it to material that can be used as landfill or fertilizer after chemical sterilization. Liquid wastes are treated separately from the sludge (Figure 14.15).

We are just beginning to explore the great potential that prokaryotes offer for bioremediation. Certain bacteria that occur naturally on ocean beaches can decompose petroleum and are useful in cleaning up oil spills

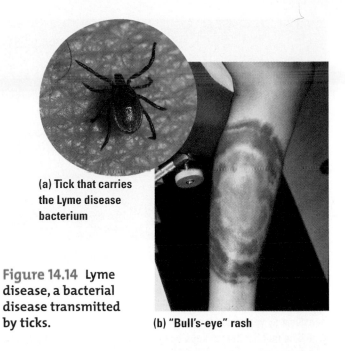

(a) Tick that carries the Lyme disease bacterium

Figure 14.14 Lyme disease, a bacterial disease transmitted by ticks.

(b) "Bull's-eye" rash

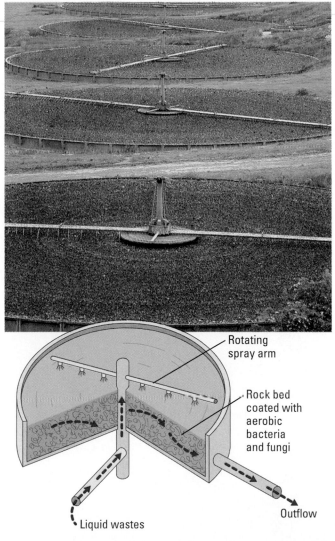

Rotating spray arm

Rock bed coated with aerobic bacteria and fungi

Outflow

Liquid wastes

Figure 14.15 Putting prokaryotes to work in sewage treatment facilities. This is a trickling filter system, one type of mechanism for treating liquid wastes after sludge is removed. The long horizontal pipes rotate slowly, spraying liquid wastes through the air onto a thick bed of rocks. Bacteria and fungi growing on the rocks remove much of the organic material dissolved in the waste. Outflow from the rock bed is sterilized and then released, usually into a river or ocean.

Figure 14.16 Treatment of an oil spill in Alaska. The workers are spraying fertilizers onto an oil-soaked beach. The fertilizers stimulate growth of naturally occurring bacteria that initiate the breakdown of the oil. This technique is the fastest and least expensive way yet devised to clean up spills on beaches. Of course, it would be much better to keep oil off the beaches in the first place!

(Figure 14.16). Genetically engineered bacteria may be able to degrade oil more rapidly than the naturally occurring oil-eaters. Bacteria may also help us clean up old mining sites. The water that drains from mines is highly acidic and is also laced with poisons—often compounds of arsenic, copper, zinc, and the heavy metals lead, mercury, and cadmium. Contamination of our soils and groundwater by these toxic substances poses a widespread threat, and cleaning up the mess is extremely expensive. Although there are no simple solutions to the problem, prokaryotes may be able to help. Bacteria called *Thiobacillus* thrive in the acidic waters that drain from mines. Some mining companies use these microbes to extract copper and other valuable metals from low-grade ores. While obtaining energy by oxidizing sulfur or sulfur-containing compounds, the bacteria also accumulate metals from the mine waters. Unfortunately, their use in cleaning up mine wastes is limited because their metabolism also adds sulfuric acid to the water. If this problem is solved, perhaps through genetic engineering, *Thiobacillus* and other prokaryotes may help us overcome some environmental dilemmas that seem intractable today. One current research focus is a bacterium that tolerates radiation doses thousands of times stronger than those that would kill people. This species may help clean up toxic dump sites that include radioactive wastes.

It is the nutritional diversity of prokaryotes that makes such benefits as chemical recycling and bioremediation possible. The various modes of nutrition and metabolic pathways we find throughout life are all variations on themes that prokaryotes "invented" during their long reign as Earth's exclusive inhabitants. Prokaryotes are at the foundation of life in both the ecological sense and the evolutionary sense. The subsequent breakthroughs in evolution were mostly structural, including the origin of the eukaryotic cell and the diversification of the organisms we call protists.

CheckPoint

1. As different as archaea and bacteria are, both groups are characterized by _____ cells, which lack nuclei and other membrane-bounded organelles.

2. Upon microscopic examination, how do you think you could distinguish the cocci that cause "staph" infections from those that cause "strep" throat?

3. A species of bacterium requires only the amino acid methionine as an organic nutrient and lives in very deep caves where no light penetrates. Based on its mode of nutrition, this species would be classified as a _____.

4. Why are some archaea referred to as "extremophiles"?

5. Why do microbiologists autoclave their laboratory instruments and glassware?

6. Contrast exotoxins with endotoxins.

7. How do bacteria help restore the atmospheric CO_2 required by plants for photosynthesis?

8. What is the pitfall in believing that genetically engineered bacteria will solve all our pollution problems?

Answers: 1. prokaryotic **2.** By the arrangement of the cell aggregates: grapelike clusters for staphylococcus and chains of cells for streptococcus **3.** chemoheterotroph **4.** Because they can thrive in extreme environments too hot, too salty, or too acidic for other organisms **5.** To kill bacterial endospores, which can survive boiling water **6.** Exotoxins are poisons secreted by pathogenic bacteria; endotoxins are components of the cell walls of pathogenic bacteria. **7.** By decomposing the organic molecules of dead organisms and organic refuse such as leaf litter, the metabolism of bacteria releases carbon from the organic matter in the form of CO_2. **8.** They won't, and falsely believing in such an elixir might increase our public tolerance for pollution.

Protists

"No more pleasant sight has met my eye than this of so many thousands of living creatures in one small drop of water," wrote Anton van Leeuwenhoek after his discovery of the microbial world more than three centuries ago. It is a world every biology student should have the opportunity to rediscover by peering through a microscope into a droplet of pond water filled with diverse creatures we call **protists.**

Protists are eukaryotic, and thus even the simplest are much more complex than the prokaryotes. The first eukaryotes to evolve from prokaryotic ancestors were protists. The very word implies great antiquity (from the Greek *protos,* first). The primal eukaryotes were not only the predecessors of the great variety of modern protists, but were also ancestral to all other eukaryotes—plants, fungi, and animals. Two of the most significant chapters in the history of life—the origin of the eukaryotic cell and the subsequent emergence of multicellular eukaryotes—unfolded during the evolution of protists.

The Origin of Eukaryotic Cells

The many differences between prokaryotic and eukaryotic cells far outnumber the differences between plant and animal cells. The fossil record indicates that eukaryotes evolved from prokaryotes more than 1.7 billion years ago. One of biology's most engaging questions is how this happened—in particular, how the membrane-enclosed organelles of eukaryotic cells arose. A widely accepted theory is that eukaryotic cells evolved through a combination of two processes. In one process, the eukaryotic cell's endomembrane system—all the membrane-enclosed organelles except mitochondria and chloroplasts (see Chapter 4)—evolved from inward folds of the plasma membrane of a prokaryotic cell (Figure 14.17a). A second, very different process, called **endosymbiosis,** generated mitochondria and chloroplasts.

Symbiosis is a close association between organisms of two or more species. The word *symbiosis* is from the Greek for "living together," and endosymbiosis refers to one species living within another, called the host. Chloroplasts and mitochondria evolved from small symbiotic prokaryotes that established residence within other, larger host prokaryotes (Figure 14.17b). The ancestors of mitochondria may have been aerobic bacteria that were able to use oxygen to release large amounts of energy from organic molecules by cellular respiration. At some point, such a prokaryote might have been an internal parasite of a larger heterotroph, or an ancestral host cell may have ingested some of these aerobic cells for food. If some of the smaller cells were indigestible, they might have remained alive and continued to perform respiration in the host cell. In a similar way, photosynthetic bacteria ancestral to chloroplasts may have come to live inside a larger host cell. Because almost all eukaryotes have mitochondria but only some have chloroplasts, it is likely that mitochondria evolved first.

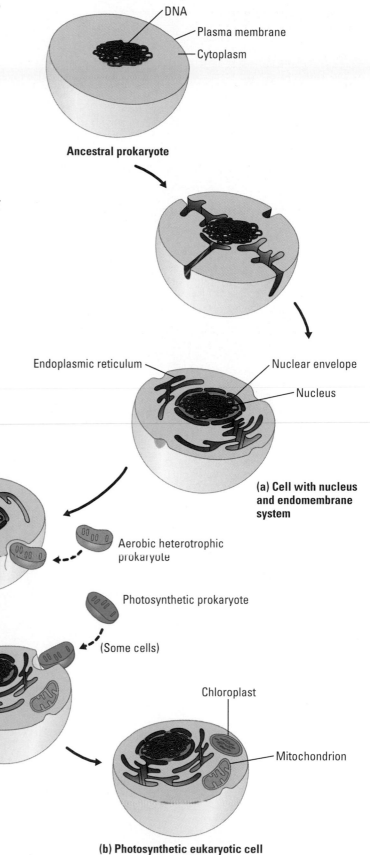

Ancestral prokaryote

- DNA
- Plasma membrane
- Cytoplasm

Endoplasmic reticulum — Nuclear envelope — Nucleus

(a) Cell with nucleus and endomembrane system

Aerobic heterotrophic prokaryote

Photosynthetic prokaryote

(Some cells)

Chloroplast

Mitochondrion

(b) Photosynthetic eukaryotic cell

Figure 14.17 How did eukaryotic cells evolve? (a) Origin of the endomembrane system. **(b)** Origin of mitochondria and chloroplasts.

By whatever means the relationships began, it is not hard to imagine the symbiosis eventually becoming mutually beneficial. In a world that was becoming increasingly aerobic, a cell that was itself an anaerobe would have benefited from aerobic endosymbionts that turned the oxygen to advantage. And a heterotrophic host could derive nourishment from photosynthetic endosymbionts. In the process of becoming more interdependent, the host and endosymbionts would have become a single organism, its parts inseparable.

Developed most extensively by Lynn Margulis, of the University of Massachusetts, the endosymbiosis theory is supported by extensive evidence. Present-day mitochondria and chloroplasts are similar to prokaryotic cells in a number of ways. For example, both types of organelles contain small amounts of DNA, RNA, and ribosomes. This equipment resembles prokaryotic versions more than eukaryotic ones. These components enable chloroplasts and mitochondria to exhibit some autonomy in their activities. The organelles transcribe and translate their DNA into polypeptides, contributing to some of their own enzymes. They also replicate their own DNA and reproduce within the cell by a process resembling the binary fission of prokaryotes.

The origin of the eukaryotic cell made more complex organisms possible, and a vast variety of protists evolved.

The Diversity of Protists

All protists are eukaryotes, but they are so diverse that few other general characteristics can be cited. In fact, protists vary in structure and function more than any other group of organisms. Most protists are unicellular, but there are some colonial and multicellular species. Because most protists are unicellular, they are justifiably considered the simplest eukaryotic organisms. But at the cellular level, many protists are exceedingly complex—the most elaborate of all cells. We should expect this of organisms that must carry out, within the boundaries of a single cell, all the basic functions performed by the collective of specialized cells that make up the bodies of plants and animals. Each unicellular protist is not at all analogous to a single cell from a human, but is itself an organism as complete as any whole animal or plant.

For our survey of these diverse organisms, we'll look at four major categories of protists, grouped more by lifestyle than by their evolutionary relationships: protozoans, slime molds, unicellular algae, and seaweeds.

Protozoans Protists that live primarily by ingesting food, a mode of nutrition that is animal-like, are called **protozoans** (from the Greek *protos,* first, and *zoion,* animal). Protozoans thrive in all types of aquatic environments, including wet soil and the watery environment inside animals. Most species eat bacteria or other protozoans, but some can absorb nutrients dissolved in the water. Protozoans that live as parasites in animals, though in the minority, cause some of the world's most harmful human diseases. We'll examine five groups of protozoans: flagellates, amoebas, forams, apicomplexans, and ciliates.

Flagellates are protozoans that move by means of one or more flagella. Most species are free-living (nonparasitic). However, there are also some nasty parasites that make humans sick. An example is *Giardia,* a flagellate that infects the human intestine and can cause abdominal cramps and severe diarrhea. People become infected mainly by drinking water contami-

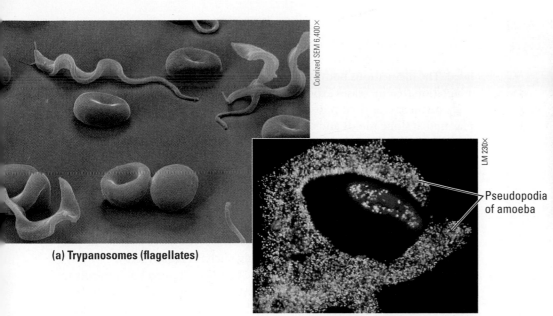

(a) Trypanosomes (flagellates)

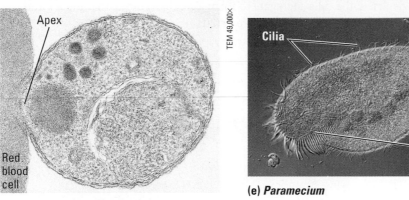

Pseudopodia of amoeba

(b) An amoeba ingesting a smaller protozoan

Apex

Red blood cell

(d) An apicomplexan: *Plasmodium*

Cilia

Oral groove

(e) *Paramecium*

(c) A foram

Figure 14.18 Examples of protozoans.
(a) Trypanosomes are flagellates that live as parasites in the bloodstream of vertebrate animals. The squiggles among these human red blood cells are trypanosomes that cause sleeping sickness, a debilitating disease common in parts of Africa. Trypanosomes escape being killed by their host's defenses by being quick-change artists. They alter the molecular structure of their coats frequently, thus preventing immunity from developing in the host. **(b)** This amoeba is ingesting a smaller protozoan as food. The amoeba's pseudopodia arch around the prey and engulf it into a food vacuole (also see Chapter 4). **(c)** Forams are almost all marine. The foram cell secretes a porous, multichambered shell made of organic material hardened with calcium carbonate, the same mineral that makes up limestone. Thin strands of cytoplasm (pseudopodia) extend through the pores, functioning in swimming, shell formation, and feeding. The shells of fossilized forams are major components of the limestone rocks that are now land formations. **(d)** *Plasmodium,* the apicomplexan that causes malaria, uses its apical complex to enter red blood cells of its human host. The parasite feeds on the host cell from within, eventually destroying it. **(e)** The ciliate *Paramecium* uses its cilia to move through pond water. Cilia also line an indentation called the oral groove, and their beating keeps a current of water containing bacteria and small protists moving toward the cell "mouth" at the base of the groove.

nated with feces from infected animals. *Giardia* can ruin a camping trip. Another group of dangerous flagellates is the trypanosomes, including a species that causes sleeping sickness in humans (Figure 14.18a).

Amoebas are characterized by great flexibility and the absence of permanent locomotor organelles. Most species move and feed by means of **pseudopodia** (singular, *pseudopodium*), which are temporary extensions of the cell (Figure 14.18b). Amoebas can assume virtually any shape as they creep over rocks, sticks, or mud at the bottom of a pond or ocean. Other protozoans with pseudopodia include the **forams** (Figure 14.18c).

Apicomplexans are all parasitic, and some cause serious human diseases. They are named for an apparatus at their apex that is specialized for penetrating host cells and tissues. This group of protozoans includes *Plasmodium,* the parasite that causes malaria (Figure 14.18d). Spread by mosquitoes, malaria is one of the most debilitating and widespread human diseases. Each year in the tropics, more than 200 million people become infected, and at least a million die in Africa alone.

Ciliates are protozoans that use locomotor structures called cilia to move and feed. Nearly all ciliates are free-living (nonparasitic). The best-known example is the freshwater ciliate *Paramecium* (Figure 14.18e).

Figure 14.19 A plasmodial slime mold. Pseudopodia of the huge cell engulf small food particles in mulch or moist soil. The weblike form is an adaptation that enlarges the organism's surface area, increasing its contact with food, water, and oxygen. Within the fine channels of the plasmodium, cytoplasm streams first one way and then the other, in pulses that are beautiful to watch with a microscope. The cytoplasmic streaming helps distribute nutrients and oxygen within the giant cell.

Slime Molds These protists are more attractive than their name. Slime molds resemble fungi in appearance and lifestyle, but the similarities are due to convergent evolution; slime molds and fungi are not at all closely related. The filamentous body of a slime mold, like that of a fungus, is an adaptation that increases exposure to the environment. This suits the role of these organisms as decomposers. The two main groups of these protists are plasmodial slime molds and cellular slime molds.

Plasmodial slime molds are named for the feeding stage in their life cycle, an amoeboid mass called a plasmodium (not to be confused with *Plasmodium*, the parasite that causes malaria). You can find plasmodial slime molds among the leaf litter and other decaying material on a forest floor, and you won't need a microscope to see them. A plasmodium can measure several centimeters across, with its network of fine filaments taking in bacteria and bits of dead organic matter amoeboid style. Large as it is, the plasmodium is actually a single cell with many nuclei (Figure 14.19).

Cellular slime molds pose a semantic question about what it means to be an individual organism. The feeding stage in the life cycle of a cellular slime mold consists of solitary amoeboid cells. They function individually, using their pseudopodia to feed on decaying organic matter. But when food is depleted, the cells aggregate to form a sluglike colony that moves and functions as a single unit (Figure 14.20).

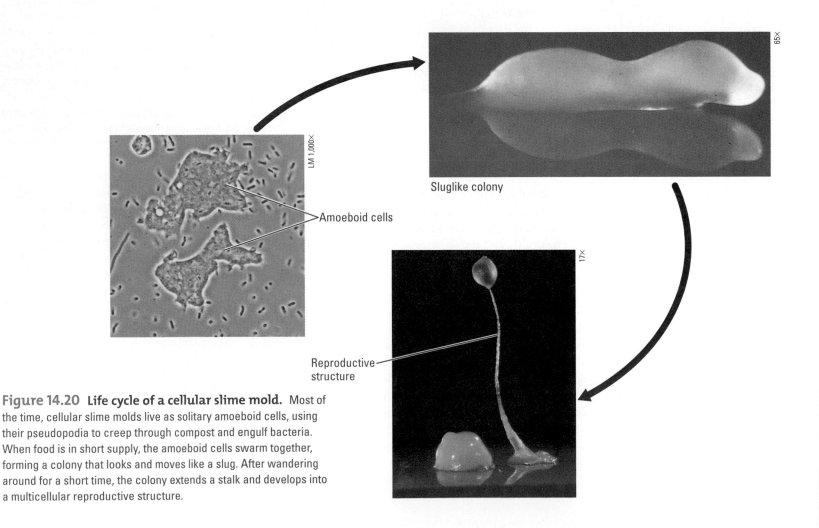

Amoeboid cells

Sluglike colony

Reproductive structure

Figure 14.20 Life cycle of a cellular slime mold. Most of the time, cellular slime molds live as solitary amoeboid cells, using their pseudopodia to creep through compost and engulf bacteria. When food is in short supply, the amoeboid cells swarm together, forming a colony that looks and moves like a slug. After wandering around for a short time, the colony extends a stalk and develops into a multicellular reproductive structure.

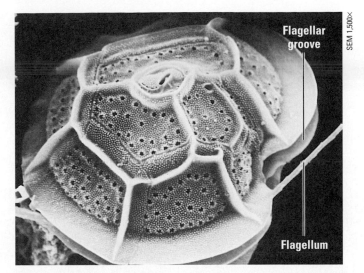

(a) A dinoflagellate

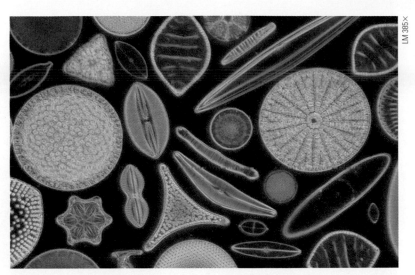

(b) Diatoms

Unicellular Algae Photosynthetic protists are called **algae** (singular, *alga*). Their chloroplasts support food chains in freshwater and marine ecosystems. Many unicellular algae are components of **plankton** (from the Greek *planktos,* wandering), the communities of organisms, mostly microscopic, that drift or swim weakly near the surfaces of ponds, lakes, and oceans. More specifically, planktonic algae are referred to as phytoplankton. We'll look at three groups of unicellular algae: dinoflagellates, diatoms, and green algae (a group that also includes colonial and truly multicellular species).

Dinoflagellates are abundant in the vast aquatic pastures of phytoplankton. Each dinoflagellate species has a characteristic shape reinforced by external plates made of cellulose (Figure 14.21a). The beating of two flagella in perpendicular grooves produces the spinning movement for which these organisms are named (Greek *dinos,* whirling). Dinoflagellate blooms—population explosions—sometimes cause warm coastal waters to turn pinkish orange, a phenomenon known as a red tide. Toxins produced by some red-tide dinoflagellates have caused massive fish kills, especially in the tropics, and are poisonous to humans as well.

Diatoms have glassy cell walls containing silica, the mineral used to make glass (Figure 14.21b). The cell wall consists of two halves that fit together like the bottom and lid of a shoebox. Diatoms store their food reserves in the form of an oil that provides buoyancy, keeping diatoms floating as phytoplankton near the sunlit surface. Massive accumulations of fossilized diatoms make up thick sediments known as diatomaceous earth, which is mined for its use as both a filtering material and an abrasive.

Green algae are named for their grass-green chloroplasts. Unicellular green algae flourish in most freshwater lakes and ponds. Some species are flagellated (Figure 14.21c). The green algal group also includes colonial forms, such as the *Volvox* in Figure 14.21d. Each *Volvox* colony is a ball of flagellated cells (the small green dots in the photo) that are very similar to certain unicellular green algae. The balls within the balls in Figure 14.21d are "daughter" colonies that will be released when the parent colonies rupture. Of all photosynthetic protists, green algae are the most closely related to true plants. (We'll examine the evidence of this evolutionary relationship in the next chapter.)

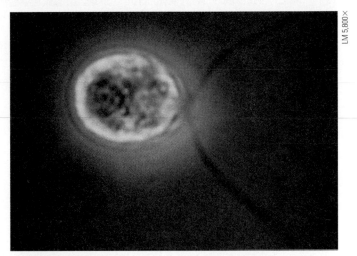

(c) *Chlamydomonas*

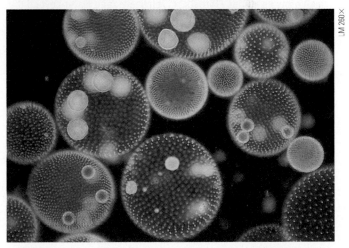

(d) *Volvox*

Figure 14.21 Unicellular and colonial algae. (a) A dinoflagellate, with its wall of protective plates. **(b)** A sample of diverse diatoms, which have glassy walls. **(c)** *Chlamydomonas,* a unicellular green alga with a pair of flagella. **(d)** *Volvox,* a colonial green alga.

(b) Red algae

(a) Green algae

(c) Brown algae

Figure 14.22 The three major groups of seaweeds.
(a) Green algae. This sea lettuce is an edible species that inhabits the intertidal zone. In addition to seaweeds, the green algal group includes unicellular and colonial species, such as those in Figures 14.21c and d. **(b)** Red algae. These seaweeds are most abundant in the warm coastal waters of the tropics. Of all the seaweeds, red algae can generally live in the deepest water. Their chloroplasts have special pigments that absorb the blue and green light that penetrates best through water. The species in this photo is an example of corraline algae, which contribute to the architecture of some coral reefs. The cell walls are hardened by a mineral. **(c)** Brown algae. This group includes the largest seaweeds, known as kelp, which grow as marine "forests" in relatively deep water beyond the intertidal zone. Some species grow to a length of over 60 meters in a single season, the fastest linear growth of any organism. Kelp is a renewable resource reaped by special boats that cut and collect the tops of the algae. More importantly, kelp forests provide habitat for many animals, including a great diversity of fishes. If you have walked on a beach covered with kelp that has washed ashore after a storm, you may have noticed the organs called floats, which keep the photosynthetic blades of the kelp in the light near the water's surface. Maybe you even picked up and popped some of those floats, the way you do those irresistible packing-material bubbles.

Seaweeds Defined as large, multicellular marine algae, **seaweeds** grow on rocky shores and just offshore beyond the zone of the pounding surf. Their cell walls have slimy and rubbery substances that cushion their bodies against the agitation of the waves. Some seaweeds are as large and complex as many plants. Even the word *seaweed* implies plant-like appearance, but the similarities between these algae and true plants are a consequence of convergent evolution. In fact, the closest relatives of seaweeds are certain unicellular algae, which is why many biologists include seaweeds with the protists. Seaweeds are classified into three different groups, based partly on the types of pigments present in their chloroplasts: green algae, red algae, and brown algae (Figure 14.22).

Coastal people, particularly in Asia, harvest seaweeds for food. For example, in Japan and Korea, some seaweed species are ingredients in soups. Other seaweeds are used to wrap sushi. Marine algae are rich in iodine and other essential minerals. However, much of their organic material consists of unusual polysaccharides that humans cannot digest, which prevents seaweeds from becoming staple foods. They are ingested mostly for their rich tastes and unusual textures. The gel-forming substances in the cell walls of seaweeds are widely used as thickeners for such processed foods as puddings, ice cream, and salad dressing. And the seaweed extract called agar provides the gel-forming base for the media microbiologists use to culture bacteria in Petri dishes.

Identify protists collected from different habitats in Web/CD Activity 14F.

Evolution Link:
The Origin of Multicellular Life

An orchestra can play a greater variety of musical compositions than a violin soloist can. Put simply, increased complexity makes more variations possible. Thus, the origin of the eukaryotic cell led to an evolutionary radiation of new forms of life. Unicellular protists, which are organized on the complex eukaryotic plan, are much more diverse in form than the simpler prokaryotes. The evolution of multicellular bodies broke through another threshold in structural organization.

Multicellular organisms are fundamentally different from unicellular ones. In a unicellular organism, all of life's activities occur within a single cell. In contrast, a multicellular organism has various specialized cells that perform different functions and are dependent on each other. For example, some cells procure food, while others transport materials or provide movement.

The evolutionary links between unicellular and multicellular life were probably colonial forms, in which unicellular protists stuck together as loose federations of independent cells (Figure 14.23). The gradual transition from colonies to truly multicellular organisms involved the cells becoming increasingly interdependent as a division of labor evolved. We can see one level of specialization and cooperation in the colonial green alga *Volvox* (see Figure 14.21d). *Volvox* produces gametes (sperm and ova), which depend on nonreproductive cells, or somatic cells, while developing. Cells in truly multicellular organisms are specialized for many more nonreproductive functions, including feeding, waste disposal, gas exchange, and protection, to name a few.

Multicellularity evolved many times among the ancestral stock of protists, leading to new waves of biological diversification. The diverse seaweeds are examples of the descendants, and so are plants, fungi, and animals. In the next chapter, we'll trace the long evolutionary movement of plants and fungi onto land. Then we'll follow the threads of animal evolution in Chapter 16.

Learn more about the origin of multicellular life in the Web Evolution Link.

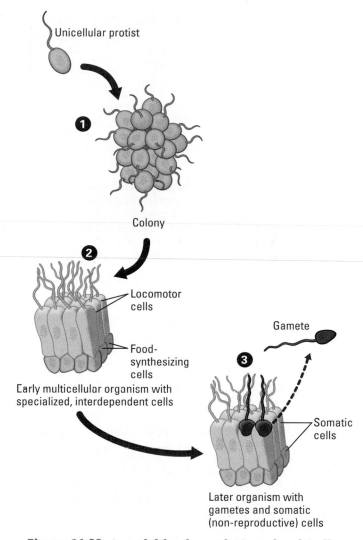

Figure 14.23 A model for the evolution of multicellular organisms from unicellular protists. ❶ An ancestral colony may have formed, as colonial protists do today, when a cell divided and its offspring remained attached to one another. ❷ The cells in the colony may have become somewhat specialized and interdependent, with different cell types becoming more and more efficient at performing specific, limited tasks. Cells that retained a flagellum may have become specialized for locomotion, while others that lost their flagellum could have assumed functions such as ingesting or synthesizing food. ❸ Additional specialization among the cells in the colony may have led to distinctions between sex cells (gametes) and non-reproductive cells (somatic cells).

Chapter Review

Summary of Key Concepts

Overview: Major Episodes in the History of Life

- Prokaryotes appeared about 3.5 billion years ago. About 1.7 billion years ago, single-celled eukaryotic organisms evolved from prokaryotes. About a billion years ago, multicellular eukaryotic organisms evolved. All the major phyla of animals evolved by the end of the Cambrian explosion, about 570 million years ago. Plants and fungi colonized the land about 475 million years ago. As amphibians evolved from fish, vertebrate life moved onto land.
- Web/CD Activity 14A *The History of Life*

The Origin of Life

- **Resolving the Biogenesis Paradox** All life today arises only by the reproduction of preexisting life. However, most biologists now think it is possible that chemical and physical processes in Earth's primordial environments produced very simple cells through a sequence of stages.

- **A Four-Stage Hypothesis for the Origin of Life** One scenario suggests that the first organisms were products of chemical evolution in four stages: the abiotic synthesis of small organic molecules, the joining of monomers into polymers, the origin of self-replicating molecules, and the packaging of these molecules into membranous pre-cells.
- Web/CD Activity 14B *Origin of Life Lab*

- **From Chemical Evolution to Darwinian Evolution** Over millions of years, natural selection favored the most efficient pre-cells, which evolved into the first prokaryotic cells.

Prokaryotes

- **They're Everywhere!** Prokaryotes are found wherever there is life. They outnumber all eukaryotes combined. Prokaryotes thrive in habitats where eukaryotes cannot live. A few prokaryotic species cause serious diseases, but most are either benign or beneficial.

- **The Two Main Branches of Prokaryotic Evolution: Bacteria and Archaea** Prokaryotes are classified into two branches, bacteria and archaea. Although they have prokaryotic cell organization in common, bacteria and archaea differ in many structural, biochemical, and physiological characteristics. The Archaea includes "extremophiles," which inhabit extreme environments.
- Web/CD The Process of Science *Investigating the Archaea: Introduction to the Archaea, Life's Third Domain*

- **The Structure, Function, and Reproduction of Prokaryotes** Prokaryotic cells lack nuclei and other membrane-enclosed organelles. Most have cell walls. Prokaryotes come in several shapes, most commonly spheres (cocci), rods (bacilli), and curves or spirals (the largest called spirochetes). About half of all prokaryotic species are motile, most of these using flagella to move. Some prokaryotes can survive extended periods of harsh conditions by forming endospores. Most prokaryotes can reproduce by binary fission at phenomenal rates if conditions are favorable, but growth is usually restricted by limited resources.
- Web/CD Activity 14C *Prokaryotic Cell Structure and Function*

- **The Nutritional Diversity of Prokaryotes** Prokaryotes exhibit four major modes of nutrition. Photoautotrophs obtain energy from sunlight and carbon from CO_2; chemoautotrophs extract energy from inorganic substances and get their carbon from CO_2; photoheterotrophs use light to generate ATP but must obtain their carbon in organic form; and chemoheterotrophs consume organic molecules for both energy and carbon.
- Web/CD Activity 14D *Modes of Nutrition in Prokaryotes*

- **The Ecological Impact of Prokaryotes** Most pathogenic bacteria cause disease by producing exotoxins or endotoxins. Sanitation, antibiotics, and education are the best defenses against bacterial disease. Prokaryotes help recycle chemical elements between the biological and physical components of ecosystems. Humans can use bacteria to remove pollutants from water, air, and soil in the process called bioremediation.
- Web/CD Activity 14E *Diversity of Prokaryotes*

Protists

- **The Origin of Eukaryotic Cells** The nucleus and endomembrane system of eukaryotes probably evolved from infoldings of the plasma membrane of ancestral prokaryotes. Mitochondria and chloroplasts probably evolved from symbiotic prokaryotes that took up residence inside larger prokaryotic cells.

- **The Diversity of Protists** Protists are unicellular eukaryotes and their closest multicellular relatives. **Protozoans** primarily live by ingesting food. They include flagellates, amoebas, apicomplexans, and ciliates. Most protozoans thrive in all types of aquatic environments, but a few live in animals and cause some of the world's most harmful diseases. **Slime molds** resemble fungi in appearance and lifestyle as decomposers, but are not at all closely related. Plasmodial slime molds are multinuclear organisms named for the plasmodium feeding stage in their life cycle. Cellular slime molds have unicellular and multicellular life stages. **Unicellular algae** are photosynthetic protists that support food chains in freshwater and marine ecosystems. They include the dinoflagellates, diatoms, and unicellular green algae. **Seaweeds** are large, multicellular marine algae that grow on and near rocky shores. Green, red, and brown algae are classified partly by the types of pigments present in their chloroplasts.
- Web/CD Activity 14F *Diversity of Protists*

Evolution Link: The Origin of Multicellular Life

- Unlike unicellular organisms, multicellular organisms have specialized cells that perform different functions and are dependent on each other. Multicellularity evolved many times among the ancestral protists. The evolutionary links between unicellular and multicellular life were probably colonial forms.
- Web Evolution Link *The Origin of Multicellular Life*

Self-Quiz

1. In terms of nutrition, autotrophs are to heterotrophs as
   a. algae are to slime molds.
   b. archaea are to bacteria.
   c. slime molds are to fungi.
   d. kelp are to diatoms.
   e. pathogenic bacteria are to harmless bacteria.

2. The bacteria that cause tetanus can be killed only by prolonged heating at temperatures considerably above boiling. This suggests that tetanus bacteria
   a. have cell walls.
   b. protect themselves by secreting antibiotics.
   c. secrete endotoxins.
   d. are autotrophic.
   e. produce endospores.

3. Which of the following steps has not yet been accomplished by scientists studying the origin of life?
   a. abiotic synthesis of small RNA polymers
   b. abiotic synthesis of polypetides
   c. formation of molecular aggregates with selectively permeable membranes
   d. formation of pre-cells that use DNA to direct the polymerization of amino acids
   e. abiotic synthesis of organic monomers

4. Part of the explanation for the absence of spontaneous generation of life on Earth today is that
   a. there is not sufficient lightning to provide an energy source.
   b. the abundance of O_2 in our modern atmosphere would destroy complex organic molecules that are not inside organisms.
   c. much less visible light is reaching Earth to serve as an energy source.
   d. there are no molten surfaces on which weak solutions of organic molecules would polymerize.
   e. all habitable places are already filled.

5. Penicillins function as antibiotics mainly by inhibiting the ability of some bacteria to
   a. form spores.
   b. replicate DNA.
   c. synthesize normal cell walls.
   d. produce functional ribosomes.
   e. synthesize ATP.

6. The two main evolutionary branches of prokaryotic life are
   a. bacteria and cyanobacteria.
   b. protozoans and algae.
   c. archaea and bacteria.
   d. autotrophs and heterotrophs.
   e. cocci and bacilli.

7. Which algal group is most closely related to plants?
   a. diatoms
   b. green algae
   c. dinoflagellates
   d. brown algae
   e. red algae

8. Of the following, which describes protists most inclusively?
   a. multicellular eukaryotes
   b. protozoans
   c. eukaryotes that are not plants, fungi, or animals
   d. organisms with the simplest eukaryotic cells
   e. single-celled organisms closely related to bacteria

9. Which of the following is an autotrophic protist?
   a. amoeba
   b. slime mold
   c. plant
   d. diatom
   e. cyanobacterium

10. The probable ancestor of mitochondria was
   a. a unicellular alga.
   b. an aerobic bacterium.
   c. a thermophilic archaean.
   d. a parasitic protist.
   e. a photosynthetic prokaryote.

• Go to the website or CD-ROM for more self-quiz questions.

The Process of Science

1. Imagine you are on a team designing a moon base that will be self-contained and self-sustaining. Once supplied with building materials, equipment, and organisms from Earth, the base will be expected to function indefinitely. One of the members of your team has suggested that everything sent to the base be chemically treated or irradiated so that no bacteria of any kind are present. Do you think this is a good idea? Predict some of the consequences of eliminating all bacteria from an environment.

2. Learn more about the archaea in The Process of Science activity on the website and CD-ROM.

Biology and Society

Many local newspapers publish a weekly list of restaurants that have been cited by inspectors for poor sanitation. Locate such a report and highlight the cases that are likely associated with potential food contamination by pathogenic prokaryotes.

CHAPTER 15

Plants, Fungi, and the Move onto Land

Some giant sequoia trees weigh more than a

dozen space shuttles.

Flowering plants such as corn, rice, and wheat pro-

vide nearly all our food. A mushroom is probably

more closely related to humans than it is to any

plant. Every two seconds humans destroy an area

of tropical rain forest equal to the area of 3 football

fields.

When you view a lush landscape, it is difficult to picture the land barren, totally uninhabited by macroscopic life. But that is how we must imagine Earth for almost the first 90% of the time life has existed. Life was cradled in the seas and ponds, and there it evolved in confinement for 3 billion years. There is fossil evidence that photosynthetic bacteria may have coated moist soil with a slime of green and other hues about 1.2 billion years ago. However, the long evolutionary pilgrimage of more complex organisms onto land did not begin until about 475 million years ago. That is when plants began to colonize the land, and the terrestrial communities they founded transformed the biosphere. Consider, for example, that there would be no humans had it not been for the chain of evolutionary events that began when certain descendants of green algae adapted to living on land.

It was not an easy evolutionary venture, this move onto land, but plants had help in the form of fungi (Figure 15.1). From the beginning, plants and fungi were partners. Plants made food, and certain symbiotic fungi helped the roots of plants absorb water and minerals from the soil. Other fungi functioned as decomposers in the chemical recycling that sustains ecosystems. Still others adapted as parasites of plants and the animals that followed them onto land.

This chapter begins by tracing how plants and fungi colonized land. Once that historical context is set, we will explore the diversity of modern plants and fungi.

Figure 15.1 A terrestrial community with a diversity of plants and fungi.

Overview: Colonizing Land

Plants are terrestrial (land-dwelling) organisms. True, some, such as water lilies, have returned to the water, but they evolved secondarily from terrestrial ancestors, making them the porpoises of the plant kingdom.

You know plants when you see them: Grasses and trees are examples. But what exactly is a plant? A plant is a multicellular eukaryote that makes organic molecules by photosynthesis. Photosynthesis distinguishes plants from the animal and fungal kingdoms. But what about large algae, including seaweeds, which we classified as protists in the preceding chapter? They, too, are multicellular, eukaryotic, and photosynthetic. It is a set of terrestrial adaptations that distinguishes plants from algae.

Terrestrial Adaptations of Plants

Structural Adaptations Living on land poses very different problems from living in water (Figure 15.2). In terrestrial habitats, the resources that a photosynthetic organism needs are found in two very different places: Light and carbon dioxide are mainly available aboveground, while water and mineral nutrients are found mainly in the soil. Thus, the complex bodies of plants show varying degrees of structural specialization into subterranean and aerial organs—**roots** and leaf-bearing **shoots,** respectively.

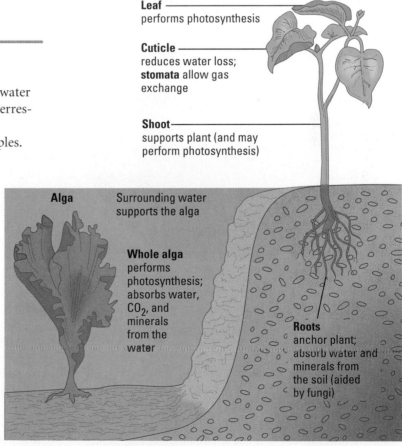

Plant

Leaf — performs photosynthesis

Cuticle — reduces water loss; **stomata** allow gas exchange

Shoot — supports plant (and may perform photosynthesis)

Alga

Surrounding water supports the alga

Whole alga performs photosynthesis; absorbs water, CO_2, and minerals from the water

Roots anchor plant; absorb water and minerals from the soil (aided by fungi)

Figure 15.2 Contrasting environments for algae and plants.

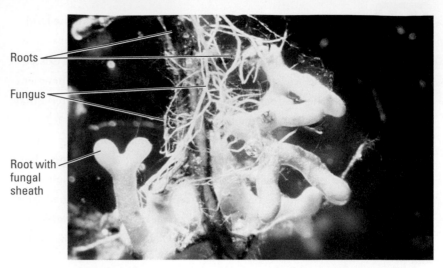

Roots

Fungus

Root with
fungal
sheath

Figure 15.3 Mycorrhizae: symbiotic associations of fungi and roots.
The finely branched filaments of the fungus provide an extensive surface area for absorption of water and minerals from the soil. The fungus provides some of those materials to the plant and benefits in turn by receiving sugars and other organic products of the plant's photosynthesis.

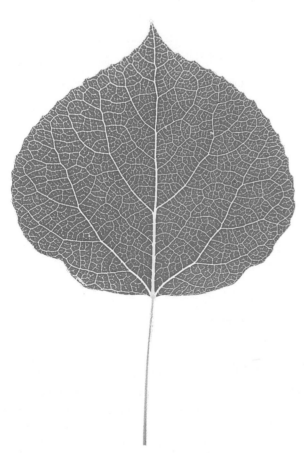

Figure 15.4 Network of veins in a leaf. The vascular tissue of the veins delivers water and minerals absorbed by the roots and carries away the sugars produced in the leaves.

Most plants have symbiotic fungi associated with their roots. These root-fungus combinations are called **mycorrhizae,** meaning "fungus root." For their part, the fungi absorb water and essential minerals from the soil and provide these materials to the plant. The sugars produced by the plant nourish the fungi. Mycorrhizae are evident on some of the oldest plant fossils. They are key adaptations that made it possible to live on land (Figure 15.3).

Leaves are the main photosynthetic organs of most plants. Exchange of carbon dioxide and oxygen between the atmosphere and the photosynthetic interior of a leaf occurs via **stomata,** the microscopic pores through the leaf's surface (see Figure 6.3). A waxy layer called the **cuticle** coats the leaves and other aerial parts of most plants, helping the plant body retain its water. (Think of the waxy surface of a cucumber or unpolished apple.)

Differentiation of the plant body into root and shoot systems solved one problem but created new ones. For the shoot system to stand up straight in the air, it must have support. This is not a problem in the water: Huge seaweeds need no skeletons because the surrounding water buoys them. An important terrestrial adaptation of plants is **lignin,** a chemical that hardens the cell walls. Imagine what would happen to you if your skeleton were to disappear or suddenly turn mushy. A tree would also collapse if it were not for its "skeleton," its framework of lignin-rich cell walls.

Specialization of the plant body into roots and shoots also introduced the problem of transporting vital materials between the distant organs. The terrestrial equipment of most plants includes **vascular tissue,** a system of tube-shaped cells that branch throughout the plant (Figure 15.4). The vascular tissue actually has two types of tissues specialized for transport: **xylem,** consisting of dead cells with tubular cavities for transporting water and minerals from roots to leaves; and **phloem,** consisting of living cells that distribute sugars from the leaves to the roots and other nonphotosynthetic parts of the plant.

Reproductive Adaptations Adapting to land also required a new mode of reproduction. For algae, the surrounding water ensures that gametes (sperm and eggs) and developing offspring stay moist. The aquatic environment also provides a means of dispersing the gametes and offspring. Plants, however, must keep their gametes and developing offspring from drying out in the air. Plants (and some algae) produce their gametes in protective structures called **gametangia** (singular, *gametangium*). A gametangium has a jacket of protective cells surrounding a moist chamber where gametes can develop without dehydrating.

In most plants, sperm reach the eggs by traveling within pollen, which is carried by wind or animals. The egg remains within tissues of the mother plant and is fertilized there. In plants, but not algae, the zygote (fertilized egg) develops into an embryo while still contained within the female parent, which protects the embryo and keeps it from dehydrating (Figure 15.5). Most plants rely on wind or animals, such as fruit-eating birds or mammals, to disperse their offspring, which are in the form of embryos contained in seeds.

> Quiz yourself and see animations of the terrestrial adaptations of plants in Web/CD Activity 15A.

The reproductive "strategy" of plants is analogous to how mammals manage to reproduce on land. As in plants, mammalian fertilization is internal (within the mother's body). And in most mammals, embryonic development also occurs within the mother's body, as it does in plants.

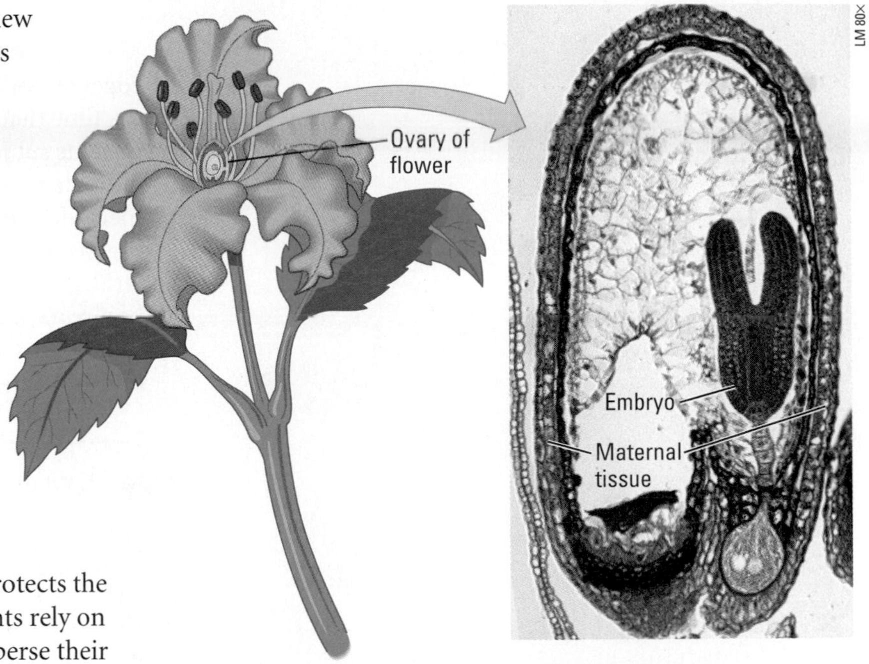

Figure 15.5 The protected embryo of a plant. Internal fertilization, with sperm and egg combining within a moist chamber on the mother plant, is an adaptation for living on land. The female parent continues to nurture and protect the plant embryo, which develops from the zygote.

The Origin of Plants from Green Algae

The move onto land and the spread of plants to diverse terrestrial environments was incremental. It paralleled the gradual accumulation of terrestrial adaptations, beginning with populations that descended from algae. Green algae are the protists most closely related to plants. More specifically, molecular comparisons and other evidence place a group of multicellular green algae called **charophytes** closest to plants (Figure 15.6).

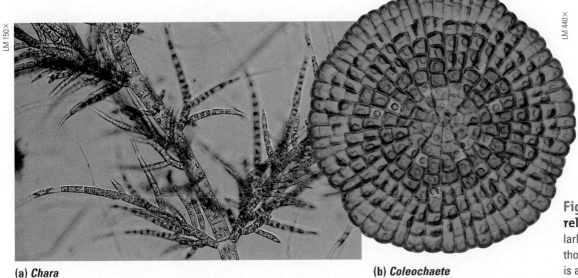

(a) Chara

(b) Coleochaete

Figure 15.6 Charophytes, closest algal relatives of plants. (a) *Chara* is a particularly elaborate green alga. **(b)** *Coleochaete*, though less plantlike than *Chara* in appearance, is actually more closely related to plants.

The evolutionary "walk" onto land was more like adaptive baby steps. Many species of modern charophytes are found in shallow water around the edges of ponds and lakes. Some of the ancient charophytes that lived about the time that land was first colonized may have inhabited shallow-water habitats subject to occasional drying. Natural selection would have favored individual algae that could survive through periods when they were not submerged. The protection of developing gametes and embryos within jacketed organs (gametangia) on the parent is one adaptation to living in shallow water that would also prove essential on land. We know that by about 475 million years ago, the vintage of the oldest plant fossils, an accumulation of adaptations allowed permanent residency above water. The plants that color our world today diversified from those early descendants of green algae.

Plant Diversity

As we survey the diversity of modern plants, remember that the past is the key to the present. The history of the plant kingdom is a story of adaptation to diverse terrestrial habitats.

Highlights of Plant Evolution

The fossil record chronicles four major periods of plant evolution, which are also evident in the diversity of modern plants (Figure 15.7). Each stage is marked by the evolution of structures that opened new opportunities on land.

The first period of evolution was the origin of plants from their aquatic ancestors, the green algae called charophytes. The first terrestrial adaptations included gametangia, which protected gametes and embryos. This made it possible for the plants known as **bryophytes,** including the mosses, to diversify from early plants. Vascular tissue also evolved relatively early in plant history. However, most bryophytes lack vascular tissue, which is why they are categorized as nonvascular plants.

The second period of plant evolution was the diversification of vascular plants (plants with vascular tissue that conducts water and nutrients). The earliest vascular plants lacked seeds. Today, this seedless condition is retained by **ferns** and a few other groups of vascular plants.

The third major period of plant evolution began with the origin of the seed. Seeds advanced the colonization of land by further protecting plant embryos from desiccation (drying) and other hazards. A seed consists of an embryo packaged along with a store of food within a protective covering. The seeds of early seed plants were not enclosed in any specialized chambers. These plants gave rise to many types of **gymnosperms** (from the Greek *gymnos*, naked, and *sperma*, seed). Today, the most widespread and diverse gymnosperms are the conifers, which are the pines and other plants with cones.

The fourth major episode in the evolutionary history of plants was the emergence of flowering plants, or **angiosperms** (from the Greek, *angion*, container, and *sperma*, seed). The flower is a complex reproductive structure that bears seeds within protective chambers (containers) called ovaries. This contrasts with the bearing of naked seeds by gymnosperms. The great majority of modern-day plants are angiosperms.

Review plant evolution in Web/CD Activity 15B.

With these highlights as our framework, we are now ready to survey the four major groups of modern plants: bryophytes, ferns, gymnosperms, and angiosperms.

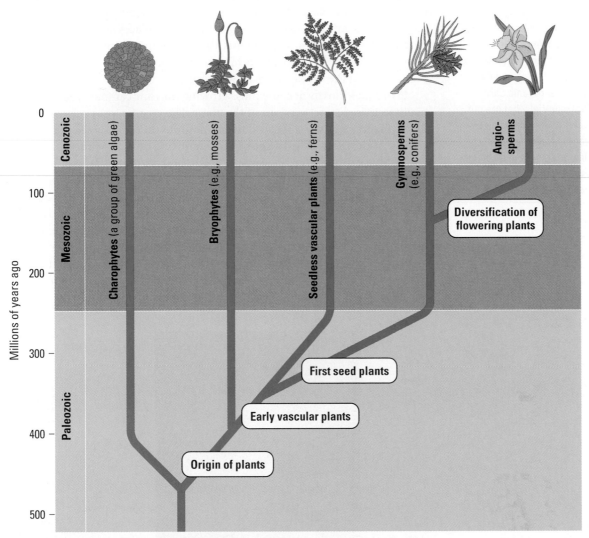

Figure 15.7 Highlights of plant evolution. Modern representatives of the major evolutionary branches are illustrated at the top of this phylogenetic tree. As we survey the diversity of plants, miniature versions of this tree will help you place each plant group in its evolutionary context.

Figure 15.8 A peat moss bog in Norway. Although mosses are short in stature, their collective impact on Earth is huge. For example, peat mosses, or *Sphagnum,* carpet at least 3% of Earth's terrestrial surface, with greatest density in high northern latitudes. The accumulation of "peat," the thick mat of living and dead plants in wetlands, ties up an enormous amount of organic carbon because peat has an abundance of chemical materials that are not easily degraded by microbes. That explains why peat makes an excellent fuel as an alternative to coal and wood. More importantly, the carbon storage by peat bogs plays an important role in stabilizing Earth's atmospheric carbon dioxide concentrations, and hence climate, through the CO_2-related greenhouse effect (see Chapter 6).

Bryophytes

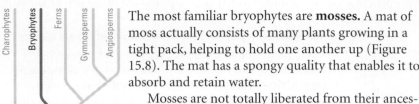

The most familiar bryophytes are **mosses.** A mat of moss actually consists of many plants growing in a tight pack, helping to hold one another up (Figure 15.8). The mat has a spongy quality that enables it to absorb and retain water.

Mosses are not totally liberated from their ancestral aquatic habitat. They do display two of the key terrestrial adaptations that made the move onto land possible: a waxy cuticle that helps to prevent dehydration; and the retention of developing embryos within the mother plant's gametangium. However, mosses need water to reproduce. Their sperm are flagellated, like those of most green algae. These sperm must swim through water to reach eggs. (A film of rainwater or dew is often enough moisture for the sperm to travel.) In addition, most mosses have no vascular tissue to carry water from soil to aerial parts of the plant. This explains why damp, shady places are the most common habitats of mosses. These plants also lack lignin, the wall-hardening material that enables other plants to stand tall. Mosses may sprawl as mats over acres, but they always have a low profile.

If you look closely at some moss growing on your campus, you may actually see two distinct versions of the plant. The greener, spongelike plant that is the more obvious is called the **gametophyte.** You may see the other version of the moss, called a **sporophyte,** growing out of a gametophyte as a stalk with a capsule at its tip (Figure 15.9). The cells of the gametophyte are haploid (one set of chromosomes; see Chapter 7). In contrast, the sporophyte is made up of diploid cells (two chromosome sets). These two different stages of the plant life cycle are named for the types of reproductive cells they produce. Gametophytes produce gametes (sperm and eggs), while sporophytes produce spores. As reproductive cells, **spores** differ from gametes in two ways: A spore can develop into a new organism without fusing with another cell (two gametes must fuse to form a zygote); and spores usually have tough coats that enable them to resist harsh environments.

Figure 15.9 The two forms of a moss. The feathery plant we generally know as a moss is the gametophyte. The stalk with the capsule at its tip is the sporophyte. The photographer caught the capsule releasing its spores, reproductive cells that can develop into new gametophytes.

The gametophyte and sporophyte are alternating generations that take turns producing each other. Gametophytes reproduce sexually, their gametes uniting to form zygotes, which develop into new sporophytes. And the

Go to Web/CD Activity 15C to help you understand the moss life cycle.

sporophytes reproduce asexually, their spores giving rise to new gametophytes. This type of life cycle, called **alternation of generations,** occurs only in plants and certain algae (Figure 15.10). Among plants, mosses and other bryophytes are unique in having the gametophyte as the dominant generation—the larger, more obvious plant. As we continue our survey of plants, we'll see an increasing dominance of the sporophyte as the more highly developed generation.

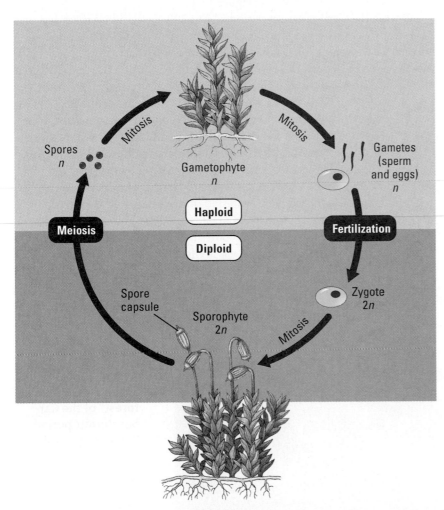

Figure 15.10 Alternation of generations. Plants have life cycles very different from ours. Each of us is a diploid individual; the only haploid stages in the human life cycle, as for nearly all animals, are sperm and eggs. By contrast, plants have alternating generations: Diploid ($2n$) individuals (sporophytes) and haploid (n) individuals (gametophytes) generate each other in the life cycle. In the case of mosses, the gametophyte is the dominant stage. In fact, the moss sporophyte remains attached to the gametophyte, depending on its parent for water and nutrients. In other plant groups, this balance is reversed, with the sporophyte being the more developed of the two generations.

Ferns

Ferns took terrestrial adaptation to the next level with the evolution of vascular tissue. However, the sperm of ferns, like those of mosses, are flagellated and must swim through a film of water to fertilize eggs. Ferns are also seedless, which helps explain why they do not dominate most modern terrestrial landscapes. However, of all seedless vascular plants, ferns are by far the most diverse today, with more than 12,000 species. Most of those species inhabit the tropics, although many species are found in temperate forests, such as most woodlands of the United States (Figure 15.11).

During the Carboniferous period, about 290–360 million years ago, ancient ferns were among a much greater diversity of seedless plants that formed vast, swampy forests that covered much of what is now Eurasia and North America (Figure 15.12). At that time, these continents were close to the equator and had tropical climates. The tropical swamp forests of the Carboniferous period generated great quantities of organic matter. As the plants died, they fell into stagnant wetlands and did not decay completely. Their remains formed thick deposits of organic rubble, or peat. Later, seawater flooded the swamps, marine sediments covered the peat, and pressure and heat gradually converted the peat to coal. Coal is black sedimentary rock made up of fossilized plant material. It formed during several geological periods, but the most extensive coal beds are derived from Carboniferous deposits. (The name Carboniferous comes from the Latin *carbo,* coal, and *fer-,* bearing.) Coal, oil, and natural gas are **fossil fuels**—fuels formed from the remains of extinct organisms. Fossil fuels are burned to generate much of our electricity. As we deplete our oil and gas reserves, the use of coal is likely to increase.

Learn more about the fern life cycle in Web/CD Activity 15D.

Figure 15.11 Ferns (seedless vascular plants). This species grows on the forest floor in the eastern United States. The "fiddleheads" in the inset on the left are young fronds (leaves) ready to unfurl. The fern generation familiar to us is the sporophyte generation. The inset on the right is the underside of a sporophyte leaf specialized for reproduction. The yellow dots consist of spore capsules that can release numerous tiny spores. The spores develop into gametophytes. However, you would have to crawl on the forest floor and explore with careful hands and sharp eyes to find fern gametophytes, tiny plants growing on or just below the soil surface.

Figure 15.12 A "coal forest" of the Carboniferous period. This painting, based on fossil evidence, reconstructs one of the great seedless forests. Most of the large trees with straight trunks are seedless plants called lycophytes. On the left, the tree with numerous feathery branches is another type of seedless plant called a horsetail. The plants near the base of the trees are ferns. Note the giant bird-sized dragonfly, which would have made quite a buzz.

Gymnosperms

"Coal forests" dominated the North American and Eurasian landscapes until near the end of the Carboniferous period. At that time, global climate turned drier and colder, and the vast swamps began to disappear. The climatic change provided an opportunity for seed plants, which can complete their life cycles on dry land and withstand long, harsh winters. Of the earliest seed plants, the most successful were the gymnosperms, and several kinds grew along with the seedless plants in the Carboniferous swamps. Their descendants include the conifers, or cone-bearing plants.

Conifers Perhaps you have had the fun of hiking or skiing through a forest of conifers, the most common gymnosperms. Pines, firs, spruces, junipers, cedars, and redwoods are all conifers.

Conifers are among the tallest, largest, and oldest organisms on Earth. Redwoods, found only in a narrow coastal strip of northern California, grow to heights of more than 110 meters; only certain eucalyptus trees in Australia are taller. The largest (most massive) organisms alive are the giant sequoias, relatives of redwoods that grow in the Sierra Nevada mountains of California. One, known as the General Sherman tree, has a trunk with a circumference of 26 meters and weighs more than the combined weight of a dozen space shuttles (Figure 15.13). Bristlecone pines, another species of California conifer, are among the oldest organisms alive. One bristlecone, named Methuselah, is more than 4600 years old; it was a young tree when humans invented writing.

Nearly all conifers are evergreens, meaning they retain leaves throughout the year. Even during winter, a limited amount of photosynthesis occurs on sunny days. And when spring comes, conifers already have fully developed leaves that can take advantage of the sunnier days. The needle-shaped leaves of pines and firs are also adapted to survive dry seasons. A thick cuticle covers the leaf, and the stomata are located in pits, further reducing water loss.

We get most of our lumber and paper pulp from the wood of conifers. What we call wood is actually an accumulation of vascular tissue with lignin, which gives the tree structural support.

Figure 15.13 Conifers, the most common gymnosperms. The large tree is the General Sherman tree, found in Sequoia National Park, California.

Terrestrial Adaptations of Seed Plants Compared to ferns, conifers and most other gymnosperms have three additional adaptations that make survival in diverse terrestrial habitats possible: (1) further reduction of the gametophyte; (2) the evolution of pollen; and (3) the advent of the seed.

The first adaptation is an even greater development of the diploid sporophyte compared to the haploid gametophyte generation (Figure 15.14).

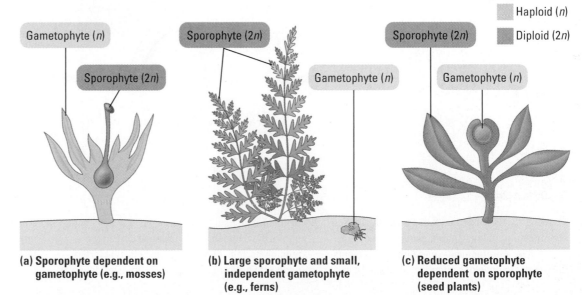

Key
| | Haploid (*n*) |
| | Diploid (2*n*) |

(a) Sporophyte dependent on gametophyte (e.g., mosses)

(b) Large sporophyte and small, independent gametophyte (e.g., ferns)

(c) Reduced gametophyte dependent on sporophyte (seed plants)

Figure 15.14 Three variations on alternation of generations in plants.

A pine tree or other conifer is actually a sporophyte with tiny gameto-phytes living in cones (Figure 15.15). The gametophytes, though multicellu-lar, are totally dependent on and protected by the tissues of the parent sporophyte. Some plant biologists speculate that the shift toward diploidy in land plants was related to the harmful impact of the sun's ionizing radiation, which causes mutations. This damaging radiation is more intense on land than in aquatic habitats, where organisms are somewhat protected by the light-filtering properties of water. Of the two generations of land plants, the diploid form (sporophyte) may cope better with mutagenic radiation. A diploid organism homozygous for a particular essential allele has a "spare tire" in the sense that one copy of the allele may be sufficient for survival if the other is damaged.

A second adaptation of seed plants to dry land was the evolution of **pollen.** A pollen grain is actually the much-reduced male gametophyte. It houses cells that will develop into sperm. In the case of conifers, wind car-ries the pollen from male to female cones, where eggs develop within female gametophytes (see Figure 15.15). This mechanism for sperm transfer con-trasts with the swimming sperm of mosses and ferns. In seed plants, this use of resistant, airborne pollen to bring gametes together is a terrestrial adapta-tion that led to even greater success and diversity of plants on land.

The third important terrestrial adaptation of seed plants is, of course, the seed itself. A **seed,** remember, consists of a plant embryo packaged along with a food supply within a protective coat. Seeds develop from structures called **ovules** (Figure 15.16). In conifers, the ovules are located on the scales of female cones. Conifers and other gymnosperms, lacking ovaries, bear their seeds "naked" on the cone scales (though the seeds do have protective coats, of course). Once released from the parent plant, the resistant seed can remain dormant for days, months, or even years. Under favorable condi-tions, the seed can then **germinate,** its embryo emerging through the seed coat as a seedling. Some seeds drop close to their parents. Others are carried far by the wind or animals.

Follow the life cycle of a pine tree in Web/CD Activity 15E.

Figure 15.15 A pine tree, a conifer. The tree bears two types of cones. The hard, woody ones we usually notice are female cones. Each scale of the female cone (upper left inset) is actually a modified leaf bearing a pair of structures called ovules on its upper surface. An ovule contains the egg-producing female gametophyte. The smaller male cones (lower right inset) produce the male game-tophytes, which are pollen grains. Mature male cones release clouds of millions of pollen grains. You may have seen yellowish conifer pollen covering car tops or floating on ponds in the spring. Some of the windblown pollen manages to land on female cones on trees of the same species. The female cones generally develop on the higher branches, where they are unlikely to be dusted with pollen from the same tree. Sperm released by pollen fertilizes eggs in the ovules of the female cones. The ovules eventually develop into seeds.

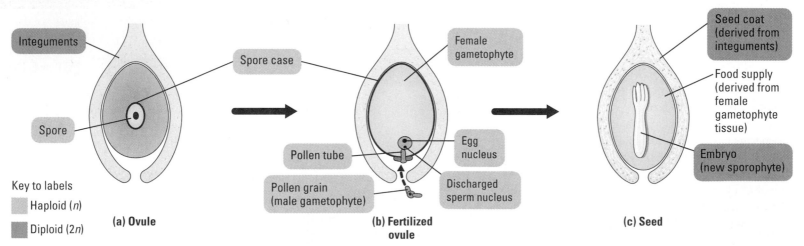

Key to labels
◻ Haploid (*n*)
◼ Diploid (2*n*)

(a) Ovule

(b) Fertilized ovule

(c) Seed

Integuments
Spore
Spore case
Female gametophyte
Pollen tube
Pollen grain (male gametophyte)
Egg nucleus
Discharged sperm nucleus
Seed coat (derived from integuments)
Food supply (derived from female gametophyte tissue)
Embryo (new sporophyte)

Figure 15.16 From ovule to seed. **(a)** The sporophyte produces spores within a tissue sur-rounded by a protective layer called integuments. **(b)** The spore develops into a female gametophyte, which produces one or more eggs. If a pollen grain enters the ovule through a special pore in the in-teguments, it discharges sperm cells that fertilize eggs. **(c)** Fertilization initiates the transformation of ovule to seed. The fertilized egg (zygote) develops into an embryo; the rest of the gametophyte forms a tissue that stockpiles food; and the integuments of the ovule harden to become the seed coat.

Coniferous Forests as a Natural Resource A broad band of coniferous forests covers much of northern Eurasia and North America and extends southward in mountainous regions (Figure 15.17a). Today, about 190 million acres in the United States, mostly in the western states and Alaska, are designated national forests. Some of these areas are set aside as wilderness; most are managed by the U.S. Forest Service for harvesting lumber, grazing, mining, public recreation, and wildlife habitat. Balancing these diverse, often conflicting uses while trying to sustain coniferous forests for future generations is a formidable challenge.

Coniferous forests provide much of our lumber. They are also harvested for wood pulp used to make paper. Currently, our demand for wood and paper is so great that clear-cut areas like the one in Figure 15.17b have become commonplace. In many areas, only about 10% of the original forest remains intact. Some forests have been replanted and are regrowing, but the rate of cutting in many coniferous forests of the northwestern United States and Canada exceeds the reforestation rate. Many scientists also predict that an increase in global temperatures, which now seems to be occurring, poses an additional threat to coniferous forests. An increase of only 1–2°C may adversely affect reproduction by some tree species.

The loss of coniferous forests also threatens other species. The original forests of North America were more biologically diverse—that is, they contained many more different kinds of trees, animals, and other organisms—than the forests that are now regrowing.

The average U.S. citizen now consumes about 50 times more paper than the average person in less developed nations. Reducing paper waste, increasing recycling, and expanding the use of electronic media can help slow the disappearance of conifers.

(a) A coniferous forest in Alberta, Canada

Angiosperms

The photograph of the coniferous forest in Figure 15.17a could give us a somewhat distorted view of today's plant life. Conifers do cover much land in the northern parts of the globe, but it is the angiosperms, or flowering plants, that dominate most other regions. There are about 250,000 angiosperm species versus about 700 species of conifers and other gymnosperms. Whereas gymnosperms supply most of our lumber and paper, angiosperms supply nearly all our food and much of our fiber for textiles. Cereal grains, including wheat, corn, oats, and barley, are flowering plants, as are citrus and other fruit trees, garden vegetables, cotton, and flax. Fine hardwoods from flowering plants such as oak, cherry, and walnut trees supplement the lumber we get from conifers.

Learn to identify trees in The Process of Science activity available on the website and CD-ROM.

Several unique adaptations account for the success of angiosperms. For example, refinements in vascular tissue make water transport even more efficient in angiosperms than in gymnosperms. Of all terrestrial adaptations, however, it is the flower that accounts for the unparalleled success of the angiosperms.

Flowers, Fruits, and the Angiosperm Life Cycle No organisms make a showier display of their sex lives than angiosperms. From roses to dandelions, flowers display a plant's male and female parts. For most angiosperms, insects and other animals transfer pollen from the male parts of one flower

(The cladogram labels, from left to right: Charophytes, Bryophytes, Ferns, Gymnosperms, Angiosperms)

(b) A clear-cut area in a coniferous forest

Figure 15.17 Coniferous forests.

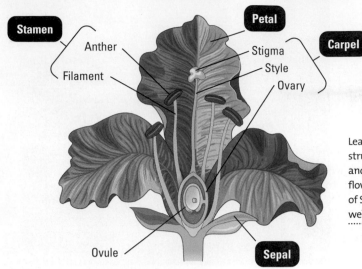

Figure 15.18 Structure of a flower.

to the female sex organs of another flower. This targets the pollen rather than relying on the capricious winds to blow the pollen between plants of the same species.

A **flower** is actually a short stem with four whorls of modified leaves: sepals, petals, stamens, and carpels (Figure 15.18). At the bottom of the flower are the **sepals,** which are usually green. They enclose the flower before it opens (think of a rosebud). Above the sepals are the **petals,** which are usually the most striking part of the flower and are often important in attracting insects and other pollinators. The actual reproductive structures are multiple stamens and one or more carpels. Each **stamen** consists of a stalk bearing a sac called an **anther,** the male organ in which pollen grains develop. The **carpel** consists of a stalk, the **style,** with an ovary at the base and a sticky tip known as the **stigma,** which traps pollen. The **ovary** is a protective chamber containing one or more ovules, in which the eggs develop.

Figure 15.19 highlights key stages in the angiosperm life cycle. The plant familiar to us is the sporophyte, which is true of all vascular plants. As in gymnosperms, the pollen grain is the male gametophyte of angiosperms. The female gametophyte is located within an ovule, which in turn resides

> Learn more about the structure, function, and significance of flowers in The Process of Science on the website and CD-ROM.

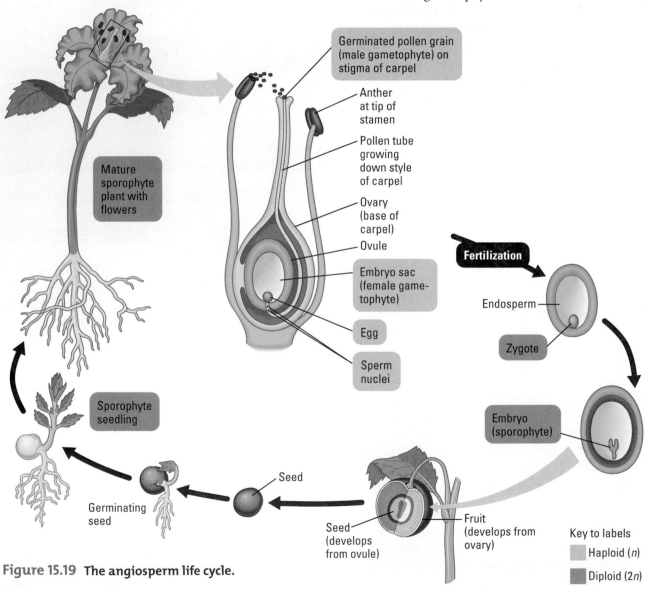

Figure 15.19 The angiosperm life cycle.

(a) Dandelion fruit dispersed by the wind

(b) Cockleburs (fruits) carried by animal fur

Figure 15.20 Fruits and seed dispersal. **(a)** Some angiosperms depend on wind for seed dispersal. For example, the dandelion fruit acts like a kite, carrying a tiny seed far away from its parent plant. **(b)** Some fruits are adapted to hitch free rides on animals. The cockleburs attached to the fur of this dog are fruits that may be carried miles before opening and releasing seeds. **(c)** Many angiosperms produce fleshy, edible fruits that are attractive to animals as food. When a mouse eats a berry, it digests the fleshy part of the fruit, but most of the tough seeds pass unharmed through the mouse's digestive tract. The mouse later deposits the seeds, along with a fertilizer supply, some distance from where it ate the fruit. Many types of garden produce are the edible fruits from plants we have domesticated. Examples include tomatoes, squash, melons, strawberries, apples, oranges, and cherries.

(c) A mouse eating fruit containing seeds that will be dispersed later with the animal's feces

within a chamber of the ovary. Pollen that lands on the sticky stigma of a carpel extends a tube down to an ovule and deposits two sperm nuclei within the female gametophyte. This **double fertilization** is an angiosperm characteristic. One sperm cell fertilizes an egg in the female gametophyte. This produces a zygote, which develops into an embryo. The second sperm cell fertilizes another female-gametophyte cell, which then develops into a nutrient-storing tissue called **endosperm.** Double fertilization thus synchronizes the development of the embryo and food reserves within an ovule. The whole ovule develops into a seed. The seed's enclosure within an ovary is what distinguishes angiosperms from the naked-seed condition of gymnosperms.

> Quiz yourself on the angiosperm life cycle in Web/CD Activity 15F.

A **fruit** is the ripened ovary of a flower. As seeds are developing from ovules, the ovary wall thickens, forming the fruit that encloses the seeds. A pea pod is an example of a fruit, with seeds (mature ovules, the peas) encased in the ripened ovary (the pod). Fruits protect and help disperse seeds (Figure 15.20).

Interdependence of Angiosperms and Animals Flowering plants and land animals have had mutually beneficial relationships throughout their evolutionary history. Most angiosperms depend on insects, birds, or

Figure 15.21 A busy bee. The Scottish broom has a tripping mechanism that dusts pollen onto the back of a visiting bee that is feeding on nectar. Some of the pollen the bee picks up will rub off onto the female parts of the next flower it visits.

mammals for pollination and seed dispersal. And most land animals depend on angiosperms for food. We can see evidence of this interdependence in the diversity of flowers and fruits.

Many angiosperms produce flowers that attract pollinators that rely mostly on the flowers' nectar and pollen for food (Figure 15.21). Nectar is a high-energy fluid that is of use to the plant only for attracting pollinators. The color and fragrance of a flower are usually keyed to a pollinator's sense of sight and smell. Many flowers also have markings that attract pollinators, leading them past pollen-bearing organs on the way to nectar. For example, flowers that are pollinated by bees often have markings that reflect ultraviolet light. Such markings are invisible to us, but vivid to bees. Many flowers pollinated by birds are red or pink, colors to which bird eyes are especially sensitive. The shape of the flower also facilitates pollination. Flowers that depend largely on hummingbirds, for example, typically have their nectar located deep in a floral tube, where only the long, thin beak and tongue of the bird are likely to reach.

Relationships between angiosperms and animals are also evident in the edible fruits. Fruits that are not yet ripe are usually green, hard, and distasteful (at least to humans). This helps the plant retain its fruit until the seeds are mature and ready for dispersal. As it ripens, the fruit becomes softer and its sugar content increases. Many fruits also become fragrant and brightly colored, advertising their ripeness to animals. One of the most common colors for ripe fruit is red, which vertebrates can probably distinguish better than insects can. Thus, most of the fruit is saved for birds and mammals, animals large enough to disperse the seeds (see Figure 15.20).

Angiosperms and Agriculture Flowering plants provide nearly all our food. All of our fruit and vegetable crops are angiosperms. Corn, rice, wheat, and the other grains are grass fruits. Grains are also the main food source for domesticated animals, such as cows and chickens. We also grow angiosperms for fiber, medications, perfumes, and decoration.

Like other animals, early humans probably collected wild seeds and fruits. Agriculture was gradually invented as humans began sowing seeds and cultivating plants to have a more dependable food source. As they domesticated certain plants, humans began to intervene in plant evolution by selective breeding designed to improve the quantity and quality of the foods. Agriculture is a unique kind of evolutionary relationship between plants and animals.

Plant Diversity as a Nonrenewable Resource

The exploding human population, with its demand for space and natural resources, is extinguishing plant species at an unprecedented rate. The problem is especially critical in the tropics, where more than half the human population lives and population growth is fastest. Tropical rain forests are being destroyed at a frightening pace. The most common cause of this destruction is slash-and-burn clearing of the forest for agricultural use. Fifty million acres, an area about the size of the state of Washington, are cleared each year, a rate that would completely eliminate Earth's tropical forests within 25 years. As the forest disappears, so do thousands of plant species. Insects and other rain forest animals that depend on these plants are also vanishing. In all, researchers estimate that the destruction of habitat in the rain forest and other ecosystems is claiming hundreds of species each year. The toll is greatest in the tropics because that is where most species live; but environmental assault is a generically human tendency. Europeans elimi-

Table 15.1	A Sampling of Medicines Derived from Plants		
Compound	**Example of Source**		**Example of Use**
Atropine	Belladonna plant		Pupil dilator in eye exams
Digitalin	Foxglove		Heart medication
Menthol	Eucalyptus tree		Ingredient in cough medicines
Morphine	Opium poppy		Pain reliever
Quinine	Quinine tree		Malaria preventive
Taxol	Pacific yew		Ovarian cancer drug
Tubocurarine	Curare tree		Muscle relaxant during surgery
Vinblastine	Periwinkle		Leukemia drug

Source: Adapted from Randy Moore et al., *Botany*, 2nd ed. Dubuque, IA: Brown, 1998. Table 2.2, p. 37.

nated most of their forests centuries ago, and habitat destruction is now endangering many species in North America. Extinction is irrevocable; plant diversity is a nonrenewable resource.

Many people have ethical concerns about contributing to the extinction of living forms. But there are also practical reasons to be concerned about the loss of plant diversity. We depend on plants for thousands of products, including food, building materials, and medicines (Table 15.1). So far, we have explored the potential uses of only a tiny fraction of the 300,000 known plant species. For example, almost all our food is based on the cultivation of only about two dozen species. More than 120 prescription drugs are extracted from plants. However, researchers have investigated fewer than 5000 plant species as potential sources of medicine. And pharmaceutical companies were led to most of these species by local peoples who use the plants in preparing their traditional medicines.

The tropical rain forest may be a medicine chest of healing plants that could become extinct before we even know they exist. This is only one reason to value what is left of plant diversity and to search for ways to slow the loss. The solutions we propose must be economically realistic. If the goal is only profit for the short term, then we will continue to slash and burn until the forests are gone. If, however, we begin to see rain forests and other ecosystems as living treasures that can regenerate only slowly, we may learn to harvest their products at sustainable rates. What else can we do to preserve plant diversity? Few questions are as important.

We have seen in our survey of plants, especially the angiosperms, how entangled the botanical world is with other terrestrial life. We switch our attention now to that other group of organisms that moved onto land with plants, the kingdom Fungi.

1. Which of the following structures is common to all four major plant groups: vascular tissue, flowers, seeds, cuticle, pollen?

2. Gametophyte is to _____ as _____ is to diploid.

3. Why are coal, oil, and natural gas called "fossil" fuels?

4. How does the evergreen nature of pines and other conifers adapt the plants for living where the growing season is very short?

5. Contrast the mode of sperm delivery in ferns with sperm delivery in conifers.

6. **(a)** In what way are forests renewable resources? **(b)** Under what conditions does reality belie the "renewable" designation?

7. _____ are to conifers as flowers are to _____.

8. What are the four main organs of a flower?

9. What is a fruit?

10. How are an angiosperm and its pollinators mutually rewarded by their relationship?

Answers: 1. Cuticle **2.** haploid; sporophyte **3.** Because they are derived from ancient organisms that did not decay completely after dying **4.** Because the plants do not lose their leaves during autumn and winter, the leaves are already fully developed for photosynthesis when the short growing season begins in spring. **5.** The flagellated sperm of ferns must swim through water to reach eggs. In contrast, the airborne pollen of conifers brings gametes together without need of an aqueous route, and the pollen releases sperm near the egg after pollination occurs. **6. (a)** Forests are renewable in the sense that new trees can grow where old growth has been removed by logging. **(b)** The current situation is the harvesting of wood at a much faster rate than new growth can replace, and thus "renewable" is an unrealistic designation for forest resources. **7.** Cones; angiosperms **8.** Sepals, petals, stamens, and carpels **9.** A ripened ovary of a flower that protects and aids in the dispersal of seeds contained in the fruit **10.** The pollinators obtain food and the plant has its pollen targeted much more efficiently than if it were carried by the wind.

Fungi

The word *fungus* often evokes some unpleasant images. Fungi rot timbers, spoil food, and afflict humans with athlete's foot and worse maladies. However, ecosystems would collapse without fungi to decompose dead organisms, fallen leaves, feces, and other organic materials, thus recycling vital chemical elements back to the environment in forms other organisms can assimilate. And you have already learned that nearly all plants have mycorrhizae, fungus-root associations that absorb minerals and water from the soil. In addition to these ecological roles, fungi have been used by humans in various ways for centuries. We eat some fungi (mushrooms, for instance), culture fungi to produce antibiotics and other drugs, add them to dough to make bread rise, culture them in milk to produce a variety of cheeses, and use them to ferment beer and wine.

Fungi are eukaryotes, and most are multicellular. They were once grouped with plants. But in fact, molecular studies indicate that fungi and animals probably arose from a common ancestor. In other words, a mushroom is probably more closely related to you than it is to any plant! However, fungi are actually a form of life so distinctive that they are now accorded their own kingdom, the kingdom Fungi (Figure 15.22).

(a) Fly agaric mushrooms

(b) A "fairy ring"

(c) *Pilobolus*

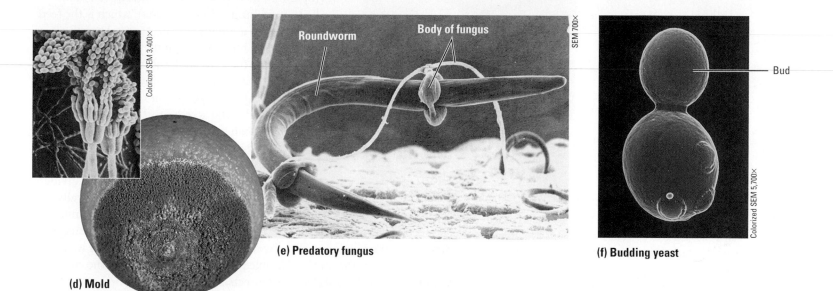

Colorized SEM 3,400×

Roundworm

Body of fungus

SEM 700×

Bud

Colorized SEM 5,700×

(d) Mold

(e) Predatory fungus

(f) Budding yeast

Figure 15.22 A gallery of diverse fungi.
(a) These mushrooms are the reproductive structures of a fungus that absorbs nutrients as it decomposes compost on a forest floor. **(b)** Some mushroom-producing fungi poke up "fairy rings," which can appear on a lawn overnight. The legendary explanation of these circles is that mushrooms spring up where fairies have danced in a ring on moonlit nights. Afterward, the tired fairies sit down on some of the mushrooms, but toads use other mushrooms as stools; hence the name toadstools. Biology offers an alternative explanation. A ring develops at the edge of the main body of the fungus, which consists of an underground mass of tiny filaments within the ring. The filaments secrete enzymes that digest soil compost. As the underground fungal mass grows outward from its center, the diameter of the fairy rings produced at its expanding perimeter increases annually. **(c)** This fungus, *Pilobolus,* decomposes animal dung. The bulbs at the tips of the stalks are sacs of spores, which are reproductive cells. *Pilobolus* can actually aim these spore sacs. The stalks bend toward light, where grass is likely to be growing, and then shoot their spore sacs like cannonballs. Grazing animals eat the spore sacs and scatter the spores in feces, where the spores grow into new fungi. **(d)** The fungi we call molds grow rapidly on their food sources, often on *our* food sources. The mold on this orange reproduces asexually by producing chains of microscopic spores (inset) that are dispersed via air currents. **(e)** This predatory fungus traps and feeds on tiny roundworms in the soil. The fungus is equipped with hoops that can constrict around a worm in less than a second. **(f)** Yeasts are unicellular fungi. This yeast cell is reproducing asexually by a process called budding. For centuries, humans have domesticated yeasts and put their metabolism to work in breweries and bakeries.

Characteristics of Fungi

Fungal Nutrition **Fungi** are heterotrophs that acquire their nutrients by **absorption.** In this mode of nutrition, small organic molecules are absorbed from the surrounding medium. A fungus digests food outside its body by secreting powerful hydrolytic enzymes into the food. The enzymes decompose complex molecules to the simpler compounds that the fungus can absorb. For example, fungi that are decomposers absorb nutrients from nonliving organic material, such as fallen logs, animal corpses, or the wastes of live organisms. Parasitic fungi absorb nutrients from the cells or body fluids of living hosts. Some of these fungi, such as certain species infecting the lungs of humans, are pathogenic. In other cases of symbiosis, such as mycorrhizae, the relationships between fungi and their hosts are mutually beneficial.

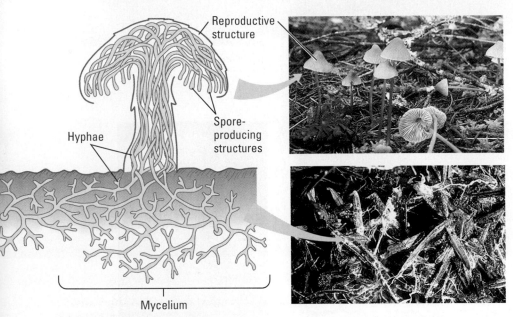

Figure 15.23 The fungal mycelium. The mushroom we see is like the tip of an iceberg. It is a reproductive structure consisting of tightly packed hyphae that extend upward from a much more massive mycelium of hyphae growing underground. The photos show mushrooms and the mycelium of cottony threads that decompose organic litter.

Fungal Structure Fungi are structurally adapted for their absorptive nutrition. The bodies of most fungi are constructed of structures called **hyphae** (singular, *hypha*). Hyphae are minute threads composed of tubular walls surrounding plasma membranes and cytoplasm. The hyphae form an interwoven mat called a **mycelium** (plural, *mycelia*), which is the feeding network of a fungus (Figure 15.23). Fungal mycelia can be huge, although they usually escape our notice because they are often subterranean. In 2000, scientists discovered the mycelium of one humongous fungus in Oregon that is 3.4 miles in diameter and spreads through 2200 acres of forest (equivalent to over 1600 football fields). This fungus is at least 2400 years old and hundreds of tons in weight, qualifying it among Earth's oldest and largest organisms.

Most fungi are multicellular, with hyphae divided into cells by cross-walls. The cross-walls generally have pores large enough to allow ribosomes, mitochondria, and even nuclei to flow from cell to cell. The cell walls of fungi differ from the cellulose walls of plants. Most fungi build their cell walls mainly of chitin, a strong but flexible polysaccharide similar to the chitin found in the external skeletons of insects.

Mingling with the organic matter it is decomposing and absorbing, a mycelium maximizes contact with its food source. Ten cubic centimeters of rich organic soil may contain as much as a kilometer of hyphae. And a fungal mycelium grows rapidly, adding as much as a kilometer of hyphae each day as it branches within its food. Fungi are nonmotile organisms; they cannot run, swim, or fly in search of food. But the mycelium makes up for the lack of mobility by swiftly extending the tips of its hyphae into new territory.

Fungal Reproduction Fungi reproduce by releasing spores that are produced either sexually or asexually. The output of spores is mind-boggling. For example, puffballs, which are the reproductive structures of certain fungi, can puff out clouds containing trillions of spores (see Figure 12.12). Carried by wind or water, spores germinate to produce mycelia if they land in a moist place where there is food. Spores thus function in dispersal and account for the wide geographic distribution of many

Watch an animation of fungal reproduction and nutrition in Web/CD Activity 15G.

species of fungi. The airborne spores of fungi have been found more than 160 km (100 miles) above Earth. Closer to home, try leaving a slice of bread out for a week or two and you will observe the furry mycelia that grow from the invisible spores raining down from the surrounding air. It's a good thing those particular molds cannot grow in our lungs.

The Ecological Impact of Fungi

Fungi have been major players in terrestrial communities ever since they moved onto land in the company of plants. Let's examine a few examples of how fungi continue to have an enormous ecological impact.

Fungi as Decomposers Fungi and bacteria are the principal decomposers that keep ecosystems stocked with the inorganic nutrients essential for plant growth. Without decomposers, carbon, nitrogen, and other elements would accumulate in organic matter. Plants and the animals they feed would starve because elements taken from the soil would not be returned (see Chapter 20).

Fungi are well adapted as decomposers of organic refuse. Their invasive hyphae enter the tissues and cells of dead organic matter and hydrolyze polymers, including the cellulose of plant cell walls. A succession of fungi, in concert with bacteria and, in some environments, invertebrate animals, is responsible for the complete breakdown of organic litter. The air is so loaded with fungal spores that as soon as a leaf falls or an insect dies, it is covered with spores and is soon infiltrated by fungal hyphae.

We may applaud fungi that decompose forest litter or dung, but it is a different story when molds attack our fruit or our shower curtains. Between 10% and 50% of the world's fruit harvest is lost each year to fungal attack. A wood-digesting fungus does not distinguish between a fallen oak limb and the oak planks of a boat. During the Revolutionary War, the British lost more ships to fungal rot than to enemy attack. And soldiers stationed in the tropics during World War II watched as their tents, clothing, boots, and binoculars were destroyed by molds. Some fungi can even decompose certain plastics.

Parasitic Fungi Of the 100,000 known species of fungi, about 30% make their living as parasites, mostly on or in plants. In some cases, fungi that infect plants have literally changed landscapes. One species, for example, has eliminated most American elm trees (Figure 15.24a). Fungi are also serious agricultural pests. Some species infect grain crops and cause tremendous economic losses each year (Figure 15.24b).

Animals are much less susceptible to parasitic fungi than are plants. Only about 50 species of fungi are known to be parasitic in humans and other animals. However, their effects are significant enough to make us take them seriously. Among the diseases that fungi cause in humans are yeast infections of the lungs, some of which can be fatal, and vaginal yeast infections. Other fungal parasites produce a skin disease called ringworm, so named because it appears as circular red areas on the skin. The ringworm fungi can infect virtually any skin surface. Most commonly, they attack the feet and cause intense itching and sometimes blisters. This condition, known as athlete's foot, is highly contagious but can be treated with various fungicidal preparations.

(b) Ergots on rye

(a) American elm trees killed by a fungus

Figure 15.24 Parasitic fungi that cause plant disease. **(a)** This photo shows American elm trees after the arrival of the parasitic fungus that causes Dutch elm disease. The fungus evolved with European species of elm trees, and it is relatively harmless to them. But it is deadly to American elms. The fungus was accidentally introduced into the United States on logs sent from Europe to pay World War I debts. Insects called bark beetles carried the fungus from tree to tree. Since then, the disease has destroyed elm trees all across North America. **(b)** The seeds of some kinds of grain, including rye, wheat, and oats, are sometimes infected with fungal growths called ergots, the dark structures on this seed head of rye. Consumption of flour made from ergot-infested grain can cause gangrene, nervous spasms, burning sensations, hallucinations, temporary insanity, and death. One epidemic in Europe in the year A.D. 944 killed more than 40,000 people. During the Middle Ages, the disease (ergotism) became known as Saint Anthony's fire because many of its victims were cared for by a Catholic nursing order dedicated to Saint Anthony. Several kinds of toxins have been isolated from ergots. One called lysergic acid is the raw material from which the hallucinogenic drug LSD is made. Certain other chemical extracts are medicinal in small doses. An ergot compound is useful in treating high blood pressure, for example.

(b) Bleu cheese

Figure 15.25 **Feeding on fungi.** **(a)** Truffles (the fungal kind, not the chocolates) are the reproductive structures of fungi that grow with tree roots as mycorrhizae. Truffles release strong odors that attract mammals and insects that excavate the fungi and disperse their spores. In some cases, the odors mimic sex attractants of certain mammals. Truffle hunters traditionally used pigs to locate their prizes. However, dogs are now more commonly used because they have the nose for the scent without the fondness for the flavor. Gourmets describe the complex flavors of truffles as nutty, musky, cheesy, or some combination of those tastes. At about $400 per pound for truffles, you probably won't get a chance to do a taste test of your own in the campus cafeteria. **(b)** The turquoise streaks in bleu cheese and Roquefort are the mycelia of a specific fungus.

(a) Truffles

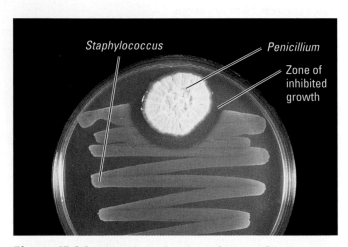

Figure 15.26 **Fungal production of an antibiotic.** The first antibiotic discovered was penicillin, which is made by the common mold called *Penicillium*. In this Petri dish, the clear area between the mold and the bacterial colony is where the antibiotic produced by *Penicillium* inhibits the growth of the bacteria, a species of *Staphylococcus*.

Commercial Uses of Fungi It would not be fair to fungi to end our discussion with an account of diseases. In addition to their positive global impact as decomposers, fungi also have a number of practical uses for humans.

Most of us have eaten mushrooms, although we may not have realized that we were ingesting the reproductive extensions of subterranean fungi. Mushrooms are often cultivated commercially in artificial caves in which cow manure is piled (be sure to wash your store-bought mushrooms thoroughly). Edible mushrooms also grow wild in fields, forests, and backyards, but so do poisonous ones. There are no simple rules to help the novice distinguish edible from deadly mushrooms. Only experts in mushroom taxonomy should dare to collect the fungi for eating.

Mushrooms are not the only fungi we eat. The fungi called truffles are highly prized by gourmets (Figure 15.25a). And the distinctive flavors of certain kinds of cheeses come from the fungi used to ripen them (Figure 15.25b). Particularly important in food production are unicellular fungi, the yeasts. As discussed in Chapter 5, yeasts are used in baking, brewing, and winemaking.

Fungi are medically valuable as well. Some fungi produce antibiotics that are used to treat bacterial diseases. In fact, the first antibiotic discovered was penicillin, which is made by the common mold called *Penicillium* (Figure 15.26).

As sources of antibiotics and food, as decomposers, and as partners with plants in mycorrhizae, fungi play vital roles in life on Earth.

Evolution Link: Mutual Symbiosis

Evolution is not just about the origin and adaptation of individual species. Interdependence between species is also an evolutionary product. This interdependence is called symbiosis when two organisms are in direct contact. Parasitism is a symbiotic relationship in which the symbiont, the parasite, benefits while harming its host in the process. Our focus here, however, is on **mutualism,** symbiosis that benefits both species.

We have seen many examples of mutualism over the past two chapters. Eukaryotic cells evolved from mutual symbiosis among prokaryotes. And today, bacteria living in the roots of certain plants provide nitrogen compounds to their host and receive food in exchange. We have our own mutually symbiotic bacteria that help keep our skin healthy and produce certain vitamins in our intestines. Plants and their specific pollinators are yet another example of mutualism. Particularly relevant to this chapter is the symbiotic association of fungi and plant roots—mycorrhizae—which made life's move onto land possible.

> Discover more about mutual symbiosis in the Web Evolution Link.

Lichens, symbiotic associations of fungi and algae, are striking examples of how two species can become so merged that the cooperative is essentially a new life-form. At a distance, it is easy to mistake lichens for mosses or other simple plants growing on rocks, rotting logs, trees, roofs, or gravestones (Figure 15.27). In fact, lichens are not mosses or any other kind of plant, nor are they even individual organisms. A lichen is a symbiotic association of millions of tiny algae embraced by a mesh of fungal hyphae. The photosynthetic algae feed the fungi. The fungal mycelium, in turn, provides a suitable habitat for the algae, helping to absorb and retain water and minerals. The mutualistic merger of partners is so complete that lichens are actually named as species, as though they are individual organisms. Mutualisms such as lichens and mycorrhizae showcase the web of life that has evolved on Earth.

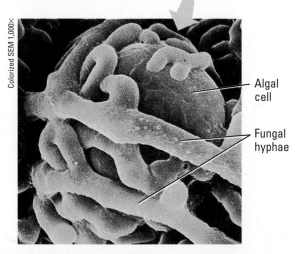

Colorized SEM 1,000×

Algal cell

Fungal hyphae

Figure 15.27 Lichens: symbiotic associations of fungi and algae. Lichens generally grow very slowly, sometimes in spurts of less than a millimeter per year. You can date the oldest lichens you see here by the engraving on the gravestone. Elsewhere, there are lichens that are thousands of years old, rivaling the oldest plants as Earth's elders. The close relationship between the fungal and algal partners is evident in the microscopic blow-up of a lichen.

Chapter Review

Summary of Key Concepts

Overview: Colonizing Land

- **Terrestrial Adaptations of Plants** Plants are multicellular photosynthetic eukaryotes. Most plants have roots, stems, and leaves that function in absorption, support, and photosynthesis. Root-fungus associations called mycorrhizae help plants absorb water and essential minerals from the soil. Leaves are the main photosynthetic organs. Gas exchange between the plant and the atmosphere occurs via stomata in the leaves. The waxy cuticle covering leaves and other aerial plant parts helps the plant retain its water. Vascular tissues transport water and minerals from roots to the leaves, distributes sugars from the leaves to the roots and other parts of the plant, and provide internal support. Plants produce their gametes in gametangia, structures that protect developing gametes from dehydration. The female gametangia produce eggs and protect the developing embryos after the eggs are fertilized.
- Web/CD Activity 15A *Terrestrial Adaptations of Plants*

- **The Origin of Plants from Green Algae** Plants probably evolved gradually from green algae. Molecular comparisons and other evidence place a group of multicellular green algae called charophytes closest to plants.

Plant Diversity

- **Highlights of Plant Evolution** Four major periods of plant evolution are marked by terrestrial adaptations. The first adaptations included gametangia, which protected gametes and embryos. Next came the diversification of vascular plants that lacked seeds. Ferns are seedless vascular plants. The third period began with the origin of seeds, protecting plant embryos from desiccation and other hazards, characteristics of modern gymnosperms. Finally, angiosperms emerged with flowers that bear ovules within protective chambers called ovaries.
- Web/CD Activity 15B *Highlights of Plant Evolution*

- **Bryophytes** The most familiar bryophytes are mosses. Mosses display two key terrestrial adaptations: a waxy cuticle that prevents dehydration and the retention of developing embryos within the mother plant's gametangium. Mosses are most common in moist environments because their sperm must swim to the eggs and because they lack vascular tissues. Mosses also lack lignin in their cell walls and thus cannot stand tall. Bryophytes are unique among plants in having the gametophyte as the dominant generation in the life cycle.
- Web/CD Activity 15C *Moss Life Cycle*

- **Ferns** Ferns are seedless plants that have vascular tissues but still use flagellated sperm to fertilize eggs. During the Carboniferous period, many giant ferns were among the plants that formed thick deposits of organic matter that were gradually converted into coal.
- Web/CD Activity 15D *Fern Life Cycle*

- **Gymnosperms** A drier and colder global climate near the end of the Carboniferous favored the evolution of the first seed plants. The most successful were the gymnosperms, represented by conifers. Needle-shaped leaves with thick cuticles and sunken stomata are adaptations to dry conditions. Conifers and most other gymnosperms have three additional terrestrial adaptations: (1) further reduction of the gametophyte generation and greater development of the diploid sporophyte; (2) the evolution of pollen, which doesn't require water for transport; and (3) the advent of the seed, which consists of a plant embryo packaged along with a food supply within a protective coat. Coniferous forests are widely used for lumbering, grazing, mining, public recreation, and wildlife habitat.
- Web/CD Activity 15E *Pine Life Cycle*

- **Angiosperms** Angiosperms supply nearly all our food and much of our fiber for textiles. The evolution of the flower and more efficient water transport help account for the success of the angiosperms. The dominant stage is a sporophyte with gametophytes in its flowers. The female gametophyte is located within an ovule, which in turn resides within a chamber of the ovary. Fertilization of an egg in the female gametophyte produces a zygote, which develops into an embryo. The whole ovule develops into a seed. The seed's enclosure within an ovary is what distinguishes angiosperms from the naked-seed condition of gymnosperms. A fruit is the ripened ovary of a flower. Fruits protect and help disperse seeds. Interactions with animals influenced the evolution of flowers and fruit. Angiosperms are a major food source for animals, while animals aid plants in pollination and seed dispersal. Agriculture constitutes a unique kind of evolutionary relationship among humans, plants, and animals.
- Web/CD The Process of Science *Using Leaf Structure to Identify Trees*
- Web/CD The Process of Science *Spring Wildflowers: An Introduction to Floral Structure, Function, and Significance*
- Web/CD Activity 15F *Angiosperm Life Cycle*

- **Plant Diversity as a Nonrenewable Resource** The exploding human population, with its demand for space and natural resources, is causing the extinction of plant species at an unprecedented rate.

Fungi

- **Characteristics of Fungi** Fungi are unicellular or multicellular eukaryotes that are probably more closely related to animals than to plants. Fungi are heterotrophs that digest their food externally and absorb the nutrients. A fungus usually consists of a mass of threadlike hyphae forming a mycelium. The cell walls of fungi are mainly composed of chitin. Although most fungi are nonmotile, the mycelium can swiftly extend the tips of its hyphae into new territory. Fungi reproduce and disperse by releasing spores that are produced either sexually or asexually.
- Web/CD Activity 15G *Fungal Reproduction and Nutrition*

- **The Ecological Impact of Fungi** Fungi and bacteria are the principal decomposers of ecosystems. Many molds destroy fruit, wood, and human-made materials. About 50 species of fungi are known to be parasitic in humans and other animals. Fungi are also commercially important as food and in baking, beer and wine production, and the manufacture of antibiotics.

Evolution Link: Mutual Symbiosis

- Lichens exemplify mutualism, a symbiotic relationship that benefits both partners.
- Web Evolution Link *Mutual Symbiosis*

Self-Quiz

1. Angiosperms are different from all other plants because only angiosperms have
   - a. a vascular system.
   - b. flowers.
   - c. a life cycle that involves alternation of generations.
   - d. seeds.
   - e. a sporophyte phase.

2. Ovule is to _____ as ovary is to _____.

   a. sporophyte; gametophyte
   b. gymnosperm; angiosperm
   c. seed; fruit
   d. carpel; stamen
   e. moss; seed plant

3. Under a microscope, a piece of a mushroom would look most like

   a. jelly.
   b. a tangle of string.
   c. grains of sugar or salt.
   d. a piece of glass.
   e. foam.

4. During the Carboniferous period, the dominant plants, which later formed the great coal beds, were mainly

   a. mosses and other bryophytes.
   b. ferns and other seedless vascular plants.
   c. charophytes and other green algae.
   d. conifers and other gymnosperms.
   e. angiosperms.

5. A plant produces flagellated sperm, and its dominant generation has diploid cells. The plant is most likely a

   a. moss.
   b. fern.
   c. pine.
   d. dandelion.
   e. seed plant.

6. Which term below includes all others in the list?

   a. angiosperm
   b. vascular plant
   c. seed plant
   d. fern
   e. gymnosperm

7. Plant diversity is greatest in

   a. tropical forests.
   b. deserts.
   c. salt marshes.
   d. the temperate forests of Europe.
   e. farmlands.

8. Gymnosperms and angiosperms have the following in common except

   a. seeds.
   b. pollen.
   c. vascular tissue.
   d. ovaries.
   e. ovules.

9. The photosynthetic symbiont of a lichen is most commonly a (an)

   a. moss.
   b. fungus.
   c. mold.
   d. alga.
   e. small vascular plant.

10. All fungi share which one of the following characteristics?

   a. heterotrophic nutrition
   b. alternation of generations
   c. multicellular body plan
   d. mutual symbiosis with plants
   e. production of mushrooms

• Go to the website or CD-ROM for more self-quiz questions.

The Process of Science

1. In April 1986, an accident at a nuclear power plant in Chernobyl, Ukraine, scattered radioactive fallout for hundreds of miles. In assessing the biological effects of the radiation, researchers found mosses to be especially valuable as organisms for monitoring the damage. Radiation damages organisms by causing mutations. Explain why it is faster to observe the genetic effects of radiation on mosses than on other types of plants. Imagine that you are conducting tests shortly after a nuclear accident. Using potted moss plants as your experimental organisms, design an experiment to test the hypothesis that the frequency of mutations decreases with the organism's distance from the source of radiation.

2. Learn about tree identification and spring wildflowers in The Process of Science activities on the website and CD-ROM.

Biology and Society

Why are tropical rain forests being destroyed at such an alarming rate? What kinds of social, technological, and economic factors are responsible? Most forests in developed Northern Hemisphere countries have already been cut. Do the developed nations have a right to pressure the developing nations in the Southern Hemisphere to slow or stop the destruction of their forests? Defend your answer. What kinds of benefits, incentives, or programs might slow the assault on the rain forests?

CHAPTER 16

The Evolution of Animals

Zoologists estimate that about a billion billion

(10^{18}) individual arthropods populate the Earth.

Tapeworms can reach lengths of 20 meters in the human

intestine. The blue whale, an endangered

species that grows to lengths of nearly 30 meters, is the

largest animal that has ever existed.

A reptile can survive on less than 10% of the calories

required by a mammal of equivalent size.

ome of us can remember our parents or teachers yelling, "Stop behaving like animals!" Hard to do, since animals are exactly what we are. Humans belong to one of about 35 phyla (major groups) in the kingdom Animalia. In this chapter, you will learn about the evolution of animals as diverse as sponges and people (Figure 16.1).

Figure 16.1 One animal checking out a diversity of others. The diver, the fish, and the many species of invertebrates (animals without backbones) of this coral reef all belong to the animal kingdom. In all, there are about a million known animal species.

Overview: The Origins of Animal Diversity

Animal life began in Precambrian seas with the evolution of multicellular creatures that ate other organisms. We are among their descendants.

What Is an Animal?

Animals are eukaryotic, multicellular, heterotrophic organisms that obtain nutrients by ingestion. *That's* a mouthful. And speaking of mouthfuls, **ingestion** means eating food. This mode of nutrition contrasts animals with fungi, which obtain nutrients by absorption after digesting the food outside the body (see Chapter 15). Animals digest their food within their bodies after ingesting other organisms, dead or alive, whole or by the piece (Figure 16.2).

A few key features of life history also distinguish animals. Most animals reproduce sexually. The zygote (fertilized egg) develops into an early embryonic stage called a **blastula,** which is usually a hollow ball of cells. The next embryonic stage in most animals is a gastrula, which has layers of cells that will eventually form the adult body parts. The gastrula also has a primitive gut, which will develop into the animal's digestive compartment. Continued development, growth, and maturation transform some animals directly from the embryo into an adult. However, the life histories of many animals include larval stages. A **larva** is a sexually immature form of an animal. It is anatomically distinct from the adult form, usually eats different food, and may even have a different habitat. Think of the contrast between

Figure 16.2 Nutrition by ingestion, the animal way of life. Most animals ingest relatively large pieces of food, though rarely as large as the prey in this case. In this amazing scene, a rock python is beginning to ingest a gazelle. The snake will spend two weeks or more in a quiet place digesting its meal.

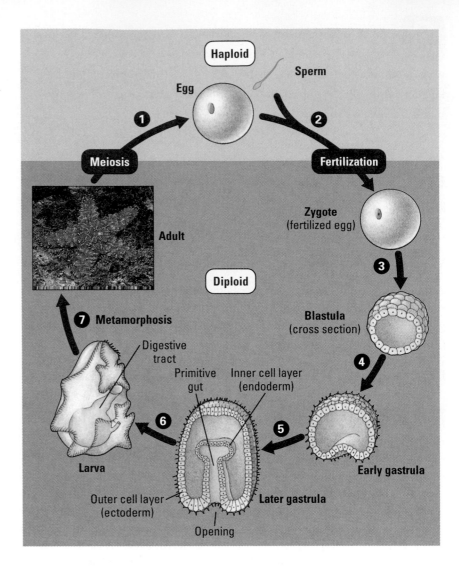

Figure 16.3 Life cycle of a sea star as an example of animal development. ❶ Male and female adult animals produce haploid gametes (eggs and sperm) by meiosis. ❷ An egg and a sperm fuse to produce a diploid zygote. ❸ Early mitotic divisions lead to an embryonic stage called a blastula, common to all animals. Typically, the blastula consists of a ball of cells surrounding a hollow cavity. ❹ Later, in the sea star and many other animals, one side of the blastula cups inward, forming an embryonic stage called a gastrula. ❺ The gastrula develops into a saclike embryo with a two-layered wall and an opening at one end. Eventually, the outer layer (ectoderm) develops into the animal's epidermis (skin) and nervous system. The inner layer (endoderm) forms the digestive tract. Still later in development, in most animals, a third layer (mesoderm) forms between the other two and develops into most of the other internal organs (not shown in the figure). ❻ Following the gastrula, many animals continue to develop and then mature directly into adults. But others, including the sea star, develop into one or more larval stages first. ❼ The larva undergoes a major change of body form, called metamorphosis, in becoming an adult—a mature animal capable of reproducing sexually.

a frog and its larval form, which we call a tadpole. A change of body form, called **metamorphosis,** eventually remodels the larva into the adult form (Figure 16.3).

Most animals have muscle cells, as well as nerve cells that control the muscles. The evolution of this equipment for coordinated movement enhanced feeding, even enabling some animals to search for or chase their food. The most complex animals, of course, can use their muscular and nervous systems for many functions other than eating. Some species even use massive networks of nerve cells called brains to think.

Early Animals and the Cambrian Explosion

Animals probably evolved from a colonial, flagellated protist that lived in Precambrian seas (Figure 16.4). By the late Precambrian, about 600–700 million years ago, a diversity of animals had already evolved. Then came the Cambrian explosion. At the beginning of the Cambrian period, 545 million years ago, animal diversity exploded. In fact, during a span of only about 10 million years, all the major animal body plans we see today evolved. It is an evolutionary episode so boldly marked in the fossil record that geologists

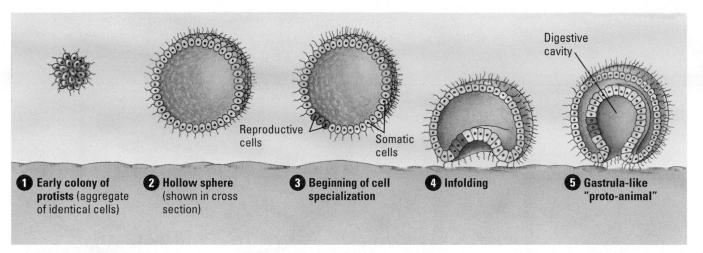

① **Early colony of protists** (aggregate of identical cells)   ② **Hollow sphere** (shown in cross section)   ③ **Beginning of cell specialization**   ④ **Infolding**   ⑤ **Gastrula-like "proto-animal"**

Figure 16.4 One hypothesis for a sequence of stages in the origin of animals from a colonial protist. ❶ The earliest colonies may have consisted of only a few cells, all of which were flagellated and basically identical. ❷ Some of the later colonies may have been hollow spheres—floating aggregates of heterotrophic cells—that ingested organic nutrients from the water. ❸ Eventually, cells in the colony may have specialized, with some cells adapted for reproduction and others for somatic (nonreproductive) functions, such as locomotion and feeding. ❹ A simple multicellular organism with cell layers may have evolved from a hollow colony, with cells on one side of the colony cupping inward, the way they do in the gastrula of a animal embryo (see Figure 16.3). ❺ A layered body plan would have enabled further division of labor among the cells. The outer flagellated cells would have provided locomotion and some protection, while the inner cells could have specialized in reproduction or feeding. With its specialized cells and a simple digestive compartment, the proto-animal shown here could have fed on organic matter on the seafloor.

use the dawn of the Cambrian period as the beginning of the Paleozoic era (see Table 13.1). Many of the Cambrian animals seem bizarre compared to the versions we see today, but most zoologists now agree that the Cambrian fossils can be classified as ancient representatives of the familiar animal phyla (Figure 16.5).

What ignited the Cambrian explosion? Hypotheses abound. Most researchers now believe that the Cambrian explosion simply extended animal diversification that was already well under way during the late Precambrian. But what caused the radiation of animal forms to accelerate so dramatically during the early Cambrian? One hypothesis emphasizes increasingly complex predator-prey relationships that led to diverse adaptations for feeding, motility, and protection. This would help explain why most Cambrian animals had shells or hard outer skeletons, in contrast to Precambrian animals, which were mostly soft-bodied. Another hypothesis focuses on the evolution of genes that control the development of animal form, such as the placement of body parts in embryos (see Chapter 10). At least some of these genes are common to diverse animal phyla. However, variation in how, when, and where these genes are expressed in an embryo can produce some of the major differences in body form that distinguish the phyla. Perhaps this developmental plasticity was partly responsible for the relatively rapid diversification of animals during the early Cambrian.

Figure 16.5 A Cambrian seascape. This drawing is based on fossils collected at a site called the Burgess Shale in British Columbia, Canada.

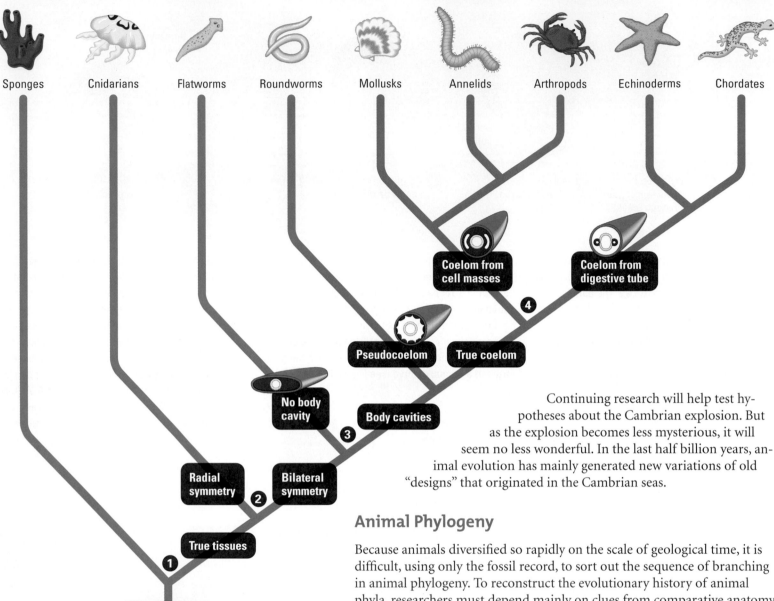

Sponges **Cnidarians** **Flatworms** **Roundworms** **Mollusks** **Annelids** **Arthropods** **Echinoderms** **Chordates**

Coelom from cell masses

Coelom from digestive tube

❹

Pseudocoelom **True coelom**

No body cavity

Body cavities

❸

Radial symmetry ❷ **Bilateral symmetry**

True tissues

❶

Multicellularity

Ancestral protist

Figure 16.6 An overview of animal phylogeny. The circled numbers key four main branch points to the text discussion. Only 9 of the more than 30 animal phyla are included in the tree.

Continuing research will help test hypotheses about the Cambrian explosion. But as the explosion becomes less mysterious, it will seem no less wonderful. In the last half billion years, animal evolution has mainly generated new variations of old "designs" that originated in the Cambrian seas.

Animal Phylogeny

Because animals diversified so rapidly on the scale of geological time, it is difficult, using only the fossil record, to sort out the sequence of branching in animal phylogeny. To reconstruct the evolutionary history of animal phyla, researchers must depend mainly on clues from comparative anatomy and embryology (see Chapter 12). Molecular methods are now providing additional tools for testing hypotheses about animal phylogeny. Figure 16.6 represents one set of hypotheses about the evolutionary relationships among nine major animal phyla. The circled numbers on the tree highlight four key evolutionary branch points, and these numbers are keyed to the following discussion.

❶ The first branch point distinguishes sponges from all other animals based on structural complexity. Sponges, though multicellular, lack the true tissues, such as epithelial (skin) tissue, that characterize more complex animals.

❷ The second major evolutionary split is based partly on body symmetry: radial versus bilateral. To understand this difference, imagine a pail and shovel. The pail has **radial symmetry,** identical all around a central axis. The shovel has **bilateral symmetry,** which means there's only one way to split it into two equal halves—right down the midline. Figure 16.7 contrasts a radial animal with a bilateral one.

The symmetry of an animal generally fits its lifestyle. Many radial animals are sessile forms (attached to a substratum) or plankton (drifting or

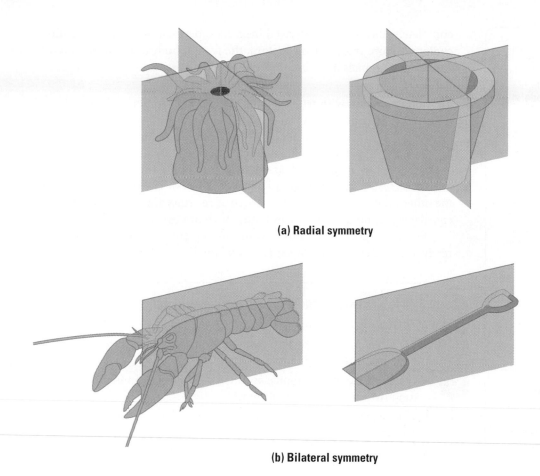

(a) Radial symmetry

(b) Bilateral symmetry

Figure 16.7 Body symmetry. (a) The parts of a radial animal, such as this sea anemone, radiate from the center. Any imaginary slice through the central axis would divide the animal into mirror images. **(b)** A bilateral animal, such as this lobster, has a left and right side. Only one imaginary cut would divide the animal into mirror-image halves.

weakly swimming aquatic forms). Their symmetry equips them to meet the environment equally well from all sides. Most animals that move actively from place to place are bilateral. A bilateral animal has a definite "head end" that encounters food, danger, and other stimuli first when the animal is traveling. In most bilateral animals, a nerve center in the form of a brain is at the head end, near a concentration of sense organs such as eyes. Thus, bilateral symmetry is an adaptation for movement, such as crawling, burrowing, or swimming.

❸ The evolution of body cavities led to more complex animals. A **body cavity** is a fluid-filled space separating the digestive tract from the outer body wall. A body cavity has many functions. Its fluid cushions the suspended organs, helping to prevent internal injury. The cavity also enables the internal organs to grow and move independently of the outer body wall. If it were not for your body cavity, exercise would be very hard on your internal organs. And every beat of your heart or ripple of your intestine would deform your body surface. It would be a scary sight. In soft-bodied animals such as earthworms, the noncompressible fluid of the body cavity is under pressure and functions as a hydrostatic skeleton against which muscles can work. In fact, body cavities may have first evolved as adaptations for burrowing.

Among animals with a body cavity, there are differences in how the cavity develops. In all cases, the cavity is at least partly lined by a middle layer of tissue, called mesoderm, which develops between the inner (endoderm) and outer (ectoderm) layers of the gastrula embryo. If the body cavity is not

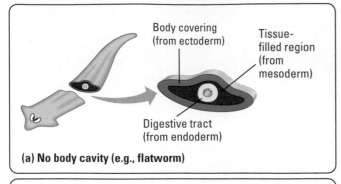

(a) No body cavity (e.g., flatworm)

Body covering (from ectoderm)
Tissue-filled region (from mesoderm)
Digestive tract (from endoderm)

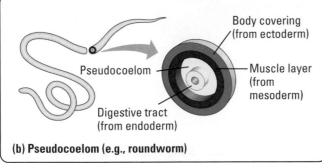

(b) Pseudocoelom (e.g., roundworm)

Body covering (from ectoderm)
Muscle layer (from mesoderm)
Pseudocoelom
Digestive tract (from endoderm)

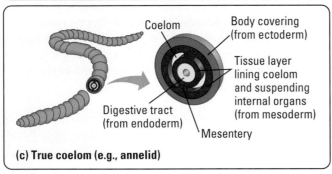

(c) True coelom (e.g., annelid)

Coelom
Body covering (from ectoderm)
Tissue layer lining coelom and suspending internal organs (from mesoderm)
Digestive tract (from endoderm)
Mesentery

Figure 16.8 Body plans of bilateral animals. The various organ systems of these animals develop from the three tissue layers that form in the embryo. **(a)** Flatworms are examples of animals that lack body cavities. **(b)** Roundworms have a pseudocoelom, a body cavity only partially lined by mesoderm, the middle tissue layer. **(c)** Earthworms and other annelids are examples of animals with true coeloms. A coelom is a body cavity completely lined by mesoderm. Mesenteries, also derived from mesoderm, suspend the organs in the fluid-filled coelom.

completely lined by tissue derived from mesoderm, it is termed a **pseudocoelom** (Figure 16.8). A true **coelom,** the type of body cavity humans and many other animals have, is completely lined by tissue derived from mesoderm.

❹ Among animals with true coeloms, there are two main evolutionary branches. They differ in several details of embryonic development, including the mechanism of coelom formation. One branch includes mollusks (such as clams, snails, and squids), annelids (such as earthworms), and arthropods (such as crustaceans, spiders, and insects). The two major phyla of the other branch are echinoderms (such as sea stars and sea urchins) and chordates (including humans and other vertebrates).

Practice classifying animals in Web/CD Activity 16A.

With the overview of animal evolution in Figure 16.6 as our guide, we're ready to take a closer look at some animal phyla.

CheckPoint

1. What embryonic stage is common to the development of all animals?
2. What is metamorphosis?
3. In key features of embryonic development, humans and other chordates are most like which other animal group?
4. What mode of heterotrophic nutrition distinguishes animals from fungi, which are also heterotrophs?
5. Why is animal evolution during the early Cambrian referred to as an "explosion"?
6. A round pizza is to _____ symmetry as a fork is to _____ symmetry.
7. The fully lined cavity between your outer body wall and your digestive tract is an example of a true _____.

Answers: **1.** Blastula **2.** The transformation from a larval form of an animal to an adult form **3.** Echinoderms **4.** Ingestion **5.** Because a great diversity of animals evolved in a relatively short time span **6.** radial; bilateral **7.** coelom

Major Invertebrate Phyla

Living as we do on land, our sense of animal diversity is biased in favor of vertebrates, which are animals with backbones. Vertebrates are well represented on land in the form of such animals as reptiles, birds, and mammals. However, vertebrates make up only one subphylum within the phylum Chordata, or less than 5% of all animal species. If we were to sample the animals in an aquatic habitat, such as a pond, tide pool, or coral reef, we would find ourselves in the realm of **invertebrates.** These are the animals *without* backbones. It is traditional to divide the animal kingdom into vertebrates and invertebrates, but this makes about as much zoological sense as sorting animals into flatworms and nonflatworms. We give special attention to the vertebrates only because we humans are among the backboned ones. However, by exploring the other 95% of the animal kingdom—the invertebrates—we'll discover an astonishing diversity of beautiful creatures that too often escape our notice.

Sponges

Sponges (Phylum **Porifera**) are sessile animals that appear so sedate to the human eye that the ancient Greeks believed them to be plants (Figure 16.9). The simplest of all animals, sponges probably evolved very early from colonial protists. Sponges range in height from about 1 centimeter to 2 meters. Sponges have no nerves or muscles, but the individual cells can sense and react to changes in the environment. The cell layers of sponges are loose federations of cells, not really tissues, because the cells are relatively unspecialized. Of the 9000 or so species of sponges, only about 100 live in fresh water; the rest are marine.

The body of a sponge resembles a sac perforated with holes (the name of the sponge phylum, Porifera, means "pore bearer"). Water is drawn through the pores into a central cavity, then flows out of the sponge through a larger opening (Figure 16.10). Most sponges feed by collecting bacteria from the water that streams through their porous bodies. Flagellated cells called **choanocytes** trap bacteria in mucus and then engulf the food by phagocytosis (see Chapter 4). Cells called **amoebocytes** pick up food from the choanocytes, digest it, and carry the nutrients to other cells. Amoebocytes are the "do-all" cells of sponges. Moving about by means of pseudopodia, they digest and distribute food, transport oxygen, and dispose of wastes. Amoebocytes also manufacture the fibers that make up a sponge's skeleton. In some sponges, these fibers are sharp and spur-like. Other sponges have softer, more flexible skeletons; we use these pliant, honeycombed skeletons as bathroom sponges.

Figure 16.9 A sponge.

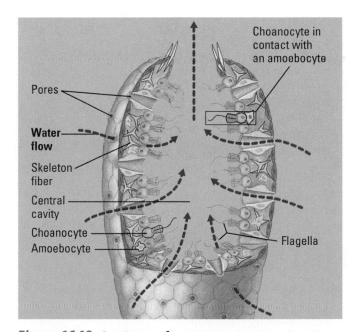

Figure 16.10 Anatomy of a sponge. Feeding cells called choanocytes have flagella that sweep water through the sponge's body. Choanocytes trap bacteria and other food particles, and amoebocytes distribute the food to other cells. To obtain enough food to grow by 100 g (about 3 ounces), a sponge must filter 1000 kg (about 275 gallons) of seawater.

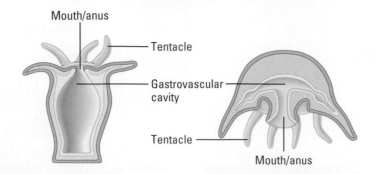

Mouth/anus

Tentacle

Gastrovascular cavity

Tentacle

Mouth/anus

(a) **Sea anemone: a polyp**

(b) **Jelly: a medusa**

Figure 16.11 Polyp and medusa forms of cnidarians. Note that cnidarians have two tissue layers, distinguished in the diagrams by blue and yellow. The gastrovascular cavity is a digestive sac, meaning that it has only one opening, which functions as both mouth and anus. **(a)** Sea anemones are examples of the polyp form of the basic cnidarian body plan. **(b)** Jellies are examples of the medusa form.

Cnidarians

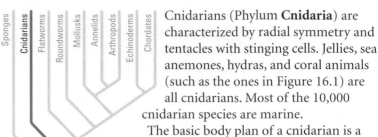

Cnidarians (Phylum **Cnidaria**) are characterized by radial symmetry and tentacles with stinging cells. Jellies, sea anemones, hydras, and coral animals (such as the ones in Figure 16.1) are all cnidarians. Most of the 10,000 cnidarian species are marine.

The basic body plan of a cnidarian is a sac with a central digestive compartment, the **gastrovascular cavity.** A single opening to this cavity functions as both mouth and anus. This basic body plan has two variations: the sessile **polyp** and the floating **medusa** (Figure 16.11). Polyps adhere to the substratum and extend their tentacles, waiting for prey. Examples of the polyp form are hydras, sea anemones, and coral animals (Figure 16.12). A medusa is a flattened, mouth-down version of the polyp. It moves freely in the water by a combination of passive drifting and contractions of its bell-shaped body. The animals we generally call jellies are medusas (jellyfish is another common name, though these animals are not really fishes, which are vertebrates). Some cnidarians exist only as polyps, others only as medusas, and still others pass sequentially through both a medusa stage and a polyp stage in their life cycle.

Cnidarians are carnivores that use tentacles arranged in a ring around the mouth to capture prey and push the food into the gastrovascular cavity, where digestion begins. The undigested remains are eliminated through the mouth/anus. The tentacles are armed with batteries of **cnidocytes,** which translates as "stinging cells," unique structures that function in defense and in the capture of prey (Figure 16.13). The phylum Cnidaria is named for these stinging cells.

Figure 16.12 Coral animals. Each polyp in this colony is about 3 mm in diameter. Coral animals secrete hard external skeletons of calcium carbonate (limestone). Each polyp builds on the skeletal remains of earlier generations to construct the "rocks" we call coral. Though individual coral animals are small, their collective construction accounts for such biological wonders as Australia's Great Barrier Reef, which Apollo astronauts were able to identify from the moon. Tropical coral reefs are home to an enormous variety of invertebrates and fishes (see Figure 16.1).

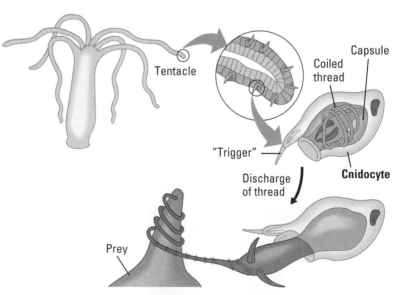

Tentacle

Coiled thread

Capsule

"Trigger"

Cnidocyte

Discharge of thread

Prey

Figure 16.13 Cnidocyte action. Each cnidocyte contains a fine thread coiled within a capsule. When a trigger is stimulated by touch, the thread shoots out. Some cnidocyte threads entangle prey, while others puncture the prey and inject a poison.

Flatworms

Flatworms (Phylum **Platyhelminthes,** from the Greek for "flat worm") are the simplest bilateral animals. True to their name, these worms are leaflike or ribbonlike, ranging from about 1 millimeter to about 20 meters in length. There are about 20,000 species of flatworms living in marine, freshwater, and damp terrestrial habitats. Planarians are examples of free-living (nonparasitic) flatworms (Figure 16.14). The phylum also includes many parasitic species, such as flukes and tapeworms.

Parasitic flatworms called blood flukes are a huge health problem in the tropics. These worms have suckers that attach to the inside of the blood vessels near the human host's intestines. This causes a long-lasting disease with such symptoms as severe abdominal pain, anemia, and dysentery. About 250 million people in 70 countries suffer from blood fluke disease. Flukes generally have complex life cycles that require more than one host species (Figure 16.15).

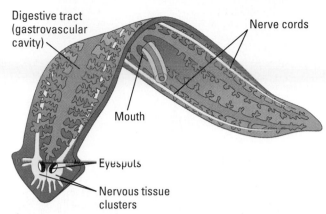

Figure 16.14 Anatomy of a planarian. This free-living flatworm has a head with two light-detecting eyespots and a flap at each side that detects certain chemicals in the water. Dense clusters of nervous tissue form a simple brain. The digestive tract of the worm is highly branched, which provides an extensive surface area for the absorption of nutrients. When the animal feeds, a muscular tube projects through the mouth (pictured here) and sucks food in. The digestive tract, like that of cnidarians, is a gastrovascular cavity, which means that it has only a single opening that functions as both mouth and anus. Planarians live on the undersurfaces of rocks in freshwater ponds and streams.

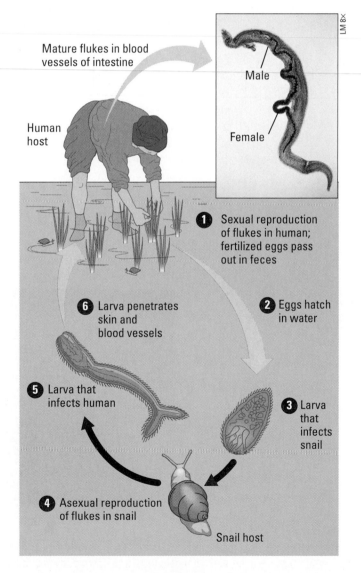

Figure 16.15 Life cycle of a blood fluke. Note in the micrograph that the female fits into a groove running the length of the larger male's body. This enables the mates to copulate frequently, producing over a thousand eggs per day. People are most commonly exposed to blood flukes while working in irrigated fields contaminated with human feces. ❶ Blood flukes living in a human host reproduce sexually, and fertilized eggs pass out in the host's feces. ❷ If an egg lands in a pond or stream, a motile larva hatches. ❸ This larva can enter a snail, the next host. ❹ Asexual reproduction in the snail eventually produces ❺ other larvae that can infect humans. ❻ A person becomes infected when these larvae penetrate the skin.

Tapeworms parasitize many vertebrates, including humans. In contrast to planarians and flukes, most tapeworms have a very long, ribbonlike body with repeated parts (Figure 16.16). They also differ from other flatworms in not having any digestive tract at all. Living in partially digested food in the intestines of their hosts, tapeworms simply absorb nutrients across their body surface. Like parasitic flukes, tapeworms have complex life cycles, usually involving more than one host. Humans can become infected with tapeworms by eating rare beef containing the worm's larvae. The larvae are microscopic, but the adults can reach lengths of 20 meters in the human intestine. Such large tapeworms can cause intestinal blockage and rob enough nutrients from the human host to cause nutritional deficiencies. An orally administered drug called niclosamide kills the adult worms.

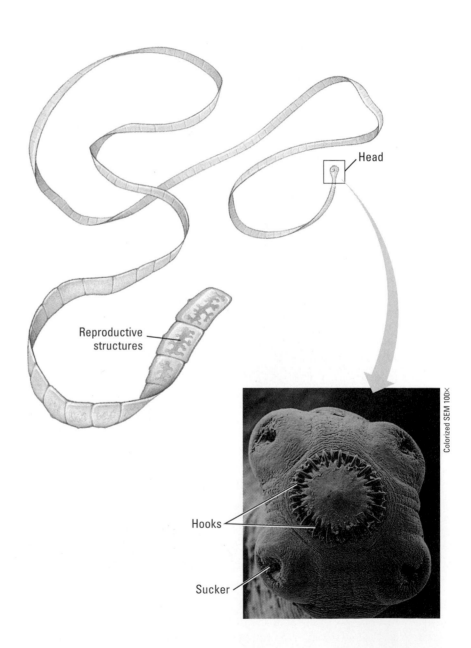

Head

Reproductive structures

Colorized SEM 100×

Hooks

Sucker

Figure 16.16 Anatomy of a tapeworm. Humans acquire larvae of these parasites by eating undercooked meat that is infected. The head of a tapeworm is armed with suckers and menacing hooks that lock the worm to the intestinal lining of the host. Behind the head is a long ribbon of units that are little more than sacs of sex organs. At the back of the worm, mature units containing thousands of eggs break off and leave the host's body with the feces.

Roundworms

Roundworms (Phylum **Nematoda**) get their common name from their cylindrical bodies, which are usually tapered at both ends (Figure 16.17a). Roundworms are among the most diverse (in species number) and widespread of all animals. About 90,000 species of roundworms are known, and perhaps ten times that number actually exist. Roundworms range in length from about a millimeter to a meter. They are found in most aquatic habitats, in wet soil, and as parasites in the body fluids and tissues of plants and animals. Free-living roundworms in the soil are important decomposers. Other species are major agricultural pests that attack the roots of plants. Humans host at least 50 parasitic roundworm species, including pinworms, hookworms, and the parasite that causes trichinosis (Figure 16.17b).

Roundworms exhibit two evolutionary innovations not found in flatworms. First, roundworms have a **complete digestive tract,** which is a digestive tube with two openings, a mouth and an anus. This anatomy contrasts with the digestive sacs, or gastrovascular cavities, of cnidarians and flatworms, which use a single opening as both mouth and anus. All of the remaining animals in our survey of the phyla have complete digestive tracts. A complete digestive tract can process food and absorb nutrients as a meal moves in one direction from one specialized digestive organ to the next. In humans, for example, the mouth, stomach, and intestines are examples of digestive organs. A second evolutionary innovation we see for the first time in roundworms is a body cavity, which in this case is a pseudocoelom (see Figure 16.8).

Colorized SEM 200×

(a) A free-living roundworm

Trichinella juvenile   Muscle tissue

LM 350×

(b) Parasitic roundworms

Figure 16.17 Roundworms. (a) This species has the classic roundworm shape: cylindrical with tapered ends. You can see the mouth at the end that is more blunt. Not visible is the anus at the other end of a complete digestive tract. This worm looks like it's wearing a corduroy coat, but the ridges actually indicate muscles that run the length of the body. **(b)** The disease called trichinosis is caused by these roundworms, encysted here in human muscle tissue. Humans acquire the parasite by eating undercooked pork or other meat that is infected. The worms then burrow into the human intestine and eventually travel to other parts of the body, encysting in muscles and other organs.

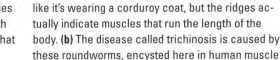

Mollusks

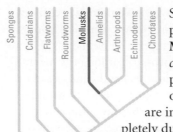

Snails and slugs, oysters and clams, and octopuses and squids are all mollusks (Phylum **Mollusca**). Mollusks (from the Latin *molluscus,* soft) are soft-bodied animals, but most are protected by a hard shell. Slugs, squids, and octopuses have reduced shells, most of which are internal, or they have lost their shells completely during their evolution. Many mollusks feed by using a straplike rasping organ called a **radula** to scrape up food. Garden snails use their radulas like tiny saws to cut pieces out of leaves.

Most of the 150,000 known species of mollusks are marine, though some inhabit fresh water, and there are land-dwelling mollusks in the form of snails and slugs. (Students at the University of California at Santa Cruz were so taken with a local species of giant slug called the banana slug, so named for the fruit it matches in color and shape, that they adopted it as their university mascot. Go Slugs!)

Despite their apparent differences, all mollusks have a similar body plan (Figure 16.18). The body has three main parts: a muscular foot, usually used for movement; a visceral mass containing most of the internal organs; and a fold of tissue called the mantle. The **mantle** drapes over the visceral mass and secretes the shell (if one is present).

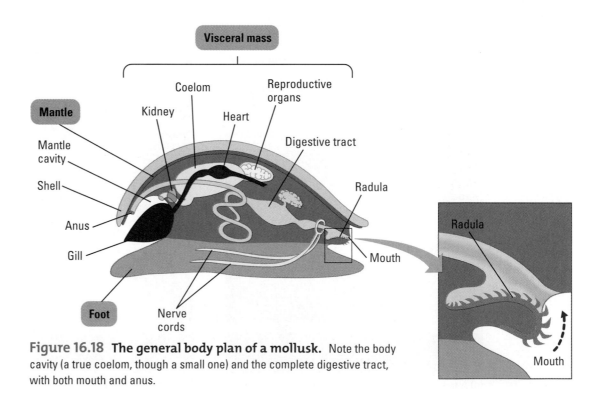

Figure 16.18 The general body plan of a mollusk. Note the body cavity (a true coelom, though a small one) and the complete digestive tract, with both mouth and anus.

(a) Gastropod shells

(b) A bivalve (a scallop)

Figure 16.19 Mollusks. **(a)** Shell collectors are delighted by the variety of gastropods. **(b)** This scallop, a bivalve, has many eyes peering out between the two halves of the hinged shell. **(c)** An octopus is a cephalopod without a shell. All cephalopods have large brains and sophisticated sense organs, which contribute to the success of these animals as mobile predators. This octopus lives on the seafloor, where it scurries about in search of crabs and other food. Its brain is larger and more complex, proportionate to body size, than that of any other invertebrates. Octopuses are very intelligent and have shown remarkable learning abilities in laboratory experiments.

(c) A cephalopod (an octopus)

The three major classes of mollusks are gastropods (including snails and slugs), bivalves (including clams and oysters), and cephalopods (including squids and octopuses). Most **gastropods** (from the Greek *gaster,* belly, and *pous,* foot) are protected by a single, spiraled shell into which the animal can retreat when threatened (Figure 16.19a). Many gastropods have a distinct head with eyes at the tips of tentacles (think of a garden snail). **Bivalves** (from the Latin *bi-,* double, and *valva,* leaf of a folding door), including numerous species of clams, oysters, mussels, and scallops, have shells divided into two halves hinged together (Figure 16.19b). Most bivalves are sedentary, living in sand or mud in marine and freshwater environments. They use their muscular foot for digging and anchoring. **Cephalopods** (from the Greek *kephale,* head, and *pous,* foot) generally differ from gastropods and sedentary bivalves in being built for speed and agility. A few cephalopods have large, heavy shells, but in most the shell is small and internal (as in squids) or missing altogether (as in octopuses). Cephalopods are marine predators that use beaklike jaws and a radula to crush or rip prey apart. The cephalopod mouth is at the base of the foot, which is drawn out into several long tentacles for catching and holding prey (Figure 16.19c).

Annelids

Sponges Cnidarians Flatworms Roundworms Mollusks Annelids Arthropods Echinoderms Chordates

Annelids (Phylum **Annelida**) are worms with body **segmentation,** which is the division of the body along its length into a series of repeated parts. The phylum name is derived from a Latin word meaning "ring," referring to body segmentation that looks like a series of fused rings. Look closely at an earthworm, an annelid you have all encountered, and you'll see why these creatures are also called segmented worms (Figure 16.20).

In all, there are about 15,000 annelid species, ranging in length from less than 1 millimeter to the 3-meter giant Australian earthworm. Annelids live in the sea, most freshwater habitats, and damp soil. The three main classes of annelids are the earthworms and their relatives, the polychaetes, and the leeches.

Earthworms eat their way through the soil, extracting nutrients as the soil passes through the digestive tube (Figure 16.21a). Undigested material, mixed with mucus secreted into the digestive tract, is eliminated as castings through the anus. Farmers value earthworms because the animals till the earth, and the castings improve the texture of the soil. Charles Darwin estimated that each acre of British farmland had about 50,000 earthworms that produced 18 tons of castings per year.

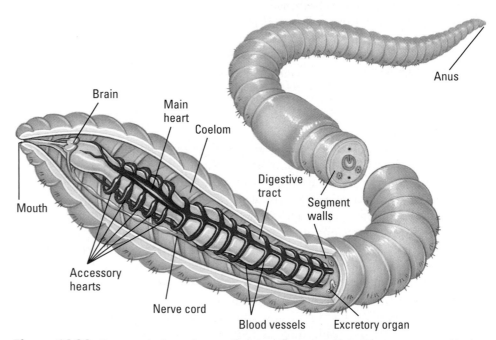

Figure 16.20 Segmented anatomy of an earthworm. Annelids are segmented both externally and internally. Many of the internal structures are repeated, segment by segment. The coelom (body cavity) is partitioned by walls (only two segment walls are fully shown here). The nervous system (yellow) includes a nerve cord with a cluster of nerve cells in each segment. Excretory organs (green), which dispose of fluid wastes, are also repeated in each segment. The digestive tract, however, is not segmented; it passes through the segment walls from the mouth to the anus. Segmental blood vessels connect continuous vessels that run along the top (dorsal location) and bottom (ventral location) of the worm. The segmental vessels include five pairs of accessory hearts. The main heart is simply an enlarged region of the dorsal blood vessel near the head end of the worm.

In contrast to earthworms, most polychaetes are marine, mainly crawling or burrowing in the seafloor. Segmental appendages with hard bristles help the worm wriggle about in search of small invertebrates to eat. The appendages also increase the animal's surface area for taking up oxygen and disposing of metabolic wastes, including carbon dioxide (Figure 16.21b).

The third group of annelids, leeches, are notorious for the bloodsucking habits of some species. However, most species are free-living carnivores that eat small invertebrates such as snails and insects. The majority of leeches live in fresh water, but a few terrestrial species inhabit moist vegetation in the tropics. Until this century, bloodsucking leeches were frequently used by physicians for bloodletting, the removal of what was considered "bad blood" from sick patients. Some leeches have razorlike jaws that cut through the skin, and they secrete saliva containing a strong anesthetic and an anticoagulant into the wound. The anesthetic makes the bite virtually painless, and the anticoagulant keeps the blood from clotting. Leech anticoagulant is now being produced commercially by genetic engineering and may find wide use in human medicine. Tests show that it prevents blood clots that can cause heart attacks. Leeches are also still occasionally used to remove blood from bruised tissues and to help relieve swelling in fingers or toes that have been sewn back on after accidents (Figure 16.21c). Blood tends to accumulate and cause swelling in a reattached finger or toe until small veins have a chance to grow back into it. Leeches are applied to remove the excess blood.

(a) A giant Australian earthworm

(b) Polychaetes

(c) A leech

Figure 16.21 Annelids. **(a)** Giant Australian earthworms are bigger than most snakes. Perhaps you've slipped on slimy worms, but imagine actually *tripping* over one! **(b)** Polychaetes have segmental appendages that function in movement and as gills. On the left is a sandworm. The beautiful polychaete on the right is an example of a fan worm, which lives in a tube it constructs by mixing mucus with bits of sand and broken shells. Fan worms use their feathery headdresses as gills and to extract food particles from the seawater. This species is called a Christmas tree worm. **(c)** A nurse applied this medicinal leech (*Hirudo medicinalis*) to a patient's sore thumb to drain blood from a hematoma (abnormal accumulation of blood around an internal injury).

Arthropods

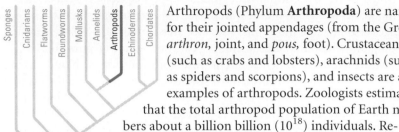

Arthropods (Phylum **Arthropoda**) are named for their jointed appendages (from the Greek *arthron*, joint, and *pous*, foot). Crustaceans (such as crabs and lobsters), arachnids (such as spiders and scorpions), and insects are all examples of arthropods. Zoologists estimate that the total arthropod population of Earth numbers about a billion billion (10^{18}) individuals. Researchers have identified about a million arthropod species, mostly insects. In fact, two out of every three species of life that have been described are arthropods. And arthropods are represented in nearly all habitats of the biosphere. On the criteria of species diversity, distribution, and sheer numbers, arthropods must be regarded as the most successful of all animal phyla.

General Characteristics of Arthropods Arthropods are segmented animals. In contrast to the matching segments of annelids, however, arthropod segments and their appendages have become specialized for a great variety of functions. This evolutionary flexibility contributed to the great diversification of arthropods. Specialization of segments, or fused groups of segments, also provides for an efficient division of labor among body regions. For example, the appendages of different segments are variously modified for walking, feeding, sensory reception, copulation, and defense (Figure 16.22).

The body of an arthropod is completely covered by an **exoskeleton** (external skeleton). This coat is constructed from layers of protein and a polysaccharide called chitin. The exoskeleton can be a thick, hard armor over some parts of the body yet paper-thin and flexible in other locations, such as the joints. The exoskeleton protects the animal and provides points of attachment for the muscles that move the appendages. There are, of course,

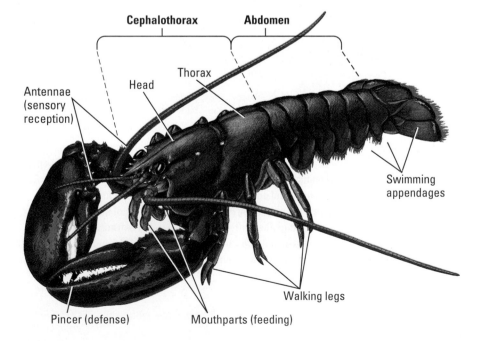

Figure 16.22 Arthropod characteristics of a lobster. The whole body, including the appendages, is covered by an exoskeleton. The two distinct regions of the body are the cephalothorax (consisting of the head and thorax) and the abdomen. The head bears a pair of eyes, each situated on a movable stalk. The body is segmented, but this characteristic is only obvious in the abdomen. The animal has a tool kit of specialized appendages, including pincers, walking legs, swimming appendages, and two pairs of sensory antennae. Even the multiple mouthparts are modified legs, which is why they work form side to side rather than up and down (as our jaws do).

(a) A scorpion (about 8 cm long)

(b) A garden spider (about 1 cm wide)

(c) A dust mite

Figure 16.23 Arachnids. (a) Scorpions are nocturnal hunters. Their ancestors were among the first terrestrial carnivores, preying on herbivorous arthropods that fed on the early land plants. Scorpions have a pair of appendages modified as large pincers that function in defense and food capture. The tip of the tail bears a poisonous stinger. Scorpions eat mainly insects and spiders. They will sting people only when prodded or stepped on. (If you camp in the desert and leave your shoes on the ground when you go to bed, make sure there are no scorpions in those shoes before putting them on in the morning.) **(b)** Spiders are usually most active during the daytime, hunting insects or trapping them in webs. Spiders spin their webs of liquid silk, which solidifies as it comes out of specialized glands. Each spider engineers a style of web that is characteristic of its species, getting the web right on the very first try. Besides building their webs of silk, spiders use the fibers in many other ways: as droplines for rapid escape; as cloth that covers eggs; and even as "gift wrapping" for food that certain male spiders offer to seduce females. **(c)** This magnified house dust mite is a ubiquitous scavenger in our homes. Each square inch of carpet and every one of those dust balls under a bed are like cities to thousands of dust mites. Unlike some mites that carry pathogenic bacteria, dust mites are harmless except to people who are allergic to the mites' feces.

advantages to wearing hard parts on the outside. Our own skeleton is interior to most of our soft tissues, an arrangement that doesn't provide much protection from injury. But our skeleton does offer the advantage of being able to grow along with the rest of our body. In contrast, a growing arthropod must occasionally shed its old exoskeleton and secrete a larger one. This process, called **molting,** leaves the animal temporarily vulnerable to predators and other dangers.

Arthropod Diversity The four main groups of arthropods are the arachnids, the crustaceans, the millipedes and centipedes, and the insects.

Most **arachnids** live on land. Scorpions, spiders, ticks, and mites are examples (Figure 16.23). Arachnids usually have four pairs of walking legs and a specialized pair of feeding appendages. In spiders, these feeding appendages are fanglike and equipped with poison glands. As a spider uses these appendages to immobilize and chew its prey, it spills digestive juices onto the torn tissues and sucks up its liquid meal.

Crustaceans are nearly all aquatic. Crabs, lobsters, crayfish, shrimps, and barnacles are all crustaceans (Figure 16.24). They all exhibit the crustacean hallmark of multiple pairs of specialized appendages (see Figure

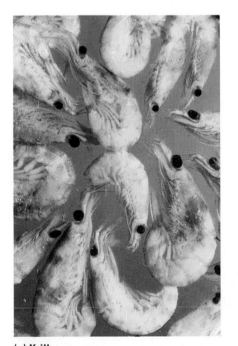

(a) Krill

(b) Barnacles

Figure 16.24 Crustaceans. (a) Planktonic crustaceans called krill are consumed by the ton by whales. **(b)** Easily confused with bivalve mollusks, barnacles are actually sessile crustaceans with exoskeletons hardened into shells by calcium carbonate (lime). The jointed appendages projecting from the shell capture small plankton.

Figure 16.25 A millipede.

16.22). One group of crustaceans, the isopods, is represented on land by pill bugs, which you have probably found on the undersides of moist leaves and other organic debris.

Millipedes and **centipedes** have similar segments over most of the body and superficially resemble annelids, but their jointed legs give them away as arthropods. Millipedes are landlubbers that eat decaying plant matter (Figure 16.25). They have two pairs of short legs per body segment. Centipedes are terrestrial carnivores, with a pair of poison claws used in defense and to paralyze prey, such as cockroaches and flies. Each of their body segments bears a single pair of long legs.

In species diversity, **insects** outnumber all other forms of life combined. They live in almost every terrestrial habitat and in fresh water, and flying insects fill the air. Insects are rare, though not absent, in the seas, where crustaceans are the dominant arthropods. There is a whole big branch of biology, called **entomology,** that specializes in the study of insects.

The oldest insect fossils date back to about 400 million years ago, during the Paleozoic era (see Table 13.1). Later, the evolution of flight sparked an explosion in insect variety (Figure 16.26). A widely held hypothesis is that the greatest diversification of insects paralleled the evolutionary radiation of flowering plants during the late Mesozoic era, about 65 million years ago. This view is challenged by new research suggesting that insects diversified extensively before the angiosperm radiation. Thus, during the evolution of flowering plants and the herbivorous insects that pollinated them, insect diversity may have been as much a cause of the angiosperm radiation as an effect.

Like the grasshopper in Figure 16.26, most insects have a three-part body: head, thorax, and abdomen. The head usually bears a pair of sensory antennae and a pair of eyes. Several pairs of mouthparts are adapted for particular kinds of eating—for example, for biting and chewing plant material in grasshoppers; for lapping up fluids in houseflies; and for piercing skin and sucking blood in mosquitoes and other biting flies. Most adult in-

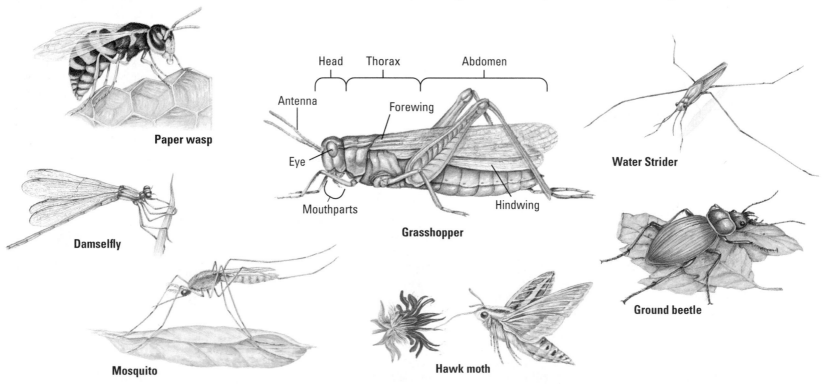

Figure 16.26 A small sample of insect diversity.

(a) Larva (caterpillar)

(b) Pupa

(c) Pupa

(d) Emerging adult

(e) Adult

sects have three pairs of legs and one or two pairs of wings, all borne on the thorax.

Flight is obviously one key to the great success of insects. An animal that can fly can escape many predators, find food and mates, and disperse to new habitats much faster than an animal that must crawl about on the ground. Because their wings are extensions of the exoskeleton and not true appendages, insects can fly without sacrificing any walking legs. By contrast, the flying vertebrates—birds and bats—have one of their two pairs of walking legs modified for wings, which explains why these vertebrates are generally not very swift on the ground.

Many insects undergo metamorphosis in their development. In the case of grasshoppers and some other insect groups, the young resemble adults but are smaller and have different body proportions. The animal goes through a series of molts, each time looking more like an adult, until it reaches full size. In other cases, insects have distinctive larval stages specialized for eating and growing that are known by such names as maggots, grubs, or caterpillars (including the swallowtail butterfly caterpillar that graces the cover of this textbook). The larval stage looks entirely different from the adult stage, which is specialized for dispersal and reproduction. Metamorphosis from the larva to the adult occurs during a pupal stage (Figure 16.27).

Animals so numerous, diverse, and widespread as insects are bound to affect the lives of all other terrestrial organisms, including humans. On one hand, we depend on bees, flies, and many other insects to pollinate our crops and orchards. On the other hand, insects are carriers of the microorganisms that cause many diseases, including malaria and African sleeping sickness. Insects also compete with humans for food. In parts of Africa, for instance, insects claim about 75% of the crops. Trying to minimize their losses, farmers in the United States spend billions of dollars each year on pesticides, spraying crops with massive doses of some of the deadliest poisons ever invented. Try as they may, not even humans have challenged the preeminence of insects and their arthropod kin. As Cornell University's Thomas Eisner puts it: "Bugs are not going to *inherit* the Earth. They own it now. So we might as well make peace with the landlord."

Explore insect diversity in The Process of Science activity, available on the website and CD-ROM.

Figure 16.27 Metamorphosis of a monarch butterfly. **(a)** The larva (caterpillar) spends its time eating and growing, molting as it grows. **(b)** After several molts, the larva encases itself in a cocoon and becomes a pupa. **(c)** Within the pupa, the larval organs break down and adult organs develop from cells that were dormant in the larva. **(d)** Finally, the adult emerges from the cocoon. **(e)** The butterfly flies off and reproduces, nourished mainly from the calories it stored when it was a caterpillar.

(a) A sea star

Echinoderms

Sponges · Cnidarians · Flatworms · Roundworms · Mollusks · Annelids · Arthropods · Echinoderms · Chordates

The echinoderms (Phylum **Echinodermata**) are named for their spiny surfaces (from the Greek *echino*, spiny, and *derma*, skin). Sea urchins, the porcupines of the invertebrates, are certainly echinoderms that live up to the phylum name. Among the other echinoderms are sea stars, sand dollars, and sea cucumbers (Figure 16.28).

Echinoderms are all marine. Most are sessile or slow moving. Echinoderms lack body segments, and most have radial symmetry as adults. Both the external and the internal parts of a sea star, for instance, radiate from the center like spokes of a wheel. In contrast to the adult, the larval stage of echinoderms is bilaterally symmetrical. This supports other evidence that echinoderms are not closely related to cnidarians or other radial animals that never show bilateral symmetry. Most echinoderms have an **endoskeleton** (interior skeleton) constructed from hard plates just beneath the skin. Bumps and spines of this endoskeleton account for the animal's rough or prickly surface. Unique to echinoderms is the **water vascular system,** a network of water-filled canals that circulate water throughout the echinoderm's body, facilitating gas exchange and waste disposal. The water vascular system also branches into extensions called tube feet. A sea star or sea urchin pulls itself slowly over the seafloor using its suction-cup-like tube feet. Sea stars also use their tube feet to grip prey during feeding (see Figure 16.28a).

Looking at sea stars and other adult echinoderms, you may think they have little in common with humans and other vertebrates. But if

(b) A sea urchin

Figure 16.28 Echinoderms. (a) The mouth of a sea star, not visible here, is located in the center of the undersurface. The inset shows how the tube feet function in feeding. When a sea star encounters an oyster or clam, its favorite foods, it grips the mollusk's shell with its tube feet and positions its mouth next to the narrow opening between the two halves of the prey's shell. The sea star then pushes its stomach out through its mouth and through the crack in the mollusk's shell. The predator then digests the soft tissue of its prey. **(b)** In contrast to sea stars, sea urchins are spherical and have no arms. If you look closely, you can see the long tube feet projecting among the spines. Unlike sea stars, which are mostly carnivorous, sea urchins mainly graze on seaweed and other algae. **(c)** On casual inspection, sea cucumbers do not look much like other echinoderms. Sea cucumbers lack spines, and the hard endoskeleton is much reduced. However, closer inspection reveals many echinoderm traits, including five rows of tube feet.

(c) A sea cucumber

you return to Figure 16.6, you'll see that echinoderms share an evolutionary branch with chordates, the phylum that includes vertebrates. Analysis of embryonic development reveals this relationship. The mechanism of coelom formation and many other details of embryology differentiate the echinoderms and chordates from the evolutionary branch that includes mollusks, annelids, and arthropods. With this phylogenetic context, we're ready now to make the transition from invertebrates to vertebrates.

Construct a table to review invertebrates in Web/CD Activity 16B.

CheckPoint

1. In what main way does the digestive tract of a jelly differ from that of a roundworm?

2. In what fundamental way does the structure of a sponge differ from that of all other animals?

3. A sea anemone is to the phylum _____ as a blood fluke is to the phylum _____.

4. Why is it a mistake for someone who eats pork chops to order them "rare" in a restaurant?

5. As representatives of classes of mollusks, a garden snail is an example of a _____; a clam is an example of a _____; and a squid is an example of a _____.

6. The phylum Arthropoda is named for its members' _____.

7. Which major arthropod group is mainly aquatic?

8. Contrast the skeleton of an echinoderm with that of an arthropod.

Answers: 1. The digestive tract of a jelly is a sac with one opening; the roundworm tract is a tube with a separate mouth and anus. **2.** A sponge has no true tissues. **3.** Cnidaria; Platyhelminthes **4.** Incomplete cooking doesn't kill nematodes and other parasites that might be present in the meat. **5.** gastropod; bivalve; cephalopod **6.** jointed appendages **7.** Crustaceans **8.** An echinoderm has an endoskeleton; an arthropod has an exoskeleton.

The Vertebrate Genealogy

Most of us are curious about our genealogy. On the personal level, we wonder about our family ancestry. As biology students, we are interested in tracing human ancestry within the broader scope of the entire animal kingdom. In this quest, we ask three questions: What were our ancestors like? How are we related to other animals? and What are our closest relatives?

In this section, we trace the evolution of the vertebrates, the group that includes humans and their closest relatives. Mammals, birds, reptiles, amphibians, and the various classes of fishes are all classified as **vertebrates.** Among the unique vertebrate features are the cranium and backbone, a series of vertebrae for which the group is named (Figure 16.29). Our first step in tracing the vertebrate genealogy is to determine where vertebrates fit in the animal kingdom.

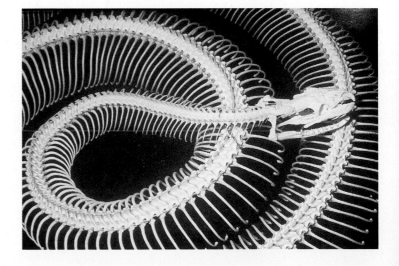

Figure 16.29 Backbone, extra long. Vertebrates are named for their backbone, which consists of a series of vertebrae. The vertebrate hallmark is apparent in this snake skeleton. You can also see the skull, the bony case protecting the brain. The backbone and skull are parts of an endoskeleton, a skeleton inside the animal rather than covering it.

(a) A lancelet

Characteristics of Chordates

Vertebrates make up one subphylum of the phylum **Chordata.** Our phylum also includes two subphyla of invertebrates, animals lacking backbones: **lancelets** and **tunicates** (Figure 16.30). These invertebrate chordates and vertebrates all share four key features that appear in the embryo and sometimes in the adult. These four chordate hallmarks are (1) a **dorsal, hollow nerve cord** (the chordate brain and spinal cord); (2) a **notochord,** which is a flexible, longitudinal rod located between the digestive tract and the nerve cord; (3) **pharyngeal slits,** which are gill structures in the pharynx, the region of the digestive tube just behind the mouth; and (4) a **post-anal tail,** which is a tail to the rear of the anus (Figure 16.31). Though these chordate characteristics are often difficult to recognize in their adult forms, they are always present in chordate embryos. For example, the notochord, for which our phylum is named, persists in adult humans only in the form of the cartilage disks that function as cushions between the vertebrae. (Back injuries described as "ruptured disks" or "slipped disks" refer to these structures.)

Body segmentation is another chordate characteristic, though not a unique one. The chordate version of segmentation probably evolved independently of the segmentation we observe in annelids and arthropods. Chordate segmentation is apparent in the backbone of vertebrates (see Figure 16.29) and in the segmental muscles of all chordates (see the chevron-shaped— >>>> —muscles in the lancelet of Figure 16.30a). Segmental musculature is not so obvious in adult humans unless one is motivated enough to sculpture those washboard "abs of steel."

Vertebrates retain the basic chordate characteristics, but have additional features that are unique, including, of course, the backbone (see Figure 16.29). Figure 16.32 is an overview of chordate and vertebrate evolution that will provide a context for our survey of the vertebrate classes.

(b) A tunicate

Figure 16.30 Invertebrate chordates. (a) Lancelets owe their name to their bladelike shape. Marine animals only a few centimeters long, lancelets wiggle backward into sand, leaving only their heads exposed. The animal filters tiny food particles from the seawater. **(b)** This adult tunicate, or sea squirt, is a sessile filter feeder that bears little resemblance to other chordates. However, a tunicate goes through a larval stage that is unmistakably chordate.

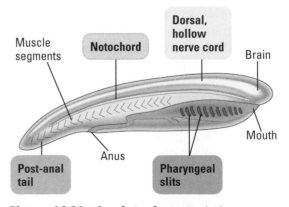

Figure 16.31 Chordate characteristics.

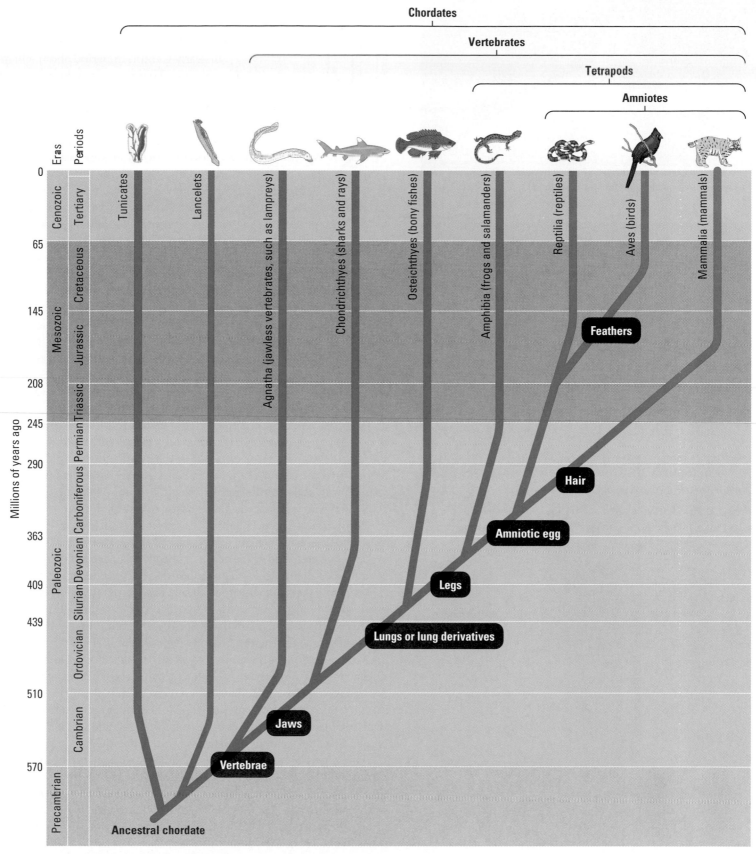

Figure 16.32 The vertebrate genealogy. "Tetrapods" refers to the four legs of terrestrial vertebrates. "Amniotes" refers to the evolution of the amniotic egg, a shelled egg that made it possible for vertebrates to reproduce on land (a chicken egg is one example).

(a) A cartilaginous fish (a blacktip reef shark)

Operculum   Lateral line

(b) A bony fish (a yellow perch)

Figure 16.33 Two classes of fishes. (a) A member of the class Chondrichthyes, the cartilaginous fishes. (b) A member of the class Osteichthyes, the bony fishes.

Fishes

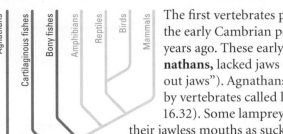

The first vertebrates probably evolved during the early Cambrian period about 540 million years ago. These early vertebrates, the **agnathans,** lacked jaws (*agnatha* means "without jaws"). Agnathans are represented today by vertebrates called lampreys (see Figure 16.32). Some lampreys are parasites that use their jawless mouths as suckers to attach to the sides of large fishes and draw blood. In contrast, most vertebrates have jaws, which are hinged skeletons that work the mouth. We know from the fossil record that the first jawed vertebrates were fishes that replaced most agnathans by about 400 million years ago. In addition to jaws, these fishes had two pairs of fins, which made them maneuverable swimmers. Some of those fishes were more than 10 meters long. Sporting jaws and fins, some of the early fishes were active predators that could chase prey and bite off chunks of flesh. Even today, most fishes are carnivores. The two major groups of living fishes are the class **Chondrichthyes** (cartilaginous fishes—the sharks and rays) and the class **Osteichthyes** (the bony fishes, including such familiar groups as tuna, trout, and goldfish).

Cartilaginous fishes have a flexible skeleton made of cartilage. Most sharks are adept predators—fast swimmers with streamlined bodies, acute senses, and powerful jaws (Figure 16.33a). A shark does not have keen eyesight, but its sense of smell is very sharp. In addition, special electrosensors on the head can detect minute electrical fields produced by muscle contractions in nearby animals. Sharks also have a **lateral line system,** a row of sensory organs running along each side of the body. Sensitive to changes in water pressure, the lateral line system enables a shark to detect minor vibrations caused by animals swimming in its neighborhood. There are fewer than 1000 living species of cartilaginous fishes, nearly all of them marine.

Bony fishes (Figure 16.33b) have a skeleton reinforced by hard calcium salts. They also have a lateral line system, a keen sense of smell, and excellent eyesight. On each side of the head, a protective flap called the **operculum** (plural, opercula) covers a chamber housing the gills. Movement of the operculum allows the fish to breathe without swimming. (By contrast, sharks lack opercula and must generally swim to pass water over their gills.) Bony fishes also have a specialized organ that helps keep them buoyant—the **swim bladder,** a gas-filled sac. Thus, many bony fishes can conserve energy by remaining almost motionless, in contrast to sharks, which sink to the bottom if they stop swimming. Some bony fishes have a connection between the swim bladder and the digestive tract that enables them to gulp air and extract oxygen from it when the dissolved oxygen level in the water gets too low. In fact, swim bladders evolved from simple lungs that augmented gills in absorbing oxygen from the water of stagnant swamps, where the first bony fishes lived.

The largest class of vertebrates (about 30,000 species), bony fishes are common in the seas and in freshwater habitats. Most bony fishes, including trout, bass, perch, and tuna, are **ray-finned fishes.** Their fins are supported by thin, flexible skeletal rays (see Figure 16.33b). A second evolutionary branch of bony fishes includes the **lungfishes** and **lobe-finned fishes.** Lungfishes live today in the Southern Hemisphere. They inhabit stagnant ponds and swamps, surfacing to gulp air into their lungs. The lobe-fins are named for their muscular fins supported by stout bones. Lobe-fins are extinct except for one species, a deep-sea dweller that may use its fins to waddle along the seafloor. Ancient freshwater lobe-finned fishes with lungs played a key role in the evolution of amphibians, the first terrestrial vertebrates.

Amphibians

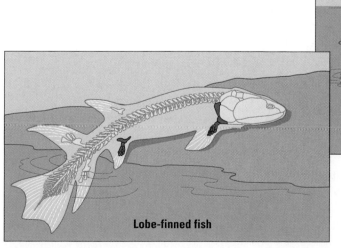

In Greek, the word *amphibios* means "living a double life." Most members of the class **Amphibia** exhibit a mixture of aquatic and terrestrial adaptations. Most species are tied to water because their eggs, lacking shells, dry out quickly in the air. The frog in Figure 16.34 spends much of its time on land, but it lays its eggs in water. An egg develops into a larva called a tadpole, a legless, aquatic algae-eater with gills, a lateral line system resembling that of fishes, and a long finned tail. In changing into a frog, the tadpole undergoes a radical metamorphosis. When a young frog crawls onto shore and begins life as a terrestrial insect-eater, it has four legs, air-breathing lungs instead of gills, a pair of external eardrums, and no lateral line system. Because of metamorphosis, many amphibians truly live a double life. But even as adults, amphibians are most abundant in damp habitats, such as swamps and rain forests. This is partly because amphibians depend on their moist skin to supplement lung function in exchanging gases with the environment. Thus, even those frogs that are adapted to relatively dry habitats spend much of their time in humid burrows or under piles of moist leaves. The amphibians of today, including frogs and salamanders, account for only about 8% of all living vertebrates, or about 4000 species.

Amphibians were the first vertebrates to colonize land. They descended from fishes that had lungs and fins with muscles and skeletal supports strong enough to enable some movement, however clumsy, on land (Figure 16.35). The fossil record chronicles the evolution of four-limbed amphibians from fishlike ancestors. Terrestrial vertebrates—amphibians, reptiles, birds, and mammals—are collectively called **tetrapods,** which means "four legs." Had our amphibian ancestors had three pairs of legs on their undersides instead of just two, we might be hexapods. This image seems silly, but serves to reinforce the point that evolution, as descent with modification, is constrained by history.

(a) Tadpole

(b) Tadpole undergoing metamorphosis

(c) Adult

Figure 16.34 The "dual life" of an amphibian. Though not all amphibians have aquatic larval stages, this class of vertebrates is named for the familiar tadpole-to-frog metamorphosis of many species.

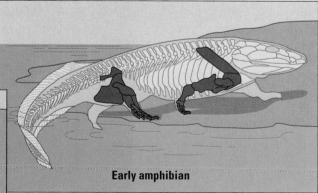

Lobe-finned fish

Early amphibian

Figure 16.35 The origin of tetrapods. Fossils of some lobe-finned fishes have skeletal supports extending into their fins. Early amphibians left fossilized limb skeletons that probably functioned in movement on land.

Figure 16.36 Terrestrial equipment of reptiles. This bull snake displays two reptilian adaptations to living on land: a waterproof skin with keratinized scales; and amniotic eggs, with shells that protect a watery, nutritious internal environment where the embryo can develop on land. Snakes evolved from lizards that adapted to a burrowing lifestyle.

Reptiles

Agnathans
Cartilaginous fishes
Bony fishes
Amphibians
Reptiles
Birds
Mammals

Class **Reptilia** includes snakes, lizards, turtles, crocodiles, and alligators. The evolution of reptiles from an amphibian ancestor paralleled many additional adaptations for living on land. Scales containing a protein called keratin waterproof the skin of a reptile, helping to prevent dehydration in dry air. Reptiles cannot breathe through their dry skin and obtain most of their oxygen with their lungs. Another breakthrough for living on land that evolved in reptiles is the **amniotic egg,** a water-containing egg enclosed in a shell (Figure 16.36). The amniotic egg functions as a "self-contained pond" that enables reptiles to complete their life cycle on land. With adaptations such as waterproof skin and amniotic eggs, reptiles broke their ancestral ties to aquatic habitats. There are about 6500 species of reptiles alive today.

Reptiles are sometimes labeled "cold-blooded" animals because they do not use their metabolism extensively to control body temperature. But reptiles do regulate body temperature by using behavioral adaptations. For example, many lizards regulate their internal temperature by basking in the sun when the air is cool and seeking shade when the air is too warm. Because they absorb external heat rather than generating much of their own, reptiles are said to be **ectotherms,** a term more accurate than *cold-blooded*. By heating directly with solar energy rather than through the metabolic breakdown of food, a reptile can survive on less than 10% of the calories required by a mammal of equivalent size.

As successful as reptiles are today, they were far more widespread, numerous, and diverse during the Mesozoic era, which is sometimes called the "age of reptiles." Reptiles diversified extensively during that era, producing a dynasty that lasted until the end of the Mesozoic about 65 million years ago. Dinosaurs, the most diverse group, included the largest animals ever to inhabit land. Some were gentle giants that lumbered about while browsing vegetation. Others were voracious carnivores that chased their larger prey on two legs (Figure 16.37).

The age of reptiles began to wane about 70 million years ago. During the Cretaceous, the last period of the Mesozoic era, global climate became cooler and more variable (see Chapter 13). This was a period of mass extinctions that claimed all the dinosaurs by about 65 million years ago, except for one lineage. That lone surviving lineage is represented today by birds.

Figure 16.37 A Mesozoic feeding frenzy. Hunting in packs, *Deinonychus* (meaning "terrible claw") probably used its sickle-shaped claws to slash at larger prey.

Birds

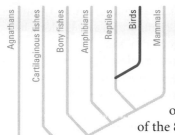

Birds (Class **Aves**) evolved during the great reptilian radiation of the Mesozoic era. Amniotic eggs and scales on the legs are just two of the reptilian features we see in birds. But modern birds look quite different from modern reptiles because of their feathers and other distinctive flight equipment. Almost all of the 8600 living bird species are airborne. The few flightless species, including the ostrich, evolved from flying ancestors. Appreciating the avian world is all about understanding flight.

Almost every element of bird anatomy is modified in some way that enhances flight. The bones have a honeycombed structure that makes them strong but light (the wings of airplanes have the same basic construction). For example, a huge seagoing species called the frigate bird has a wingspan of more than 2 meters, but its whole skeleton weighs only about 113 grams (4 ounces). Another adaptation that reduces the weight of birds is the absence of some internal organs found in other vertebrates. Female birds, for instance, have only one ovary instead of a pair. Also, modern birds are toothless, an adaptation that trims the weight of the head (no uncontrolled nosedives). Birds do not chew food in the mouth but grind it in the gizzard, a chamber of the digestive tract near the stomach.

Flying requires a great expenditure of energy from an active metabolism. In contrast to the ectothermic reptiles, birds are **endotherms.** That means they use their own metabolic heat to maintain a warm, constant body temperature.

A bird's most obvious flight equipment is its wings. Bird wings are airfoils that illustrate the same principles of aerodynamics as the wings of an airplane (Figure 16.38). A bird's flight motors are its powerful breast muscles, which are anchored to a keel-like breastbone. It is mainly these flight muscles that we call "white meat" on a turkey or chicken. Some birds, such as eagles and hawks, have wings adapted for soaring on air currents and flap their wings only occasionally. Other birds, including hummingbirds, excel at maneuvering but must flap continuously to stay aloft. In either case, it is the shape and arrangement of the feathers that form the wing into an airfoil. Feathers are made of keratin, the same protein that forms our hair and fingernails as well as the scales of reptiles. Feathers may have functioned first as insulation, helping birds retain body heat, only later being co-opted as flight gear.

In tracing the ancestry of birds back to the Mesozoic era, we must search for the oldest fossils with feathered wings that could have functioned in flight. Fossils of an ancient bird named *Archaeopteryx* have been found in Bavarian limestone in Germany dating back some 150 million years into the Jurassic period (see Figure 13.15). *Archaeopteryx* is not considered the ancestor of modern birds, and paleontologists place it on a side branch of the avian lineage. Nonetheless, *Archaeopteryx* probably was derived from ancestral forms that also gave rise to modern birds. Its skeletal anatomy indicates that it was a weak flyer, perhaps mainly a tree-dwelling glider. A combination of gliding downward and jumping into the air from the ground may have been the earliest mode of flying in the bird lineage.

Figure 16.38 A bald eagle in flight. Bird wings are airfoils, which have shapes that create lift by altering air currents. Air passing over a wing must travel farther in the same amount of time than air passing under the wing. This expands the air above the wing relative to the air below the wing. And this makes the air pressure pushing upward against the lower wing surface greater than the pressure of the expanded air pushing downward on the wing. The wings of birds and airplanes owe their "lift" to this pressure differential.

Mammals

Mammals (Class **Mammalia**) evolved from reptiles about 225 million years ago, long before there were any dinosaurs. During the peak of the age of reptiles, there were a number of mouse-sized, nocturnal mammals, which lived on a diet of insects. Mammals became much more diverse after the downfall of the dinosaurs. Most mammals are terrestrial. However, there are nearly 1000 species of winged mammals, the bats. And about 80 species of dolphins, porpoises, and whales are totally aquatic. The blue whale, an endangered species that grows to lengths of nearly 30 meters, is the largest animal that has ever existed.

Two features—hair and mammary glands that produce milk that nourishes the young—are mammalian hallmarks. The main function of hair is to insulate the body and help maintain a warm, constant internal temperature; mammals, like birds, are endotherms.

There are three major groups of mammals: the monotremes, the marsupials, and the eutherians. The duck-billed platypus is one of only three existing species of **monotremes,** the egg-laying mammals. The platypus lives along rivers in eastern Australia and on the nearby island of Tasmania. It eats mainly small shrimps and aquatic insects. The female usually lays two eggs and incubates them in a leaf nest. After hatching, the young nurse by licking up milk secreted onto the mother's fur. The animals called echidnas are also monotremes (Figure 16.39a).

(a) A monotreme (an echidna)

(b) A young marsupial in its mother's pouch (a brushtail opossum)

(c) Eutherians (placental mammals) (zebras)

Figure 16.39 The three major groups of mammals. (a) Monotremes, such as this echidna, are the only mammals that lay eggs (inset). (b) The young of marsupials, such as this brushtail opossum, are born very early in their development. They finish their growth while nursing from a nipple in their mother's pouch. (c) In eutherians (placentals), such as these zebras, young develop within the uterus of the mother. There they are nurtured by the flow of blood though the dense network of vessels in the placenta. The reddish portion of the afterbirth clinging to the newborn zebra in this photograph is the placenta.

Most mammals are born rather than hatched. During gestation in marsupials and eutherians, the embryos are nurtured inside the mother by an organ called the **placenta.** Consisting of both embryonic and maternal tissues, the placenta joins the embryo to the mother within the uterus. The embryo is nurtured by maternal blood that flows close to the embryonic blood system in the placenta. The embryo is bathed in fluid contained by an amniotic sac, which is homologous to the fluid compartment within the amniotic eggs of reptiles.

Marsupials are the so-called pouched mammals, including kangaroos and koalas. These mammals have a brief gestation and give birth to tiny, embryonic offspring that complete development while attached to the mother's nipples. The nursing young are usually housed in an external pouch, called the marsupium, on the mother's abdomen (Figure 16.39b). Nearly all marsupials live in Australia, New Zealand, and Central and South America. Australia has been a marsupial sanctuary for much of the past 60 million years. Australian marsupials have diversified extensively, filling terrestrial habitats that on other continents are occupied by eutherian mammals.

Eutherians are also called placental mammals because their placentas provide more intimate and long-lasting association between the mother and her developing young than do marsupial placentas (Figure 16.39c). Eutherians make up almost 95% of the 4500 species of living mammals. Dogs, cats, cows, rodents, rabbits, bats, and whales are all examples of eutherian mammals. One of the eutherian groups is the order **Primates,** which includes monkeys, apes, and humans.

Test your knowledge of chordates in Web/CD Activity 16C.

CheckPoint

1. What four features do we share with invertebrate chordates such as lancelets?

2. What feature of your skeleton is an example of segmentation?

3. A shark has a _____ skeleton, while a tuna has a _____ skeleton.

4. The oldest class of tetrapods are the _____.

5. What is an amniotic egg?

6. Birds and reptiles differ in their main source of body heat, with birds being _____ and reptiles being _____.

7. What are two main hallmarks of mammals?

Answers: 1. (1) Dorsal, hollow nerve chord; (2) notochord; (3) gill structures at some time during development; (4) post-anal tail at some time during development **2.** The vertebrae of your backbone **3.** cartilaginous; bony **4.** amphibians **5.** A shelled egg with the embryo contained in a fluid-filled sac **6.** endotherms; ectotherms **7.** Hair and mammary glands

The Human Ancestry

We have now traced the animal genealogy to the mammalian group that includes *Homo sapiens* and its closest kin. We are primates. To understand what that means, we must trace our ancestry back to the trees, where some of our most treasured traits originated as arboreal adaptations.

The Evolution of Primates

Primate evolution provides a context for understanding human origins. The fossil record supports the hypothesis that primates evolved from insect-eating mammals during the late Cretaceous period, about 65 million years ago. Those early primates were small, arboreal (tree-dwelling) mammals. Thus, the order Primates was first distinguished by characteristics that were shaped, through natural selection, by the demands of living in the trees. For example, primates have limber shoulder joints, which make it possible to brachiate (swing from one branch to another). The dexterous hands of primates can hang on to branches and manipulate food. Nails have replaced claws in many primate species, and the fingers are very sensitive. The eyes of primates are close together on the front of the face. The overlapping fields of vision of the two eyes enhance depth perception, an obvious advantage when brachiating. Excellent eye-hand coordination is also important for arboreal maneuvering. Parental care is essential for young animals in the trees. Mammals devote more energy to caring for their young than most other vertebrates, and primates are among the most attentive parents of all mammals (Figure 16.40). Most primates have single births and nurture their offspring for a long time. Though humans do not live in trees, we retain in modified form many of the traits that originated there.

Taxonomists divide the primates into two main groups: prosimians and anthropoids. The oldest primate fossils are **prosimians.** Modern prosimians include the lemurs of Madagascar and the lorises, pottos, and tarsiers that live in tropical Africa and southern Asia (Figure 16.41a). **Anthropoids** include monkeys, apes, and humans. All monkeys in the New World (the Americas) are arboreal and are distinguished by prehensile tails that function as an extra appendage for brachiating (Figure 16.41b). (If you see a monkey in a zoo swinging by its tail, you know it's from the New World.) Although some Old World monkeys are also arboreal, their tails are not prehensile. And many Old World monkeys, including baboons, macaques, and mandrills, are mainly ground-dwellers (Figure 16.41c).

Figure 16.40 The arboreal athleticism of primates.
This orangutan mother, toting her baby, displays primate adaptations for living in the trees: limber shoulder joints, manual dexterity, stereo vision due to eyes on the front of the face, and parental care.

(a) A prosimian (a brown mouse lemur)

Figure 16.41 Primate diversity.
(a) A prosimian. (b)–(c) Monkeys.
(d)–(g) Apes. (h) A human.

(b) A New World monkey (a white-faced capucin)

(c) Old World monkeys (pig-tailed macaques)

Our closest anthropoid relatives are the apes: gibbons, orangutans, gorillas, and chimpanzees (Figure 16.41d-g). Modern apes live only in tropical regions of the Old World. With the exception of some gibbons, apes are larger than monkeys, with relatively long arms and short legs and no tail. Although all the apes are capable of brachiation, only gibbons and orangutans are primarily arboreal. Gorillas and chimpanzees are highly social. Apes have larger brains proportionate to body size than monkeys, and their behavior is consequently more adaptable.

Sort out the primates in Web/CD Activity 16D.

(d) A gibbon

(e) An orangutan

(g) A chimpanzee

(f) A gorilla

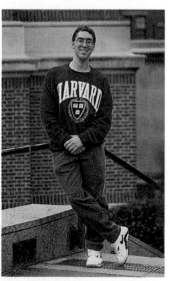

(h) A human

The Emergence of Humankind

Humanity is one very young twig on the vertebrate branch, just one of many branches on the tree of life. In the continuum of life spanning 3.5 billion years, humans and apes have shared a common ancestry for all but the last 5–7 million years (Figure 16.42). Put another way, if we compressed the history of life to a year, humans and chimpanzees diverged from a common ancestor less than 18 hours ago. The fossil record and molecular systematics concur in that vintage for the human lineage. **Paleoanthropology,** the study of human evolution, focuses on this very thin slice of biological history.

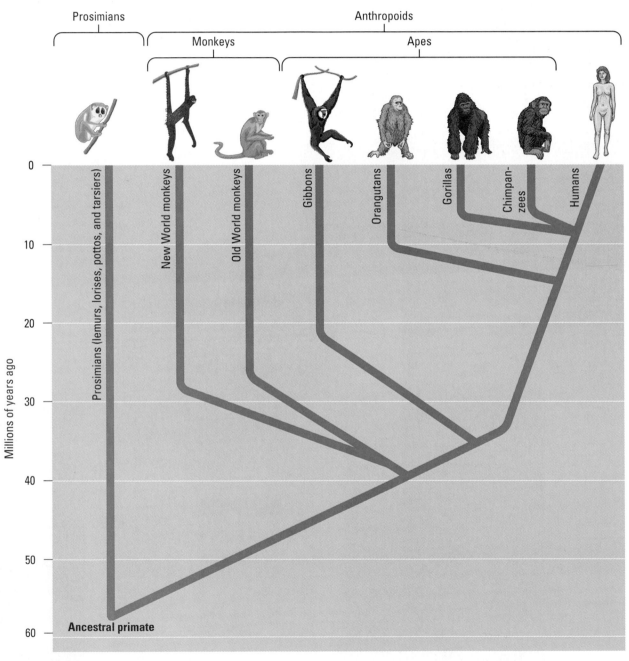

Figure 16.42 Primate phylogeny.

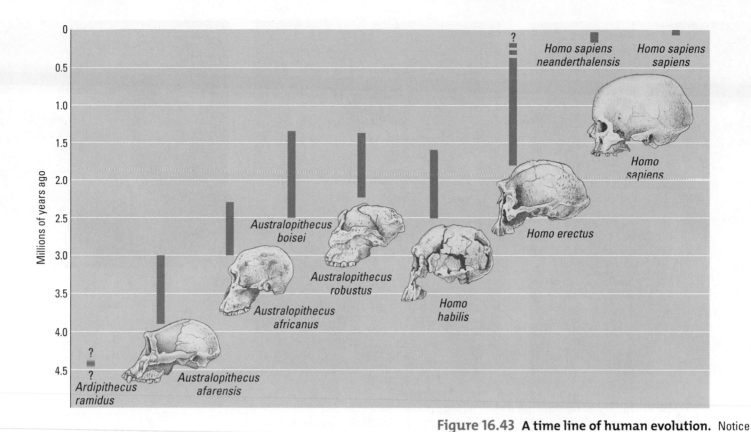

Figure 16.43 A time line of human evolution. Notice that there have been times when two or more hominid species coexisted. The skulls are all drawn to the same scale to enable you to compare the sizes of craniums and hence brains.

Some Common Misconceptions Certain misconceptions about human evolution that were generated during the early part of the twentieth century still persist today in the minds of many, long after these myths have been debunked by the fossil evidence.

Let's first dispose of the myth that our ancestors were chimpanzees or any other modern apes. Chimpanzees and humans represent two divergent branches of the anthropoid tree that evolved from a common, less specialized ancestor. Chimps are not our parent species, but more like our phylogenetic siblings or cousins.

Another misconception envisions human evolution as a ladder with a series of steps leading directly from an ancestral anthropoid to *Homo sapiens*. This is often illustrated as a parade of fossil **hominids** (members of the human family) becoming progressively more modern as they march across the page. If human evolution is a parade, then it is a disorderly one, with many splinter groups having traveled down dead ends. At times in hominid history, several different human species coexisted (Figure 16.43). Human phylogeny is more like a multibranched bush than a ladder, with our species being the tip of the only twig that still lives.

One more myth we must bury is the notion that various human characteristics, such as upright posture and an enlarged brain, evolved in unison. A popular image is of early humans as half-stooped, half-witted cave-dwellers. In fact, we know from the fossil record that different human features evolved at different rates, with erect posture, or bipedalism, leading the way. Our pedigree includes ancestors who walked upright but had ape-sized brains.

After dismissing some of the folklore on human evolution, however, we must admit that many questions about our ancestry have not yet been resolved.

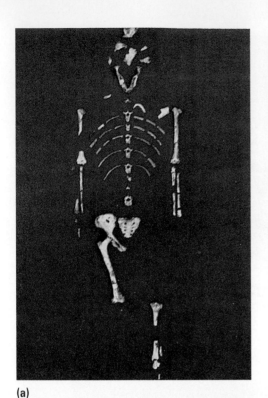

(a)

(b)

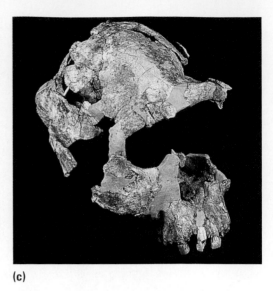

(c)

Figure 16.44 The antiquity of upright posture. (a) Lucy, a 3.18-million-year-old skeleton, represents the hominid species *Australopithecus afarensis*. Fragments of the pelvis and skull put *A. afarensis* on two feet. **(b)** Some 3.7 million years ago, several bipedal (upright-walking) hominids left footprints in damp volcanic ash in what is now Tanzania in East Africa. The prints fossilized and were discovered by British anthropologist Mary Leakey in 1978. The footprints are part of the strong evidence that bipedalism is a very old human trait. **(c)** This *A. afarensis* skull, 3.9 million years old, articulated with a vertical backbone. The upright posture of humans is at least that old.

***Australopithecus* and the Antiquity of Bipedalism** Before there was *Homo*, several hominid species of the genus *Australopithecus* walked the African savanna (grasslands with clumps of trees). Paleoanthropologists have focused much of their attention on *A. afarensis*, an early species of *Australopithecus*. Fossil evidence now pushes bipedalism in *A. afarensis* back to at least 4 million years ago (Figure 16.44). Doubt remains whether an even older hominid, *Ardipithecus ramidus*, which dates back at least 4.4 million years, was bipedal.

One of the most complete fossil skeletons of *A. afarensis* dates to about 3.2 million years ago in East Africa. Nicknamed Lucy by her discoverers, the individual was a female, only about 3 feet tall, with a head about the size of a softball (see Figure 16.44a). Lucy and her kind lived in savanna areas and may have subsisted on nuts and seeds, bird eggs, and whatever animals they could catch or scavenge from kills made by more efficient predators such as large cats and dogs.

All *Australopithecus* species were extinct by about 1.4 million years ago. Some of the later species overlapped in time with early species of our own genus, *Homo* (see Figure 16.43). Much debate centers on the evolutionary relationships of the *Australopithecus* species to each other and to *Homo*. Were the *Australopithecus* species all evolutionary side branches? Or were some of them ancestors to later humans? Either way, these early hominids show us that the fundamental human trait of bipedalism evolved millions of years before the other major human trait—an enlarged brain. As evolutionary biologist Stephen Jay Gould puts it, "Mankind stood up first and got smart later."

Homo habilis and the Evolution of Inventive Minds Enlargement of the human brain is first evident in fossils from East Africa dating to the latter part of the era of *Australopithecus,* about 2.5 million years ago. Anthropologists have found skulls with brain capacities intermediate in size between those of the latest *Australopithecus* species and those of *Homo sapiens.* Simple handmade stone tools are sometimes found with the larger-brained fossils, which have been dubbed *Homo habilis* ("handy man"). After walking upright for about 2 million years, humans were finally beginning to use their manual dexterity and big brains to invent tools that enhanced their hunting, gathering, and scavenging on the African savanna.

Homo erectus and the Global Dispersal of Humanity The first species to extend humanity's range from its birthplace in Africa to other continents was *Homo erectus,* perhaps a descendant of *H. habilis.* The global dispersal began about 1.8 million years ago. But don't picture this migration as a mad dash for new territory or even as a casual stroll. If *H. erectus* simply expanded its range from Africa by about a mile per year, it would take only about 12,000 years to populate many regions of Asia and Europe. The gradual spread of humanity may have been associated with a change in diet to include a larger proportion of meat. In general, animals that hunt require more geographic territory than animals that feed mainly on vegetation.

Homo erectus was taller than *H. habilis* and had a larger brain capacity. During the 1.5 million years the species existed, the *H. erectus* brain increased to as large as 1200 cubic centimeters (cm^3), a brain capacity that overlaps the normal range for modern humans. Intelligence enabled humans to continue succeeding in Africa and also to survive in the colder climates of the north. *Homo erectus* resided in huts or caves, built fires, made clothes from animal skins, and designed stone tools that were more refined than the tools of *H. habilis.* In anatomical and physiological adaptations, *H. erectus* was poorly equipped for life outside the tropics but made up for the deficiencies with cleverness and social cooperation.

Some African, Asian, European, and Australasian (from Indonesia, New Guinea, and Australia) populations of *H. erectus* gave rise to regionally diverse descendants that had even larger brains. Among these descendants of *H. erectus* were the Neanderthals, who lived in Europe, the Middle East, and parts of Asia from about 130,000 years ago to about 35,000 years ago. (They are named Neanderthals because their fossils were first found in the Neander Valley of Germany.) Compared with us, Neanderthals had slightly heavier browridges and less pronounced chins, but their brains, on average, were slightly larger than ours. Neanderthals were skilled toolmakers, and they participated in burials and other rituals that required abstract thought. Much current research on the Neanderthal skull addresses an intriguing question: Did Neanderthals have the anatomical equipment necessary for speech?

The Origin of Homo sapiens The oldest known post–*H. erectus* fossils, dating back over 300,000 years, are found in Africa. Many paleoanthropologists group these African fossils along with Neanderthals and various Asian and Australasian fossils as the earliest forms of our species, *Homo sapiens.* These regionally diverse descendants of *H. erectus* are sometimes referred to as "archaic *Homo sapiens.*" The oldest fully modern fossils of *H. sapiens*—skulls and other bones that look essentially like those of today's humans— are about 100,000 years old and are located in Africa. Similar fossils almost as ancient have also been discovered in caves in Israel. The famous fossils of

Test your knowledge of human evolution in Web/CD Activity 16E.

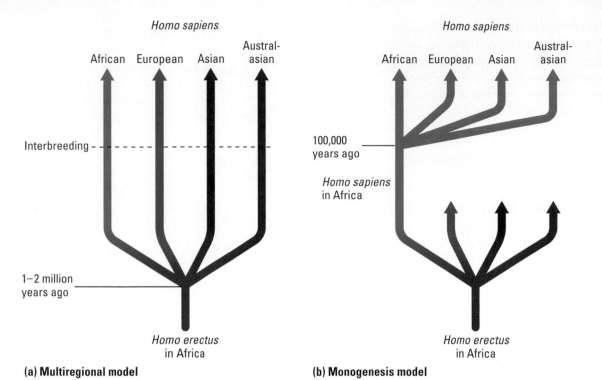

(a) Multiregional model

Homo sapiens

African European Asian Austral-asian

Interbreeding -

1–2 million
years ago

Homo erectus
in Africa

(b) Monogenesis model

Homo sapiens

African European Asian Austral-asian

100,000
years ago

Homo sapiens
in Africa

Homo erectus
in Africa

Figure 16.45 "Out of Africa"—but when? Two models for the origin of modern *Homo sapiens*. There is no question that Africa is the cradle of humanity; we can all trace our ancestry back to that continent. But how recently was the ancestor common to all the world's modern humans still in Africa? **(a)** According to the multiregional model, modern humans throughout the world evolved from the *Homo erectus* populations that spread to different regions beginning almost 2 million years ago. Interbreeding between populations (dashed line in the diagram) kept humans very similar genetically. **(b)** In contrast, the monogenesis model postulates that all regional descendants of *H. erectus* became extinct except in Africa. And according to this model, it was a *second* dispersal out of Africa, just 100,000 years ago, that populated the world with modern humans, which had evolved from *H. erectus* in Africa.

modern humans from the Cro-Magnon caves of France date back about 35,000 years.

The relationship of the Cro-Magnons and other modern humans to archaic *Homo sapiens* is a question that will continue to engage paleoanthropologists for decades. What was the fate of the various descendants of *Homo erectus* who populated different parts of the world? In the view of some anthropologists, these archaic *H. sapiens* populations gave rise to modern humans. According to this hypothesis, called the **multiregional model,** modern humans evolved simultaneously in different parts of the world (Figure 16.45a). If this view is correct, then the geographic diversity of humans originated relatively early, when *H. erectus* spread from Africa into the other continents between 1 and 2 million years ago. This model accounts for the great genetic similarity of all modern people by pointing out that interbreeding among neighboring populations has always provided corridors for gene flow throughout the entire geographic range of humanity.

In sharp contrast to the multiregional model is a hypothesis that modern *H. sapiens* arose from a single archaic group in Africa. According to this **monogenesis model,** the Neanderthals and other archaic peoples outside Africa were evolutionary dead ends (Figure 16.45b). Proponents of this hypothesis argue that modern humans spread out of Africa about 100,000 years ago and completely replaced the archaic *H. sapiens* in other regions. This hypothesis is based mainly on molecular data—specifically, analyses of DNA found in the mitochondria of human cells (mtDNA). The mtDNA in today's global human population seems to be very uniform, and supporters of monogenesis argue that such uniformity could only stem from a single ancestral stock, not from diverse, geographically isolated populations. Proponents of the multiregional model counter that interbreeding between populations can explain our great genetic similarity. Debate about which hypothesis is correct—a multiregional or a monogenetic origin of modern humans—centers mainly on differences in interpreting the fossil record and molecular data. As research continues, perhaps anthropologists will come closer to a consensus on our origins.

Cultural Evolution An erect stance was the most radical anatomical change in our evolution; it required major remodeling of the foot, pelvis, and vertebral column. Enlargement of the brain was a secondary alteration made possible by prolonging the growth period of the skull and its contents (see Figure 13.17). The primate brain continues to grow after birth, and the period of growth is longer for a human than for any other primate. The ex-

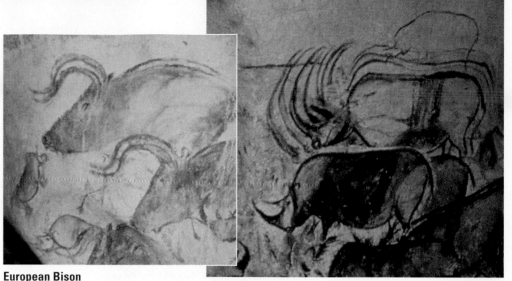

European Bison

Rhinoceroses

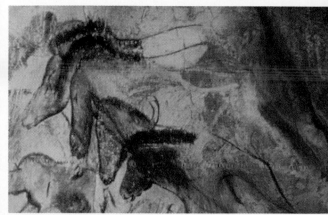

Horses

Figure 16.46 Art history goes way back—and so does our fascination with and dependence on animal diversity. Cro-Magnon wildlife artists created these remarkable paintings beginning about 30,000 years ago. Three cave explorers found this prehistoric art gallery on Christmas eve, 1994, when they ventured into a cavern near Vallon-Pont d'Arc in southern France.

Owl

tended period of human development also lengthens the time parents care for their offspring, which contributes to the child's ability to benefit from the experiences of earlier generations. This is the basis of **culture**—the transmission of accumulated knowledge over generations. The major means of this transmission is language, written and spoken.

Cultural evolution is continuous, but there have been three major stages. The first stage began with nomads who hunted and gathered food on the African grasslands 2 million years ago. They made tools, organized communal activities, and divided labor. Beautiful ancient art is just one example of our cultural roots in these early societies (Figure 16.46). The second main stage of cultural evolution came with the development of agriculture in Africa, Eurasia, and the Americas about 10,000 to 15,000 years ago. Along with agriculture came permanent settlements and the first cities. The third major stage in our cultural evolution was the Industrial Revolution, which began in the eighteenth century. Since then, new technology has escalated exponentially; a single generation spanned the flight of the Wright brothers and Neil Armstrong's walk on the moon. It took less than a decade for the Internet to transform commerce, communication, and education. Through all this cultural evolution, from simple hunter-gatherers to high-tech societies, we have not changed biologically in any significant way. We are probably no more intelligent than our cave-dwelling ancestors. The same toolmaker who chipped away at stones now designs microchips and software. The know-how to build skyscrapers, computers, and spaceships is stored not in our genes but in the cumulative product of hundreds of generations of human experience, passed along by parents, teachers, books, and electronic media.

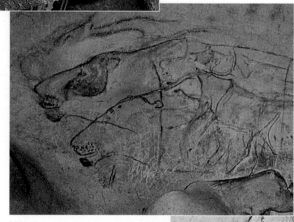

Lions

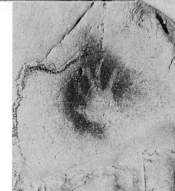

Human hand

1. To which mammalian order do we belong? What are the two main subgroups of this order?

2. Based on the fossil evidence in Figure 16.43, how many hominid species existed 1.7 million years ago?

3. Humans first evolved on which continent?

4. What role does a long period of parental care play in culture?

5. Why is *Homo habilis* known as "handy man"?

6. Lucy is an example of an early hominid belonging to the species _____ _____.

7. Which hypothesis—multiregional or monogenesis—postulates that all *Homo erectus* populations outside of Africa became extinct?

Answers: 1. Primates; prosimians and anthropoids **2.** Four **3.** Africa **4.** Extends opportunity for parents to transmit the lessons of the past to offspring **5.** Because it was the earliest hominid that definitely created and used stone tools **6.** *Australopithecus afarensis* **7.** Monogenesis

Evolution Link: Earth's New Crisis

Evolution of the human brain may have been anatomically simpler than acquiring an upright stance, but the global consequences of cerebral expansion have been enormous. Cultural evolution made *Homo sapiens* a new force in the history of life—a species that could defy its physical limitations and shortcut biological evolution. We do not have to wait to adapt to an environment through natural selection; we simply change the environment to meet our needs. We are the most numerous and widespread of all large animals, and everywhere we go, we bring environmental change. There is nothing new about environmental change. The history of life is the story of biological evolution on a changing planet. But it is unlikely that change has ever been as rapid as in the age of humans. Cultural evolution outpaces biological evolution by orders of magnitude. We are changing the world faster than many species can adapt; the rate of extinctions in the twentieth century was 50 times greater than the average for the past 100,000 years.

This rapid rate of extinction is mainly a result of habitat destruction, which is a function of human cultural changes and overpopulation. Feeding, clothing, and housing 6 billion people imposes an enormous strain on Earth's capacity to sustain life. If all these people suddenly assumed the high standard of living enjoyed by many people in developed nations, it is likely that Earth's support systems would be overwhelmed. Already, for example, current rates of fossil fuel consumption, mainly by developed nations, are so great that waste carbon dioxide may be causing the temperature of the atmosphere to increase enough to alter world climates. Today, it is not just individual species that are endangered, but entire ecosystems, the global atmosphere, and the oceans. Tropical rain forests, which play a vital role in moderating global weather, are being cut down at a startling rate. Scientists have hardly begun to study these ecosystems, and many species in them may become extinct before they are even discovered.

Learn more about Earth's crises in the Web Evolution Link.

Of the many crises in the history of life, the impact of one species, *Homo sapiens,* is the latest and potentially the most devastating.

Summary of Key Concepts

Overview: The Origins of Animal Diversity

- **What Is an Animal?** Animals are eukaryotic, multicellular, heterotrophic organisms that obtain nutrients by ingestion. Most animals reproduce sexually and develop from a zygote to a blastula, gastrula, and perhaps a larval stage before becoming an adult.

- **Early Animals and the Cambrian Explosion** Animals probably evolved from a colonial, flagellated protist more than 700 million years ago. At the beginning of the Cambrian period, 545 million years ago, animal diversity exploded.

- **Animal Phylogeny** Major branches of animal evolution are defined by four key evolutionary differences: the presence or absence of true tissues; radial versus bilateral body symmetry; presence or absence of a body cavity at least partly lined by mesoderm; and details of embryonic development.
- **Web/CD Activity 16A** *Overview of Animal Phylogeny*

Major Invertebrate Phyla

- **Sponges** Sponges (Phylum Porifera) are sessile animals with porous bodies and choanocytes but no true tissues. They filter-feed by drawing water through pores in the sides of the body.

- **Cnidarians** Cnidarians (Phylum Cnidaria) have radial symmetry, a gastrovascular cavity, and tentacles with cnidocytes. The body is either a sessile polyp or floating medusa.

- **Flatworms** Flatworms (Phylum Platyhelminthes) are the simplest bilateral animals. They may be free-living or parasitic in or on other animals.

- **Roundworms** Roundworms (Phylum Nematoda) are unsegmented and cylindrical with tapered ends. They have a complete digestive tract and a pseudocoelom. They may be free-living or parasitic in other animals.

- **Mollusks** Mollusks (Phylum Mollusca) are soft-bodied animals often protected by a hard shell. The body has three main parts: a muscular foot, a visceral mass, and a fold of tissue called the mantle.

- **Annelids** Annelids (Phylum Annelida) are segmented worms. They may be free-living or parasitic on other animals.

- **Arthropods** Arthropods (Phylum Arthropoda) are segmented animals with an exoskeleton and specialized, jointed appendages. Arthropods consist of four main groups: Arachnids, crustaceans, millipedes and centipedes, and insects. In species diversity, insects outnumber all other forms of life combined.
- **Web/CD The Process of Science** *Exploring Insect Diversity*

- **Echinoderms** Echinoderms (Phylum Echinodermata) are sessile or slow-moving marine animals that lack body segments and possess a unique water vascular system. Bilaterally symmetrical larvae usually change to radially symmetrical adults.
- **Web/CD Activity 16B** *Characteristics of Invertebrates*

The Vertebrate Genealogy

- **Characteristics of Chordates** Chordates (Phylum Chordata) are defined by a dorsal, hollow nerve cord; a flexible notochord; pharyngeal slits; and a post-anal tail. Vertebrates are chordates that possess a cranium and backbone.

- **Fishes** Agnathans are jawless vertebrates. Cartilaginous fishes (Class Chondrichthyes), such as sharks, are mostly predators with powerful jaws and a flexible skeleton made of cartilage. Bony fishes (Class Osteichthyes) have a stiff skeleton reinforced by hard calcium salts. Bony fishes are further classified into ray-finned fishes, lungfishes, and lobe-finned fishes.

- **Amphibians** Amphibians (Class Amphibia) are tetrapod vertebrates that usually deposit their eggs (lacking shells) in water. Aquatic larvae typically undergo a radical metamorphosis into the adult stage. Their moist skin requires that amphibians spend much of their adult life in moist environments.

- **Reptiles** Reptiles (Class Reptilia) are terrestrial ectotherms with lungs and waterproof skin covered by scales. Their amniotic eggs enhanced reproduction on land.

- **Birds** Birds (Class Aves) are endothermic vertebrates with amniotic eggs, wings, feathers, and other adaptations for flight.

- **Mammals** Mammals (Class Mammalia) are endothermic vertebrates with hair and mammary glands. There are three major groups: Monotremes lay eggs; marsupials use a placenta but give birth to tiny offspring that usually complete development while attached to nipples inside the mother's pouch; and eutherians, or placental mammals, use their placenta in a longer-lasting association between the mother and her developing young.
- **Web/CD Activity 16C** *Characteristics of Chordates*

The Human Ancestry

- **The Evolution of Primates** The first primates were small, arboreal mammals that evolved from insect-eating mammals about 65 million years ago. Two subgroups of primates are the prosimians and anthropoids. Anthropoids consist of New World monkeys (with prehensile tails), Old World monkeys (without prehensile tails), apes, and humans. Modern apes are confined to tropical regions of the Old World.
- **Web/CD Activity 16D** *Primate Diversity*

- **The Emergence of Humankind** Chimpanzees and humans evolved from a common, less specialized ancestor about 5–7 million years ago. Upright posture evolved in several hominid species of the genus *Australopithecus* at least 4 million years ago. Enlargement of the human brain came later, about 2.5 million years ago, when *Homo habilis* coexisted with smaller-brained hominids of the genus *Australopithecus*. *Homo erectus* was the first species to extend humanity's range from its birthplace in Africa to other continents. *Homo erectus* gave rise to regionally diverse descendants, such as the Neanderthals. According to the multiregional model, modern humans evolved in several locations from archaic *Homo sapiens*. In contrast, the monogenesis model views all but the African archaic *Homo sapiens* as evolutionary dead ends. According to this model, a relatively recent dispersal (100,000 years ago) of modern Africans gave rise to today's human diversity. Human cultural evolution began in Africa with wandering hunter-gatherers, then progressed to the development of agriculture and eventually to the Industrial Revolution, which today continues with accelerating technological change.
- **Web/CD Activity 16E** *Human Evolution*

Evolution Link: Earth's New Crisis

• Cultural evolution made *Homo sapiens* a new force in the history of life—a species that could defy its physical limitations and shortcut biological evolution. The exploding human population now threatens Earth's ecosystems.
• **Web Evolution Link** *Earth's New Crisis*

Self-Quiz

1. Reptiles are much more extensively adapted to life on land than amphibians in that reptiles
   a. have a complete digestive tract.
   b. lay eggs that are enclosed in shells.
   c. are endothermic.
   d. have legs.
   e. go through a larval stage.

2. A small group of worms classified in the phylum Gnathostomulida (not discussed in this chapter) live between sand grains on ocean beaches. These animals are unsegmented and lack a body cavity, but some have a complete digestive tract. These characteristics suggest that gnathostomulid worms might branch from the main trunk of the animal phylogenetic tree between
   a. flatworms and nematodes.
   b. annelids and arthropods.
   c. mollusks and annelids.
   d. cnidarians and flatworms.
   e. nematodes and annelids.

3. Which of the following categories includes all others in the list?
   a. arthropod   d. arachnid
   b. insect   e. butterfly
   c. invertebrate

4. The two major groups of primates are
   a. monkeys and anthropoids.
   b. prosimians and apes.
   c. monkeys and apes.
   d. prosimians and anthropoids.
   e. Old World monkeys and New World monkeys.

5. Which of the following correctly lists probable ancestors of modern humans, from the oldest to the most recent?
   a. *Homo erectus, Australopithecus, Homo habilis*
   b. *Australopithecus, Homo habilis, Homo erectus*
   c. *Australopithecus, Homo erectus, Homo habilis*
   d. *Homo erectus, Homo habilis, Australopithecus*
   e. *Homo habilis, Homo erectus, Australopithecus*

6. Fossils suggest that the first major trait distinguishing human primates from other primates was
   a. forward-facing eyes with depth perception.
   b. a larger brain.
   c. erect posture.
   d. grasping hands.
   e. toolmaking.

7. Bilateral symmetry in the animal kingdom is best correlated with
   a. an ability to sense equally in all directions.
   b. the presence of a skeleton.
   c. motility and active predation and escape.
   d. development of a true coelom.
   e. adaptation to terrestrial environments.

8. Which of the following is an *incorrect* match of animal with phylum?
   a. human—Chordata   d. sponge—Porifera
   b. leech—Arthropoda   e. earthworm—Annelida
   c. sea anemone—Cnidaria

9. Which of the following types of animals is not included in the human ancestry?
   a. a reptile   d. a bird
   b. a bony fish   e. an amphibian
   c. a primate

10. Our phylum is named for which of the following structures?
    a. nerve cord   d. mammary glands
    b. notochord   e. blastula stage of embryonic development
    c. backbone

• **Go to the website or CD-ROM for more self-quiz questions.**

The Process of Science

1. You have dredged up an unknown animal from the seafloor. Describe some of the characteristics you should look at to determine the animal phylum to which the creature should be assigned.

2. **Explore insect diversity in The Process of Science activity on the website and CD-ROM.**

Biology and Society

Give examples from at least three phyla of animals that humans consume as food.

Ecology

The Biosphere: Earth's Diverse Environments

The oceans cover about 75% of Earth's surface. Seasons result from the permanent tilt  of Earth on its axis as it orbits the sun. As the Earth rotates, a person standing at the equator is moving much faster than a person standing in New  York City. Humans currently use about 60% of Earth's land, mostly as cropland, forest, or  rangeland.

All organisms, including humans, interact continuously with their environments. We breathe, exchanging gases with the atmosphere. We eat other organisms, taking in second-hand energy that entered plants as sunlight. We add urea and other wastes to water and soil. We absorb and radiate heat. We are inextricably connected to the outside world, as are all of Earth's creatures.

The scientific study of the interactions between organisms and their environments is called **ecology** (from the Greek *oikos,* home, and *logos,* to study). This straightforward definition masks an enormously complex and exciting area of biology that is also of crucial practical importance. Photographs of Earth, taken by Apollo astronauts over 30 years ago, remind us that our planet is a finite home in the vastness of space, not an unlimited frontier that we humans can abuse indefinitely (Figure 17.1). The science of ecology provides a basic understanding of how natural processes and organisms interact, giving us the tools we need to manage the planet's limited resources over the long term. This chapter begins our study of ecology with an emphasis on Earth's diverse environments and the adaptations that have evolved in these varied ecological arenas.

Figure 17.1 An earthrise photographed from the moon. Apollo astronaut Rusty Schweickart, reflecting on such a view, once remarked: "On that small blue-and-white planet below is everything that means anything to you. National boundaries and human artifacts no longer seem real. Only the biosphere, whole and home of life."

Overview: The Scope of Ecology

Let's take a closer look at three key words in our definition of ecology: the *scientific* study of the *interactions* between organisms and their *environments.*

Ecology as Scientific Study

Humans have always had an interest in other organisms and their environments. As hunters and gatherers, prehistoric people had to learn where game and edible plants could be found in greatest abundance. And naturalists, from Aristotle to Darwin, made the process of observing and describing organisms in their natural habitats an end in itself rather than simply a means of survival. Extraordinary insight can still be gained from this descriptive approach of watching nature and recording its structure and processes. (As Yogi Berra once put it, "You can observe a lot just by watching.") Thus, natural history as a "discovery science" (see Chapter 1) remains fundamental to ecology (Figure 17.2). But in the past few decades, ecology has become increasingly experimental. In spite of the difficulty of

(a)

(b)

Figure 17.2 Ecological research in a rain forest canopy. (a) This hot-air dirigible is placing a giant rubber raft on the treetops of a tropical rain forest in French Guiana, in northeastern South America. (b) Living and working on this field station 30 meters above the forest floor, an international research team began describing the canopy environment and its life in the 1990s. French biologist Pierre Grard and two other scientists are cataloging plants and the insects that feed on those plants. Among Earth's diverse environments, the canopy (upper tier) of tropical rain forests is especially rich in its diversity of insects, birds, and other animals.

Figure 17.3 Experimental ecology on a grand scale. In this classic experiment, David Schindler (University of Alberta) and his colleagues tested the hypothesis that the availability of mineral nutrients such as nitrogen and phosphorus can promote the growth of algae in lakes. The researchers used a plastic curtain to partition this lake into an "experimental lake" (top) and a "control lake." The team added certain mineral nutrients to the experimental lake, but not to the control, which served as a basis of comparison. Within two months after the scientists added phosphorus, a bloom (population explosion) of algae gave the experimental lake the cloudy, whitish appearance you see in the photograph. The experiment helped ecologists understand how the pollution of lakes with phosphorus from sewage and runoff from fertilized farmland can cause algal blooms. In such "overfertilized" lakes, bacteria and other decomposers of the algal mass sometimes consume all the oxygen, leading to the death of fish and other aerobic organisms. Scientific evidence has resulted in better controls on water quality, including the banning of phosphate-containing detergents in many states. (Reprinted with permission from D. W. Schindler, "Eutrophication and Recovery in Experimental Lakes: Implications for Lake Management," *Science* vol. 184, page 897, Figure 1.49 (1974). Copyright © 1974 American Association for the Advancement of Science.)

conducting experiments that often involve large amounts of time and space, many ecologists are testing hypotheses in the laboratory and in the field. Ecologists also complement descriptive and experimental studies with computer simulations of large-scale experiments that might be impossible to conduct in the field. Whatever the approach, ecology employs that most basic of scientific processes, the posing of hypotheses and the use of observations and experiments to test those hypotheses. In the experiment depicted in Figure 17.3, for example, scientists tested the hypothesis that mineral nutrients can promote the growth of algae in lakes.

Major Components of the Environment

The environment includes two major components. The **abiotic component** consists of nonliving chemical and physical factors, such as temperature, light, water, minerals, and air. The **biotic component** includes the living factors—all the other organisms that are part of an individual's environment. Other organisms may compete with an individual for food and other resources, prey upon it, or change its physical and chemical environment.

A Hierarchy of Interactions

When we study the interactions between organisms and their environments, it is convenient to divide ecology into four increasingly comprehensive levels: organismal ecology, population ecology, community ecology, and ecosystem ecology (see Figure 2.1).

Organismal ecology is concerned with the evolutionary adaptations that enable individual organisms to meet the challenges posed by their abiotic environments. For example, an organismal ecologist might be interested in the special equipment that enables diving whales to stay under water for so long (Figure 17.4a). The distribution of organisms is limited by the abiotic conditions they can tolerate. Earthworms, for instance, are restricted to moist environments because their skin does not prevent dehydration.

(b) Population ecology: What factors limit the number of striped mice that can inhabit a particular area?

(c) Community ecology: What factors influence the diversity of tree species that make up a particular forest?

(a) Organismal ecology: How do diving whales stay under water for so long?

Figure 17.4 Examples of questions at different levels of ecology.

(d) Ecosystem ecology: What processes recycle vital chemical elements such as nitrogen within a savanna ecosystem?

The next level of organization in ecology is the **population,** a group of individuals of the same species living in a particular geographic area. **Population ecology** concentrates mainly on factors that affect population density and growth. What factors, for example, limit the number of striped mice that can inhabit a particular area (Figure 17.4b)?

A **community** consists of all the organisms that inhabit a particular area; it is an assemblage of populations of different species. Questions in **community ecology** focus on how interactions between species, such as predation, competition, and symbiosis, affect community structure and organization. For example, what factors influence the diversity of tree species that make up a particular forest (Figure 17.4c)?

An **ecosystem** includes all the abiotic factors in addition to the community of species in a certain area. For example, a forest ecosystem includes not only the organisms, such as diverse plants and animals, but also the soil, water sources, sunlight, and other abiotic factors of the environment. In **ecosystem ecology,** questions concern energy flow and the cycling of chemicals among the various biotic and abiotic factors. What processes, for instance, recycle vital chemical elements such as nitrogen within a savanna ecosystem (Figure 17.4d)?

The **biosphere** is the global ecosystem—the sum of all the planet's ecosystems, or all of life and where it lives. The most complex level in ecology, the biosphere includes the atmosphere to an altitude of several kilometers, the land down to water-bearing rocks about 1500 meters deep, lakes and streams, caves, and the oceans to a depth of several kilometers. Isolated in space, the biosphere is self-contained, or closed, except that its photosynthetic producers derive energy from sunlight, and it loses heat to space.

Figure 17.5 Rachel Carson and *Silent Spring*. Although her book, seminal to the modern environmental movement, focused on the biosphere's hangover from the pesticide DDT, Carson's message was much broader: "The 'control of nature' is a phrase conceived in arrogance, born of the Neanderthal age of biology and philosophy, when it was supposed that nature exists for the convenience of man."

Ecology and Environmentalism

The science of ecology should be distinguished from the informal use of the word ecology to refer to environmental concerns. And yet, we need to understand the complicated and delicate relationships between organisms and their environments to address environmental problems.

Our current awareness of the biosphere's limits stems mainly from the 1960s, a time of growing disillusionment with environmental practices of the past. In the 1950s, technology seemed poised to free humankind from several age-old bonds. New chemical fertilizers and pesticides, for example, showed great promise for increasing agricultural productivity and eliminating insect-borne diseases. Fertilizers were applied extensively, and pests were attacked with massive aerial spraying of pesticides, including a poison called DDT. The immediate results were astonishing: Increases in farm productivity enabled developed nations such as the United States to grow surplus food and market it overseas, and the worldwide incidence of malaria and several other insect-borne diseases was markedly reduced. The pesticide DDT was hailed as a miracle weapon to wield anywhere insects caused problems.

Our enthusiasm for chemical fertilizers and pesticides began to wane as some of the side effects of DDT and other widely used poisons began appearing in the late 1950s. One of the first to perceive the global dangers of pesticide abuse was the late Rachel Carson (Figure 17.5). We can trace our current environmental awareness to Carson's 1962 book, *Silent Spring*. Her warnings were underscored when scientists began reporting that DDT was threatening the survival of predatory birds and was showing up in human milk. Another serious problem was the genetic resistance to pesticides that evolved in an increasing number of pest populations (see Chapter 12). At the same time, some of the ill effects of other technological developments began to be widely publicized. By the early 1970s, disillusionment with the overuse of chemicals and a realization that our finite biosphere could not tolerate unlimited exploitation had developed into widespread concern about the environment. The environmental movement that Carson helped catalyze continues, with students of your generation sustaining a tradition of activism on behalf of the biosphere's future health.

Today, it's clear that no part of the biosphere is untouched by the abusive impact of human populations and their technology. Depletion of natural resources, localized famine aggravated by land misuse and expanding population, the growing list of species extinguished or endangered by loss of habitat, the poisoning of soil and streams with toxic wastes, and global warming caused by deforestation and combustion of fossil fuels—these are just a few of the problems that we have created and now must solve. Analyzing environmental issues and planning for better practices start with a basic understanding of ecology, which should be part of every student's education.

Analyze the pros and cons of DDT in Web/CD Activity 17A.

CheckPoint

1. A (an) _____ consists of a biological _____, or all the biotic factors in the area, along with the nonliving, or _____, factors.

2. Why is it more accurate to define the biosphere as the global *ecosystem* rather than the global *community*?

3. Rachel Carson's *Silent Spring* focused on the destructive consequences of toxic pollutants, especially _____.

Answers: 1. ecosystem; community; abiotic **2.** Because the biosphere includes both abiotic and biotic factors **3.** the pesticide DDT

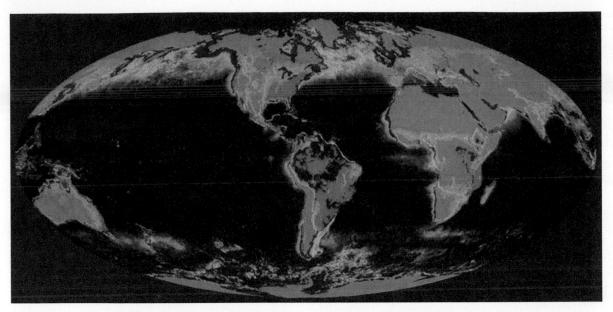

Figure 17.6 Regional distribution of life in the biosphere. In this image of Earth, based on data sent from satellites, colors are keyed to the relative abundance of chlorophyll, which correlates with the regional densities of life. Green areas on land are dense forests, including the tropical forests of South America and Africa. Orange areas on land are relatively barren regions, such as the Sahara of Africa. The patchy productivity of the oceans is also evident, with green regions having a greater abundance of phytoplankton (algae) than darker regions. The patchiness of the biosphere is mainly due to regional variations in abiotic factors such as temperature and the availability of water and mineral nutrients.

Abiotic Factors of the Biosphere

The biosphere is patchy. We can see this environmental patchwork on several levels. On a global scale, ecologists have long recognized striking regional patterns in the distribution of terrestrial and aquatic life (Figure 17.6). These patterns mainly reflect regional differences in climate and other abiotic factors. Figure 17.7 shows patchiness on a local scale; we can see a mixture of forest, small lakes, a meandering river, and open meadows. If we moved even closer, into any one of these different environments, we would find patchiness on yet a smaller scale. For example, we would find that each lake has several different **habitats**—environmental situations in which organisms live. And each habitat has a characteristic community of organisms. As with the global patchiness of the biosphere, smaller-scale environmental variation is based mainly on differences in abiotic factors, such as temperature and the availability of water and light.

Major Abiotic Factors

Sunlight Solar energy powers nearly all ecosystems. In aquatic environments, the availability of sunlight has a significant effect on the growth and distribution of algae. Because water itself and the microorganisms in it absorb light and keep it from penetrating very far, most photosynthesis occurs near the surface of the water. In terrestrial environments, light is often not the most important factor limiting plant growth. In many forests, however, shading by trees creates intense competition for light at ground level.

Figure 17.7 Patchiness of the environment in an Alaskan wilderness.

Figure 17.8 Home of hot prokaryotes. Heat loving prokaryotes thrive in the Grand Prismatic Pool at Yellowstone National Park, where temperatures may exceed 80°C (176°F).

Figure 17.9 Wind as an abiotic factor that shapes trees. The prevailing winds gradually caused the "flagging" of these fir trees on a ridge on Oregon's Mt. Hood. The mechanical disturbance of the wind inhibits limb growth on the windward side of the trees, while limbs on the leeward side grow normally. This growth response is an evolutionary adaptation that reduces the number of limbs that are broken during strong winds.

Water Aquatic organisms have a seemingly unlimited supply of water, but they face problems of water balance if their own solute concentration does not match that of their surroundings. (Figure 4.26 shows water balance in cells.) For a terrestrial organism, the main water problem is the threat of drying out. Many land species have watertight coverings that reduce water loss. For example, a waxy coating on the leaves and other aerial parts of most plants helps prevent dehydration. (Rubbing a cucumber is a good way to sample this wax.) And humans and other mammals have a layer of dead outer skin containing a waterproofing protein. Moreover, the ability of our kidneys to excrete a very concentrated urine is an evolutionary adaptation that enables us to rid our body of urea, a waste product, with minimal water loss.

Temperature Environmental temperature is an important abiotic factor because of its effect on metabolism. Few organisms can maintain a sufficiently active metabolism at temperatures close to 0°C (32°F), and temperatures above 50°C (122°F) destroy the enzymes of most organisms. Still, extraordinary adaptations enable some species to live outside this temperature range. For example, some of the frogs and turtles living in the northern United States and Canada can freeze during the winter months and still survive, and prokaryotes living in hot springs have enzymes that function optimally at extremely high temperatures (Figure 17.8). Mammals and birds can remain considerably warmer than their surroundings and can be active in a fairly wide range of temperatures, but even these animals function best at certain temperatures.

Wind Some organisms—for example, bacteria, protists, and many insects that live on snow-covered mountain peaks—depend on nutrients blown to them by winds. Many plants depend on wind to disperse their pollen and seeds. Local wind damage often creates openings in forests, contributing to patchiness in ecosystems. Wind also increases an organism's rate of water loss by evaporation. The consequent increase in evaporative cooling can be advantageous on a hot summer day, but can cause dangerous wind chill in the winter. Wind can also affect the pattern of a plant's growth (Figure 17.9).

Investigate the effects of changing abiotic factors in Web/CD Activity 17B/The Process of Science.

Rocks and Soil The physical structure and chemical composition of rocks and soil limit the distribution of plants and of the animals that feed on the vegetation. Soil variation contributes to the patchiness we see in terrestrial landscapes such as in Figure 17.7. In streams and rivers, the composition of the substrate can affect water chemistry, which in turn influences the resident plants and animals. In marine environments, the structure of the substrates in the intertidal zone and on seafloors determines the types of organisms that can attach or burrow in those habitats.

Periodic Disturbances

Catastrophic disturbances, such as fires, hurricanes, tornadoes, and volcanic eruptions, can devastate biological communities. After the disturbance, the area is recolonized by organisms or repopulated by survivors, but the structure of the community undergoes a succession of changes during the rebound (Figure 17.10). Some disturbances, such as volcanic eruptions, are so infrequent and irregular over space and time that organisms have not acquired evolutionary adaptations to them. Fire, on the other hand, although unpredictable over the short term, recurs frequently in some communities, and many plants have adapted to this periodic disturbance. In fact, several communities actually depend on periodic fire to maintain them, as we will see later in this chapter.

Figure 17.10 Recovery after a forest fire. Just a few months after a 1988 forest fire swept through Yellowstone National Park, wildflowers and other small plants were already colonizing the area. The increased sunlight and soil nutrients released from the trees that burned were among the abiotic factors contributing to the regreening of the scorched land.

Global Climate Patterns

When we ask what determines whether a particular organism or community of organisms lives in a certain area, the climate of the region—especially temperature and rainfall—is often a large part of the answer. Earth's global climate patterns are largely determined by the input of solar energy and the planet's movement in space.

Solar Energy and Latitude Because of its curvature, Earth receives an uneven distribution of solar energy (Figure 17.11). The sun's rays strike equatorial areas most directly (perpendicularly). At latitudes away from the equator, the rays strike Earth's surface at oblique angles. As a result, the same amount of solar energy is spread over a larger area. Thus, any particular area of land or ocean near the equator absorbs more heat than comparable areas in the more northern or southern latitudes. The uneven heating of Earth's surface is a major factor driving air movements and water currents.

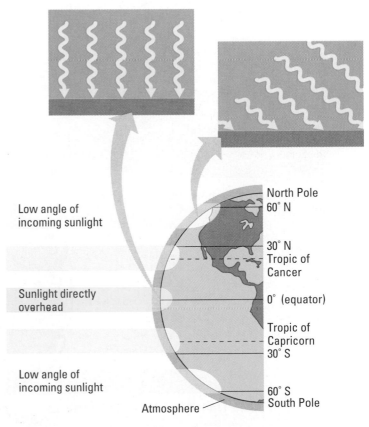

Figure 17.11 Uneven heating of Earth's surface. The equator receives the greatest intensity of solar radiation. At higher latitudes, sunlight has a longer path through the atmosphere and strikes Earth's surface at an oblique angle.

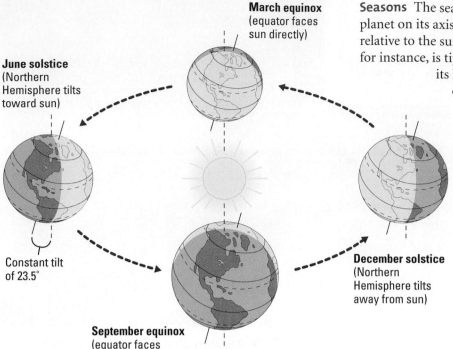

March equinox
(equator faces
sun directly)

June solstice
(Northern
Hemisphere tilts
toward sun)

Constant tilt
of 23.5°

September equinox
(equator faces
sun directly)

December solstice
(Northern
Hemisphere tilts
away from sun)

Figure 17.12 Earth's tilts and the seasons.

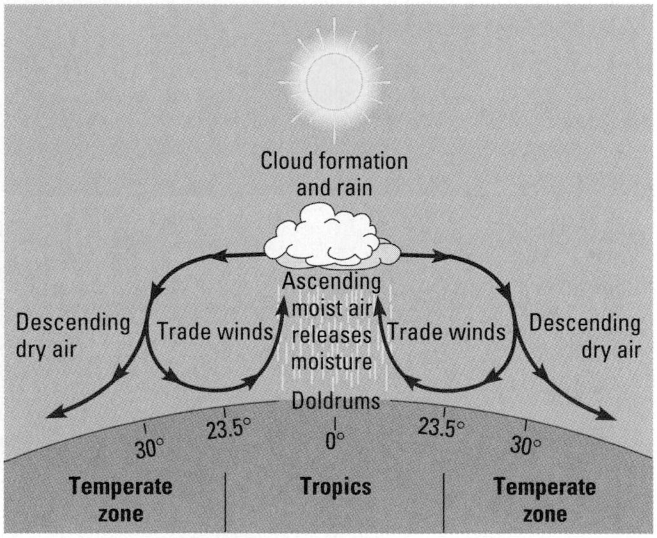

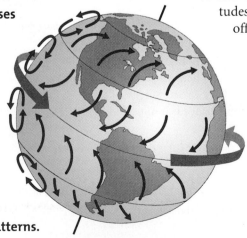

Cloud formation
and rain

Ascending
moist air
releases
moisture

Descending
dry air

Trade winds

Trade winds

Descending
dry air

Doldrums

30°   23.5°   0°   23.5°   30°

**Temperate
zone**

Tropics

**Temperate
zone**

**Figure 17.13 How uneven heating of Earth causes
rain and wind.**

Seasons The seasons of the year result from the permanent tilt of the planet on its axis as it orbits the sun (Figure 17.12). The globe's orientation relative to the sun changes throughout the year. The Northern Hemisphere, for instance, is tipped most toward the sun in June, creating summer, with its long, warm days, in that hemisphere. At the same time, days are short and it's winter in the Southern Hemisphere. Conversely, the Southern Hemisphere is tipped most to ward the sun in December, creating summer there and winter in the Northern Hemisphere.

Global Air Circulation, Precipitation, Winds, and Ocean Currents Figure 17.13 shows some of the effects of uneven warming on global wind patterns and rainfall. When air is warmed, it rises and tends to absorb moisture; when air is cooled, it falls and tends to lose moisture. Heated by the direct rays of the sun, air at the equator rises, creating an area of calm or of very light winds known as the doldrums. As warm equatorial air rises, it cools, forms clouds, and drops rain. Warm temperatures throughout the year and heavy rainfall largely explain why rain forests are concentrated near the equator.

After losing moisture over equatorial zones, high-altitude air masses spread away from the equator until they cool and descend again at latitudes of about 30° north and south. Many of the world's great deserts—the Sahara in North Africa and the Arabian on the Arabian Peninsula, for example—are centered at these latitudes because of the dry air they receive from the equator. As the dry air descends, some of it spreads back toward the equator. This movement creates the **trade winds,** which dominate the **tropics** (latitudes between 23.5° north and 23.5° south). As the air moves back toward the equator, it warms and picks up moisture until it is uplifted again.

Latitudes between the tropics and the Arctic Circle in the north and the Antarctic Circle in the south are called **temperate zones.** Generally, these regions have milder climates than the tropics or the polar regions. Notice in Figure 17.13 that some of the descending dry air heads into the latitudes above 30°. At first these air masses pick up moisture, but they tend to drop it as they cool at higher latitudes. This is why the north and south temperate zones, especially latitudes between about 40° and 60°, tend to be relatively wet. New York City, for example, sits at a latitude of 40.4° north. Deserts and arid regions in higher latitudes usually result from mountain ranges that cut off the flow of moist air, as we'll see.

The rising and falling of air masses combines with Earth's rotation to produce **prevailing winds** (Figure 17.14). Because Earth is spherical, its surface moves faster at the equator than at other latitudes. (A point on the equator must travel much farther in 24 hours than a point nearer the poles.) In the tropics, Earth's rapidly moving surface deflects vertically circulating air, making the trade winds blow

Figure 17.14 Prevailing wind patterns.

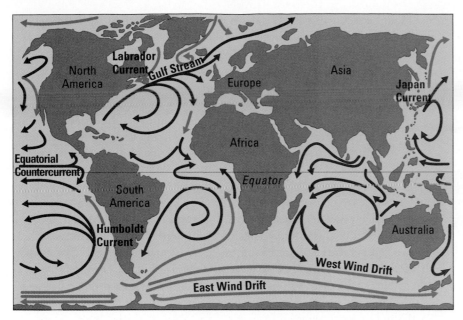

Figure 17.15 **Ocean currents.**

from the northeast in the Northern Hemisphere and from the southeast in the Southern Hemisphere. In temperate zones, the slower-moving surface produces the westerlies, winds that blow from west to east. Earth's prevailing winds had a huge impact on human civilizations, as people with goods sailed across the oceans for centuries.

Ocean currents are riverlike flow patterns in the oceans (Figure 17.15). They result from a combination of the prevailing winds, the planet's rotation, unequal heating of surface waters, and the locations and shapes of the continents. Ocean currents have a profound effect on regional climates. For instance, the Gulf Stream circulates warm water northward from the Gulf of Mexico and makes the climate on the west coast of Great Britain warmer than many areas farther south. In contrast, the northward-flowing Humboldt Current cools the west coast of South America.

Local Climate

On a good March day in Southern California, it is possible to spend the morning snowboarding in freezing temperatures in the mountains and then go to the beach in the afternoon to surf and bask under 65°F (18°C) skies. And in August, the 75°F (24°C) beach offers welcome relief to those who live just 40 or so miles inland in valleys that often cook at 100°F (38°C) or higher (Figure 17.16). Such local variations in climate are superimposed on the global patterns.

Proximity to large bodies of water and the presence of landforms such as mountain ranges are the two most important factors affecting local variations in climate. Oceans and large lakes moderate climate by absorbing heat when the air is warm and releasing heat to cold air (see Chapter 2 to review this property of water).

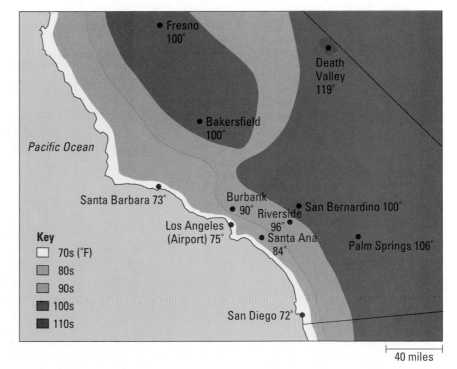

Figure 17.16 **Local high temperatures for August 6, 2000 in Southern California.**

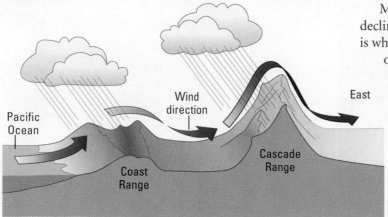

Figure 17.17 How mountains affect rainfall. This drawing represents major landforms across the state of Washington, but mountain ranges cause similar effects elsewhere. Washington is a temperate area in which the prevailing winds are westerlies. As moist air moves in off the Pacific Ocean and encounters the westernmost mountains (the Coast Range), it flows upward, cools at higher altitudes, and drops a large amount of water. The biological community in this wet region is a temperate rain forest. Some of the world's tallest trees, the Douglas firs, thrive here. Farther inland, precipitation increases again as the air moves up and over higher mountains (the Cascade Range). On the eastern side of the Cascades, there is little precipitation. As a result of this rain shadow, much of central Washington is very arid, almost qualifying as desert.

Mountains affect local climate in two major ways. First, air temperature declines by about 6°C with every 1000-meter increase in elevation, which is why it is so much cooler in the mountains than in nearby lowlands. Second, mountains can block the flow of cool, moist air from a coast, causing radically different climates on opposite sides of a mountain range (Figure 17.17).

Climate also varies on a very fine scale, called **microclimate.** For example, ecologists often refer to the microclimate on a forest floor or under a rock. Many features in the environment influence microclimates by casting shade, reducing evaporation from soil, and minimizing the effects of wind. If you have ever lifted a log or large stone in the woods, you have observed that there are organisms (such as salamanders, worms, and some insects) that live in the shelter of this microenvironment, buffered from the extremes of temperature and moisture. Every environment on Earth is similarly characterized by a mosaic of small-scale differences in the abiotic factors that influence the distribution of organisms. For now, however, let's return to the global scale to see how climate patterns determine the distribution of Earth's major ecosystems.

Terrestrial Biomes

Major types of ecosystems that cover large geographic regions are called **biomes.** You can think of them as types of landscapes. Tropical forests and deserts are examples. Figure 17.18 shows a map of Earth's major terrestrial biomes.

The distribution of the biomes largely depends on climate, with temperature and rainfall often the key factors determining the kind of biome that exists in a particular region. Because there are latitudinal patterns of climate over Earth's surface, there are also latitudinal patterns of biome distribution.

If the climate in two geographically separate areas is similar, the same type of biome may occur in both places. Coniferous forests, for instance, extend in a broad band across North America, Europe, and Asia. Note, however, that a biome is characterized by a *type* of biological community, not a collection of certain species of organisms. For example, the groups of species living in the Sahara Desert of Africa and the Gobi Desert of eastern

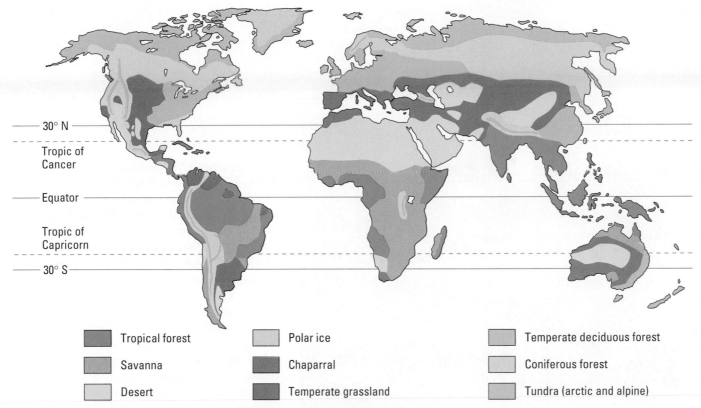

Figure 17.18 Map of the major terrestrial biomes. Although this map has sharp boundaries, biomes actually grade into one another. The tropics are the low-latitude regions bordered by the Tropic of Cancer and the Tropic of Capricorn. We'll use smaller versions of this map, highlighted by color coding, during our closer look at the terrestrial biomes on pages 404–407.

Asia are different, but the species in both regions are adapted to desert conditions. Widely separated biomes may look alike because of convergence, the evolution of similar adaptations in independently evolved species living in similar environments (see Chapter 13). We'll return to this point at the end of the chapter.

Most biomes are named for major physical or climatic features and for their predominant vegetation. For example, temperate grasslands are dominated by various grass species and are generally found in middle latitudes, where the climate is more moderate than in the tropics or polar regions. Each biome is also characterized by microorganisms, fungi, and animals adapted to that particular environment. Temperate grasslands, for example, are more likely than forests to be populated by large grazing mammals.

In most biomes today, extensive human activities have radically altered the natural patterns. Most of the eastern United States, for example, is classified as temperate deciduous forest, but human activity has eliminated all but a tiny percentage of the original forest. In fact, humans have altered much of Earth's surface, replacing original biomes with urban and agricultural ones.

Review the major terrestrial biomes in Web/CD Activity 17C.

Figure 17.19, spread over the next four pages, surveys the major terrestrial biomes, beginning near the equator and generally approaching the poles.

Figure 17.19 Major terrestrial biomes.

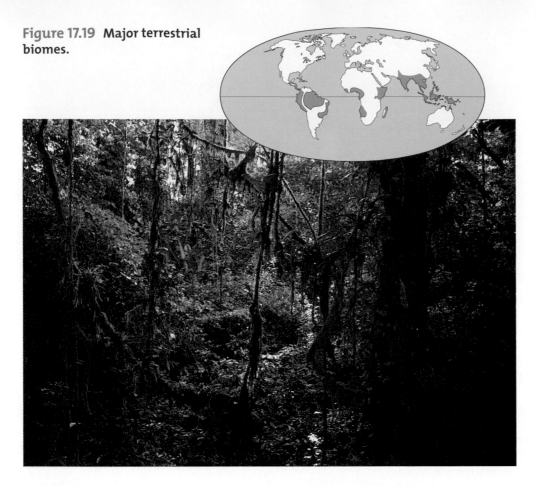

(a) Tropical forest. The photograph shows a tropical rain forest in Costa Rica. Tropical rain forests have pronounced vertical stratification, or layering. Trees in the canopy make up the topmost stratum. The canopy is often closed, so that little light reaches the ground below. When an opening does occur, perhaps because of a fallen tree, other trees and large woody vines grow rapidly, competing for light and space as they fill the gap. Many of the trees are covered with epiphytes (plants that grow on other plants rather than in soil), such as orchids and bromeliads. Rainfall, which is quite variable in the tropics, is the prime determinant of the vegetation growing in an area. In lowland areas that have a prolonged dry season or scarce rainfall at any time, tropical dry forests predominate. The plants found there are a mixture of thorny shrubs and trees and succulents. In regions with distinct wet and dry seasons, tropical deciduous trees are common.

(b) Savanna. Savannas are grasslands with scattered trees. This Kenyan savanna is a showcase of large herbivores and their predators. Actually, the dominant herbivores here and in other savannas are insects, especially ants and termites. Fire is an important abiotic component, and the dominant plant species are fire adapted. The luxuriant growth of grasses and forbs (small broadleaf plants) during the rainy season provides a rich food source for animals. However, large grazing mammals must migrate to greener pastures and scattered watering holes during regular periods of seasonal drought.

(c) Desert. Sparse rainfall (less than 30 centimeters per year) largely determines that an area will be a desert. Some deserts have soil surface temperatures above 60°C (140°F) during the day. Other deserts, such as those west of the Rocky Mountains and in central Asia, are relatively cold. The Sonoran Desert of southern Arizona (shown here) is characterized by giant saguaro cacti and deeply rooted shrubs. Evolutionary adaptations of desert plants and animals include a remarkable array of mechanisms that store water. The "pleated" structure of saguaro cacti enables the plants to expand when they absorb water during wet periods. Some desert mice never drink, deriving all their water from the metabolic breakdown of the seeds they eat. Protective adaptations that deter feeding by mammals and insects, such as spines on cacti and poisons in the leaves of shrubs, are common in desert plants.

(d) Chaparral. This biome occurs in midlatitude coastal areas with mild, rainy winters and long, hot, dry summers. Dense, spiny, evergreen shrubs dominate chaparral. Plants of the chaparral, such as those in this California scrubland, are adapted to and dependent on periodic fires. The dry, woody shrubs are frequently ignited by lightning and by careless human activities, creating summer and autumn brushfires in the densely populated canyons of Southern California and elsewhere. Some of the shrubs produce seeds that will germinate only after a hot fire. Food reserves stored in their fire-resistant roots enable them to resprout quickly and use mineral nutrients released by fires.

Figure 17.19 Major terrestrial biomes, continued.

(e) Temperate grassland. The puszta of Hungary, the pampas of Argentina and Uruguay, the steppes of Russia, and the plains and prairies of central North America are all temperate grasslands. The key to the persistence of grasslands is seasonal drought, occasional fires, and grazing by large mammals, all of which prevent establishment of woody shrubs and trees. Temperate grasslands such as the tallgrass prairie in Kansas (shown here) once covered much of central North America. Because grassland soil is both deep and rich in nutrients, these habitats provide fertile land for agriculture. Most grassland in the United States has been converted to farmland, and very little natural prairie exists today.

(f) Temperate deciduous forest. Dense stands of deciduous trees are trademarks of temperate deciduous forests, such as this one in Great Smoky Mountains National Park in North Carolina. Temperate deciduous forests occur throughout midlatitudes where there is sufficient moisture to support the growth of large trees. Deciduous forest trees drop their leaves before winter, when temperatures are too low for effective photosynthesis and water lost by evaporation is not easily replaced from frozen soil. Many temperate deciduous forest mammals also enter a dormant winter state called hibernation, and some bird species migrate to warmer climates. Virtually all the original deciduous forests in North America were destroyed by logging and land clearing for agriculture and urban development. In contrast to drier biomes, these forests tend to recover after disturbance, and today we see deciduous trees dominating undeveloped areas over much of their former range.

(g) Coniferous forest. Cone-bearing evergreen trees such as pine, spruce, fir, and hemlock dominate coniferous forests. Coastal coniferous forests of the U.S. Pacific Northwest , such as the one in Olympic National Park in western Washington shown here, are actually temperate rain forests. Warm, moist air from the Pacific Ocean supports these unique communities, which like most coniferous forests are dominated by one or a few tree species. Extending in a broad band across northern North America and Eurasia to the southern border of the arctic tundra, the northern coniferous forest, or taiga, is the largest terrestrial biome on Earth (see Figure 17.18). Taiga receives heavy snowfall during winter. The conical shape of many conifers prevents too much snow from accumulating on and breaking their branches. Coniferous forests are being logged at an alarming rate, and the old-growth stands of these trees may soon disappear.

(h) Tundra. Permafrost (permanently frozen subsoil), bitterly cold temperatures, and high winds are responsible for the absence of trees and other tall plants in this arctic tundra in central Alaska (photographed in autumn). Although the arctic tundra receives very little annual rainfall, water cannot penetrate the underlying permafrost and accumulates in pools on the shallow topsoil during the short summer. Tundra covers expansive areas of the Arctic, amounting to 20% of Earth's land surface. High winds and cold temperatures create similar plant communities, called alpine tundra, on very high mountaintops at all latitudes, including the tropics.

CheckPoint

1. How do fires help maintain savannas as grassland ecosystems?

2. Why is homeowners insurance relatively expensive for people who live in the chaparral?

3. How do humans now use most of the North American land that was once temperate grasslands?

4. How does the loss of leaves function as an adaptation of deciduous trees to cold winters?

5. What two abiotic factors account for the rarity of trees in arctic tundra?

Answers: 1. By repeatedly preventing the spread of trees and other woody plants **2.** High fire risk **3.** For farming **4.** By reducing loss of water from the trees when that water cannot be replaced because of frozen soil **5.** Long, very cold winters (short growing season) and permafrost

Aquatic Biomes

Aquatic biomes, consisting of freshwater and marine ecosystems, occupy the largest part of the biosphere.

Freshwater Biomes

Lakes and Ponds Standing bodies of water range from small ponds only a few square meters in area to large lakes, such as North America's Great Lakes, that are thousands of square kilometers (Figure 17.20a).

In lakes and large ponds, the communities of plants, algae, and animals are distributed according to the depth of the water and its distance from shore. Shallow water near shore and the upper stratum of water away from shore make up the **photic zone,** so named because light is available for photosynthesis. **Phytoplankton** (microscopic algae and cyanobacteria) grow in the photic zone, joined by rooted plants and floating plants such as water lilies in photic regions near shore. If a lake or pond is deep enough or murky enough, it has an **aphotic zone,** where light levels are too low to support photosynthesis.

At the bottom of all aquatic biomes, the substrate is called the **benthic zone.** Made up of sand and organic and inorganic sediments ("ooze"), the benthic zone is occupied by communities of organisms, including a diversity of bacteria, collectively called **benthos** (from the Greek *benthos,* depth of the sea). (If you have ever waded barefoot into a pond or lake, you have felt the benthos squish between your toes.) A major source of food for the benthos is dead organic matter called **detritus.** Detritus "rains" down from the productive surface waters of the photic zone.

Temperature has a profound effect on freshwater communities, especially in temperate areas. During the summer, lakes often have a distinct upper layer of water that has been warmed by the sun. The warm-water layer is less dense than underlying, cooler water and does not mix with it. Fish spend much of their time in the deep, cool waters of a lake unless oxygen levels there become depleted by decomposers such as the aerobic bacteria of the benthos.

Nitrogen and phosphorus are the mineral nutrients that usually limit the amount of phytoplankton growth in a lake or pond. When there are

(a) Satellite view of the Great Lakes

Figure 17.20 Freshwater biomes.

(b) A stream flowing into the Snake River, Idaho

(c) A freshwater wetland in Pennsylvania

temperature layers in a lake, for instance, nutrients released by decomposers can become trapped near the bottom, out of reach of the phytoplankton. During the summer months, this may limit the growth of algae in the photic zone. As autumn approaches, the surface water becomes denser as it cools; it then tends to mix with the deeper water, allowing nutrients to return to the surface, where phytoplankton can again use them. Seasonal mixing also restores oxygen to the depths.

Today, many lakes and ponds are affected by large inputs of nitrogen and phosphorus from sewage and runoff from fertilized lawns and agricultural fields. These nutrients often produce blooms, or population explosions, of algae. Heavy algal growth reduces light penetration into the water, and when the algae die and decompose, a pond or lake can suffer serious oxygen depletion. Fish and other aerobic organisms may then die (see Figure 17.3).

Rivers and Streams Bodies of water flowing in one direction, rivers and streams generally support quite different communities of organisms than lakes and ponds (Figure 17.20b). A river or stream changes greatly between its source (perhaps a spring or snowmelt) and the point at which it empties into a lake or the ocean. Near a source, the water is usually cold, low in nutrients, and clear. The channel is often narrow, with a swift current that does not allow much silt to accumulate on the bottom. The current also inhibits the growth of phytoplankton; most of the organisms found here are supported by the photosynthesis of algae attached to rocks or organic material (such as leaves) carried into the stream from the surrounding land. The most abundant benthic animals are usually insects that eat algae, leaves, or one another. Trout are often the predominant fishes, locating their food, including insects, mainly by sight in the clear water.

Downstream, a river or stream generally widens and slows. Marshes, ponds, and other wetlands are common in downstream areas (Figure 17.20c). There, the water is usually warmer and may be murkier because of sediments and phytoplankton suspended in it. Worms and insects that burrow into mud are often abundant, as are waterfowl, frogs, and catfish and other fishes that find food more by scent and taste than by sight.

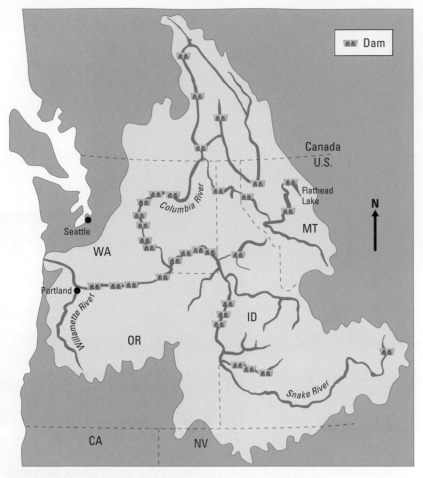

Figure 17.21 Damming the Columbia River Basin. If Lewis and Clark lived today, they would have a hard time navigating the Columbia. This map shows only the largest of the 250 dams that have altered freshwater ecosystems throughout the Pacific Northwest. The great concrete obstacles make it difficult for salmon to swim upriver to their breeding streams, though many dams now have "fish ladders" that provide detours.

Many streams and rivers have been affected by pollution from human activities, by stream channelization to speed water flow, and by dams that hold water. For centuries, humans used streams and rivers as depositories of waste, thinking that these materials would be diluted and carried downstream. While some pollutants are carried far from their source, many settle to the bottom, where they can be taken up by aquatic organisms. Even the pollutants that are carried away contribute to ocean and lake pollution. In many cases, dams have completely changed the downstream ecosystems, altering the intensity and volume of water flow and affecting fish and invertebrate populations (Figure 17.21).

Marine Biomes

Life originated in the sea and evolved there for almost 3 billion years before plants and animals began colonizing land. Covering about 75% of the planet's surface, oceans have always had an enormous impact on the biosphere. Their evaporation provides most of Earth's rainfall, and as you learned earlier in the chapter, ocean temperatures have a major effect on climate and wind patterns. Photosynthesis by marine algae supplies a substantial portion of the biosphere's oxygen. Let us begin our exploration of marine biomes with estuaries, environments where freshwater and seawater mix.

Estuaries The area where a freshwater stream or river merges with the ocean is called an **estuary.** Most estuaries are bordered by extensive coastal wetlands called mudflats and saltmarshes (Figure 17.22). Salinity (salt concentration) varies spatially within estuaries, from nearly that of fresh water to that of the ocean; it also varies over the course of a day with the rise and fall of the tides. Nutrients from the river enrich estuarine waters, making estuaries one of the most biologically productive environments on Earth.

Saltmarsh grasses and algae are the major producers in estuaries. This environment also supports a variety of worms, oysters, crabs, and many of the fish species that humans consume. A diversity of marine invertebrates and fishes use estuaries as a breeding ground or migrate through them to freshwater habitats upstream. Estuaries are also crucial feeding areas for a great diversity of birds.

Although estuaries support a wide variety of commercially valuable species, areas around estuaries are also prime locations for commercial and residential developments. In addition, estuaries are unfortunately at the receiving end for pollutants dumped upstream. Very little undisturbed estuarine habitat remains, and a large percentage has been totally eliminated by landfill and development. Many states have now—rather belatedly—taken steps to preserve their remaining estuaries.

Figure 17.22 An estuary on the edge of Chesapeake Bay, Maryland.

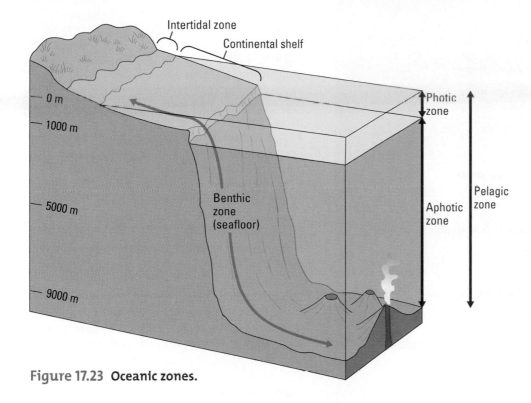

Intertidal zone
Continental shelf
0 m
1000 m
5000 m
9000 m
Benthic zone (seafloor)
Photic zone
Aphotic zone
Pelagic zone

Figure 17.23 Oceanic zones.

Major Oceanic Zones Similar to what we find in freshwater lakes, marine life is distributed according to depth of the water, degree of light penetration, distance from shore, and open water versus bottom (Figure 17.23).

The area where land meets sea is called the **intertidal zone.** This environment is alternately submerged and exposed by the twice-daily cycle of tides. Intertidal organisms are therefore subject to huge daily variations in the availability of seawater (and the nutrients it carries) and in temperature. Also, intertidal organisms are subject to the mechanical forces of wave action, which can dislodge them from their habitats. The organisms that are adapted to this agitation attach to rocks or vegetation or burrow into mud or sand. Examples are certain seaweeds (large algae), sessile crustaceans such as barnacles, and echinoderms such as sea stars and sea urchins (Figure 17.24).

The open ocean itself, called the **pelagic zone** (from the Greek *pelagos,* sea), supports communities dominated by motile animals such as fishes, squids, and marine mammals, including whales and dolphins. Microscopic algae and cyanobacteria—the phytoplankton—drift passively in the pelagic zone. Phytoplankton are the ocean's main photosynthetic producers, making most of the organic food molecules on which other ocean-dwellers depend (see Chapter 14). **Zooplankton** are animals that drift in the pelagic zone either because they are too small to resist ocean currents or because they don't swim. Zooplankton eat phytoplankton and in turn are consumed by other animals, including fishes.

The seafloor, like the lake bottom, is called the **benthic zone.** Depending on depth and light penetration, the benthic community consists of attached algae, fungi, bacteria, sponges, burrowing worms, sea anemones, clams, crabs, and fishes.

Note in Figure 17.23 that marine biologists often group the illuminated regions of the benthic and pelagic communities together, calling them the

Figure 17.24 A sampling of organisms in a tide pool.

Figure 17.25 Exploring hydrothermal vent communities. Research vessels such as *Alvin* have enabled scientists to explore the seafloor to depths of 2500 meters, more than a mile deeper than sunlight penetrates. In the late 1970s, researchers discovered seafloor life that does not depend on photosynthesis. On these expeditions, humans first encountered a diversity of organisms living around hydrothermal vents, where molten rock and hot gases escape from Earth's interior. Nearby lives a community of sea anemones, giant clams, shrimps, worms, and a few fish species. The food chain is supported by sulfur bacteria, which use the oxidation of hydrogen sulfide (H_2S) in the vent's effluent to power the synthesis of organic molecules from CO_2. Some of the animals eat the bacteria. Others, such as the tube worms in the inset, harbor the food-producing bacteria as symbionts. Unlike all the other ecosystems we know, life at the hydrothermal vents on the seafloor is driven by energy from Earth itself rather than from the sun.

photic zone, as we did for lakes. Underlying the photic zone is the vast, dark aphotic zone. This is the most extensive part of the biosphere. Without light, there are no photosynthetic organisms, but life is still diverse in the aphotic zone. Many kinds of invertebrates, such as sea urchins and polychaete worms, and some fishes scavenge organic matter that sinks from the lighted waters above. The seafloor also includes **hydrothermal vent communities,** which are powered by chemical energy from Earth's interior rather than sunlight (Figure 17.25).

> Learn more about organisms from different freshwater and oceanic zones in Web/CD Activity 17D.

Coral Reefs In warm tropical waters, coral reefs are a conspicuous and distinctive biome. Coral reefs are dominated by the structure of the coral itself, formed mainly by cnidarians that secrete hard external skeletons made of calcium carbonate (see Figure 16.12). These skeletons vary in shape, forming a substrate on which other corals, sponges, and algae grow. The coral animals themselves feed on microscopic organisms and particles of organic debris. They also obtain organic molecules from the photosynthesis of symbiotic algae that live in their tissues. In addition to the coral creatures themselves, reefs support a diversity of invertebrates and fishes (Figure 17.26).

Some coral reefs cover enormous expanses of shallow ocean, but this delicate biome is easily degraded by pollution and development, as well as by souvenir hunters who gather the coral skeletons. Corals are also subject to damage from both native and introduced predators, such as the crown-of-thorns sea star, which has undergone a population explosion in many regions and actually destroyed coral reefs in parts of the western Pacific Ocean. Reef communities are very old and grow very slowly, and they may not be able to withstand continued human encroachment. No biome seems safe from our intrusions.

Figure 17.26 A coral reef in Fiji.

1. The _____ , small photosynthetic organisms inhabiting the _____ zone of the pelagic zone, provide most of the food for oceanic life.

2. Why does sewage cause algal blooms in lakes?

3. What is the energy source for hydrothermal vent communities?

Answers: 1. phytoplankton; photic **2.** The sewage adds minerals that stimulate growth of the algae. **3.** Chemicals that spew up from Earth's interior

Ecology and Evolutionary Adaptation

In surveying the biomes, we have seen that the abiotic conditions of the biosphere largely determine the distribution of life. We now turn to the organisms themselves and look at some of the evolutionary adaptations that allow species to meet the challenges of their environment.

The fields of ecology and evolutionary biology are tightly linked. Charles Darwin was an ecologist (although he predated the word *ecology*). It was the geographic distribution of organisms and their exquisite adaptations to specific environments that provided Darwin with evidence for evolution. Evolutionary adaptation via natural selection results from the interaction of organisms with their environments, which brings us back to our definition of ecology. Thus, events that occur in the time frame of what is sometimes called **ecological time** translate into effects over the longer scale of **evolutionary time.** For instance, hawks feeding on field mice have an impact on the gene pool of the prey population by curtailing the reproductive success of certain individuals. One long-term effect of such a predator-prey interaction may be the prevalence in the mouse population of fur coloration that camouflages the animals.

In our brief survey of organismal ecology here, we'll focus on three types of adaptations—physiological, anatomical, and behavioral—that enable plants and animals to adjust to changes in their environments. Note that these changes occur during the lifetime of an individual, so they do not qualify as evolution, which is change in a population over time (see Chapter 12). But an individual organism's abilities to adjust to environmental change during ecological time are themselves adaptations refined by natural selection during evolutionary time.

Physiological Responses

You've seen how a cat's fur fluffs up on a cold day, a response that helps insulate the animal's body. The mechanism for this response is the contraction of tiny muscles attached to the hairs. (Our own muscles do this, too, but as almost hairless mammals, we just get "goose bumps" instead of a furry insulation.) The blood vessels in the cat's skin also constrict, which slows the loss of body heat. (And this works for humans, too.) These are examples of physiological responses to environmental change. In these mechanisms of temperature regulation (thermoregulation), the response occurs in just seconds.

Physiological response that is longer term, though still reversible, is called **acclimation.** For example, suppose you moved from Boston, which is essentially at sea level, to the mile-high city of Denver. One physiological response to the lower oxygen supply in your new environment would be a

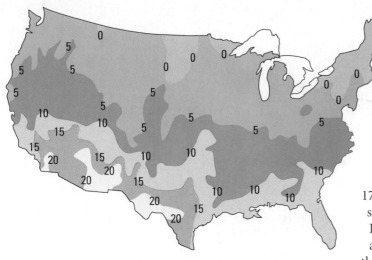

Figure 17.27 The number of lizard species in different regions of the contiguous United States. The northward decrease in the diversity of reptiles reflects their ectothermic physiology, which depends on environmental heat for keeping the body warm enough for the animal to be active.

gradual increase in the number of your red blood cells, which transport O_2 from your lungs to other parts of your body. Acclimation can take days or weeks.

The ability to acclimate is generally related to the range of environmental conditions the species naturally experiences. Species that live in very warm climates, for example, usually do not acclimate to extreme cold. Among vertebrates, birds and mammals can generally tolerate the greatest temperature extremes because, as endotherms, they use their metabolism to regulate internal temperature (see Chapter 16). In contrast, reptiles, which are ectotherms, are more limited in the climates they can tolerate (Figure 17.27). Contrary to popular legend, it wasn't Saint Patrick who chased the snakes out of Ireland; the ice ages did. Even now, there are no snakes in Ireland. They have a hard time getting there across the sea, and Ireland is apparently too cool for snake populations to become established even if they manage to reach the island from warmer parts of the world.

Anatomical Responses

Many organisms respond to environmental challenge with some type of change in body shape or anatomy (structure). In some cases, these responses are examples of acclimation, since they are reversible. Many mammals and birds, for example, grow a heavier coat of fur or feathers in winter; sometimes coat color changes seasonally as well, camouflaging the animal against winter snow and summer vegetation.

Other anatomical changes are irreversible over the lifetime of an individual. In some cases, environmental variation can affect growth and development so much that there are remarkable differences in body shape within a population. You can see an example in Figure 17.9, the "flagging" that wind causes in certain trees. In general, plants are more anatomically plastic than animals. Rooted and unable to move to a better location, plants rely entirely on their anatomical and physiological responses to survive environmental fluctuations.

Behavioral Responses

In contrast to plants, most animals can respond to an unfavorable change in the environment by moving to a new location. Such movement may be fairly localized. For example, many desert ectotherms, including reptiles, maintain a reasonably constant body temperature by shuttling between sun and shade. Some animals, however, are capable of migrating great distances in response to such environmental cues as the changing seasons. Many migratory birds overwinter in Central and South America, returning to northern latitudes to breed in the summer. And humans, with their large brains and technology, have an especially rich range of behavioral responses available to them (Figure 17.28).

Explore evolutionary adaptations of desert and forest organisms in Web/CD Activity 17E.

Figure 17.28 Behavioral responses have expanded the geographic range of humans. Dressing for the weather, illustrated by these Siberian children, is a thermoregulatory behavior unique to humans.

CheckPoint

1. Why does the set of evolutionary adaptations characterizing a species tend to limit the geographic distribution of that species?

2. What is acclimation?

3. Contrast "ecological time" with "evolutionary time."

Answers: 1. To the extent that the species is adapted to particular environmental conditions, it is not so well equipped to survive and reproduce where there are different conditions to which the species is not adapted. **2.** A gradual, reversible change in anatomy or physiology in response to an environmental change **3.** Ecological time is the temporal scale for the present interactions between organisms and their environments. Evolutionary time is the longer-term consequence of those interactions in the adaptations that evolve via natural selection.

Figure 17.29 Riding through the "Old West" in a spaghetti western. The vegetation in certain parts of Italy and Spain is superficially similar to the landscape of the American Southwest because adaptations of the plants have been shaped by similar climates.

Evolution Link: Spaghetti Westerns and Evolutionary Convergence

In the late 1960s, Sergio Leone and other European movie directors crafted a genre of blood-and-guts films that have become cult favorites. Known as "spaghetti westerns" because many were filmed in Italy (others in Spain), the titles included *The Hills Run Red, Death Rides a Horse, A Fistful of Dollars* and its sequel *For a Few Dollars More,* and the classic *The Good, the Bad, and the Ugly.* For English-language audiences, most of the characters' voices were dubbed, as the actors were mainly Italian and Spanish. However, as the main character in several spaghetti westerns, a young American actor named Clint Eastwood, always playing a nameless stranger, rode to stardom on horseback through hills that looked to all the world like the "Old West."

Why did the locations for shooting the spaghetti westerns work so well as backdrops for plots that were supposedly unfolding along the border of the United States and Mexico? The answer is that climate and other abiotic factors in certain parts of Italy and Spain are very similar to those in the southwestern United States (Figure 17.29). Though thousands of miles apart, the hills of the spaghetti westerns and the hills where outlaws roamed the Old West are the same biome, a type of arid brushland. A biome, remember, is a kind of place, not a specific place. If two distant locations have similar climates, the landscapes may also appear similar because the vegetation in the two locales is adapted to a common set of abiotic factors. The plants that make up that vegetation, however, may not be closely related. Their similarity is due to convergent evolution, the independent evolution of similar adaptations in similar environments. Just as the actors in the spaghetti westerns were pretending to be outlaws, the plants in the Mediterranean landscape were substituting for completely different species in the American Southwest. Figure 17.30 illustrates another example of evolutionary convergence in plants. Convergent evolution highlights the central role that interactions between organisms and their environments play in shaping life on Earth.

Learn more about convergent evolution in the Web Evolution Link.

Figure 17.30 An example of convergent evolution. The ocotillo of the southwestern United States and northwestern Mexico (left) looks remarkably similar to the allauidia (right) of Madagascar. However, these two plants are not closely related. They owe their superficial resemblance to analogous adaptations that evolved independently in similar environments.

Summary of Key Concepts

Overview: The Scope of Ecology

- **Ecology as Scientific Study** Ecology is the scientific study of interactions between organisms and their environments. Ecologists use observation, experiments, and computer models to test hypothetical explanations of these interactions.

- **Major Components of the Environment** The environment includes abiotic (nonliving) and biotic (living) components.

- **A Hierarchy of Interactions** Ecologists study interactions at four increasingly complex levels: organismal ecology, population ecology, community ecology, and ecosystem ecology.

- **Ecology and Environmentalism** Human activities have impacted all parts of the biosphere. Ecology provides the basis for understanding and addressing these environmental problems.
• Web/CD Activity 17A *Case Studies of DDT Use*

Abiotic Factors of the Biosphere

- **Major Abiotic Factors** The biosphere is an environmental patchwork in which several abiotic factors affect the distribution and abundance of organisms. These include the availability of sunlight and water, temperature, wind, rock and soil characteristics, and catastrophic disturbances such as fires, hurricanes, tornadoes, and volcanic eruptions.
• Web/CD Activity 17B/The Process of Science *Investigating Abiotic Factors*

- **Global Climate Patterns** Global climates and seasonality are largely determined by the uneven input of solar energy. Latitudinal variation in precipitation, winds, and ocean currents account for the uneven geographical distribution of major ecosystems.

- **Local Climate** Proximity to large bodies of water and the presence of land forms such as mountains are the most important factors affecting local variations in climate. Wind, shade, and evaporation are key factors affecting microclimates.

Terrestrial Biomes

- The geographical distribution of terrestrial biomes is based mainly on regional variations in climate. If the climate in two geographically separate areas is similar, the same type of biome may occur in them. Most biomes are named for major physical or climatic features and for their predominant vegetation. The major terrestrial biomes include the tropical forest, savanna, desert, chaparral, temperate grassland, temperate deciduous forest, coniferous forest, and tundra.
• Web/CD Activity 17C *Major Terrestrial Biomes*

Aquatic Biomes

- Aquatic biomes occupy the largest part of the biosphere.

- **Freshwater Biomes** Freshwater biomes include lakes, ponds, rivers, streams, and wetlands. Lakes are often stratified vertically with regard to light penetration, temperature, nutrients, oxygen levels, and community structure. Rivers change greatly from their source to the point at which they empty into a lake or an ocean.

- **Marine Biomes** Estuaries, located where a freshwater river or stream merges with the ocean, are one of the most biologically productive environments on Earth. As in freshwater environments, marine life is distributed into distinct zones (intertidal, pelagic, and benthic) according to the depth of the water, degree of light penetration, distance from shore, and open water versus bottom. Hydrothermal vent communities are a marine deepwater biome powered by chemical energy from the Earth's interior instead of sunlight. Coral reefs are a fragile biome that supports a broad diversity of invertebrates and fishes.
• Web/CD Activity 17D *Aquatic Biomes*

Ecology and Evolutionary Adaptation

- **Physiological Responses** Most organisms adjust their physiological conditions in response to changes in the environment. Acclimation is a longer term response that can take days or weeks. The ability to acclimate is generally related to the range of environmental conditions that a species naturally experiences.

- **Anatomical Responses** Many organisms respond to environmental change with reversible or irreversible changes in anatomy. Unable to move to a better location, plants are generally more anatomically plastic than animals.

- **Behavioral Responses** Able to travel about, animals frequently adjust to poor environmental conditions by moving to a new location. These may be small adjustments such as shuttling between sun and shade or seasonal migrations to new biomes.
• Web/CD Activity 17E *Evolutionary Adaptations*

Evolution Link: Spaghetti Westerns and Evolutionary Convergence

- The types of organisms living in the same biome throughout the world may not be related yet appear similar due to convergent evolution. The common environmental conditions favor similar adaptations in similar environments.
• Web Evolution Link *Spaghetti Westerns and Evolutionary Convergence*

Self-Quiz

1. Imagine some cosmic catastrophe that jolts Earth so that it is no longer tilted. Instead its axis is perpendicular to the line between the sun and Earth. The most predictable effect of this change would be
   a. no more night and day.
   b. a big change in the length of the year.
   c. a cooling of the equator.
   d. a loss of seasonal variations at northern and southern latitudes.
   e. the elimination of ocean currents.

2. Imagine that you are a passenger on a plane flying over temperate deciduous forest, then grassland and desert, and finally landing at an airport in chaparral. The route of your flight was between
   a. New York and Denver.
   b. Philadelphia and Los Angeles.
   c. Denver and Los Angeles.
   d. Washington, D.C., and Phoenix.
   e. Seattle and Washington, D.C.

3. What makes the Gobi Desert of Asia a desert?
   a. The growing season there is very short.
   b. Its vegetation is sparse.
   c. It is hot.
   d. Temperatures vary little from summer to winter.
   e. It is dry.

4. Which of the following sea creatures might be described as a pelagic animal of the aphotic zone?
   a. a coral reef fish
   b. a giant clam near a deep-sea hydrothermal vent
   c. an intertidal snail
   d. a deep-sea squid
   e. a harbor seal

5. We are on a coastal hillside on a hot, dry summer day among evergreen shrubs that are adapted to fire. We are most likely standing in
   a. desert.
   b. taiga.
   c. tundra.
   d. chaparral.
   e. temperate deciduous forest.

6. The growing season would generally be shortest in which biome?
   a. tropical rain forest
   b. savanna
   c. taiga (coniferous forest)
   d. temperate deciduous forest
   e. temperate grassland

7. While climbing in the Rocky Mountains, one observes transitions in biological communities that are analogous to the changes one encounters
   a. in biomes at different latitudes.
   b. at different depths in the ocean.
   c. in a community through different seasons.
   d. in an ecosystem as it evolves over time.
   e. traveling across the United States from east to west.

8. If we compare plants found in the deciduous forests of Australia with plants found in the deciduous forests of the eastern United States, similarities between trees are probably due to
   a. the communities being populated by the same tree species.
   b. a common origin of the tree species from ancestors that existed before the continents separated.
   c. evolutionary convergence during adaptation to similar environments.
   d. selection for similar characteristics by the browsing of forest animals.
   e. the two regions being about the same distance north of the equator.

9. Tide pool animals have adaptations that help them hang onto their substratum. You would expect to find similar adaptations among animals in
   a. estuaries.
   b. upstream regions of streams and rivers.
   c. lakes.
   d. ponds.
   e. the benthic zone of the deep sea.

10. In *volume*, which biome is largest?
    a. deserts
    b. tropical forests
    c. the intertidal zone
    d. the photic zone of the oceans
    e. the pelagic zone of the oceans

• Go to the website or CD-ROM for more self-quiz questions.

The Process of Science

1. Design a laboratory procedure to measure the effect of water temperature on the population growth of a certain phytoplankton species from a pond.

2. **Investigate how changes in sunlight, water, temperature, and wind affect the environment in The Process of Science activity on the website and CD-ROM.**

Biology and Society

During the summer of 1988, huge forest fires burned a large portion of Yellowstone National Park (see Figure 17.10). The National Park Service has a natural-burn policy: Fires that start naturally are allowed to burn unless they endanger human settlements. Lightning ignited the Yellowstone fires, so they were allowed to spread and burn themselves out as much as possible; firefighters primarily protected people. This drew a lot of public criticism; the Park Service was accused of letting a national treasure go up in flames. Park Service scientists stuck with the natural-burn policy. Do you think this was the best decision? Support your position. More recently, in the spring of 2000, park officials authorized a controlled burn of a forested area near Los Alamos, New Mexico. The strategy for such a controlled burn is to clear away brush and dead wood in order to reduce the severity of future fires. Unfortunately, a weather warning was ignored and the Los Alamos fire escaped control, claiming over 200 homes and thousands of acres of forest. What impact do you think the Los Alamos fire will have on public opinion about controlled burning? Assuming that the basic concept of controlled burning is scientifically sound, how would you justify future burns to a public that remembers the Los Alamos fiasco?

Population Ecology

The human population doubled from

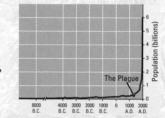

3 billion to 6 billion in just 40 years. Every

three years the world's population increases by the popu-

lation equivalent of the United States. On average,

one species of plant or animal becomes extinct every

20 minutes. A female Pacific salmon produces

millions of eggs in a single reproductive event

and then dies.

ometime during the summer of 1999, Earth's human population passed the 6 billion mark—and kept right on exploding into the new millennium. The population had doubled from 3 billion in just 40 years. At the present growth rate of 78 million people per year, it takes only three days to add the population equivalent of a San Francisco, or just over three years to add the population equivalent of the United States. We are by far the most abundant large animals, and given our technological prowess, we have a disproportionately high impact on the environment (Figure 18.1).

Every day, we hear about local and global problems that threaten our well-being or provoke disputes between individuals or nations—global warming, toxic waste, conflicts over oil in the Middle East, the declining health of the oceans, civil strife aggravated by depressed economics. Contributing to all these apparently unrelated problems is a common factor: the continued increase of the human population in the face of limited resources. The human population explosion is now Earth's most significant biological phenomenon. Our species requires vast amounts of materials and space, including places to live, land to grow our food, and places to dump our waste. Incessantly expanding our presence on Earth, we have devastated the environment for many other species and now threaten to make it unfit for ourselves.

To understand the problem of human population growth on more than a superficial level, we must understand the principles of population ecology that apply to all species. After considering these key concepts of population ecology, we'll return to our discussion of the human population.

Figure 18.1 Making way for a shopping center.

Overview: What Is Population Ecology?

No population can grow indefinitely. Species other than humans sometimes exhibit population explosions, but their populations inevitably crash. In contrast to these radical booms and busts, many populations are relatively stable over time, with only minor increases or decreases in population size. **Population ecology** focuses on the factors that influence a population's size (number of individuals), growth rate (rate of change in population size), density (number of individuals per unit area or volume), and features of population structure (such as relative numbers of individuals of different ages).

Look at global human population ecology in Web/CD Activity 18A.

Before we go on, it's important to understand what biologists mean by a population. A **population** is a group of individuals of the same species living in a given area at a given time. A population's geographic boundaries may be natural, as with certain species of trout in an isolated lake. But ecologists often define a population's boundaries in more arbitrary ways that fit their research questions. For example, an ecologist studying the contribution of asexual reproduction to the population growth of sea anemones might define a population as all the anemones of one species in a tide pool. Another researcher studying the effects of hunting on deer might define a population as all the deer within a particular state. Yet another researcher, attempting to determine which segment of the human population will be most affected by the AIDS epidemic, might study the HIV infection rate of the human population in one nation or throughout the world. Whatever the scale of the population we're studying, there are some common principles of population structure and growth that will guide our analysis.

Figure 18.2 An indirect census of a prairie dog population. We could estimate the number of prairie dogs in this colony in South Dakota by counting the number of mounds constructed by the rodents. The estimate is rough because the animals are social, and the number of individuals that cohabit a system of tunnels under each mound varies.

The Structure and Growth of Populations

Population Density

Population density is the number of individuals of a species per unit area or volume—the number of oak trees per square kilometer (km^2) in a forest, for example, or the number of earthworms per cubic meter (m^3) in the forest's soil.

How do we measure population density? In rare cases, it is possible to actually count all individuals within the boundaries of the population. For example, we could count the total number of oak trees (say, 200) in a forest covering 50 km^2. The population density would be the total number of trees divided by the area, or 4/km^2.

In most cases, it is impractical or impossible to count all individuals in a population. Instead, ecologists use a variety of sampling techniques to estimate population densities. For example, they might estimate the density of alligators in the Florida Everglades based on a count of individuals in a few sample plots of 1 km^2 each. The larger the number and size of sample plots, the more accurate the estimates. In some cases, population densities are estimated not by counts of organisms but by indirect indicators, such as number of bird nests or rodent burrows (Figure 18.2).

Another sampling technique commonly used to estimate wildlife populations is the **mark-recapture method.** The scientist places traps within the boundaries of the population under study. The researcher then marks the captured animals with tags, collars, bands, or spots of dye. The marked animals are released. After a few days or weeks—enough time for the marked individuals to mix randomly with unmarked members of the population—traps are set again. This second capture will yield both marked and unmarked individuals in a proportion that gives an estimate of the number of individuals (N) in the population:

$$N = \frac{\text{Marked individuals} \times \text{total catch second time}}{\text{Recaptured marked individuals}}$$

Figure 18.3 shows this method of estimating *N* for a bird population. The mark-recapture method assumes that each marked individual has the same probability of being trapped as each unmarked individual. This is not always a safe assumption, because an animal that has been trapped once may be wary of traps in the future.

Try your hand at estimating population density in Web/CD Activity 18B/The Process of Science.

Patterns of Dispersion

The **dispersion pattern** of a population is the way individuals are spaced within the population's geographical range. A **clumped** pattern, in which individuals are aggregated in patches, is the most common in nature. Clumping often results from an unequal distribution of resources in the environment. For instance, cottonwood trees are usually clumped along a streamside in patches of moist and sandy soil. Clumping of animals is often associated with uneven food distribution or with mating or other social behavior. For instance, mosquitoes often swarm in great numbers, which increases their chances for mating. Schooling fishes are another example (Figure 18.4a).

A **uniform** pattern of dispersion often results from interactions among the individuals of a population. For instance, creosote bushes in the desert tend to be uniformly spaced because their roots compete for water and dissolved nutrients. Animals often exhibit uniform dispersion as a result of social interactions. Examples are birds nesting in large numbers on small islands (Figure 18.4b).

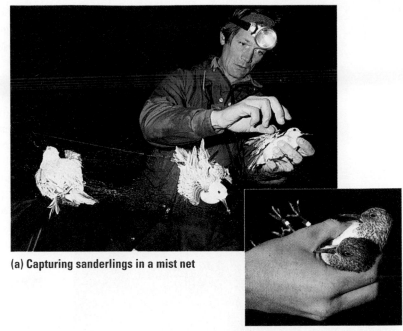

(a) Capturing sanderlings in a mist net

(b) Marking birds with leg bands so they can be identified among birds trapped during a second capture

Figure 18.3 A mark-recapture estimate of population size. This biologist is using the mark-recapture method to estimate *N* (number of individuals) in a population of birds called sanderlings. **(a)** Let's say that he has captured 50 sanderlings in a harmless trap called a mist net. **(b)** He marks the birds with leg bands and then releases them back to the whole population. A second capture two weeks later yields a total of 100 sanderlings, of which 10 are marked birds that have been recaptured. We can estimate that 10% of the total sanderling population is marked. Since the biologist marked 50 birds, our estimate for the entire population is about 500 birds.

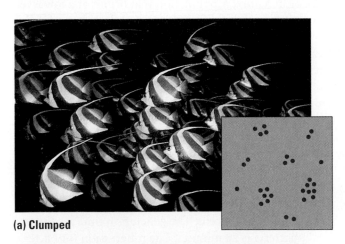

(a) Clumped

(b) Uniform

(c) Random

Figure 18.4 Patterns of dispersion within a population's geographical range. (a) Butterfly fish, like many fishes, often clump into what are called schools. Schooling may enhance the hydrodynamic efficiency of swimming, reduce predation ("safety in numbers"), and increase feeding efficiency. **(b)** Birds nesting on small islands, such as these king penguins on South Georgia Island in the South Atlantic, often exhibit uniform spacing. Territorial behavior helps stabilize the distance between individuals. **(c)** Trees of the same species are often randomly distributed among trees of other species in tropical rain forests.

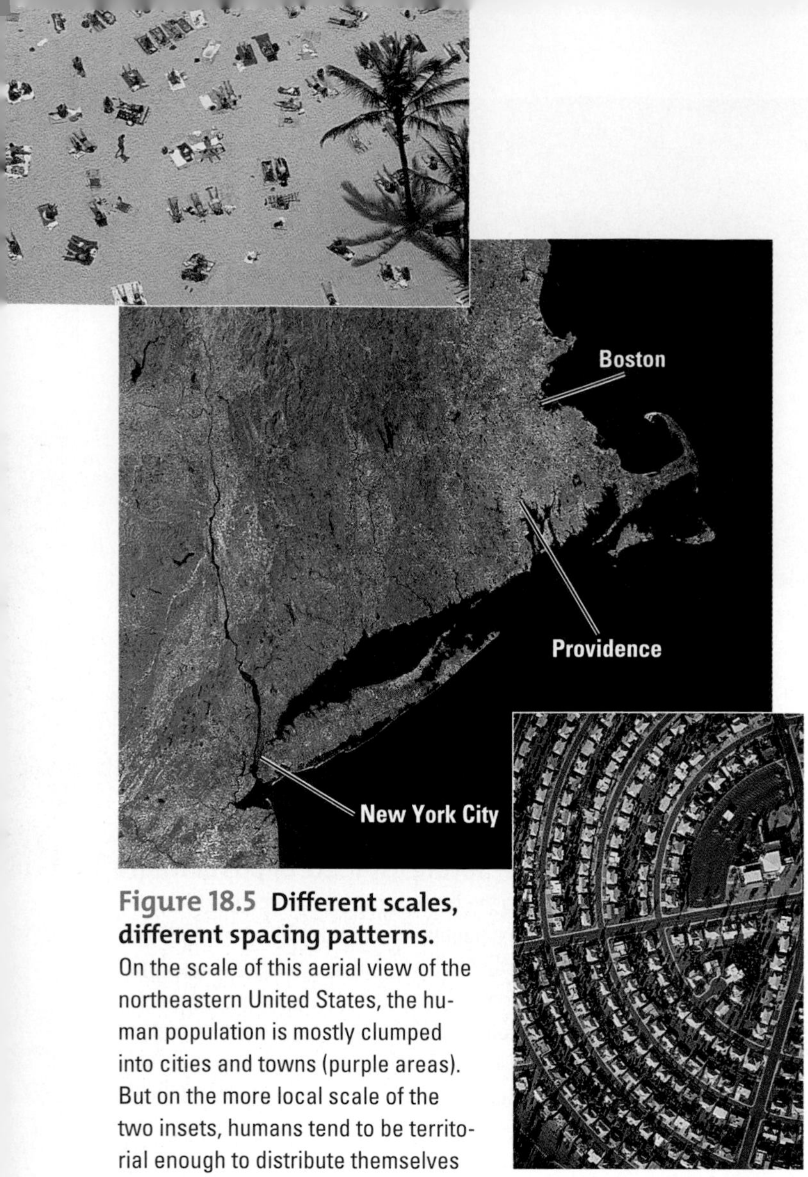

In a third type of dispersion, a **random** dispersion, individuals in a population are spaced in a patternless, unpredictable way. This only occurs in the absence of strong attractions or repulsions among individuals in a population. Clams living in a coastal mudflat, for instance, might be randomly dispersed at times of the year when they are not breeding and when resources are plentiful and do not affect their distribution. Forest trees are also randomly distributed in some cases (see Figure 18.4c). However, environmental conditions and social interactions make random dispersion rare.

Some populations exhibit both clumped and uniform dispersion patterns, but on different scales. For instance, if you studied dispersion patterns of the human population in the northeastern United States, you would find most of the population clumped in metropolitan areas such as New York City, Providence, and Boston. Within each clump, however, individuals or family groups might be more or less uniformly dispersed—in housing tracts of fixed lot size, for example (Figure 18.5).

Population Growth Models

To appreciate the explosive potential for population increase, consider a single bacterium that can reproduce by fission every 20 minutes under ideal laboratory conditions. At the end of this time, there would be two bacteria, four after 40 minutes, and so on. If this continued for only a day and a half—a mere 36 hours—there would be bacteria enough to form a layer a foot deep over the entire Earth. At the other extreme, elephants may produce only six young in a 100-year life span. Still, Darwin calculated that it would take only 750 years for a single mating pair of elephants to give rise to a population of 19 million. Obviously, indefinite increase does not occur, either in the laboratory or in nature. A population that begins at a low level in a favorable environment may increase rapidly for a while, but eventually the numbers must, as a result of limited resources and other factors, stop growing. We'll take a look at two mathematical models that will help us understand the raw potential and limitations for population growth.

The Exponential Growth Model: The Ideal of an Unlimited Environment

The rate of expansion of a population under ideal, unregulated conditions is described by the **exponential growth model,** in which the whole population multiplies by a constant factor during constant time intervals (the generation time). For example, the constant factor for the bacterial population represented in Figure 18.6 is 2, because each parent cell splits to produce two daughter cells. The generation time for these bacteria is 20 minutes. The progression for bacterial growth—2, 4, 8, 16, and so on—is the number 2 raised to a successively higher power (exponent) each generation (that is, 2^1, 2^2, 2^3, 2^4, and so on).

Suppose you have a summer job in a microbiology research lab, and you are asked to monitor the growth of a bacterial population being cultured in a Petri dish. You would measure the population by counting the number of bacterial cells (N) in a small sample of the colony at regular intervals of time (t). You could then plot the numbers you obtained for N against t. The graph in Figure 18.6b shows the type of curve you would obtain if you plotted the number of cells in a bacterial popula-

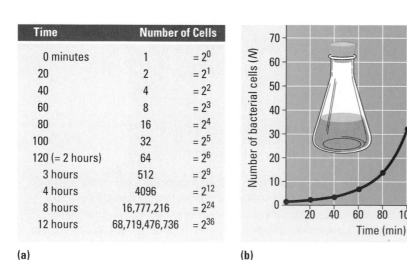

Figure 18.5 **Different scales, different spacing patterns.** On the scale of this aerial view of the northeastern United States, the human population is mostly clumped into cities and towns (purple areas). But on the more local scale of the two insets, humans tend to be territorial enough to distribute themselves uniformly.

Time	Number of Cells	
0 minutes	1	$= 2^0$
20	2	$= 2^1$
40	4	$= 2^2$
60	8	$= 2^3$
80	16	$= 2^4$
100	32	$= 2^5$
120 (= 2 hours)	64	$= 2^6$
3 hours	512	$= 2^9$
4 hours	4096	$= 2^{12}$
8 hours	16,777,216	$= 2^{24}$
12 hours	68,719,476,736	$= 2^{36}$

(a)

(b)

$G = rN$

Number of bacterial cells (N)

Time (min)

Figure 18.6 Exponential growth of a bacterial colony. (a) A table of data for the colony. (b) Plotting the data as a graph.

tion that was expanding exponentially. We can describe the J-shaped curve, which is typical of exponential growth, with this equation:

$$G = rN$$

The G stands for the growth rate of the population; N stands for the population size (the number of individuals in the population); and r stands for the **intrinsic rate of increase,** an organism's inherent capacity to reproduce in an ideal environment that provides unlimited space and resources. The value for r depends on the kind of organism and is influenced by such factors as generation span. We can estimate the value of r by subtracting the death rate (the number of individuals dying in a given unit of time) from the birth rate (the number of individuals produced in that same time interval).

Here's a key feature of the exponential growth equation: The rate at which a population grows depends on the number of individuals already in the population. It's like compound interest on a long-term savings account at a fixed annual yield (analogous to the constant r in our equation). At 7% per annum, you make only $70 the first year on a $1000 deposit. But leave it in the bank until retirement age, and you'll be supplementing your pension with about $2200 per year on an account that has grown to over $30,000. Similarly, in our equation for exponential population growth, the bigger the value of N, the faster the population increases. As time goes by, N gets bigger faster and faster. On the graph in Figure 18.6b, the lower part of the J results from the relatively slow growth when the population is small. The steep, upper part of the J results from N being large.

The exponential growth model gives an idealized picture of the unregulated growth of a population. For bacteria, unregulated growth means there is no restriction on the ability of the cells to live, grow, and reproduce. Given a few days of unregulated growth, bacteria would smother every other living thing. Obviously, long periods of exponential increases are not common in the real world, or life could not continue on Earth. Where we do observe exponential growth in nature, it is generally a short-lived consequence of organisms being introduced to a new or underexploited environment.

The Logistic Growth Model: The Reality of a Limited Environment In nature, a population may grow exponentially for a while, but eventually, one or more environmental factors will limit its growth. Population size then stops increasing or may even crash. Environmental factors that restrict population growth are called **population-limiting factors.** The graph for the seal population in Figure 18.7 resembles what is called the **logistic growth model,** a description of idealized population growth that is slowed by limiting factors. Figure 18.8 contrasts the logistic growth model with the exponential growth model. Note that the logistic curve is J-shaped at first, but gradually levels off to resemble a lazy S.

Compared with the equation for exponential growth, the logistic growth equation looks a bit more complicated:

$$G = rN\frac{(K-N)}{K}$$

You can think of the logistic equation as the exponential equation ($G = rN$) modified by the term $(K - N)/K$. This term represents the overall effect of population-limiting factors. Notice that the only new letter in the equation is K, which stands for carrying capacity.

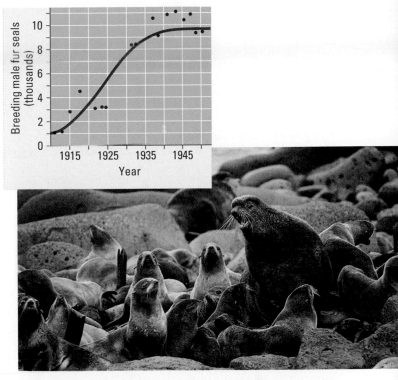

Figure 18.7 Effect of population-limiting factors on the growth of an animal population. The graph profiles the growth of a population of fur seals on Saint Paul Island, off the coast of Alaska. (For simplicity, only the mated bulls were counted. Each has a harem of females, as shown in the photograph.) Before 1925, the seal population on the island remained low because of uncontrolled hunting, although it changed from year to year. After hunting was controlled, the population increased rapidly until about 1935, when it leveled off and began fluctuating around a population size of about 10,000 bull seals. At this point, a number of population-limiting factors, including some hunting and the amount of space suitable for breeding, restricted population growth.

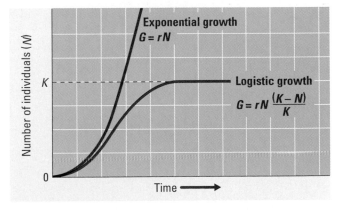

Figure 18.8 Logistic growth and exponential growth contrasted.

Carrying capacity is the number of individuals in a population that the environment can just maintain ("carry") with no net increase or decrease. For the fur seal population in Figure 18.7, for instance, K is about 10,000 mated males. The value of K varies, depending on species and habitat. Carrying capacity might be considerably less than 10,000 for a fur seal population on a smaller island with fewer breeding sites.

Let's see how the term $(K - N)/K$ produces the S-shaped logistic curve. When the population first starts growing, N is close to zero—very small compared with the carrying capacity (K). At this time, N has little effect on the $(K - N)/K$ term; in fact, the term nearly equals K/K, or 1. When this is the case, population growth G is close to rN—that is, exponential growth. However, as the population increases and N gets close to carrying capacity, it has a large effect on the term $(K - N)/K$. In fact, the term becomes an increasingly smaller fraction. And the value rN is multiplied by that fraction, with population growth slowing down more and more. At carrying capacity, the population is as big as it can theoretically get in its environment. At this point, $N = K$, so $(K - N)/K = 0$, and the population growth rate (G) is zero.

The logistic model predicts that a population's growth rate will be low when the population size is either small or large, and highest when the population is at an intermediate level relative to the carrying capacity. At a low population level, resources are abundant, and the population is able to grow nearly exponentially. At this point, however, the increase is small because N is small. In contrast, at a high population level, population-limiting factors strongly oppose the population's potential to increase. In nature, there might be less food available per individual or fewer breeding sites, nest sites, or shelters. The limiting factors make the birth rate decrease, the death rate increase, or both. Eventually, the population stabilizes at the carrying capacity (K), when the birth rate equals the death rate.

Both the logistic growth model and the exponential growth model are mathematical ideals. No natural populations fit either one perfectly. However, these models are useful starting points for studying population growth. Ecologists use them to predict how populations will grow in certain environments and as a basis for constructing more complex models. The models have stimulated research leading to a greater understanding of populations in nature.

Regulation of Population Growth

Let's take a closer look at the population-limiting factors that contribute to carrying capacity. Ecologists classify these factors into two categories: density-dependent factors and density-independent factors.

Density-Dependent Factors The major biological implication of the logistic model is that increasing population density reduces the resources available for individual organisms, ultimately limiting population growth. The logistic model is actually a description of **intraspecific competition**—competition between individuals of the same species for the same limited resources. As population size increases, competition becomes more intense and the growth rate (G) declines in proportion to the intensity of competition. Thus, population growth rate is density-dependent. A **density-dependent factor** is a population-limiting factor whose effects intensify as the population increases in size. Put another way, density-dependent factors affect a greater percentage of individuals in a population as the number of individuals increases. Limited food supply and the buildup of poisonous

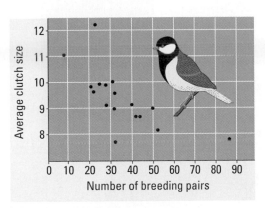

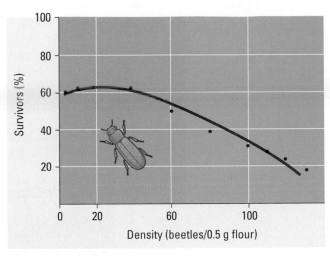

(a) Decreased birth rate

(b) Increased death rate

Figure 18.9 **Density-dependent regulation of population growth.** (a) Decreased birth rate. This graph relates the clutch size (a "litter" of eggs a female bird lays) to population density for a forest population of a species called the great tit. Note that clutch size decreases as population density increases. (b) Increased death rate. In this laboratory culture of an insect called the flour beetle, the percentage of individuals that survive from the egg stage to reproductive maturity decreases as the population becomes denser.

wastes are examples of density-dependent factors. Such factors depress a population's growth rate by increasing the death rate, decreasing the birth rate, or both (Figure 18.9).

We often see density-dependent regulation in laboratory populations. For example, when a pair of fruit flies is placed in a jar with a limited amount of food added each day, population growth fits the logistic model. After a rapid increase, population growth levels off as the flies become so numerous that they outstrip their limited food supply. Each individual in a large population has a smaller share of the limited food than it would in a small population. Also, the more flies, the more concentrated the poisonous wastes become in the jar.

Laboratory populations are one thing; natural populations are another. We do not often see clear-cut cases of density-dependent factors regulating populations in nature. To test whether such factors are operating, it is necessary to change the density of individuals in the population while keeping other factors constant. This is sometimes done in managing game populations. For instance, state agencies often allow hunters to reduce populations of white-tailed deer to levels that keep the animals from permanently damaging the plants they use for food. White-tailed deer are browsers, preferring the highly nutritious parts of woody shrubs—young stems, leaves, and buds. When deer populations are kept low and high-quality food is therefore abundant, a high percentage of females become pregnant and bear offspring; in fact, many of them produce twins (Figure 18.10). On the other hand, when populations are high and food quality is poor, many females fail to reproduce at all. These observations support the hypothesis that food supply is a density-dependent factor regulating white-tailed deer populations.

Figure 18.10 **Increased birth rates in times of plenty.** In deer populations, the birth of twin fawns is much more common when population densities are low.

Density-Independent Factors A population-limiting factor whose intensity is unrelated to population density is called a **density-independent factor.** Examples are such abiotic factors as unfavorable changes in the weather. A freeze in the fall, for example, may kill a certain percentage of insects in a population. The date and severity of the first freeze obviously are not affected by the density of the insect population. Density-independent factors such as a killing frost affect the same percentage of individuals regardless of population size. (In larger populations, of course, greater numbers will die.)

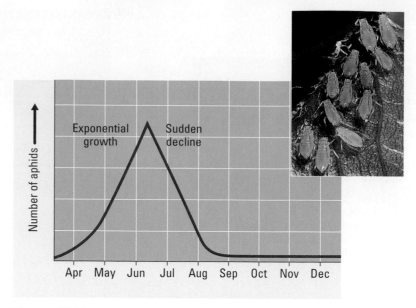

Figure 18.11 **Weather change as a density-independent factor limiting growth of an aphid population.**

In many natural populations, density-independent factors limit population size well before resources or other density-dependent factors become important. In such cases, the population may decline suddenly. If we look at the growth curve of such a population, we see something like exponential growth followed by a rapid decline rather than a leveling off. Ecologists have observed such history, for example, in certain populations of aphids, insects that feed on the sap of plants (Figure 18.11). Aphids and many other insects often show virtually exponential growth in the spring and then rapid die-offs when it becomes hot and dry in the summer. A few individuals may remain, and these may allow population growth to resume again if favorable conditions return. Some insect populations—many mosquitoes and grasshoppers, for instance—will die off entirely, leaving only eggs, which will initiate population growth the following year. In addition to seasonal changes in the weather, abrupt environmental trauma, such as fire, floods, storms, and habitat disruption by human activity, can affect populations in a density-independent manner.

Over the long term, most populations are probably regulated by a mixture of density-independent and density-dependent factors. Many populations remain fairly stable in size and are presumably close to a carrying capacity that is determined by density-dependent factors. In addition, however, many show short-term fluctuations due to density-independent factors. In some cases, the distinction between density-dependent and density-independent factors is not clear. In the case of white-tailed deer, for example, many individuals may starve to death in very cold, snowy areas. The severity of this effect is related to the harshness of the winter; cold temperatures increase energy requirements (and therefore the need for food), while deeper snow makes it harder to find food. But the severity of the effect is also density dependent, because the larger the population, the less food available per individual.

Population Cycles Some populations of insects, birds, and mammals have regular boom-and-bust cycles. Perhaps the most striking is that of the periodical cicadas, grasshopper-like insects that complete their life cycle every

13 or 17 years, emerging from the ground at phenomenal densities (as high as 600 individuals per m²). This long life cycle may be an adaptation that reduces predation; few predators can wait 13 or 17 years for their prey to appear.

As a case study of population cycles, let's examine the boom-and-bust cycles of the snowshoe hare and one of its predators, the lynx. Both animals inhabit the taiga of North America (Figure 18.12). About every 10 years, both hare and lynx populations have a rapid increase (a "boom") followed by a sharp decline (a "bust"). What causes these boom-and-bust cycles? The ups and downs in the two populations seem to almost match each other on the graph. Does this mean that changes in one directly affect the other? In other words, does predation by the lynx make the hare population fluctuate, and do the ups and downs of the hare population cause the changes in the lynx population? For the lynx and many other predators that depend heavily on a single species of prey, the availability of prey can influence population changes. Thus, the 10-year cycles in the lynx population probably result at least in part from the 10-year cycles in the hare population. But we cannot conclude from the graph that lynx predation alone causes the cyclical changes in the hare population. In fact, researchers have discovered that hare populations will cycle about every 10 years, whether or not lynx are present. Another hypothesis is that fluctuations in hare populations are tied to fluctuations in the populations of plants eaten by the hares. There is evidence that when certain plants are damaged by herbivores, the nutrient content of the plants decreases. Experimental studies performed in the field support the hypothesis that the 10-year cycles of snowshoe hares result from the combined effects of predation and fluctuations in the hare's food sources.

Populations of many rodents, including lemmings, also exhibit boom-and-bust changes, often cycling every three to five years. For such short-term cycles, some researchers postulate that either predation or food supply alone may be the underlying cause. Another hypothesis, based on laboratory studies of mice and other small rodents, is that stress from crowding may alter hormonal balance and reduce fertility. The causes of cycles probably vary among species and maybe even among populations of the same species.

Figure 18.12 **Population cycles of the snowshoe hare and the lynx.**

CheckPoint

1. An aquarium population of guppies has reached a stable population size. We decide to add twice as much guppy food per day to the aquarium, but this turns out to have no effect on population size. What is the most likely explanation for this observation?

2. Of the following factors with the potential to limit the growth of a human population, which one is most density independent? **(a)** lowering of fertility due to hormonal changes in very crowded conditions; **(b)** a famine; **(c)** mass drowning caused by hurricane floods; **(d)** epidemic of a highly contagious disease; **(e)** freezing deaths due to a shortage of housing

3. What new research question arises from evidence that the population cycle of snowshoe hares is caused at least partly by a cycle in the availability or nutritional value of the plants eaten by the hares?

4. What causes a population's growth to level off if its behavior approximates the logistic model?

Answers: 1. The population was already at carrying capacity before we increased food supply, and the key limiting factor was something other than food availability. **2. (c) 3.** What causes the cyclical changes in the plants? **4.** The population size reaches the environment's carrying capacity.

(a)

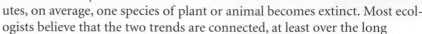

Human Population Growth

Now that we have examined some general concepts of population dynamics, let's return to the specific case of the human population. Every 20 minutes, our population increases by 3500 individuals. In the same 20 minutes, on average, one species of plant or animal becomes extinct. Most ecologists believe that the two trends are connected, at least over the long term—a connection between the exploding human population and the mass extinction the biosphere is now experiencing. The most important factor is our encroachment and destruction of habitat as we spread out and "tame" more land to satisfy our growing demand for space, food, shelter, fuel, water, and other resources (Figure 18.13).

The History of Global Population Growth

The human population has been growing almost exponentially for centuries. In fact, if we compare the history of human population growth in Figure 18.14 with the exponential growth model in Figure 18.6b, it almost looks as if we've been multiplying like bacteria. Of course, our generation span is about 20 years instead of the mere 20 minutes for bacteria, so our population explosion has been stretched out in time. Still, when we compare the two graphs, it seems as though we've been proliferating into the space and resources of the biosphere as though it were an enormous Petri dish. And like bacterial growth in a real Petri dish, exponential growth cannot continue forever.

You can see in Figure 18.14 that the human population increased relatively slowly until about 1650, when approximately 500 million people inhabited Earth. The population doubled to 1 billion within the next two centuries, doubled again to 2 billion between 1850 and 1930, and doubled still again by 1975 to more than 4 billion. If the present growth rate persists, the population will reach 8 billion people by the year 2017. Recall that this hockey-stick-like arch, from a relatively horizontal to an almost vertical curve, is characteristic of exponential growth.

Human population growth is based on the same two general parameters that affect other animal and plant populations: birth rates and death rates. Birth rates increased and death rates decreased when agricultural societies replaced a lifestyle of hunting and gathering about 10,000 years ago. Since the Industrial Revolution, virtually exponential growth has resulted mainly from a drop in death rates, especially infant mortality, even in the least developed countries. Improved nutrition, better medical care, and sanitation have all contributed to an increased percentage of newborns that survive long enough to leave offspring of their own. A decrease in mortality cou-

(b)

Figure 18.13 "Subduing the Earth." (a) Clear-cutting of an old-growth forest in Oregon to harvest lumber to support a growing demand for housing. (b) Slash-and-burn clearing of a tropical rain forest in the Amazon Basin to provide temporary farmland that will become unproductive in just a few years because of depletion of soil nutrients.

Figure 18.14 The history of human population growth.

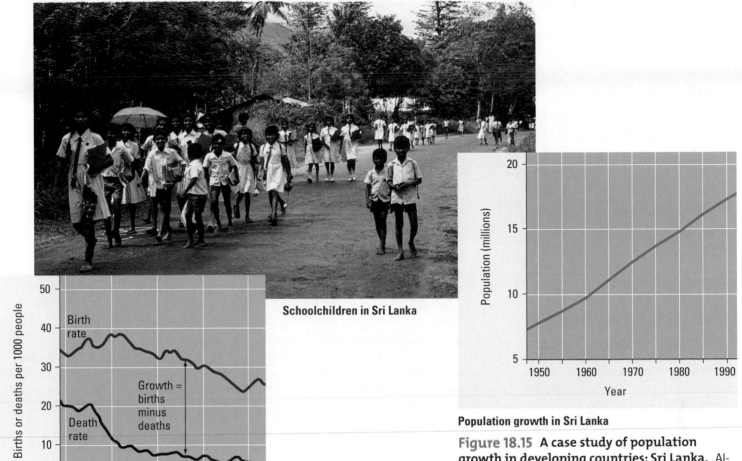

Schoolchildren in Sri Lanka

Changes in birth rate and death rate in Sri Lanka

Population growth in Sri Lanka

Figure 18.15 A case study of population growth in developing countries: Sri Lanka. Although family-planning education and better medical care reduced both birth rates and death rates in Sri Lanka, the population continued to grow because of the large difference between birth rates and death rates.

pled with birth rates that are still relatively high in most developing countries results in an actual increase in population growth rates. In Sri Lanka, for example, the birth rate has been decreasing over the past 50 years, but it has never declined enough to offset the drop in the death rate. Thus, the Sri Lankan population continues to grow rapidly (Figure 18.15).

Age Structure and Population Growth

Worldwide, human population growth is a mosaic of various rates of growth in different countries. Some developed countries, such as Sweden, have stable populations because birth rates and death rates balance. In sharp contrast to Sweden, most developing nations have burgeoning populations in which birth rates greatly exceed death rates. Partly as a result of such unchecked growth, many people in such countries face serious housing, water, and food shortages, as well as severe pollution problems.

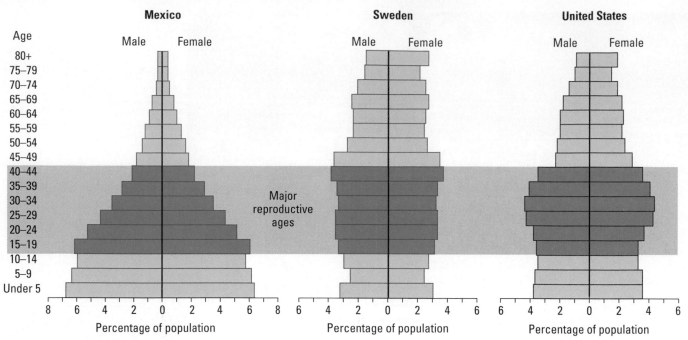

Figure 18.16 Age structures of three nations. The proportion of individuals in different age-groups has a significant impact on the potential for future population growth. Mexico, like many developing countries, has a disproportionate number of young people, and their future reproduction will sustain a steep population increase for many years. In contrast, Sweden's population is distributed more evenly over all age classes, with a high proportion of individuals past their prime reproductive years. The United States has a fairly even age distribution except for the bulge corresponding to the post–World War II "baby boom." (The graphs are based on 1990 data because the numbers from the U.S. 2000 census were not yet available.)

A population characteristic called age structure can help us predict the future growth of populations in different countries. The **age structure** of a population is the proportion of individuals in different age-groups. The relatively uniform age distribution in Sweden, for instance, contributes to that country's stable population size; individuals of reproductive age or younger are not disproportionately represented in the population (Figure 18.16). In contrast, Mexico has an age structure that is bottom-heavy, skewed toward young individuals who will grow up and sustain the explosive growth with their own reproduction.

Notice in Figure 18.16 that the age structure for the United States is relatively even except for a bulge that corresponds to the "baby boom" that lasted for about two decades after the end of World War II. Even though couples born during those years had an average of fewer than two children, population continued to increase during the past few decades because there were so many "boomers" of reproductive age. Because the boomers now range in age from 35 to 55, they are having less impact on population growth. The U.S. population is still growing, however, because the combination of the birth rate and immigration exceeds the death rate. Immigration (legal and illegal combined) now contributes about 40% of the current growth of the U.S. population. And while the average woman in the United States tends to have only two children, the total number of women having children remains high enough for the birth rate to exceed the death rate. Population researchers predict that the U.S. population will continue to grow well into the twenty-first century, perhaps increasing from about 280 million today to about 390 million by the year 2050.

The Sociology, Economics, and Politics of Population Growth

Age-structure diagrams not only reveal a population's growth trends; they also relate to social conditions. Based on the diagrams in Figure 18.16, we

Analyze age-structure diagrams of different countries and make predictions about future conditions in Web/CD Activity 18C.

can predict, for instance, that employment for an increasing number of working-age people will continue to be a significant problem for Mexico in the foreseeable future. For Sweden and the United States, a decreasing proportion of working-age people—mostly those of college age today—will be supporting an increasing proportion of retired people. Programs such as the U.S. Social Security system and Medicare, which are crucial to many older citizens, will become severely strained as the proportion of senior citizens swells.

Predictions of future trends in global human population growth vary widely. The gloomiest models predict a continuing high rate of growth through the twenty-first century, with a doubling to about 12 billion as early as 2050. Perhaps more realistic is a computer model, based on the almost-global trend toward smaller families, which predicts that by about 2080, the human population will peak at about 10.6 billion and then begin a slight decline to about 10.4 billion by the end of the twenty-first century. Either way, there will be a lot more people consuming resources and dumping pollutants, bruising a biosphere that already ails (Figure 18.17).

A unique feature of human population growth is our ability to control it with voluntary contraception and government-sponsored family planning. Social change and the rising educational and career aspirations of women in many cultures encourage them to delay marriage and postpone reproduction. Delayed reproduction dramatically decreases population growth rates. You can get a sense of this phenomenon by imagining two populations in which women each produce three children but begin reproduction at different ages. In one population, females first give birth at age 15, and in the other, at age 30. If we start with a group of newborn girls, then after 30 years the women in the first population will already begin to have grandchildren, whereas women in the second population will be giving birth to their first children. After 60 years, women in the first population will have a large number of great-great-grandchildren (who will themselves begin to reproduce 15 years later), but women in the second population will just begin to see their grandchildren being born.

There is a great deal of heated disagreement among leaders in almost every country as to how much support should be provided for family planning. The issue is certainly socially charged, but there are also political and economic threads to the debate. For example, the United States and many other developed countries, as well as some developing ones, currently face labor shortages, a problem cited by some opponents to policies that encourage family planning. But the debate is also related in part to the difficulty of answering the question, "How many people are too many?" The problem of defining carrying capacity for humans is confounded by the observation that carrying capacity has changed with human cultural evolution (see Chapter 16). The advent of agricultural and industrial technology has significantly increased carrying capacity at least twice during human history, and opponents of population control are counting on some new, as yet unidentified technological breakthrough that will allow our population to grow and plateau at some higher level.

(a)

(b)

Figure 18.17 Two very crowded places. (a) This scene captures the stifling, unsanitary conditions in a slum that has arisen within a garbage dump in Manila, in the Philippines. One of the world's largest cities, Manila has well over 10 million people. **(b)** Commuters who crawl along on the Los Angeles freeways to go to work each day take more than their share of our growing population's toll on the biosphere. Although developing countries are generally growing in population faster than developed ones, it is unfair to assess the problem of human population size in terms of numbers alone. On a per capita basis, people in developed countries have much more impact on the environment than people in developing countries with lower standards of living. In the United States, for example, each person, on average, consumes ten times as much energy (mostly as fossil fuels) as each person in a developing country.

Technology has undoubtedly increased Earth's carrying capacity for humans, but no population can continue to grow indefinitely. Ideally, human populations would reach carrying capacity smoothly and then level off. This will occur when birth rates and death rates are equal, and a decrease in birth rate is more desirable than an increase in death rate. If, however, the population fluctuated about carrying capacity, we would expect periods of increase followed by mass death, as has occurred during plagues, localized famines, and international military conflicts. In any case, the human population must eventually stop growing. Unlike other organisms, we can decide whether zero population growth will be attained through social changes involving individual choice or government intervention or through increased mortality due to resource limitation and environmental degradation. For better or worse, we have the unique responsibility to decide the fate of our species and the rest of the biosphere.

CheckPoint

1. **(a)** How does the age structure of the U.S. population explain the current surplus in the Social Security fund? **(b)** If the system is not changed, why will the surplus give way to a deficit sometime in the next few decades?

2. This question will tell you how well you interpret a graph. What was the approximate percentage increase in the Sri Lankan population over the 20-year span from 1960 to 1980? (See Figure 18.15.)

Answers: 1. (a) The largest population segment, the boomers, are currently in the work force in their peak earning years, paying into the system. **(b)** The boomers will retire over the next few decades and begin drawing from the system at a time when there will be fewer employees paying into Social Security. **2.** 50%, an increase from about 10 million to 15 million people

Life Histories and Their Evolution

In Chapter 17, we looked at some of the ways the responses of organisms to environmental variation increase their chances of survival. However, natural selection does not act only on traits that increase survival; organisms that survive but do not reproduce are not at all "fit" in the Darwinian sense (see Chapter 12). Clearly, an organism can pass along its genes only if it survives long enough to reproduce. But how long is long enough? In many cases there are trade-offs between survival traits and traits that enhance reproductive output—traits such as frequency of reproduction, investment in parental care, and the number of offspring per reproductive episode (usually called seed crop for seed plants and litter size or clutch size for animals).

The traits that affect an organism's schedule of reproduction and death make up its **life history.** Of course, a particular life history, like most characteristics of an organism, is the result of natural selection operating over evolutionary time. In this section, we'll see how life history traits affect population growth.

Life Tables and Survivorship Curves

When the life insurance industry was established about a century ago, insurance companies based their business plans on some of the early scientific

Table 18.1 — Life Table for the U.S. Population in 1980

Age Interval	Number Living at Start of Age Interval (N)	Number Dying During Interval (D)	Mortality (Death Rate) During Interval (D/N)	Chance of Surviving Interval (1 − D/N)
0–10	10,000,000	121,678	0.012	0.988
10–20	9,878,322	124,163	0.013	0.987
20–30	9,754,159	174,161	0.018	0.982
30–40	9,579,998	202,773	0.021	0.979
40–50	9,377,225	410,607	0.044	0.956
50–60	8,966,618	882,352	0.098	0.902
60–70	8,084,266	1,810,106	0.224	0.776
70–80	6,274,160	2,999,619	0.478	0.522
80–90	3,274,541	2,628,753	0.803	0.197
90–100	645,788	645,788	1.000	0.000

studies of human populations. Needing to determine how long, on average, an individual of a given age could be expected to live, the insurance companies began using what are called life tables.

A **life table** tracks survivorship and mortality (death) in a population. For example, Table 18.1 arranges the survivorship/mortality data for a sample of 10 million U.S. citizens over a ten-year period, ending in 1980. Using this table, an insurance agent could predict that a 21-year-old has about a 0.982 (98.2%) chance of surviving to age 30. Borrowing the basic idea from the insurance industry, population ecologists construct life tables for plants and nonhuman animals.

A graphic way of representing some of the data in a life table is to draw a **survivorship curve,** a plot of the number of people still alive at each age (Figure 18.18). We can classify survivorship curves for diverse organisms into three general types. A Type I curve is relatively flat at the start, reflecting low death rates during early and middle life and dropping steeply as death rates increase among older age-groups. Humans and many other large mammals that produce relatively few offspring but provide them with good care often exhibit this kind of curve. In contrast, a Type III curve indicates high death rates for the very young and then a period when death rates are much lower for those few individuals who survive to a certain age. Species with this type of survivorship curve usually produce very large numbers of offspring but provide little or no care for them. An oyster, for instance, may release millions of eggs, but most offspring die as larvae from predation or other causes. A Type II curve is intermediate, with mortality more constant over the life span. This type of survivorship has been observed in several invertebrates, including hydras, and in certain rodents, such as the gray squirrel.

A population's pattern of mortality is certainly a key feature of life history. But we have defined life history as the set of traits that affect an organism's schedule of *reproduction* and death. Let's take a closer look now at how natural selection affects the reproductive strategies that evolve in populations.

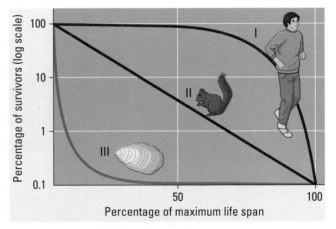

Figure 18.18 **Three idealized types of survivorship curves.**

Figure 18.19 The "big bang" reproduction of century plants. A century plant, a type of agave, grows for many years (a century in some cases) without flowering or reproducing. Then one spring it grows a floral stalk that may be as tall as a telephone pole. That season, the plant produces many seeds and then withers and dies, its food reserves and water spent in the formation of its massive bloom. The century plant's life history is an example of what ecologists call the "big bang" strategy of reproduction. It's an evolutionary adaptation to the organism's environmental situation. Century plants grow in arid climates with sparse and unpredictable rainfall. Their shallow roots catch water after rain showers but are dry during droughts. This unpredictable water supply may prevent seed production for several years at a time. By growing and storing nutrients until an unusually wet year and then putting all its resources into one grand burst of seed production, the plant maximizes its reproductive success.

Life History Traits as Evolutionary Adaptations

Some key life history traits affecting population growth are the age at which reproduction first occurs, the number of offspring produced for each reproductive episode, the amount of parental care committed to offspring, and the overall energy cost of reproduction. For a given population living in a particular environmental context, natural selection will favor the combination of life history traits that maximizes an individual's output of viable, fertile offspring. In other words, life history traits, like anatomical features, are shaped by adaptive evolution (Figure 18.19).

Life history strategies vary with species and may change even for a single population as the environmental context changes. However, it is useful to contrast two extreme types of life history: opportunistic life history and equilibrial life history.

We can find examples of **opportunistic life histories** among certain populations of small-bodied species. Individuals reproduce when young, and they produce many offspring. The population tends to grow exponentially when conditions are favorable (hence the term *opportunistic*). Such a population typically lives in an unpredictable environment and is controlled by density-independent factors such as the weather. For example, dandelions and many other annual weeds grow quickly in open, disturbed areas, producing a large number of seeds in a brief time when the weather is favorable. Although most of the seeds will not produce mature plants, their large number and ability to disperse to new habitats ensure that at least some will grow and eventually produce seeds themselves. Many insects, including locusts, also exhibit opportunistic life histories, maximizing reproductive output whenever environmental opportunity knocks. For such species, natural selection has reinforced quantity of reproduction more than individual survivorship. In general, populations with an opportunistic life history exhibit a Type III survivorship curve (see Figure 18.18).

In contrast, some populations, mostly larger-bodied species, exhibit an **equilibrial life history,** which generally results in a Type I survivorship curve. Individuals usually mature later and produce few offspring but care for their young. The population size may be quite stable, held near the carrying capacity by density-dependent factors (stabilization around carrying capacity accounts for the term *equilibrial*). Natural selection has resulted in the production of better-endowed offspring that can become established in the well-adapted population into which they are born. The life histories of many large terrestrial vertebrates fit this model. Among polar bears, for instance, a female has only one or two offspring every three years, but the cubs remain in her protective custody for over two years. In the plant kingdom, the coconut palm tree also fits the equilibrial model. Compared with most other trees, it produces relatively few, very large seeds, which provide nutrients (including the coconut "milk") for the embryo; this is a plant's version of parental care.

Table 18.2 contrasts some life history traits between opportunistic and equilibrial strategies. Most populations probably fall between the opportunistic and equilibrial extremes. Thus, these two life history strategies are only hypothetical models, starting points for studying the complex interplay of the forces of natural selection on reproductive characteristics.

Investigate the life histories of various organisms in Web/CD Activity 18D.

Table 18.2	Some Life History Characteristics of Opportunistic and Equilibrial Populations	

Characteristic	Opportunistic Populations (e.g., many wildflowers)	Equilibrial Populations (e.g., many large mammals)
Climate	Relatively unpredictable	Relatively predictable
Maturation time	Short	Long
Life span	Short	Long
Death rate	Often high	Usually low
Number of offspring produced per reproductive episode	Many	Few
Number of reproductions per lifetime	Usually one	Often several
Timing of first reproduction	Early in life	Later in life
Size of offspring or eggs	Small	Large
Parental care	None	Often extensive

Source: Adapted from E. R. Pianka, *Evolutionary Ecology,* 6th ed. (San Francisco, CA: Benjamin/Cummings, 2000), p. 186.

CheckPoint

1. How do the terms *opportunistic* and *equilibrial* contrast the key characteristics of these life history strategies?

2. What is the key feature of the mortality column in the life table for a population exhibiting a Type II survivorship curve?

Answers: 1. Opportunistic life histories are characterized by an ability to produce a large number of offspring very rapidly when the environment affords a temporary opportunity for exponential growth; equilibrial life histories are characterized by a population size that fluctuates only slightly from carrying capacity. **2.** The mortality is about the same for every age interval.

Evolution Link: Testing a Darwinian Hypothesis

A body of evidence supports the Darwinian view that a population's life history traits are evolutionary adaptations shaped by natural selection. As a case study of the process of science, let's examine a classic set of experiments on the evolution of life history.

For many years, David Reznick of the University of California, Riverside, and John Endler of the University of California, Santa Barbara, have been investigating the life histories of guppy populations in Trinidad, a Caribbean island. Guppies are small freshwater fish you probably recognize as popular aquarium pets. In the Aripo River system of Trinidad, guppies live in small pools as populations that are relatively isolated from one another. In some cases, two populations inhabiting the same stream live less than 100 meters apart, but they are separated by a waterfall that impedes the migration of guppies between the two ponds.

Figure 18.20 David Reznick conducting field experiments on guppy evolution in Trinidad.

Early in their research, Reznick and Endler recognized that certain life history traits among guppy populations correlated with the main type of predator in a stream pool. Certain guppy populations live in pools where the predator is the killifish, which eats mainly small, immature guppies. Other guppy populations live where larger fish, called pike-cichlids, eat mostly large, mature guppies. Where preyed on by pike-cichlids, guppies tend to be smaller, mature earlier, and produce more offspring each time they give birth than those in areas without pike-cichlids. If the differences between the populations result from natural selection, the life history traits should be heritable. To test for heritability, the scientists raised guppies from both types of populations in the laboratory without predators. The two populations retained their life history differences when followed through several generations, indicating that the differences were indeed inherited. A reasonable hypothesis is that the selective predation of larger versus smaller guppies results in the life history adaptations that the researchers observed. Apparently, when predators such as pike-cichlids prey mainly on reproductively mature adults, the chance that a guppy will survive to reproduce several times is relatively low. The guppies with the greatest reproductive success should then be the individuals that mature at a young age and small size and reproduce at least once before growing to a size preferred by the local predator.

Reznick and Endler tested their hypothesis with field experiments (Figure 18.20). Here is the scientific reasoning, based on the "*If . . . then*" logic you learned about in Chapter 1:

> **Hypothesis:** *If* the feeding preferences of different predators cause contrasting life histories in different guppy populations by natural selection,
>
> **Experiment:** and guppies are transplanted from locations with pike-cichlids (predators of mature guppies) to guppy-free sites inhabited by killifish (predators of juvenile guppies),
>
> **Predicted Result:** *then* the transplanted guppy populations should show a generation-to-generation trend toward later maturation and larger size—life history traits typical of natural populations that coexist with killifish.

Reznick and Endler performed these guppy transplant experiments and studied the populations for 11 years, measuring age and size at maturity and other life history traits (Figure 18.21). Over the 11 years, or 30 to 60 guppy generations, the average weight at maturity for guppies in the transplanted (experimental) populations increased by about 14% compared with control populations. Other life history traits also changed in the direction predicted by the hypothesis of evolutionary adaptation due to selective predation.

Without a control group for comparison, there would be no way to tell whether it was the killifish or some other factor that caused the transplanted guppy populations to change. But because control sites and experimental sites were usually nearby pools of the same stream, the main variable was probably the presence of different predators. And these careful scientists observed similar results when guppy populations were reared in artificial streams that were identical except for the type of predator. Reznick and Endler had in fact tested a Darwinian hypothesis and documented evolution in a natural setting over a relatively short time.

Evolutionary biology qualifies as science because its questions stimulate scientists to articulate hypotheses they can test by further observation and experiments.

Learn more about evolutionary population studies in the Web Evolution Link.

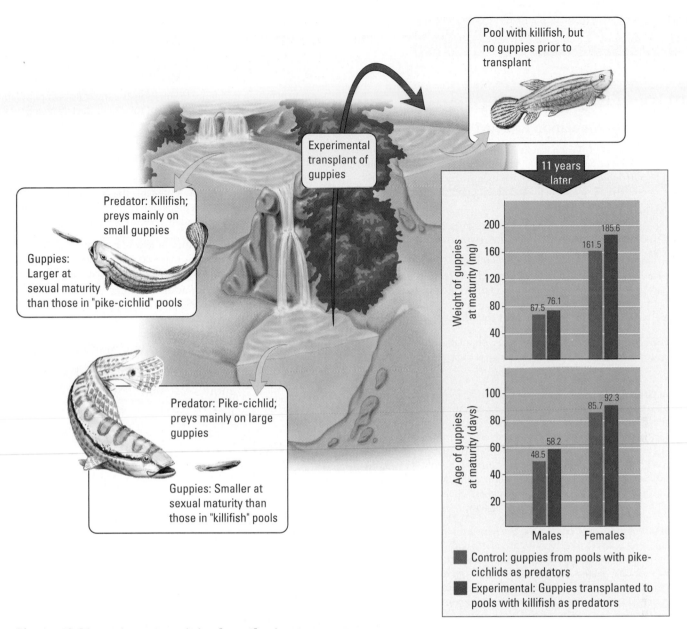

Figure 18.21 Testing a Darwinian hypothesis. This illustration represents three pools along streams of Trinidad's Aripo River system. In one pool (upper left), the predators are killifish, which prey on relatively small immature guppies. In another pool (lower left), the predators are pike-cichlids, which eat primarily large mature guppies. Guppy populations living in pools where the predator is the pike-cichlid tend to be smaller at sexual maturity than those in the "killifish pools." Researchers tested the hypothesis that selective predation accounts for the life-history differences between guppy populations. The scientists transplanted guppies from pike-cichlid pools to pools that contained killifish but had no natural guppy populations (upper right). The biologists then tracked the evolution of weight and age at sexual maturity in the experimental guppy populations for 11 years, comparing their measurements to guppies from control pools inhabited by pike-cichlids. As shown in the graphs, the researchers observed that the average weight and age at sexual maturity of the transplanted populations increased significantly as compared to the control populations.

Summary of Key Concepts

Overview: What Is Population Ecology?

• The human population explosion is now Earth's most significant biological phenomenon. Principles of population ecology help us understand the problem of human population growth. Population ecology focuses on the factors that influence a population's size, growth rate, density, and characteristics.
• **Web/CD Activity 18A** *Human Population Ecology*

The Structure and Growth of Populations

• **Population Density** Population density, the number of individuals of a species per unit area or volume, is usually estimated by a variety of sampling techniques. These include counting the number of individuals in a sample plot or the mark-recapture method.
• **Web/CD Activity 18B/The Process of Science** *Population Density Lab*

• **Patterns of Dispersion** Dispersal patterns of a population range from clumped, to uniform, to random, as determined by various environmental or social factors.

• **Population Growth Models** The exponential model of population growth describes an idealized population in an unlimited environment. This model predicts that the larger a population becomes, the faster it grows. Exponential growth in nature is generally a short-lived consequence of organisms being introduced to a new or under-exploited environment. The logistic model of population growth describes an idealized population that is slowed by limiting factors. This model predicts that a population's growth rate will be small when the population size is either small or large, and highest when the population is at an intermediate level relative to the carrying capacity.

• **Regulation of Population Growth** Over the long term, most population growth is limited by a mixture of density-independent and density-dependent factors. Density-dependent factors intensify as a population increases in size, increasing the death rate, decreasing the birth rate, or both in a population. Density-independent factors affect the same percentage of individuals regardless of population size. Some populations have regular boom and bust cycles.

Human Population Growth

• **The History of Global Population Growth** The human population has been growing almost exponentially for centuries. Human population growth is based on the same two general parameters that affect other animal and plant populations: birth rates and death rates. Birth rates increased and death rates decreased when agricultural societies replaced a lifestyle of hunting and gathering.

• **Age Structure and Population Growth** The age structure of the population is a major factor in the different growth rates of different countries. Population researchers predict that the U.S. population will continue to grow well into the twenty-first century.

• **The Sociology, Economics, and Politics of Population Growth** Age-structure diagrams also predict social predicaments. Predictions of future trends in global human population growth vary widely. The human species is unique in having the ability to consciously control its own population growth, the fate of our species, and the fate of the rest of the biosphere.
• **Web/CD Activity 18C** *Analyzing Age-Structure Diagrams*

Life Histories and Their Evolution

• **Life Tables and Survivorship Curves** A population's pattern of mortality is a key feature of life history. A life table tracks survivorship and mortality in a population. Survivorship curves can be classified into three general types, depending on the rate of mortality over the entire life span.

• **Life History Traits as Evolutionary Adaptations** Life history traits are shaped by adaptive evolution, may vary within a species, and may change as the environmental context changes. Most populations probably fall between the extreme opportunistic strategies of many insects and equilibrial strategies of many larger-bodied species.
• **Web/CD Activity 18D** *Investigating Life Histories*

Evolution Link: Testing a Darwinian Hypothesis

• Scientists have conducted field tests of the hypothesis that natural selection shapes life history.
• **Web Evolution Link** *Testing a Darwinian Hypothesis*

Self-Quiz

1. Which of the following shows the effects of a density-dependent limiting factor?
   a. A forest fire kills all the pine trees in a patch of forest.
   b. Early rainfall triggers the explosion of a locust population.
   c. Drought decimates a wheat crop.
   d. Silt from logging kills half the young salmon in a stream.
   e. Rabbits multiply, and their food supply begins to dwindle.

2. With regard to its percent increase, a population that is growing logistically
   a. grows fastest when density is lowest.
   b. has a high intrinsic rate of increase.
   c. grows fastest at an intermediate population density.
   d. grows fastest as it approaches carrying capacity.
   e. is always slowed by density-independent factors.

3. Pine trees in a forest tend to shade and kill pine seedlings that sprout nearby. This causes the pine trees to
   a. increase exponentially.
   b. grow in a clumped pattern.
   c. grow in a uniform pattern.
   d. exceed their carrying capacity.
   e. grow in a random pattern.

4. To figure out the human population density of your community, you would need to know the number of people living there and
   a. the land area in which they live.
   b. the birth rate of the population.
   c. whether population growth is logistic or exponential.
   d. the dispersion pattern of the population.
   e. the carrying capacity.

5. Skyrocketing growth of the human population since the beginning of the Industrial Revolution appears to be mainly a result of

    a. migration to thinly settled regions of the globe.

    b. better nutrition boosting the birth rate.

    c. a drop in the death rate due to better nutrition and health care.

    d. the concentration of humans in cities.

    e. social changes that make it desirable to have more children.

6. A uniform dispersion pattern for a population may indicate that

    a. the population is spreading out and increasing its range.

    b. resources are heterogeneously distributed.

    c. individuals of the population are competing for some resource, such as water and minerals for plants or nesting sites for animals.

    d. there is an absence of strong attractions or repulsions among individuals.

    e. the density of the population is low.

7. A Type III survivorship curve would be expected in a species in which

    a. mortality occurs at a constant rate over the life span.

    b. parental care is extensive.

    c. a large number of offspring are produced but parental care is minimal.

    d. mortality rate is quite low for the young.

    e. an equilibrial life history prevails.

8. The example of the population cycles of the snowshoe hare and its predator, the lynx, illustrates that

    a. predators are the major factor in controlling the size of prey populations and are in turn regulated in their numbers by the oscillating supply of prey.

    b. the two species must have evolved in close contact with each other because their life histories are intertwined.

    c. one should not conclude a cause-and-effect relationship when viewing population patterns without careful observation and experimentation.

    d. both populations are controlled by density-independent factors.

    e. the hare population has an opportunistic life history, whereas the lynx population has an equilibrial life history.

9. Consider five human populations that differ in their age structures. The population that will grow the most in the next 30 years is probably going to be the one with the greatest fraction of people in which age group?

    a. 10–20   d. 40–50

    b. 20–30   e. 50–60

    c. 30–40

10. All these characteristics are typical of human populations in most developed countries *except*

    a. relatively small family size.

    b. relatively slow population growth.

    c. opportunistic life history.

    d. Type I survivorship curve.

    e. relatively even age structure.

• **Go to the website or CD-ROM for more self-quiz questions.**

The Process of Science

1. We estimate the size of a population of small mice in a particular field by the mark-recapture method. Our estimate is $N = 350$. Later, we learn from experiments on the behavior of these mice that they can locate a baited trap faster if they have already been rewarded with food by visiting that trap once before. Does this mean that our original estimate of 350 individuals was (a) too low or (b) too high? Explain your answer in terms of the equation for the mark-recapture method.

2. **Try your hand at estimating population density in The Process of Science activity available on the website and CD-ROM.**

Biology and Society

Many people regard the rapid population growth of developing countries as our most serious environmental problem. Others think that the population growth in developed countries, though smaller, is actually a greater threat to the environment. What kinds of problems result from population growth in (a) developing countries and (b) industrialized countries? Which do you think is the greater threat, and why?

Community Ecology

Humans intentionally introduced the pests

kudzu and the European starling to the United States.

Many insects use chemical sensors on their feet to

find food. 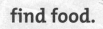 Some plants recruit parasitic

wasps that lay their eggs in caterpillars that eat the

plants. The United States has 50,000

introduced species, with a cost to the economy of $130

billion in damage and control efforts.

On your next walk through a field or woodland, or even across campus or through a park or your backyard, try to observe some of the interactions among the species present. You may see birds using trees as nesting sites, bees pollinating flowers, shelf fungi growing on trees, caterpillars feeding on leaves, spiders trapping insects in their webs, cats stalking small rodents, ferns growing in shade provided by trees—a sample of the many interactions that occur in any ecological theater. In addition to the physical and chemical factors (abiotic factors) we discussed in Chapter 17, an organism's environment includes other individuals in its population and populations of other species living in the same area. Such an assemblage of species living close enough together for potential interaction is called a **community.** In Figure 19.1, the lion, the zebra, the hyena, the vultures, and the grasses and other plants are all members of a community in Kenya.

In this chapter, we examine the diverse interactions among organisms and how those relationships determine the species composition and other features of communities. Throughout the chapter, biology's core theme of evolution will be apparent in the adaptations that fit organisms to their ecological roles in communities.

Figure 19.1 Diverse species interacting in a Kenyan savanna community.

Overview: Key Properties of Communities

Four key properties of a community are its diversity, its prevalent form of vegetation, its stability, and its trophic structure.

Diversity

The diversity of a community—the variety of different kinds of organisms that make up the community—has two components. One is species richness, or the total number of different species in the community. The other is the relative abundance of the different species. For example, imagine two forest communities, each with 100 individuals distributed among four different tree species (A, B, C, and D) as follows:

Community 1: 25A, 25B, 25C, 25D

Community 2: 80A, 10B, 5C, 5D

The species richness is the same for both communities because they both contain four species, but the relative abundance is very different (Figure 19.2). You would easily notice the four different types of trees in community 1, but without looking carefully, you might see only the abundant species A in the second forest. Most observers would intuitively describe community 1 as the more diverse of the two communities. Indeed, the term **species diversity,** as used by ecologists, considers *both* diversity factors: richness and relative abundance.

Analyze species richness and relative abundance in Web/CD Activity 19A/The Process of Science.

Community 1

Community 2

Figure 19.2 Which forest is more diverse? With the same four tree species in both forests, the two communities are equal in their species richness of trees. But if we factor in the relative abundance of species, then community 1 is the more diverse because of the more equitable representation of the different tree species.

Figure 19.3 Trophic structure: feeding relationships. After spending four to six years feeding in the open ocean, this salmon is attempting to return to the stream of its birth. Along the way, however, it may become a meal for a grizzly bear.

Prevalent Form of Vegetation

The second property of a community, its prevalent form of vegetation, applies mainly to terrestrial situations. For example, deciduous trees are the prevalent components of the community in a temperate deciduous forest. When we look more closely at such a community, we see not only which plants are dominant, but also how the plants are arranged, or "structured." For instance, a deciduous forest has a pronounced vertical structure: The treetops form a top layer, or canopy, under which there is a subcanopy of lower branches, and small shrubs and herbs carpet the forest floor. The types and structural features of plants largely determine the kinds of animals that live in a community.

Stability

The third property of a community, **stability,** refers to the community's ability to resist change and return to its original species composition after being disturbed. Stability depends on both the type of community and the nature of disturbances. For example, a forest dominated by cedar and hemlock trees is a highly stable community in that it may last for thousands of years with little change in species composition. Large cedars and hemlocks even withstand most lightning-caused fires, which kill small trees and shrubs growing in forest openings or in the shade of the dominant trees. However, when a fire does kill the dominant trees, a cedar/hemlock forest might seem less stable than, say, a grassland, because it will take much longer for the forest to return to its original species composition.

Trophic Structure

The fourth property of a community is its **trophic structure** (from the Greek *trophe,* nourishment), the feeding relationships among the various species making up the community. A community's trophic structure determines the passage of energy and nutrients from plants and other photosynthetic organisms to herbivores and then to carnivores. Trophic relationships are certainly implied by the scene in Figure 19.3, and we'll see many more examples as we take a closer look at community interactions.

Interspecific Interactions in Communities

With the four main properties of a community in mind, we turn next to the various kinds of interactions between species—what ecologists call **interspecific interactions.** The three main types of interspecific interactions we'll explore are competition, predation, and symbiosis. In each case, we'll see how these community relationships function as environmental factors in the adaptive evolution of organisms through natural selection.

Competition Between Species

When populations of two or more species in a community rely on similar limiting resources, they may be subject to **interspecific competition.** You already learned in Chapter 18 that as a population's density increases and nears carrying capacity, every individual has access to a smaller share of some limiting resource, such as food. As a result, mortality rates increase, birth rates decrease, and population growth is curtailed. In interspecific competition, however, the population growth of a species may be limited by the density of competing species as well as by the density of its own population. For example, if several bird species in a forest feed on a limited population of insects, the density of each species may have a negative impact on population growth in the other species. Similarly, species may compete for nesting sites, shelters, or any other resource that is in short supply.

Competitive Exclusion Principle In 1934, Russian ecologist G. F. Gause studied the effects of interspecific competition in laboratory experiments with two closely related species of protists, *Paramecium aurelia* and *Paramecium caudatum* (Figure 19.4). Gause cultured the protists under stable conditions with a constant amount of food added every day. When he grew the two species in separate cultures, each population grew rapidly and then leveled off at what was apparently the carrying capacity of the culture. But when Gause cultured the two species together, *P. aurelia* apparently had a competitive edge in obtaining food, and *P. caudatum* was driven to extinction in the culture. Gause concluded that two species so similar that they compete for the same limiting resources cannot coexist in the same place. One will use the resources more efficiently and thus reproduce more rapidly. Even a slight reproductive advantage will eventually lead to local elimination of the inferior competitor. Ecologists called Gause's concept the **competitive exclusion principle.** Numerous tests of the hypothesis include classic field experiments with two species of barnacles that attach to intertidal rocks on the North Atlantic coast (Figure 19.5).

The Ecological Niche The sum total of a species' use of the biotic and abiotic resources in its environment is called the species' **ecological niche.** One way to grasp the concept is through an analogy made by ecologist Eugene Odum: If an organism's habitat is its address, the niche is that habitat plus the organism's occupation. Put another way, an organism's niche is its ecological role—how it "fits into" an ecosystem. The niche of a population of

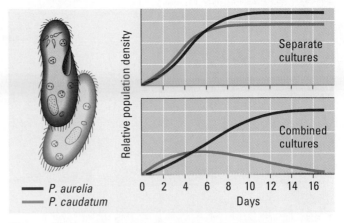

Figure 19.4 Competitive exclusion in laboratory populations of *Paramecium*. Cultured separately (upper graph) with constant amounts of food (bacteria) added daily, populations of each of the two *Paramecium* species grow to carrying capacity. But when the two species are cultured together (lower graph), *P. aurelia* has a competitive edge in obtaining food, an advantage that drives *P. caudatum* to extinction in the microcosm of the culture jar.

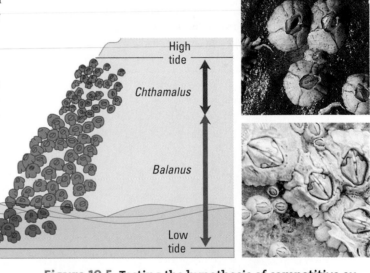

Figure 19.5 Testing the hypothesis of competitive exclusion in the field. These two types of barnacles, *Balanus* and *Chthamalus*, grow on rocks that are exposed during low tide. Both types are attached as adults, but have free-swimming larvae that may settle and begin to develop on virtually any rock surface. *Chthamalus* occupies the upper parts of the rocks, which are out of the water longer during low tides. *Balanus* fails to survive as high on the rocks as *Chthamalus*, apparently because *Balanus* dries out in the air. When ecologists experimented by removing *Balanus* from the lower rocks, *Chthamalus* spread lower, colonizing the unoccupied rocks. However, when both species colonize the same rock, *Balanus* eventually displaces *Chthamalus* on the lower part of the rock. The researchers concluded that the upper limit of *Balanus*'s distribution is set mainly by the availability of water, whereas the lower limit of *Chthamalus*'s distribution is set by competition.

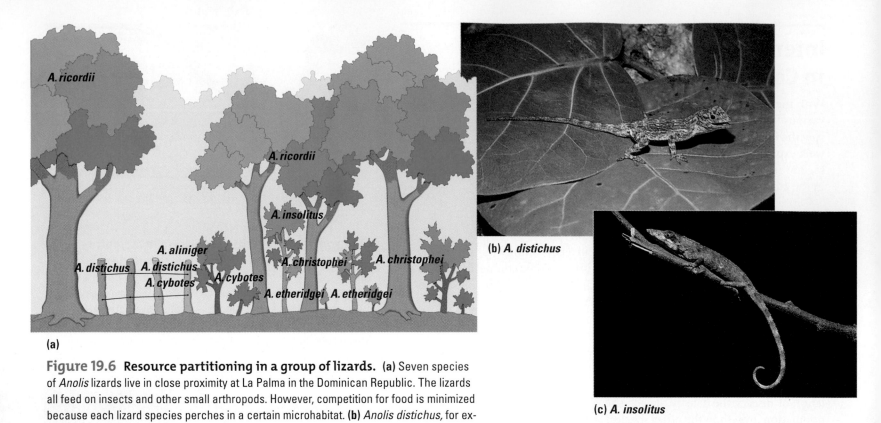

(a)

(b) *A. distichus*

(c) *A. insolitus*

Figure 19.6 **Resource partitioning in a group of lizards.** **(a)** Seven species of *Anolis* lizards live in close proximity at La Palma in the Dominican Republic. The lizards all feed on insects and other small arthropods. However, competition for food is minimized because each lizard species perches in a certain microhabitat. **(b)** *Anolis distichus,* for example, perches on fence posts and other sunny surfaces (such as this leaf), whereas **(c)** *A. insolitus* usually perches on shady branches.

tropical tree lizards, for example, consists of, among many other components, the temperature range it tolerates, the size of trees on which it perches, the time of day in which it is active, and the size and type of insects it eats.

We can now restate the competitive exclusion principle to say that two species cannot coexist in a community if their niches are identical. However, ecologically similar species can coexist in a community if there are one or more significant differences in their niches.

Resource Partitioning There are two possible outcomes of competition between species having identical niches: Either the less competitive species will be driven to local extinction, or one of the species may evolve enough to use a different set of resources. This differentiation of niches that enables similar species to coexist in a community is called **resource partitioning.** Figure 19.6 illustrates an example. We can think of such cases of resource partitioning within a community as "the ghost of competition past"—circumstantial evidence of earlier interspecific competition resolved by the evolution of niche differences.

Predation

In everyday usage, the term *community* is benign, maybe even implying the warmness of cooperation, as in *community spirit.* In contrast, an ecological community exhibits the Darwinian realities of competition and predation, where organisms eat other organisms. In the interspecific interaction called **predation,** the consumer is the **predator** and the food species is the **prey.**

We include herbivory, the eating of plants by animals, as a form of predation, even in cases such as grazing, where the animal does not kill the whole plant.

It won't surprise you that predation is a potent factor in adaptive evolution. Eating and avoiding being eaten are prerequisite to reproductive success. Natural selection refines the adaptations of both predators and prey.

Predator Adaptations Many important feeding adaptations of predators are both obvious and familiar. Most predators have acute senses that enable them to locate and identify potential prey. In addition, many predators have adaptations such as claws, teeth, fangs, stingers, or poison that help catch and subdue the organisms on which they feed. Rattlesnakes and other pit vipers, for example, locate their prey with special heat-sensing organs located between each eye and nostril, and they kill small birds and mammals by injecting them with toxins through their fangs. Similarly, many herbivorous insects locate appropriate food plants by using chemical sensors on their feet, and their mouthparts are adapted for shredding tough vegetation. Predators that pursue their prey are generally fast and agile, whereas those that lie in ambush are often camouflaged in their environments. In our own case, perhaps it is valid to view our capacity for learning and teaching to be adaptations contributing to our success as predators; the invention and refinement of agriculture over the centuries has certainly increased Earth's carrying capacity for humans (see Chapter 16).

Plant Defenses Against Herbivores Plants cannot run away from herbivores. Chemical toxins, often in combination with various kinds of antipredator spines and thorns, are plants' main arsenals against being eaten to extinction. Among such chemical weapons are the poison strychnine, produced by a tropical vine called *Strychnos toxifera;* morphine, from the opium poppy; nicotine, produced by the tobacco plant; mescaline, from peyote cactus; and tannins, from a variety of plant species. Other defensive compounds that are not toxic to humans but may be distasteful to other herbivores are responsible for the familiar flavors of cinnamon, cloves, and peppermint. Some plants even produce chemicals that imitate insect hormones and cause abnormal development in some insects that eat them.

Animal Defenses Against Predators Animals can avoid being eaten by using passive defenses, such as hiding, or active defenses, such as escaping or defending themselves against predators. Fleeing is a common antipredator response, though it can be very costly in terms of energy. Many animals flee into a shelter and avoid being caught without expending the energy required for a prolonged flight. Active self-defense is less common, though some large grazing mammals will vigorously defend their young from predators such as lions. Other behavioral defenses include alarm calls, which often bring in many individuals of the prey species that mob the predator. Mobbing can involve either harassment at a safe distance or direct attack (Figure 19.7). Distraction displays direct the attention of the predator away from a vulnerable prey, such as a bird chick, to another potential prey that is more likely to escape, such as the chick's parent (Figure 19.8).

Many other defenses rely on adaptive coloration, which has evolved repeatedly among animals. Camouflage, called **cryptic coloration,** is passive defense that makes potential prey difficult to spot against its background (Figure 19.9).

Figure 19.7 Mobbing, a behavioral defense against predators. Many prey species turn the tables and attack their predators. Here, two crows mob a barn owl, a predator that often kills and eats crow eggs and nestlings.

Figure 19.8 Protecting offspring by faking an injury. This killdeer uses deception to defend her nest against predators and human disturbance. When danger threatens, she leaves the nest and fakes a broken wing. This behavior distracts and draws a potential predator away from the nest, and then the mother just flies away. The trickery often saves the lives of a killdeer's offspring.

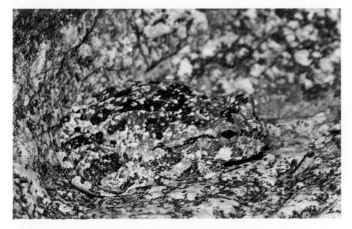

Figure 19.9 Camouflage: a canyon tree frog disappearing into a background of granite.

Figure 19.10 Warning coloration of a poison-arrow frog. The vivid markings of this tree frog, an inhabitant of rain forests in Costa Rica, warn of noxious chemicals in the frog's skin; predators learn about this as soon as they touch the frog. In some parts of South America, human hunters in the rain forest tip their arrows with poisons from similar frogs to bring down large mammals.

(a) Hawk moth larva

(b) Snake

Figure 19.11 Batesian mimicry. When disturbed, **(a)** the hawk moth larva resembles **(b)** a snake.

Some animals have mechanical or chemical defenses against would-be predators. Most predators are strongly discouraged by the familiar defenses of skunks and porcupines. Some animals, such as poisonous toads and frogs, can synthesize toxins. Others acquire chemical defense passively by accumulating toxins from the plants they eat. For example, monarch butterflies store poisons from the milkweed plants they eat as larvae, which makes the butterflies distasteful to some potential predators. Animals with chemical defenses are often brightly colored, a caution to predators known as **warning coloration** (Figure 19.10).

A species of prey may gain significant protection through mimicry, a "copycat" adaptation in which one species mimics the appearance of another. In **Batesian mimicry,** a palatable or harmless species mimics an unpalatable or harmful model. In one intriguing example, the larva of the hawkmoth puffs up its head and thorax when disturbed, looking like the head of a small poisonous snake, complete with eyes (Figure 19.11). The mimicry even involves behavior; the larva weaves its head back and forth and hisses like a snake. (The swallowtail butterfly larva on the cover of this book is another example of a snake imitator.) In **Müllerian mimicry,** two or more unpalatable species resemble each other. Presumably, each species gains an additional advantage, because the pooling of numbers causes predators to learn more quickly to avoid any prey with a particular appearance (Figure 19.12).

Predators also use mimicry in a variety of ways. For example, some snapping turtles have tongues that resemble a wriggling worm, thus luring small fish; any fish that tries to eat the "bait" is itself quickly consumed as the turtle's strong jaws snap closed.

(a) Cuckoo bee

Figure 19.12 Müllerian mimicry. Both **(a)** the cuckoo bee and **(b)** the yellow jacket wasp have stingers that release toxins. The cross-mimicry in appearance presumably benefits both species because predators learn more quickly to avoid any prey with these distinctive markings.

(b) Yellow jacket

Predation and Species Diversity in Communities You might think that organisms eating other organisms would always reduce species diversity, but field experiments reveal that predator-prey relationships can actually preserve diversity. Experiments by American ecologist Robert Paine in the 1960s were among the first to provide such evidence. Paine removed the dominant predator, a sea star of the genus *Pisaster,* from experimental areas within the intertidal zone of the Washington coast (Figure 19.13). The result was that *Pisaster*'s main prey, a mussel of the genus *Mytilus,* outcompeted many of the other shoreline organisms (barnacles and snails, for instance) for the important resource of space on the rocks. The number of species dropped.

The experiments of Paine and others led to the concept of keystone predators. A **keystone predator** is a species, such as *Pisaster,* that reduces the density of the strongest competitors in a community. In so doing, the predator helps maintain species diversity by preventing competitive exclusion of weaker competitors. By checking the population growth of mussels, the most successful (and most abundant) species, the sea star *Pisaster* helped maintain populations of several less competitive species in Paine's experimental tide pools. Predation has its constructive side.

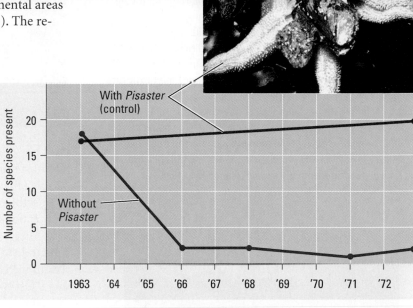

Figure 19.13 Effect of a keystone predator on species diversity. Ecologist Robert Paine removed the sea star *Pisaster* (inset, eating its favorite food, a mussel) from a tide pool in 1963. Mussels then gradually took over the rock space and eliminated other invertebrates and algae.

Symbiotic Relationships

The term *symbiosis* is derived from the Greek meaning "living together." A **symbiotic relationship** is an interspecific interaction in which one species, the **symbiont,** lives in or on another species, the **host.** The two types of symbiotic relationships most important in structuring ecological communities are parasitism and mutualism.

Parasitism A symbiotic relationship is called **parasitism** if one organism benefits at the expense of the other, which is harmed in the process. The **parasite,** usually the smaller of the two organisms, obtains its nutrients by living on or in its **host** organism. (The word *parasite* comes from the Greek meaning "near food.") We can actually think of parasitism as a specialized form of predation. Tapeworms and the protozoans that cause malaria are examples of internal parasites (see Figures 16.16 and 14.18d). External parasites include mosquitoes, which suck blood from animals, and aphids, which tap into the sap of plants. The term *host* is misleading in its implication that the welcome mat is out. In everyday usage, we sometimes use the term *parasite* for a person who is always taking without ever giving anything in return. Biological parasitism is similarly one-sided.

Natural selection has refined the relationships between parasites and their hosts. Many parasites, particularly microorganisms, have adapted to specific hosts, often a single species. In any parasite population, reproductive success is greatest for individuals that are best at locating and feeding on their hosts. For example, some aquatic leeches first locate a host by detecting movement in the water and then confirm its identity based on temperature and chemical cues on the host's skin. Natural selection has also favored the evolution of host defenses. In humans and other vertebrates, an elaborate immune system helps defend the body against specific internal parasites. With natural selection working on both host and parasite, the eventual outcome is usually a relatively stable relationship that does not kill

Figure 19.14 A case study in the evolution of host-parasite relations.
In the 1940s, Australia was overrun by hundreds of millions of rabbits, all descended from just 12 pairs imported a century earlier onto an estate for sport. The rabbits destroyed vegetation in huge expanses of Australia and threatened the sheep and cattle industries. In 1950, in an effort to control the exploding rabbit population, biologists deliberately introduced a myxoma virus that specifically parasitizes rabbits. The virus spread rapidly and killed almost all (99.8%) the infected rabbits. However, a second exposure to the virus killed only 90% of the rabbit population derived from survivors of the first application. And a third exposure eliminated only about 50% of the rabbit population. Apparently, viral infection selected for host genotypes that were better able to resist the parasite. Over the generations, the myxoma virus had less and less effect, and the rabbit population rebounded. The Australian government is now having more success with a different virus, which was introduced into the rabbit population in 1995.

the host quickly, an excess that would eliminate the parasite as well. An example of how rapidly natural selection can temper a host-parasite relationship is the evolution of resistance to a viral parasite in Australian rabbit populations (Figure 19.14).

Mutualism In contrast to the one-sidedness of parasitism, **mutualism** (from the Latin *mutualis,* reciprocal) is a symbiosis that benefits both partners. Just two of the examples we encountered in earlier chapters are the root-fungus associations called mycorrhizae and the specific interactions between certain pollinators and flowering plants (see Figures 15.3 and 15.21). Figure 19.15 illustrates another case of mutual symbiosis, the relationship between acacia trees and the ants that protect these trees from herbivorous insects and competing plants.

Many mutualistic relationships may have evolved from predator-prey or host-parasite interactions. Certain angiosperm plants, for example, have adaptations that attract animals that function in pollination or seed dispersal; these adaptations may represent an evolutionary response to a history of the herbivores' feeding on pollen and seeds. In many cases, pollen is spared when the pollinator is able to consume nectar instead. Indigestible seeds within fruits are dispersed by animals, which are nourished by the fruits. Any plants that could derive some benefit by sacrificing organic materials other than pollen or seeds would increase their reproductive success, and the adaptations for mutualistic interactions would spread through the plant population over the generations.

The Complexity of Community Networks

So far, we have reduced the networks known as biological communities to interactions between species: competition, predator-prey interactions, and symbiosis. It is the branching of these interactions that makes communities so complex. For example, some plants recruit parasitic wasps that lay their eggs in caterpillars that eat the plants (Figure 19.16). Both the wasps and caterpillars are also eaten by predators such as spiders and birds, and all of these organisms are hosts to specific parasites. Ecologists are only beginning to sort out the complex networks of just a few biological communities. Contributing to the challenges of community ecology is the fact that the structure of a community may change, sometimes over relatively short periods of time, due to a variety of disturbances. The next section examines some of these changes.

Figure 19.15 Mutualism between acacia trees and ants.
Certain species of Central and South American acacia trees have hollow thorns that house stinging ants. The ants feed on sugar and proteins from specialized glands on the trees (the orange swellings at the tips of the leaflets). The acacia also benefits from housing and feeding this population of ants, for the ants will attack anything that touches the tree. The pugnacious ants sting other insects, remove fungal spores, and clip surrounding vegetation that happens to grow close to the foliage of the acacia.

Categorize various interactions between species in Web/CD Activity 19B.

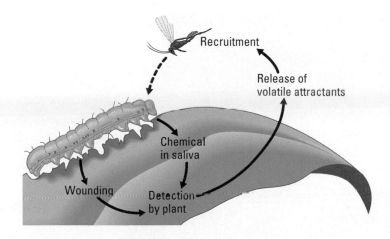

Recruitment

Release of
volatile attractants

Chemical
in saliva

Wounding

Detection
by plant

Figure 19.16 A three-way community interaction. Some plants recruit wasps that lay their eggs in caterpillars feeding on the plants. A leaf releases a wasp-attracting vapor in response to saliva from the caterpillar and injury of the plant's cells. The wasp stings the leaf-munching caterpillar and then lays its eggs within the caterpillar. The wasp larvae that hatch from these eggs will eat their way out of the caterpillar. (Adapted with permission from Edward Farmer, "Plant Biology: New Fatty Acid-Based Signals: A Lesson from the Plant World," *Science* vol. 276, page 912 (May 9, 1997). Copyright © 1997 American Association for the Advancement of Science.)

CheckPoint

1. Reexamine the distribution of the two barnacle species in Figure 19.5. What experiment could you perform to test the hypothesis that it is mainly susceptibility to drying, not competitive exclusion, that keeps *Balanus* from populating the upper parts of the rock?

2. What is the advantage to a keystone predator of being specialized to feed mainly on those prey species that are otherwise the most successful among potential prey species?

3. Which type of symbiotic relationship is represented by bacteria in the human colon (large intestine) that produce vitamin K, a nutrient that can be utilized by the human host?

Answers: 1. You could remove *Chthamalus* from the upper parts of the rocks to see whether *Balanus* still failed to spread upward, even in the absence of a competitor. **2.** The most competitive prey species probably represent the most abundant and dependable food source for the predator. **3.** Mutualism

Disturbance of Communities

Disturbances are episodes that damage biological communities, at least temporarily, by destroying organisms and altering the availability of resources such as mineral nutrients and water. Examples of disturbances are storms, fires, floods and the severe erosion they can cause, droughts, and human activities such as deforestation and the introduction of domesticated animals that graze.

Decades ago, most ecologists viewed communities as being relatively stable, with a more or less constant "balance of nature" due mainly to interactions among organisms. According to that traditional model, when a community is disturbed by such factors as storms, fires, floods, and droughts, it tends to return to an original, balanced condition. But further research challenged that view. Today, ecologists generally agree that disturbances affect all communities, which rarely have time to settle into a constant state before another disturbance occurs. Put another way, most communities spend much of their time in various stages of recovery from various disturbances. Change seems to characterize most communities more than constancy and balance.

❶ Retreating glacier

❷ Barren landscape after glacial retreat

Figure 19.17 Succession after retreat of glaciers. Ecologists sometimes deduce the process of succession by studying several locations that are at different successional stages. For example, these photographs, taken at different sites where glaciers are retreating (melting), represent about 200 years of succession.

❸ Moss and lichen stage

❹ Alders and cottonwoods covering the hillsides

❺ Spruce coming into the alder and cottonwood forest

❻ Spruce and hemlock forest

Ecological Succession

Communities may change drastically after a flood, fire, glacial advance and retreat, or volcanic eruption strips away their vegetation. A variety of species may colonize the disturbed area. Later these may be replaced as yet other species colonize the area. This process of community change is called **ecological succession.**

When a community arises in a virtually lifeless area with no soil, the change is called **primary succession.** Examples of such areas are new volcanic islands or the rubble left by a retreating glacier. Often the only life-forms initially present are autotrophic microorganisms. Lichens and mosses, which grow from windblown spores, are commonly the first large photosynthetic organisms to colonize the barren ground. Soil develops gradually as organic matter accumulates from the decomposed remains of the early colonizers. Once soil is present, the lichens and mosses are overgrown by grasses, shrubs, and trees that sprout from seeds blown in from nearby areas or carried in by animals. Eventually, the area may be colonized by plants that will become the community's prevalent form of vegetation. Primary succession from barren soil to a community such as a deciduous forest can take hundreds or thousands of years (Figure 19.17).

> Practice putting the steps of primary succession in order in Web/CD Activity 19C.

Secondary succession occurs where a disturbance has destroyed an existing community but left the soil intact. For instance, forested areas in the eastern United States that are cleared for farming will, if abandoned, undergo secondary succession and may eventually return to forest. The earliest plants to recolonize an area are often species with opportunistic life histories (see Chapter 18). Many are herbaceous (nonwoody) species that grow from windblown or animal-borne seeds. These plants thrive where there is little competition from other plants. Woody shrubs may eventually replace most of the herbaceous species. Later yet, trees may replace most of the shrubs.

A Dynamic View of Community Structure

As part of the "balance of nature" ideal that prevailed among biologists in the early 1900s, ecological succession was viewed as a linear sequence leading to some predictable, stable end point—a *climax community*. According to this hypothesis, the balanced state is reached when interactions between species are so intricate that the community is essentially saturated. No additional species can "fit" into the community because all ecological roles are filled. In this view, the particular type of climax community that develops —grassland or deciduous forest, for example—depends only on such abiotic factors as climate and soil type. Indeed, some communities, such as temperate forests dominated by oak and hickory trees, seem to be mature stages that will persist indefinitely. In many areas, however, what appear to be climax communities may not be stable over long periods, and many communities are routinely disturbed and never reach a climax stage. For instance, historically, prairie grasslands were often swept by fire. Without fire, some grassland communities would eventually become forests. In this case, we might say that forest is the climax community, but that makes little sense if the forest community never develops. In effect, periodic disturbances (fires) maintain the community at a stage that does not fit the traditional idea of a climax community. More generally, disturbances keep communities in a continual flux, making them mosaics of patches at various successional stages and preventing them from ever reaching a state of equilibrium or complete balance.

We tend to think of disturbances as having negative impacts, but this is only part of the story, at least in the case of natural disturbances. Small-scale disturbances often have positive effects, such as creating new opportunities for species. For example, when a tree falls, it disturbs the immediate surroundings; however, the fallen log fosters new habitats, and the depression left by its roots may fill with water and be used as egg-laying sites by frogs, salamanders, and numerous insects (Figure 19.18).

If most communities never really stabilize, what is the effect of this constant change on species diversity? When disturbance is severe and relatively frequent, the community may include only good colonizers typical of early stages of succession. Temperate grasslands maintained by periodic fires are an example. At the other end of the disturbance scale, when disruptions are mild and rare in a particular location, then the late-successional species that are most competitive will make up the community to the exclusion of other species. Between these two extremes, in an area where disturbance is moderate in both severity and frequency, species diversity may be greatest. Studies of species diversity in tropical rain forests support this **intermediate disturbance hypothesis.** Scattered throughout these forests are gaps where trees have fallen. In these disturbed areas, species of various successional stages coexist within a relatively small space.

Figure 19.18 A small-scale disturbance. When this tree fell in a windstorm, its root system and the surrounding soil uplifted, resulting in a depression that filled with water. The dead tree, the root mound, and the water-filled depression created new habitats in this forest in Michigan.

CheckPoint

1. What is the main abiotic factor that distinguishes primary from secondary succession?

2. Why are trees unlikely to be among the first colonizers of a landscape cleared by the advance and retreat of a glacier?

Answers: 1. Absence of soil (primary succession) versus presence of soil (secondary succession) at the onset of succession **2.** Trees can't grow until earlier colonizers form soil.

Figure 19.19 A large-scale human disturbance. This half-mile-deep open-pit mine in Butte, Montana is known as the Berkeley Pit. The mine has not operated since 1983, when its high-quality copper ore was used up. Water in the pit is highly acidic and loaded with heavy metal ions. It threatens groundwater, soils, adjacent rivers, and communities for hundreds of miles downstream. The Berkeley Pit is one of the most troublesome cleanup sites in the United States, and the federal government has spent millions to initiate waste removal at this site.

Human Impact on Biological Communities

We have seen that natural disturbances that are intermediate in scale can be constructive in enhancing species diversity in a community. Unfortunately, human disturbance of biological communities is almost always destructive, reducing species diversity.

Human Disturbance of Communities

Of all animals, humans have the greatest impact on communities worldwide (Figure 19.19). Logging and clearing for urban development, mining, and farming have reduced large tracts of forests to small patches of disconnected woodlots in many parts of the United States and throughout Europe. Similarly, agricultural development has disrupted what were once the vast grasslands of the North American prairie.

Much of the United States is now a hodgepodge of early successional growth where more mature communities once prevailed. After forests are clear-cut and abandoned, weedy and shrubby vegetation often colonizes the area and dominates it for many years. This type of vegetation is also found extensively in agricultural fields that are no longer under cultivation and in vacant lots and construction sites that are periodically cleared.

Human disturbance of communities is by no means limited to the United States and Europe; nor is it a recent problem. Tropical rain forests are quickly disappearing as a result of clear-cutting for lumber and pastureland. Centuries of overgrazing and agricultural disturbance have contributed to the current famine in parts of Africa by turning seasonal grasslands into great barren areas.

Human disturbance usually reduces species diversity in communities. We currently use about 60% of Earth's land in one way or another, mostly as cropland and rangeland. Most crops are grown in monocultures, intensive cultivations of a single plant variety over large areas. Even forests that are used to produce pulpwood and lumber are often replanted in single-species stands. And the effects of intensive grazing on rangelands often include the removal of several native plant species and replacement with only a few introduced species. Let's take a closer look at the problem of introduced species.

Introduced Species

Sometimes called exotic species, **introduced species** are those that humans move from the species' native locations to new geographic regions. In some cases, the introductions are intentional. Australia's rabbit fiasco is an example (see Figure 19.14). In other cases, humans transplant species accidentally. For instance, the fungus that causes Dutch elm disease stowed away with logs shipped from Europe to the United States after World War I (see Figure 15.24a). Whether purposeful or unintentional, introduced species that gain a foothold usually disrupt their adopted community, often by preying on native organisms or outcompeting native species that use some of the same resources.

Throughout human history, moving plants and animals from place to place was considered "improving on nature." This was especially true of domesticated species. As Europeans began colonizing the Americas in the fif-

(b) European starlings

(a) Kudzu

Figure 19.20 A few of America's 50,000 introduced species. (a) Kudzu.
(b) European starlings. (Figure 19.20 is continued on page 454.)

teenth and sixteenth centuries, plants and animals that provide food were
transported back and forth across the Atlantic in what we now call the
"Columbian Exchange." Corn, potatoes, tomatoes, beans, and certain kinds
of squash are just a few examples of New World species that became European staples.

Humans have also introduced many nonfood species with the best of
intentions. For example, the U.S. Department of Agriculture encouraged
the import of a Japanese plant called kudzu to the American South in the
1930s to help control erosion, especially along irrigation canals. At first, the
government paid farmers to plant kudzu vines. The enthusiasm for the new
vines led to kudzu festivals in southern towns, complete with the crowning
of kudzu queens. But kudzu celebrations ended decades ago as the invasive
plant took over vast expanses of the southern landscape (Figure 19.20a).
Another introduced plant called purple loosestrife is claiming over 200,000
acres of wetlands per year, crowding out native plants and the animals that
feed on the native flora. The story is similar for the introduction to the
United States of a bird called the European starling. A citizens group intent
on introducing all plants and animals mentioned in Shakespeare's plays imported 120 starlings to New York's Central Park in 1890 (the starling is
mentioned in just one line of Shakespeare's *Henry IV*). From that foothold,
starlings spread rapidly across North America. In less than a century, the
population increased to about 100 million, displacing many of the native
songbird species in the United States and Canada (Figure 19.20b).

(c) Argentine ants

(d) Zebra mussels

(e) The seaweed *Caulerpa*

Figure 19.20 Introduced species, continued. (c) Argentine ants ganging up on a red ant native to California. **(d)** Zebra mussels. (The shopping cart, covered with the mussels, was retrieved from a lake.) **(e)** An aquarium-bred, hypervigorous variety of the seaweed *Caulerpa*. In this underwater photo, you can see the algae (foreground) crowding out the native eelgrass.

The ease of travel by ships and airplanes has accelerated the transplant of species, especially unintentional introductions. For example, fire ants, which can inflict very painful beelike stings, reached the southeastern United States in the early 1900s from South America, probably in the hold of a produce ship. Fire ants have been extending their range northward and westward ever since. In Texas, for example, fire ants have apparently managed to eliminate about two-thirds of the native ant species. Another accidentally introduced ant species, the Argentine ant, is decimating populations of native ants in California (Figure 19.20c). Still another recent troublesome invader in the United States is the zebra mussel, a mollusk that entered the St. Lawrence Seaway in the mid-1980s in ballast water released by a cargo ship that had traveled from the mussels' native Caspian Sea (Figure 19.20d). By the summer of 1993, zebra mussels were found in the Mississippi River as far south as New Orleans. Of the many problems caused by the mussels, those that have provoked the most uproar have resulted from their competition with humans—by clogging reservoir intake pipes, for example. However, zebra mussels also compete with native shellfish for space and with fish for the plankton used for nourishment. The full extent to which this new competitor is altering community structure has yet to be determined.

Learn more about fire ants in Web/CD Activity 19D.

An even more recent example of introduced species is the appearance in 2000 of an alga called *Caulerpa* in a California lagoon (Figure 19.20e). The small seaweed was probably introduced by someone dumping a home saltwater aquarium. Native to Caribbean waters, the California invader is a variety of the alga that has been domesticated and selectively bred as an aquarium "plant" for its vigor and resistance to disease and herbivores. An earlier invasion of the Mediterranean Sea by this super seaweed is displacing many of the native algae there, and the same thing could happen now all along the Pacific coast of North America.

All told, the United States has at least 50,000 introduced species, with a cost to the economy of over $130 billion in damage and control efforts. And that does not include the priceless loss of native species. Fortunately, most introduced species fail to thrive in their new homes. But for those exotic species that succeed, elimination of native species is a common consequence. In fact, introduced species rank second only to habitat destruction as a cause of extinctions and loss of Earth's biodiversity. Why should some introduced species be able to outcompete native organisms? In many cases, the answer is that there are relatively few pathogens, parasites, and predators to hold population explosions of the introduced species in check once the globetrotters arrive. In contrast, the native species of a biological community fit into a network of interactions shaped by the adaptive evolution of multiple species. We close the chapter by connecting to this evolutionary context in community ecology.

CheckPoint

1. What is an "exotic" or "introduced" species?

2. If an introduced species gains a foothold in its new home, why does its success often contribute to loss of native species?

Answers: 1. A species that humans have transferred from the species' native location to another location **2.** The introduced species may prey on some native species and outcompete others due to the absence of the introduced species' natural predators and parasites.

Evolution Link: Coevolution in Biological Communities

We have seen many examples in this chapter of adaptations that evolve in populations as a result of interactions with other species in a community. Defenses of plants against herbivores, animal defenses against predators, mutual symbiosis, and parasite-host relationships are all examples. Ecologists use the term **coevolution** when the adaptations of two species are closely connected—that is, when an adaptation in one species leads to a counteradaptation in a second species. Let's examine one such case of coevolution, the reciprocal adaptations between certain butterfly species and passionflowers, plant species on which the caterpillars of the butterflies feed.

Passionflower vines of the genus *Passiflora* are protected against most herbivorous insects by their production of toxic compounds in young leaves and shoots. However, the larvae (caterpillars) of butterflies of the genus *Heliconius* can tolerate these defensive chemicals. This counteradaptation has enabled *Heliconius* larvae to become specialized feeders on plants that few other insects can eat (Figure 19.21). Survival of the larvae is further enhanced by a behavioral adaptation of the butterflies. The eggs that female *Heliconius* butterflies lay on the leaves of passionflower vines are bright yellow, and other females generally avoid laying eggs on leaves marked by these yellow dots. This behavior presumably reduces competition among the larvae for food.

An infestation of *Heliconius* larvae can devastate a passionflower vine, and these poison-resistant insects are likely to be a strong selection force favoring the evolution of more defenses in the plants. In some species of *Passiflora*, the leaves have conspicuous yellow spots that mimic *Heliconius eggs*, an adaptation that may divert the butterflies to other plants in their search for egg-laying sites.

On closer inspection, such seemingly clear-cut examples of coevolution between two species usually turn out to be more complicated. The yellow "spots" on the passionflower vine are actually nectar-secreting glands, which attract ants and wasps that prey on *Heliconius* eggs or larvae. And there is evidence that the mere presence of ants on a leaf will discourage a *Heliconius* butterfly from laying eggs there. (We saw a similar example in the mutualism of ants and acacia trees.)

Explore coevolution further in the Web Evolution Link.

We see once again the complexity of communities, in which the evolution of each species reflects its interactions with many other species. In the next chapter, we explore an even higher level of complexity: ecosystems, which are defined by their community interactions plus the interface of the community with soil, water, and other abiotic factors in the environment.

(a) *Heliconius* caterpillar

(b) *Heliconius* eggs

Sugar deposits

(c) Passionflower leaves with sugar-secreting nectar glands

Figure 19.21 Coevolution of a plant and insect. **(a)** Passionflower vines produce toxic chemicals that help protect leaves from most herbivorous insects. A counteradaptation has evolved in *Heliconius* butterflies; their larva can feed on passionflower leaves because they have digestive enzymes that break down the plant's toxins. **(b)** The female butterflies avoid laying eggs on passionflower leaves with egg clusters deposited by other females. This behavior ensures an adequate food supply when eggs hatch and the larvae begin feeding on the leaves. **(c)** Some passionflower species have yellow nectaries (sugar-secreting glands) that mimic *Heliconius* eggs. The butterflies avoid laying eggs on leaves with these yellow spots. The nectaries also attract ants that prey on the butterfly eggs.

Chapter Review

Summary of Key Concepts

Overview: Key Properties of Communities

- Four key properties of a community are its diversity, its prevalent form of vegetation, its stability, and its trophic structure.

- **Diversity** Community diversity includes the species richness and relative abundance of different species.
- Web/CD Activity 19A/The Process of Science *Community Diversity*

- **Prevalent Form of Vegetation** The types and structural features of plants largely determine the kinds of animals that live in a community.

- **Stability** A community's stability is its ability to resist change and return to its original species composition after being disturbed. Stability depends on the type of community and the nature of the disturbances.

- **Trophic Structure** The trophic structure consists of the feeding relationships among the various species making up the community.

Interspecific Interactions in Communities

- **Competition Between Species** When populations of two or more species in a community rely on similar limiting resources, they may be subject to interspecific competition. Two species cannot coexist in a community if their niches are identical. Resource partitioning is the differentiation of niches that enables similar species to coexist in a local community.

- **Predation** Natural selection refines the adaptations of both predators and prey. Most predators have acute senses that enable them to locate and identify potential prey. Plants mainly use chemical toxins, spines, and thorns to defend against predators. Animals may defend themselves by using passive defenses, such as cryptic coloration (camouflage), or active defenses, such as escaping, alarm calls, mobbing, or distraction displays. Animals with chemical defenses are often brightly colored, a warning to potential predators. A species of prey may also gain protection through mimicry. In Batesian mimicry, a palatable species mimics an unpalatable model. In Müllerian mimicry, two or more unpalatable species resemble each other. Predator-prey relationships can preserve diversity. A keystone predator is a species that reduces the density of the strongest competitors in a community. This predator helps maintain species diversity by preventing competitive exclusion of weaker competitors.

- **Symbiotic Relationships** A symbiotic relationship is an interspecific interaction in which a symbiont species lives in or on a host species. Parasitism is a one-sided relationship in which one organism benefits at the expense of the other. Natural selection has refined the relationships between parasites and their hosts. Mutualism is a symbiosis that benefits both partners. Many mutualistic relationships may have evolved from predator-prey or host-parasite interactions.

- **The Complexity of Community Networks** The branching of interactions between species makes communities complex.
- Web/CD Activity 19B *Interactions Between Species*

Disturbance of Communities

- Disturbances are episodes that damage biological communities, at least temporarily, by destroying organisms and altering the availability of resources such as mineral nutrients and water. Most communities spend much of their time in various stages of recovery from disturbances.

- **Ecological Succession** The sequence of changes in a community after a disturbance is called ecological succession. Primary succession occurs where a community arises in a virtually lifeless area with no soil. Secondary succession occurs where a disturbance has destroyed an existing community but left the soil intact.
- Web/CD Activity 19C *Primary Succession*

- **A Dynamic View of Community Structure** In general, disturbances keep communities in a continual flux, making them mosaics of patches at various successional stages and preventing them from reaching a completely stable state. Small-scale disturbances often have positive effects, such as creating new opportunities for species. Species diversity may be greatest in places where disturbances are moderate in severity and frequency.

Human Impact on Biological Communities

- **Human Disturbance of Communities** Human disturbance usually reduces species diversity in communities. Much of the United States is now a hodgepodge of early successional growth where more mature communities once prevailed.

- **Introduced Species** Introduced species are those that humans intentionally or accidentally move from the species' native locations to new geographic regions. Introduced species rank second only to habitat destruction as a cause of extinctions and loss of Earth's biodiversity.
- Web/CD Activity 19D *Fire Ants*

Evolution Link: Coevolution in Biological Communities

- Coevolution occurs when an adaptation in one species leads to a counter-adaptation in a second species.
- Web Evolution Link *Coevolution in Biological Communities*

Self-Quiz

1. Which of the following best illustrates ecological succession?
   a. A mouse eats seeds, and an owl eats the mouse.
   b. Decomposition in soil releases nitrogen that plants can use.
   c. Grass grows on a sand dune, then shrubs, and then trees.
   d. Imported pheasants increase, while local quail disappear.
   e. Overgrazing causes a loss of nutrients from soil.

2. A bat locates insect prey in the dark by bouncing high-pitched sounds off them. One species of moth escapes predation by diving to the ground when it hears "sonar" of a particular bat species. This illustrates _____ between the bat and moth.
   a. mutualism   d. interspecific competition
   b. competitive exclusion   e. coevolution
   c. ecological succession

3. The concept of trophic structure of a community emphasizes the
   a. prevalent form of vegetation.
   b. keystone predator.
   c. feeding relationships within a community.
   d. effects of coevolution.
   e. species richness of the community.

4. According to the concept of competitive exclusion,
   a. two species cannot coexist in the same habitat.
   b. extinction or emigration is the only possible result of competitive interactions.
   c. intraspecific competition results in the success of the best-adapted individuals.
   d. two species cannot share the same niche in a community.
   e. resource partitioning will allow a species to utilize all the resources of its niche.

5. The effect of a keystone predator within a community may be to
   a. competitively exclude other predators from the community.
   b. maintain species diversity by preying on the prey species that is the dominant competitor.
   c. increase the relative abundance of the most competitive prey species.
   d. encourage the coevolution of predator and prey adaptations.
   e. create a climax community.

6. An example of camouflage coloration is
   a. the green color of a plant.
   b. the bright markings of a poisonous tropical frog.
   c. the stripes of a skunk.
   d. the mottled coloring of moths that rest on lichens.
   e. the bright colors of an insect-pollinated flower.

7. An example of Müllerian mimicry is
   a. a butterfly that resembles a leaf.
   b. two poisonous frogs that resemble each other in coloration.
   c. a minnow with spots that look like large eyes.
   d. a beetle that resembles a scorpion.
   e. a carnivorous fish with a wormlike tongue that lures prey.

8. To be certain that two species had coevolved, one would ideally need to establish that
   a. the two species originated about the same time.
   b. local extinction of one species dooms the other species.
   c. each species affects the population density of the other species.
   d. one species has adaptations that specifically tracked evolutionary change in the other species, and vice versa.
   e. the two species are adapted to a common set of environmental conditions.

9. In terms of how it affects the populations of two interacting species, parasitism is most like
   a. mutualism.
   b. competition.
   c. predation.
   d. ecological succession.
   e. coevolution.

10. The disturbances that are most likely to enhance species diversity in a community are those that are
   a. caused by humans.
   b. moderate in severity and frequency.
   c. severe and rare.
   d. very frequent and moderate in severity.
   e. caused by introduced species.

• Go to the website or CD-ROM for more self-quiz questions.

The Process of Science

1. An ecologist studying desert plants performed the following experiment. She staked out two identical plots that included a few sagebrush plants and numerous small annual wildflowers. She found the same five wildflower species in similar numbers in both plots. Then she enclosed one of the plots with a fence to keep out kangaroo rats, the most common herbivores in the area. After two years, four species of wildflowers were no longer present in the fenced plot, but one wildflower species had increased dramatically. The control plot had not changed significantly in species composition. Using the concepts discussed in the chapter, what do you think happened?

2. Analyze species richness and relative abundance in The Process of Science activity on the website and CD-ROM.

Biology and Society

By 1935, hunting and trapping had eliminated wolves from the United States outside Alaska. Since wolves have been protected as an endangered species, they have moved south from Canada and have become reestablished in the Rocky Mountains and northern Great Lakes. Conservationists who would like to speed up this process have reintroduced wolves into Yellowstone National Park. Local ranchers are opposed to bringing back the wolves because they fear predation on their cattle and sheep. What are some reasons for reestablishing wolves in Yellowstone Park? What effects might the reintroduction of wolves have on the ecological communities in the region? What might be done to mitigate the conflicts between ranchers and wolves?

Ecosystems and Conservation Biology

Every year, humans destroy an area of tropical

rain forest equal to the size of West Virginia.

Concentrations of carbon dioxide in the atmosphere

have increased about 14% over the last 40 years.

In the United States, we consume more energy than

do the total populations of Central and South America, Africa,

India, and China, combined.   About 11% of the known

bird species are endangered.

A n ecological system, or **ecosystem,** is the highest level of biological organization. It consists of all the organisms in a given area plus the physical environment, including soil, water, and air. In other words, an ecosystem is a biological community and the abiotic factors with which the community interacts. In this chapter, we'll explore the dynamics of individual ecosystems, such as forests, ponds, and oceanic zones. We'll also widen our scope to view the biosphere as the global ecosystem, the sum of all ecosystems. And we'll see how human activities such as deforestation and pollution are putting ecosystems and their biological diversity at risk, both on the local level and throughout the whole biosphere. This last chapter of *Essential Biology* closes with the science of conservation biology and the challenge to do what we can as individuals and a society to slow the loss of biodiversity.

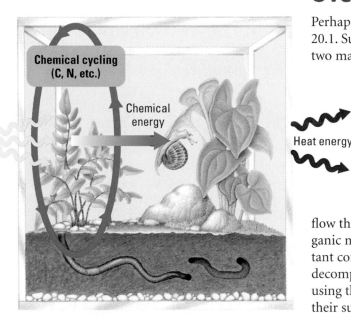

Figure 20.1 A terrarium ecosystem. Small and artificial as it is, this microcosm illustrates the two major ecosystem processes: energy flow and chemical cycling.

Overview: Ecosystem Dynamics

Perhaps you have seen or even made a terrarium such as the one in Figure 20.1. Such a microcosm qualifies as an ecosystem, because it exhibits the two major processes that sustain all ecosystems, from terrariums to forests and oceans all the way to the entire biosphere. Those two key processes of ecosystem dynamics are energy flow and chemical cycling. **Energy flow** is the passage of energy through the components of the ecosystem. **Chemical cycling** is the use and reuse of chemical elements such as carbon and nitrogen within the ecosystem.

Energy reaches most ecosystems in the form of sunlight. Plants and other photosynthetic producers convert the light energy to the chemical energy of food. Energy continues its flow through an ecosystem when animals and other consumers acquire organic matter by feeding on plants and other producers. Especially important consumers are the soil bacteria and fungi that obtain their energy by decomposing the dead remains of plants, animals, and other organisms. In using their chemical energy for work, all organisms dissipate heat energy to their surroundings. Thus, energy cannot be recycled within an ecosystem, but must flow continuously through the ecosystem, entering as light and exiting as heat (see Figure 20.1).

In contrast to the flow of energy through an ecosystem, chemical elements can be recycled between an ecosystem's living community and the abiotic environment. The plants and other producers acquire their carbon, nitrogen, and other chemical elements in inorganic form from the air and soil. Photosynthesis then enables plants to incorporate these elements into organic compounds such as carbohydrates and proteins. Animals acquire these elements in organic form by eating. The metabolism of all organisms returns some of the chemical elements to the abiotic environment in inorganic form. Cellular respiration, for example, breaks organic molecules down to carbon dioxide and water. This job of recycling is finished by microorganisms that decompose dead organisms and their organic wastes, such as feces and leaf litter. These decomposers restock the soil, water, and air with chemical elements in the inorganic form that plants and other producers can again build into organic matter. And the chemical cycles continue.

Note again the key distinction between energy *flow* and chemical *cycling*. Because energy, unlike matter, cannot be recycled, an ecosystem must be powered by a continuous influx of energy from an external source, usually the sun. Notice also that an ecosystem's energy flow and chemical cycling are closely related. Both depend on transfer of substances in the feeding relationships, or **trophic structure,** of the ecosystem.

Increase your understanding of energy flow and chemical cycling in Web/CD Activity 20A.

The Trophic Structure of Ecosystems

Trophic relationships determine an ecosystem's routes of energy flow and chemical cycling. In analyzing these feeding relationships, ecologists divide the species of an ecosystem into different **trophic levels** based on their main sources of nutrition. Let's see how these trophic levels connect as the living components of an ecosystem's energy flow and chemical cycling.

Trophic Levels and Food Chains

The sequence of food transfer from trophic level to trophic level is called a **food chain.** Figure 20.2 compares a terrestrial food chain and a marine food chain. Starting at the bottom of such diagrams, the trophic level that supports all others consists of the **producers,** the autotrophic organisms (see Chapter 6). Photosynthetic producers are organisms that use light energy to power the synthesis of organic compounds. Plants are the main producers on land. In aquatic ecosystems, the producers are mainly photosynthetic protists and bacteria, collectively known as phytoplankton. Multicellular algae and aquatic plants are also important producers in shallow waters.

All organisms in trophic levels above the producers are consumers, the heterotrophs directly or indirectly dependent on the output of producers (see Chapters 5 and 6). **Herbivores,** which eat plants, algae, or autotrophic bacteria, are the **primary consumers** of an ecosystem. Primary consumers on land include grasshoppers and many other insects, snails, and certain vertebrates, such as grazing mammals and birds that eat seeds and fruit. In aquatic ecosystems, primary consumers include a variety of zooplankton (mainly protists and microscopic animals such as small shrimps) that prey on the phytoplankton.

Above the primary consumers, the trophic levels are made up of **carnivores,** which eat the consumers from the levels below. On land, **secondary consumers** include many small mammals, such as rodents that eat herbivorous insects, and a great variety of small birds, frogs, and spiders, as well as lions and other large carnivores that eat grazers. In aquatic ecosystems,

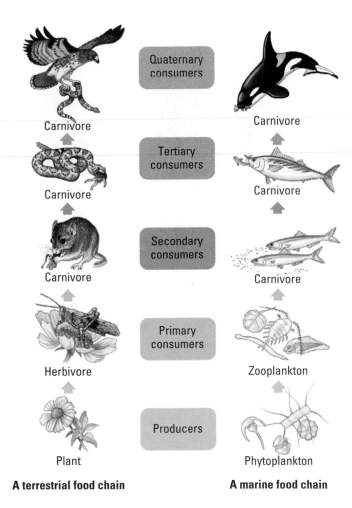

A terrestrial food chain **A marine food chain**

Figure 20.2 Examples of food chains. The arrows trace the transfer of food from producers through various levels of consumers. Detritivores (decomposers), important consumers in all ecosystems, are not included in these simplified diagrams of terrestrial and marine ecosystems.

Figure 20.3 Fungi decomposing a dead log.

secondary consumers are mainly small fishes that eat zooplankton and bottom-dwelling invertebrates. Higher trophic levels include **tertiary consumers,** such as snakes that eat mice and other secondary consumers. Some ecosystems even have **quaternary consumers,** such as hawks in terrestrial ecosystems and killer whales in marine environments (see Figure 20.2).

Most diagrams of food chains are incomplete by not showing a critical trophic level of consumers called **detritivores,** another name for **decomposers.** Detritivores derive their energy from detritus, the dead material left by all trophic levels. Detritus includes animal wastes, plant litter, and all sorts of dead organisms. Most organic matter eventually becomes detritus and is consumed by detritivores. A great variety of animals, often called scavengers, eat detritus. For instance, earthworms, many rodents, and insects eat fallen leaves and other detritus. Other scavengers include crayfish, catfish, and vultures. But an ecosystem's main detritivores are the prokaryotes and fungi (Figure 20.3). Enormous numbers of microorganisms in the soil and in mud at the bottoms of lakes and oceans recycle most of the ecosystem's organic materials to inorganic compounds that plants and other producers can reuse.

Food Webs

Actually, few ecosystems are so simple that they are characterized by a single unbranched food chain. Several types of primary consumers usually feed on the same plant species, and one species of primary consumer may eat several different plants. Such branching of food chains occurs at the other trophic levels as well. For example, adult frogs, which are secondary consumers, eat several insect species that may also be eaten by various birds. In addition, some consumers feed at several different trophic levels. An owl, for instance, may eat mice, which are mainly primary consumers that may also eat some invertebrates; but an owl may also feed on snakes, which are strictly carnivorous. **Omnivores,** including humans, eat producers as well as consumers of different levels. Thus, the feeding relationships in an ecosystem are usually woven into elaborate **food webs**.

Though more realistic than a food chain, a food web diagram, such as the one in Figure 20.4, is still a highly simplified model of the feeding relationships in an ecosystem. An actual food web would involve many more organisms at each trophic level, and most of the animals would have a more diverse diet than shown in Figure 20.4. Indicating "who eats whom," the arrows in a food web diagram outline an ecosystem's overall dynamics, including the flow of energy.

Practice interpreting food webs in Web/CD Activity 20B.

CheckPoint

1. I'm eating a cheese pizza. At which trophic level(s) am I feeding?
2. In the "who eats whom" dynamics of a food web, even consumers of the highest level in the ecosystem eventually become food for _____.

Answers: 1. Primary consumer (flour and tomato sauce) and secondary consumer (cheese, a product from cows, which are primary consumers) **2.** detritivores, or decomposers

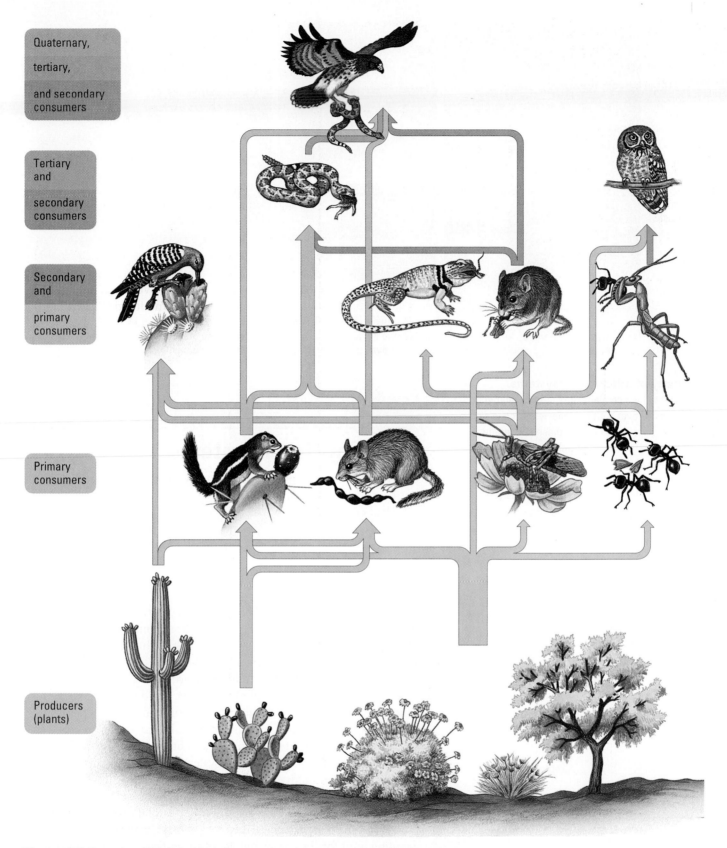

Figure 20.4 A simplified food web. As in the food chain of Figure 20.2, the arrows in this web indicate the direction of nutrient transfers. We also continue the color coding for the trophic levels and food transfers introduced in Figure 20.2. Note that in a food web, such as this one for a Sonoran desert ecosystem, some species feed at more than one trophic level. The other branching factor that turns food chains into webs is the consumption of the same species by more than one consumer.

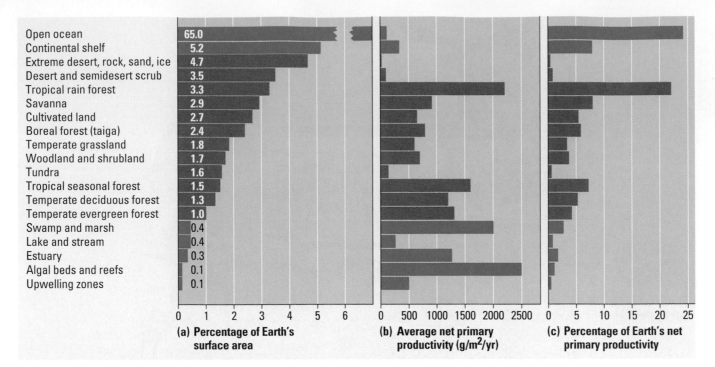

Open ocean	65.0	
Continental shelf	5.2	
Extreme desert, rock, sand, ice	4.7	
Desert and semidesert scrub	3.5	
Tropical rain forest	3.3	
Savanna	2.9	
Cultivated land	2.7	
Boreal forest (taiga)	2.4	
Temperate grassland	1.8	
Woodland and shrubland	1.7	
Tundra	1.6	
Tropical seasonal forest	1.5	
Temperate deciduous forest	1.3	
Temperate evergreen forest	1.0	
Swamp and marsh	0.4	
Lake and stream	0.4	
Estuary	0.3	
Algal beds and reefs	0.1	
Upwelling zones	0.1	

(a) Percentage of Earth's surface area

(b) Average net primary productivity (g/m²/yr)

(c) Percentage of Earth's net primary productivity

Figure 20.5 Productivity of different ecosystems.
Primary productivity is the rate at which plants and other producers store chemical energy in biomass over a unit of time, in this case a year. Aquatic ecosystems are color coded blue in these histograms; terrestrial ecosystems are brown. **(a)** The geographic extent and **(b)** the productivity per unit area of different ecosystems determine **(c)** their total contribution to worldwide primary productivity. Open ocean, for example, contributes a lot to the planet's productivity despite its low productivity per unit area because of its large size. In contrast, tropical rain forests contribute a lot to the biosphere's energy budget mainly because of their high productivity per unit area.

Energy Flow in Ecosystems

All organisms require energy for growth, maintenance, reproduction, and, in some species, locomotion (see Chapter 5). In this section, we take a closer look at energy flow through ecosystems. Along the way, we will answer two key questions: What limits the length of food chains? and How do lessons about energy flow apply to human nutrition?

Productivity and the Energy Budgets of Ecosystems

Each day, Earth is bombarded by about 10^{19} kilocalories of solar radiation. This is the energy equivalent of about 100 million atomic bombs the size of the one that devastated Hiroshima in 1945. Most of this solar energy is absorbed, scattered, or reflected by the atmosphere or by Earth's surface. Of the visible light that reaches plants and other producers, only about 1% is converted to chemical energy by photosynthesis. But on a global scale, this is enough to produce about 170 billion tons of organic material per year.

Ecologists call the amount, or mass, of organic material in an ecosystem the **biomass.** And the rate at which plants and other producers build biomass, or organic matter, is called the ecosystem's **primary productivity.** Thus, the primary productivity of the entire biosphere is 170 billion tons of organic material per year. Different ecosystems vary considerably in their productivity as well as in their contribution to the total productivity of the biosphere (Figure 20.5). Whatever the ecosystem, primary productivity sets the spending limit for the energy budget of the entire ecosystem because consumers must acquire their organic fuels from producers. Now let's see how this energy budget is divided among the different trophic levels in an ecosystem's food web.

Energy Pyramids

When energy flows as organic matter through the trophic levels of an ecosystem, much of it is lost at each link in a food chain. Consider the transfer of organic matter from plants (producers) to herbivores (primary consumers). In most ecosystems, herbivores manage to eat only a fraction of the plant material produced, and they cannot digest all the organic compounds they do ingest. Of the organic compounds an herbivore *can* use, much of it is consumed as fuel for cellular respiration. Only the remainder is available as raw material for growth of the herbivore. For example, in the case of a caterpillar consuming leaves, only about 15% of the biomass of the food is transformed into caterpillar biomass (Figure 20.6). In general, an average of only about 10% of energy in the form of organic matter at each trophic level is stored as biomass in the next level of the food chain. The rest of the energy is accounted for by the heat released by all the working organisms in the ecosystem, including the detritivores decomposing feces and other organic wastes (see Figure 20.1).

The cumulative loss of energy from a food chain can be represented in a diagram called an **energy pyramid.** The trophic levels are stacked in blocks, with the block representing producers forming the foundation of the pyramid (Figure 20.7). The size of each block is proportional to the productivity of each trophic level, the rate of energy storage as biomass at that level. The pyramid owes its steep slopes to the loss of 90% or so of the energy with each food transfer in the chain.

An important implication of this stepwise decline of energy along a food chain is that the amount of energy available to top-level consumers is small compared with that available to lower-level consumers. Only a tiny fraction of the energy stored by photosynthesis flows through a food chain to a tertiary consumer, such as a snake feeding on a mouse (see Figure 20.7). This explains why top-level consumers such as lions and hawks require so much geographic territory; it takes a lot of vegetation to support trophic levels so many steps removed from photosynthetic production. You can also understand now why most food chains are limited to three to five levels; there is

Build an energy pyramid for an East African savanna ecosystem in Web/CD Activity 20C.

simply not enough energy at the very top of an energy pyramid to support another trophic level. There are, for example, no nonhuman predators of lions, eagles, and killer whales; the biomass in populations of these top-level consumers is insufficient to support yet another trophic level with a reliable source of nutrition.

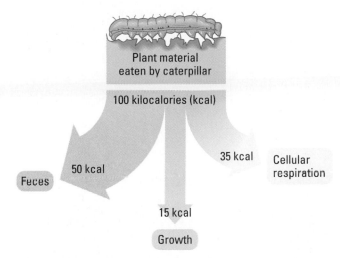

Figure 20.6 What becomes of a caterpillar's food? Only about 15% of the calories of plant material this herbivore consumes becomes stored as biomass available to the next link in the food chain.

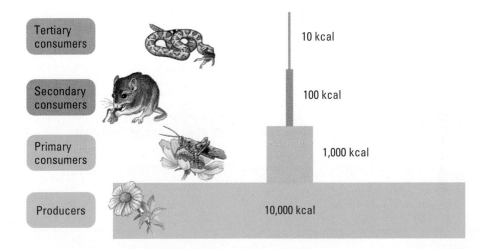

Figure 20.7 An idealized energy pyramid. In this case, 10% of the chemical energy available at each trophic level is converted to new biomass in the trophic level above it.

Ecosystem Energetics and Human Nutrition

The dynamics of energy flow apply to the human population as much as to other organisms. Like other consumers, we depend entirely on productivity by plants for our food. As omnivores, most of us eat both plant material and meat. When we eat grain or fruit, we are primary consumers; when we eat beef or other meat cut from herbivores, we are secondary consumers. When we eat fish like salmon (which eat insects and other small animals) and tuna (which eat smaller fish), we are tertiary or quaternary consumers.

Figure 20.8 applies the concept of energy pyramids to show why consuming a high proportion of our calories in the form of meat is a relatively inefficient way of tapping photosynthetic productivity. The pyramids are based on the 10% assumption for energy transfer between trophic levels. About 90% of the food energy in a crop such as corn is lost to us if that corn is first "processed" by primary consumers such as hogs or cattle. Put still another way, it takes at least 10 pounds of feed corn to produce 1 pound of bacon or beef steak. And the 10% estimate is actually high for a flow of photosynthetic energy to humans via primary consumers that are mammals or birds. As endotherms, mammals, such as cattle and hogs, and birds, such as chickens and turkeys, expend a great deal of energy to control body temperature—much more than do ectotherms, such as grasshoppers (see Chapter 16). Accounting for this consumption of food energy for thermoregulation, it may actually take closer to 100 times more photosynthetic product to feed us on cattle (and other mammals and birds) than directly on plants.

This analysis of energy flow to human consumers is not meant as a plea for vegetarianism or a switch from hamburgers to grasshopper sandwiches. The main point is that eating meat of any kind is an expensive luxury, both economically and environmentally. In many developing countries, people are mainly vegetarian by necessity because they cannot afford to buy much meat or their country cannot afford to produce it. Wherever meat is on the menu, producing it requires that more land be cultivated, more water be used for irrigation, and more chemical fertilizers and pesticides be applied to croplands used for growing grain. It is likely that as the human population continues to expand, meat consumption will become even more of a luxury than it is today.

Trophic level

Secondary consumers

Primary consumers

Producers

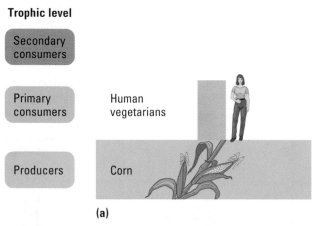

Human vegetarians

Corn

(a)

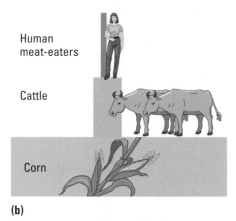

Human meat-eaters

Cattle

Corn

(b)

Figure 20.8 Food energy available to the human population at different trophic levels. Most humans have a diet between these two extremes. The point here is the greater proportion of photosynthetic energy that reaches us when we feed as primary consumers directly on plants than when we are nourished by photosynthesis indirectly by feeding as secondary consumers on animals.

CheckPoint

1. In a shrubland ecosystem, which is likely to have the greatest total biomass: the sum of all insects or the sum of all birds that feed on the insects?

2. Why is a pound of bacon so much more expensive than a pound of corn?

Answers: 1. The insects **2.** Because it took at least 10 pounds of feed corn to produce that pound of bacon

Chemical Cycles in Ecosystems

The sun supplies ecosystems with energy, but there are no extraterrestrial sources of the chemical elements life requires. Ecosystems depend on a recycling of these chemical elements. Even while an individual organism is alive, much of its chemical stock is rotated continuously as nutrients are acquired and waste products are released. Atoms present in the complex molecules of an organism at the time of its death are returned in simpler compounds to the atmosphere, water, or soil by the action of bacterial and fungal detritivores. This decomposition replenishes the pools of inorganic nutrients that plants and other producers use to build new organic matter. In a sense, we only borrow an ecosystem's chemical elements, returning what is left in our bodies after we die. "Ashes to ashes, dust to dust" is one metaphor for this fact of life. Let's take a closer look at how chemicals cycle between the organisms and the abiotic (nonliving) components of ecosystems.

The General Scheme of Chemical Cycling

Because chemical cycles in an ecosystem involve both biotic and abiotic components, they are also called **biogeochemical cycles.** Figure 20.9 is a general scheme for these cycles. Here are three key points to note:

- Each circuit, whether for carbon, nitrogen, or some other chemical material required for life, has an abiotic reservoir through which the chemical cycles. Carbon's main **abiotic reservoir,** for example, is the atmosphere, where carbon atoms are stocked mainly in the form of CO_2 gas (carbon dioxide). Carbon makes the abiotic→biotic transition when it is incorporated from CO_2 into food by plants and other producers. And the carbon is returned to the abiotic reservoir by the cellular respiration of the ecosystem's organisms, including detritivores.

- A portion of chemical cycling can bypass the biotic components and rely completely on geological processes. Water, for example, can exit lakes and oceans by evaporation and recycle to those abiotic reservoirs by precipitation, such as rain.

- Some chemical elements require "processing" by certain microorganisms before they are available to plants as inorganic nutrients. The element nitrogen is an example. Its main abiotic reservoir is the atmosphere, which is almost 80% nitrogen in the form of N_2 gas. However, plants can utilize nitrogen only in the forms of ammonium (NH_4^+) and nitrate (NO_3^-), which roots absorb from the soil. Certain soil bacteria convert atmospheric N_2 to ammonium and nitrate in the soil, making nitrogen available to plants (see Chapter 14). In fact, microorganisms, mainly prokaryotes, play a critical role in all biogeochemical cycles.

Examples of Biogeochemical Cycles

Like most generalizations, the basic scheme of Figure 20.9 can go only so far in helping us understand biogeochemical cycles. A chemical's specific route through an ecosystem varies with the particular element and the trophic structure of the ecosystem. Figure 20.10 on the next two pages will help you apply the basic principles of chemical cycling to four key materials: carbon, nitrogen, phosphorus, and water (the only one that's a compound, not an element).

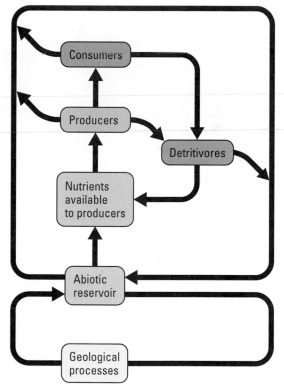

Figure 20.9 Generalized scheme for biogeochemical cycles.

Figure 20.10 Examples of biogeochemical cycles.

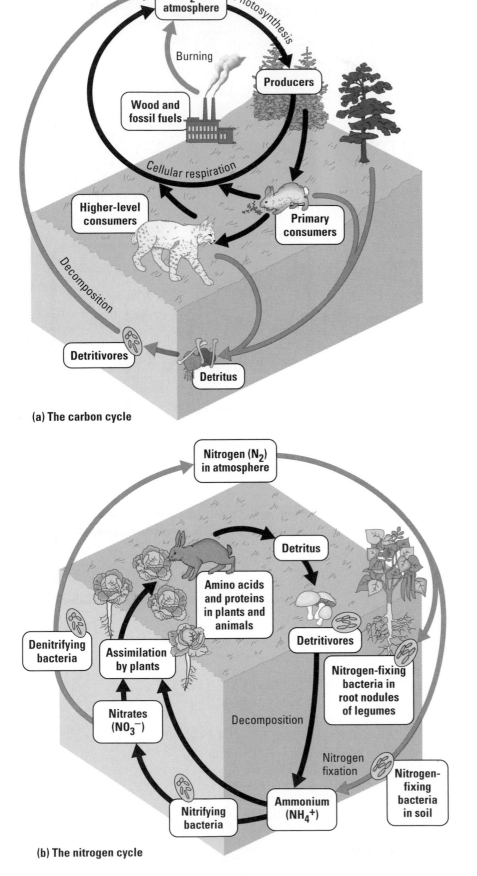

(a) The carbon cycle

(a) The carbon cycle. Carbon is a major ingredient of all organic molecules. With the atmosphere as its abiotic reservoir, carbon cycles globally. The reciprocal metabolic processes of photosynthesis and respiration are mainly responsible for the cycling of carbon between the biotic and abiotic world (black arrows in diagram). On a global scale, the return of CO_2 to the atmosphere by respiration closely balances its removal by photosynthesis. However, the increased burning of wood (during slash-and-burn deforestation) and fossil fuels (coal and petroleum) is steadily raising the level of CO_2 in the atmosphere. This may be causing significant environmental problems, such as global warming, which we discuss later in the chapter.

(b) The nitrogen cycle

(b) The nitrogen cycle. As an ingredient of amino acids and other organic compounds essential to life, nitrogen is a key chemical element in all ecosystems. Earth's atmosphere is almost 80% N_2 gas, but that form of nitrogen is unavailable to plants. Soil bacteria called nitrogen fixers convert the N_2 to ammonium, and nitrifying bacteria convert the ammonium to nitrates. Ammonium and nitrates are the soil minerals the roots of plants absorb as a nitrogen source for synthesis of amino acids and other organic molecules. Notice in the diagram that there are two groups of nitrogen-fixing bacteria: free-living bacteria and those that live as symbiotic organisms in nodules of beans and other legumes (see Chapter 14). Note that other types of bacteria play key roles at various junctures in the nitrogen cycle. For example, denitrifying bacteria complete the nitrogen cycle by converting soil nitrate to atmospheric N_2. Much of the nitrogen cycling by bacteria in many ecosystems involves the inner cycle in the diagram (black arrows). The outer cycle (the gray arrows) often moves only a tiny fraction of nitrogen into and out of natural ecosystems.

(c) The phosphorus cycle. Organisms require phosphorus as an ingredient of ATP and nucleic acids (such as DNA) and (in vertebrates) as a major component of bones and teeth. In contrast to carbon and nitrogen, phosphorus does not have an atmospheric presence and tends to recycle only locally. The main abiotic reservoir is rock, which, upon weathering, releases phosphorus mainly in the form of the mineral phosphate (PO_4^{3-}). Plants absorb the dissolved phosphate ions in the soil and build them into organic compounds. Consumers obtain phosphorus in organic form from plants. Detritivores return phosphates to the soil. Some phosphates also precipitate out of solution at the bottom of deep lakes and oceans. The phosphates in this form may eventually become part of new rocks and will not cycle back into living organisms until geological processes uplift the rocks and expose them to weathering. Thus, phosphorus actually cycles on two time scales: through a local ecosystem on the scale of ecological time (black arrows in diagram) and through strictly geological processes on the much longer scale of geological time (gray arrows).

(d) The water cycle. Water is essential to life, both because organisms are made mostly of water and because water's unique properties make environments on Earth habitable (see Chapter 2). In this diagram of the global water cycle, the numbers indicate water movements in billion billion (10^{18}) grams per year. That's a lot of water, but it's the relative comparisons that are important here. Those relative quantities are also indicated by the widths of the arrows. Three major processes driven by solar energy—precipitation, evaporation, and transpiration from plants (evaporation from leaves)—continuously move water between the land, the oceans, and the atmosphere. Note that over the oceans, evaporation exceeds precipitation. The result is a net movement of this excess amount of water vapor in clouds that are carried by winds from the oceans across the land. On land, precipitation exceeds evaporation and transpiration. The excess precipitation forms systems of surface water (such as lakes and rivers) and groundwater, all of which flow back to the sea, completing the water cycle. The water cycle has a global character because there is a large reservoir of water in the atmosphere. Thus, water molecules that have evaporated from the Pacific Ocean, for instance, may appear in a lake or in an animal's body far inland in North America.

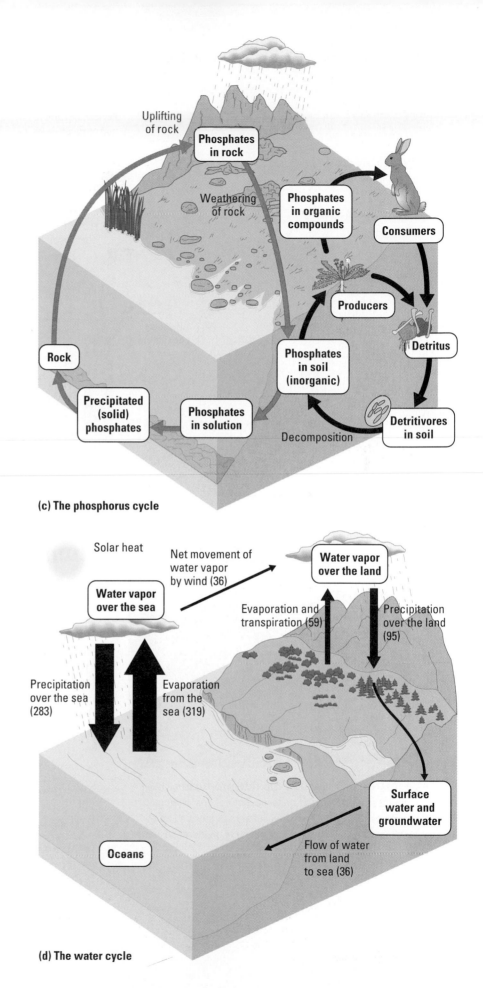

(c) The phosphorus cycle

(d) The water cycle

The carbon, nitrogen, phosphorus, and water cycles (and others) can be classified into two main categories based on how mobile the chemicals are within environments. Some chemicals, including phosphorus, are not very mobile and thus cycle almost entirely locally, at least over the short term. Soil is the main reservoir for these nutrients, which are absorbed by plant roots and eventually returned to the soil by detritivores living in the same vicinity. In contrast, for those materials that spend part of their time in gaseous form—carbon, nitrogen, and water are examples—the cycling is essentially global. For instance, some of the carbon atoms a plant acquires from the air as CO_2 may have been released to the atmosphere by the respiration of some animal living far away.

For both local and global cycling of chemicals, human activities have intruded, as we will now see by examining our impact on the dynamics of ecosystems and the whole biosphere.

Review the carbon and nitrogen cycles in Web/CD Activities 20D and 20E.

Review the carbon and nitrogen cycles in Web/CD Activities 20D and 20E.

CheckPoint

1. What is the main abiotic reservoir for carbon?
2. What would happen to the carbon cycle if all the detritivores suddenly went on "strike" and stopped working?
3. Over the short term, why does phosphorus cycling tend to be more localized than carbon, nitrogen, or water cycling?

Answers: 1. The atmospheric stock of CO_2 **2.** Carbon would accumulate in organic mass, the atmospheric reservoir of carbon would decline, and plants would eventually be starved for CO_2. **3.** Because phosphorus is cycled almost entirely within the soil rather than being transferred over long distance via the atmosphere

Human Impact on Ecosystems

Human population growth plus technology add up to a badly bruised biosphere. We have managed to intrude in one way or another into the dynamics of most ecosystems. Even where we have not completely destroyed a natural system, our actions have disrupted the trophic structure, energy flow, and chemical cycling of ecosystems in most areas of the world. The effects are sometimes local or regional, but the ecological impact of humans can be far-reaching or even global. For example, acid precipitation may be carried by prevailing winds and fall as rain hundreds or thousands of miles from the smokestacks emitting the chemicals that produce it (see Chapter 2). And the CO_2 exhaust of our machinery is probably causing a global warming that will affect all life on Earth, including humans (Figure 20.11). In this section, we'll examine a few of our ecological footprints on both local ecosystems and the biosphere, the global ecosystem.

Impact on Chemical Cycles

Human activities often intrude in biogeochemical cycles by removing nutrients from one location and adding them to another. This may result in the depletion of key nutrients in one area, excesses in another place, and the disruption of the natural chemical cycling in both locations. For example,

Figure 20.11 Carbon dioxide producers. Burning fossil fuels, such as gasoline, natural gas, and coal, produces CO_2. The rising level of carbon dioxide in the atmosphere is probably contributing to an increase in the average global temperature.

nutrients in the soil of croplands soon appear in the wastes of humans and livestock and then appear in streams and lakes through runoff from stockyards and discharge as sewage. Someone eating a salad in Washington, D.C., is consuming nutrients that only days before might have been in the soil in California. And a short time later, some of these nutrients will be in the Potomac River on their way to the sea, having passed through an individual's digestive system and the local sewage facilities.

Humans have altered chemical cycles to such an extent that it is no longer possible to understand any cycle without taking the human impact into account. Let's consider just a few examples.

Impact on the Carbon Cycle The increased burning of fossil fuels (coal and petroleum) as well as wood from deforested areas is steadily raising the level of CO_2 in the atmosphere. This is leading to significant environmental problems, such as global warming (discussed later in the chapter).

Impact on the Nitrogen Cycle Sewage treatment facilities typically empty large amounts of dissolved inorganic nitrogen compounds into rivers or streams. Farmers routinely apply large amounts of inorganic nitrogen fertilizers, mainly ammonium and nitrates, to croplands. Lawns and golf courses also receive sizable doses of fertilizer. Crop and lawn plants take up some of the nitrogen compounds, and soil bacteria convert some into atmospheric N_2 (see Figure 20.9b). However, chemical fertilizers usually exceed the soil's natural recycling capacity. The excess nitrogen compounds often enter streams, lakes, and groundwater. In lakes and streams, these nitrogen compounds continue to fertilize, causing heavy growth of algae. Groundwater pollution by nitrogen fertilizers is also a serious problem in many agricultural areas. In the human digestive tract, nitrates in drinking water are converted to nitrites, which can be toxic.

> Explore the problem of water pollution from nitrates in Web/CD Activity 20F.

Impact on the Phosphorus Cycle Like nitrogen compounds, phosphates are a major component of sewage outflow. They are also used extensively in agricultural fertilizers and are a common ingredient in pesticides. Phosphate pollution of lakes and rivers, like nitrate pollution, stimulates a heavy algal growth. This leads to population explosions of detritivores that can eventually suffocate all the aerobic life, including fish, in a lake or pond. You can see the results of such "overfertilization," or **eutrophication,** of lakes in Figure 17.3.

Impact on the Water Cycle One of the main sources of atmospheric water is transpiration (evaporation) from the dense vegetation of tropical rain forests. The destruction of these forests, which is occurring rapidly today, will change the amount of water vapor in the air (Figure 20.12). This, in turn, will probably alter local, and perhaps global, weather patterns. Another change in the water cycle caused by humans results from pumping large amounts of groundwater to the surface to use for crop irrigation. This practice can increase the rate of evaporation from soil, and unless this loss is balanced by increased rainfall over land, groundwater supplies are depleted. Large areas in the midwestern United States, the southwestern American desert, parts of California, and areas bordering the Gulf of Mexico currently face this problem.

Figure 20.12 Deforestation and the water cycle. Destruction of tropical forests, such as this one in Belize, reduces return of water to the atmosphere via transpiration.

(b) Logged watersheds in the Hubbard Brook Forest

(a) A dam at the Hubbard Brook study site

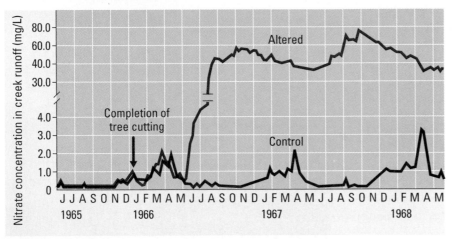

(c) The loss of nitrate from a deforested watershed

Figure 20.13 Chemical cycling in an experimental forest: the Hubbard Brook study. (a) A dam at a water-sampling station in the Hubbard Brook Forest. **(b)** Logged ("experimental group") and unlogged ("control group") watersheds in the Hubbard Brook Forest. **(c)** The loss of nitrate from a deforested watershed. Nitrate levels in the runoff from the deforested area began to rise markedly about 5 months after the plants were killed. Within 8 months, the nitrate loss was about 60 times greater in the altered watershed than in the control area. Without plants to take up and hold nitrate, this mineral nutrient drained out of the ecosystem.

Deforestation and Chemical Cycles: A Case Study

Since 1963, a team of scientists has been conducting a long-term study of chemical cycling in a forest ecosystem both under natural conditions and after vegetation is removed. The study site is the Hubbard Brook Experimental Forest in the White Mountains of New Hampshire. It is a deciduous forest with several valleys, each drained by a small creek that is a tributary of Hubbard Brook. Bedrock impenetrable to water is close to the surface of the soil, and each valley constitutes a watershed that can drain only through its creek.

The research team first determined the mineral budget for each of six valleys by measuring the inflow and outflow of several key nutrients. They collected rainfall at several sites to measure the amount of water and dissolved minerals added to the ecosystem. To monitor the loss of water and minerals, the scientists constructed small concrete dams, each with a V-shaped spillway, across the creek at the bottom of each valley (Figure 20.13a). About 60% of the water added to the ecosystem as rainfall and snow exits through the streams, and the remaining 40% is lost by transpiration from plants and evaporation from the soil.

Measurements confirmed that local cycling within a terrestrial ecosystem conserves most of the mineral nutrients. Mineral inflow and outflow balanced and were relatively small compared with the quantity of minerals being recycled within the forest ecosystem.

In 1966, one of the valleys, with an area of 15.6 hectares, was completely logged and then sprayed with herbicides for three years to prevent regrowth of plants (Figure 20.13b). All the original plant material was left in place to decompose. The inflow and outflow of water and minerals in the experimentally altered watershed were compared with those in a control watershed for three years. Water runoff from the altered watershed increased by 30–40%, apparently because there were no plants to absorb and transpire water from the soil. Net losses of minerals from the altered watershed were huge. Most remarkable was the loss of nitrate, which increased in concentration in the creek 60-fold (Figure 20.13c). Not only was this vital mineral nutrient drained from the ecosystem, but nitrate in the creek reached a level considered unsafe for drinking water.

The Hubbard Brook research is demonstrating that the amount of nutrients leaving an intact forest ecosystem is controlled mainly by plants. When plants are not present to retain them, nutrients are lost from the system. These effects are almost immediate, occurring within a few months of logging, and continue as long as plants are absent.

While scientists designed the Hubbard Brook experiments to assess natural ecosystem dynamics, the results are also providing important insights into the mechanisms by which human activities such as deforestation affect these processes.

Release of Toxic Chemicals to Ecosystems

In addition to transporting vital chemical elements from one location to another, we have added entirely new materials, many of them toxic, to ecosystems. Humans produce an immense variety of these toxic chemicals, including thousands of synthetics previously unknown in nature. Many of these poisons cannot be degraded by microorganisms and consequently persist in the environment for years or even decades. In other cases, chemicals released into the environment may be relatively harmless but are converted to more toxic products by reaction with other substances or by the metabolism of microorganisms. For example, mercury, a by-product of plastic production, was once routinely expelled into rivers and the sea in an insoluble form. Bacteria in the bottom mud converted the waste to methyl mercury, an extremely toxic soluble compound that then accumulated in the tissues of organisms, including humans who consumed fish from the contaminated waters.

Organisms acquire toxic substances from the environment along with nutrients and water. Some of the poisons are metabolized or excreted, but others accumulate in specific tissues, especially fat. Examples of industrially synthesized compounds that act in this manner are the chlorinated hydrocarbons (which include many pesticides, such as DDT) as well as the industrial chemicals called PCBs (polychlorinated biphenols).

One of the reasons the toxins we add to ecosystems are such ecological disasters is that they become more concentrated in successive trophic levels of a food web, a process called **biological magnification.** Magnification occurs because the biomass at any given trophic level is accumulated from a much larger toxin-containing biomass ingested from the level below (see Figure 20.7 to review energy pyramids). Thus, top-level carnivores are usually the organisms most severely damaged by toxic compounds that have been released into the environment.

A classic example of biological magnification involves DDT, the poisonous pollutant that Rachel Carson warned about almost 40 years ago (see Chapter 17). DDT was used to control insects such as mosquitoes and agricultural pests. But DDT persisted in the environment and was transported by water to areas far from where it was sprayed, rapidly becoming a global menace. Because the compound is soluble in lipids, it collects in the fatty tissues of animals, and its concentration is magnified in higher trophic levels (Figure 20.14). Traces of DDT have been found in nearly every organism tested; it has even been detected in human breast milk throughout the world. One of the first signs that DDT was a serious environmental problem was a decline in the populations of pelicans, ospreys, and eagles, birds that feed at the tops of food chains. DDT was banned in the United States in 1971, and a dramatic recovery in populations of the affected bird species followed. The pesticide is still used in many developing countries, however. And in the United States, we have replaced DDT with other pesticides that may also be biologically magnified in the food chains of ecosystems. For example, in 1999, several areas of New York City were sprayed with insecticides called pyrethroids. The spraying was a precaution against a pathogen called the West Nile virus, which is carried by mosquitoes. Within months, there was a massive die-off of lobsters in Long Island Sound, and there is evidence that the insecticides became magnified in these commercially important animals. Scientists hypothesize that the poisons reached the ocean in rainstorm runoff from the city.

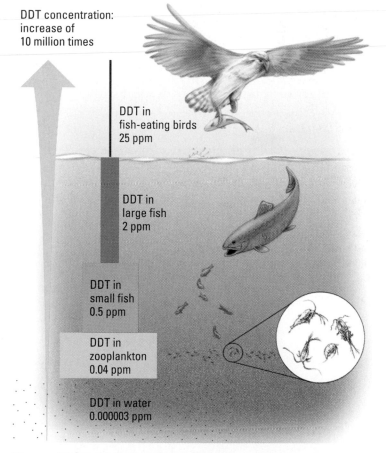

DDT concentration: increase of 10 million times

DDT in fish-eating birds 25 ppm

DDT in large fish 2 ppm

DDT in small fish 0.5 ppm

DDT in zooplankton 0.04 ppm

DDT in water 0.000003 ppm

Figure 20.14 Biological magnification of DDT in a food chain. DDT concentration in a Long Island Sound food chain was magnified by a factor of 10 million ppm. (The abbreviation stands for "parts per million," a unit of measurement commonly used for toxins.)

Figure 20.15 **Increase in atmospheric CO₂ in the past four decades.**

Human Impact on the Atmosphere and Climate

We are causing radical changes in the composition of the atmosphere and, consequently, in the global climate. Our activities release a variety of gaseous waste products. We once thought that the vastness of the atmosphere could absorb these materials without significant consequences, but an astronaut's view of our little planet squashes such naive notions (see Figure 17.1). One pressing problem that relates directly to one of the chemical cycles we examined is the rising level of carbon dioxide in the atmosphere.

Carbon Dioxide Emissions, the Greenhouse Effect, and Global Warming Since the Industrial Revolution, the concentration of CO_2 in the atmosphere has been increasing as a result of the combustion of fossil fuels and the burning of enormous quantities of wood removed by deforestation. Various methods have estimated that the average carbon dioxide concentration in the atmosphere before 1850 was about 274 parts per million (ppm). When a monitoring station on Hawaii's Mauna Loa peak began making very accurate measurements in 1958, the CO_2 concentration was 316 ppm (Figure 20.15). Today, the concentration of CO_2 in the atmosphere exceeds 360 ppm, an increase of about 14% since the measurements began just over 40 years ago. If CO_2 emissions continue to increase at the present rate, by the year 2075, the atmospheric concentration of this gas will be double what it was at the start of the Industrial Revolution. It is difficult to predict the multiple ways this intrusion in the carbon cycle will affect the biosphere and its various ecosystems.

One factor that complicates predictions about the long-term effects of rising atmospheric CO_2 concentration is its possible influence on Earth's heat budget. Much of the solar radiation that strikes the planet is reflected back into space. Although CO_2 and water vapor in the atmosphere are transparent to visible light, they intercept and absorb much of the reflected heat radiation, bouncing it back toward Earth. This process, called the **greenhouse effect**, retains some of the solar heat. Figure 20.16 adds some details to the illustration of the greenhouse effect you saw in Chapter 6.

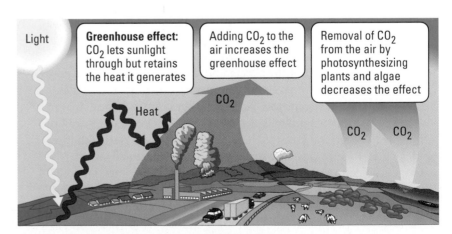

Figure 20.16 **Factors influencing the greenhouse effect.**

The marked increase in atmospheric CO_2 concentrations during the last 150 years concerns ecologists because of its potential effect on global temperature through the greenhouse effect. A number of studies predict that a doubling of CO_2 concentration by the end of the twenty-first century will cause an average global temperature increase of about 2°C (Figure 20.17). An increase of only 1.3°C would make the world warmer than at any time in the past 100,000 years. A worst-case scenario suggests that the warming would be greatest near the poles. Melting of polar ice might raise sea level by an estimated 100 meters, gradually flooding coastal areas 150 kilometers (or more) inland from the current coastline. New York, Miami, Los Angeles, and many other cities would then be underwater. A warming trend would also alter the geographic distribution of precipitation, making major agricultural areas of the central United States much drier. Most of Earth's natural ecosystems would also be affected, with boundaries between systems such as forests and grasslands shifting. However, the various mathematical models disagree about the details of how warming on a global level will change the climate in each region.

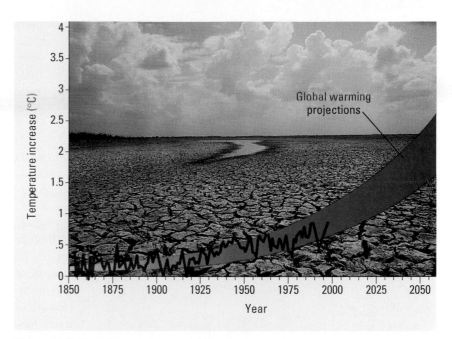

Figure 20.17 **Average atmospheric temperatures and projections for global warming.**

By studying how prehistoric periods of global warming and cooling affected plant communities, ecologists are using another strategy to help predict the consequences of future temperature changes. Records from fossilized pollen provide evidence that plant communities are altered dramatically by climate change. However, past climate changes occurred gradually, and plant and animal populations could spread into areas where conditions allowed them to survive. A major concern about the global warming under way now is that it is so rapid that many species may not be able to survive.

Relatively few scientists still doubt the evidence that human activity is heating up our planet. But what can be done to lessen the chances of greenhouse disaster? The burning of trees after deforestation to clear land in the tropics accounts for about 20% of the excess CO_2 released into the atmosphere. The burning of fossil fuels is the cause of the other 80%. International cooperation and national and individual action are needed to decrease fossil fuel consumption and to reduce the destruction of forests. Many nations have called for efforts to cap the level of CO_2 output from fossil fuel combustion. Because fossil fuels currently power much of our industry and economic growth, meeting the terms of any international agreement will require strong individual commitment and acceptance of some major lifestyle changes.

Those of us in developed countries have the greatest responsibility to reduce energy consumption. People in the United States consume more energy than do the total populations of Central and South America, Africa, India, and China combined—about 3.7 billion people compared to 270 million in the United States. Thus, the moderation of global warming depends mainly on the richest countries reducing their use of fossil fuels by conserving energy and by developing alternative energy sources, such as wind, solar, and geothermal energy. We can help individually by becoming more energy efficient at home and reducing our reliance on the automobile.

Investigate the problem of global warming further in The Process of Science activity on the website and CD-ROM.

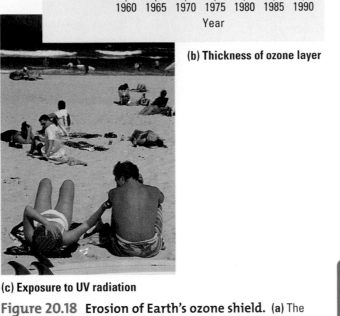

(a) Ozone hole

(b) Thickness of ozone layer

Ozone layer thickness (Dobson units) — Year axis: 1960, 1965, 1970, 1975, 1980, 1985, 1990 — Monthly averages for October

(c) Exposure to UV radiation

Figure 20.18 Erosion of Earth's ozone shield. **(a)** The ozone hole over the Antarctic is visible as the blue patch in this image based on atmospheric data. **(b)** This graph tracks the thickness of the ozone layer in units called Dobsons. **(c)** Australia, home to these sun lovers, already has the world's highest rate of skin cancer. Increased UV radiation through an eroded ozone shield won't help.

Depletion of Atmospheric Ozone Life on Earth is protected from the damaging effects of ultraviolet (UV) radiation by a very thin protective layer of ozone molecules (O_3) located in the atmosphere between 17 and 25 kilometers above Earth's surface. This **ozone layer** absorbs UV radiation, preventing much of it from contacting organisms in the biosphere. Measurements by atmospheric scientists document that the ozone layer has been gradually thinning since 1975.

The destruction of atmospheric ozone probably results mainly from the accumulation of chlorofluorocarbons, chemicals used in refrigeration, as propellants in aerosol cans, and in certain manufacturing processes. When the breakdown products from these chemicals rise in the atmosphere, the chlorine they contain reacts with ozone, converting it to O_2. Subsequent chemical reactions liberate the chlorine, allowing it to react with other ozone molecules in a catalytic chain reaction. The effect is most apparent over Antarctica, where cold winter temperatures facilitate these atmospheric reactions (Figure 20.18). Scientists first described the "ozone hole" over Antarctica in 1985. Since then, the size of the ozone hole has increased, sometimes extending as far as the southernmost portions of Australia, New Zealand, and South America. And at the more heavily populated middle latitudes, ozone levels have decreased 2–10% during the past 20 years.

The consequences of ozone depletion may be quite severe for all life on Earth, including humans. Some scientists expect the growing intensity of UV radiation to increase the incidence of skin cancer and cataracts among humans. It is likely that there will also be damaging effects on crops and natural communities, especially the phytoplankton that are responsible for a large proportion of the biosphere's primary productivity. The danger posed by ozone depletion is so great that many nations agreed in 1987 to end the production of chlorofluorocarbons by the year 2010. (The United States and other developed nations have already substituted safer compounds for chlorofluorocarbons, but a grace period was allowed for developing countries.) Unfortunately, even if all chlorofluorocarbons were banned today, the chlorine molecules already in the atmosphere will continue to influence atmospheric ozone levels for at least a century. It is just one more example of how far our technological tentacles reach in disrupting the dynamics of ecosystems and the entire biosphere.

CheckPoint

1. How can clear-cutting a forest (removing all trees) damage the water quality of nearby lakes?

2. How does excessive addition of mineral nutrients to a lake eventually result in the loss of most fish in the lake?

3. How is biological magnification relevant to the health of most humans in developed countries?

4. Here's one of those questions that requires interpreting data and graphs (see Figure 20.15): What was the approximate percentage increase in atmospheric CO_2 between 1960 and 1990?

Answers: 1. Without the growing trees to assimilate minerals from the soil, more of the minerals run off and end up polluting water resources. **2.** The eutrophication (overfertilization) initially causes population explosions of algae and the organisms that feed on them. The respiration of so much life, including the detritivores working on all the organic refuse, consumes most of the lake's oxygen, which the fish require. **3.** People in developed countries generally eat more meat than do people in developing countries. As secondary or tertiary consumers in a food chain, meat-eaters acquire a greater dose of some toxic chemicals than if they fed exclusively on plants as primary consumers. **4.** About 12%, from 315 ppm to about 352 ppm ($352 - 315 = 37$, and $37/315 \times 100 = 11.7\%$)

The Biodiversity Crisis

As a result of our numbers and technology and incessant intrusions in the world's ecosystems, we are now presiding over an alarming **biodiversity crisis,** a precipitous decline in Earth's great variety of life.

The Three Levels of Biodiversity

Biodiversity, short for biological diversity, has three main components. The first is the diversity of ecosystems. Each ecosystem, be it a rain forest, desert, or coral reef, has a unique biological community and characteristic patterns of energy flow and chemical cycling. And each ecosystem has a unique impact on the entire biosphere. For example, the productive "pastures" of phytoplankton in the oceans help moderate the greenhouse effect by consuming massive quantities of atmospheric CO_2 for photosynthesis and shell building (many microscopic protists in plankton secrete shells of bicarbonate, a derivative of CO_2). Some ecosystems are being erased from the biosphere at an astonishing rate. For example, the cumulative area of all tropical rain forests is only about the size of the 48 contiguous United States, and we lose an area equal to the state of West Virginia each year.

The second component of biodiversity is the variety of species that make up the biological community of any ecosystem (Figure 20.19). And the third component is the genetic variation within each species. You learned in Chapter 12 that the loss of genetic diversity—by a severe reduction in population size, for example—can hasten the demise of a dwindling species.

Though human impact reaches all three levels of biodiversity, most of the research focus so far has been on species extinction.

Figure 20.19 **A coral reef is a showcase of biodiversity.**

The Loss of Species

The seventh mass extinction in the history of life is well under way. Previous episodes, including the Cretaceous crunch that claimed the dinosaurs and many other groups, pale by comparison. The current mass extinction is both broader and faster, extinguishing species at a rate at least 50 times faster than just a few centuries ago. And unlike past poundings of biodiversity, which were triggered mainly by physical processes, such as climate change caused by volcanism or asteroid crashes, this latest mass extinction is due to the evolution of a single species—a big-brained, manually dexterous, environment-manipulating toolmaker that has named itself *Homo sapiens.*

We do not know the full scale of the biodiversity crisis in terms of a species "body count," for we are undoubtedly losing species that we didn't even know existed; the 1.5 million species that have been identified probably represent less than 10% of the true number of species. However, there are already enough signs to know that the biosphere is in deep trouble:

- About 11% of the 9040 known bird species in the world are endangered. In the past 40 years, population densities of migratory songbirds in the mid-Atlantic United States dropped 50%.

- Of the approximately 20,000 known plant species in the United States, over 600 are very close to extinction.

- Throughout the world, 970 tree species have been classified as critically endangered. At least 5 of those species are down to fewer than a half dozen surviving individuals.

- About 20% of the known freshwater fishes in the world have either become extinct during historical times or are seriously threatened. The toll on amphibians and reptiles has been almost as great.
- Harvard biologist Edward O. Wilson, a renowned scholar of biodiversity, has compiled what he grimly calls the Hundred Heartbeat Club. The species that belong are those animals that number fewer than 100 individuals and so are only that many heartbeats away from extinction (Figure 20.20).
- Several researchers estimate that at the current rate of destruction, over half of all plant and animal species will be gone by the end of this new century.

(a) Philippine eagle

(b) Chinese river dolphin

(c) Javan rhinoceros

Figure 20.20 A hundred heartbeats from extinction.
These are just three of the many members of what E.O. Wilson calls the Hundred Heartbeat Club, species with fewer than 100 individuals remaining on Earth.

The modern mass extinction is different from earlier biodiversity shakeouts in still another important way. The prehistoric crashes were all followed by rebounds in diversity as the survivors radiated and adapted to ecological niches left vacant by the extinctions. But as long as we humans are around to destroy habitats and degrade biodiversity at the ecosystem level, there can be no rebound in the evolutionary diversification of life. In fact, the trend is toward increased geographic range and prevalence of "disaster species," those life-forms such as house mice, kudzu and other weeds, cockroaches, and fire ants that seem to thrive in environments disrupted by human activities. Unless we can reverse the current trend of increasing loss of biodiversity, we will leave our children and grandchildren a biosphere that is much less interesting and much more biologically impoverished.

The Three Main Causes of the Biodiversity Crisis

Habitat Destruction Human alteration of habitats poses the single greatest threat to biodiversity throughout the biosphere (Figure 20.21a). Assaults on diversity at the ecosystem level result from the expansion of agriculture to feed the burgeoning human population, urban development, forestry, mining, and environmental pollution. The amount of human-altered land surface is approaching 50%, and we use over half of all accessible surface fresh water. Some of the most productive aquatic habitats in estuaries and intertidal wetlands are also prime locations for commercial and residential developments. The loss of marine habitats is also severe, especially in coastal areas and coral reefs.

Introduced Species Ranking second behind habitat loss as a cause of the biodiversity crisis is human introduction of exotic (non-native) species that eliminate native species through predation or competition (see Chapter 19). For example, if your campus is in an urban setting, there is a good chance that the birds you see most often as you walk between classes are starlings, rock doves (often called "pigeons"), and house sparrows—all introduced species that have replaced native birds in many areas of North America. One of the largest rapid-extinction events yet recorded is the loss of freshwater fishes in Lake Victoria in East Africa. About 200 of the 300 species of native fishes, found nowhere else but in this lake, have become extinct since Europeans introduced a non-native predator, the Nile perch, in the 1960s (Figure 20.21b).

Overexploitation As a third major threat to biodiversity, overexploitation of wildlife often compounds problems of shrinking habitat and introduced species. Animal species whose numbers have been drastically reduced by excessive commercial harvest or sport hunting include whales, the American bison, Galápagos tortoises, and numerous fishes. Many fish stocks in the ocean have been overfished to levels that cannot sustain further human exploitation (Figure 20.21c). In addition to the commercially important species, members of many other species are often killed by harvesting methods; for example, dolphins, marine turtles, and seabirds are caught in fishing nets, and countless numbers of invertebrates are killed by marine trawls (big nets). An expanding, often illegal world trade in wildlife products, including rhinocerous horns, elephant tusks, and grizzly bear gallbladders, also threatens many species.

Lake Victoria

(a) Habitat destruction: clearing a rain forest

(b) Introduced species: Nile perch

Figure 20.21 The three main causes of the biodiversity crisis.
(a) Habitat destruction. This is an all-too-common scene in the tropics: the clearing of a rain forest for lumber, agriculture, housing projects, or, in this case, a road. **(b) Introduced species.** One of the largest freshwater fishes (up to 2 meters long and weighing up to 450 kilograms), the Nile perch was introduced to Lake Victoria in East Africa to provide high-protein food for the growing human population. Unfortunately, the perch's main effect has been to wipe out about 200 smaller native species, reducing its own food supply to a critical level. **(c) Overexploitation.** Until the past few decades, the North Atlantic bluefin tuna was considered a sport fish of little commercial value—just a few cents per pound as cat food. Then, beginning in the 1980s, wholesalers began airfreighting fresh, iced bluefin to Japan for sushi and sashimi. In that market, the fish now brings up to $100 per pound! With that kind of demand, the results are predictable. It took just ten years to reduce the North Atlantic bluefin population to less than 20% of its 1980 size. In spite of quotas, the high price that bluefin tuna brings probably dooms the species to extinction.

(c) Overexploitation: North Atlantic bluefin tuna being auctioned in a Japanese fish market

Figure 20.22 Madagascar's rosy periwinkle, a source of anticancer drugs.

Why Biodiversity Matters

Why should we care about the loss of biodiversity? First of all, we depend on many other species for food, clothing, shelter, oxygen, soil fertility—the list goes on and on. In the United States, 25% of all prescriptions dispensed from pharmacies contain substances derived from plants. For instance, two drugs effective against Hodgkin's disease and certain other forms of cancer come from the rosy periwinkle, a flowering plant native to the island of Madagascar (Figure 20.22). Madagascar alone harbors some 8000 species of flowering plants, 80% of which occur only there. Among these unique plants are several species of wild coffee trees, some of which yield beans lacking caffeine (naturally "decaffeinated"). With an estimated 200,000 species of plants and animals, Madagascar is among the top five most biologically diverse countries in the world. Unfortunately, most of Madagascar's species are in serious trouble. People have lived on the island for only about 2000 years, but in that time, Madagascar has lost 80% of its forests and about 50% of its native species. Madagascar's dilemma represents that of much of the developing world. The island is home to over 10 million people, most of whom are desperately poor and hardly in a position to be concerned with environmental conservation. Yet the people of Madagascar as well as others around the globe could derive vital benefits from the biodiversity that is being destroyed.

> Explore the biodiversity crisis in Madagascar further in Web/CD Activity 20G.

Who knows which of the species disappearing from the biosphere could provide new sources of food or medicine? And who understands the dynamics of ecosystems well enough to know which species are least essential to biological communities? The great American naturalist Aldo Leopold explained it this way: "If the biota, in the course of aeons, has built something we like but do not understand, then who but a fool would discard seemingly useless parts? To keep every cog and wheel is the first precaution of intelligent tinkering."

Another reason to be concerned about the changes that underlie the biodiversity crisis is that the human population itself is threatened by large-scale alterations in the biosphere. Like all other species, we evolved in Earth's ecosystems, and we are dependent on the living and nonliving components of these systems. By allowing the extinction of species and the degradation of habitats to continue, we are taking a risk with our own species' survival.

In an attempt to counter what they see as a tendency of policymakers and governments to undervalue life-sustaining features of the biosphere, a team of ecologists and economists recently estimated the cost of losing ecosystem "services." For example, they estimated part of the value of a wetland from the cost of flood damage that occurred because of the loss of the wetland's ability to hold floodwater. For a single year in the late 1990s, the scientists estimated the average annual value of ecosystem dynamics in the biosphere at 33 trillion U.S. dollars. In contrast, the global gross national product for the same year was 18 trillion U.S. dollars. Although rough, these estimates help make the important point that we cannot afford to continue to take ecosystems for granted.

CheckPoint

1. What are the three main levels of biodiversity?

2. What are the three main causes of the biodiversity crisis?

Answers: 1. Ecosystem diversity, species diversity, and the genetic diversity of populations
2. Habitat destruction, introduced species, and overexploitation

Conservation Biology

Conservation biology is a goal-oriented science that seeks to counter the loss of biodiversity. It began to take form in the late 1970s as an interdisciplinary collaboration of scientists working toward the long-term maintenance of functional ecosystems and a reduction of the rate of species extinction. Promoting research on biodiversity and the means to save it, an international group of scientists and educators founded the Society for Conservation Biology in 1985. Today, with about 5000 members, the society helps unite the conservation efforts of biologists, anthropologists, sociologists, economists, government and industry officials, and private citizens. Conservation biologists recognize that biodiversity can be sustained only if the evolutionary mechanisms that have given rise to species and communities of organisms continue to operate. Thus, the goal is not simply to preserve individual species but to sustain ecosystems, where natural selection can continue to function, and to maintain the genetic variability on which natural selection acts. The frontlines for conservation biology are geographic areas that are especially rich in endangered biodiversity.

Biodiversity "Hot Spots"

A **biodiversity hot spot** is a relatively small area with an exceptional concentration of species (Figure 20.23). Many of the organisms in biodiversity hot spots are **endemic species,** meaning they are found nowhere else. For example, nearly 30% of all bird species are confined to only about 2% of Earth's land area. And about 50,000 plant species, or 20% of all known plant species, inhabit 18 hot spots making up only about 0.5% of the global land surface. Overall, the "hottest" of the biodiversity hot spots, including rain forests and dry shrublands (such as California's chaparral), total less than 1.5% of Earth's land but are home to a third of all species of plants and vertebrates. Conservation biologists have also identified aquatic ecosystems, including certain river systems and coral reefs, that are biodiversity hot spots.

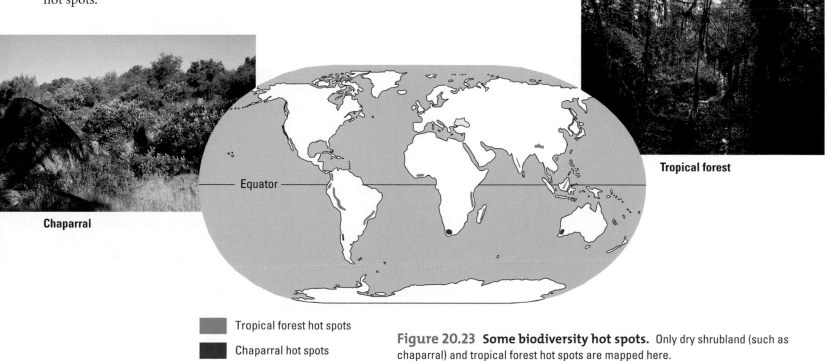

Tropical forest

Chaparral

Equator

Tropical forest hot spots

Chaparral hot spots

Figure 20.23 Some biodiversity hot spots. Only dry shrubland (such as chaparral) and tropical forest hot spots are mapped here.

Because endemic species are limited to specific areas, they are highly sensitive to habitat degradation. To date, 6 of the 18 hot spot areas shown in Figure 20.23 have lost nearly 90% of their original habitats to human development. At the current rate of habitat destruction, the rest could lose similar amounts in the next two decades. Conservation biologists estimate that this loss of habitat will cause the extinction of about half the species in the hot spots. Thus, biodiversity hot spots are also hot spots of extinction and rank high on the list of areas demanding strong global conservation efforts. In the United States, the greatest numbers of endangered species (terrestrial and freshwater) occur in Hawaii, southern California, the southern Appalachians, and the southeastern coastal states (especially Florida), areas with the highest numbers of endemic species.

With so much biodiversity concentrated in a relatively small portion of the biosphere, there is reason for optimism that conservation biologists can accomplish much by focusing on these hot spots. The goal is to keep as many of these hot spots as wild as possible. On the other hand, the biodiversity crisis is a global problem, and focusing on hot spots should not detract from efforts to conserve habitats and species diversity in other areas.

Conservation at the Population and Species Levels

Much of the popular and political discussion of the biodiversity crisis centers on species. The U.S. Endangered Species Act (ESA) defines an **endangered species** as one that is "in danger of extinction throughout all or a significant portion of its range." Also defined for protection by the ESA, **threatened species** are those that are likely to become endangered in the foreseeable future throughout all or a significant portion of their geographic range.

Habitat Fragmentation and Subdivided Populations The focus of scientific studies concerned with sustaining species is on the dynamics of populations that have been reduced in numbers and fragmented by human activities. Because habitats are patchy, the populations of many species were subdivided into groups with varying degrees of isolation from one another before humans began altering habitats significantly. Gene flow among population subgroups varied according to the degree of isolation (see Chapter 12).

Today, severe **population fragmentation,** the splitting and consequent isolation of portions of populations by habitat degradation, is one of the most harmful effects of habitat loss due to human activities (Figure 20.24). We'll refer to these fragments of a population as subpopulations. A decrease in the overall size of populations and a reduction in gene flow among subpopulations usually accompany fragmentation.

In most cases, subpopulations are separated into habitat patches that vary in quality. For instance, in a fragmented forest, the presence of some large, dead trees can make the difference between a high-quality patch and a low-quality one for squirrels, owls, and many birds that require rot cavities for their nests. Patches with abundant high-quality resources tend to have stable, persistent subpopulations. Reproductive individuals in a high-quality patch tend to produce more offspring than the patch can sustain. Such an area of habitat where a subpopulation's reproductive success exceeds its death rate is called a **source habitat.** Source habitats produce enough new individuals that some disperse to other areas, often in search

Northern spotted owl

Figure 20.24 Fragmentation of a forest ecosystem. This aerial photograph illustrates fragmentation of a coniferous forest in the Mt. Hood National Forest in northwestern Oregon. The forest was originally contiguous. The open areas in the photo were logged, creating forest fragments, some of which are islands within clear-cut areas. A common result of human activities, this kind of habitat alteration has reduced and fragmented the populations of many species. An example is the northern spotted owl (inset), which inhabits coniferous forests of the U.S. Pacific Northwest. Owl populations declined markedly and were fragmented after these forests were logged.

of food or a place to reproduce. In contrast, a habitat where a subpopulation's death rate exceeds its reproductive success is called a **sink habitat.** The persistence of many subpopulations in sink habitats depends on individuals dispersing from source habitats.

Today, because of habitat loss, more sink habitats exist than were present for most species historically, and the dispersal of individuals to sinks can sometimes threaten the survival of subpopulations in source habitats. For example, the remaining source habitats of the northern spotted owl are relatively small fragments of old-growth rain forests. The owls tend to disperse from small fragments of source habitats into larger, neighboring areas where trees are regrowing after logging. Such areas are sink habitats for the owls, and sustaining the owl populations depends in part on keeping enough reproductive individuals as robust breeding stock in the source habitats. Some researchers have suggested surrounding old-growth fragments with distinct boundaries—such as clear-cut areas—that the owls will not enter. The spotted owl situation illustrates the importance of identifying source and sink habitats and of protecting source habitats.

(a) Red-cockaded woodpecker

(b) Forest that can sustain red-cockaded woodpeckers

What Makes a Good Habitat? What factors make some habitats sources and others sinks? Identifying the specific combination of habitat factors that is critical for a species is fundamental in conservation biology.

As a case study in identifying critical habitat factors, we'll consider the red-cockaded woodpecker (*Picoides borealis*), an endangered, endemic species originally found throughout the southeastern United States. This species requires mature pine forests, preferably ones dominated by the longleaf pine. Most woodpeckers nest in dead trees, but the red-cockaded woodpecker drills its nest holes in mature, living pine trees (Figure 20.25a). The heartwood (deep wood) of mature longleaf pines is usually rotted and softened by fungi, allowing the woodpeckers adequate space for nesting once they excavate into the heartwood. Red-cockaded woodpeckers also drill small holes around the entrance to their nest cavity, which causes resin from the tree to ooze down the trunk. The resin seems to repel certain predators, such as corn snakes, that eat bird eggs and nestlings. Another critical habitat factor for this woodpecker is a low growth of plants among the mature pine trees (Figure 20.25b). Biologists have found that a habitat becomes a sink, with breeding birds tending to abandon nests, when vegetation among the pines is thick and higher than about 15 feet (Figure 20.25c). Apparently, the birds require a clear flight path between their home trees and the neighboring feeding grounds. Historically, periodic fires swept through longleaf pine forests, keeping the undergrowth low.

The recent recovery of the red-cockaded woodpecker from near-extinction to sustainable populations is largely due to recognizing the key habitat factors and protecting some longleaf pine forests that support viable numbers of the birds. The use of controlled fires to reduce forest undergrowth helps maintain mature pine trees as well as the woodpeckers.

(c) Forest that cannot sustain red-cockaded woodpeckers

Figure 20.25 Habitat requirements of the red-cockaded woodpecker. **(a)** A red-cockaded woodpecker at the entrance to its nest site in a longleaf pine tree. **(b)** Forest habitat with low undergrowth that sustains red-cockaded woodpeckers. **(c)** High, dense undergrowth that impedes the woodpeckers' access to feeding grounds.

Conserving Species in the Context of Conflicting Demands Determining habitat requirements is only one aspect of the effort to save species. It is usually necessary to weigh a species' biological and ecological needs against the conflicting demands of our complex culture. Thus, conservation biology often highlights the relationships between biology and society. For example, an ongoing, sometimes bitter debate in the U.S. Pacific Northwest pits saving habitats for populations of the northern spotted owl, timber wolf, grizzly bear, and bull trout against demands for jobs in the timber, mining, and other resource-extraction industries. Programs to restock wolves and to bolster the populations of grizzly bears and other large carnivores are opposed by some recreationists concerned about safety and by many ranchers concerned with potential losses of livestock.

Large, high-profile vertebrates are not always the focal point in conflicts involving conservation biology. It is habitat use that is almost always at issue. Should work proceed on a new highway bridge if it destroys the only remaining habitat of a species of freshwater mussel? If you were the owner of a coffee plantation growing varieties that thrive in bright sunlight, do you think you would be willing to change to shade-tolerant coffee varieties that are less productive and less profitable but support large numbers of songbirds?

Conservation at the Ecosystem Level

Most conservation efforts in the past have focused on saving individual species, and this work continues. More and more, however, conservation biology aims at sustaining the biodiversity of entire communities and ecosystems. On an even broader scale, conservation biology considers the biodiversity of whole landscapes. Ecologically, a landscape is a regional assemblage of interacting ecosystems, such as an area with forest, adjacent fields, wetlands, streams, and streamside habitats. **Landscape ecology** is the application of ecological principles to the study of land-use patterns. Its goal is to make ecosystem conservation a functional part of the planning for land use.

Edges and Corridors Edges between ecosystems are prominent features of landscapes, whether natural or altered by humans (Figure 20.26). Such edges have their own sets of physical conditions, such as soil type and surface features that differ from either side. Edges also may have their own type and amount of disturbance. For instance, the edge of a forest often has more blown-down trees than a forest interior because the edge is less protected from strong winds. Because of their specific physical features, edges also have their own communities of organisms. Some organisms depend on edges because they require resources of the two adjacent areas. For instance, white-tailed deer thrive in edge habitats where they can browse on woody shrubs, and their populations often expand when forests are logged.

Edges can have both positive and negative effects on biodiversity. A recent study in a tropical rain forest in western Africa indicated that natural edge communities are important sites of speciation (the origin of new species). On the other hand, landscapes where human activities have produced edges often have fewer species and are dominated by species that are

(a) Natural edges between ecosystems

(b) Edges created by human activities

Figure 20.26 Landscape edges between ecosystems.
(a) This relatively natural landscape in Australia includes a dry forest, a rocky area with grassy islands, and a flat, grass-covered lakeshore. **(b)** Human activities, such as logging and road building, often create edges that are more abrupt than those delineating natural landscapes. Sharp edges surround clear-cuts in this photograph of a heavily logged rain forest in Malaysia.

Figure 20.27 An artificial corridor. This highway underpass allows movement between protected areas for the few remaining Florida panthers. High fences along the highway reduce road kills of panthers and other species.

adapted to edges. An example is the brown-headed cowbird, an edge-adapted species that is currently expanding its populations in many areas of North America. Cowbirds forage in open fields on insects disturbed by or attracted to cattle and other large herbivores. But the cowbirds also need forests, where they lay their eggs in the nests of other bird species. The "host" bird feeds the cowbird young as though they were the host's own off-spring. This is a form of parasitism. Cowbird numbers are burgeoning where forests are being heavily cut and fragmented, creating more forest-edge habitats and open land for cattle, horses, and sheep. Increasing cow-bird populations and loss of habitats are correlated with declining popula-tions of several songbirds, such as warblers, that the cowbirds parasitize.

Another important landscape feature, especially where habitats have been severely fragmented, is the **movement corridor,** a narrow strip or se-ries of small clumps of quality habitat connecting otherwise isolated patches. Streamside habitats often serve as natural corridors, and govern-ment policy in some nations prohibits destruction of these areas. In places where there is extremely heavy human impact, government agencies some-times construct artificial corridors (Figure 20.27). Corridors can promote dispersal and help sustain populations, and they are especially important to species that migrate between different habitats seasonally. On the other hand, a corridor can be harmful—as, for example, in the spread of diseases, especially among small subpopulations in closely situated habitat patches. A movement corridor that connects a source habitat to a sink habitat may also reduce population numbers in the source. The effects of movement corri-dors have not been thoroughly studied, and researchers tend to evaluate the potential effects of corridors on a case-by-case basis.

Zoned Reserves In an attempt to slow the disruption of ecosystems, a number of countries are setting up what they call zoned reserves. A **zoned reserve** is an extensive region of land that includes one or more areas undisturbed by humans. The undisturbed areas are surrounded by lands that have been changed by human activity and are used for economic gain. The key factor of the zoned reserve concept is the development of a social and economic climate in the surrounding lands that is compatible with ecosystem conservation. These surrounding areas continue to be used to support the human population, but they are protected from extensive alter-ation. As a result, they serve as buffer zones against further intrusion into the undisturbed areas.

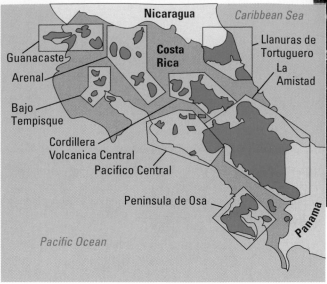

(a)

(b)

Figure 20.28 Zoned reserves in Costa Rica. (a) The green areas on the map are national park lands, core areas where human disruption is minimized. Surrounding these conservation cores, gold areas map buffer zones. These are transition areas, mainly privately owned, where most of the human population live and work. Ideally, the most destructive practices—industries such as mining, large-scale monoculture (growth of a single type of crop over a large area), and urban development—are confined to the outermost fringes of the buffer zones. Within the buffer zones, the trend is toward sustainable agriculture and forestry, activities that can provide comfortable economic support for local residents without drastically altering habitats. **(b)** Local students marvel at the diversity of life in one of Costa Rica's reserves.

The small Central American nation of Costa Rica has become a world leader in establishing zoned reserves. In exchange for reducing its international debt, the Costa Rican government established eight zoned reserves, called "conservation areas" (Figure 20.28). Costa Rica is making progress toward managing its zoned reserves so that the buffer zones provide a steady, lasting supply of forest products, water, and hydroelectric power and also support sustainable agriculture and tourism. An important goal is providing a stable economic base for people living there. Destructive practices that are not compatible with long-term ecosystem conservation, and from which there is often little local profit, are gradually being discouraged. Such destructive practices include massive logging, large-scale single-crop agriculture, and extensive mining. Costa Rica looks to its zoned reserve system to maintain at least 80% of its native species.

The Goal of Sustainable Development

With conservation progress in countries such as Costa Rica as a model, many nations, scientific organizations, and private foundations are embracing the concept of **sustainable development.** Balancing human needs with the health of the biosphere, the goal of sustainable development is the long-term prosperity of human societies and the ecosystems that support them. The significance of that responsibility was nicely phrased by former Norwegian Prime Minister G. H. Brundtland: "We must consider our planet to be on loan from our children, rather than being a gift from our ancestors."

Sustainable development will depend on the continued research and applications of basic ecology and conservation biology. It will also require a cultural commitment to conserve ecosystem processes and biodiversity. That is a priority that, so far, relatively few nations have placed at the top of their political, economic, and social agendas. Those of us living in affluent developed nations are responsible for the greatest amount of environmental degradation. Reality demands that we rearrange some of our priorities, learn to revere the natural processes that sustain us, and temper our orientation toward short-term personal gain.

The current state of the biosphere demonstrates that we are treading precariously on uncharted ecological ground. But despite the uncertainties, now is not the time for gloom and doom, but a time to meet the challenges to pursue more knowledge about life and to work as individuals and a society toward long-term sustainability. Along the way, we will reap the bonus of appreciating our connections to the biosphere and its diversity of life.

Review the concepts of conservation biology in Web/CD Activity 20H.

CheckPoint

1. What is a biodiversity hot spot?

2. Why is the fragmentation of populations of endangered species increasing?

3. Are critical habitat factors generally more favorable in a source habitat or a sink habitat?

4. How is a landscape different from an ecosystem?

5. How can "living on the edge" be a good thing for some species, such as white-tailed deer and cowbirds?

6. Why is a concern for the well-being of future generations essential for progress toward sustainable development?

Answers: 1. A relatively small area with a disproportionate number of species, including endangered species **2.** By destroying habitat, human activities are fragmenting the populations of many species into subpopulations that live in the separate patches of habitat that remain. **3.** Source habitat **4.** A landscape is more inclusive in that it consists of several interacting ecosystems in the same region. **5.** They use a combination of resources from the two ecosystems on either side of the edge. **6.** Sustainable development is a long-term goal—longer than a human lifetime. Concern only with personal gain in the here and now is an obstacle to sustainable development because it discourages behavior that benefits future generations.

Evolution Link: Biophilia and an Environmental Ethic

Not many people today live in truly wild environments or even visit such places often. Our modern lives are very different from those of early humans, who hunted and gathered and painted wildlife murals on cave walls (see Figure 16.46). But our behavior reflects remnants of our ancestral attachment to nature and the diversity of life. People keep pets, nurture houseplants, invite avian visitors with backyard birdhouses, and visit zoos, gardens, and nature parks. These pleasures are examples of what Harvard biologist Edward O. Wilson calls *biophilia,* a human desire to affiliate with other life in its many forms. (You met Wilson earlier in the context of the Hundred Heartbeat Club.) Wilson extends biophilia to include our attraction to pristine landscapes with clean water and lush vegetation (Figure 20.29). We evolved in natural environments rich in biodiversity, and we still have an affinity for such settings. Wilson makes the case that our biophilia is innate, an evolutionary product of natural selection acting on a brainy species whose survival depended on a close connection to the environment and a practical appreciation of plants and animals.

It will come as no surprise that most biologists have embraced the concept of biophilia. After all, these are people who have turned their passion for nature into careers. But biophilia strikes a harmonic chord with biologists for another reason. If biophilia is evolutionarily embedded in our genomes, then there is hope that we can become better custodians of the biosphere. If we all pay more attention to our biophilia, a new environmental ethic could catch on among individuals and societies. And that ethic is a resolve never to knowingly allow a single species to become extinct or any

Figure 20.29 Doing what comes naturally. One of modern biology's greatest naturalists, Edward O. Wilson has helped teach scientists and the general public a greater respect for Earth's biodiversity (two of his books have won Pulitzer Prizes). He coined the term *biophilia* to denote humans' inherent passion for nature. This photograph finds biophiliac Wilson in the woods near Massachusetts's Walden Pond, a landscape immortalized by another great naturalist and writer, Henry David Thoreau.

Figure 20.30 The face of biophilia. Biologist Carlos Rivera Gonzales, who is participating in a biodiversity survey in a remote region of Peru, could not resist a closer look at a tiny tree frog. You can also see biophilia in the faces of the children in Figure 20.28b.

ecosystem to be destroyed as long as there are reasonable ways to prevent such ecological violence. It is an environmental ethic that balances out another human trait—our tendency to "subdue the Earth." Yes, we should be motivated to preserve biodiversity because we depend on it for food, medicine, building materials, fertile soil, flood control, habitable climate, drinkable water, and breathable air. But maybe we can also work harder to prevent the extinction of other forms of life just because it is the ethical thing for us to do as the most thoughtful species in the biosphere. Again, Wilson sounds the call: "Right now, we're pushing the species of the world through a bottleneck. We've got to make it a major moral principle to get as many of them through this as possible. It's the challenge now and for the next century. And there's one good thing about our species: We like a challenge!"

Learn more about biophilia and environmental ethics in the Web Evolution Link.

Biophilia is a fitting capstone for this book. Modern biology is the scientific extension of our human tendency to feel connected to and curious about all forms of life. We hope that *Essential Biology* has deepened your biophilia and broadened your education.

Chapter Review

Summary of Key Concepts

Overview: Ecosystem Dynamics

- An ecosystem is a biological community and the abiotic factors with which the community interacts. Energy flow and chemical cycling are two closely related and key processes of ecosystem dynamics. Energy must flow continuously through an ecosystem, from producers to consumers and decomposers, entering an ecosystem as light and exiting as heat. Chemical elements can be recycled between an ecosystem's living community and the abiotic environment.
- Web/CD Activity 20A *Energy Flow and Chemical Cycling*

Trophic Structure of Ecosystems

- **Trophic Levels and Food Chains** Trophic relationships determine an ecosystem's routes of energy flow and chemical cycling. Producers are autotrophic organisms that ultimately support all others in an ecosystem. Consumers are heterotrophs directly or indirectly dependent on the output of producers. Herbivores, which eat producers, are primary consumers. Higher trophic levels are made up of carnivores, which eat consumers from lower levels. Detritivores are consumers that decompose organic materials into inorganic compounds that plants and other producers can reuse.

- **Food Webs** The feeding relationships in an ecosystem are usually woven into elaborate food webs.
- Web/CD Activity 20B *Food Webs*

Energy Flow in Ecosystems

- **Productivity and the Energy Budgets of Ecosystems** The rate at which plants and other producers build biomass is the ecosystem's primary productivity. Ecosystems vary considerably in their productivity. Primary productivity sets the spending limit for the energy budget of the entire ecosystem because consumers must acquire their organic fuels from producers.

- **Energy Pyramids** In general, about 10% of the energy in the form of organic matter at each trophic level is stored as biomass in the next level of the food chain. This cumulative loss of energy can be represented in a diagram called an energy pyramid.
- Web/CD Activity 20C *Energy Pyramids*

- **Ecosystem Energetics and Human Nutrition** Eating producers instead of consumers requires less photosynthetic productivity and reduces the impact to the environment.

Chemical Cycles in Ecosystems

- **The General Scheme of Chemical Cycling** Biogeochemical cycles involve biotic and abiotic components. Each circuit has an abiotic reservoir through which the chemical cycles. Some chemical elements require "processing" by certain microorganisms before they are available to plants as inorganic nutrients.

- **Examples of Biogeochemical Cycles** A chemical's specific route through an ecosystem varies according to the trophic structure. Phosphorus is not very mobile and is cycled locally. Carbon, nitrogen, and water spend part of their time in gaseous form and are cycled globally.
- Web/CD Activity 20D *The Carbon Cycle*
- Web/CD Activity 20E *The Nitrogen Cycle*

Human Impact on Ecosystems

- **Impact on Chemical Cycles** Human activities often intrude in biogeochemical cycles by removing nutrients from one location and adding them to another. For example, the increased burning of fossil fuels is steadily raising CO_2 in the atmosphere and sewage-treatment facilities and fertilizers add large amounts of nitrogen and phosphorus to aquatic systems. Deforestation and extensive removal of ground water change the water cycle.
- Web/CD Activity 20F *Water Pollution from Nitrates*

- **Deforestation and Chemical Cycles: A Case Study** Research at the Hubbard Brook Experimental Forest is demonstrating that the amount of nutrients leaving an intact forest ecosystem is controlled by plants.

- **Release of Toxic Chemicals to Ecosystems** Humans have added entirely new materials, many of them toxic, to ecosystems. These toxins often become more concentrated in successive trophic levels of a food web through biological magnification. One classic example was the accumulation of DDT in aquatic birds.

- **Human Impact on the Atmosphere and Climate** Deforestation and the burning of fossil fuels have increased concentrations of CO_2 in the atmosphere, contributing to the greenhouse effect and global warming with potentially disastrous consequences. Developed countries, which have the greatest energy consumption, have the greatest responsibility to reduce their use of fossil fuels by conserving energy and by developing alternative energy sources. In addition, the protective ozone layer has been gradually thinning since 1975 because of the accumulation of chlorofluorocarbons. Scientists expect that a thinning ozone layer will result in an increase of skin cancer and cataracts among humans.
- Web/CD The Process of Science *Global Warming: Evaluating Alternatives*

The Biodiversity Crisis

- **Three Levels of Biodiversity** Biodiversity consists of the diversity of ecosystems, the diversity of species that make up the biological community of an ecosystem, and the genetic variation within each species.

- **The Loss of Species** The current mass extinction, caused by human activities, is broader and faster than the Cretaceous extinction that claimed dinosaurs and many other groups. Species now go extinct at a rate at least 50 times faster than just a few centuries ago.

- **The Three Main Causes of the Biodiversity Crisis** The main causes of the biodiversity crisis, in order of impact, are habitat destruction, the introduction of non-native species, and the exploitation of wildlife.

- **Why Biodiversity Matters** Humans have relied upon biodiversity for food, clothing, shelter, oxygen, soil fertility, and medicinal substances. The loss of diversity limits the potential for new discoveries and reflects large-scale changes in the biosphere that could have catastrophic consequences.
- Web/CD Activity 20G *Madagascar and the Biodiversity Crisis*

Conservation Biology

- Conservation biology is a goal-oriented science that seeks to counter the loss of biodiversity.

- **Biodiversity "Hot Spots"** The front lines for conservation biology are relatively small geographic areas that are especially rich in endangered biodiversity.

- **Conservation at the Population and Species Levels** Severe population fragmentation is one of the most harmful effects of habitat loss due to human activities. Fragmentation usually results in a decrease in the overall size of populations and a reduction in gene flow among subpopulations. In addition, because of habitat loss, the survival of subpopulations in source habitats is threatened by the dispersal of individuals to an increasing number of sink habitats. Identifying the specific combination of habitat factors that is critical for a species is fundamental in conservation biology. Conservation biology often highlights the relationships between biology and society. Competing demands for habitat use is almost always at issue.

- **Conservation at the Ecosystem Level** Increasingly, conservation biology aims at sustaining the biodiversity of entire communities, ecosystems, and landscapes. Edges between ecosystems are prominent features of landscapes, with positive and negative effects on biodiversity. Natural edge communities are important sites of speciation. But human-produced edges often have fewer species and are dominated by species adapted to edges, such as cowbirds. Corridors can promote dispersal and help sustain populations. But corridors can also promote the spread of diseases and connect source and sink habitats. Zoned reserves are now used to slow the disruptions of ecosystems.

- **The Goal of Sustainable Development** Balancing human needs with the health of the biosphere, the goal of sustainable development is the long-term prosperity of human societies and the ecosystems that support them.
- Web/CD Activity 20H *Conservation Biology Review*

Evolution Link: Biophilia and an Environmental Ethic

- We evolved in natural environments rich in biodiversity, and we still have an affinity for such settings. E.O. Wilson argues that our biophilia is innate, an evolutionary product of natural selection acting on an intelligent species whose survival depended on a close connection to the environment and a practical appreciation of plants and animals.
- Web Evolution Link *Biophilia and an Environmental Ethic*

Self-Quiz

1. Local conditions, such as heavy rainfall or the removal of plants, may limit the amount of nitrogen, phosphorus, or calcium available to a particular ecosystem, but the amount of carbon available to the system is seldom a problem. Why?
   a. Organisms do not need very much carbon.
   b. Plants can make their own carbon using water and sunlight.
   c. Plants are much better at absorbing carbon from the soil.
   d. Many nutrients come from the soil, but carbon comes from the air.
   e. Symbiotic bacteria help plants capture carbon.

2. One of the lessons from an energy pyramid is that
   a. only one-half of the energy in one trophic level is passed on to the next level.
   b. most of the energy from one trophic level is incorporated into the biomass of the next level.
   c. the energy lost as heat or in cellular respiration is 10% of the available energy of each trophic level.
   d. food chains are essentially unlimited in the number of trophic levels possible.
   e. eating grain-fed beef is a relatively inefficient means of obtaining the energy trapped by photosynthesis.

3. The Hubbard Brook Experimental Forest study demonstrated all of the following *except*:
   a. Most minerals were recycled within the forest ecosystem.
   b. Mineral inflow and outflow within a natural watershed were nearly balanced.
   c. Deforestation resulted in an increase in water runoff.
   d. The nitrate concentration in waters draining the deforested area became dangerously high.
   e. Deforestation caused a large increase in the density of soil bacteria.

4. The recent increase in atmospheric CO_2 concentration is mainly a result of an increase in
   a. primary productivity.
   b. the biosphere's biomass.
   c. the absorption of infrared radiation escaping from Earth.
   d. the burning of fossil fuels and wood.
   e. cellular respiration by the exploding human population.

5. Falcons and other top predators in food chains are most severely affected by pesticides such as DDT because
   a. their systems are especially sensitive to chemicals.
   b. they have rapid reproductive rates.
   c. the pesticides become concentrated in their prey.
   d. they cannot store the pesticides in their tissues.
   e. they are directly exposed to pesticides in the air.

6. Which of the following statements most comprehensively addresses what conservation biologists mean by the "biodiversity crisis"?
   a. Worldwide extinction rates are currently 50 times greater than at any time during the past 100,000 years.
   b. Introduced species, such as house sparrows and starlings, have rapidly expanded their ranges.
   c. Harvests of marine fish, such as cod and bluefin tuna, are declining.
   d. Many pest species have developed resistance and are no longer effectively controlled by insecticide applications.
   e. The rate of patent applications for pharmaceuticals developed from living organisms is currently declining.

7. Which of these ecosystems has the lowest primary productivity per square meter? (See Figure 20.5.)
   a. a salt marsh
   b. an open ocean
   c. a coral reef
   d. a grassland
   e. a tropical rain forest

8. Which of the following organisms is *incorrectly* paired with its trophic level?
   a. alga—producer
   b. grasshopper—primary consumer
   c. zooplankton—secondary consumer
   d. eagle—tertiary consumer
   e. fungi—detritivore

9. Which of the following statements concerning the water cycle is *incorrect*? (See Figure 20.10d.)
   a. There is a net movement of water vapor from oceans to terrestrial environments.
   b. Precipitation exceeds evaporation on land.
   c. Most of the water that evaporates from oceans is returned by runoff from land.
   d. Transpiration makes a significant contribution to evaporative water loss from terrestrial ecosystems.
   e. Evaporation exceeds precipitation over the seas.

10. Currently, the number one cause of biodiversity loss is
    a. excessive hunting of wildlife.
    b. destruction of habitat.
    c. introduction of non-native species.
    d. biological magnification.
    e. UV radiation.

Go to the website or CD-ROM for more self-quiz questions.

The Process of Science

1. Imagine that you have been chosen as the biologist for the design team for a self-contained space station to be assembled in orbit. It will be stocked with organisms you choose to create, an ecosystem that will support you and five other people for two years. Describe the main functions you expect the organisms to perform. List the types of organisms you would select, and explain why you chose them.

2. Investigate the problem of global warming in The Process of Science activity on the website and CD-ROM.

Biology and Society

Some organizations are starting to envision a sustainable society—one in which each generation inherits sufficient natural and economic resources and a relatively stable environment. The Worldwatch Institute, an environmental policy organization, estimates that we must reach sustainability by the year 2030 to avoid economic and environmental collapse. To get there, we must begin shaping a sustainable society during the next ten years or so. In what ways is our current system not sustainable? What might we do to work toward sustainability, and what are the major roadblocks to achieving it? How would your life be different in a sustainable society?

Metric Conversion Table

Measurement	Unit and Abbreviation	Metric Equivalent	Approximate Metric-to-English Conversion Factor	Approximate English-to-Metric Conversion Factor
Length	1 kilometer (km)	$= 1000 \ (10^3)$ meters	1 km = 0.6 mile	1 mile = 1.6 km
	1 meter (m)	$= 100 \ (10^2)$ centimeters	1 m = 1.1 yards	1 yard = 0.9 m
		= 1000 millimeters	1 m = 3.3 feet	1 foot = 0.3 m
			1 m = 39.4 inches	
	1 centimeter (cm)	$= 0.01 \ (10^{-2})$ meter	1 cm = 0.4 inch	1 foot = 30.5 cm
				1 inch = 2.5 cm
	1 millimeter (mm)	$= 0.001 \ (10^{-3})$ meter	1 mm = 0.04 inch	
	1 micrometer (μm)	$= 10^{-6}$ meter $(10^{-3}$ mm$)$		
	1 nanometer (nm)	$= 10^{-9}$ meter $(10^{-3}$ μm$)$		
	1 angstrom (Å)	$= 10^{-10}$ meter $(10^{-4}$ μm$)$		
Area	1 hectare (ha)	= 10,000 square meters	1 ha = 2.5 acres	1 acre = 0.4 ha
	1 square meter (m^2)	= 10,000 square centimeters	$1 \ m^2$ = 1.2 square yards	1 square yard = 0.8 m^2
			$1 \ m^2$ = 10.8 square feet	1 square foot = 0.09 m^2
	1 square centimeter (cm^2)	= 100 square millimeters	$1 \ cm^2$ = 0.16 square inch	1 square inch = 6.5 cm^2
Mass	1 metric ton (t)	= 1000 kilograms	1 t = 1.1 tons	1 ton = 0.91 t
	1 kilogram (kg)	= 1000 grams	1 kg = 2.2 pounds	1 pound = 0.45 kg
	1 gram (g)	= 1000 milligrams	1 g = 0.04 ounce	1 ounce = 28.35 g
			1 g = 15.4 grains	
	1 milligram (mg)	$= 10^{-3}$ gram	1 mg = 0.02 grain	
	1 microgram (μg)	$= 10^{-6}$ gram		
Volume (Solids)	1 cubic meter (m^3)	= 1,000,000 cubic centimeters	$1 \ m^3$ = 1.3 cubic yards	1 cubic yard = 0.8 m^3
			$1 \ m^3$ = 35.3 cubic feet	1 cubic foot = 0.03 m^3
	1 cubic centimeter (cm^3 or cc)	$= 10^{-6}$ cubic meter	$1 \ cm^3$ = 0.06 cubic inch	1 cubic inch = 16.4 cm^3
	1 cubic millimeter (mm^3)	$= 10^{-9}$ cubic meter $(10^{-3}$ cubic centimeter$)$		
Volume (Liquids and Gases)	1 kiloliter (kL or kl)	= 1000 liters	1 kL = 264.2 gallons	1 gallon = 3.79 L
	1 liter (L)	= 1000 milliliters	1 L = 0.26 gallon	1 quart = 0.95 L
			1 L = 1.06 quarts	
	1 milliliter (mL or ml)	$= 10^{-3}$ liter	1 mL = 0.03 fluid ounce	1 quart = 946 mL
		= 1 cubic centimeter	1 mL = approx. $\frac{1}{4}$ teaspoon	1 pint = 473 mL
			1 mL = approx. 15–16 drops	1 fluid ounce = 29.6 mL
				1 teaspoon = approx. 5 mL
Volume (Liquids and Gases)	1 microliter (μl or μL)	$= 10^{-6}$ liter $(10^{-3}$ milliliters$)$		
Time	1 second (s)	$= \frac{1}{60}$ minute		
	1 millisecond (ms)	$= 10^{-3}$ second		
Temperature	Degrees Celsius (°C)		$°F = \frac{9}{5}°C + 32$	$°C = \frac{5}{9}(°F - 32)$

Answers to Self-Quiz Questions

CHAPTER 2

1. d
2. element
3. c
4. b
5. a
6. electron transfer; ionic bond
7. d
8. e
9. e
10. solvent, solute, acidic

CHAPTER 3

1.

2. These digestive enzymes *hydrolyze* the polymers in food.

3. b
4. c
5. b
6. nitrogen
7. e
8. e
9. d
10. a

CHAPTER 4

1. b
2. e
3. c
4. a
5. d
6. d
7. b
8. b
9. a
10. a

CHAPTER 5

1. c
2. c
3. b
4. d
5. b
6. a
7. c
8. d
9. c
10. b

CHAPTER 6

1. b
2. b
3. c
4. a
5. c
6. d
7. c
8. a
9. e
10. e

CHAPTER 7

1. c
2. a
3. b
4. c
5. e
6. c
7. b
8. c
9. b
10. d
11. a. metaphase
    b. anaphase
    c. telophase
    d. prophase

CHAPTER 8

1. c
2. d
3. d
4. e
5. e

More Genetics Problems

1. The parental gametes are WS and ws. Recombinant gametes are Ws and wS, produced by crossing over.

2. Height appears to be a quantitative trait, resulting from polygenic inheritance, like human skin color. See Figure 8.24.

3. The brown allele appears to be dominant, the white allele recessive. The brown parent appears to be homozygous dominant, *BB*, and the white mouse is homozygous recessive, *bb*. The F_1 mice are all heterozygous, *Bb*. If two of the F_1 mice are mated, ¾ of the F_2 mice will be brown.

4. The best way to find out whether an F_2 mouse is homozygous dominant or heterozygous is to do a testcross: Mate the brown mouse with a white mouse. If the brown mouse is homozygous, all the offspring will be brown. If the brown mouse is heterozygous, you would expect half the offspring to be brown and half to be white.

5. Freckles is dominant, so Tim and Jan must both be heterozygous. There is a ¾ chance that they will produce a child with freckles, a ¼ chance they will produce a child without freckles. The probability that the next two children will have freckles is ¾ × ¾ = 9⁄16.

6. Half their children will be heterozygous and have elevated cholesterol levels. There is a ¼ chance that their next child will be homozygous, *hh*, and have an extremely high cholesterol level, like Zoe.

7. The bristle-shape alleles are sex-linked, carried on the X chromosome. Normal bristles is dominant (*F*) and forked is recessive (*f*). The genotype of the female parent is $X^f X^f$. The genotype of the male parent is $X^F Y$. Their female offspring are $X^F X^f$; their male offspring $X^f Y$.

8. ¼ will be boys suffering from hemophilia; ¼ will be female carriers. (The mother is a heterozygous carrier, and the father is normal.)

9. In order for a woman to be color-blind, she must inherit X chromosomes bearing the color blindness allele from both parents. Her father has only one X chromosome, which he passes on to all his daughters, so he must be color-blind. A male only needs to inherit the color blindness allele from a carrier mother; both his parents are usually phenotypically normal.

CHAPTER 9

1. e
2. e
3. b
4. c

5. A gene is the polynucleotide sequence with information for making one polypeptide. Each codon—a triplet of bases in DNA or RNA—codes for one amino acid. Transcription occurs when RNA polymerase produces mRNA using one strand of DNA as a template. A ribosome is the site of translation, or polypeptide synthesis, and tRNA molecules serve as interpreters of the genetic code. Each tRNA molecule has an amino acid attached at one end, and a three-base anticodon at the other end. Beginning at the start codon, mRNA moves relative to the ribosome a codon at a time. A tRNA with a complementary anticodon pairs with each codon, adding its amino acid to the polypeptide chain. The amino acids are linked by peptide bonds. Translation stops at a stop codon, and the finished polypeptide is released. The polypeptide folds to form a functional protein, sometimes in combination with other polypeptides.

CHAPTER 10

1. b
2. c
3. b
4. a
5. b
6. e

7. The nucleus of a differentiated tadpole intestine cell can shape the development of an entire embryo. Salamander cells can redifferentiate and regenerate a lost leg. A carrot plant can grow from a single root cell.

8. A mutation in a single gene can influence the actions of many other genes if the mutated gene is a control gene, such as a homeotic gene. A single control gene may encode a protein that affects (activates or represses) the expression of a number of other genes. In addition, some of the affected genes may themselves be control genes that, in turn, affect other batteries of genes. Cascades of gene expression are common in embryonic development.

CHAPTER 11

1. d
2. e
3. c
4. b
5. c
6. b
7. d

CHAPTER 12

1. b	6. c
2. d	7. b
3. d	8. b
4. c	9. b
5. c	10. d

CHAPTER 13

1. b	6. e
2. d	7. b
3. c	8. d
4. c	9. c
5. d	10. a

CHAPTER 14

1. a	6. c
2. e	7. b
3. d	8. c
4. b	9. d
5. c	10. b

CHAPTER 15

1. b	6. b
2. c	7. a
3. b	8. d
4. b	9. d
5. b	10. a

CHAPTER 16

1. b	6. c
2. a	7. c
3. c	8. b
4. d	9. d
5. b	10. b

CHAPTER 17

1. d	6. c
2. b	7. a
3. e	8. c
4. d	9. b
5. d	10. e

CHAPTER 18

1. e	6. c
2. c	7. c
3. c	8. c
4. a	9. a
5. c	10. c

CHAPTER 19

1. c	6. d
2. e	7. b
3. c	8. d
4. d	9. c
5. b	10. b

CHAPTER 20

1. d	6. a
2. e	7. b
3. e	8. c
4. d	9. c
5. c	10. b

Photographic Credits

Front matter: Pages i, v, and xvi © Jeff Lepore/Photo Researchers, Inc. Page viii, see Chapter 3 opening photos. Page x © Jack Fields/Photo Researchers, Inc. Page xii, © Dr. Tudor Parfitt, University of London. Page xvii from top, © G. I. Bernard/Animals Animals; Rendered by Graham Johnston. Page xviii © M. Schliwa/Visuals Unlimited. Page xix from top, © Volker Steger/Peter Arnold, Inc.; © Tom Brakefield/Planet Earth Pictures. Page xx © Heather Angel/Natural Visions. Page xxi from top, © David M. Phillips/Photo Researchers, Inc.; © Ed Reschke. Page xxii from top, Bettmann Archive; Barry Runk/Grant Heilman Photography, Inc. Page xxiii Courtesy of the Roslin Institute, Edinburgh. Page xxiv © Hank Morgan/Photo Researchers, Inc. Page xxv from top, © Steve Miller/AP/Wide World Photos; Larry Burgess, *Life* Magazine, © Time Inc. Page xxvi from top, © Tui de Roi/Bruce Coleman, Inc.; © Fred Bavendam/Minden Pictures. Page xxvii from top, © Barbara Gerlach/Tom Stack & Associates; © Manfred Kage/Peter Arnold. Page xxviii from top, © Bill Coster/NHPA; © Therisa Stack/Tom Stack & Associates. Page xxix from top, © Ken Lucas/Planet Earth Pictures; © Tom Brakefield/Planet Earth Pictures; © David Lazenby/Planet Earth Pictures. Page xxx from top, © David Hall, 1984/Photo Researchers, Inc.; © Ernest Manewa/Visuals Unlimited; © Frans Lanting/Minden Pictures. Page xxxi from top © Kevin Schafer/Tom Stack & Associates; © Daniel Heuclin/NHPA. Page xxxii from top, © Frans Lanting/Minden Pictures; NASA/Goddard Space Flight Center.

Unit openers: Unit I © Volker Steger/Peter Arnold, Inc. **Unit II** © Andrew Bajer. **Unit III** © Jim Zipp/Ardea London Ltd. **Unit IV** © David Lazenby/Planet Earth Pictures

Chapter opening background photos: Chapter 1 © David Lazenby/Planet Earth Pictures **Chapters 2–6** © Volker Steger/Peter Arnold, Inc. **Chapters 7–11** © Andrew Bajer **Chapters 12–16** © Jim Zipp/Ardea London Ltd. **Chapters 17–20** © David Lazenby/Planet Earth Pictures. **Evolution Link** photo on pages 37, 55, 80, 101, 116, 139, 168, 195, 216, 245, 271, 300, 323, 347, 388, 415, 435, 455, 487 © Tom Bean/CORBIS.

Chapter 1: Chapter opening photos, top to bottom © Jeff Lepore/Photo Researchers, Inc.; © H. Reinhard/Okapia/Photo Researchers, Inc.; N. L. Max, University of California/BPS; © 2000 The Economist Newspaper Group, Inc.; © Pete Seaward/Stone **1.1 counterclockwise from top** © *U.S. News & World Report*; TimePix; Copyright © 2000 by The New York Times Co. Reprinted by permission; © 2000 The Economist Newspaper Group, Inc.; TimePix; From *Newsweek*, September 4, 2000, © 2000 Newsweek, Inc. All rights reserved. Reprinted by permission; Copyright © 2000 *Los Angeles Times*. Reprinted with permission. **1.2, from top** © Tom Van Sant/Geosphere Project, Santa Monica/Science Photo Library/Photo Researchers, Inc.; © CNES/Spot Image Corporation/Photo Researchers, Inc.; © Rafael Macia/Photo Researchers, Inc.; © Jeff Lepore/Photo Researchers, Inc.; Michael Adams and Steve Irving, Dept. of Entomology, UC Riverside; N. L. Max, University of California/BPS **1.6** © Hank Morgan/Photo Researchers, Inc. **1.7** © Charles H. Phillips **1.8 top left** © A.B. Dowsett/Science Source/Photo Researchers, Inc.; **top right** © Ralph Robinson/Visuals Unlimited; **middle left** © D.P. Wilson/Science Source/Photo Researchers, Inc.; **middle left and bottom left and right** © 1998 Photodisc

1.9a © Manfred Kage/Peter Arnold, Inc. **1.9b** David M. Phillips/Visuals Unlimited **1.9c** Courtesy W. L. Dentler, University of Kansas/BPS **1.10** © François Gohier/Photo Researchers, Inc. **1.12** Courtesy of Richard Milner **1.13 left** Larry Burrows, *Life* Magazine © Time Inc.; **right** Courtesy of the National Maritime Museum, London **1.14** © William Paton/NHPA. **1.17** © Jack Wilburn/Earth Scenes/Animals Animals **1.18 top** © Breck P. Kent/Animals Animals; **bottom** © Breck P. Kent/Animals Animals **1.19** Mary DeChirico/Benjamin Cummings **1.22 (both)** © Bernard Roitberg/Simon Fraser University **1.25** Courtesy of NYU/Strongin **1.26** © Peter Menzel

Chapter 2: Chapter opening photos, top to bottom © Rod Planck/Photo Researchers, Inc.; © Anne Dowie; © Michael Newman/PhotoEdit; © G. I. Bernard/Animals Animals; © Harry How/Allsport USA; © James King-Holmes/Science Photo Library/Photo Researchers, Inc. **2.1 top to bottom** © Nigel J. Dennis/NHPA; © Frank Krahmer/Planet Earth Pictures; N. L. Max, University of California/BPS **2.3a** © Ivan Polunin/Bruce Coleman, Inc. **2.3b** © Anne Dowie **2.4** © Anne Dowie **2.5a** © Hank Morgan/Science Source/Photo Researchers, Inc. **2.5b** Courtesy of M. E. Raichle **2.6** © Stuart Isett/Corbis Sygma **2.10** Courtesy of NASA **2.12 (both)** © Dr. E. R. Degginger **2.13** © G. I. Bernard/Animals Animals **2.14** © Harry How/Allsport USA **2.18** © Maresa Pryor/Earth Scenes **2.19** © Jagga Gudlaugnd

Chapter 3: Chapter opening photos, top to bottom © Anne Dowie; Cary Groner; © Alain McGlaughlin; © Dr. E. R. Degginger; N. L. Max, University of California/BPS; © Volker Steger/Peter Arnold, Inc. **3.1** Rendered by Graham Johnson **3.4 (both)** © Anne Dowie **3.8** © Scott Camazine/Photo Researchers, Inc. **3.12** © Anne Dowie **3.13a** © Biophoto Associates/Photo Researchers, Inc. **3.13b** © L. M. Beidler, Florida State University **3.13c** © Biophoto Associates/Photo Researchers, Inc. **3.15** © The Kobal Collection **3.16** © Mike Neveux **Page 47** © Anne Dowie **3.17a** © Bob Torrez/Stone **3.17b** © Garry D. McMichael/Photo Researchers, Inc. **3.17c** © Paul J. Sutton/Duomo **3.17d** © R. M. Motta and S. Correr/Science Photo Library/Photo Researchers, Inc. **3.21 (both)** © Stanley Flegler/Visuals Unlimited **3.28** © Bill Davilla/Retna Ltd.

Chapter 4: Chapter opening photos, top to bottom © Anne Dowie; © Volker Steger/Peter Arnold, Inc.; © Anne Dowie; © R. M. Motta and S. Correr/Science Photo Library/Photo Researchers, Inc.; © Dr. Rodolfo Llinas/Peter Arnold, Inc.; © Fred Felleman/Tony Stone Images **4.1** © M. Schliwa/Visuals Unlimited **4.2a** © M. Abbey/Photo Researchers, Inc. **4.2b** © CNRI/Science Photo Library/Photo Researchers, Inc. **4.2c** © David M. Phillips/Visuals Unlimited **4.4 left** CNRI/Science Photo Library; **right** CNRI/Science Photo Library **4.9** © Barry King/Biological Photo Service **4.11** © Garry Cole/Biological Photo Service **Page 66** © James King-Holmes/Science Photo Library **4.13a** © Roland Birke/Peter Arnold, Inc. **4.13b** W. P. Wergin, courtesy of E. H. Newcomb, University of Wisconsin, Madison **4.15** Courtesy of W. P. Wergin and E. H. Newcomb, University of Wisconsin/Biological Photo Service **4.16** Courtesy of Daniel S. Friend, Harvard Medical School **4.17a** Courtesy of Drs. Frank Solomon and J. Dinsmore, MIT **4.17b** Mark S. Ladinsky and J. Richard McIntosh, University of Colorado **4.17c** Courtesy of Dr. Mary Osborn, Max Planck Institute **4.18a** © Richard Kessel/Visuals Unlimited **4.18b** © 1990 Dennis Kunkel **4.18c** © Eddy Gray/Science Photo Library/Photo Researchers,

Inc. **4.19a (both)** W. L. Dentler, University of Kansas/Biological Photo Service **4.29** © Mike Abbey/Visuals Unlimited **4.31** © David Gonzales **4.32** © Peter Armstrong, University of California, Davis

Chapter 5: Chapter opening photos, top to bottom © Anne Dowie (first three photos); © Paul J. Sutton/Duomo; © Anne Dowie **5.1** © Anne Dowie **5.2** © Tom Brakefield/Planet Earth Pictures **5.4 (all)** Russell Chun and Maureen Kennedy **5.10** © Alain McGlaughlin **5.19** © Davis Barber/PhotoEdit **5.22** © Steve Welsh/Liaison

Chapter 6: Chapter opening photos, top to bottom © Felix Labhardt/Bruce Coleman Ltd.; © Heather Angel/Natural Visions; © Marge Lawson; © Kevin Schafer **6.1** © Staffan Widstrand/Bruce Coleman Ltd. **6.2a** © Renee Lynn/Photo Researchers, Inc. **6.2b** © Bob Evans/Peter Arnold, Inc. **6.2c** © Dwight Kuhn **6.2d** © J. R. Waaland, University of Washington/BPS **6.2e** © Paul Johnson/Biological Photo Service **6.3 middle** © M. Eichelberger/Visuals Unlimited; **bottom** Courtesy of W. P. Wergin and E. H. Newcomb, University of Wisconsin/Biological Photo Service **6.4** © Felix Labhardt/Bruce Coleman Ltd. **6.7** © Heather Angel/Natural Visions **Page 108** © Wally Eberhart/Visuals Unlimited **6.8** © Christine L. Case **6.14a** © C.F. Miescke/Biological Photo Service **6.14b** © Phil Degginger **6.16** © Tom and Pat Leeson

Chapter 7: Chapter opening photos, top to bottom © Andy Walker, Midland Fertility Services/Science Photography Library/Photo Researchers, Inc.; © Science Photo Library/Photo Researchers, Inc.; © David M. Phillips/Photo Researchers, Inc.; © K. G. Murti/Visuals Unlimited; © G. Schatten/Science Photo Library/Photo Researchers, Inc. **7.1** © Andy Walker, Midland Fertility Services/Science Photography Library/Photo Researchers, Inc. **7.2** © Biophoto Associates/Photo Researchers, Inc. **7.3** © Andrew Bajer/University of Oregon **7.4** © Biophoto Associates/Photo Researchers, Inc. **7.7** © Ed Reschke **7.8a** © David M. Phillips/Visuals Unlimited **7.8b** © Carolina Biological Supply/Phototake NYC **7.10** © James King-Holmes/Science Photo Library/Photo Researchers, Inc. **7.11** © Chris Pizzello/AP/Wide World Photos **7.12** © CNRI/SPL/Photo Researchers, Inc. **7.19 left** © CNRI/Science Photo Library/Photo Researchers, Inc.; **right** © Mike Greenlar/The Image Works **7.26** © Dr. Milton Gallardo, Universidad Austral de Chile **Page 142** © Carolina Biological Supply/Phototake

Chapter 8: Chapter opening photos, top to bottom Bettmann Archive; © Barry Runk/Grant Heilman Photography, Inc.; © Bill Longcore/Photo Researchers, Inc.; AKG London **8.1** © Hans Reinhard/Bruce Coleman Photography **8.3** Bettmann Archive **Page 146** © Barry Runk/Grant Heilman Photography, Inc. **8.10** Science Education Resources Pty. Ltd., Victoria, Australia **Page 151** © D. Boone/Corbis **8.13 top left** © Boisvieux/Explorer/Photo Researchers, Inc.; **top right** Anthony Loveday/Benjamin Cummings; **middle left** © Llewellyn/Uniphoto Picture Agency; **middle right** © Eliane Sulle/The Image Bank; **bottom (both)** Anthony Loveday/Benjamin Cummings **8.16** © Dick Zimmerman/Shooting Star **8.18a** © Blair Seitz/Photo Researchers, Inc. **8.18b** © Howard Sochurek **8.22 left** © Dr. Tony Brain/Science Photo Library/Photo Researchers, Inc.; **right** © Bill Longcore/Photo Researchers, Inc. **8.23** © Custom Medical Stock Photos **8.28** From *Thomas Hunt Morgan: The Man and His Science*, Garland Allan (Princeton University Press, 1978). Photo by Dr. Tore Mohr, Fredrikstad, Norway **8.31a** © Jean Claude Revy/Phototake, NYC **8.31b** © Carolina Biological Supply/Phototake, NYC **8.34** © Culver Pictures, Inc. **8.35** © Dr. Tudor Parfitt, University of London **Page 171** © Norma Jubinville

Chapter 9: Chapter opening photos, top to bottom © Manfred Kage/Peter Arnold, Inc.; © Keith V. Wood; Richard Wagner, UCSF Graphics; © CDC/Phototake, NYC **9.2** © Oliver Meckes/E.O.S./Photo Researchers **9.8a** Cold Spring Harbor Archives **9.8b** Photo Researchers, Inc. **9.10c** Richard Wagner, UCSF Graphics **9.12** © Bob Daemmrich/The Image Works **9.15** © Christine Case **9.18** © Keith V. Wood **Page 190** © A. Witte/C. Mahaney/Tony Stone Images **9.29** © Francis Leroy, Biocosmos/Science Photo Library/Photo Researchers, Inc. **9.31** © Holt Studios/Jurgen Dielenschneider/Photo Researchers, Inc.

9.34c © Lennart Nilsson, Boehringer Ingelheim International **9.35a** © CDC/Phototake, NYC **9.35b** © Keith V. Wood/Photo Researchers, Inc.

Chapter 10: Chapter opening photos, top to bottom © Anne Dowie; © Biophoto Associates/Science Source/Photo Researchers; © Jack Dermid/Bruce Coleman Ltd.; © Oliver Meckes/Photo Researchers, Inc.; © Robert Dowling/Corbis; © Ed Reschke **10.1 (both)** © Edward Lewis, California Institute of Technology **10.2 (all)** © Ed Reschke **Page 201** © Anne Dowie **10.8** Courtesy of the Roslin Institute, Edinburgh **10.9b** Courtesy Mergen Ltd. **10.10 top** © A. L. Olins, University of Tennessee/ Biological Photo Service; **bottom** GF Bahr, Armed Forces Institute of Pathology **10.11** © Grant Heilman/Grant Heilman Photography **10.21** Courtesy of University of Washington **Page 215** © Lee Snider/The Image Works **10.22 (both)** © Oliver Meckes/Photo Researchers, Inc.

Chapter 11: Chapter opening photos, top to bottom © Scala/Art Resource; Courtesy Department of Energy, Joint Genome Institute. Photograph by Michael Anthony; © Anne Dowie; © C. Dauguet/Institut Pasteur/Photo Researchers, Inc.; © Arthur C. Smith III/Grant Heilman Photography, Inc. **11.1** © Arthur C. Smith/Grant Heilman Photography, Inc. **11.2** © Nita Winter **11.5** Courtesy of Huntington Potter, University of South Florida and David Dressler, Oxford University; **inset** © S. Cohen/Science Photo Library/Photo Researchers, Inc. **11.14** Courtesy Department of Energy, Joint Genome Institute. Photograph by Michael Anthony **11.19** © Peter Lansdorp, Terry Fox Laboratory, Vancouver, B.C. **11.20** © Associated Press/World Wide Photos **11.21** Courtesy of Virginia Walbot, Stanford University **11.23** Courtesy of PE Biosystems **11.24** © Steve Miller/AP/Wide World Photos **11.25** Courtesy of Cellmark Diagnostics Inc., Germantown, Maryland **Pages 236 & 237** © Hank Morgan/Photo Researchers, Inc. **11.26** © Hank Morgan/Photo Researchers, Inc. **11.28** © Dr. John C. Sanford/Cornell University **11.29** © PPL Therapeutics **Page 242** © Alfred Wolf/Explorer/Science Source/Photo Researchers, Inc. **11.31** © Bill Beatty/Visuals Unlimited **11.32** Courtesy of University of California, San Francisco **11.33** Robin Heyden/Benjamin Cummings **11.34** Courtesy of Dr. Nancy Wexler, Columbia University **11.35** Davis Freeman, University of Washington

Chapter 12: Chapter opening photos, top to bottom © Terry Wild Studio; © Leonard Lessin/Peter Arnold, Inc.; Larry Burrows, *Life* Magazine © Time Inc. **12.1** © James L. Amos/Photo Researchers, Inc. **12.3a** © E. S. Ross, California Academy of Sciences **12.3b** © K. G. Preston Mafham/Animals Animals **12.3c** © P. and W. Ward/Animals Animals **12.5** © Jules Cowan/Bruce Coleman Ltd. **12.6** Philip Gingerich © 1991. Reprinted with permission of *Discover* Magazine **12.7 left** © WorldSat International and J. Knighton/Photo Researchers, Inc.; **middle** © Peter Skinner/Photo Researchers, Inc.; **right** © John Cancalosi/Peter Arnold, Inc. **12.9a** © Dwight Kuhn **12.9b** © Lennart Nilsson/Albert Bonniers Forlag AB, *A Child Is Born*, Dell Publishing, 1990 **12.11a & c** © Tui de Roy /Bruce Coleman, Inc. **12.11b** © Mike Putland/Ardea London Ltd. **12.12** © Adrian Davies/Bruce Coleman, Inc. **Page 260** Photo Courtesy Department of Library Services/American Museum of Natural History **12.14** © Jack Fields/Photo Researchers, Inc. **12.15a** © David Cavagnaro **12.15b** USAF, NOAA/NESDIS at University of Colorado, CIRES/National Snow and Ice Data Center **12.16** © Edmund Brodie, University of Indiana **12.18** © Anne Dowie **12.21** © Gunter Ziesler/Peter Arnold, Inc. **12.22** © Hulton-Deutsch Collection/Corbis **12.23** © 1993 *Time* Magazine **12.24** © Bob and Clara Calhoun/Bruce Coleman **12.26** © Bill Longcore/Photo Researchers, Inc. **12.28** © Megan Moore

Chapter 13: Chapter opening photos, top to bottom © Fred Bavendam/Minden Pictures; © Gary W. Carter/Visuals Unlimited; Auguste Rodin, Le Penseur, private collection/The Bridgeman Art Library; © George Bernard/Science Photo Library/Photo Researchers, Inc. **13.1** © Heather Angel/Natural Visions **13.3** Courtesy of Ernst Mayr, Museum of Comparative Zoology, Harvard University **13.4a (both)** © John Shaw/Tom Stack & Associates **13.4b top left & bottom right** © Photodisc; **top right** © David Young Wolff/Stone; **bottom left**

© Brian Vikander/Corbis **13.6** © Barbara Gerlach/Tom Stack & Associates **13.7 left** © Ralph A. Reinhold/Animals Animals; **middle** © Breck P. Kent/Animals Animals; **right** © E. R. Degginger/Animals Animals **13.9** © Richard Sisk/Panoramic Images; **left inset** © John Shaw/Bruce Coleman, Inc.; **right inset** © Michael Fogden/Bruce Coleman **13.11** Courtesy of the University of Amsterdam **13.15** © Chip Clark **13.16** © Jane Burton/Bruce Coleman **13.17 top** © Tom McHugh/Photo Researchers, Inc.; **bottom** © Michael Newman/PhotoEdit **13.18a** © Georg Gerster/Photo Researchers, Inc. **13.18b** © Margo Crabtree (deceased) **13.18c** © Tom Till **13.18d** © Manfred Kage/Peter Arnold, Inc. **13.18e** © Walter H. Hodge/Peter Arnold, Inc. **13.18f** © Dr. Martin Lockley, University of Colorado **13.18g** Courtesy Dr. David A. Grimald. Photo by Jacklyn Beckett/The American Museum of Natural History, N.Y. **13.18h** © F. Latreille/Cerpolex **13.24** © Gamma Liaison/Hanny Paul

Chapter 14: Chapter opening photos, top to bottom © P. Motta/Science Photo Library/Photo Researchers, Inc.; © Cleve Bryant/PhotoEdit; © USDA/Science Source/Photo Researchers, Inc.; © David M. Frazier/Photo Researchers, Inc.; Frederick P. Mertz/Visuals Unlimited **14.1** © Roland Birke/Peter Arnold **Page 306** Artist: Peter Sawyer © NMNH Smithsonian Inst. **14.3** © Roger Ressmeyer/Corbis **14.6a** © Sidney Fox, University of Miami/BPS **14.6b** Courtesy of F. M. Menger and Kurt Gabrielson, Emory University **14.7a** Stanley Awramik/Biological Photo Service **14.7b** Dr. Tony Brain and David Parker/Science Photo Library/Photo Researchers, Inc. **14.8** Helen E. Carr/BPS **14.9a & b** © David M. Phillips/Visuals Unlimited **14.9c** CNRI/SPL/Photo Researchers Inc. **14.10a** © David M. Phillips/Science Source/Photo Researchers **14.10b** P. W. Johnson and Jon M. Sieburth. University of Rhodes Island University/BPS **14.10c** © Heide Schulz, Max Planck Institute for Marine Microbiology **14.11** © Lee D. Simon/Science Source/Photo Researchers, Inc. **14.12** © H. S. Pankratz, T. C. Beaman/Biological Photo Service **14.13** © Dr. Tony Brain/Science Photo Library/Photo Researchers, Inc. **14.14a** © Larry West/Bruce Coleman, Inc. **14.14b** Centers for Disease Control **14.15** © Martin Bond/Science Photo Library/Photo Researchers, Inc. **14.16** Courtesy of Exxon Corporation **14.18a** © Oliver Meckes/Photo Researchers, Inc. **14.18b** © Peter Parks/Oxford Scientific Films/Animals Animals **14.18c** © Manfred Kage/Peter Arnold, Inc. **14.18d** © Dr. Masamichi Aikawa **14.18e** © M. Abbey/Photo Researchers, Inc. **14.19** © Robert Calentine/Visuals Unlimited **14.20 left** Courtesy of Matt Springer, Stanford University; **right (both)** Courtesy of Robert Kay, MRC Cambridge **14.21a** © Biophoto Associates/Photo Researchers, Inc. **14.21b** © Kent Wood/Photo Researchers, Inc. **14.21c** © Eric Grave/Phototake **14.21d** © Manfred Kage/Peter Arnold **14.22a** © Laurie Campbell/NHPA **14.22b** © Gary Robinson/Visuals Unlimited **14.22c** © W. Lewis Trusty/Animals Animals

Chapter 15: Chapter opening photos, top to bottom National Park Service; NASA; © Larry Lefever/Grant Heilman Photography, Inc.; © Anne Dowie; © G. Prance/Visuals Unlimited **15.1** © Bill Coster/NHPA **15.3** © Dana Richter/Visuals Unlimited **15.4** © David Middleton/NHPA **15.5** © Graham Kent **15.6a** © E. R. Degginger/Photo Researchers, Inc. **15.6b** © L. E. Graham **15.8** Courtesy of J. Shaw, Biology Department, Duke University **15.9** © Dwight Kuhn **15.11** © John Shaw/Tom Stack & Associates; **inset bottom** © Glenn Oliver/Visuals Unlimited; **inset right** © Milton Rand/Tom Stack & Associates **15.12** © The Field Museum **15.13** National Park Service **15.15** © Doug Sokell/Visuals Unlimited; **inset left** © Derrick Ditchburn/Visuals Unlimited; **inset right** © Gerald & Buff Corsi/Visuals Unlimited **15.17a** © François Gohier/Ardea London Ltd. **15.17b** © Lara Hartley **15.20a** © Jane Burton/Bruce Coleman, Inc. **15.20b** © Scott Camazine/Photo Researchers, Inc. **15.20c** © Dwight R. Kuhn **15.21** © D. Wilder **15.22a** © H. Reinhard/Okapia/Photo Researchers, Inc. **15.22b** © Rob Simpson/Visuals Unlimited **15.22c** © G. L. Barron, University of Guelph/Biological Photo Service **15.22d left** M. F. Brown/Visuals Unlimited; **right** Jack Bostrack/Visuals Unlimited **15.22e** N. Allin and G. L. Barron, University of Guelph/Biological Photo Service **15.22f** © J. Forsdyke/Gene Cox/Science Photo Library/Photo Researchers, Inc. **15.23 (both)** Fred Rhoades/Mycena Consulting **15.24a** © Stu-

art Bebb/Oxford Scientific Films/Animals Animals/Earth Scenes **15.24b** © David Cavagnaro/Visuals Unlimited **15.25a** © Robb Walsh **15.25b** © MG3/Vision/ Photo Researchers **15.26** © Christine Case **15.27a** © Dr. Jeremy Burgess/Science Photo Library/Photo Researchers, Inc. **15.27b** © V. Ahmadijian/Visuals Unlimited

Chapter 16: Chapter opening photos, top to bottom © Steve Hopkin/Planet Earth Pictures; © Dwight Kuhn; © Robert Calentine/Visuals Unlimited; © François Gohier/Photo Researchers, Inc.; © C. Allan Morgan **16.1** © Charles & Sandra Hood/Bruce Coleman Ltd. **16.2** © Gunter Ziesler/Peter Arnold, Inc. **16.3** © Anthony Mercieca/Photo Researchers, Inc. **16.9** © Therisa Stack/Tom Stack & Associates **16.11a** © Ken Lucas/Planet Earth Pictures **16.11b** © Claudia Mills/Friday Harbor Labs **16.12** © Chris Huss/The Wildlife Collection **16.15** Centers for Disease Control. Public Domain **16.16** © Stanley Fleger/Visuals Unlimited **16.17a** Reprinted with permission from A. Eizinger and R. Sommer, Max Planck Institut für Entwicklungsbiologie, Tübingen. Copyright 2000, American Association for the Advancement of Science **16.17b** © Andrew Syred/Science Photo Library/Photo Researchers **16.19a** Tony Craddock/Science Photo Library/Photo Researchers, Inc. **16.19b** H. W. Pratt/BPS **16.19c** © Charles R. Wyttenbach/Biological Photo Service **16.21a** © A.N.T./NHPA **16.21b left** © R. DeGoursey/Visuals Unlimited; **right** © Kjell Sandved **16.21c** © Astrid & Hanns-Frieder Michler/Science Photo Library/Photo Researchers, Inc. **16.23a** © Frans Lanting/Minden Pictures **16.23b** © Paul Skelcher/Rainbow **16.23c** © David Scharf **16.24a** © Tom McHugh/Photo Researchers **16.24b** © C. R. Wyttenbach/University of Kansas/BPS **16.25** © Robert and Linda Mitchell **16.27a–e** © John Shaw/Tom Stack & Associates **16.28a** © Gerald Corsi/Visuals Unlimited; **inset** © Gary Milburn/Tom Stack **16.28b** © David Wrobel **16.28c** © Fred Bavendam/Peter Arnold, Inc. **16.29** © Biophoto Associates/Photo Researchers, Inc. **16.30a** © Runk/Schoenberger/Grant Heilman Photography, Inc. **16.30b** © Robert Brons/Biological Photo Service **16.33a** Steinhart Aquarium /Photo Researchers. Tom McHugh/Photo Researchers **16.33b** © J. M. Labat/Jacana/Photo Researchers, Inc. **16.34a & b** © Hans Pfletschinger/Peter Arnold Inc. **16.34c** © Dr. Eckart Pott/Bruce Coleman Ltd. **16.36** © Robert and Linda Mitchell **16.37** The Natural History Museum, London **16.38** © Stephen J. Kraseman/DRK Photo **16.39a** Courtesy of Mervyn Griffiths/CSIRO; **inset** © D. Parer and E. Parer Cook/Auscape **16.39b** © Dan Hadden/Ardea London Ltd. **16.39c** © Mitch Reardon/Photo Researchers, Inc. **16.40** © D. Robert Franz/Planet Earth Pictures **16.41a** © Konrad Wothe/Minden Pictures **16.41b** © Kevin Schafer/NHPA **16.41c** © Mickey Gibson/Animals Animals **16.41d–f** Frans Lanting/Minden Pictures **16.41g** © Tom Brakefield/Planet Earth Pictures **16.41h** © Bill Horsman/Stock, Boston **16.44a** Cleveland Museum of Natural History **16.44b** © John Reader/SPL/Photo Researchers, Inc. **16.44c** © Institute of Human Origins, photo by Donald Johanson **16.46 clockwise from left** © Jean Clottes/Corbis Sygma (three photos); © Chauvet/Le Seuil/Corbis Sygma; © Jean Clottes/Corbis Sygma; © Jean-Marie Chauvet/Corbis Sygma

Chapter 17: Chapter opening photos, top to bottom Courtesy of NASA; © Jean-Paul Ferrero/Auscape; © Carr Clifton; © Bob Abraham/Stock Market **17.1** Courtesy of NASA–LBJ Space Center **17.2a & b** © Raphael Gaillarde/Liaison **17.3** Reprinted with permission from D. W. Schindler, *Science* 184 (1974): 897, Figure 1.49. ©1974 American Association for the Advancement of Science **17.4a** © François Gohier/Photo Researchers, Inc. **17.4b** © Ingrid Van Den Berg/Animals Animals **17.4c** © David Lazenby/Planet Earth Pictures **17.4d** © Nigel J. Dennis/NHPA **17.5** © Erich Hartmann/Magnum **17.6** © Copyright 1995, *Los Angeles Times.* Reprinted with permission **17.7** © Stephen Krasemann/Photo Researchers, Inc. **17.8** © Larry Ultich/DRK Photo **17.9** © David Cavagnaro **17.10 left** © Robert Bower; **right** © Arthur C. Smith III/Grant Heilman Photography **17.19a** © Peter Ward/Bruce Coleman **17.19b** © Jonathan Scott/Planet Earth Pictures **17.19c** © Charlie Ott/Photo Researchers, Inc. **17.19d** © John D. Cunningham/Visuals Unlimited **17.19e** © Frank Oberle **17.19f** © Carr Clifton **17.19g** © Michio Hoshino/Minden

Pictures **17.19h** © J. Warden/Superstock, Inc. **17.20a** © WorldSat International/Science Source/Photo Researchers, Inc. **17.20b** © Williams H. Mullins/Photo Researchers, Inc. **17.20c** © Michael Gadomski/Bruce Coleman, Inc. **17.22** © M. E. Warren/Photo Researchers **17.24** © Ron Sefton/Bruce Coleman, Inc. **17.25** © Emory Kristof/National Geographic Image Collection; **inset** © Robert Hessler/Planet Earth Pictures **17.26** © David Hall, 1984/Photo Researchers, Inc. **17.28** © Sovfoto/Eastfoto **17.29** "Sabata" © 1970 United Artists Corporation/Photofest **17.30 (both)** © Tom McHugh/Photo Researchers, Inc.

Chapter 18: Chapter opening photos, top to bottom © Alain Evrard/Photo Researchers, Inc.; © Pete Seaward/Stone; © John D. Cunningham/Visuals Unlimited; © Henry Ausloos/NHPA **18.1** © Will and Deni McIntyre/Photo Researchers, Inc. **18.2** © Jim Brandenburg/Minden Pictures **18.3a & b** © Frans Lanting/Minden Pictures; **18.4a** Sophie de Wilde/Jacana/Photo Researchers, Inc. **18.4b** © Frans Lanting/Minden Pictures **18.4c** © Will & Deni McIntyre/Photo Researchers, Inc. **18.5 top** © Tim Thompson/Stone; **middle** © Earth Satellite Corporation; **bottom** © Randy Wells/Stone **18.7** © Erwin and Peggy Bauer/Natural Selection **18.10** © Runk/Schoenberger/Grant Heilman **18.11** © Nigel Cattlin/Holt Studios International/Photo Researchers, Inc. **18.12** Alan Carey/Photo Researchers, Inc. **18.13a** © Martin Miller/Visuals Unlimited **18.13b** © G. Prance/Visuals Unlimited **Page 428** © Rob Crandall/The Image Works **18.15** © Arvind Garg **18.17a** © Alain Evrard/Photo Researchers, Inc. **18.17b** © Pete Seaward/Stone **18.19** © François Gohier/Ardea London Ltd. **Table 18.2 left** © François Gohier/Ardea London Ltd; **right** © Ernest Manewa/Visuals Unlimited **18.20** Courtesy of David Reznick

Chapter 19: Chapter opening photos, top to bottom © David Dennis/Animals Animals; © Laurie Campbell/NHPA; © Scott Camazine/Photo Researchers, Inc **19.1** © Richard D. Estes/Photo Researchers, Inc. **19.3** © Henry Ausloos/NHPA **19.5 (both)** © Heather Angel/Biofotos **19.6b** © 1985 Joseph T. Collins/Photo Researchers, Inc. **19.6c** © Kevin de Queiroz, National Museum of Natural History **19.7** Arthur Morris/© VIREO **19.8** © Jeff Foott/Bruce Coleman, Inc. **19.9** © C. Allan Morgan/Peter Arnold, Inc. **19.10**

© Kevin Schafer/Tom Stack & Associates **19.11a** © Lincoln Brower, Sweet Briar College **19.11b** © Peter J. Mayne **19.12a** © E. S. Ross **19.12b** © Runk/Schoenberger/Grant Heilman **19.13** © William E. Townsend/Photo Researchers, Inc. **19.14** © Jack Cameron/ANT Photo Library **19.15** © Robert and Linda Mitchell **19.17 (all)** © Tom Bean **19.18** © John Sohlden/Visuals Unlimited **19.19** Courtesy of Susan Barnett **19.20a** © David Dennis/Animals Animals **19.20b** © Laurie Campbell/NHPA **19.20c** © Marc Dantzker **19.20d** © J. Lubner/Wisconsin Sea Grant; **inset** © Scott Camazine/Photo Researchers, Inc. **19.20e** © Rachel Woodfield, Merkel + Associates **19.21a–c** © Lawrence E. Gilbert/Biological Photo Service

Chapter 20: Chapter opening photos, top to bottom © David Lazenby/Planet Earth Pictures; © Juan Manuel Renjifo/Earth Scenes; © P. G. Adam/Publiphoto/Photo Researchers, Inc.; © Dana White/PhotoEdit; © Daniel Heuclin/NHPA **20.3** © Gregory G. Dimijian/Photo Researchers, Inc. **20.11** © Dana White/PhotoEdit **20.12** © Nigel Tucker/Planet Earth Pictures **20.13a** © John D. Cunningham/Visuals Unlimited **20.13b** Provided by the Northeastern Forest Experiment Station, Forest Service, United States Department of Agriculture **20.15** © P. G. Adam/Publiphoto/Photo Researchers, Inc. **20.17** © D. MacDonald/Animals Animals **20.18a** NASA/Goddard Space Flight Center **20.18c** © Bill Bachman **20.19** Tony Stone Images/Chicago, Inc. **20.20a** © Daniel Heuclin/NHPA **20.20b** © Mark Carwardine/Still Pictures/Peter Arnold, Inc. **20.20c** © Dieter & Mary Plage/Bruce Coleman Inc. **20.21a** © Juan Manuel Renjifo/Earth Scenes **20.21b top** © Wendy Stone/Liaison; **bottom** © Gary Kramer **20.21c** © Richard Vogel/Liaison **20.22** © Richard Shiell/Earth Scenes **20.23 left** © John D. Cunningham/Visuals Unlimited, **right** © Peter Ward/Bruce Coleman **20.24** © Gary Braasch/Woodfin Camp & Associates; **inset** © Greg Vaughn/Tom Stack & Associates **20.25a** © Rob Curtis **20.25b** © Raymond K. Gehman/National Geographic Society **20.25c** © Blanche Haning **20.26a** © David Hosking/Photo Researchers, Inc. **20.26b** © James P. Blair/National Geographic Image Collection **20.27** Courtesy Florida Department of Transportation **20.28b** © Frans Lanting/Minden Pictures **20.29 & 20.30** © Frans Lanting/Minden Pictures

Illustration and Text Credits

The following figures are adapted from Lawrence G. Mitchell, John A. Mutchmor, and Warren D. Dolphin, *Zoology* (Menlo Park, CA: Benjamin/Cummings, 1988) © 1988 The Benjamin/Cummings Publishing Company: **16.2, 16.29**

The following figures are adapted from Gerard J. Tortora, Berdell R. Funke, and Christine L. Case, *Microbiology: An Introduction,* 5th ed. (Menlo Park, CA: Benjamin/Cummings, 1998) © 1998 The Benjamin/Cummings Publishing Company: **9.3, 11.6**

The following figures are adapted from C. K. Mathews and K. E. van Holde, *Biochemistry* (Menlo Park, CA: Benjamin/Cummings, 1996) © 1996 The Benjamin/Cummings Publishing Company: **5.18, 10.10**

The following figures are adapted from W. M. Becker, L. J. Kleinsmith, and J. Hardin, *The World of the Cell* (San Francisco, CA: Benjamin Cummings, 2000) © 2000 The Benjamin Cummings Publishing Company: **4.16, 11.17**

Figure 1.23: Data from Monica H. Mather and Bernard D. Roitberg, "A Sheep in Wolf's Clothing: Tephritid Flies Mimic Spider Predators," *Science*, vol. 236, page 309 (April 17, 1987). Copyright © 1987 American Association for the Advancement of Science.

Figure 1.24: Adapted with permission from Erick Greene, Larry J. Orsak, and Douglas W. Whitman, "A Tephritid Fly Mimics the Territorial Displays of Its Jumping Spider Predators," *Science,* vol. 236, p. 310 (April 17, 1987). Copyright © 1987 American Association for the Advancement of Science.

Table 5.6: Data from C. M. Taylor and G. M. McLeod, *Rose's Laboratory Handbook for Dietetics,* 5th ed. (New York: Macmillan, 1949), p. 18; J. V. G. A. Durnin and R. Passmore, 1967, *Energy and Protein Requirements* in *FAO/WHO Technical Report* No. 522, 1973; W. D. McArdle, F. I. Katch, and V. L. Katch, 1981, *Exercise Physiology* (Philadelphia, PA: Lea & Feibiger, 1981); R. Passmore and J. V. G. A. Durnin, *Physiological Reviews* 35 (1955): 801–840; USDA, McDonald's, Kentucky Fried Chicken, www.cyberdiet.com

Figure 5.12: Adapted from Alberts et al., *Molecular Biology of the Cell,* 2nd ed., fig. 7.17, p. 351 (New York: Garland Publishing, 1989). Copyright © 1989. Reproduced by permission of Taylor & Francis, Inc., http://www.routledge-ny.com

Figure 6.11: From Richard and David Walker, *Energy, Plants and Man,* fig. 4.1, p. 69 (Sheffield: University of Sheffield). Copyright David and Richard Walker. Reprinted by permission.

Figure 7.23: Adapted from F. Vogel and A. G. Motulsky, *Human Genetics* (New York: Springer-Verlag, 1982). Copyright 1982 Springer-Verlag.

Figure 8.14: Adapted from *Everyone Here Spoke Sign Language* by Nora Ellen Groce. Copyright 1985 by Nora Ellen Groce. Reprinted by permission of Harvard University Press.

Figure 8.22: From *Introduction to Genetic Analysis,* 4th ed. by Suzuki, Griffiths, Miller, and Lewontin. Copyright 1976, 1981, 1986, 1989, 1993, 1996 by W. H. Freeman and Company. Used with permission.

Chapter 9, page 195: Text quotation by Joshua Lederberg, from Barbara J. Culliton, "Emerging Viruses, Emerging Threat," *Science*, vol. 247, p. 279 (19 January 1990). Copyright © 1990 American Association for the Advancement of Science.

Table 10.1: Data from *1999 Facts and Figures: Estimated New Cancer Cases and Deaths*, published by the American Cancer Society.

Figure 10.23: Adapted from an illustration by William McGinnis, UCSD.

Chapter 11, page 244: Excerpted with permission from James Watson, *Science*, April 6, 1990. Copyright © 1990 American Association for the Advancement of Science.

Chapter 11, pages 244 & 245: Adapted from an interview with Nancy Wexler in Neil A. Campbell, *Biology*, 2nd ed. (Menlo Park, CA: Benjamin/Cummings, 1990).

Chapter 11, page 245: Text quotation from Leroy Hood. Reprinted with permission from *Science News,* copyright 1989, Science Service, Inc.

Figure 12.27: Adapted from A. C. Allison, "Abnormal Hemoglobin and Erythrovute Enzyme-Deficiency Traits," in *Genetic Variation in Human Populations* by G. A. Harrison, ed. (Oxford: Elsevier Science, 1961).

Table 15.1: Adapted from Randy Moore et al., *Botany,* 2nd ed. Dubuque, IA: Brown, 1998. Table 2.2, p. 37.

Figure 16.45: Adapted from an illustration from Laurie Grace, "The Recent African Genesis of Humans," by A. C. Wilson, R. I. Cann, *Scientific American,* 1992:73.

Table 18.1: Data from F. Black and H. D. Skipper, Jr., *Life Insurance,* 12th ed. (Englewood Cliffs, NJ: Prentice-Hall, 1988).

Table 18.2: Adapted from Pianka, E. R., *Evolutionary Ecology,* 6th ed. (San Francisco, CA: Benjamin Cummings, 2000), p. 186. © 2000 The Benjamin Cummings Publishing Company.

Figure 19.16: Adapted with permission from Edward Farmer, *Science*, vol. 276 (1997): 912. Copyright © 1997 American Association for the Advancement of Science.

Figure 20.13c: From C. E. Likens, et al., "Effects of Forest Cutting and Herbicide Treatment on Nutrient Budgets in the Hubbard Brook Watershed Ecosystem," *Ecological Monographs,* 1966, 50: 22–32. Copyright 1966 The Ecological Society of America. Reprinted by permission.

Figure 20.14: From G. Tyler Miller, *Living in the Environment,* 2nd. ed., p. 87 (Belmont, CA: Wadsworth Publishing, 1979). © 1979 Wadsworth Publishing Company.

Figure 20.17: Data from "Global Warming Trends," by P. D. Jones and T. M. L. Wigley, *Scientific American,* August 1990.

Figure 20.18b: Data from University of Cambridge Centre for Atmospheric Science website http://www.atm.ch.oam.ao.uk/tour/part2.html

Glossary

A

abiotic factor (ā′-bī-ot′-ik) A nonliving component of an ecosystem, such as air, water, or temperature.

abiotic reservoir The part of an ecosystem where a chemical, such as carbon or nitrogen, accumulates or is stockpiled outside of living organisms.

ABO blood groups Genetically determined classes of human blood that are based on the presence or absence of carbohydrates A and B on the surface of red blood cells. The ABO blood group phenotypes, also called blood types, are A, B, AB, and O.

absorption The uptake of small nutrient molecules by an organism's own body.

acclimation (ak′-li-mā′-shun) Long-term but reversible physiological response to an environmental change.

achondroplasia (uh-kon′-druh-plā′-zhuh) A form of human dwarfism caused by a single dominant allele; the homozygous condition is lethal.

acid A substance that increases the hydrogen ion (H⁺) concentration in a solution.

acid precipitation Rain, snow, sleet, hail, drizzle, and so on, with a pH below 5.6; can damage or destroy organisms by acidifying lakes, streams, and possibly land habitats.

acid rain Rain, snow, sleet, hail, drizzle, and so on, with a pH below 5.6; can damage or destroy organisms by acidifying lakes, streams, and possibly land habitats.

activation energy The amount of energy that reactants must absorb before a chemical reaction will start.

activator A protein that switches on a gene or group of genes.

active site The part of an enzyme molecule where a substrate molecule attaches; typically, a pocket or groove on the enzyme's surface.

active transport The movement of a substance across a biological membrane against its concentration gradient, aided by specific transport proteins and requiring the input of energy (often as ATP).

adaptation An inherited characteristic that enhances an organism's ability to survive and reproduce in a particular environment.

adenine (A) (ad′-uh-nēn) A double-ring nitrogenous base found in DNA and RNA.

aerobic (ār-ō′-bik) Containing or requiring oxygen (O₂).

age structure The proportion of individuals in different age-groups in a population.

agnathan (ag-nā′-thun) A vertebrate animal that is superficially fishlike but lacks jaws and paired fins.

AIDS Acquired immune deficiency syndrome; the late stages of HIV infection, characterized by a reduced number of T cells; usually results in death caused by other diseases.

algae (al′-jē) (singular, *alga*) Photosynthetic protists; may be unicellular, colonial, or multicellular, such as seaweeds.

allele (uh-lēl′) An alternative form of a gene.

allopatric speciation (al′-ō-pat′-rik) The formation of a new species as a result of an ancestral population's becoming isolated by a geographic barrier.

alpha helix (al′-fuh hē′-liks) The spiral shape resulting from the coiling of a polypeptide in a protein's secondary structure.

alternation of generations A life cycle in which there is both a multicellular diploid form, the sporophyte, and a multicellular haploid form, the gametophyte; a characteristic of plants and multicellular green algae.

amino acid (uh-mēn′-ō) An organic molecule containing a carboxyl group, an amino group, a hydrogen atom, and a variable side group bonded to a central carbon atom; serves as the monomer of proteins.

amino group In an organic molecule, a functional group consisting of a nitrogen atom bonded to two hydrogen atoms.

amniocentesis (am′-nē-ō-sen-tē′-sis) A technique for diagnosing genetic defects while a fetus is in the uterus. A sample of amniotic fluid, obtained via a needle inserted into the amnion, is analyzed for telltale chemicals and defective fetal cells.

amniotic egg (am′-nē-ot′-ik) A shelled egg in which an embryo develops within a fluid-filled amniotic sac and is nourished by yolk; produced by reptiles, birds, and egg-laying mammals, it enables them to complete their life cycles on dry land.

amoeba (uh-mē′-buh) A type of protist characterized by great flexibility and the presence of pseudopodia.

amoebocyte (uh-mē′-buh-sīt) An amoeba-like cell that moves by pseudopodia; in sponges, they digest and distribute food, transport oxygen, dispose of wastes, and make skeletal fibers.

Amphibia (am-fib′-ē-uh) A class of vertebrate animals that consists of the amphibians, such as frogs, toads, and salamanders.

anabolic steroid (an′-uh-bol′-ik stār′-oid) A synthetic variant of the male hormone testosterone that mimics some of its effects.

anaerobic (an′-ār-ō′-bik) Lacking or not requiring oxygen (O₂).

analogy Anatomical similarity due to convergent evolution.

anaphase The third stage of mitosis, beginning when sister chromatids separate from each other and ending when a complete set of daughter chromosomes have arrived at each of the two poles of the cell.

anchoring junction A junction that connects tissue cells to each other (or to an extracellular matrix) and allows materials to pass from cell to cell.

angiosperm (an′-jē-ō-sperm) A flowering plant, which forms seeds inside a protective chamber called an ovary.

Animalia (an′-uh-mōl′-ē-uh) The kingdom that contains the animals.

Annelida (a-nel′-lid-uh) The phylum that contains the segmented worms, or annelids, characterized by uniform segmentation; includes earthworms, polychaetes, and leeches.

anther A sac in which pollen grains develop, located at the tip of a flower's stamen.

anthropoid (an′-thruh-poid) A member of a primate group made up of the monkeys, apes (gibbons, orangutans, gorillas, and chimpanzees), and humans.

anticodon (an′-tī-kō′-don) On a tRNA molecule, a specific sequence of three nucleotides that is complementary to a codon triplet on mRNA.

antisense nucleic acid (an′-tī-sens) Single-stranded DNA or RNA created by scientists to bind to, and prevent translation of, particular mRNA molecules (for example, mRNA encoding disease-causing proteins).

aphotic zone (ā-fō′-tik) The region of an aquatic ecosystem beneath the photic zone, where light does not penetrate enough for photosynthesis to take place.

apicomplexan (ap′-ē-com-pleks′-un) One of a group of parasitic protozoans, some of which cause human diseases.

aqueous solution (ak′-wē-us) A mixture of two or more substances, one of which is water.

arachnid A member of a major arthropod group that includes spiders, scorpions, ticks, and mites.

Archaea (ar′-kē-uh) One of two prokaryotic domains of life, the other being Bacteria.

Arthropoda (ar-throp′-uh-duh) The most diverse phylum in the animal kingdom; includes arachnids (for example, spiders, ticks, scorpions, and mites), crustaceans (for example, crayfish, lobsters, crabs, barnacles), millipedes, centipedes, and insects. Arthropods are characterized by a chitinous exoskeleton, molting, jointed appendages, and a body formed of distinct groups of segments.

asexual reproduction The creation of offspring by a single parent, without the participation of sperm and egg.

atom The smallest unit of matter that retains the properties of an element.

atomic number The number of protons in each atom of a particular element.

ATP (a-den′-ō-sēn trī-fos′-fāt) Adenosine triphosphate, the main energy source for cells.

ATP synthase A complex (cluster) of several proteins that synthesizes ATP using the energy of hydrogen ions moving through the ATP synthase; found in the inner membrane of mitochondria and the thylakoid membrane of chloroplasts.

autosome A chromosome not directly involved in determining the sex of an organism; in mammals, for example, any chromosome other than X or Y.

autotroph (ot′-ō-trōf) An organism that makes all its own organic matter from inorganic nutrients, using an energy source such as the sun. Plants, algae, and photosynthetic bacteria are autotrophs.

Aves (ā′-vēz) A class of vertebrate animals that consists of the birds.

B

bacillus (buh-sil′-us) (plural, *bacilli*) A rod-shaped prokaryotic cell.

Bacteria One of two prokaryotic domains of life, the other being Archaea.

bacterial chromosome The single, circular DNA molecule found in bacteria.

bacteriophage (bak-tēr′-ē-ō-fāj) A virus that infects bacteria; also called a phage.

basal body (bā′-sul) A eukaryotic cell organelle consisting of nine sets of microtubule triplets; may organize the microtubule assembly of a cilium or flagellum; structurally identical to a centriole.

base A substance that decreases the hydrogen ion (H^+) concentration in a solution.

Batesian mimicry (bāt′-zē-un mi′-muh-krē) A type of mimicry in which a species that a predator can eat looks like a species that is poisonous or otherwise harmful to the predator.

behavioral isolation A type of pre-zygotic barrier between species; two species remain isolated because individuals of neither species are sexually attracted to individuals of the other species.

benign tumor An abnormal mass of cells that remains at its original site in the body.

benthic zone A seafloor, or the bottom of a freshwater lake, pond, river, or stream.

benthos (ben′-thōz) The communities of organisms living in the benthic zone of an aquatic biome.

bilateral symmetry An arrangement of body parts such that an organism can be divided equally by a single cut passing longitudinally through it. A bilaterally symmetrical organism has mirror-image right and left sides.

binary fission A means of asexual reproduction in which a parent organism, often a single cell, divides into two individuals of about equal size.

binomial A two-part, latinized name of a species; for example, *Homo sapiens*.

biodiversity All of the variety of life; usually refers to the variety of species that make up a community; concerns both species richness (the total number of different species) and the relative abundance of the different species.

biodiversity crisis The current rapid decline in the variety of life on Earth, largely due to the effects of human culture.

biodiversity hot spot A small geographic area with an exceptional concentration of species, especially endemic species (those found nowhere else).

biogenesis The principle that all life arises by the reproduction of pre-existing life.

biogeochemical cycle Any of the various chemical circuits occurring in an ecosystem, involving both biotic and abiotic components of the ecosystem.

biogeography The study of the geographic distribution of species.

biological community All the organisms living together and potentially interacting in a particular area.

biological magnification The accumulation of persistent chemicals in the living tissues of consumers in food chains.

biological species concept The definition of a species as a population or group of populations whose members have the potential in nature to interbreed and produce fertile offspring.

biomass The amount, or mass, of organic material in an ecosystem.

biome (bī′-ōm) A major type of ecosystem that covers a large geographic region.

bioremediation The use of living organisms to detoxify (remove pollutants from) water, air, and soil.

biosphere The global ecosystem; that portion of Earth that is alive; all of life and where it lives.

biotechnology The use of living organisms (often microbes) to perform useful tasks; today, usually involves DNA technology.

biotic factor (bī-ot′-ik) A living component (organism) of a biological community.

bivalve A member of a group of mollusks that includes clams, mussels, scallops, and oysters.

blastula (blas′-tyū-luh) An embryonic stage that marks the end of cleavage during animal development; a hollow ball of cells in many species.

body cavity A fluid-filled space between the digestive tract and the body wall.

bony fish A member of the vertebrate class Osteichthyes; for example, trout and goldfish.

bottleneck effect Genetic drift resulting from a drastic reduction in population size.

brown algae One of several groups of marine, multicellular, autotrophic protists; the most common and largest type of seaweed; includes the kelps.

bryophyte (brī′-uh-fīt) One of a group of plants that lack vascular tissue. Bryophytes include mosses and their close relatives.

buffer A chemical substance that resists changes in pH by accepting hydrogen ions from a solution (when there is an excess of H^+) or donating hydrogen ions to a solution (when H^+ is depleted).

C

calorie The amount of energy that raises the temperature of 1 gram of water by 1°C.

Calvin cycle The second of two stages of photosynthesis; a cyclic series of chemical reactions that occur in the stroma of a chloroplast, using the carbon in CO_2 and the ATP and NADPH produced by the light reactions to make the energy-rich sugar molecule G3P.

cancer cell A cell that is not subject to normal cell cycle control mechanisms and that divides continuously, invading other parts of the body and often killing the organism if unchecked.

cap Extra nucleotides added to the beginning of an RNA transcript in the nucleus of a eukaryotic cell.

capsule A sticky layer that surrounds the cell walls of some bacteria, protecting the cell surface and sometimes helping to glue the cell to surfaces.

carbohydrate A class of biological molecules that includes simple single-monomer sugars (monosaccharides), two-monomer sugars (disaccharides), and polymers made of long chains of sugar units (polysaccharides).

carbon skeleton The chain of carbon atoms that forms the structural backbone of an organic molecule.

carbonyl group (kar′-buh-nēl′) In an organic molecule, a functional group consisting of a carbon atom linked by a double bond to an oxygen atom.

carboxyl group (kar-bok′-sil) In an organic molecule, a functional group consisting of an oxygen atom double-bonded to a carbon atom that is also bonded to a hydroxyl group.

carboxylic acid (kar′-bok-sil′-ik) An organic compound containing a carboxyl group.

carcinogen (kar-sin′-uh-jin) A cancer-causing agent, either high-energy radiation (such as X-rays or UV light) or a chemical.

carcinoma (kar′-si-nō′-muh) Cancer that originates in one of the coverings of the body, such as the skin or the lining of the intestinal tract.

carnivore An animal that eats other animals.

carpel (kar′-pul) The female part of a flower, consisting of a stalk with an ovary at the base and a stigma, which traps pollen, at the tip.

carrier An individual who is heterozygous for a recessively inherited disorder and who therefore does not show symptoms of that disorder.

carrying capacity The number of individuals in a population that an environment can sustain.

cartilaginous fish (kar′-ti-laj′-uh-nus) A member of the class Chondrichthyes; for example, sharks and rays.

cell The fundamental structural unit of life; a basic unit of living matter separated from its environment by a plasma membrane.

cell cycle An orderly sequence of events (including interphase and the mitotic phase) from the time a cell first arises from cell division until it itself divides.

cell division The reproduction of a cell.

cell junction A structure that connects cells within a tissue to one another.

cell plate A membranous disk in the middle of a dividing plant cell; the site where the new cell wall forms during cytokinesis.

cell theory The theory that all living things are composed of cells and all cells come from other cells.

cellular differentiation The specialization in the structure and function of cells that occurs during the development of an organism; results from selective activation and deactivation of the cells' genes.

cellular respiration The aerobic harvesting of energy from food molecules; the energy-releasing chemical breakdown of food molecules, such as glucose, and the storage of potential energy as ATP, a form that cells can use to perform work; involves glycolysis, the Krebs cycle, and electron transport chains.

cellular slime mold A type of protist that has unicellular amoeboid cells during its feeding stage but also functions as a multicellular colony.

cellulose A large polysaccharide composed of many glucose monomers linked into cable-like fibrils that provide structural support in plant cell walls.

cell wall A protective layer external to the plasma membrane in plant cells, bacteria, fungi, and some protists; protects the cell and helps maintain its shape.

centipede A carnivorous terrestrial arthropod that has one pair of long legs for each of its numerous body segments, with the front pair modified as poison claws.

central vacuole (vak′-ū-ōl) A membrane-enclosed sac occupying most of the interior of a mature plant cell, having diverse roles in reproduction, growth, and development.

centriole (sen′-trē-ōl) A structure in an animal cell composed of nine sets of microtubule triplets. An animal cell usually has a pair of centrioles within each of its centrosomes.

centromere (sen′-trō-mēr) The region of a chromosome where two sister chromatids are joined and where spindle microtubules attach during mitosis and meiosis. The centromere divides at the onset of anaphase during mitosis and anaphase II of meiosis.

centrosome (sen′-trō-sōm) A cloud of cytoplasmic material in a eukaryotic cell that gives rise to microtubules; important in mitosis and meiosis. An animal cell usually has a pair of centrioles within each of its centrosomes.

cephalopod (sef′-uh-luh-pod) A member of a group of mollusks that includes squids and octopuses.

chaparral (shap'-uh-ral') A biome dominated by spiny evergreen shrubs adapted to periodic drought and fires; found where cold ocean currents circulate off-shore, creating mild, rainy winters and long, hot, dry summers.

charophyte (kār'-ō-fīt) A member of a group of green algae closely related to plants. Species of charophytes living today probably represent the ancestors of the plant kingdom.

chemical bond An attraction between two atoms resulting from a sharing of outer-shell electrons or the presence of opposite charges on the atoms. The bonded atoms gain complete outer electron shells.

chemical cycling The use and reuse of chemical elements such as carbon within an ecosystem.

chemical energy Energy stored in the chemical bonds of molecules; a form of potential energy.

chemical reaction A process leading to chemical changes in matter; involves the making and/or breaking of chemical bonds.

chemoautotroph (kē'-mō-ot'-ō-trōf) An organism that obtains both energy and carbon from inorganic chemicals; makes its own organic compounds from CO_2 without using light energy.

chemoheterotroph (kē'-mō-het'-er-ō-trōf) An organism that obtains energy and carbon from organic molecules.

chemotherapy (kē'-mo-thār'-uh-pē) Treatment for cancer in which drugs are administered to disrupt cell division of the cancer cells.

chiasma (plural, *chiasmata*) (kī-az'-muh, kī-az'-muh-tuh) The microscopically visible site where crossing over has occurred between chromatids of homologous chromosomes during prophase I of meiosis.

chlorophyll *a* (klor'-ō-fil ā) A green pigment in chloroplasts that participates directly in the light reactions.

chloroplast (klor'-ō-plast) An organelle found in plants and photosynthetic protists. Enclosed by two concentric membranes, a chloroplast absorbs sunlight and uses it to power the synthesis of organic food molecules (sugars).

choanocyte (kō-an'-uh-sīt) A flagellated cell that traps bacteria in mucus, then engulfs the food by phagocytosis.

Chondrichthyes (kon-drik'-thēz) A class of cartilaginous fishes that includes sharks and rays.

Chordata (kor-dā'-tuh) The phylum of the chordates; characterized by a dorsal, hollow nerve cord, a notochord, gill structures, and a post-anal tail; includes lancelets, tunicates, and vertebrates.

chorionic villus sampling (CVS) (kor'-ē-on'-ik) A technique for diagnosing genetic defects while the fetus is in the uterus. A small sample of the fetal portion of the placenta is removed and analyzed.

chromatin (krō'-muh-tin) The combination of DNA and proteins that constitutes eukaryotic chromosomes; often refers to the diffuse, very extended form taken by the chromosomes when a eukaryotic cell is not dividing.

chromosome (krō'-muh-sōm) A gene-carrying, thread-like structure found in the nucleus of a eukaryotic cell and most visible during mitosis and meiosis; also, the main gene-carrying structure of a prokaryotic cell. Chromosomes consist of chromatin.

chromosome theory of inheritance A basic principle in biology stating that genes are located on chromosomes and that the behavior of chromosomes during meiosis and fertilization accounts for inheritance patterns.

ciliate (sil'-ē-it) A type of protozoan that moves by means of cilia.

cilium (plural, *cilia*) (sil'-ē-um, sil'-ē-uh) A short appendage that propels some protists through water and moves fluids across the surface of many tissue cells in animals. In common with eukaryotic flagella, cilia have a 9 + 2 arrangement of microtubules covered by the cell's plasma membrane.

cladistic analysis (kluh-dis'-tik) The scientific search for clades, taxonomic groups composed of an ancestor and all its descendants.

class In classification, the taxonomic category above order.

cleavage furrow The first sign of cytokinesis during cell division in an animal cell; an indentation at the equator of the cell.

clone As a verb, to produce genetically identical copies of a cell, organism, or DNA molecule. As a noun, the collection of cells, organisms, or molecules resulting from cloning; also (colloquially), a single organism that is genetically identical to another because it arose from the cloning of a somatic cell.

clumped Describing a dispersion pattern in which individuals are aggregated in patches.

Cnidaria (nī-dār'-ē-uh) The phylum that contains the hydras, jellies, sea anemones, corals, and related animals characterized by cnidocytes, radial symmetry, a gastrovascular cavity, polyps, and medusae.

cnidocyte (nī'-duh-sīt) A specialized cell for which the phylum Cnidaria is named; consists of a capsule containing a fine coiled thread, which, when discharged, functions in defense and prey capture.

coccus (plural, *cocci*) (kok'-us, kok'-sē) A spherical prokaryotic cell.

codominance The expression of two different alleles of a gene in a heterozygote.

codon (kō'-don) A three-nucleotide sequence in mRNA that specifies a particular amino acid or polypeptide termination signal; the basic unit of the genetic code.

coelom (sē'-lōm) A body cavity completely lined with mesoderm.

coevolution The reciprocal evolutionary influences between two species.

cohesion The attraction between molecules of the same kind.

communicating junction A channel between adjacent tissue cells through which water and other small molecules pass freely.

community All the organisms living together and potentially interacting in a particular area.

community ecology The study of how interactions between species affect community structure and organization.

comparative anatomy The comparison of body structures between different species.

comparative embryology (em'-brē-ol'-ō-jē) The comparison of structures that appear during the development of different organisms.

competitive exclusion principle The concept that populations of two species cannot coexist in a community if their niches are nearly identical. Using resources more efficiently and having a reproductive advantage, one of the populations will eventually outcompete and eliminate the other.

complete digestive tract A digestive tube with two openings, a mouth and an anus.

compound A substance containing two or more elements in a fixed ratio; for example, table salt (NaCl) consists of one atom of the element sodium (Na) for every atom of chlorine (Cl).

coniferous forest (kō-nif'-er-us) A biome characterized by conifers, cone-bearing evergreen trees.

conjugation The union (mating) of two bacterial cells or protist cells and the transfer of DNA between the two cells.

conservation biology The science of species preservation; the scientific study of ways to counter the loss of biodiversity.

conservation of energy The principle that energy can neither be created nor destroyed.

consumer An organism that obtains its food by eating plants or by eating animals that have eaten plants.

convergent evolution Adaptive change resulting in non-homologous (analogous) similarities among organisms. Species from different evolutionary lineages come to resemble each other (evolve analogous structures) as a result of living in very similar environments.

covalent bond (kō-vā'-lent) An attraction between atoms that share one or more pairs of outer-shell electrons; symbolized by a single line between the atoms.

creatine phosphate (krē'-uh-tin) A compound in muscle cells that provides energy.

crista (plural, *cristae*) (kris'-tuh, kris'-tē) A fold of the inner membrane of a mitochondrion. Enzyme molecules embedded in cristae make ATP.

cross The cross-fertilization of two different varieties of an organism or of two different species; also called hybridization.

cross-fertilization The fusion of sperm and egg derived from two different individuals.

crossing over The exchange of corresponding segments between homologous chromosomes during prophase I of meiosis; also, the exchange of segments between DNA molecules in prokaryotes.

crustacean (krus-tā'-shē-un) A member of a major arthropod group that includes lobsters, crayfish, crabs, shrimps, and barnacles.

cryptic coloration A type of camouflage that makes potential prey difficult to spot against its background.

culture The transmission of accumulated knowledge over generations.

cuticle (1) In animals, a tough, nonliving outer layer of the skin. (2) In plants, a waxy coating on the surface of stems and leaves that helps retain water.

cyanobacteria (sī-an'-ō-bak-tēr'-ē-uh) Photosynthetic, oxygen-producing bacteria.

cystic fibrosis (sis'-tik fī-brō'-sis) A genetic disease that occurs in people with two copies of a certain recessive allele; characterized by an excessive secretion of mucus from the lungs and other organs; fatal if untreated.

cytokinesis (sī'-tō-ki-nē-sis) The division of the cytoplasm to form two separate daughter cells. Cytokinesis usually occurs during telophase of mitosis, and the two processes make up the mitotic (M) phase of the cell cycle.

cytoplasm (sī'-tō-plaz'-um) Everything inside a cell between the plasma membrane and the nucleus; consists of a semifluid medium and organelles.

cytosine (C) (sī'-tuh-sin) A single-ring nitrogenous base found in DNA and RNA.

cytoskeleton A network of fine fibers that provides structural support for a eukaryotic cell.

cytosol (sī'-tuh-sol) The semifluid medium of a cell's cytoplasm.

D

Darwinian fitness The contribution an individual makes to the gene pool of the next generation relative to the contributions of other individuals in the population.

Darwinian medicine The study of health problems in an evolutionary context.

decomposer An organism that derives its energy from organic wastes and dead organisms; also called a detritivore.

decomposition The breakdown of organic materials into inorganic ones.

dehydration synthesis A chemical process in which a polymer forms as monomers are linked by the removal of water molecules. One molecule of water is removed for each pair of monomers linked.

deletion The loss of one or more nucleotides from a gene by mutation; the loss of a fragment of a chromosome.

denaturation (dē-nā′-chur-ā′-shun) A process in which a protein unravels, losing its specific shape and hence function; can be caused by changes in pH or salt concentration or high temperature; also refers to the separation of the two strands of the DNA double helix, caused by similar factors.

density-dependent factor A population-limiting factor whose effects intensify as the population increases in size.

density-independent factor A population-limiting factor whose occurrence and effects are not affected by population density.

deoxyribonucleic acid (DNA) (dē-ok′-sē-rī′-bo-nū-klā′-ik) The genetic material that organisms inherit from their parents; a double-stranded helical macromolecule consisting of nucleotide monomers with deoxyribose sugar and the nitrogenous bases adenine (A), cytosine (C), guanine (G), and thymine (T). *See also* gene.

derived characters Homologous features that have changed from a primitive (ancestral) condition and that are unique to an evolutionary lineage; features found in members of a lineage but not found in ancestors of the lineage.

desert A biome characterized by sparse rainfall (less than 30 centimeters per year).

detritivore (de-trī′-tuh-vōr) An organism that derives its energy from organic wastes and dead organisms; also called a decomposer.

detritus (de-trī′-tus) Nonliving organic matter, including animal wastes, plant litter, and dead organisms.

diatom (dī′-uh-tom) A unicellular photosynthetic alga with a unique glassy cell wall containing silica.

diffusion The tendency of molecules of any kind to move from where they are more concentrated to where they are less concentrated.

dihybrid cross (dī′-hī′-brid) An experimental mating of individuals in which the inheritance of two traits is tracked.

dinoflagellate (dī′-nō-flaj′-uh-let) A unicellular photosynthetic alga with two flagella situated in perpendicular grooves in cellulose plates covering the cell.

diploid (dip′-loid) Containing two homologous sets of chromosomes in each cell, one set inherited from each parent; referring to a 2*n* cell.

directional selection Natural selection that acts in favor of the individuals at one end of a phenotypic range.

disaccharide (dī-sak′-uh-rīd) A sugar molecule consisting of two monosaccharides linked by dehydration synthesis.

dispersion pattern The manner in which individuals in a population are spaced within their area. Three types of dispersion patterns are clumped (individuals are aggregated in patches), uniform (individuals are evenly distributed), and random (unpredictable distribution).

disturbance A force that changes a biological community and usually removes organisms from it. Disturbances, such as fire and storms, play pivotal roles in structuring many biological communities.

diversifying selection Natural selection that favors extreme over intermediate phenotypes.

DNA Deoxyribonucleic acid; the genetic material that organisms inherit from their parents; a double-stranded helical macromolecule consisting of nucleotide monomers with deoxyribose sugar and the nitrogenous bases adenine (A), cytosine (C), guanine (G), and thymine (T). *See also* gene.

DNA fingerprint An individual's unique collection of DNA restriction fragments, detected by electrophoresis and nucleic acid probes.

DNA ligase (lī′-gās) An enzyme, essential for DNA replication, that catalyzes the covalent bonding of adjacent DNA nucleotides; used in genetic engineering to paste a specific piece of DNA containing a gene of interest into a bacterial plasmid or other vector.

DNA polymerase (puh-lim′-er-ās) An enzyme that assembles DNA nucleotides into polynucleotides using a preexisting strand of DNA as a template.

DNA technology Methods used to study and/or manipulate DNA, including recombinant DNA technology. Recombinant DNA technology uses techniques for combining genes from different sources into a single DNA molecule and then transferring the new molecule into cells, where it can be replicated and expressed; also known as genetic engineering.

domain A taxonomic category above the kingdom level; the three domains of life are Archaea, Bacteria, and Eukarya.

dominant allele In a heterozygote, the allele that determines the phenotype with respect to a particular gene.

dorsal Pertaining to the back of a bilaterally symmetrical animal.

double bond A type of covalent bond in which two atoms share two pairs of electrons; symbolized by a pair of lines between the bonded atoms.

double fertilization In flowering plants, the formation of both a zygote and a cell with a triploid nucleus, which develops into the endosperm.

double helix The two complementary strands of DNA that form the characteristic helical (spiral) structure.

Down syndrome A human genetic disorder resulting from the presence of an extra chromosome 21 (trisomy 21) and characterized by heart and respiratory defects and varying degrees of mental retardation.

Duchenne muscular dystrophy (duh-shen′, dis′-truh-fē) A human genetic disease caused by a sex-linked recessive allele; characterized by progressive weakening and a loss of muscle tissue.

duplication Repetition of part of a chromosome resulting from fusion with a fragment from a homologous chromosome; can result from an error in meiosis or from mutagenesis.

dynein arm (dī-nin) A protein extension from a microtubule doublet in a cilium or flagellum; involved in energy conversions that drive the bending of cilia and flagella.

E

Echinodermata (ē-kī′-nō-der-ma′-tuh) The phylum of echinoderms, including sea stars, sea urchins, and sand dollars; characterized by a rough or spiny skin, a water vascular system, an endoskeleton, and radial symmetry in adults.

ecological niche A population's role in its community; the sum total of a species' use of the biotic and abiotic resources of its habitat.

ecological succession The process of biological community change resulting from disturbance; transition in the species composition of a biological community, often following a flood, fire, or volcanic eruption. *See also* primary succession; secondary succession.

ecological time The time scale for the present interactions between organisms and their environments.

ecology The scientific study of how organisms interact with their environments.

ecosystem All the organisms in a given area, along with the nonliving (abiotic) factors with which they interact; a biological community and its physical environment.

ecosystem ecology The study of energy flow and the cycling of chemicals among the various biotic and abiotic factors in an ecosystem.

ectoderm (ek′-tō-derm) The outer layer of three embryonic cell layers in a gastrula; forms the skin of the gastrula and gives rise to the epidermis and nervous system in the adult.

ectotherm (ek′-tō-therm) An animal that warms itself mainly by absorbing heat from its surroundings.

electromagnetic energy Solar energy, or radiation, which travels in space as rhythmic waves and also behaves as discrete packets of energy called photons.

electromagnetic spectrum The full range of radiation, from the very short wavelengths of gamma rays to the very long wavelengths of radio signals.

electron A subatomic particle with a single negative electrical charge. One or more electrons move around the nucleus of an atom.

electron microscope (EM) An instrument that focuses an electron beam through a specimen or onto its surface. An electron microscope achieves a thousand times more resolving power than a light microscope; the most powerful EM can distinguish objects as small as 0.2 nanometer.

electron transport chain A series of electron-carrier molecules that shuttle electrons during the redox reactions that release energy used to make ATP; located in the inner membrane of mitochondria, the thylakoid membrane of chloroplasts, and the plasma membrane of prokaryotes.

electrophoresis (ē-lek′-trō-for-ē′-sis) Gel electrophoresis; a technique for separating and purifying macromolecules. A mixture of molecules is placed on a gel between a positively charged electrode and a negatively charged one; negative charges on the molecules are attracted to the positive electrode, and the molecules migrate toward that electrode; the molecules separate in the gel according to their rates of migration.

element A substance that cannot be broken down to other substances by chemical means. Scientists recognize 92 chemical elements occurring in nature.

embryo sac (em′-brē-ō) The female gametophyte contained in the ovule of a flowering plant.

endangered species A species in danger of extinction throughout all or a significant portion of its range.

endemic species (en-dem′-ik) A species of organism whose distribution is limited to a specific geographic area.

endocytosis (en′-dō-sī-tō′-sis) The movement of materials into the cytoplasm of a cell via membranous vesicles or vacuoles.

endoderm (en′-dō-derm) The innermost of three embryonic cell layers in a gastrula; forms the primitive gut in the gastrula and gives rise to the innermost linings of the digestive tract and other hollow organs in the adult.

endomembrane system A network of membranous organelles that partition the cytoplasm of eukaryotic cells into functional compartments. Some of the organelles are structurally connected to each other, whereas others are structurally separate but functionally connected by the traffic of membranous vesicles between them.

endoplasmic reticulum (ER) (en′-dō-plaz′-mik re-tik′-ū-lum) An extensive membranous network in a eukaryotic cell, continuous with the outer nuclear membrane and composed of ribosome-studded (rough) and ribosome-free (smooth) regions. *See also* rough ER; smooth ER.

endoskeleton A hard skeleton located within the soft tissues of an animal; includes the skeletons of some sponges, the hard plates of echinoderms, and the cartilage and bony skeletons of many vertebrates.

endosperm In flowering plants, a nutrient-rich mass formed by the union of a sperm cell with two polar nuclei during double fertilization; provides nourishment to the developing embryo in the seed.

endospore A thick-coated, protective cell produced within a bacterial cell exposed to harsh conditions.

endosymbiosis (en′-dō-sim′-bē-ō′-sis) A process by which the mitochondria and chloroplasts of eukaryotic cells probably evolved from symbiotic associations between small prokaryotic cells living inside larger ones.

endotherm An animal that derives most of its body heat from its own metabolism.

endotoxin A poisonous component of the cell walls of certain bacteria.

energy The capacity to perform work, or to move matter in a direction in which it would not move if left alone.

energy coupling In cellular metabolism, the transfer of energy from processes that yield energy to processes that consume energy.

energy flow The passage of energy through the components of an ecosystem.

energy pyramid A diagram depicting the cumulative loss of energy from a food chain.

enhancer A eukaryotic DNA sequence that helps stimulate the transcription of a gene at some distance from it. An enhancer functions by means of a transcription factor called an activator, which binds to it and then to the rest of the transcription apparatus. *See also* silencer.

entomology (en′-tuh-mol′-ō-jē) The branch of biology that specializes in the study of insects.

entropy (en′-truh-pē) A measure of disorder, or randomness. One form of disorder is heat, which is random molecular motion.

enzyme (en′-zīm) A protein that serves as a biological catalyst, changing the rate of a chemical reaction without itself being changed in the process.

equilibrial life history (ē′-kwi-lib′-rē-ul) Often seen in larger-bodied species, the pattern of reproducing when mature and producing few offspring but caring for the young.

estuary (es′-tyū-ār′-ē) An area where fresh water merges with seawater.

Eukarya (ū-kār′-ē-uh) The domain of eukaryotes, organisms made of eukaryotic cells; includes all of the protists, plants, fungi, and animals.

eukaryotic cell (ū′-kār-ē-ot′-ik) A type of cell that has a membrane-enclosed nucleus and other membrane-enclosed organelles. All organisms except bacteria and archaea are composed of eukaryotic cells.

eutherian (ū-thēr′-ē-um) A placental mammal; a mammal whose young complete their embryonic development within the uterus, joined to the mother by the placenta.

eutrophication (ū′-truh-fi-kā′-shun) An increase in productivity of an aquatic ecosystem.

evaporative cooling Surface cooling that results when a substance evaporates. This occurs because the "hottest" molecules vaporize first.

evolution Genetic change in a population or species over generations; all the changes that transform life on Earth; the heritable changes that have produced Earth's diversity of organisms.

evolutionary adaptation An inherited characteristic that enhances an organism's ability to survive and reproduce in a particular environment.

evolutionary time The time scale of evolution.

exaptation (ek′-sap-tā-shun) A structure that has evolved in one context and later becomes adapted for a different function in a different context.

exocytosis (ek′-sō-sī-tō′-sis) The movement of materials out of the cytoplasm of a cell via membranous vesicles or vacuoles.

exon (ek′-son) In eukaryotes, a coding portion of a gene. *See also* intron.

exoskeleton A hard, external skeleton that protects an animal and provides points of attachment for muscles.

exotoxin A poisonous protein secreted by bacterial cells.

exponential growth model A mathematical description of idealized, unregulated population growth.

extracellular matrix A sticky coat secreted by most animal cells.

F

facultative anaerobe (fak′-ul-tā′-tiv an′-uh-rōb) A microorganism that makes ATP by aerobic respiration if oxygen is present, but that switches to fermentation when oxygen is absent.

family In classification, the taxonomic category above genus.

fat A large lipid molecule made from an alcohol called glycerol and three fatty acids; a triglyceride. Most fats function as energy-storage molecules.

feedback regulation A method of metabolic control in which the end product of a metabolic pathway acts as an inhibitor of an enzyme within that pathway.

fermentation The anaerobic harvest of food by some cells.

fern One of a group of seedless vascular plants.

fertilization The union of a sperm cell with an egg cell, producing a zygote.

fetoscopy (fē-tos′-kuh-pē) A technique for examining a fetus for anatomical deformities. A needle-thin tube containing a viewing scope is inserted into the uterus, giving a direct view of the fetus.

F₁ generation The offspring of two parental (P generation) individuals; F_1 stands for first filial.

F₂ generation The offspring of the F_1 generation; F_2 stands for second filial.

fitness The contribution an individual makes to the gene pool of the next generation relative to the contributions of other individuals in the population.

five-kingdom system A system of taxonomic classification based on the five kingdoms Monera, Plantae, Fungi, Animalia, and Protista.

flagellate (flaj′-uh-let) A protist (protozoan) that moves by means of one or more flagella.

flagellum (plural, *flagella*) (fluh-jel′-um) A long appendage that propels protists through water and moves fluids across the surface of many tissue cells in animals. A cell may have one or more flagella. Like cilia, flagella have a 9 + 2 arrangement of microtubules covered by the cell's plasma membrane.

flatworm A member of the phylum Platyhelminthes.

flower In angiosperms, a short stem with four sets of modified leaves (sepals, petals, stamens, and carpels) that bear structures that function in sexual reproduction.

fluid mosaic A description of membrane structure, depicting a cellular membrane as a mosaic of diverse protein molecules embedded in a fluid bilayer made of phospholipid molecules.

fluke One of a group of parasitic flatworms.

food chain The sequence of food transfer from producers through several levels of consumers in an ecosystem.

food vacuole (vak′-ū-ōl) The simplest type of digestive cavity, found in single-celled organisms.

food web A network of interconnecting food chains.

foram A marine protozoan that secretes a shell and extends pseudopodia through pores in its shell.

fossil A preserved remnant or impression of an organism that lived in the past.

fossil fuel An energy deposit formed from the remains of extinct organisms; includes coal, oil, and natural gas.

fossil record The chronicle of evolution over millions of years of geological time engraved in the order in which fossils appear in rock strata.

founder effect Random change in the gene pool that occurs in a small colony of a population.

fruit The ripened, thickened ovary of a flower, which protects dormant seeds and aids in their dispersal.

functional group The atoms that form the chemically reactive part of an organic molecule.

Fungi (fun′-jē) The kingdom that contains the fungi.

fungus (plural, *fungi*) A heterotrophic eukaryote that digests its food externally and absorbs the resulting small nutrient molecules. Most fungi consist of a netlike mass of filaments called hyphae. Molds, mushrooms, and yeasts are examples of fungi.

G

gametangium (plural, *gametangia*) (gam′-uh-tan′-jē-um) A reproductive organ that houses and protects the gametes of a plant.

gamete (gam′-ēt) A sex cell; a haploid egg or sperm. The union of two gametes of opposite sex (fertilization) produces a zygote.

gametic isolation (guh-mē′-tik) A type of pre-zygotic barrier between species; the species remain isolated because male and female gametes of the different species cannot fuse, or they die before they unite.

gametophyte (guh-mē′-tō-fīt′) The multicellular haploid form in the life cycle of organisms undergoing alternation of generations; mitotically produces haploid gametes that unite and grow into the sporophyte generation.

gastropod A member of a group of mollusks that includes snails and slugs.

gastrovascular cavity A digestive compartment with a single opening that functions as both mouth and anus.

gastrula (gas′-trū-luh) The embryonic stage resulting from gastrulation in animal development. Most animals have a gastrula made up of three layers of cells: ectoderm, endoderm, and mesoderm.

gastrulation (gas′-trū-lā′-shun) The phase of embryonic development that transforms the blastula into a gastrula. Gastrulation adds more cells to the embryo and sorts the cells into distinct cell layers.

gel electrophoresis (jel ē-lek′-trō-for-ē′-sis) A technique for separating and purifying macromolecules. A mixture of molecules is placed on a gel between a positively charged electrode and a negatively charged one; negative charges on the molecules are attracted to the positive electrode, and the molecules migrate toward that electrode; the molecules separate in the gel according to their rates of migration.

gene A discrete unit of hereditary information consisting of a specific nucleotide sequence in DNA (or RNA, in some viruses). Most of the genes of a eukaryote are located in its chromosomal DNA; a few are carried by the DNA of mitochondria and chloroplasts.

gene cloning The production of multiple copies of a gene.

gene expression The process whereby genetic information flows from genes to proteins; the flow of genetic information from the genotype to the phenotype.

gene flow The gain or loss of alleles in a population by the movement of individuals or gametes into or out of the population.

gene pool All the genes in a population at any one time.

genetically modified (GM) organism An organism that has acquired one or more genes by artificial means; also known as a transgenic organism.

genetic code The set of rules relating nucleotide sequence to amino acid sequence.

genetic drift A change in the gene pool of a small population due to chance.

genetic marker A segment of chromosomal DNA whose inheritance can be tracked in a genetic study.

genetic recombination The production, by crossing over and/or independent assortment of chromosomes during meiosis, of offspring with allele combinations different from those in the parents. The term may also be used more specifically to mean the production by crossing over of eukaryotic or prokaryotic chromosomes with gene combinations different from those in the original chromosomes.

genetics The science of heredity.

genome (jē′-nōm) A complete (haploid) set of an organism's genes; an organism's genetic material.

genomic library (juh-nō′-mik) A set of DNA segments representing the entire genome of an organism. Each segment is usually carried by a plasmid or phage.

genotype (jē′-nō-tīp) The genetic makeup of an organism.

genus (plural, *genera*) (jē′-nus) In classification, the taxonomic category above species; the first part of a species' binomial; for example, *Homo*.

geological time scale A time scale devised by geologists that reflects a consistent sequence of geological periods.

germination The beginning of growth.

glycogen (glī′-kuh-jen) A complex, extensively branched polysaccharide consisting of many glucose monomers; serves as an energy-storage molecule in liver and muscle cells.

glycolysis (glī-kol′-uh-sis) The multi-step chemical breakdown of a molecule of glucose into two molecules of pyruvic acid; the first stage of cellular respiration in all organisms; occurs in the cytoplasmic fluid.

Golgi apparatus (gol′-jē) An organelle in eukaryotic cells consisting of stacks of membranous sacs that modify, store, and ship products of the endoplasmic reticulum.

granum (plural, *grana*) (gran′-um) A stack of disk-like membranous sacs called thylakoids in a chloroplast. Grana are the sites where light energy is trapped by chlorophyll and converted to chemical energy during the light reactions of photosynthesis.

green algae Photosynthetic protists that include unicellular, colonial, and multicellular species with grass-green chloroplasts; closely related to true plants.

greenhouse effect The warming of the atmosphere caused by CO_2 and other gases that absorb infrared radiation and slow its escape from Earth's surface.

growth factor A protein secreted by certain body cells that stimulates other cells to divide.

guanine (G) (gwan′-ēn) A double-ring nitrogenous base found in DNA and RNA.

gymnosperm (jim′-nō-sperm) A naked-seed plant. Its seed is said to be naked because it is not enclosed in a fruit.

H

habitat A place where an organism lives; an environmental situation in which an organism lives.

habitat isolation A type of pre-zygotic barrier between species. The species remain isolated because they breed in different habitats.

haploid Containing a single set of chromosomes; referring to an *n* cell.

Hardy-Weinberg equilibrium The condition describing a nonevolving population (one that is in genetic equilibrium).

Hardy-Weinberg formula A formula for calculating the frequencies of genotypes in a gene pool from the frequencies of alleles, and vice versa.

heat The amount of energy associated with the movement of the atoms and molecules in a body of matter.

hemophilia (hē-mō-fil′-ē-uh) A human genetic disease caused by a sex-linked recessive allele; characterized by excessive bleeding following injury.

herbivore An animal that eats plants, algae, or autotrophic bacteria.

heterotroph (het′-er-ō-trōf′) An organism that cannot make its own organic food molecules and must obtain them by consuming other organisms or their organic products; a consumer or decomposer in a food chain.

heterozygous (het′-er-ō-zī′-gus) Having two different alleles for a given gene.

histone A small basic protein molecule associated with DNA and important in DNA packing in the eukaryotic chromosome.

HIV Human immunodeficiency virus, the retrovirus that attacks the human immune system and causes AIDS.

homeobox (hō′-mē-ō-boks′) A 180-nucleotide sequence within a homeotic gene encoding the part of the protein that binds to the DNA of the genes regulated by the protein.

homeotic gene (hō′-mē-ot′-ik) A master control gene that regulates batteries of other genes that actually create the anatomical identity of parts of a developing organism.

hominid (hom′-uh-nid) A member of the family Hominidae, including *Homo sapiens* and our ancestors.

homologous chromosomes (hō-mol′-uh-gus) The two chromosomes that make up a matched pair in a diploid cell. Homologous chromosomes are of the same length, centromere position, and staining pattern and possess genes for the same characteristics at corresponding loci. One homologous chromosome is inherited from the organism's father, the other from the mother.

homologous structures Structures that are similar in different species of common ancestry.

homology (hō-mol′-uh-jē) Anatomical similarity due to common ancestry.

homozygous (hō′-mō-zī′-gus) Having two identical alleles for a given gene.

host The larger participant in a symbiotic relationship, serving as home and feeding ground to the symbiont.

Human Genome Project An international collaborative effort to map and sequence the DNA of the entire human genome.

Huntington's disease A human genetic disease caused by a dominant allele; characterized by uncontrollable body movements and degeneration of the nervous system; usually fatal 10 to 20 years after the onset of symptoms.

hybrid The offspring of parents of two different species or of two different varieties of one species; the offspring of two parents that differ in one or more inherited traits; an individual that is heterozygous for one or more pairs of genes.

hybrid inviability A type of post-zygotic barrier between species; the species remain isolated because hybrid zygotes do not develop or hybrids do not become sexually mature.

hybridization The cross-fertilization of two different varieties of an organism or of two different species; also called a cross.

hybrid sterility A type of post-zygotic barrier between species; the species remain isolated because hybrids fail to produce functional gametes.

hydrocarbon An organic molecule composed only of the elements carbon and hydrogen.

hydrogen bond A type of weak chemical bond formed when the partially positive hydrogen atom in one molecule is attracted to the partially negative atom of a neighboring molecule.

hydrolysis (hī-drol′-uh-sis) A chemical process in which macromolecules are broken down by the chemical addition of water molecules to the bonds linking their monomers; an essential part of digestion.

hydrophilic (hī′-drō-fil′-ik) "Water-loving"; pertaining to polar or charged molecules (or parts of molecules) that are soluble in water.

hydrophobic (hī'-drō-fō'-bik) "Water-fearing"; pertaining to nonpolar molecules (or parts of molecules) that do not dissolve in water.

hydrothermal vent community A seafloor community powered by chemical energy from Earth's interior rather than by sunlight.

hydroxyl group (hī-drok'-sul) In an organic molecule, a functional group consisting of a hydrogen atom bonded to an oxygen atom.

hypercholesterolemia An inherited human disease characterized by an excessively high level of cholesterol in the blood.

hypertonic In comparing two solutions, referring to the one with the greater concentration of solutes.

hypha (plural, *hyphae*) (hī'-fuh, hī'-fē) One of many filaments making up the body of a fungus.

hypothesis (plural, *hypotheses*) A tentative explanation a scientist proposes for a specific phenomenon that has been observed.

hypotonic In comparing two solutions, referring to the one with the lower concentration of solutes.

I

incomplete dominance A type of inheritance in which the phenotype of a heterozygote *(Aa)* is intermediate between the phenotypes of the two homozygotes *(AA* and *aa).*

induced fit The interaction between a substrate molecule and the active site of an enzyme, which changes shape slightly to embrace the substrate and catalyze the reaction.

ingestion The act of eating; the first main stage of food processing.

in group In a cladistic study of evolutionary relationships among taxa of organisms, the group of taxa that is actually being analyzed. *See also* out-group.

insect An arthropod that usually has three body segments (head, thorax, and abdomen), three pairs of legs, and one or two pairs of wings.

intermediate disturbance hypothesis The idea that species diversity is greatest in places where disturbance is moderate in both severity and frequency.

intermediate filament An intermediate-sized protein fiber that is one of the three main kinds of fibers making up the cytoskeleton of a eukaryotic cell; a ropelike rod made of fibrous proteins.

interphase The period in the eukaryotic cell cycle when the cell is not actually dividing. *See* mitosis.

interspecific competition The competition between populations of two or more species that require similar limited resources.

interspecific interaction Any interaction between members of different species.

intertidal zone A shallow zone where the waters of an estuary or ocean meet land.

intraspecific competition Competition between individuals of the same species for a limited resource.

intrinsic rate of increase An organism's inherent capacity to reproduce in an ideal environment.

introduced species A species that humans move from the species' native location to a new geographic region; sometimes called an exotic species.

intron (in'-tron) In eukaryotes, a nonexpressed (noncoding) portion of a gene that is excised from the RNA transcript. *See also* exon.

inversion A change in a chromosome resulting from reattachment of a chromosome fragment to the original chromosome, but in the reverse direction. Mutagens and errors during meiosis can cause inversions.

invertebrate An animal that lacks a backbone.

ion An atom or molecule that has gained or lost one or more electrons, thus acquiring an electrical charge.

ionic bond (ī-on'-ik) An attraction between two ions with opposite electrical charges. The electrical attraction of the opposite charges holds the ions together.

isomers (ī'-suh-merz) Organic compounds with the same molecular formula but different structures and therefore different properties.

isotonic (ī'-sō-ton'-ik) Having the same solute concentration as another solution.

isotope (ī'-suh-tōp) A variant form of an atom. Isotopes of an element have the same number of protons but different numbers of neutrons.

K

karyotype (kār'-ē-uh-tīp) A display of micrographs of the metaphase chromosomes of a cell, arranged by size and centromere position.

kelp A giant brown alga, up to 100 meters long, that forms extensive undersea forests.

keystone predator A predator species that reduces the density of the strongest competitors in a community, thereby helping maintain species diversity.

kilocalorie (kcal) A quantity of heat equal to 1000 calories. Used to measure the energy content of food, it is usually called a "calorie."

kinetic energy (kuh-net'-ik) The energy of motion.

kingdom In classification, the broad taxonomic category above phylum.

Krebs cycle The metabolic cycle that is fueled by the acetyl CoA formed after glycolysis in cellular respiration. Chemical reactions in the Krebs cycle complete the metabolic breakdown of glucose molecules to carbon dioxide. The Krebs cycle occurs in the mitochondria and supplies most of the NADH molecules that carry energy to the electron transport chains.

L

lancelet One of a group of invertebrate chordates.

landscape ecology The application of ecological principles to the study of land-use patterns; the scientific study of the biodiversity of interacting ecosystems.

larva (plural, *larvae*) A sexually immature form of an animal that is structurally and often ecologically very different from the adult.

lateral line system A row of sensory organs along each side of a fish's body. Sensitive to changes in water pressure, it enables a fish to detect minor vibrations in the water.

LDL Low-density lipoprotein; a cholesterol-carrying particle in the blood, made up of cholesterol and other lipids surrounded by a single layer of phospholipids in which proteins are embedded.

leukemia (lū-kē'-mē-uh) A type of cancer of the blood-forming tissues, characterized by an excessive production of white blood cells and an abnormally high number of them in the blood; cancer of the bone marrow cells that produce leukocytes.

lichen (lī'-ken) A mutualistic association between a fungus and an alga or between a fungus and a cyanobacterium.

life cycle The entire sequence of stages in the life of an organism, from the adults of one generation to the adults of the next.

life history The traits that affect an organism's schedule of reproduction and death.

life table A listing of survivals and deaths in a population in a particular time period and predictions of how long, on average, an individual of a given age will live.

light microscope (LM) An optical instrument with lenses that bend visible light to magnify images and project them into a viewer's eye or onto photographic film.

light reactions The first of two stages in photosynthesis; the steps in which solar energy is absorbed and converted to chemical energy in the form of ATP and NADPH. The light reactions power the sugar-producing Calvin cycle but produce no sugar themselves.

lignin (lig'-nin) A chemical that hardens the cell walls of plants.

linkage map A map of a chromosome showing the relative positions of genes.

linked genes Genes located close enough together on a chromosome to be usually inherited together.

lipid A biological molecule that does not mix with water.

lobe-finned fish A bony fish with strong, muscular fins supported by bones. Lobe-fins are extinct except for one species.

locus (plural, *loci*) The particular site where a gene is found on a chromosome. Homologous chromosomes have corresponding gene loci.

logistic growth model A mathematical description of idealized population growth that is restricted by limiting factors.

low-density lipoprotein (LDL) A cholesterol-carrying particle in the blood, made up of cholesterol and other lipids surrounded by a single layer of phospholipids in which proteins are embedded.

lungfish A bony fish with lungs that comes to the surface of the water to gulp air.

Lyme disease A debilitating human disease caused by a spirochete bacterium; characterized at first by a red rash at the site of the tick bite and, if not treated, by heart disease, arthritis, and nervous disorders.

lymphoma (lim-fō'-muh) Cancer of the tissues that form white blood cells.

lysogenic cycle (lī'-sō-jen'-ik) A type of bacteriophage replication cycle in which the viral genome is incorporated into the bacterial host chromosome as a prophage. New phages are not produced, and the host cell is not killed or lysed unless the viral genome leaves the host chromosome.

lysosomal storage disease (lī'-suh-sō'-mul) A hereditary disorder associated with abnormal lysosomes, where the sufferer is missing one of the lysosomal digestive enzymes.

lysosome (lī'-suh-sōm) A digestive organelle in eukaryotic cells; contains enzymes that digest the cell's food and wastes.

lytic cycle (lit'-ik) A type of viral replication cycle resulting in the release of new viruses by lysis (breaking open) of the host cell.

M

macroevolution The main events in the evolutionary history of life on Earth.

macromolecule A giant molecule in a living organism: a protein, polysaccharide, or nucleic acid.

magnification An increase in the apparent size of an object.

malignant tumor An abnormal tissue mass that can spread into neighboring tissue and to other parts of the body; a cancerous tumor.

Mammalia (muh-mā′-lē-uh) The vertebrate class of mammals, characterized by body hair and mammary glands that produce milk to nourish the young.

mantle In mollusks, the outgrowth of the body surface that drapes over the animal. The mantle produces the shell and forms the mantle cavity.

mark-recapture method A sampling technique used to estimate wildlife populations.

marsupial (mar-sū′-pē-ul) A pouched mammal, such as a kangaroo, opossum, or koala. Marsupials give birth to embryonic offspring that complete development while housed in a pouch and attached to nipples on the mother's abdomen.

mass number The sum of the number of protons and neutrons in an atom's nucleus.

matter Anything that occupies space and has mass.

mechanical isolation A type of pre-zygotic barrier between species; the species remain isolated because structural differences between them prevent fertilization.

medusa (plural, *medusae*) (muh-dū′-suh, muh-dū′-sē) One of two types of cnidarian body forms; an umbrella-like body form. Jellies (jellyfish) have a medusa body form.

meiosis (mī-ō′-sis) In a sexually reproducing organism, the division of a single diploid nucleus into four haploid daughter nuclei. Meiosis and cytokinesis produce haploid gametes from diploid cells in the reproductive organs of the parents.

mesoderm (mez′-ō-derm) The middle layer of the three embryonic cell layers in a gastrula; gives rise to muscles, bones, the dermis of the skin, and most other organs in the adult.

mesophyll (mez′-ō-fil) The green tissue in the interior of a leaf; the main site of photosynthesis.

messenger RNA (**mRNA**) The type of ribonucleic acid that encodes genetic information from DNA and conveys it to ribosomes, where the information is translated into amino acid sequences.

metabolism (muh-tab′-uh-liz-um) The many chemical reactions that occur in organisms.

metamorphosis (met′-uh-mor′-fuh-sis) The transformation of a larva into an adult.

metaphase (met′-uh-fāz) The second stage of mitosis, during which all the cell's duplicated chromosomes are lined up at an imaginary plane equidistant between the poles of the mitotic spindle.

metastasis (muh-tas′-tuh-sis) The spread of cancer cells beyond their original site.

microclimate Climate on a small scale.

microevolution A change in a population's gene pool over a succession of generations; evolutionary changes in species over relatively brief periods of geological time.

microfilament The thinnest of the three main kinds of protein fibers making up the cytoskeleton of a eukaryotic cell; a solid, helical rod composed of the globular protein actin. Microfilaments help some cells change shape and move by assembling at one end while disassembling at the other.

microtubule The thickest of the three main kinds of fibers making up the cytoskeleton of a eukaryotic cell; a straight, hollow tube made of globular proteins called tubulins. Microtubules form the basis of the structure and movement of cilia and flagella.

millipede A terrestrial arthropod that has two pairs of short legs for each of its numerous body segments and that eats decaying plant matter.

mitochondrion (plural, *mitochondria*) (mī-tō-kon′-drē-on) An organelle in eukaryotic cells where cellular respiration occurs. Enclosed by two concentric membranes, it is where most of the cell's ATP is made.

mitosis (mī-tō′-sis) The division of a single nucleus into two genetically identical daughter nuclei. Mitosis and cytokinesis make up the mitotic (M) phase of the cell cycle.

mitotic phase (mī-tot′-ik) The part of the cell cycle when the cell is actually dividing.

mitotic spindle (mī-tot′-ik) A spindle-shaped structure formed of microtubules and associated proteins that is involved in the movements of chromosomes during mitosis and meiosis. (A spindle is shaped roughly like a football.)

modern synthesis A comprehensive theory of evolution that incorporates genetics and includes most of Darwin's ideas, focusing on populations as the fundamental units of evolution.

molecular biology The study of the molecular basis of genes and gene expression; molecular genetics.

molecule A group of two or more atoms held together by covalent bonds.

Mollusca (mol-lus′-kuh) The phylum that contains the mollusks; characterized by a muscular foot, mantle, mantle cavity, and radula; includes gastropods (snails and slugs), bivalves (clams, oysters, and scallops), and cephalopods (squids and octopuses).

molting In arthropods, the process of shedding an old exoskeleton and secreting a new, larger one.

monogenesis model (mon′-ō-jen′-uh-sis) The hypothesis that humans arose from a single archaic group in Africa.

monohybrid cross An experimental mating of individuals in which the inheritance of a single characteristic is tracked.

monomer (mon′-uh-mer) A chemical subunit that serves as a building block of a polymer.

monosaccharide (mon′-ō-sak′-uh-rīd) The smallest kind of sugar molecule; a single-unit sugar. Monosaccharides are the building blocks of more complex sugars and polysaccharides.

monotreme (mon′-uh-trēm) An egg-laying mammal, such as the duck-billed platypus.

moss One of a group of seedless nonvascular plants.

movement corridor A series of small clumps or a narrow strip of quality habitat (usable by organisms) that connects otherwise isolated patches of quality habitat.

mRNA Messenger RNA; the type of ribonucleic acid that encodes genetic information from DNA and conveys it to ribosomes, where the information is translated into amino acid sequences.

Müllerian mimicry (myū-lār′-ē-un) A mutual mimicry by two species, both of which are poisonous or otherwise harmful to a predator.

multiregional model The hypothesis that humans evolved simultaneously in different parts of the world.

mutagen (myū′-tuh-jen) A chemical or physical agent that interacts with DNA and causes a mutation.

mutagenesis (myū′-tuh-jen′-uh-sis) The creation of a mutation.

mutation A change in the nucleotide sequence of DNA; the ultimate source of genetic diversity.

mutualism A symbiotic relationship in which both partners benefit.

mycelium (plural, *mycelia*) (mī-sē′-lē-um) The densely branched network of hyphae in a fungus.

mycorrhiza (plural, *mycorrhizae*) (mī′-kō-rī′-zuh) A symbiotic association between plant roots and fungi.

N

NADH A molecule that carries electrons from glucose and other fuel molecules and deposits them at the top of an electron transport chain. NADH is generated during glycolysis and the Krebs cycle.

NADPH An electron carrier involved in photosynthesis. Light drives electrons from chlorophyll to $NADP^+$, forming NADPH, which provides the high-energy electrons for the reduction of carbon dioxide to sugar in the Calvin cycle.

natural selection The idea that a population of organisms can change over the generations if individuals having certain heritable traits leave more offspring than other individuals, resulting in a change in the population's genetic composition over time.

Nematoda (nē′-muh-tō′-duh) The phylum that contains the roundworms, or nematodes; characterized by a pseudocoelom, a complete digestive tract, and a cylindrical, wormlike body form.

nerve cord An elongated bundle of axons and dendrites, usually extending longitudinally from the brain or anterior ganglia. One or more nerve cords and the brain make up the central nervous system in many animals.

neutron A subatomic particle that is electrically neutral (has no electrical charge); found in the nucleus of an atom.

niche (nich) A population's role in its community; the sum total of a population's use of the biotic and abiotic resources of its habitat.

nitrogen fixation The conversion of atmospheric nitrogen (N_2) into nitrogen compounds (NH_4^+, NO_3^-) that plants can absorb and use.

nitrogenous base (nī-troj′-uh-nus) An organic molecule that is a base and that contains the element nitrogen; DNA and RNA contain nitrogenous bases.

nondisjunction An accident of meiosis or mitosis in which a pair of homologous chromosomes or a pair of sister chromatids fail to separate at anaphase.

notochord (nō′-tuh-kord) A flexible, longitudinal rod located between the digestive tract and nerve cord in chordate animals; present only in embryos in many species.

nuclear envelope A double membrane, perforated with pores, that encloses the nucleus and separates it from the rest of the eukaryotic cell.

nucleic acid (nū-klā′-ik) A polymer consisting of many nucleotide monomers; serves as a blueprint for proteins and, through the actions of proteins, for all cellular structures and activities. The two types of nucleic acids are DNA and RNA.

nucleic acid probe (nū-klā′-ik) In DNA technology, a labeled single-stranded nucleic acid molecule used to find a specific gene or other nucleotide sequence within a mass of DNA. The probe hydrogen-bonds to the complementary sequence in the targeted DNA.

nucleoid region (nū′-klē-oid) The region in a prokaryotic cell consisting of a concentrated mass of DNA.

nucleolus (nū-klē′-ō-lus) A structure within the nucleus of a eukaryotic cell where ribosomes are made.

nucleosome (nū′-klē-ō-sōm) The beadlike unit of DNA packaging in a eukaryotic cell; consists of DNA wound around a protein core made up of eight histone molecules.

nucleotide (nū′-klē-ō-tīd) An organic monomer consisting of a five-carbon sugar covalently bonded to a nitrogenous base and a phosphate group. Nucleotides are the building blocks of nucleic acids.

nucleus (plural, *nuclei*) (1) An atom's central core, containing protons and neutrons. (2) The genetic control center of a eukaryotic cell.

O

obligate aerobe (ob′-li-get ār′-ōb) An organism that cannot survive without oxygen (O_2).

obligate anaerobe (ob′-li-get an′-uh-rōb) An organism that cannot survive in the presence of oxygen (O_2).

ocean current One of the riverlike flow patterns in the oceans.

old-growth forest An ancient forest that has never been seriously disturbed by humans. Trees dominating an old-growth forest may be thousands of years old.

omnivore An animal that eats both plants and animals.

oncogene (on′-kō-jēn) A cancer-causing gene; usually contributes to malignancy by abnormally enhancing the amount or activity of a growth factor made by the cell.

operator In prokaryotic DNA, a sequence of nucleotides near the start of an operon to which an active repressor can attach. The binding of a repressor prevents RNA polymerase from attaching to the promoter and transcribing the genes of the operon.

operculum (plural, *opercula*) (ō-per′-kyū-lum) A protective flap on each side of a fish's head that covers a chamber housing the gills.

operon (op′-er-on) A unit of genetic regulation common in prokaryotes; a cluster of genes with related functions, along with the promoter and operator that control their transcription.

opportunistic life history Often seen in small-bodied species, the pattern of reproducing when young and producing many offspring.

order In classification, the taxonomic category above family.

organ A structure consisting of several tissues adapted as a group to perform specific functions.

organelle (or-guh-nel′) A structure with a specialized function within a cell.

organic chemistry The study of carbon compounds.

organic compound A chemical compound containing the element carbon and usually synthesized by cells.

organism An individual living thing, such as a bacterium, fungus, protist, plant, or animal.

organismal ecology The study of the evolutionary adaptations that enable individual organisms to meet the challenges posed by their abiotic environments.

organ system A group of organs that work together in performing vital body functions.

osmoregulation The control of water and solute balance in an organism.

osmosis (oz-mō′-sis) The passive transport of water across a selectively permeable membrane.

Osteichthyes (os-tē-ik′-thēz) The vertebrate class of bony fishes; for example, trout and goldfish.

out-group In a cladistic study of evolutionary relationships among taxa of organisms, a taxon or group of taxa with a known relationship to, but not a member of, the taxa being studied. *See also* in-group.

ovary (1) In animals, the female gonad, which produces egg cells and reproductive hormones. (2) In flowering plants, the basal portion of a carpel in which the egg-containing ovules develop.

ovule (ō′-vyūl) A reproductive structure in a seed plant; contains the female gametophyte and the developing egg. An ovule develops into a seed.

ovum (plural, *ova*) (ō′-vum) An unfertilized egg, or female gamete.

oxidation The loss of electrons from a substance involved in a redox reaction; always accompanies reduction.

ozone layer The layer of O_3 in the upper atmosphere that protects life on Earth from the harmful ultraviolet rays in sunlight.

P

paedomorphosis (pēd′-uh-mor′-fuh-sis) The retention of juvenile body features in an adult.

paleoanthropology The study of human evolution.

parasite An organism that benefits at the expense of another organism, which is harmed in the process.

parasitism (pār′-uh-si-tiz-um) A symbiotic relationship in which the symbiont (parasite) benefits at the expense of the host by living either within the host or on its surface and deriving its food from the host.

passive transport The diffusion of a substance across a biological membrane, not requiring any input of energy.

pathogen A disease-causing organism.

pedigree A family tree representing the occurrence of heritable traits in parents and offspring across a number of generations.

pelagic zone (puh-laj′-ik) The open ocean.

peptide bond The covalent linkage between two amino acid units in a polypeptide; formed by dehydration synthesis.

petal A modified leaf of a flowering plant. Petals are the often colorful parts of a flower that advertise it to insects and other pollinators.

P generation The parent individuals from which offspring are derived in studies of inheritance; P stands for parental.

phage (fāj) Bacteriophage; a virus that infects bacteria.

phagocytosis (fag′-ō-sī-tō′-sis) "Cellular eating"; a type of endocytosis whereby a cell engulfs a particle into its cytoplasm.

pharyngeal slit (fuh-rin′-jē-ul) A gill structure in the pharynx; found in chordate embryos and some adult chordates.

phenotype (fē′-nō-tīp) The expressed traits of an organism.

phenylketonuria (PKU) (fē′-nul-kē′-tuh-nū′-rē-uh) A recessive genetic disorder characterized by an inability to properly break down the amino acid phenylalanine, resulting in an accumulation of toxins in the blood and mental retardation.

phloem (flō′-um) The portion of a plant's vascular system that distributes sugars from the leaves throughout the plant.

phosphate group A functional group consisting of a phosphorus atom covalently bonded to four oxygen atoms.

phospholipid (fos′-fō-lip′-id) A type of lipid molecule containing the element phosphorus in a hydrophilic "head," to which are attached two fatty acids that form two hydrophobic "tails." A bilayer (double layer) of phospholipids makes up the basic fabric of biological membranes.

photic zone (fō′-tik) The region of an aquatic ecosystem into which light penetrates and where photosynthesis occurs.

photoautotroph An organism that obtains energy from sunlight and carbon from CO_2 by photosynthesis.

photoheterotroph An organism that obtains energy from sunlight and carbon from organic sources.

photon (fō′-ton) A fixed quantity of light energy. The shorter the wavelength of light, the greater the energy of a photon.

photosynthesis (fō′-tō-sin′-thuh-sis) The process by which plants, autotrophic protists, and some bacteria use light energy to make sugars and other organic food molecules from carbon dioxide and water.

photosystem A light-harvesting unit of a chloroplast's thylakoid membrane; consists of several hundred antenna molecules, a reaction-center chlorophyll, and a primary electron acceptor.

pH scale A measure of the relative acidity of a solution, ranging in value from 0 (most acidic) to 14 (most basic); pH stands for potential hydrogen and refers to the concentration of hydrogen ions (H^+).

phylogenetic tree (fī′-lō-juh-net′-ik) A branching diagram that represents a hypothesis about evolutionary relationships among organisms.

phylogeny (fī-loj′-uh-nē) The evolutionary history of a species.

phylum (plural, *phyla*) (fī-lum) In classification, the taxonomic category above class and below kingdom; members of a phylum all have a similar general body plan.

phytoplankton (fī′-tō-plank′-ton) Algae and photosynthetic bacteria that drift passively in the pelagic zone of an aquatic environment.

pili (singular, *pilus*) (pī′-lī, pī′-lus) Short projections on the surface of prokaryotic cells that help prokaryotes attach to other surfaces.

pinocytosis (pīn′-ō-sī-tō′-sis) "Cellular drinking"; a type of endocytosis in which the cell takes fluid and dissolved solutes into small membranous vesicles.

placenta (pluh-sen′-tuh) In most mammals, the organ that provides nutrients and oxygen to the embryo and helps dispose of its metabolic wastes via the mother's blood vessels.

placental mammal (pluh-sen′-tul) A mammal whose young complete their embyonic development in the uterus, nourished via the mother's blood vessels in the placenta; also called a eutherian.

plankton Algae and other organisms, mostly microscopic, that drift passively in ponds, lakes, and oceans.

Plantae (plant′-tē) The kingdom that contains the plants.

plasma membrane The thin layer of lipids and proteins that sets a cell off from its surroundings and acts as a selective barrier to the passage of ions and molecules into and out of the cell; consists of a phospholipid bilayer in which are embedded molecules of protein and cholesterol.

plasmid A small ring of DNA separate from the chromosome(s); plasmids are found in prokaryotes and yeast.

plasmodesma (plural, *plasmodesmata*) (plaz′-mō-dez′-muh) An open channel in a plant cell wall, through which strands of cytoplasm connect from adjacent walls.

plasmodial slime mold (plaz-mō′-dē-ul) A type of protist that has amoeboid cells, flagellated cells, and an amoeboid plasmodial feeding stage in its life cycle.

plasmolysis (plaz-mol′-uh-sis) A phenomenon that occurs in plant cells in a hypertonic environment. The cell loses water and shrivels, and its plasma membrane pulls away from the cell wall, usually killing the cell.

Platyhelminthes (plat´-ē-hel-min´-thēz) The phylum that contains the flatworms, the bilateral animals with a thin, flat body form, gastrovascular cavity or no digestive system, and no body cavity; the free-living flatworms, flukes, and tapeworms.

pleated sheet The folded arrangement of a polypeptide in a protein's secondary structure.

pleiotropy (plī´-uh-trō-pē) The control of more than one phenotypic characteristic by a single gene.

polar covalent bond An attraction between atoms that share electrons unequally. The shared electrons are pulled closer to the atom that attracts electrons more strongly, making it partially negative and the other atom partially positive.

polar molecule A molecule that has opposite charges on opposite ends.

pollen In a seed plant, the male gametophytes that develop within the anthers of stamens.

pollination In seed plants, the delivery, by wind or animals, of pollen from the male parts of a plant to the stigma of a carpel on the female.

polygenic inheritance (pol´-ē-jen´-ik) The additive effect of two or more genes on a single phenotypic characteristic.

polymer (pol´-uh-mer) A large molecule consisting of many identical or similar smaller molecular units, called monomers, covalently joined together in a chain.

polymerase chain reaction (PCR) (puh-lim´-uh-rās) A technique used to obtain many copies of a DNA molecule or part of a DNA molecule. When a small amount of DNA is mixed with the enzyme DNA polymerase along with DNA nucleotides and a few other ingredients, the DNA replicates repeatedly in a test tube.

polymorphic (pol´-ē-mor´-fik) Existing in different forms; may pertain to a population in which two or more morphs are present in readily noticeable frequencies.

polynucleotide (pol´-ē-nū´-klē-uh-tīd) A polymer made up of many nucleotides covalently bonded together.

polyp (pol´-ip) One of two types of cnidarian body forms; a columnar, hydra-like body.

polypeptide A chain of amino acids linked by peptide bonds.

polyploid (pol´-ē-ploid) Containing more than two complete sets of homologous chromosomes in each somatic cell.

polysaccharide (pol´-ē-sak´-uh-rīd) A carbohydrate consisting of hundreds to thousands of monosaccharides (sugars) linked by covalent bonds.

population A group of interacting individuals belonging to one species and living in the same geographic area.

population density The number of individuals of a species per unit area or volume.

population ecology The study of how members of a population interact with their environment, focusing on factors that influence population density and growth.

population fragmentation The splitting and consequent isolation of a biological population, usually by human-caused habitat degradation.

population genetics The study of genetic changes in populations; the science of microevolutionary changes in populations.

population-limiting factor An environmental factor that restricts population growth.

Porifera (po-rif´-er-uh) The phylum that contains the sponges, characterized by choanocytes, a porous body wall, and no true tissues.

positron-emission tomography (PET) Imaging technology that uses radioactively labeled biological molecules, such as glucose, to obtain information about metabolic processes at specific locations in the body. The labeled molecules are injected into the bloodstream, and a PET scan for radioactive emissions determines which tissues have taken up the molecules.

post-anal tail A tail posterior to the anus; found in chordate embryos and most adult chordates.

post-zygotic barrier (pōst´-zī-got´-ik) A reproductive barrier that operates should interspecies mating occur and form hybrid zygotes.

potential energy Stored energy; the capacity to perform work that an object has because of its location or arrangement. Water behind a dam possesses potential energy.

predation An interaction between species in which one species, the predator, eats the other, the prey.

predator A consumer in a biological community.

prevailing winds Winds that result from the combined effects of Earth's rotation and the rising and falling of air masses.

prey An organism eaten by a predator.

pre-zygotic barrier (prē´-zī-got´-ik) A reproductive barrier that impedes mating between species or hinders fertilization of eggs if members of different species should attempt to mate.

primary consumer An organism that eats plants, algae, or autotrophic bacteria.

primary electron acceptor A molecule that traps the light-excited electron from the reaction-center chlorophyll in a photosystem.

primary productivity The rate at which an ecosystem's plants and other producers build biomass, or organic matter.

primary structure The first level of protein structure; the specific sequence of amino acids making up a polypeptide chain.

primary succession A type of ecological succession in which a biological community arises in an area without soil. *See also* secondary succession.

Primates The mammalian order that includes prosimians, monkeys, apes, and humans.

primitive characters Homologous features found in members of a lineage and also in the ancestors of the lineage; ancestral features.

principle of independent assortment A general rule in inheritance that when gametes form during meiosis, each pair of alleles for a particular characteristic segregate independently; also known as Mendel's second law of inheritance.

principle of segregation A general rule in inheritance that individuals have two alleles for each gene and that when gametes form by meiosis, the two alleles separate, and each resulting gamete ends up with only one allele of each gene; also known as Mendel's first law of inheritance.

producer An organism that makes organic food molecules from CO_2, H_2O, and other inorganic raw materials: a plant, alga, or autotrophic bacterium.

product An ending material in a chemical reaction.

prokaryotic cell (prō´-kār-ē-ot´-ik) A type of cell lacking a membrane-enclosed nucleus and other membrane-enclosed organelles; found only in the domains Bacteria and Archaea.

prokaryotic flagellum (prō´-kār-ē-ot´-ik fluh-jel´-um) A long surface projection that propels a prokaryotic cell through its liquid environment; totally different from the flagellum of a eukaryotic cell.

promoter A specific nucleotide sequence in DNA, located at the start of a gene, that is the binding site for RNA polymerase and the place where transcription begins.

prophage (prō´-faj) Phage DNA that has inserted by genetic recombination into the DNA of a prokaryotic chromosome.

prophase The first stage of mitosis, during which duplicated chromosomes condense to form structures visible with a light microscope and the mitotic spindle forms and begins moving the chromosomes toward the center of the cell.

prosimian (prō-sim´-ē-un) A member of the primate group comprised of lorises, pottos, tarsiers, and lemurs.

protein A biological macromolecule made of one or more polypeptides.

protist (prō´-tist) A member of the kingdom Protista. Protists are unicellular eukaryotes and closely related multicellular organisms that are not plants, fungi, or animals.

Protista (prō-tis´-tuh) In the standard five-kingdom classification system, the kingdom that contains the unicellular eukaryotes and closely related multicellular organisms called protists.

proton A subatomic particle with a single positive electrical charge; found in the nucleus of an atom.

proto-oncogene (prō´-tō-on´-kō-jēn) A normal gene with the potential to become a cancer-causing gene.

protozoan (prō´-tō-zō´-un) A protist that lives primarily by ingesting food; a heterotrophic, animal-like protist.

provirus Viral DNA that inserts into a host genome.

pseudocoelom (sū´-dō-sē´-lum) A body cavity only partly lined by mesoderm.

pseudopod (plural, *pseudopodia*) A temporary extension of an amoeboid cell. Pseudopodia function in moving cells and engulfing food.

punctuated equilibrium The idea that speciation occurs in spurts followed by long periods of little change.

Punnett square A diagram used in the study of inheritance to show the results of random fertilization.

Q

quaternary consumer An organism that eats tertiary consumers.

quaternary structure The fourth level of protein structure; the shape resulting from the association of two or more polypeptide subunits.

R

radial symmetry An arrangement of the body parts of an organism like pieces of a pie around an imaginary central axis. Any slice passing longitudinally through a radially symmetrical organism's central axis divides it into mirror-image halves.

radiation therapy Treatment for cancer in which parts of the body that have cancerous tumors are exposed to high-energy radiation to disrupt cell division of the cancer cells.

radioactive isotope An isotope whose nucleus decays spontaneously, giving off particles and energy.

radiometric dating A method for determining the age of fossils and rocks based on the decay of radioactive atoms.

radula (rad′-ū-luh) A rasping organ used to scrape up or shred food; found in many mollusks.

random Describing a dispersion pattern in which individuals are spaced in a patternless, unpredictable way.

ray-finned fish A bony fish having fins supported by thin, flexible skeletal rays. Most bony fishes are rayfins. *See also* lobe-finned fish.

reactant A starting material in a chemical reaction.

reaction center In a photosystem in a chloroplast, the chlorophyll *a* molecule and the primary electron acceptor that trigger the light reactions of photosynthesis. The chlorophyll donates an electron excited by light energy to the primary electron acceptor, which passes an electron to an electron transport chain.

reading frame The way a cell's mRNA-translating machinery groups the mRNA nucleotides into codons.

receptor On or in a cell, a specific protein molecule whose shape fits that of a specific molecular messenger, such as a hormone.

receptor-mediated endocytosis (en′-dō-sī-tō′-sis) The movement of specific molecules into a cell by the inward budding of membranous vesicles. The vesicles contain proteins with receptor sites specific to the molecules being taken in.

recessive allele In a heterozygous individual, the allele that has no noticeable effect on the phenotype.

recombinant DNA A DNA molecule carrying genes derived from two or more sources.

recombinant DNA technology Techniques for combining genes from different sources into a single DNA molecule and then transferring the new molecule into cells, where it can be replicated and expressed; also known as genetic engineering.

recombination frequency With respect to two given genes, the number of recombinant progeny from a mating divided by the total number of progeny. Recombinant progeny carry combinations of alleles different from those in either of the parents as a result of independent assortment of chromosomes or crossing over.

red algae Marine, mostly multicellular, autotrophic protists; includes the reef-building coralline algae.

red-green color blindness A class of common sex-linked human disorders involving several genes on the X chromosome; characterized by a malfunction of light-sensitive cells in the eyes; affects mostly males but also homozygous females.

redox reaction Short for oxidation-reduction reaction; a chemical reaction in which electrons are lost from one substance (oxidation) and added to another (reduction). Oxidation and reduction always occur together.

reduction The gain of electrons by a substance involved in a redox reaction; always accompanies oxidation.

regeneration The regrowth of body parts from pieces of an organism.

regulatory gene A gene that codes for a protein, such as a repressor, that controls the transcription of another gene or group of genes.

repetitive DNA Nucleotide sequences that are present in many copies in the genome. The repeated sequences may be long or short and may be located next to each other or dispersed in the DNA.

repressor A protein that blocks the transcription of a gene or operon.

reproductive barrier A biological feature of a species that prevents it from interbreeding with other species even when populations of the two species live together.

Reptilia (rep-til′-ē-uh) A class of vertebrate animals that consists of the reptiles, including snakes, lizards, turtles, crocodiles, and alligators.

resolving power A measure of the clarity of an image; the ability of an optical instrument to show two objects as separate.

resource partitioning The division of environmental resources by coexisting species populations such that the niche of each species differs by one or more significant factors from the niches of all coexisting species populations.

respiration (1) Gas exchange, or breathing; the exchange of O_2 and CO_2 between an organism and its environment. An aerobic organism takes up O_2 and gives off CO_2. (2) Cellular respiration; the aerobic harvest of energy from food molecules by cells.

restriction enzyme A bacterial enzyme that cuts up foreign DNA, thus protecting bacteria against intruding DNA from phages and other organisms. Restriction enzymes are used in DNA technology to cut DNA molecules in reproducible ways.

restriction fragment One of the molecules of DNA produced from a longer DNA molecule cut up by a restriction enzyme; used in RFLP (restriction fragment length polymorphism) analysis, genome mapping, and other applications.

retrovirus (ret′-trō-vī′-rus) An RNA virus that reproduces by means of a DNA molecule; it reverse-transcribes its RNA into DNA, inserts the DNA into a cellular chromosome, and then transcribes more copies of the RNA from the viral DNA. HIV and a number of cancer-causing viruses are retroviruses.

reverse transcriptase (tran-skrip′-tās) An enzyme that catalyzes the synthesis of DNA on an RNA template.

ribonucleic acid (RNA) (rī′-bō-nū-klā′-ik) A type of nucleic acid consisting of nucleotide monomers with a ribose sugar and the nitrogenous bases adenine (A), cytosine (C), guanine (G), and uracil (U); usually single-stranded; functions in protein synthesis and as the genome of some viruses.

ribosomal RNA (rRNA) (rī′-buh-sōm′-ul) The type of ribonucleic acid that, together with proteins, makes up ribosomes; the most abundant type of RNA.

ribosome (rī′-bō-sōm) A cell organelle that functions as the site of protein synthesis in the cytoplasm. The ribosomal components are constructed in the nucleolus.

ribozyme (rī′-bō-zīm) An enzymatic RNA molecule that catalyzes chemical reactions.

RNA Ribonucleic acid; a type of nucleic acid consisting of nucleotide monomers with a ribose sugar and the nitrogenous bases adenine (A), cytosine (C), guanine (G), and uracil (U); usually single-stranded; functions in protein synthesis and as the genome of some viruses.

RNA polymerase (puh-lim′-uh-rās) An enzyme that links together the growing chain of RNA nucleotides during transcription, using a DNA strand as a template.

RNA splicing The removal of introns and joining of exons in eukaryotic RNA, forming an mRNA molecule with a continuous coding sequence; occurs before mRNA leaves the nucleus.

RNA world A hypothetical period in the evolution of life when RNA served as rudimentary genes and the sole catalytic molecules.

root A plant structure that anchors the plant in the soil, absorbs and transports minerals and water, and stores food.

rough ER (en′-dō-plaz′-mik re-tik′-ū-lum) Rough endoplasmic reticulum; a network of interconnected membranous sacs in a eukaryotic cell's cytoplasm. Rough ER membranes are studded with ribosomes that make membrane proteins and secretory proteins. The rough ER constructs membrane from phospholipids and proteins.

roundworm A member of the phylum Nematoda.

R plasmid A bacterial plasmid that carries genes for enzymes that destroy particular antibiotics, thus making the bacterium resistant to the antibiotics.

rRNA Ribosomal RNA; the type of ribonucleic acid that, together with proteins, makes up ribosomes; the most abundant type of RNA.

rule of addition A rule stating that the probability of an event occurring when there are two or more alternative ways is the sum of the separate probabilities of the different ways.

rule of multiplication A rule stating that the probability of a compound event is the product of the separate probabilities of the independent events.

S

sarcoma (sar-kō′-muh) Cancer of supportive tissue, such as bone, cartilage, or muscle.

saturated Pertaining to fats and fatty acids whose hydrocarbon chains contain the maximum number of hydrogens and therefore have no double covalent bonds. Saturated fats and fatty acids solidify at room temperature.

savanna A biome dominated by grasses and scattered trees.

scanning electron microscope (SEM) A microscope that uses an electron beam to study the surface architecture of a cell or other specimen.

seaweed A large, multicellular marine alga.

secondary consumer An organism that eats primary consumers.

secondary structure The second level of protein structure; the regular patterns of coils or folds of a polypeptide chain.

secondary succession A type of ecological succession that occurs where a disturbance has destroyed an existing biological community but left the soil intact. *See also* primary succession.

seed A plant embryo packaged with a food supply within a protective covering.

seed coat A tough outer covering of a seed, formed from the outer coat (integuments) of an ovule. In a flowering plant, it encloses and protects the embryo and endosperm.

seed dormancy The temporary suspension of growth and development of a seed.

segmentation Division of an animal body along its length into a series of repeated parts called segments.

selectively permeable (per′-mē-uh-bul) Allowing some substances to cross a biological membrane more easily than others and blocking the passage of other substances altogether.

self-fertilization The fusion of sperm and egg that are produced by the same individual organism.

sepal (sē′-pul) A modified leaf of a flowering plant. A whorl of sepals encloses and protects the flower bud before it opens.

sex chromosome A chromosome that determines whether an individual is male or female.

sex-linked gene A gene located on a sex chromosome.

sexual reproduction The creation of offspring by the fusion of two haploid sex cells (gametes), forming a diploid zygote.

shoot The stem and leaves of a plant.

sickle-cell disease A genetic disorder in which the red blood cells have abnormal hemoglobin molecules and take on an abnormal shape.

signal-transduction pathway A series of molecular changes that converts a signal on a target cell's surface into a specific response inside the cell.

silencer A eukaryotic DNA sequence that functions to inhibit the start of gene transcription by binding a repressor.

sink habitat An area of habitat where a species' death rate exceeds its reproductive success.

sister chromatid (krō′-muh-tid) One of two identical parts of a duplicated chromosome in a eukaryotic cell.

slime mold A type of protist that has a filamentous body and is a decomposer; includes cellular slime molds and plasmodial slime molds.

smooth ER (en′-dō-plaz′-mik re-tik′-ū-lum) Smooth endoplasmic reticulum; a network of interconnected membranous tubules in a eukaryotic cell's cytoplasm. Smooth ER lacks ribosomes. Enzymes embedded in the smooth ER membrane function in the synthesis of certain kinds of molecules, such as lipids.

solute (sol′-ūt) A substance that is dissolved in a solution.

solution A fluid mixture of two or more substances, consisting of a dissolving agent, the solvent, and a substance that is dissolved, the solute.

solvent The dissolving agent in a solution. Water is the most versatile known solvent.

somatic cell (sō-mat′-ik) Any cell in a multicellular organism except a sperm or egg cell or a cell that develops into a sperm or egg.

source habitat An area of habitat where a species' reproductive success exceeds its death rate and from which new individuals often disperse to other areas.

speciation The origin of new species.

species In classification, the taxonomic category just below genus. *See also* biological species concept.

species diversity The number and relative abundance of species in a biological community.

sperm A male gamete.

spirochete (spī′-rō-kēt) A large spiral-shaped (curved) prokaryotic cell.

spontaneous generation The incorrect notion that life can emerge from inanimate material.

spore (1) In plants and algae, a haploid cell that can develop into a multicellular individual without fusing with another cell. (2) In prokaryotes, protists, and fungi, any of a variety of thick-walled life cycle stages capable of surviving unfavorable environmental conditions.

sporophyte (spō′-rō-fīt) The multicellular diploid form in the life cycle of organisms undergoing alternation of generations; results from a union of gametes; meiotically produces haploid spores that grow into the gametophyte generation.

stability The tendency of a biological community to resist change and return to its original species composition after being disturbed.

stabilizing selection Natural selection that favors intermediate variants by acting against extreme phenotypes.

stamen (stā′-men) A pollen-producing male reproductive part of a flower, consisting of a stalk and an anther.

starch A storage polysaccharide found in the roots of plants and certain other cells; consists of glucose monomers.

start codon (kō′-don) On mRNA, the specific three-nucleotide sequence (AUG) to which an initiator tRNA molecule binds, starting translation of genetic information.

stem Part of a plant's shoot system that supports the leaves and reproductive structures.

steroid A type of lipid whose carbon skeleton is bent to form four fused rings: three 6-sided rings and one 5-sided ring; examples are cholesterol, testosterone, and estrogen.

stigma (plural, *stigmata*) The sticky tip of a flower's carpel, which traps pollen grains.

stoma (plural, *stomata*) (stō′-muh) A pore surrounded by guard cells in the epidermis of a leaf. When stomata are open, CO_2 enters a leaf and water and O_2 exit. A plant conserves water when its stomata are closed.

stop codon (kō′-don) In mRNA, one of three triplets (UAG, UAA, UGA) that signal gene translation to stop.

stroma (strō′-muh) A thick fluid enclosed by the inner membrane of a chloroplast. Sugars are made in the stroma by the enzymes of the Calvin cycle.

style The stalk of a flower's carpel, with the ovary at the base and the stigma at the top.

substrate (1) A specific substance (reactant) on which an enzyme acts. Each enzyme recognizes only the specific substrate of the reaction it catalyzes. (2) A surface in or on which an organism lives.

succession *See* ecological succession, primary succession, and secondary succession.

sugar-phosphate backbone The alternating chain of sugar and phosphate to which the DNA and RNA nitrogenous bases are attached.

surface tension A measure of how difficult it is to stretch or break the surface of a liquid.

survivorship curve A plot of the number of people at each age; a graphic way to represent some of the data in a life table.

sustainable development The long-term prosperity of human societies and the ecosystems that support them.

swim bladder A gas-filled internal sac that helps bony fish maintain buoyancy.

symbiont (sim′-bē-unt) The smaller participant in a symbiotic relationship, living in or on the host.

symbiosis (sim′-bē-ō′-sis) An interspecific interaction in which one species, the symbiont, lives in or on another species, the host.

symbiotic relationship (sim′-bē-ot′-ik) An interspecific interaction in which one species, the symbiont, lives in or on another species, the host.

sympatric speciation (sim-pat′-rik) The formation of a new species as a result of a genetic change that produces a reproductive barrier between the changed population (mutants) and the parent population.

systematics The scientific study of biological diversity and its classification.

T

taiga (tī′-guh) The northern (boreal) coniferous forest, which extends across North America and Eurasia, to the southern border of the arctic tundra; also found just below alpine tundra on mountainsides in temperate zones.

tail Extra nucleotides added at the end of an RNA transcript in the nucleus of a eukaryotic cell.

tapeworm A parasitic flatworm characterized by the absence of a digestive tract.

target cell A cell that responds to a regulatory signal, such as a hormone.

taxonomy (tak-son′-uh-mē) The branch of biology concerned with identifying, naming, and classifying species.

telomere (tel′-uh-mēr) The repetitive DNA at each end of a eukaryotic chromosome.

telophase (tel′-uh-fāz) The fourth and final stage of mitosis, during which daughter nuclei form at the two poles of a cell. Telophase usually occurs together with cytokinesis.

temperate deciduous forest A biome located throughout midlatitude regions where there is sufficient moisture to support the growth of large, broadleaf deciduous trees.

temperate grassland Grassland regions maintained by seasonal drought, occasional fires, and grazing by large mammals.

temperate zone Region whose latitude is between the tropics and the Arctic Circle in the north or the Antarctic Circle in the south; region with a milder climate than the tropics or polar regions.

temperature A measure of the intensity of heat, reflecting the average kinetic energy or speed of molecules.

temporal isolation A type of pre-zygotic barrier between species; the species remain isolated because they breed at different times.

terminator A special sequence of nucleotides in DNA that marks the end of a gene. It signals RNA polymerase to release the newly made RNA molecule, which then departs from the gene.

tertiary consumer An organism that eats secondary consumers.

tertiary structure The third level of protein structure; the overall three-dimensional shape of a polypeptide in a protein.

testcross The mating between an individual of unknown genotype for a particular characteristic and an individual that is homozygous recessive for that same characteristic.

testosterone An androgen hormone that stimulates an embryo to develop into a male and promotes male body features.

tetrad A paired set of homologous chromosomes, each composed of two sister chromatids. Tetrads form during prophase I of meiosis.

tetrapod A terrestrial vertebrate with two pairs of limbs. Tetrapods include amphibians, reptiles, birds, and mammals.

theory A widely accepted explanatory idea that is broad in scope and supported by a large body of evidence.

threatened species A species likely to become endangered in the foreseeable future throughout all or a significant portion of its geographic range.

three-domain system A system of taxonomic classification based on three basic groups: Bacteria, Archaea, and Eukarya.

thylakoid (thī′-luh-koid) One of a number of disk-shaped membranous sacs inside a chloroplast. Thylakoid membranes contain chlorophyll and the enzymes of the light reactions of photosynthesis. A stack of thylakoids is called a granum.

thymine (T) (thī′-min) A single-ring nitrogenous base found in DNA.

tight junction A junction that binds tissue cells together in a leakproof sheet.

tissue A cooperative unit of many similar cells that perform a specific function within a multicellular organism.

trace element An element that is essential for the survival of an organism but only in minute quantities.

trade winds The movement of air in the tropics from approximately 23.5° north and south toward the equator.

transcription The synthesis of RNA on a DNA template.

transcription factor In the eukaryotic cell, a protein that functions in initiating or regulating transcription. Transcription factors bind to DNA or to other proteins that bind to DNA.

transduction The transfer of bacterial genes from one bacterial cell to another by a phage. *See* signal-transduction pathway.

transfer RNA (tRNA) A type of ribonucleic acid that functions as an interpreter in translation. Each tRNA molecule has a specific anticodon, picks up a specific amino acid, and conveys the amino acid to the appropriate codon on mRNA.

transformation The incorporation of new genes into a cell from DNA that the cell takes up from the fluid around it.

transgenic organism An organism that has acquired one or more genes by artificial means; also known as a genetically-modified (GM) organism.

translation The synthesis of a polypeptide using the genetic information encoded in an mRNA molecule. There is a change of "language" from nucleotides to amino acids.

translocation (1) During protein synthesis, the movement of a tRNA molecule carrying a growing polypeptide chain from the A site to the P site on a ribosome. (2) A change in a chromosome resulting from a chromosomal fragment attaching to a nonhomologous chromosome; can occur as a result of an error in meiosis or from mutagenesis.

transmission electron microscope (TEM) A microscope that uses an electron beam to study the internal structure of thinly sectioned specimens.

transport protein A membrane protein that helps move substances across a cell membrane.

transport vesicle A tiny membranous sac in a cell's cytoplasm carrying molecules produced by the cell. The vesicle buds from the endoplasmic reticulum or Golgi and eventually fuses with another membranous or-

ganelle or the plasma membrane, releasing its contents.

transposon (trans-pō′-zon) A transposable genetic element, or "jumping gene"; a segment of DNA that can move from one site to another within a cell and serve as an agent of genetic change.

triglyceride (trī-glis′-uh-rīd) A fat, which consists of a molecule of glycerol linked to three fatty acids.

trisomy 21 (trī-suh-mē) The condition of having three number 21 chromosomes, which results in the human genetic disorder Down syndrome, characterized by heart and respiratory defects and varying degrees of mental retardation.

tRNA Transfer RNA; a type of ribonucleic acid that functions as an interpreter in translation. Each tRNA molecule has a specific anticodon, picks up a specific amino acid, and conveys the amino acid to the appropriate codon on mRNA.

trophic level (trō′-fik) The level in a food chain.

trophic structure (trō′-fik) The feeding relationships in an ecosystem; determines the route of energy flow and the pattern of chemical cycling in an ecosystem.

tropical forest A biome in the tropics that varies from tropical rain forests in wet areas to tropical dry forests in areas with scarce rainfall to tropical deciduous forests in areas with distinct wet and dry seasons.

tropics Latitudes between 23.5° north and south.

true-breeding variety Organisms for which sexual reproduction produces offspring with inherited trait(s) identical to those of the parents. The organisms are homozygous for the characteristic(s) under consideration.

tumor-suppressor gene A gene whose product inhibits cell division, thereby preventing uncontrolled cell growth.

tundra A biome at the northernmost limits of plant growth and at high altitudes, characterized by dwarf woody shrubs, grasses, mosses, and lichens.

tunicate (tū′-nuh-kāt) One of a group of invertebrate chordates.

U

ultrasound imaging A technique for examining a fetus in the uterus. High-frequency sound waves echoing off the fetus are used to produce an image of the fetus.

uniform Describing a dispersion pattern in which individuals are evenly distributed.

unsaturated Pertaining to fats and fatty acids whose hydrocarbon chains lack the maximum number of hydrogen atoms and therefore have one or more double covalent bonds. Unsaturated fats and fatty acids do not solidify at room temperature.

uracil (U) (ū′-ruh-sil) A single-ring nitrogenous base found in RNA.

V

vaccine A harmless variant or derivative of a pathogen used to stimulate a host organism's immune system to mount a long-term defense against the pathogen.

vacuole (vak′-ū-ōl) A membrane-enclosed sac, part of the endomembrane system of a eukaryotic cell, having any of a number of diverse functions.

vascular tissue (vas′-kyū-ler) A network of cells forming narrow tubes that extend throughout a plant. The two types of vascular tissue in a plant are xylem, which channels water and minerals upward from the roots, and phloem, which channels sugars produced in the leaves throughout the plant.

vector In molecular biology, a piece of DNA, usually a plasmid or a viral genome, that moves genes from one cell to another.

vertebrate A chordate animal with a backbone; includes agnathans, fishes, amphibians, reptiles, birds, and mammals.

W

warning coloration The often brightly colored markings of animals possessing chemical defenses. The coloration provides a caution to their predators.

water vascular system In echinoderms, a radially arranged system of water-filled canals that branch into extensions called tube feet. The system provides movement and circulates water, facilitating gas exchange and waste disposal.

wavelength The distance between the crests of adjacent waves, such as those of the electromagnetic spectrum.

wild type The phenotype most commonly found in nature.

X

X chromosome inactivation In female mammals, the inactivation of one X chromosome in each somatic cell.

xylem (zī′-lum) The nonliving portion of a plant's vascular system that provides support and transports water and minerals from the roots to the rest of the plant.

Z

zoned reserve An extensive region of land that includes one or more areas undisturbed by humans. The undisturbed areas are surrounded by lands that have been altered by human activity.

zooplankton (zō′-ō-plank′-tun) Animals that drift in the pelagic zone of an aquatic environment.

zygote (zī′-gōt) The fertilized egg, which is diploid, that results from the union of a sperm cell with an egg cell.

Index

Consumers (heterotrophs) (continued)
in food web, 462, 463*fig.*
fungi as, 344, 462, 462*fig.*
Contraception, human population growth and, 431
Contractile proteins, 48, 48*fig.*
Contractile vacuole, 67, 67*fig.*
Control sequences, in *lac* operon, 201
Convergent evolution, 295, 403, 415, 415*fig.*,
415*web/CD Evolution Link*
analogy and, 295
Cooling, evaporative, 33–34, 34*fig.*
Copy-and-paste transposon, 233
Coral animals, 358, 358*fig.*
Coral reefs, 412, 412*fig.*
Corridors, in landscape ecology, 485, 485*fig.*
Courtship rituals, as reproductive barrier, 280, 280*fig.*
Covalent bonds, 30, 30*fig.*, 30*web/CD 2F*
in water, 32, 32*web/CD 2G*
Creatine phosphate, 99
Cretaceous period, mass extinctions/explosive diversi-
fications at end of, 292–293, 293*fig.*
Crick, Francis, 177, 177*fig.*
Cri du chat ("cat cry") syndrome, 138
Crimes, DNA fingerprinting and, 22, 22*fig.*, 235–236,
235*fig.*, 236*fig.*, 236*web/CD 11F*
Cristae, 69, 69*fig.*
Cro-Magnons, 386
artwork of, 387, 387*fig.*
Crop irrigation, water cycle affected by, 471
Crops, genetically modified, 220, 220*fig.*, 239–240,
239*fig.*
controversy about, 243, 243*fig.*
Cross (hybrid), 146
dihybrid, 148–149, 149*fig.*
monohybrid, 146–148, 147*fig.*, 148*web/CD 8B and*
The Process of Science
polyploid species arising from, 283–284, 283*fig.*
Cross-fertilization, 145–146, 145*fig.*, 148*web/CD 8B*
and The Process of Science
Crossing over, 130*fig.*, 131, 134–135, 134*fig.*, 163*fig.*,
164, 164*web/CD 8D*
in bacteria, 221–222, 221*fig.*
linked genes and, 163*fig.*, 164, 164*web/CD 8D*
Crossover data, for gene mapping, 164–165, 165*fig.*
Human Genome Project and, 234
Crustaceans, 367–368, 368*fig.*
Cryptic coloration, as defense against predators,
445, 445*fig.*
Crystallin gene, gene expression pattern for, 204,
204*fig.*
CSF (colony-stimulating factor), DNA technology in
production of, 237*table*
Cultural evolution, 386–387, 387*fig.*
earth's crisis and, 388, 388*web/CD Evolution Link*
Culture of science, 20–21, 21*fig.*
Cut-and-paste transposon, 233, 233*fig.*
Cuticle, of plant, 327*fig.*, 328
CVS (chorionic villus sampling), 155, 155*fig.*
complication rate for, 156
Cyanobacteria, 312*fig.*
as autotrophs, 104*fig.*
in oxygen revolution, 116, 116*web/CD Evolution*
Link
as photoautotrophs, 313

Cystic fibrosis, 152–153, 153*table*
Cytokinesis, 123, 125–126, 125*fig.*, 126*fig.*, 131*fig.*
in animal cells, 125, 125*fig.*, 126*fig.*
in plant cells, 126, 126*fig.*
Cytoplasm, 62
Cytosine (C), 53, 176*fig.*, 177
Cytoskeleton, 62*fig.*, 63*fig.*, 69–72
fiber types in, 62*fig.*, 70, 70*fig.*
membrane proteins bonded to, 75*fig.*
Cytosol, 62

D

Dams, ecosystems affected by, 410, 410*fig.*
Damselfly, 368*fig.*
Darwin, Charles, 10, 10*fig.*, 251, 260*fig.*, 277
theories of
cultural and scientific, 10–13, 12*fig.*, 12*web/CD*
The Process of Science, 251–252, 251*web/CD*
12A, 252*fig.*, 253*fig.*
modern synthesis and, 261–265, 262*fig.*, 263*fig.*,
263*web/CD 12D*, 264*fig.*, 265*fig.*
natural selection, 13–16, 14*fig.*, 15*fig.*, 252,
253*fig.*, 258–261, 258*web/CD The Process of*
Science, 269–271, 269*web/CD 12E*
testing, 435–436, 436*fig.*, 436*web/CD Evolution*
Link, 437*fig.*
Darwinian fitness, 269, 269*fig. See also* Natural
selection
Darwinian medicine, 271–273, 273*web/CD Evolution*
Link
Daughter cells, cell division producing, 121, 122–123,
122*fig.*
DDT, 396, 396*web/CD 17A*
biological magnification of, 473, 473*fig.*
evolution of insects resistant to, 260–261, 261*fig.*
Deafness, hereditary, 152, 152*fig.*
Death rates, in human population growth, 428–429,
429*fig.*
Deciduous forest, temperate, 403*fig.*, 406*fig.*
Decomposers, 462, 462*fig.*
chemical cycling and, 467, 467*fig.*
fungi as, 345, 462, 462*fig.*
prokaryotes as, 315–316, 315*fig.*, 316*fig.*, 462
Deduction, in hypothetico-deductive reasoning, 17
Defenses
against parasites, natural selection affecting,
447–448, 448*fig.*
against predators, 445–446, 445*fig.*, 446*fig.*
Defensive proteins, 48
Deforestation, 340–341
chemical cycles affected by, 471, 472, 472*fig.*
earth's crisis and, 388
Dehydration synthesis, 42, 42*fig.*, 42*web/CD 3C*
ATP as energy source for, 89
in disaccharide formation, 44, 44*fig.*
in DNA strand formation, 53, 53*fig.*
in protein formation, 49, 49*fig.*
Deletion, chromosome, 138, 138*fig.*
Denaturation, protein, 51
Density-dependent factors, in regulation of population growth, 424–425, 425*fig.*
interspecific competition and, 443

Density-independent factors, in regulation of popula-
tion growth, 423–424
Deoxyribonucleic acid. *See* DNA
Deoxyribose, in DNA, 176, 176*fig.*
Descent with modification, 251–252, 252*fig.*, 253*fig.*
See also Evolution
Desert, 403*fig.*, 405*fig.*
Detritus, 408
Detritivores (decomposers), 462, 462*fig.*
chemical cycling and, 467, 467*fig.*
fungi as, 345, 462, 462*fig.*
prokaryotes as, 315–316, 315*fig.*, 316*fig.*, 462
Development, evolutionary biology and ("evo-devo"),
287–288, 287*fig.*, 288*fig.*, 288*web/CD 13C*
de Vries, Hugo, 283, 283*fig.*
Diatoms, 321, 321*fig.*
Diet, cancer risk affected by, 215
Differentiation, 199, 199*fig.*
gene expression patterns and, 204–205, 204*fig.*
Diffusion, 76, 76*fig.*, 76*web/CD 4K*
osmosis (water balance) and, 76–77, 76*fig.*, 77*fig.*,
77*web/CD 4L*
Digestive tract, in roundworm, 361
Dihybrid cross, 148–149, 149*fig.*
Dinoflagellates, 321, 321*fig.*
Dinosaurs, 376, 376*fig.*
Diploid cells, 129
Directional selection, 270, 270*fig.*
Disaccharides, 44, 44*fig.*
Discovery science, 16–17, 16*fig.*
Disease
animal viruses causing, 192–193, 192*fig.*, 193*fig.*
bacteria causing, 314–315, 314*fig.*, 315*fig.*
defense against, 314–315
in evolutionary context (Darwinian medicine),
271–273, 273*web/CD Evolution Link*
gene therapy in treatment of, 241–242, 241*fig.*
ethical issues and, 241–242, 244
inherited
ethical issues in DNA technology and, 244–245,
244*fig.*
genetic drift and, 268
population genetics and, 264–265
Dispersion patterns of population, 421–422, 421*fig.*,
422*fig.*
clumped, 421, 421*fig.*
mixed, 422, 422*fig.*
random, 421*fig.*, 422
uniform, 421, 421*fig.*
Disturbances, 449–451, 450*fig.*, 451*fig.*, 452*fig.*
dynamic view of community structure and, 451,
451*fig.*
ecological succession and, 450, 450*fig.*,
450*web/CD 19C*
human, 452, 452*fig.*
intermediate, 451
periodic/catastrophic, 399, 399*fig.*
small-scale, 451, 451*fig.*
Diversifying selection, 270, 270*fig.*
Diversity (biological/biodiversity), 6–8, 6*fig.*, 276–302
animal, 351–356, 351*fig.*
arthropod, 367–369, 367*fig.*, 369*web/CD The*
Process of Science

Sunlight
  as abiotic factor, 397
    air movement/water currents affected by,
      400–401, 400*fig.*
    global climate patterns and, 399, 399*fig.*
  energy for photosynthesis provided by, 106, 106*fig.*
    See also Light reactions
  exposure to (ultraviolet radiation)
    cancer risk and, 214, 476, 476*fig.*
    ozone depletion and, 476, 476*fig.*
  nature of, 108, 108*fig.*
  for photoautotrophs, 313, 314*table*
Surface tension, 33, 33*fig.*
"Survival of the fittest," 269. *See also* Natural selection
Survivorship curve, 433, 433*fig.*, 433*web/CD 18D*
Sustainable development, 486, 486*web/CD 20H*
Sweating, evaporative cooling and, 34, 34*fig.*
Swim bladder, in bony fishes, 374
Symbiont, 447
Symbiosis, 447–448, 448*fig.*
  in evolution, 317–318, 347, 347*fig.*, 347*web/CD
    Evolution Link*
  root-fungus (mycorrhizae), 328, 328*fig.*, 347
Symbiotic relationship, 447–448, 448*fig.*
  mutualism, 448, 448*fig.*
  parasitism, 447–448, 448*fig.*
Symmetry, body, animal phylogeny and, 354–355,
  354*fig.*, 355*fig.*
Sympatric speciation, 281, 281*fig.*, 283–284, 283*fig.*,
  283*web/CD 13B*, 284*fig.*
Systematics, 294. *See also* Taxonomy
  cladistic analysis and, 296–297, 296*fig.*, 297*fig.*
  molecular biology as tool in, 295–296, 296*fig.*,
    296*web/CD 13E/The Process of Science*

T

Tadpole, 375, 375 *fig.*
Tail
  post-anal, in chordates, 372, 372*fig.*
  RNA transcript, 185, 185*fig.*
Tapeworms, 360, 360*fig.*
Tatum, Edward, 181, 220
Taxol, DNA technology in production of, 237*table*
Taxonomy, 294–298, 294*fig.*, 299*fig.*
  cladistic analysis and, 296–297, 296*fig.*, 297*fig.*
  definition of, 6, 294
  hierarchical classification and, 294, 294*fig.*
  kingdoms as category in, 297–298, 299*fig.*
  molecular biology in, 295–296, 296*fig.*, 296*web/CD
    13E/The Process of Science*
  naming species and, 294
  phylogeny and, 295–297, 295*fig.*, 296*fig.*,
    296*web/CD 13E*, 297*fig.*
Tay-Sachs disease, 67, 153*table*
Technology
  DNA. *See* DNA technology
  science and society and, 22, 22*fig.*
Telomeres, 232, 232*fig.*
Telophase
  of meiosis I, 131*fig.*
  of meiosis II, 131*fig.*
  of mitosis, 125*fig.*
Temperate deciduous forest, 403*fig.*, 406*fig.*

Temperate grassland, 403*fig.*, 406*fig.*
Temperate zones, 400, 400*fig.*
Temperature, 33. *See also* Climate
  as abiotic factor, 398, 398*fig.*
  carbon dioxide emissions/greenhouse effect/global
    warming and, 115, 115*fig.*, 470, 470*fig.*, 471,
    474–475, 474*fig.*, 475*fig.*, 475*web/CD The
    Process of Science*
  freshwater biomes affected by, 408
  photosynthesis affecting, 114, 115–116, 115*fig.*,
    115*web/CD 6G and The Process of Science*
  water in moderation of, 33–34, 34*fig.*, 401, 401*fig.*
Temporal isolation, as reproductive barrier, 280,
  280*fig.*
TEM (transmission electron microscope), 59*fig.*, 60
Termination, 187, 188*fig.*
  in transcription, 184, 184*fig.*
Terminator, 184, 184*fig.*
Terrarium ecosystem, 460, 460*fig.*
Terrestrial biomes, 402–408, 403*fig.*, 403*web/CD 17C*,
  404–407*fig.*
Tertiary consumers
  in energy pyramid, 465, 465*fig.*
  in food chain, 461*fig.*, 462
  in food web, 462, 463*fig.*
Tertiary structure of protein, 50, 50*fig.*
Testcross, 150, 150*fig.*
Testosterone, 47, 47*fig.*
Tetrad, 130*fig.*
Tetrapods, evolution of, 373*fig.*, 375, 375*fig.*
Theory, scientific, 21
  evolution as, 300, 300*web/CD Evolution Link*
Thermophiles, extreme, 311
Thiobacillus, in bioremediation, 316
Threatened species, 482
Three-domain system, 298, 299*fig.*
Thylakoids, 105, 105*fig.*
  chloroplast pigments in, 108, 108*fig.*
  light reactions in, 110–111, 111*fig. See also* Light
    reactions
Thymine (T), 53, 176*fig.*, 177
Thyroid gland, iodine and, 26, 26*fig.*
Ticks, 367
Tide pools, 411, 411*fig.*
Tight junctions, 73, 73*fig.*
Ti plasmid, as vector, 239, 239*fig.*
Tissue plasminogen activator, DNA technology in
  production of, 237, 237*table*
Tobacco, cancer risk and, 214, 215*table*
Tobacco mosaic virus, 192*fig.*
Tornadoes, as abiotic factors, 399
Toxic chemicals, human release of to ecosystem, 473,
  473*fig.*
t-PA (tissue plasminogen factor), DNA technology in
  production of, 237, 237*table*
Trace elements, 26, 26*fig.*
Trade winds, 400, 400*fig.*
Transcription, 181, 181*fig.*, 182, 182*fig.*, 184, 184*fig.*,
  184*web/CD 9E*, 188, 189*fig.*
  in gene expression, 199–200, 200*fig.*, 206–207, 207*fig.*
Transcription factors, 206–207, 207*fig.*
Transduction
  bacterial, 221, 221*fig.*
  in cell signaling, 79, 79*fig.*

Transfer RNA (tRNA), 185–186, 185*fig.*
Transformation, bacterial, 221, 221*fig.*
Transforming factor, 173, 173*fig. See also* DNA
Transfusion reaction, 158–159, 159*fig.*
Transfusions, blood groups and, 158–159, 159*fig.*
Transgenic (genetically modified) organisms, 239
  animals, 240–241, 240*fig.*
  plants, 220, 220*fig.*, 239–240, 239*fig.*
  controversy about, 243, 243*fig.*
Transitional fossils, 255, 255*fig.*
Translation, 181, 181*fig.*, 182, 182*fig.*, 185–187, 185*fig.*,
  187*fig.*, 187*web/CD 9F*, 188*fig.*, 189
  in gene expression, 208
Translocation
  chromosome, 138, 138*fig.*
  in translation, 187, 187*fig.*
Transmission electron microscope (TEM), 59*fig.*, 60
Transpiration, in water cycle, 469*fig.*
  deforestation affecting, 471, 471*fig.*, 472
Transport
  active, 78, 78*fig.*, 78*web/CD 4M*
    ATP in, 78, 89, 89*fig.*
  exocytosis and endocytosis, 78–79, 78*fig.*,
    78*web/CD 4N*, 79*fig.*
  passive (diffusion), 76, 76*fig.*, 76*web/CD 4K*
    osmosis (water balance) and, 76–77, 76*fig.*, 77*fig.*,
      77*web/CD 4L*
Transport proteins, 48, 48*fig.*, 75
  selective permeability and, 75
Transport vesicles, 65, 65*fig.*, 68*fig.*
Transposons ("jumping genes"), 233, 233*fig.*
Triglycerides, 46, 46*fig.*
Triplet code, 183
Trisomy 21 (Down syndrome), 135, 135*fig.*
tRNA (transfer RNA), 185–186, 185*fig.*
Trophic levels, 461
  energy pyramids and, 465, 465*fig.*,
    465*web/CD 20C*
  food chains and, 461–462, 461*fig.*, 462*fig.*
  food webs and, 462, 463*fig.*
  human nutrition and, 466, 466*fig.*
Trophic structure, 442, 442*fig.*, 461–462, 461*fig.*,
  462*fig.*, 462*web/CD 20B*, 463*fig.*
Tropical rain forest, 403*fig.*, 404*fig.*
  destruction of, 340–341
    chemical cycles affected by, 471, 471*fig.*, 472,
      472*fig.*
  ecological research in, 393*fig.*
Tropics, 400, 400*fig.*
trp operon, 202, 202*fig.*
Truffles, 346, 346*fig.*
Trypanosomes, 319, 319*fig.*
Tubulins, in microtubules, 70, 70*fig.*
Tumor
  benign, 126
  malignant, 126, 127*fig. See also* Cancer
Tumor necrosis factor, DNA technology in produc-
  tion of, 237*table*
Tumor-suppressor genes, 212, 212*fig.*
Tundra, 403*fig.*, 407*fig.*
Tunicates, 372, 372*fig.*, 373*fig.*
Turner (XO) syndrome, 137, 137*fig.*, 137*table*
Two-kingdom system, 297–298
Tyrolean Ice Man, 296*fig.*